Structural Control

Structural Control

Proceedings of the Second International Symposium
on Structural Control, University of Waterloo, Ontario,
Canada, July 15–17, 1985

Edited by

H.H.E. Leipholz

University of Waterloo
Ontario, Canada

1987 SPRINGER-SCIENCE+BUSINESS MEDIA, B.V.

Library of Congress Cataloging in Publication Data

International Symposium on Structural Control (2nd :
    1985 : University of Waterloo)
    Structural control.

    1. Structural engineering--Congresses. 2. Control
theory--Congresses. 3. System analysis--Congresses.
I. Leipholz, H. H. E. (Horst H. E.), 1919-
II. Title.
TA630.I548 1985   624.1   87-5492

ISBN 978-94-010-8075-0   ISBN 978-94-009-3525-9 (eBook)
DOI 10.1007/ 978-94-009-3525-9

Copyright

PREFACE

The topic of "structural control", which had already experienced some attention through publications, for example by Roorda, Yao, Yang, Abdel-Rohman, Leipholz etc., mostly in journals of ASCE, was given its first international forum at the University of Waterloo, Waterloo, Ontario, Canada, via an IUTAM - Symposium held in June, 1979. This very successful event gathered experts from a variety of technical and theoretical domains in which control plays traditionally an essential role and was meant to present the new idea of structural control to a broad audience, thus triggering interest and commitment as well as cross-fertilization. However, the peculiarities of structural control were already sufficiently well pointed out and stressed by those participants of the symposium who had devoted themselves earlier to this specific topic for some time. The result of presentations and discussions are collected in a set of Proceedings entitled "Structural Control", published by North - Holland Publishing Company and the Solid Mechanics Division (SMD) of the University of Waterloo.

The stimulation following this first symposium was quite noticeable in the literature and led to the conviction of many researchers that after a reasonable period of time, a second meeting should be held to collect the fruits produced by the intermediate efforts of those working with increased emphasis on structural control.

Therefore, Professors J. T. P. Yao, Department of Civil Engineering, Purdue University and H. H. E. Leipholz, Department of Civil Engineering, University of Waterloo, serving as chairmen, and assisted by a board of scientific advisors comprising:

M. Abdel-Rohman, University of Kuwait
M. P. Gaus, National Science Foundation (U.S.A.)
N. G. Lind, University of Waterloo
L. Meirovitch, Virginia Polytech. Inst. and State University
G. Thierauf, Universität G.H. Essen
J. N. Yang, George Washington University

organized the Second International Symposium on Structural Control which took place in July, 1985, again at the University of Waterloo in Waterloo, Ontario, Canada. The three key-note speakers were:

Dr. J. L. Lions, I.N.R.I.A., France
who presented a paper entitled
Homogenization and Reinforced Structures

Dr. T. T. Soong, State University of New York
who presented a paper entitled
A Standarized Model for Structural Control Experiments and some Experimental Results (co-authors A. M. Reinhorn, J. N. Yang)

Dr. R. L. Kosut, Integrated Systems Incorporated (USA)

who presented a paper entitled
Feedback Control of Large Space Structures: A View from the Outside.

The expectation, that at this new meeting a noticeable development in structural control as far as theory and implications are concerned would be apparent, was fully confirmed. Indeed, many authors were able to report on practical realizations of structural control experimentally and as part of the full scale design of actual buildings and other civil engineering structures.

Again, experts from various fields dedicated to control were present, and the "spillover" of ideas and concrete technical suggestions into structural control took place as before to the benefit of this new area. However, as an advancement with respect to the first symposium, one could notice that the suggestions made were already more precisely pointed than before towards the specifics of structural control. For example, a strong effort was made by many contributors to adjust the general theory of control to the needs of structural control which involves in the first place distributed parameters and non-selfadjoined mathematical terms raising thus the important question how far a modal approach to control might remain feasible.

Highlights of the presentations were reports on experiments and on control systems actually added to certain buildings as part of their design. Also impressive was the fact that experts working on structural control are addressing themselves now to such problems as time lag, stochastic effects, systems identification, effects of damping and of structural changes on optimal control, and other topics. This fact indicates that the time of considering idealized situations only is being replaced by a time in which the practical implication of control is more closely considered. This requires working with systems which are more or less imperfect and bound to be exposed to secondary effects.

It became clear through the papers presented that the main objects of structural control are, for the time being, tall buildings exposed to wind and earthquake; bridges exposed to flutter causing wind and to support movement; towers, stacks, chimneys, antennas; and last not least, large space structures.

Finally, the opportunity may be taken to express thanks to all those who sponsored the symposium and contributed to its realization:

The National Science Foundation of the United States contributed essentially to the travel expenses of many participants whose valuable contributions to the symposium may otherwise have been lost,

The Natural Science and Engineering Research Council of Canada who, together with the University of Waterloo, contributed the funds needed for the local organization of the symposium,

The American Society of Civil Engineering,

The Royal Society of Canada, and

The Canadian Society for Mechanical Engineering, who jointly recommended to their members participation in the symposium,

and

Martinus Nijhoff Publishing Company who serves as the publisher and will distribute the proceedings.

Preparation of the manuscript and the wordprocessing was done by Mrs. Nancy Simpson and Mrs. Chu-Yong Yeo for which they deserve special thanks. However, without the devotion, competence, and great skill of Dr. F. Afagh, who served as the technical editor, these proceedings would never have been completed. During a critical period of collapse of the wordprocessing activities of the Solid Mechanics Division of this University, he stepped in in order to take over the preparation of the manuscript for the printing with never ceasing enthusiasm inspiring his co-workers. He not only contributed to the solution of technical problems but also worked with the two ladies mentioned above on the proofreading and excercised excellent judgement in questions related to the scientific content.

Thanks are also due to the local organizing committee and to all the participants who jointly made the symposium a great event. May the Proceedings, which serve as a testimony of their remarkable efforts, be well received by the scientific and engineering community of the world, and may a meeting of those devoted to structural control be repeated in due course in order to yield again the compilation of the results obtained over the future years in that discipline.

H. H. E. Leipholz
Waterloo, Ontario, Canada
June, 1986

Table of Contents

THE FEASIBILITY OF ACTIVE CONTROL OF TALL BUILDINGS

Mohamed Abdel-Rohman
Department of Civil Engineering
Kuwait University, Kuwait

1. Introduction

Researchers have suggested and investigated the use of three active control mechanisms for tall buildings control. These mechanisms are the active tendon mechanism [1-4], the active tuned mass damper mechanism [4-7], and the active aerodynmaic appendage mechanism [8,11]. Each mechanism was claimed to be superior to the others either by comparing the degree of control provided, or by the magnitude of the control force consumed. However, these studies have not shown how feasible and practical is the suggested control mechanism.

The feasibility of active control could be studied by considering various aspects as, for example, the time delay effect on the controlled response, the effectiveness of the control mechanism in suppressing the vibration, the effect of changing the structural parameters due to the feedback control forces, and the size of actuators required to achieve the designed control law.

This paper investigates the feasibility of using the three suggested active control mechanisms for tall buildings considering the size and behaviour of the actuators to perform the designed control law. The other aspects of feasibility have been previously studied [12-15] and shall briefly be mentioned when suggesting a practical control mechanism for tall buildings control.

1

2. Feasibility of Active Tendon Control

The tall building considered in this study is 1000 ft high, with a square cross section of 100 ft width. The building parameters are $m = 13.42 \times 10^3$ lb.sec^2/ft^2, and $\omega_1 = 1.228$ rps. The tendons are arranged as in Figure 1 [4] which provide a passive stiffness natural frequency, ω'_1, of 0.890 rps. The equation of motion of the first mode reads

$$\ddot{A} + 2\varsigma_1\omega_1\dot{A} + \omega_1^2 A(t) = \frac{F(t)}{mL} - \frac{1.32}{mL}U(t) - \omega_1'^2 A(t), \tag{1}$$

in which $\varsigma_1 = 0.01$, and $U(t)$ is the active control force.

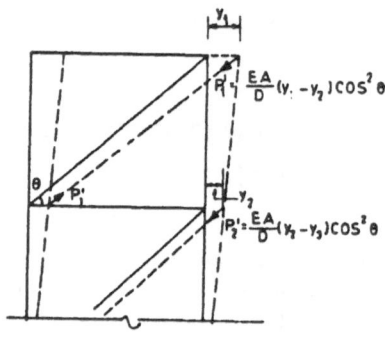

PASSIVE CONTROL FORCES

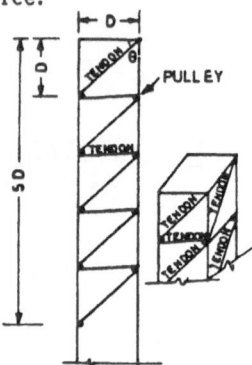

CONTROL MECHANISM

Figure 1: Active tendon control mechanism

In order to have closed-loop eigenvalues of (-0.38±J1.5175) for the first mode, the following gain matrix must be used [16]:

$$\mathbf{K} = 10^6[1.466 \; 7.477] . \tag{2}$$

The wind force $F(t)$ has been calculated from

$$F(t) = \int_0^L F_1(x,t)\phi_1(x)dx, \tag{3}$$

in which $\phi_1(x) = $ first mode shape, and $F_1(x,t)$ is given by

$$F_1(x,t) = 0.5\rho BC_D U^2(x,t), \tag{4}$$

in which $\rho = $ air density (0.076 pcf), $B = $ building width (100 ft), $C_D = $ Drag coefficient (1.3), and $U(x,t) = $ wind speed.

The wind speed has been modeled as [9,10]:

$$U(x,t) = \overline{U}(x) + u(t), \tag{5}$$

in which $\overline{U}(x) = $ mean wind speed at height x.

The fluctuating wind speed has been calculated as follows:

$$u(t) = \sqrt{2} \sum_{i=1}^{25} [2S_u(\omega_i) \Delta\omega]^{1/2} \cos(\omega_i t + \phi_i),$$
(6)

in which,

$$\omega_i = (i - 0.5)\frac{40}{25},$$
(7)

$$S_u(\omega_i) = \frac{2K\phi^2 |\omega_i|}{\pi^2 \left[1 + \left(\frac{\phi\omega_i}{\pi U_o}\right)^2\right]^{4/3}},$$
(8)

in which K = surface drag coefficient (0.04); ϕ = scale turbulence (4000 ft), U_o = mean wind speed at height 33 ft from ground, and ϕ_i = random phase angle uniformly distributed between zero and 2π.

Considering U_o = 66 fps, and according to the logarithmic law, the mean wind speed is found from

$$\bar{U}(x) = 66 Ln\left(\frac{x}{x_i}\right) / Ln\left(\frac{33}{x_o}\right),$$
(9)

in which x_o = roughness of the terrain (0.8 ft).

Substituting Eqs. (5)-(9) into Eqs. (3) and (4), the generalized wind force can be found. This force divided by (mL) is shown in Fig. 2 for a 24 seconds period.

The active control force $U(t)$ generated by the tendons is proportional with the magnitude of the actuator displacement. The control force can be expressed by

$$U(t) = \frac{y(t)}{L} EA,$$
(10)

in which E = Young's Modulus of tendons, A = cross section areas of tendons, $y(t)$ = elongation of tendons caused by actuators, and L = tendon length.

The rate of flow of the fluid in and out of the actuators is obtained from

$$Q(t) = \dot{y}(t) A_p,$$
(11)

in which A_p = cross section area of the pistons in the actuator, and $\dot{y}(t)$ = rate of flow per unit piston area.

As an index for the rate of flow per unit area one may use

$$J_1 = \int_0^T \dot{y}^2 dt = \int_0^T \dot{U}^2(t)\left(\frac{L}{EA}\right)^2 dt,$$
(12)

in which $U(t)$ is the active control force obtained from

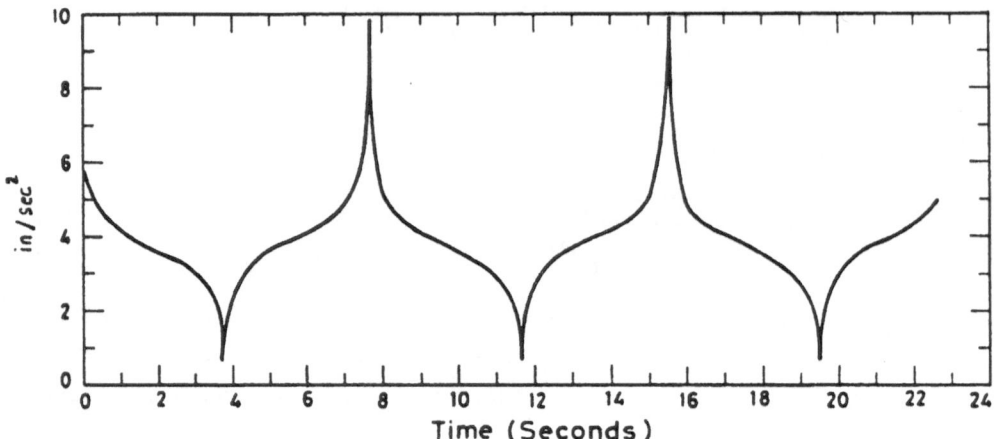

Figure 2: Generalized wind forces

$$U(t) = \mathbf{K}[A(t) \quad \dot{A}(t)]^T. \tag{13}$$

Considering the tendons cross section areas to be 1 ft^2, and its total length is 1100 ft, the rate of flow per unit piston area was calculated and plotted in Fig. 3, and the index J_1 is plotted in Fig. 4.

To facilitate the comparison between the various control schemes, the following indices are used, respectively, for deflection response of building and for the control force response:

$$J_D = \int_0^T A^2(t)\,dt, \tag{14}$$

$$J_U = \int_0^T U^2(t)\,dt. \tag{15}$$

The building deflection response, J_D, is shown in Fig. 5, and the necessary control force response, J_U, is shown in Fig. 6, using the notation of "Active Tendon 1". Another design denoted by "Active Tendon 2" is also shown in Figs. 3-6 in which the gain matrix is designed to provide eigenvalues of (-0.19 $\pm J$ 1.505) for the first mode. The gain matrix in this case is

$$\mathbf{K} = [-6.979 \times 10^2 \quad 3.614 \times 10^6]. \tag{16}$$

The control force response is shown in Fig. 7.

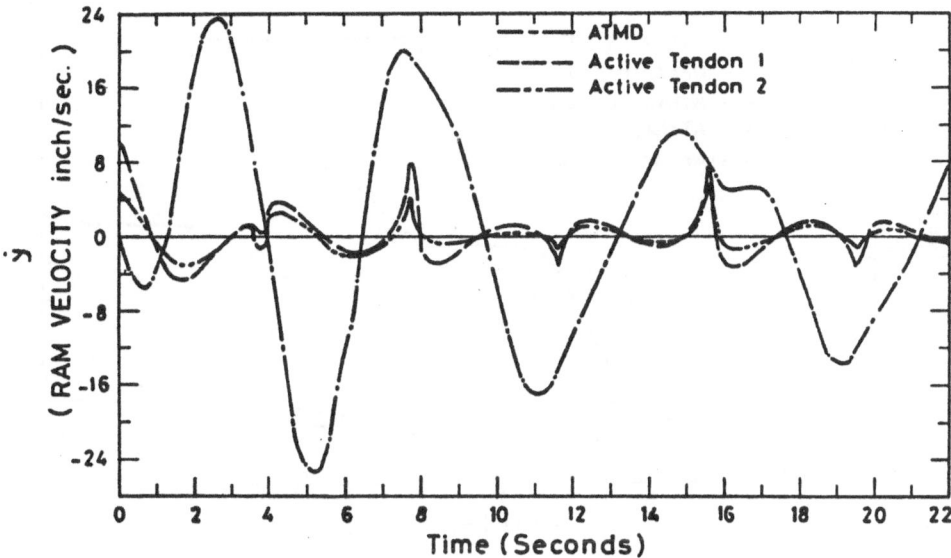

Figure 3: Actuators response

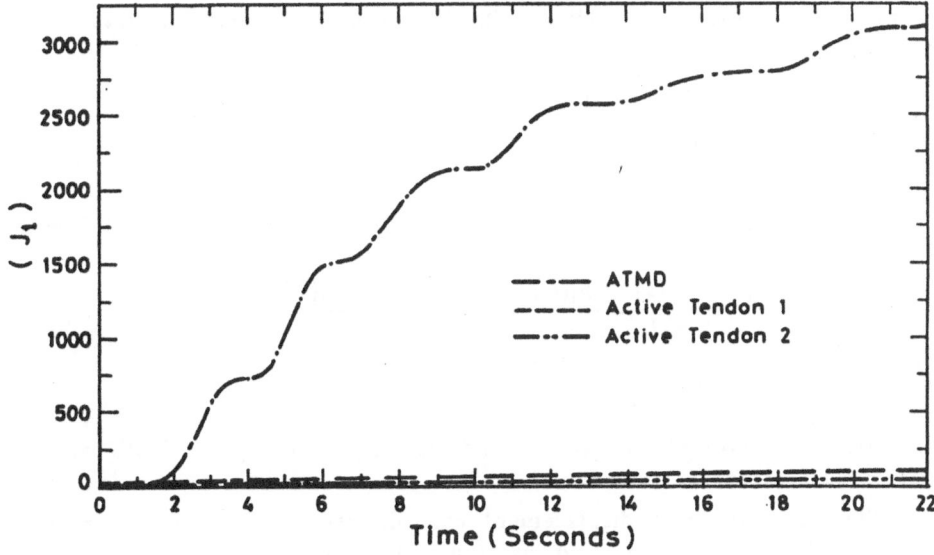

Figure 4: Rate of flow index

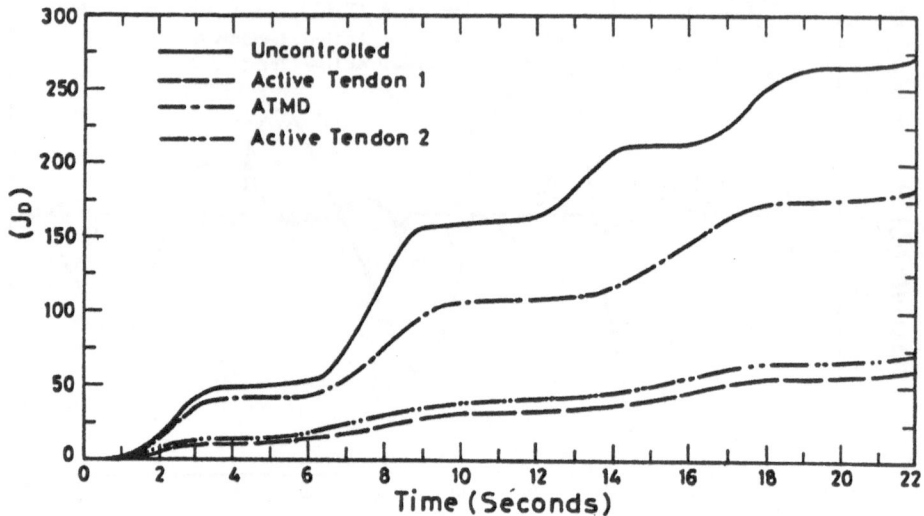

Figure 5: Deflection index

3. Feasibility of Active TMD Control

In this case, the actuator movements are represented by the relative sway between the damper and building. Thus, the ram displacement is obtained as in Fig. 8 from

$$y(t) = WT(t) - A(t)\phi_1(L), \qquad (17)$$

in which $WT(t)$ is the damper displacement, $A(t)$ is the coordinate of the first mode, and $\phi_1(L)$ is the first mode shape at height L.

The equations of motion of the first mode of the building and the damper are:

$$\ddot{A} + 2\varsigma_1\omega_1\dot{A} + \omega_1^2 A = \frac{F_1(t)}{mL} + \frac{\phi_1(L)}{mL}(C_2\dot{Z} + K_2 Z) - \frac{\phi_1(L)}{mL}U(t), \qquad (18)$$

$$\ddot{Z} + 2\varsigma_2\omega_2\dot{Z} + \omega_2^2 Z = -\ddot{A}\phi_1(L) + \frac{U(t)}{M_2}, \qquad (19)$$

in which $Z(t) = WT(t) - A(t)\phi_1(L)$, and M_2, C_2, and K_2 are TMD structural parameters.

Therefore, the ram displacement can, directly, be determined from $Z(t)$. The rate of flow of fluid per unit piston area is thus

$$\dot{y}(t) = \dot{Z}(t). \qquad (20)$$

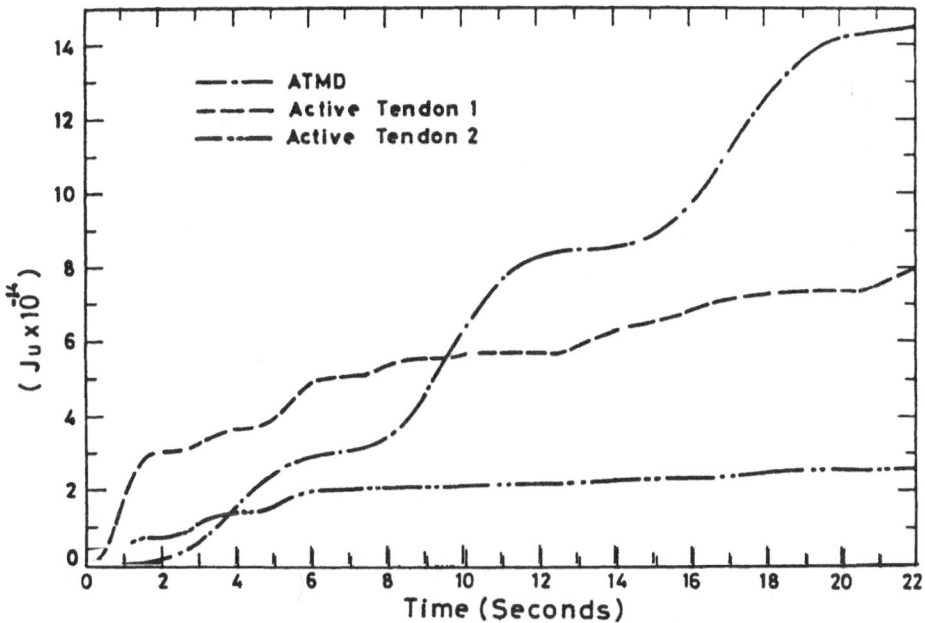

Figure 6: Control forces index

The tuned mass structural parameters were designed considering the variations due to the feedback control force [14]. The optimal design has provided
$M_2 = 0.02$ mL, $\omega_2 = 1.375$ rps and $\varsigma_2 = 0.07$. The eigenvalues of the passive controlled building are $(-0.068 \pm J \ 1.392)$ and $(-0.035 \pm J \ 1.06)$. The control force
was designed to provide eigenvalues of $(-0.35 \pm J \ 1.392)$ and $(-0.35 \pm J \ 1.06)$.
The necessary gain matrix to do that is:

$$\mathbf{K} = 10^6 [2.769 \quad -1.69 \quad 0.0505 \quad -0.353]. \tag{21}$$

For the same wind forces considered in the previous section, the rate of flow
of the fluid in the actuator per unit piston area is shown in Fig. 3, and the flow
index is given in Fig. 4. Figures 5-7 show the building deflection index, control
force index and the control force response, respectively.

It is obvious that active TMD was not feasible and not effective for forced
vibration control [13] as it consumes more energy than the tendon mechanism but
does not provide an acceptable control. This is mainly attributed to the lack of
stiffness in the building due to the coupling with the tuned mass damper [4].

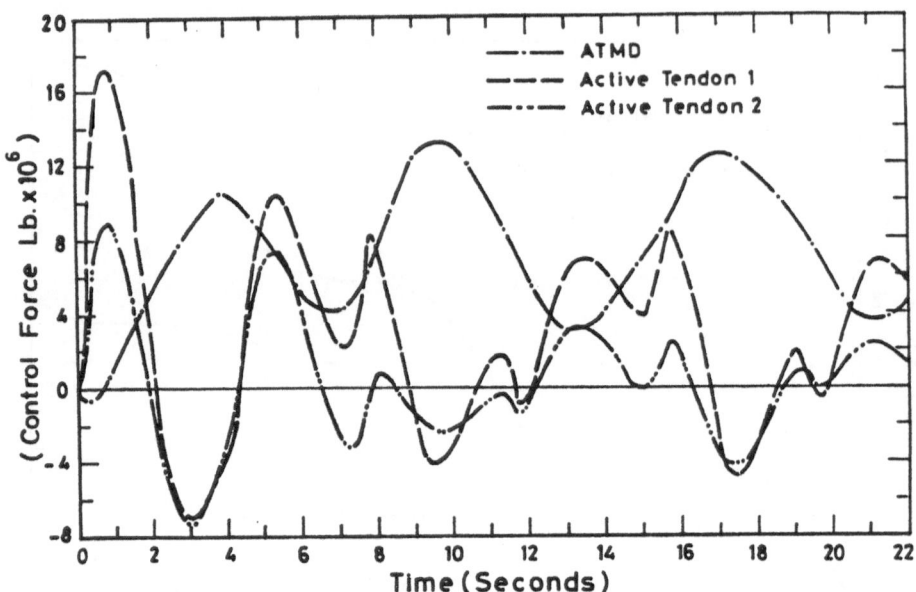

Figure 7: Response of control forces

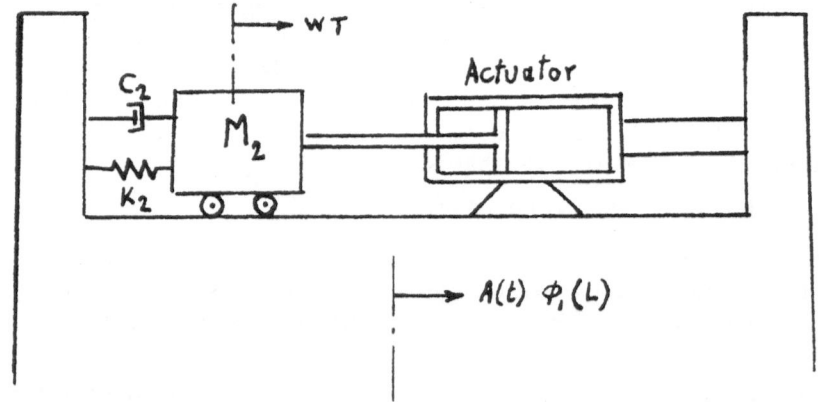

Figure 8: Active tuned mass damper mechanism

4. Feasibility of Active Control by Appendages

One main disadvantage in using appendages is that the effective control occurs only when the appendage is perpendicular to the mean wind direction, which is not permanent. in this section, the feasibility is studied assuming the appendage is always perpendicular to the mean wind direction.

The appendage used is a steel plate of dimensions $100 \times 40 \times (2/12)$ feet. The appendage is hinged along one building side and can be raised up or down using compression members and actuators [11], as shown in Fig. 9.

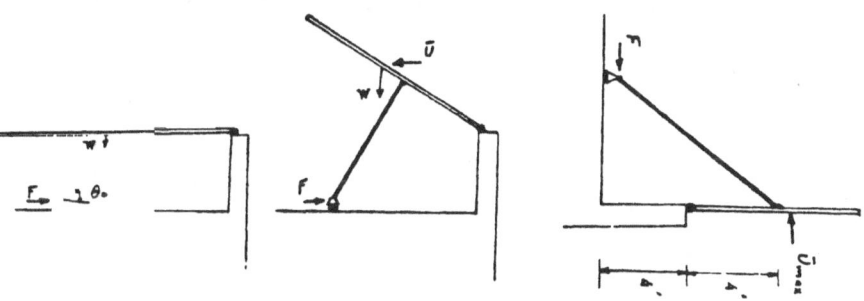

Figure 9: Aerodynamic Mechanism

The equation of motion of the building controlled by an aerodynamic appendage is given by

$$EI\frac{\partial^4 W}{\partial x^4} + C\frac{\partial W}{\partial t} + m\frac{\partial^2 W}{\partial t^2} = F_1(x,t) + F_2(L,t)U(t)\delta(x-L), \qquad (22)$$

in which $U(t) =$ deployment coefficient equals unity when the appendage is fully deployed and zero when the appendage is fully folded; $\delta(x-L) =$ Dirac delta function; and $F_2(L,t)$ is the wind force on the appendage given by

$$F_2(L,t) = 0.5\rho A_p C_D' U^2(L,t), \qquad (23)$$

where A_p is the windward area of the appendage (40×100); $C_D' =$ appendage drag coefficient (1.3), and $U(L,t)$ is the wind speed at height L.

The deployment coefficient $U(t)$ was designed to minimize the following performance index;

$$J = \frac{1}{2} \int_0^\infty (Q_{11}A^2 + Q_{22}\dot{A}^2 + \frac{U^2}{\xi})\,dt, \tag{24}$$

in which $Q_{11} = Q_{22} = 0.2 \times 10^5$ and $\xi = 0.05 \times 10^{-4}$ were chosen.

In order to determine the actuator force which operates the appendage, one has the following relations from Fig. 10:

$$\text{Sin}\,\theta' = 2.5\sin\theta - 1, \tag{25}$$

$$\bar{U}(t) = \bar{U}_{max}(t)\sin\theta', \tag{26}$$

in which θ' is the angle made by the appendage with the horizontal; θ is the angle made by the link attached to the actuator with the horizontal; $\bar{U}_{max}(t)$ is the maximum wind force generated on fully deployed appendage; and $\bar{U}(t)$ is the wind force generated on the appendage in the general operating condition.

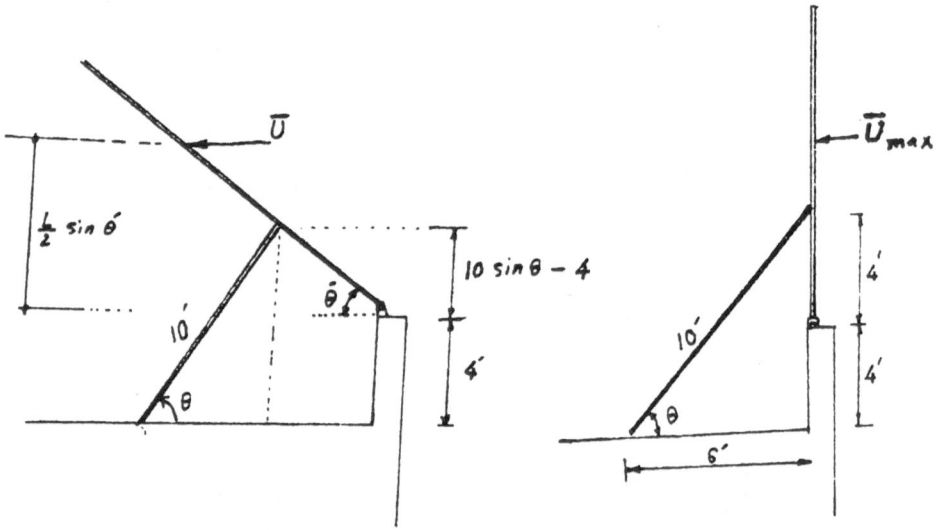

Figure 10: Dimensions of appendage mechanism

The equation of motion of the appendage can be derived using Fig. 11 as:

$$J_A\ddot{\theta}' + \bar{U}\frac{L_p}{2}\sin\theta' + W\frac{L_p}{2}\cos\theta' - 4F'\sin(\theta + \theta') = 0, \tag{27}$$

in which J_A = mass moment of inertia about A; L_p = appendage length (40 ft); W = appendage weight; and F' is the compression force as a result of the actuator force F.

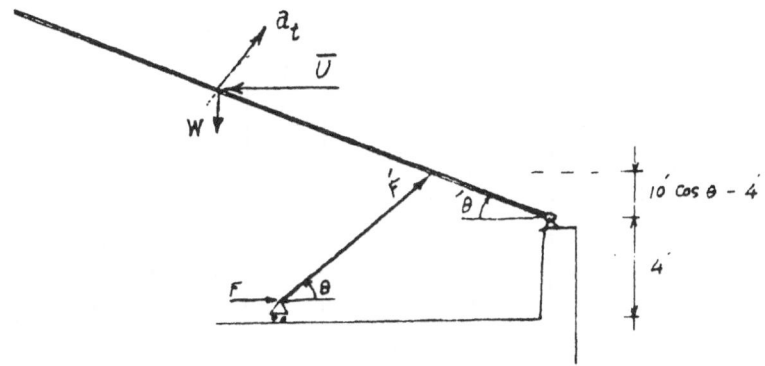

Figure 11: General operating condition

It is obvious that the angle θ' can directly be determined from Eq. (26) using the deployment coefficient U. In other words, $\sin\theta' = U(t)$, and the force F' can be found using Eq. (27). However, when $U = 0$ (i.e. $\theta' = 0$), the force F' is given by

$$F' = \frac{W(L_p/2)}{4\sin\theta}, \quad (\theta' = 0), \tag{28}$$

when $U = 1$ (i.e. $\theta' = 90°$), the force F' would be:

$$F' = \frac{\bar{U}_{max}(L_p/2)}{4\cos\theta}, \quad (\theta' = 90°). \tag{29}$$

The actuator force in any case is given by

$$F = \frac{F'}{\cos\theta}. \tag{30}$$

The control force index shall then be

$$J_U = \int_0^T F^2(t)\,dt. \tag{31}$$

For the same building parameters and wind data considered in the previous sections, the response of the actuator force F, the deflection index J_D, and the control index J_U are respectively plotted in Figs. 12-14. It is observed that the maximum actuator force would be 616000 kips which is much higher than the active tendon and ATMD mechanisms. However, these forces are applied for very small intervals of time. This indicates that the consumed energy is less than the other two mechanisms. But, the feasibility of this mechanism is still questionable as the reduction in the building response was not great. The actuator response is shown in Fig. 15.

M. Abdel-Rohman

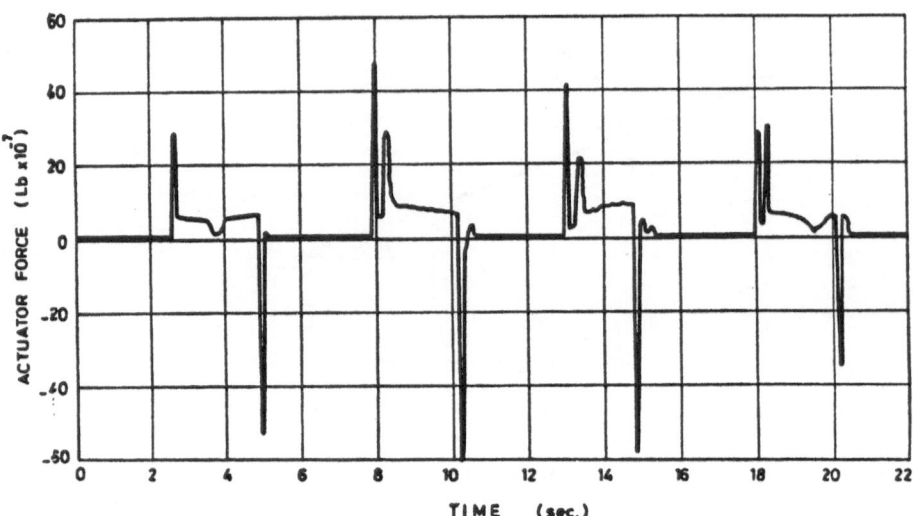

Figure 12:   Response of actuator force

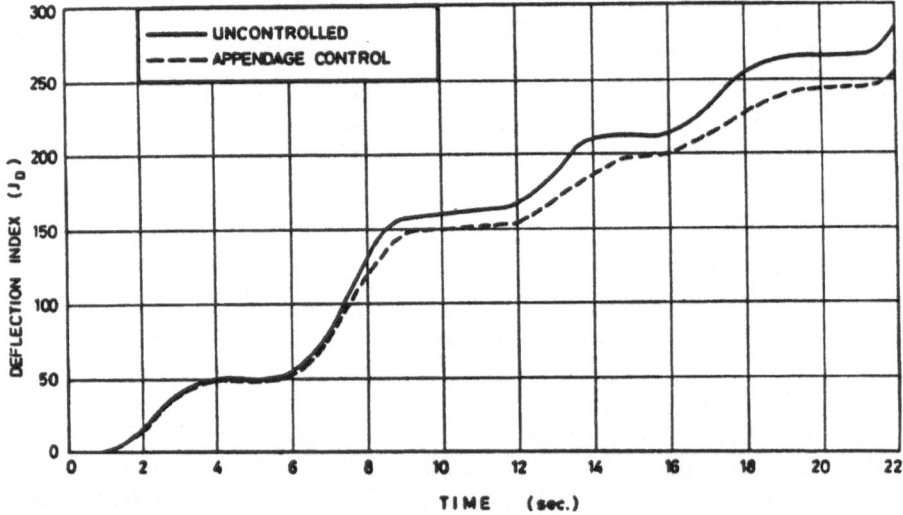

Figure 13:   Deflection index

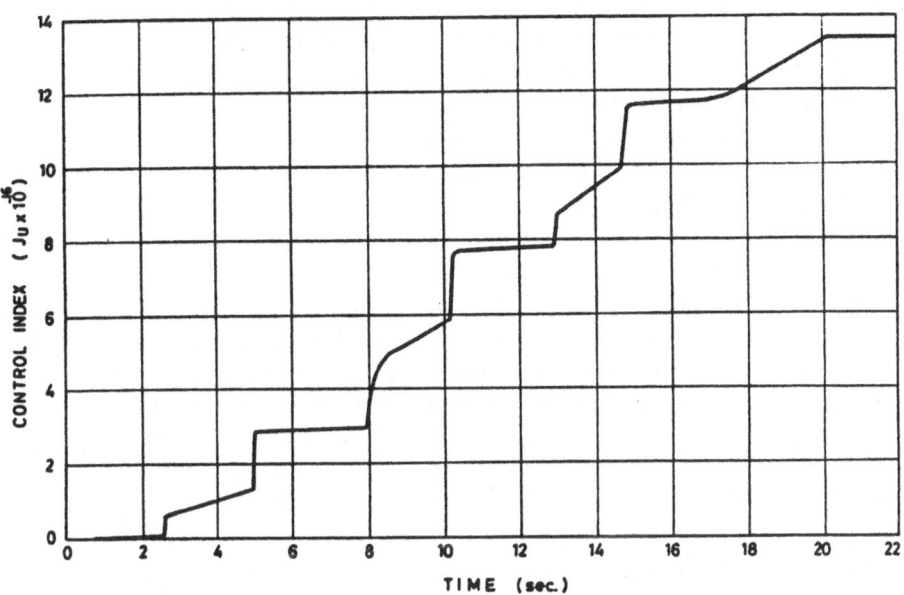

Figure 14: Control force index

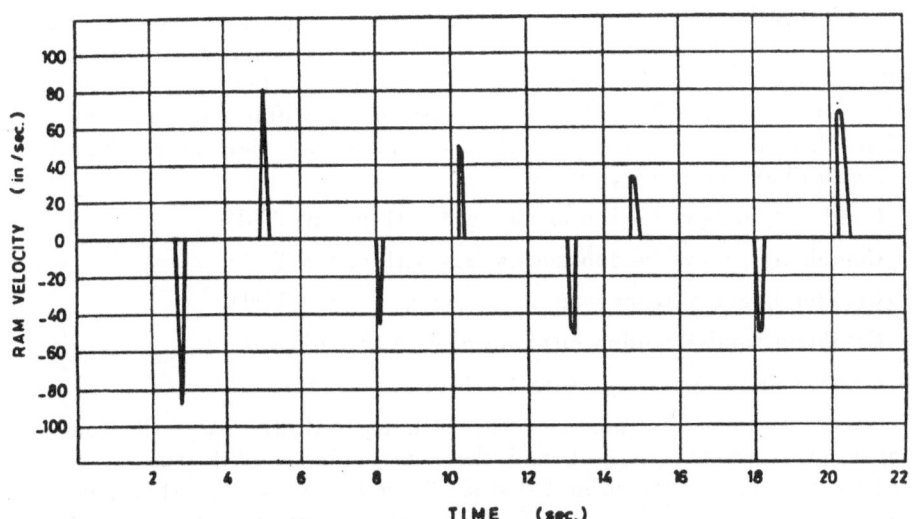

Figure 15: Actuator response

5. Determination of a Practical and Feasible Mechanism

The results which have been shown in the previous sections and the results obtained for other aspects of feasibility [12-15] can be used in proposing a practical mechanism for tall buildings.

The active tendon mechanism was very efficient in providing the active damping and active stiffness for the building. However, it was sensitive for time delay effect. It has been shown in [12] that if the time delay is taken into consideration then the consumed control energy would almost be the same as when the delay is neglected.

The active tuned mass damper was only effective for free vibration control [13] because it introduces active damping to the building. The active TMD was not effective for forced vibration control because of the lack of stiffness. The tuned mass parameters are influenced by the feedback control force and one has to design these parameters according to the desired degree of control [14]. On the other hand, active TMD was not sensitive for time delay effect [15], but the actuators need long strokes and huge fluids for their operation.

The aerodynamic appendage is only effective when the mean wind direction is perpendicular to the appendage, and when the building is very flexible. Its feasibility, however, was questionable as it needs very large actuator force for very small intervals of time.

From these results one may investigate the posssibility of combining any of the suggested control mechanisms in order to obtain a feasible and practical one able to compensate the present disadvantages by the advantages of the other mechanisms. For example, when combining a passive tendon with an active TMD one would expect a compensation for the lack in the building stiffness due to the coupling with ATMD by the increase in the stiffness provided by passive tendons, keeping the active control force the same. Several combinations of mechanisms have been investigated as:

(1) Combining passive tendon mechanism with passive TMD

(2) Combining passive tendon mechanism with active TMD

(3) Combining active tendon mechanism with passive TMD

(4) Combining active tendon mechanism with active TMD

(5) Combining an appendage with 1,2,3, or 4.

The mean square response of the building deflection, J_D, the tuned mass damper displacement, (J_{TMDF}), and the control forces (J_U) using several combined mechanisms are shown in Table 1. These results indicate that using a combined passive tendon and passive TMD is very economical but on the account of violent TMD response. A better mechanism is a combined active tendon and active TMD. If this mechanism is designed properly one would expect significant reduction in both the building and TMD responses, as in case 9 of Table 1.

Table 1: *Control mechanism*

Case		J_D	J_{TMDF}	J_U
Uncontrolled	1	265.0		
Uncontrolled + App.	2	239.99		0.56×10^{17}
Passive Tendon + Passive TMD	3	77.45	546.4	
Passive Tendon + Passive TMD + App.	4	74.87	451.9	0.8398×10^{14}
Passive Tendon + ATMD	5	69.51	4258.0	0.5287×10^{15}
Active Tendon + Passive TMD	6	50.7	179.6	0.6932×10^{15}
Active Tendon + Passive TMD + App.	7	51.01	158.3	0.7337×10^{15}
Active Tendon + ATMD(Addition)	8	49.34	3039	0.1064×10^{16}
Similar to Active Tendon + ATMD	9	61.44	105.2	0.2118×10^{15}
Similar to Active Tendon + ATMD + App.	10	61.47	99.1	0.2379×10^{15}
Similar to Passive Tendon + ATMD	11	71.23	600.2	0.324×10^{14}
Similar to Passive Tendon + ATMD'	12	72.42	299.2	0.1653×10^{14}

In general, by using a combined control mechanism one is able to get a better controlled response, consuming moderate control energy, and aleviating problems like time delay effect, feedback control effect on structural parameters, and uneffectiveness during forced vibration response.

Acknowledgement

The work of this paper is a part of a project supported by Kuwait Foundation for the Advancement of Sciences under grant No. 81-08-02.

References

[1] Zuk, W. and Clark, R.H., *Kinetic Architecture*, Van Nostrand Reinhold Co., N.Y., 1970.

[2] Roorda, J., "Tendon control in tall structures," *Journal of the Structural Division, ASCE*, Vol. 101, ST3, March, 1975, pp. 505-521.

[3] Yang, J.N. and Giannopoulas, F., "Active Tendon Control of Structures," *Journal of the Engineering Mechanics Division*, ASCE, Vol. 104, EM3, June 1978, pp. 551-568.

[4] Abdel-Rohman, M. and Leipholz, H.H., "Active Control of Tall Buildings," *Journal of the Structural Division*, ASCE, Vol. 109, ST3, March 1983, pp. 628-645.

[5] Lund, R.A., "Active Damping of Large Structures in Winds," *Proceedings of IUTAM Symposium on Structural Control*, held at the University of Waterloo, June 4-7, 1979, pp. 459-471.

[6]   Chang, J.C. and Soong, T.T., "Structural Control Using Active Tuned Mass Dampers," *Journal of the Engineering Mechanics Division*, ASCE, Vol. 106, EM6, Dec. 1980, pp. 1091-1098.

[7]   Peterson, N.R., "Design of Large Scale Tuned Mass Dampers," *Proceedings of ASCE Convention and Exposition*, April 1978, Boston, Reprint No. 3578.

[8]   Klein, R.E., Cusano, C., and Slukel, J.V., "Investigation of Method to Stabilize Wind Induced Oscillations in Large Structures," ASME Winter Annual Meeting, Paper No. 72-Wt/AUT-11, N.Y., Nov. 1972.

[9]   Chang, J.C. and Soong, T.T., "The Use of Aerodynamic Appendages for Tall Building Control," *IUTAM Proceedings*, 1980, North-Holland Publishing Co. & SM Publications, pp. 199-210.

[10] Soong, T.T. and Skinner, G.T., "Experimental Study of Active Structural Control," *Journal of the Engineering Mechanics Division, ASCE*, Vol. 107, EM6, Dec. 1981, pp. 1057-1068.

[11] Abdel-Rohman, M., "Optimal Control of Tall Buildings by Appendages," *Journal of the Structural Engineering, ASCE*, Vol. 110, May 1984, pp. 937-947.

[12] Abdel-Rohman, M., "Structural Control Considering Time Delay Effect," appears in *Transaction of CSME*.

[13] Abdel-Rohman, M., "Effectiveness of Active TMD in Buildings Control," *Transaction of CSME*, Vol. 8, No. 4, Dec. 1984, pp. 179-184.

[14] Abdel-Rohman, M., "Design of Active TMD for Tall Building Control," *International Journal of Buildings and Environment*, Pergamon Press, Vol. 19, No. 3, 1984, pp. 191-195.

[15] Abdel-Rohman, M., "Control of Tall Buildings Response Against Wind Forces," Final Report, Project KFAS 81-08-02, Kuwait Foundation for Advancement of Sciences, Kuwait.

[16] Abdel-Rohman, M. and Leipholz, H.H.E., "Structural Control by the Pole Assignment Method," *Journal of the Engineering Mechanics Division, ASCE*, Vol. 104, EM5, Oct. 1978, pp. 1159-1175.

A CONTROLLABILITY-STABILIZABILITY RESULT FOR THE NASA-IEEE SPACECRAFT CONTROL LABORATORY EXPERIMENT (S.C.O.L.E) CONFIGURATION

R. Araya
Dept. of Electrical Engineering
UCLA, LA
CA 90024

1. Introduction

A Spacecraft Control Laboratory Experiment (SCOLE) has been proposed in reference [9] to evaluate control laws for flexible spacecrafts. This configuration consists of a large antenna attached to the Space Shuttle orbiter by a flexible beam. A non-linear distributed parameter mathematical model is offered in reference [9]. A slightly different version of that model is adopted here. We follow the formulation of references [5] and [3], where elastic displacements are given relative to coordinate axes fixed to the shuttle body. Another difference is that we assume no inherent structural damping (see reference [5] for a discussion about the importance of this point). Thus the goal is to determine if the control system can be designed in such a way that the desired behavior can be obtained without having to rely on the system damping.

The main contribution of this paper is the controllability-stabilizability result for the SCOLE configuration. A rate-plus-position feedback control is proposed for this purpose. This linear control law points the line of sight of the reflector to any given target direction and at the same time stabilizes, in the strong sense, the oscillations of the entire flexible configuration.

17

2. Equations of Motion

We consider in this paper the case when the antenna is attached to the mast at its center of gravity.

Three hyperbolic partial differential equations describe the dynamics of the system. Non-linear terms and coupling between these equations are introduced at the boundary conditions.

$$
I_4(\overline{\omega} + \begin{bmatrix} -\dfrac{\partial^3 u^\phi}{\partial t^2 \partial \xi}(t,-L) \\[2mm] \dfrac{\partial^3 u^\theta}{\partial t^2 \partial \xi}(t,-L) \\[2mm] \dfrac{\partial^2 u_\psi}{\partial t^2}(t,-L) \end{bmatrix}) + (\overline{\omega} + \begin{bmatrix} -\dfrac{\partial^2 u^\phi}{\partial t \partial \xi}(t,-L) \\[2mm] \dfrac{\partial^2 u^\theta}{\partial t \partial \xi}(t,-L) \\[2mm] \dfrac{\partial u_\psi}{\partial t}(t,-L) \end{bmatrix}) \times I_4(\overline{\omega} + \begin{bmatrix} \dfrac{\partial^2 u^\phi}{\partial t \partial \xi}(t,-L) \\[2mm] \dfrac{\partial^2 u^\theta}{\partial t \partial \xi}(t,-L) \\[2mm] \dfrac{\partial u_\psi}{\partial t}(t,-L) \end{bmatrix})
$$

$$
+ \; I'\overline{\omega} + \overline{\omega} \times I'\overline{\omega} = \overline{M}_1 + \overline{M}_4 + \overline{\rho} \times \begin{bmatrix} F_\theta \\ F_\phi \\ 0 \end{bmatrix}, \quad t \geq 0, \quad (I' \equiv I - I_4, \rho \equiv \begin{bmatrix} 0 \\ 0 \\ -L \end{bmatrix}), \quad (1)
$$

$$
P\frac{\partial^2 u^\theta}{\partial t^2}(t,\xi) + EI\frac{\partial^4 u^\theta}{\partial \xi^4}(t,\xi) = 0, \qquad t \geq 0, \; \xi \epsilon [-L,0], \tag{2}
$$

$$
P\frac{\partial^2 u^\phi}{\partial t^2}(t,\xi) + EI\frac{\partial^4 u^\phi}{\partial \xi^4}(t,\xi) = 0, \qquad t \geq 0, \; \xi \epsilon [-L,0], \tag{3}
$$

$$
P\frac{\partial^2 u_\psi}{\partial t^2}(t,\xi) - G\frac{\partial^2 u_\psi}{\partial \xi^2}(t,\xi) = 0, \qquad t \geq 0, \xi \epsilon [-L,0], \tag{4}
$$

$$
m_4 \begin{bmatrix} \dfrac{\partial^2 u^\theta}{\partial t^2}(t,-L) \\[2mm] \dfrac{\partial^2 u^\phi}{\partial t^2}(t,-L) \end{bmatrix} + \begin{bmatrix} EI\dfrac{\partial^3 u^\theta}{\partial \xi^3}(t,-L) \\[2mm] EI\dfrac{\partial^3 u^\phi}{\partial \xi^3}(t,-L) \end{bmatrix} = \begin{bmatrix} F_\theta \\ F_\phi \end{bmatrix}, \qquad t \geq 0, \tag{5}
$$

$$
I_4(\overline{\omega} + \begin{bmatrix} \dfrac{-\partial^3 u^\phi}{\partial \xi \partial t^2}(t,-L) \\[2mm] \dfrac{\partial^3 u^\theta}{\partial t^2 \partial \xi}(t,-L) \\[2mm] \dfrac{\partial^2 u_\psi}{\partial t^2}(t,-L) \end{bmatrix}) + (\overline{\omega} + \begin{bmatrix} -\dfrac{\partial^2 u^\phi}{\partial t \partial \xi}(t,-L) \\[2mm] \dfrac{\partial^2 u^\theta}{\partial t \partial \xi}(t,-L) \\[2mm] \dfrac{\partial u_\psi}{\partial t}(t,-L) \end{bmatrix}) \times
$$

$$I_4\left(\overline{\omega}+\begin{bmatrix}\dfrac{-\partial^2 u^\phi}{\partial t\partial\xi}(t,-L)\\[2mm]\dfrac{\partial^2 u^\theta}{\partial t\partial\xi}(t,-L)\\[2mm]\dfrac{\partial u_\psi}{\partial t}(t,-L)\end{bmatrix}\right)+\begin{bmatrix}EI\dfrac{\partial^2 u^\phi}{\partial\xi^2}(t,-L)\\[2mm]-EI\dfrac{\partial^2 u^\theta}{\partial\xi^2}(t,-L)\\[2mm]-EI\dfrac{\partial u_\psi}{\partial\xi}(t,-L)\end{bmatrix}=M_4,\qquad t\geq 0,\tag{6}$$

$$u^\theta(t,0)=u^\phi(t,0)=\frac{\partial u^\theta}{\partial\xi}(t,0)=\frac{\partial u^\phi}{\partial\xi}(t,0)=u_\psi(t,0)=0,\quad t\geq 0.\tag{7}$$

Define $\overline{D}$ as the difference between the line of sight $\overline{R}$ and the target direction $\overline{S}$,

$$\overline{D}=-\overline{\omega}\times\overline{D}+\overline{\omega}\times\overline{R}.\tag{8}$$

(See reference [9] for a description of the symbols).

Now, we develop Liapunov's approach.

If we form:

$$\overline{\omega}\cdot(1)+\int_{-L}^0 (2)\frac{\partial u^\theta}{\partial t}+(3)\frac{\partial u^\phi}{\partial t}+(4)\frac{\partial u_\psi}{\partial t}d\xi+\lambda(8)\cdot\overline{D},$$

we get

$$\frac{d}{dt}(T(t)+V(t)+\lambda D(t))=\overline{\omega}\cdot(\overline{M}_1-\lambda\overline{D}\times\overline{R})+\left(\overline{\omega}+\begin{bmatrix}\dfrac{-\partial^2 u^\phi}{\partial t\partial\xi}(t,-L)\\[2mm]\dfrac{\partial^2 u^\theta}{\partial t\partial\xi}(t,-L)\\[2mm]\dfrac{\partial u_\psi}{\partial t}(t,-L)\end{bmatrix}\right)\cdot\overline{M}_4$$

$$+\left(\begin{bmatrix}\dfrac{\partial u^\theta}{\partial t}(t,-L)\\[2mm]\dfrac{\partial u^\phi}{\partial t}(t,-L)\\[2mm]0\end{bmatrix}+\overline{\omega}\times\overline{\rho}\right)\cdot\begin{bmatrix}F_\theta\\ F_\phi\\ 0\end{bmatrix},\tag{9}$$

where

$$T(t)=\overline{\omega}\cdot I'\overline{\omega}+(\overline{\omega}+\begin{bmatrix}-\dfrac{\partial^2 u^\phi}{\partial t\partial\xi}(t,-L)\\[2mm]\dfrac{\partial^2 u^\theta}{\partial t\partial\xi}(t,-L)\\[2mm]\dfrac{\partial u_\psi}{\partial t}(t,-L)\end{bmatrix})\cdot I_4(\overline{\omega}+\begin{bmatrix}-\dfrac{\partial^2 u^\phi}{\partial t\partial\xi}(t,-L)\\[2mm]\dfrac{\partial^2 u^\theta}{\partial t\partial\xi}(t,-L)\\[2mm]\dfrac{\partial u_\psi}{\partial t}(t,-L)\end{bmatrix})$$

$$+\ P\int_{-L}^{0}[\frac{\partial u^\theta}{\partial t}(t,\xi)]^2+[\frac{\partial u^\phi}{\partial t}(t,\xi)]^2+I_\psi[\frac{\partial u_\psi}{\partial t}(t,\xi)]^2 d\xi$$

$$+\ m_4[\frac{\partial u^\theta}{\partial t}(t,-L)]^2+m_4[\frac{\partial u^\phi}{\partial t}(t,-L)]^2$$

=Kinetic energy of the entire configuration,

$$V(t)=\int_{-L}^{0}EI[\frac{\partial^2 u^\theta}{\partial\xi^2}(t,\xi)]^2+EI[\frac{\partial^2 u^\phi}{\partial\xi^2}(t,\xi)]^2+GI[\frac{\partial u_\psi}{\partial\xi}(t,\xi)]^2 d\xi$$

=Potential energy of deformation,

$$\overline{D}(t)=\overline{D}(t)\cdot\overline{D}(t)$$

=square of the distance between the target direction

and the actual line of sight direction.

$\lambda D(t)+T(t)+V(t)$ is then a Liapunov functional.

Define:

$$\overline{r}(t,\xi)=\begin{bmatrix}u^\theta(t,\xi)\\u^\phi(t,\xi)\end{bmatrix},$$

$$\overline{r}_4(t)=\overline{r}(t,-L),$$

$$\overline{\alpha}_4(t)=\begin{bmatrix}-\dfrac{\partial u^\phi}{\partial\xi}(t,-L)\\[2mm]\dfrac{\partial u^\theta}{\partial\xi}(t,-L)\\[2mm]u_\psi(t,-L)\end{bmatrix}.$$

Note that $\dot{\overline{\alpha}}_4$ is the angular velocity of the antenna with respect to the Shuttle orbiter, $\overline{\omega}_4\underline{\triangle}\ \overline{\omega}+\dot{\overline{\alpha}}_4$ is the angular velocity of the antenna and $v_4=\dot{\overline{r}}_4+\overline{\omega}\times\overline{\rho}$ is the interial velocity of the antenna.

We write Equation (9) as:

$$\frac{d}{dt}(D(t)+T(t)+V(t))=\overline{\omega}\cdot(\overline{M}_1-\lambda\overline{D}\times\overline{R})+\overline{\omega}_4\cdot\overline{M}_4+\overline{v}_4\cdot\overline{F}. \tag{9'}$$

We propose control laws that make the right side of expression (9') negative. This procedure produces position plus rate feedback control laws that are familiar in attitude control for rigid spacecraft [10],[11]. For example:

$$\overline{M}_1=\lambda\overline{D}\times\overline{R}-K_1\overline{\omega},$$
$$\overline{M}_4=-K_4\overline{\omega}_4, \tag{10}$$
$$\overline{F}=-K\overline{v}_4,$$

where K_1, K_4, K are positive matrices.

The main result of this paper is the proof that the type of linear control law points the line of sight of the antenna to any given target direction and stabilizes the oscillations of the entire flexible configuration. That means, $D(t)$, $T(t)$ and $V(t)$ go to zero as t goes to infinite.

3. Hilbert Space Model

Define

$$D(A_o^o)\triangleq\mathbb{R}^3\times\{L_2(-L,0)\}^3\times\mathbb{R}^5$$

with the canonical inner product.

$$D(A_o)\triangleq\Big\{(\overline{D},u^\theta,u^\phi,u_\psi,r_1,r_2,\epsilon D(A_o^o),\text{ such that}$$

$$u^\theta,u^\phi\epsilon H^4(-L,0),$$

$$u_\psi\epsilon H^2(-L,0),$$

$$u^\theta(0)=u^\phi(0)=u_\psi(0)=\frac{\partial u^\theta}{\partial\xi}(0)=\frac{\partial u^\phi}{\partial\xi}(0)=0,$$

$$u^\theta(-L)=r_1,\qquad u^\phi(-L)=r_2,$$

$$\frac{\partial u^\theta}{\partial\xi}(-L)=\alpha_2,\quad -\frac{\partial u^\phi}{\partial\xi}(-L)=\alpha_1,\quad u_\psi(-L)=\alpha_3\Big\},$$

$$A_o \left\{ \begin{array}{c} \overline{D} \\ u^\theta \\ u^\phi \\ u_\psi \\ r_1 \\ r_2 \\ \alpha_1 \\ \alpha_2 \\ \alpha_3 \end{array} \right\} \triangleq \left\{ \begin{array}{c} \lambda\overline{D} \\[4pt] EI\dfrac{\partial^4 u^\theta}{\partial\xi^4} \\[8pt] EI\dfrac{\partial^4 u^\phi}{\partial\xi^4} \\[8pt] -GI\dfrac{\partial^2 u_\psi}{\partial\xi^2} \\[8pt] EI\dfrac{\partial^3 u^\theta}{\partial\xi^3}(-L) \\[8pt] EI\dfrac{\partial^3 u^\theta}{\partial\xi^3}(-L) \\[8pt] EI\dfrac{\partial^2 u^\theta}{\partial\xi^2}(-L) \\[8pt] -EI\dfrac{\partial^2 u^\theta}{\partial\xi^2}(-L) \\[8pt] -GI\dfrac{\partial u_\psi}{\partial\xi}(-L) \end{array} \right\}.$$

A_o is a closed linear operator from $D(A_o^o)$ into itself, with a dense domain, self-adjoint, positive, with compact inverse and for $x_2 \epsilon D(A_o)$ we have

$$[A_o x_2, x_2] = \lambda D + V$$

$$= \lambda \text{ (square distance) + potential energy of deformation.}$$

Define M as a linear operator from $D(A_o^o)$ into itself by:

$$M \triangleq \begin{bmatrix} I & 0 & 0 & 0 & I_4 \\ 0 & P & 0 & 0 & 0 \\ 0 & 0 & PI_\psi & 0 & 0 \\ 0 & 0 & 0 & m_4 & 0 \\ I_4 & 0 & 0 & 0 & I_4 \end{bmatrix}.$$

M is a linear bounded operator, self-adjoint and positive ,

$$[M x_2, x_2] = T = \text{ Kinetic Energy}.$$

Define

$$H \triangleq D(A_o^{1/2}) \times D(A_o^o),$$

with the inner product

$$\left[\begin{bmatrix} x_1 \\ x_2 \end{bmatrix}, \begin{bmatrix} y_1 \\ y_2 \end{bmatrix} \right] = [A_o^{1/2} x_1, A_o^{1/2} y_1] + [M x_2, y_2],$$

$$D(\widetilde{A}) \triangleq (A_o) \times D(A_o^{1/2}),$$

$$\widetilde{A} \triangleq \begin{bmatrix} 0 & I \\ -A_o & 0 \end{bmatrix}, \quad \overline{M} = \begin{bmatrix} I & 0 \\ 0 & M \end{bmatrix}, \quad u(t) = \begin{bmatrix} \overline{M}_1(t) \\ \overline{M}_4(t) \\ \overline{F}(t) \end{bmatrix} \epsilon H_u = \mathbb{R}^8,$$

$$\widetilde{K}(x) \triangleq \begin{bmatrix} -\overline{\omega} \times \overline{D} - \overline{\omega} \times \overline{R} - \overline{\omega} \\ 0 \\ 0 \\ 0 \\ 0 \\ -(\overline{\omega} + \overline{\alpha}_4) \times I_4(\overline{\omega} + \overline{\alpha}_4) + \lambda \overline{D} - \overline{\omega} \times I'\overline{\omega} \\ 0 \\ 0 \\ 0 \\ -(\overline{\omega} + \overline{\alpha}_4) \times I_4(\overline{\omega} + \overline{\alpha}_4) \end{bmatrix}, \quad B_u \triangleq \begin{bmatrix} 0 \\ 0 \\ 0 \\ 0 \\ 0 \\ \overline{M}_1 + \overline{M}_4 + \overline{\rho} \times \overline{F} \\ 0 \\ 0 \\ \overline{F} \\ \overline{M}_4 \end{bmatrix}.$$

Thus the equations of motion (1),...,(8) can be written as:

$$\widetilde{M}\dot{x} = \widetilde{A}x + \widetilde{K}(x) + Bu, \quad x(0) = x_0. \tag{11}$$

We will study the behavior of this system under the linear feedback control law:

$$\overline{M}_1 = \lambda \overline{D} \times R - K_1 \overline{\omega},$$
$$\overline{M}_4 = -K_4(\overline{\omega} + \overline{\alpha}_4), \tag{10}$$
$$\overline{F} = -K(\dot{\overline{r}}_4 + \overline{\omega} \times \overline{\rho}).$$

Therefore, we have an autonomus system:

$$\widetilde{M}x = \widetilde{A}x + J(x), \quad x(0) = x_0, \tag{12}$$

or

$$\dot{x} = \widetilde{M}^{-1}\widetilde{A}x + \widetilde{M}^{-1}J(x), \quad x(0) = x_0. \tag{12'}$$

$\widetilde{M}^{-1}\widetilde{A}$ is a closed linear operator in H with dense domain and compact inverse, and generates a continuous group $e^{\widetilde{M}^{-1}\widetilde{A}t}$ on H. $\widetilde{M}^{-1}J$ is a C^∞ non-linear operator in H. Then

$$x(t) = F(t)x_0$$

$$= e^{\widetilde{M}^{-1}\widetilde{A}t}x_0 + \int_0^t e^{\widetilde{M}^{-1}\widetilde{A}(t-s)}\widetilde{M}^{-1}\widetilde{J}(x(s))\,ds \qquad (13)$$

defines a unique local flow on H with $F(t)$ being C^∞ from H into H. Furthermore, if $x_0 \epsilon D(\widetilde{M}^{-1}\widetilde{A})$ then

$$x(t)\epsilon D(\widetilde{M}^{-1}\widetilde{A}) \qquad \forall\, t \geq 0,$$

and Equation (12') holds strongly.

Also note that

$$\|x(t)\|^2 = \lambda D(T) + V(t) + T(t),$$

therefore

$$\frac{d}{dt}\|x(t)\|^2 \leq 0,$$

then

$$\|F(t)x_0\|^2 = \|x(t)\|^2 \leq \|x_0\|^2 \quad \forall\, x_0 \epsilon D(\widetilde{M}^{-2}\widetilde{A}) \quad \forall\, t \geq 0.$$

This implies

$$\|x(t)\|^2 \leq \|x_0\|^2 \quad \forall\, x_0 \epsilon H, \qquad \forall\, t \geq 0,$$

then

$$t \rightarrow \|\widetilde{M}^{-1}J(x(s))\|$$

is integrable and we have:

Theorem 1: The flow $F(t)$ is global, C^∞. If $x_0 \epsilon D(\widetilde{M}^{-1}\widetilde{A})$ then

$$x(t) = F(t)x_0 \epsilon D(\widetilde{M}^{-1}\widetilde{A}), \quad \forall\, t \geq 0,$$

is the unique strong solution of (12').

4. Asymptotic Behavior

We study in this section the asymptotic behavior of the system. We begin by proving the precompactness of the positive orbits

$$\gamma^+(t)\underset{=}{\Delta} UF(t)x_0, \qquad t \geq 0,$$

for $x_0 \epsilon D(\widetilde{M}^{-1}\widetilde{A})$. To obtain that result we use the theory of non-linear semi-groups generated by quasi-accretive operators. We prove that the generator has compact resolvent.

Next we prove that the positive limit set belongs to $D(\widetilde{M}^{-1}\widetilde{A})$ when the initial condition belongs to this domain also.

Finally using La Salle's invariant principle and harmonic analysis we conclude that the positive limit set is reduced to 0.

Define α: $\mathbb{R} \to \mathbb{R}$, a C^{∞} function such that

$$\alpha(s) = \begin{cases} 1 & \text{for } s \leq s_1 \\ 0 & \text{for } s > s_2 \end{cases} \qquad s_2 > s_1.$$

$B : H \to H$.

$$B(x) \underline{\underline{\triangle}} \alpha(\|x\|^2) \widetilde{M}^{-1} J(x).$$

B is Lipschitz, C^{∞}, and $B(x) = \widetilde{M}^{-1} J(x)$ for $\|x\|^2 \leq s_1$

The operator

$$\widetilde{A}_B x \overset{\triangle}{=} -\widetilde{M}^{-1} \widetilde{A} - B(x),$$

with

$$D(\widetilde{A}_B) = D(\widetilde{M}^{-1} \widetilde{A})$$

is a closed quasi-accretive non-linear operator, that is $\exists \omega \in \mathbb{R}$ such that

$$\widetilde{A}_B + \omega I$$

is accretive.

Moreover, from the fact that

$$R(I - \lambda \widetilde{M}^{-1} \widetilde{A}) = H \quad \forall \lambda > 0,$$

we have

$$R(I + \lambda \widetilde{A}_B) = H$$

for all $\lambda > 0$ sufficiently small.

Then

$$S(t)x \underline{\underline{\triangle}} \lim_{n \to \infty} (I + t/n \widetilde{A}_B)^{-n} x$$

is well defined, uniformly in t belonging to bounded sets of $\mathbb{R}^+$ and is a strongly continuous non-linear semigroup satisfying

$$\|S(t)x - S(t)y\| \leq e^{\omega t} \|x - y\| \quad \forall t \in \mathbb{R}^+, \quad \forall x, y \in H$$

If

$$x_0 \in D(\widetilde{A}_B) = D(\widetilde{M}^{-1} \widetilde{A})$$

then,

$$S(t)x_0 \in D(\widetilde{M}^{-1} \widetilde{A})$$

and it is the unique strong solution of

$$\begin{cases} \dot{x}(t) = \widetilde{M}^{-1}\widetilde{A}x(t) + B(x(t)), \\ x(0) = X_0. \end{cases} \tag{14}$$

From the fact that $\widetilde{M}^{-1}\widetilde{A}$ has compact inverse we have that the generator $\widetilde{A}_B$ has compact resolvent also, that is,

$$J_\lambda \triangleq (I + \lambda\widetilde{A}_B)^{-1}$$

is compact for all $\lambda > 0$ sufficiently small.

Then if

$$x_0 \epsilon D(\widetilde{M}^{-1}\widetilde{A})$$

there exist $z \epsilon H$ such that $x_0 = J_\lambda z$, and

$$\underset{t>0}{U}S(t)x_0 = \underset{t>0}{U}\ S(t)J_\lambda z$$

$$= \underset{t>0}{U}\ J_\lambda S(t)z$$

$$= J_\lambda \underset{t>0}{U}S(t)z.$$

But

$$\underset{t>0}{U}S(t)z$$

is bounded, then

$$\underset{t>0}{U}S(t)x_0 = J_\lambda \underset{t>0}{U}S(t)z$$

is precompact.

If we take $s_1 > \|x_0\|$ then $x(t) = F(t)x_0$ is strong solution of (14) and threfore we have

$$F(t)x_0 = S(t)x_0.$$

Hence:

Proposition 2:   If $x_0 \epsilon D(\widetilde{M}^{-1}\widetilde{A})$ then the orbit $\lambda^+(x_0) = \underset{t>0}{U}F(t)x_0$ is precompact.

Define the positive limit set of x_0 by:

$$L^+(x_0) \triangleq \{z \epsilon H : F(t_n)x_0 \to z \text{ for some sequence } t_n \to +\infty\}.$$

Let $x_0 \epsilon D(\widetilde{M}^{-1}\widetilde{A})$, then $z \epsilon L^+(x_0)$ implies that there is a sequence

$$\{t_n\}\ ,\ t_n \to +\infty\ ,\ \text{such that}$$

$$S(t_n)x_0 \to z.$$

But

$$\widetilde{A}_B S(t)x = S(t)\widetilde{A}_B x \quad \forall\, t \geq 0, \quad \forall\, x \in D(\widetilde{A}_B).$$

Therefore

$$\widetilde{A}_B S(t_n)x_0 = S(t_n)\widetilde{A}_B x_0,$$

and then it is a bounded sequence.

Thus

$$\|M^{-1}A S(t_n)x_0\| \leq \|\widetilde{A}_B S(t_n)x_0\| + \|B(S(t_n)x_0\| < C \quad \forall\, n,$$

therefore

$$\widetilde{M}^{-1}\widetilde{A} S(t_n)x_0 \xrightarrow{H} v,$$

this means

$$[\widetilde{M}^{-1}\widetilde{A} S(t_n)x_0, y] \to [v, y] \quad \forall\, y \in H.$$

Let

$$y \in D(\widetilde{M}^{-1}\widetilde{A}) = D((\widetilde{M}^{-1}\widetilde{A})^*),$$

then

$$[S(t_n)x_0, \ (\widetilde{M}^{-1}\widetilde{A})^* y] \to [v, y],$$

and also we have

$$[S(t_n)x_0, \ (\widetilde{M}^{-1}\widetilde{A})^* y] \to [z, (\widetilde{M}^{-1}\widetilde{A})^* y].$$

Therefore

$$[z, -\widetilde{M}^{-1}\widetilde{A} y] = [v, y] \quad \forall\, y \in D(\widetilde{M}^{-1}\widetilde{A}),$$

and then

$$z \in D(\widetilde{M}^{-1}\widetilde{A}).$$

Therefore we have:

Proposition 3: The positive limit set of x_0, $L^+(x_0)$, is contained on $D(\widetilde{M}^{-1}\widetilde{A})$ for all $x_0 \in D(\widetilde{M}^{-1}\widetilde{A})$.

Let

$$z \in L^+(x_0) \ , \quad x_0 \in D(\widetilde{M}^{-1}\widetilde{A}),$$

$$\|z\|^2 = \lim_{t_n \to \infty} \|S(t_n)x_0\|$$

$$= \lim_{t \to \infty} \|S(t)x_0\|,$$

therefore

$$\|z(t)\| = \|S(t)z\| = \text{constant } \forall\, t \geq 0,$$

then

$$\frac{d}{dt}\|z(t)\|^2 = 0 \quad \forall\, t \geq 0,$$

and we have

$$\overline{\omega}(t) = 0\ ,\quad \dot{\overline{\alpha}}_4(t) = 0\ ,\quad \dot{\overline{r}}_4(t) = 0,\quad \forall\, t > 0,$$

therefore

$$\frac{\partial^2 u^\theta}{\partial t^2}(t,-L) = 0\ ,\quad \frac{\partial^2 u^\phi}{\partial t^2}(t,-L) = 0\ ,\quad \frac{\partial^2 u^\phi}{\partial \xi \partial t}(t,-L) = 0,$$

$$\frac{\partial^2 u^\theta}{\partial \xi \partial t}(t,-L) = 0,\quad \frac{\partial u_\psi}{\partial t}(t,-L) = 0,\quad \forall\, t > 0.$$

Then

$$\frac{\partial^3 u^\theta}{\partial \xi^3}(t,-L) = \frac{\partial^3 u^\phi}{\partial \xi^3}(t,-L) = \frac{\partial^2 u^\phi}{\partial \xi^2}(t,-L) = \frac{\partial^2 u^\theta}{\partial \xi^2}(t,-L) = \frac{\partial u_\psi}{\partial \xi}(t,-L) = 0,\ \forall\, t > 0.$$

Therefore $u^\theta(t,\xi)$ is a strong solution of

$$\mathrm{I}\ \begin{cases} P\dfrac{\partial^2 u^\theta}{\partial t^2}(t,\xi) + EI\dfrac{\partial^4 u^\theta}{\partial \xi^4}(t,\xi) = 0, & \forall\, t > 0, \\[2mm] u^\theta(t,0) = 0\ ,\quad \dfrac{\partial u^\theta}{\partial \xi}(t,0) = 0, & \forall\, t > 0, \\[2mm] \dfrac{\partial^2 u^\theta}{\partial \xi^2}(t,-L) = 0\ ,\quad \dfrac{\partial^3 u^\theta}{\partial \xi^3}(t,-L) = 0, & \forall\, t > 0, \\[2mm] \dfrac{\partial u^\theta}{\partial t}(t,-L) = 0\ ,\quad \dfrac{\partial^2 u^\theta}{\partial \xi \partial t}(t,-L) = 0, & \forall\, t > 0, \end{cases}$$

with $u^\theta(t,\xi)\epsilon H^4(-L,0)$, $\dfrac{\partial u^\theta}{\partial t}(t,\xi)\epsilon H^2(-L,0)$, $\forall\, t \geq 0.$

A similar equation is satisfied by $u^\phi(t,\xi)$, and $u_\psi(t,\xi)$ is the strong solution of

$$
\text{II} \quad
\begin{cases}
P\dfrac{\partial^2 u_\psi}{\partial t^2}(t,\xi) - G\dfrac{\partial^2 u_\psi}{\partial \xi^2}(t,\xi)=0, & \forall\, t>0, \\[2mm]
u_\psi(t,0)=0\ , \quad \dfrac{\partial u_\psi}{\partial \xi}(t,-L)=0, & \forall\, t>0, \\[2mm]
\dfrac{\dot{\partial u_\psi}}{\partial t}(t,-L)=0,
\end{cases}
$$

with $u_\psi(t,\xi)\epsilon H^2(-L,0)$, $\partial u_\psi/\partial t(t,\xi)\epsilon H^1(-L,0)$ $\forall\, t\geq 0$.
Then, harmonic analysis [2] allowed us to conclude that

$$u^\theta(t,\xi)=u^\phi(t,\xi)=u^\psi(t,\xi)=0 \quad \forall\, \xi\epsilon[-L,0],\ \ t\geq 0.$$

Then

$$\overline{M}_1(t)=0, \quad t>0,$$

therefore

$$\overline{D}(t)\times\overline{R}=0, \quad t>0;$$

but

$$\dot{\overline{D}}(t)=0\ ,$$

then $\overline{S}(t)=\overline{R}-\overline{D}(t)$ is constant and we can have either $\overline{S}=\overline{R}$ or $\overline{S}=-\overline{R}$. That means D is either 0 or 4.

Theorem 2: If the system is initially in $D(\widetilde{M}^{-1}\widetilde{A})$ and the initial direction of the line of sight is not just opposite to the target direction then, choosing λ appropriately, the proposed linear feedback control law points the line of sight of the antenna to the target and stabilizes the oscillations of the entire flexible configuration.

Proof: Initially the line of sight R is not just opposite to the target direction S. That is $\overline{S}\neq-\overline{R}$. Then the square distance at $t=0$, $D(0)$, is strictly less than 4. Choose $\lambda>[V(0)+T(0)]/[4-D(0)]$, where $V(0)$ is the initial potential energy and $T(0)$ is the kinetic energy. Then

$$\|z(t)\|^2\leq\lambda D(0)+T(0)+V(0)<4\lambda,\ \ t>0;$$

but $\|z(t)\|^2=\lambda D$, where D can be 0 or 4, then $\|z(t)\|^2=0$, $\forall\, t\geq 0$, and therefore $z=0$, or $L^+(x_0)=\{0\}$.

Finally we note that since $D(t)+V(t)+T(t)$ is decreasing, we have that $\overline{\omega}(t)$, $\overline{\omega}_4(t)$, $\overline{V}_4(t)$, $\overline{D}(t)$ are bounded. Then the controls $\overline{M}_1(t)$, $\overline{M}_4(t)$ and $\overline{F}(t)$ are also bounded.

5. Concluding Remarks

The main contribution of this paper is given in Theorem 2. We prove that the line of sight of the antenna can be pointed to any target and the oscillations induced on the flexible configuration can be stabilized by means of a linear feedback control law. This control law is very simple and requires the following data: (1) the angular velocity $\overline{\omega}$ of the space shuttle orbiter, (2) the angular velocity $\overline{\omega}_4$ of the reflector body, (3) the inertial velocity $\overline{v}_4$ of the reflector body and (4) the target direction $\overline{S}$ expressed in the shuttle body fixed frame, which is computed from the direction cosine matrix for the shuttle body.

Acknowledgement

The author would like to express his appreciation to Professor A.V. Balakrishnan for his guidance and encouragement throughout this research. Research supported in part under AFOSR grant No. 83-0318. Applied Mathematics Division.

References

[1] Balakrishnan, A.V., *Applied Functional Analysis*, 2nd Ed., Springer-Verlag, 1981.

[2] Ball, J.M., Marsden, J.E., and Slemrod, M., "Controllability for Distributed Bilinear Systems", *SIAM J. Control*, 20, 4, 1982, pp. 575-597.

[3] Biswas, S.K. and Ahmed, N.U., "Stabilization of a Flexible Spacecraft Governed by a Couple System of Ordinary and Partial Differential Equations", to appear.

[4] Dafermos, C.M., "Asymptotic Behavior of Solutions of Evolutions Equations", in *Nonlinear Evolution Equations*, M. Crandall (Ed.), Academic Press, 1977.

[5] Mackay, M.K., "Active Control of a Large Flexible Space Structure", Ph.D. Thesis, U.C.L.A., 1983.

[6] Marsden, J.E. and Hughes, T.J.R., *Mathematical Foundations of Elasticity*, Prentice-Hall, 1983.

[7] Quinn, J.P. and Russel, D.T., "Assymptotic stability and energy decay rates for solutions of hyperbolic equations with boundary damping", *Proc. Roy. Soc. Edinburg*, 77A, 1977, pp. 97-127.

[8] Saperstone, S.H., *Semidynamical Systems in Infinite Dimensional Spaces*, Springer-Verlag, 1981.

[9] Taylor, L.W. and Balakrishnan, A.V., "A Mathematical Problem and a Spacecraft Control Laboratory Experiment (SCOLE) Used to Evaluate Control Laws for Flexible Spacecraft ... NASA/IEEE Design Challenge", January, 1984.

[10] Wertz, J.R., *Spacecraft Attitude Determination and Control*, D. Reidel Publishing Company, 1980.

[11] Zoubov, V., *Théorie de la Commande*, Mir, 1978.

IDENTIFICATION OF STRUCTURAL SYSTEMS USING WALSH FUNCTIONS

A.S. Ashok, B. Sahay
Indian Institute of Technology
Kanpur, India
T. Prasad
University of Waterloo
Waterloo, Ontario
Canada N2L 3G1

1. Introduction

System theoretic approaches for analysis and design of dynamic structures has been attracting considerable attention in the recent past. There are a number of methods available for system identification [1], some of which have been applied for the identification of dynamic structures. The objective is the determination of specific parameter values of the system model such that the behavior of the system and model output are as close as possible under the same operating conditions.

Identification of dynamic systems is required mostly for two purposes: firstly for designing the system and, secondly to analyse the system properties such that its behavior due to any given input can be predicted. In the latter case we obtain a simplified parametric model of the dynamic system.

The parameters associated with the system models are to be identified using the input-output record of the system. Here, Walsh functions have been used to identify the system parameters for the input-output records of dynamic structures.

32

The system identification algorithm formulated here does not include problems such as state estimation or input process identification. But the Walsh functions method can be used to identify the initial and boundary conditions along with the parameters.

In this paper, we shall confine ourselves to the identification of constant parameter systems, although the method can be applied to systems with slowly varying parameters also.

The essential features of the Walsh functions are described first. The algorithms for identification of lumped parameter structures (LPS) and distributed parameter structures (DPS) are described next.

2. Walsh Functions

Rademacher (1922) and Walsh (1923) independently developed functions in the form of a set of rectangular waves. These are shown in Figs. 1 and 2. Walsh functions form a complete orthonormal set of rectangular waves; whereas, Rademacher functions are not complete as they include only odd functions. The Walsh and Rademacher functions are related [2] by the following formula

$$\phi_n(t) = (r_i)^{b_k}(r_{i-1})^{b_{k-1}}(r_{i-2})^{b_{k-2}}.... \tag{1}$$

where $i = largest\ integer\ of\ [\log_2 n] + 1$, and $b_k.\ b_{k-1}....b_0$ is the binary expression for the decimal number 'n'.

Orthogonality and convergence of Walsh functions and series have been treated by Paley [3] and Fine [4]. It has been shown that, if $f(t)$ is continuous in (0,1), then the Walsh series expension of $f(t)$ converges uniformly. If $f(t)$ is not continuous then there is convergence in the mean.

2.1. Walsh coefficients

A function $f(t)$, absolutely integrable in (0,1], may be expanded into a Walsh series

$$f(t) = \sum_{i=0}^{\infty} C_i \phi_i(t) \tag{2}$$

where C_i ($i=1,2,...$) are coefficients of the Walsh series for $f(t)$. Since the Walsh functions $\phi(t)$ are orthonormal in (0,1], the coefficients C_i can be determined, using ISE criterion, from

$$C_i = \int_0^1 \phi_i(t) f(t)\ dt. \tag{3}$$

If the given function is in the form of tabulated data, we can obtain the Walsh series expansion using the discrete form of Eq. (2) as

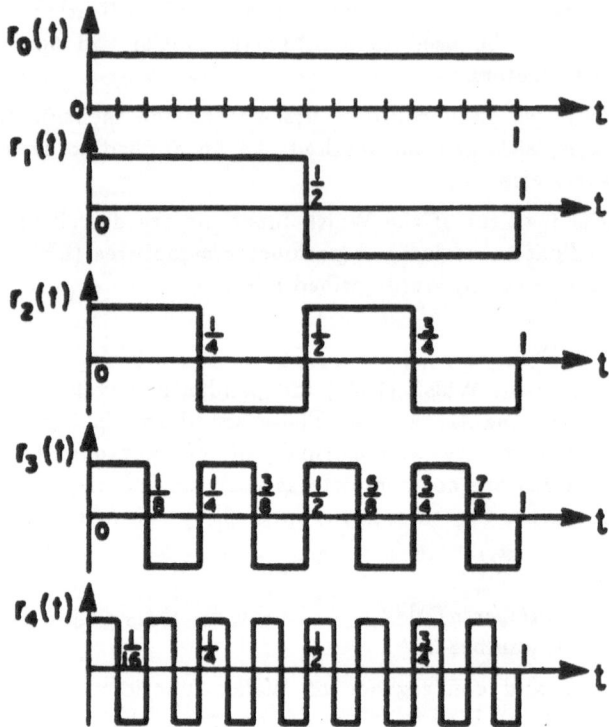

Figure 1: Rademacher functions

$$f_k = \sum_{n=0}^{m-1} \phi_n(t_k)C_n, \qquad k = 0,1,2,...,m-1, \tag{4}$$

$$C_n = \sum_{k=0}^{m-1} \phi_n(t_k)f(t_k)\frac{1}{m}, \qquad n = 0,1,2,...,m-1, \tag{5}$$

where f_k is the average value of the function in the kth subinterval, $\phi_n(t_k)$ is the value of the nth Walsh function in kth subinterval, m is the number of subintervals, and $f(t_k)$ is the value of $f(t)$ in the kth subinterval.

Equations (4) and (5), expressed in the vector matrix form, are

$$f_k = \mathbf{C}^T\phi(t_k) \tag{6}$$

and

$$\mathbf{C} = \Phi f(t_k)\frac{1}{m} \tag{7}$$

where $\mathbf{C}$ and $\phi(t_k)$ are vectors of the coefficients C_n and $\phi_n(t_k)$ respectively, and

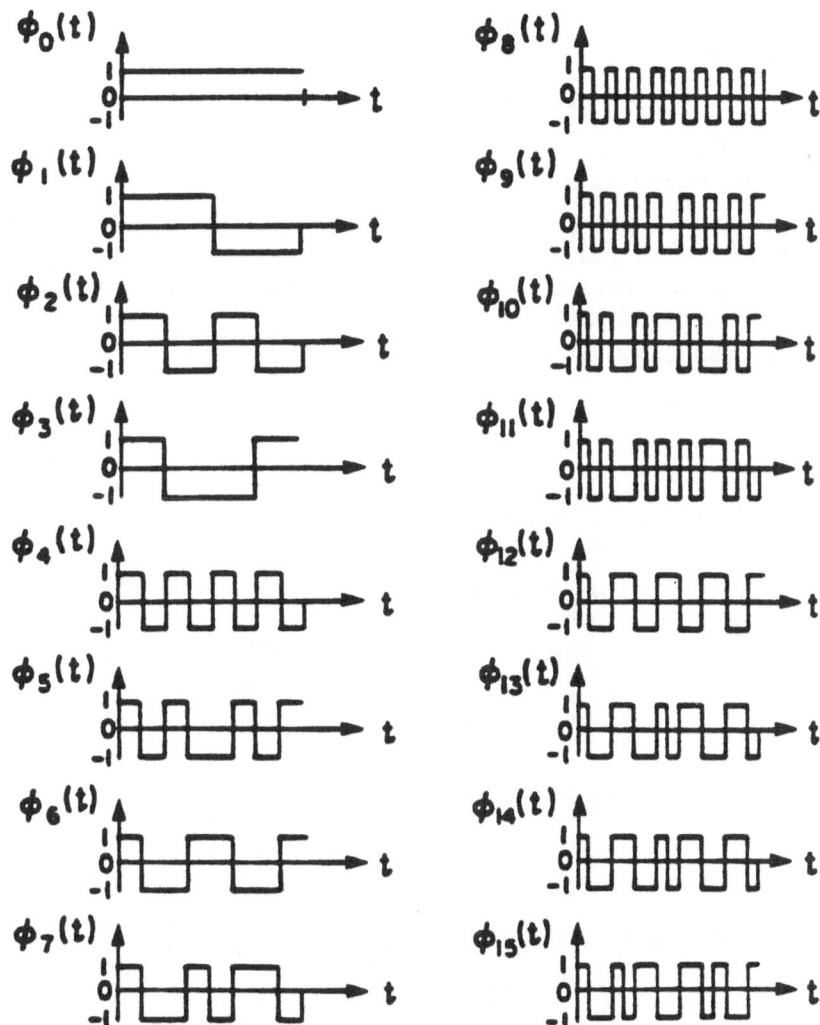

Figure 2: Walsh functions

Φ is the matrix $[\phi_n(t_k)]$. A 4×4 Walsh matrix is given by

$$\Phi_{4\times4} \begin{bmatrix} 1 & 1 & 1 & 1 \\ 1 & 1 & -1 & -1 \\ 1 & -1 & 1 & -1 \\ 1 & -1 & -1 & 1 \end{bmatrix} \tag{8}$$

2.2. The operational matrix

Figure 3 shows the first few Walsh functions and their first integrals. Integration of Walsh functions is just the multiplication by a constant matrix called the Operational Matrix, obtained by evaluating the Walsh Coefficients for the triangular waves of the Walsh function integrals.

Consider the 4-term approximation for

$$f(t) = t \tag{9}$$

Let

$$f(t) = t = \sum_{i=0}^{3} C_i \phi_i(t). \tag{10}$$

C_i can be obtained by using Eq. (7)

$$\begin{bmatrix} C_0 \\ C_1 \\ C_2 \\ C_3 \end{bmatrix} = \begin{bmatrix} 1 & 1 & 1 & 1 \\ 1 & 1 & -1 & -1 \\ 1 & -1 & 1 & -1 \\ 1 & -1 & -1 & 1 \end{bmatrix} \begin{bmatrix} \frac{1}{2}(0 + \frac{1}{4}) \\ \frac{1}{2}(\frac{1}{4} + \frac{1}{2}) \\ \frac{1}{2}(\frac{1}{2} + \frac{3}{4}) \\ \frac{1}{2}(\frac{3}{4} + 1) \end{bmatrix} \times \frac{1}{4} = \begin{bmatrix} \frac{1}{2} \\ -\frac{1}{4} \\ -\frac{1}{8} \\ 0 \end{bmatrix}. \tag{11}$$

Then, the 4-term approximation for $f(t) = t$ is given by

$$f(t) = t = \frac{1}{2}\phi_0(t) - \frac{1}{4}\phi_1(t) - \frac{1}{8}\phi_2(t) + 0.\phi_3(t). \tag{12}$$

Proceeding on similar lines, the integrals of the Walsh functions (Fig. 3) can be built up. These can be written in the form

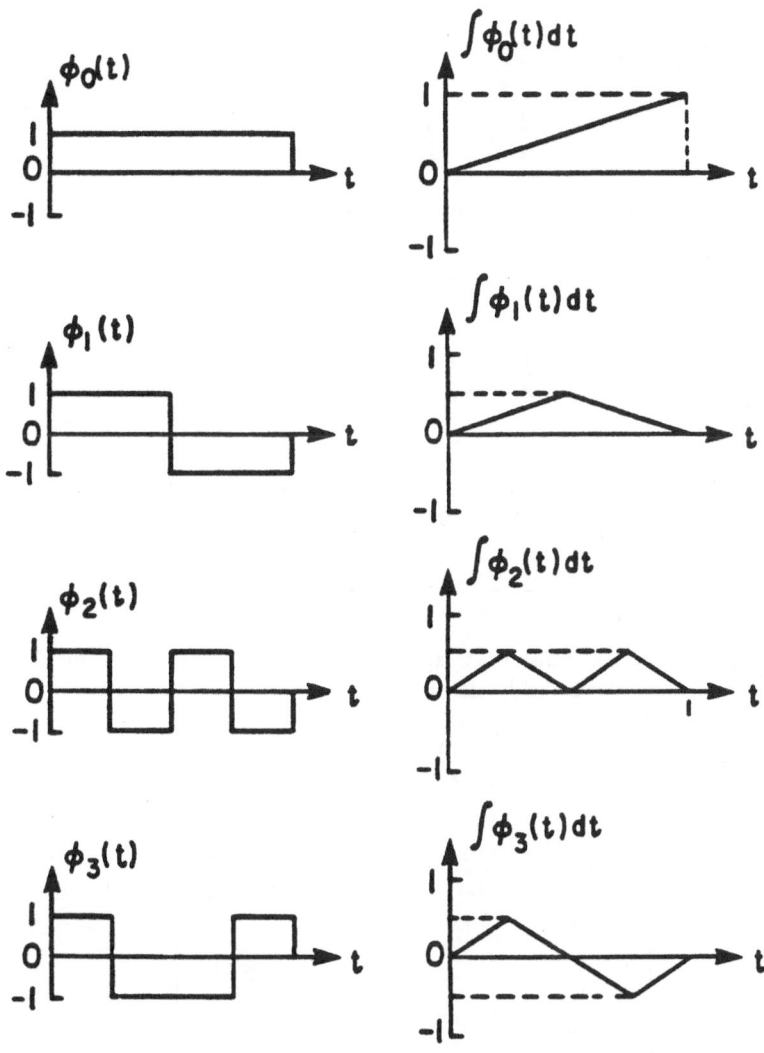

Figure 3: Walsh functions and their first integrals

$$\int_0^t \begin{bmatrix} \phi_0(t) \\ \phi_1(t) \\ \phi_2(t) \\ \phi_3(t) \end{bmatrix} dt = \begin{bmatrix} \dfrac{1}{2} & -\dfrac{1}{4} & -\dfrac{1}{8} & 0 \\[2mm] \dfrac{1}{4} & 0 & 0 & -\dfrac{1}{8} \\[2mm] \dfrac{1}{8} & 0 & 0 & 0 \\[2mm] 0 & -\dfrac{1}{8} & 0 & 0 \end{bmatrix} \begin{bmatrix} \phi_0(t) \\ \phi_1(t) \\ \phi_2(t) \\ \phi_3(t) \end{bmatrix}. \tag{13}$$

In the vector matrix form it can be written as

$$\int_0^t \phi(t) = P_{4\times4}\phi(t) \tag{14}$$

where

$$P_{4\times4} = \begin{bmatrix} \dfrac{1}{2} & -\dfrac{1}{4} & -\dfrac{1}{8} & 0 \\[2mm] \dfrac{1}{4} & 0 & 0 & -\dfrac{1}{8} \\[2mm] \dfrac{1}{8} & 0 & 0 & 0 \\[2mm] 0 & -\dfrac{1}{8} & 0 & 0 \end{bmatrix}. \tag{15}$$

P is called the operational matrix. As the number of terms in the expansion increases, the size of matrix P also imcreases.

$$P = \begin{bmatrix} \dfrac{1}{2} & \dfrac{-2}{m}\dfrac{\text{Im}}{8} & & \\[2mm] & & \dfrac{-1}{m}\dfrac{\text{Im}}{4} & \\[2mm] \dfrac{2}{m}\dfrac{\text{Im}}{8} & \dfrac{Om}{8} & & \dfrac{-2}{2m}\dfrac{\text{Im}}{2} \\[2mm] \dfrac{1}{m}\dfrac{\text{Im}}{4} & & \dfrac{Om}{4} & \\[2mm] \dfrac{1}{2m}\dfrac{\text{Im}}{2} & & & \dfrac{Om}{2} \end{bmatrix}. \tag{16}$$

Solution of identification problems using Walsh functions is highly dependent on the use of the matrices P and Φ.

2.3. Double Walsh series

A function $y(x,t)$ of two variables $0 < x \leq 1$ and $0 < t \leq 1$ can be expanded in a double Walsh Series [5], [6]:

$$y(x,t) = \sum_{j=0}^{\infty}\sum_{i=0}^{\infty} \psi_j(x)Y_{ij}\phi_i(t) = \psi(x)Y_{MN}\phi_N(t) \qquad (17)$$

where $\psi_j(x)$, $j = 1,2,...,M$ and $\phi_i(t)$ $i = 1,1,...N$ are Walsh functions of variables x and t respectively. Y_{ij} are the coefficients of the double Walsh series expansion given by

$$Y_{ij} = \int_0^1 \int_0^1 \psi_j(x)y(x,t)\phi_i(t) \, dxdt \qquad (18)$$

For the discrete form, let $y(x_\ell, t_k)$ be the value of y for ℓth observation in x and kth observation in t. This can be represented by the matrix of observations

$$y_{NM} = \begin{bmatrix} y(x_0,t_0) & Y(x_1,t_0) & \cdots & y(x_M,t_0) \\ y(x_0,t_1) & Y(x_1,t_1) & \cdots & y(x_M,t_1) \\ \cdot & \cdot & & \cdot \\ \cdot & \cdot & & \cdot \\ \cdot & \cdot & & \cdot \\ y(x_0,t_N) & Y(x_1,t_N) & \cdots & y(x_M,t_N) \end{bmatrix}. \qquad (19)$$

There are a total number of $(N + 1) \times (M + 1)$ observations and hence $M \times N$ subintervals.

Let

$$y''_{\ell,k} = \frac{1}{n}W_{NN} \cdot \frac{1}{2}[y(x_\ell,t_{k-1}) + y(x_\ell,t_k)], \qquad (20)$$

$$y'_{p,q} = \frac{1}{2}y''_{p-1,q} + y''_{p,q} \quad \begin{cases} p=1,2,...,M, \\ q=1,2,...,N \end{cases}, \qquad (21)$$

and

$$Y_{MN} = \frac{1}{M}W_{MM}[y'_{p,q}]^T, \qquad (22)$$

where W is Walsh matrix.

Equation (22) gives the Walsh Coefficient matrix for Double Walsh series.

3. Identification of Lumped Parameters Structures

In general, a lumped parameter structure will have the following mathematical form:

$$\frac{d^n y}{dt^n} + a_{n-1}\frac{d^{n-1}y}{dt^{n-1}} + \cdots + a_1\frac{dy}{dt} + a_0 y = u(t) \tag{23}$$

where $y(t)$ is the displacement and $u(t)$ is the forcing function. $Y_0, Y_1, \ldots, Y_{n-1}$ are the initial conditions of $y(t)$ and its derivatives.

Expanding $u(t)$, $y(t)$ and t^n in Walsh series,

$$\begin{aligned} y(t) &= \mathbf{C}^T \boldsymbol{\phi}(t), \\ u(t) &= \mathbf{h}^T \boldsymbol{\phi}(t), \\ t^n &= (\boldsymbol{\epsilon}^T)_n \boldsymbol{\phi}(t). \end{aligned} \tag{24}$$

It can be shown from Eqs. (23) and (24) that

$$[a_{n-1}P^T\mathbf{C},\, a_{n-2}(P^2)^T\mathbf{C},..,a_0(P^n)^T\mathbf{C},\, \psi_0\epsilon_0^T,\, \psi_1\epsilon_1^T,.., \psi_{n-1}\epsilon_{n-1}^T] = (P^n)^T\mathbf{h} - \mathbf{C} \tag{25}$$

where

$$\begin{aligned} \psi_0 &= -Y_0, \\ \psi_1 &= -t[Y_1 + a_{n-1}Y_0], \\ \psi_2 &= -t^2[Y_2 + a_{n-2}Y_1 + a_{n-3}Y_0], \\ \psi_{n-1} &= -t^{n-1}[Y_{n-1} + a_{n-1}Y_{n-2} + ...A_1Y_0]. \end{aligned} \tag{26}$$

Equation (25) can be written in a simpler form

$$A\mathbf{x} = H \tag{27}$$

where (letting $P^T = Q$),

$$A = \begin{bmatrix} (Q\ \mathbf{c})_1 (Q^2\ \mathbf{c})_1 & \cdots & (Q^n\ \mathbf{c})_1, & (\epsilon_0^T)_1, & (\epsilon_1)_1^T, & \ldots, & (\epsilon_{n-1})_1^T \\ (Q\ \mathbf{c})_2 (Q^2\ \mathbf{c})_2 & \cdots & (Q^n\ \mathbf{c})_2, & (\epsilon_0^T)_2, & (\epsilon_1)_2^T, & \ldots, & (\epsilon_{n-1})_2^T \\ (Q\ \mathbf{c})_M (Q^2\ \mathbf{c})_M & \cdots & (Q^n\ \mathbf{c})_M, & (\epsilon_0^T)_M, & (\epsilon_1)_M^T, & \ldots, & (\epsilon_{n-1})_M^T \end{bmatrix},$$

$$H = \begin{bmatrix} (Q^n\mathbf{h} - \mathbf{c})_1 \\ (Q^n\mathbf{h} - \mathbf{c})_M \end{bmatrix},$$

$$\mathbf{x} = [a_{n-1},\, a_{n-2},..,\, a_0,\, \psi_0,\, \psi_1,\, \ldots,\, \psi_{n-1}]^T.$$

By least squares method the unknown vector x can be obtained from

$$\hat{\mathbf{x}} = (A^T A)^{-1} A^T H. \tag{28}$$

The vector $\hat{\mathbf{x}}$ gives the estimate of the parameters a and initial condition functions $\boldsymbol{\psi}$. The initial conditions can be obtained from

$$
\begin{bmatrix} Y_0 \\ Y_1 \\ \cdot \\ \cdot \\ \cdot \\ Y_{n-1} \end{bmatrix} = \begin{bmatrix} 1 \\ a_{n-1} & 1 \\ \cdot \\ \cdot \\ \cdot \\ \dfrac{a_1}{(n-1)!} & \dfrac{a_2}{(n-1)!} & \cdots & \dfrac{1}{(n-1)!} \end{bmatrix}^{-1} \begin{bmatrix} \psi_0 \\ \psi_1 \\ \cdot \\ \cdot \\ \cdot \\ \psi_{n-1} \end{bmatrix}. \tag{29}
$$

4. Identification of Distributed Parameter Systems (DPS)

Let a second order system be modelled by

$$
a_5 \frac{\partial^2 y(x,t)}{\partial t^2} + a_4 \frac{\partial^2 y(x,t)}{\partial x \partial t} + a_3 \frac{\partial^2 y(x,t)}{\partial x^2} + a_2 \frac{\partial y(x,t)}{\partial t}
$$
$$
+ a_1 \frac{\partial y(x,t)}{\partial x} + a_0 y(x,t) = U(x,t). \tag{30}
$$

Equation (30) is integrated two times with respect to (w.r.t.) t and two times w.r.t. x, with initial and boundary conditions. We get

$$
\alpha(x) = -a_5 \left[\frac{\partial y(x,t)}{\partial t} \right]_{t=0} - a_2 y(x,0),
$$
$$
\beta(t) = -a_3 \left[\frac{\partial y(x,t)}{\partial x} \right]_{x=0} - a_1 y(0,t) - a_4 \frac{\partial y(0,t)}{\partial t}, \tag{31}
$$
$$
\nu(x) = -a_4 \left[\frac{\partial y(x,t)}{\partial x} \right]_{t=0}.
$$

The identification problem now consists of determining the parameters a_0 to a_5 and the initial and boundary conditions $\alpha(x)$, $\beta(t)$, $\nu(x)$, $y(0,t)$ and $y(x,0)$. Expanding $y(x,t)$ and $U(x,t)$ in double Walsh series and $\alpha(x)$, $\beta(t)$ and $\nu(x)$ in Walsh Series, and using the discrete form as in Eqs. (19 - 22), we get

$$
y(x,t) = \psi_M^T(x) Y_{MN} \phi_N(t),
$$
$$
U(x,t) = \psi_M^T(x) U_{MN} \phi_N(t), \tag{32}
$$
$$
\alpha(x) = \alpha_r^T \psi_r(x),
$$
$$
\beta(t) = \beta_s^T \phi_s(t),
$$
$$
\nu(x) = \nu_\delta^T \psi^T(x),
$$
$$
y(0,t) = C_q^T \phi_q(t),
$$

$$y(x,0) = b_p^T \phi_p(x). \tag{33}$$

Here α_r, β_s, ν_δ are coefficient vectors of the expansions above. The expansion in Eq. (33) may be converted into double Walsh series by introducing $(M \times N)$ matrix E_{ij} which has only the (ij)th element unity and the remaining elements as zero. Thus

$$\alpha(x) = \sum_{i=0}^{r-1} \alpha_i \psi_i(x) = \sum_{i=0}^{r-1} \alpha_i \psi_M^T(x) E_{i+1,1} \phi_N(t). \tag{34}$$

Other expansions in Eq. (33) can be similarly expressed in this form. Introducing these expressions in the double integrals of Eq. (30) and equating the coefficients of like Walsh function products on both sides we get $(Q_M = P_M^T)$,

$$a_r (Q_M)^2 Y_{MN} + a_4 Q_M Y_{MN} P_N + a_3 Y_{MN} P_N^2 + a_2 Q_M^2 Y_{MN} P_N$$
$$+ a_1 Q_M Y_{MN} P_N^2 + a_0 Q_M^2 Y_{MN} P_N^2 + \sum_{i=0}^{r-1} \alpha_i Q_M^2 E_{iH,1} P_N$$
$$+ \sum_{i=0}^{s-1} \beta_i Q_M E_{1,i+1} P_N^2 + \sum_{i=0}^{\delta-1} \nu_i Q_M^2 E_{i+1,1} P_N$$
$$+ \sum_{i=0}^{q-1} \hat{b}_i Q_M^2 E_{1,i+1} + \sum_{i=0}^{\phi-1} \hat{C}_i E_{i+1,1} P_N^2 = Q_M^2 U_{MN} P_N^2 \tag{35}$$

where

$$\hat{b}_i = -a_5 b_i, \qquad \hat{C}_i = -a_3 C_i.$$

Let

$$\mathbf{x}^T = [a_5, a_4, \ldots, a_1, a_0, \alpha_0, \alpha_1, \ldots, \alpha_{r-1}, \beta_0, \beta_1, \ldots, \beta_{s-1},$$
$$\nu_0, \nu_1, \ldots, \nu_{\delta-1}, b_0, b_1, \ldots, b_{q-1}, C_0, C_1, \ldots, C_{p-1}].$$

Then, Eq. (35) can be expressed as

$$A\mathbf{x} = \mathbf{h} \tag{36}$$

where A will be a matrix consisting of $Q_M^2 Y_{MN}$, $Q_M Y_{MN} P_N$, $Y_{MN} P_N^2$ etc. at data points 1,2,...., N, formed from Eq. (31), and

$$\mathbf{h} = \begin{bmatrix} (Q_M^2 U_{MN} P_N^2)_1 \\ (Q_M^2 U_{MN} P_N^2)_2 \\ \cdot \\ \cdot \\ \cdot \\ (Q_M^2 U_{MN} P_N^2)_N \end{bmatrix}. \tag{37}$$

$\mathbf{x}^T$ is the unknown parameter vector to be identified. Thus,

$$\hat{\mathbf{x}} = (A^T A)^{-1} A^T \mathbf{h}. \tag{38}$$

$\hat{\mathbf{x}}$ is the estimated value of the parameter vector.

5. Examples

5.1. Example 1: A linear lumped system

A mass M is mounted on a platform with a spring and damper as shown in Fig. 4. Under the static load the spring sags by 10 cms. The platform is moved up and down at a frequency of 3 rad/sec. and amplitude unity. The system parameters (the spring stiffness k, the damping coefficient β and the mass M) are to be obtained when the experimental data for the displacement y are as given in the Table below.

t (secs)	0	$\frac{1}{8}$	$\frac{1}{4}$	$\frac{3}{8}$	$\frac{1}{2}$	$\frac{5}{8}$	$\frac{3}{4}$	$\frac{7}{8}$	1
y (cms)	10.00	4.60	2.26	3.20	0.60	1.15	0.85	0.09	0.43
$\bar{y}$ (cms)		7.30	3.43	2.73	1.90	0.875	1.00	0.47	0.26

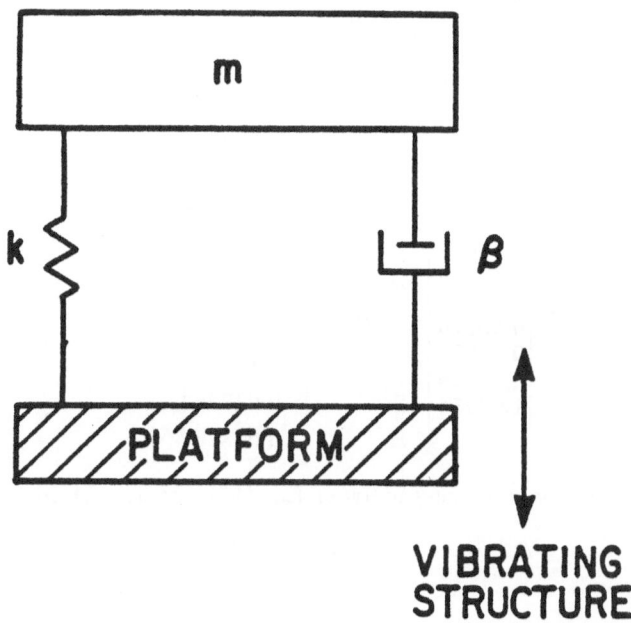

Figure 4: Spring-mass-damper system of example 1

Solution: The system is described by

$$m\ddot{y} + \beta\dot{y} + ky = Sin\,3t; \quad y(0) = 10, \quad \dot{y}(0) = 0.$$

Let $y(t) = \sum_{n=1}^{8} c_n \phi_n(t)$,

$$s = Sin\,3t = \sum_{i=1}^{8} h_i \phi_i(t).$$

The coefficient vectors c and h are given by

$$c = \frac{1}{8}W\overline{y}, \quad h = \frac{1}{8}W\overline{s}$$

where

$$W = \begin{bmatrix} 1 & 1 & 1 & 1 & 1 & 1 & 1 & 1 \\ 1 & 1 & 1 & 1 & 1 & -1 & -1 & -1 \\ 1 & 1 & -1 & -1 & -1 & 1 & 1 & -1 \\ 1 & 1 & -1 & -1 & -1 & -1 & -1 & 1 \\ 1 & -1 & 1 & -1 & -1 & 1 & -1 & -1 \\ 1 & -1 & -1 & -1 & -1 & -1 & 1 & -1 \\ 1 & -1 & -1 & 1 & 1 & 1 & -1 & 1 \\ 1 & -1 & 1 & 1 & 1 & -1 & 1 & 1 \end{bmatrix}.$$

Thus,

$$c = \begin{bmatrix} 0.67243 \\ 0.32064 \\ 1.78498 \\ 1.54039 \\ 0.65176 \\ 0.68930 \\ 0.65749 \\ 1.02954 \end{bmatrix}, \quad h = \begin{bmatrix} 0.85553 \\ -0.04330 \\ -0.01829 \\ -0.24038 \\ -0.00882 \\ -0.11586 \\ -0.04895 \\ 0.00323 \end{bmatrix}.$$

The matrix P is constructed from Eq. (16). Matrix A in Eq. (27) is then calculated as

$$A = \begin{bmatrix} 0.67243 & 0.68023 & 0.34900 & 1.0000 & 0.5000 \\ 0.32064 & 0.06752 & -0.16827 & 0.0000 & -0.2500 \\ 1.78496 & -0.04296 & -0.09200 & 0.0000 & -0.1250 \\ 1.54039 & 0.02426 & -0.01445 & 0.0000 & 0.0000 \\ 0.65176 & -0.04202 & -0.04251 & 0.0000 & -0.0625 \\ 0.58930 & -0.02004 & -0.00422 & 0.0000 & 0.0000 \\ 0.65749 & -0.11156 & 0.00268 & 0.0000 & 0.0000 \\ 1.02954 & -0.09627 & -0.00156 & 0.0000 & 0.0000 \end{bmatrix} ; \quad H = \begin{bmatrix} 0.93572 \\ -0.07765 \\ -0.03919 \\ 0.02608 \\ 0.01963 \\ 0.01257 \\ 0.00531 \\ -0.00035 \end{bmatrix}.$$

From Eq. (25),

$$\hat{\mathbf{x}}^T = [4.0625 \quad 35.712 \quad 490.98 \quad -40.62 \quad -315.324].$$

This gives,

$$M = 4.0625, \quad \beta = 35.712, \quad k = 490.09,$$

$$\psi_0 = -40.62, \quad \psi_1 = -315.324.$$

The actual differential equation was

$$5.1\frac{d^2y}{dt^2} + 32\frac{dy}{dt} + 500y = Sin\,3t.$$

5.2. Example 2: A distributed parameter system

Consider a second order system

$$\frac{1}{c^2}\frac{\partial^2 y(x,t)}{\partial t^2} - \alpha\frac{\partial^2 y(x,t)}{\partial x^2} = 5\,Sin\,x.$$

It may represent a vibrating string of length unity and a suddenly applied load of intensity $5\,Sin\,x$. The initial conditiond are all zero. The observation matrices are given by

$$y_{3\times3} = \begin{bmatrix} 0.000 & 0.0000 & 0.000 \\ 0.000 & 0.000299 & 0.01197 \\ 0.000 & 0.00525 & 0.02101 \end{bmatrix}$$

and

$$u_{3\times3} = \begin{bmatrix} 0.0000 & 0.000 & 0.000 \\ 2.3971 & 2.3971 & 2.3971 \\ 4.2073 & 4.2073 & 4.2073 \end{bmatrix}$$

Using discrete form of Eq. (32) the coefficient matrices of the double Walsh series expansion of $y(x,t)$ and $u(x,t)$ will be

$$y_{2\times2} = \begin{bmatrix} 0.00043 & -0.0035 \\ -0.00020 & 0.00016 \end{bmatrix}, \quad u_{2\times2} = \begin{bmatrix} 2.25 & 0.000 \\ -1.0518 & 0.000 \end{bmatrix}.$$

The matrices A and h given in Eq. (36) can now be formed as

$$A = \begin{bmatrix} 0.00043 & 0.00054 \\ -0.00020 & -0.00040 \\ -0.00035 & -0.00036 \\ 0.00016 & 0.00026 \end{bmatrix}; \quad h = \begin{bmatrix} 0.05446 \\ -0.04041 \\ -0.03630 \\ 0.0294 \end{bmatrix}.$$

From Eq. (34)

$$\hat{X}^T = [a_1 \quad a_0] = (A^TA)^{-1}A^Th.$$

Solving this by least squares technique, we get

$$a_1 = 102.414 \qquad a_0 = -1.6714.$$

Thus $\dfrac{1}{c^2} = 102.414$, $\alpha = 1.6714$.

The actual differential equation was

$$100\frac{\partial^2 y(x,t)}{\partial t^2} - \frac{\partial^2 y(x,t)}{\partial x^2} = 5Sinx.$$

The error is caused due to small number of terms in the expansion.

6. Conclusion

The above examples show that the Walsh function approach provides an efficient discrete time technique for the identification of structural system parameters for lumped and distributed structures. The parameter determination helps in determining the dominant eigenvalues of the structure. The method also determines the initial conditions from the current input-output data simultaneously along with the parameters. The technique provides considerable reduction in computational time because the Walsh matrix and the operatianal matrix can be in-built in the software and need not be generated again and again.

The discrete character of the Walsh function makes it possible for the algorithm to be implemented directly on microprocessors.

Acknowledgement

This research was partially supported by the NSERC (Natural Sciences and Engineering Research Council) of Canada, which the authors gratefully acknowledge.

References

[1] Astrom, K.J. and Eykhoff, P., "System Identification - A survey," *Automatica*, Vol. 7, 1971, pp. 123-162.

[2] Beauchamp, K.G., *Walsh Functions - Theory and Applications*, Academic Press, 1975.

[3] Paley, R.E.A.C., "A Remarkable Series of Orthogonal Functions," *Proc. of London Math. Soc.*, Vol. 34, 1932, pp. 241-279.

[4] Fine, N.J., "On the Walsh Functions," *Trans. Am. Math. Soc.*, Vol. 65, 1949, pp. 372-414.

[5] Paraskevopoulos, P.N. and Bounos, A.C., "Distributed Parameter Systems Identification Via Walsh Function," *Int. J. System Science*, Vol. 9, No. 1, 1978, pp. 75-83.

[6] Tzafestas, S.G., "Walsh Series Approach to Lumped and Distributed System Identification," *Journal of the Franklin Institute*, Vol. 305, 1978, pp. 192-220.

COMPATIBILITY ASPECTS OF ACTIVE CONTROL TECHNOLOGIES WITH AIRCRAFT STRUCTURE DESIGN

J. Becker, F. Weiss, O. Sensburg
Messerschmitt-Bölkow-Blohm GmbH
Helicopter and Military Aircraft Division
D-8012 Ottobrunn, West Germany

1. Introduction

The usual aircraft structural design procedure of unaugmented airplanes may strongly be altered when Active Control Technology (ACT) is implemented.

The many different applications of ACT for the purpose of stability augmentation, stabilisation of unstable configurations, for drag reduction, gust and maneuver load alleviation, ride comfort improvement, vibration alleviation, elastic mode control and flutter suppression will lead each to special structural design problems.

In general all the effects of the systems have to be investigated with respect to the different requirements of control system handling, loads, weight and flutter and vibration and with respect to specifications for transport or military airplanes. There are often contradictory effects which would limit the full use of the optimal solution for one purpose due to restrictions and requirements from the various disciplines. During structural design with ACT compatibility of different requirements should be achieved with a reasonable amount of costs. Some considerations in this direction were described in several papers [1,2,3]. In this paper typical examples will be demonstrated which illustrate problem areas arising from ACT.

48

The first example deals with active gust alleviation and ride improvement investigations on a transport (commuter) airplane [4]. Improvement of passenger comfort could be reached but some adverse effects on handling and local dynamic design and fatigue loads had to be paid for the profit.

The second example concentrates on aspects of an original unstable military aircraft with flight control system (CCV-FCS). Here the degrading effects of static elastic deformations on handling and on ride qualities are illustrated. In addition structural coupling prevention is of special interest.

It is shown in advance that a discrete gust design load calculation (dynamic loads) may be of interest also for military aircraft wing dynamic loads, although these airplanes are mainly designed for high static g levels.

2. General Remarks

The aircraft structural design philosophy is well defined and established for unaugmented versions. When dealing with ACT the procedures of defining all properties, for instance static and dynamic load cases, may be different for the various ACT applications, especially if artificial stability is introduced for drag reduction to the aircraft.

In general in the case of a controlled aircraft, reactions of the elastic aircraft structure are measured by sensor signals (Fig. 1) and introduced in the feedback loop. The sensor signals are contaminated by static elastic deformations in case of flow sensors, as for example angle of attack and sideslip α, β vanes or radar signals from the deformable front fuselage. The sensor signals are influenced as well by global rigid and local elastic aircraft motions excited by turbulent gusts, maneuvers or other excitations.

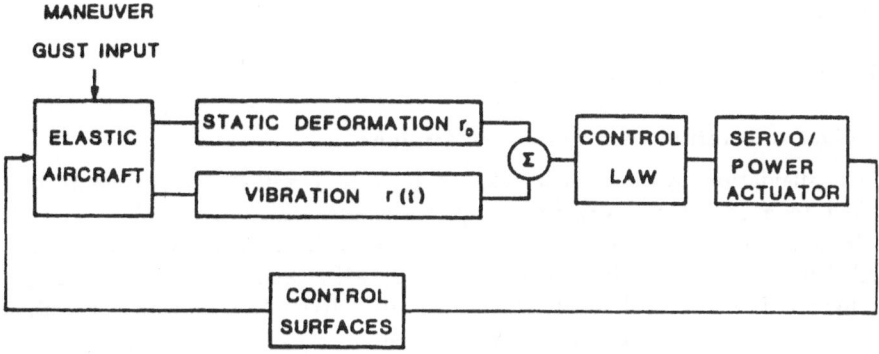

Figure 1: Block diagram

In addition the controlled aircraft may strongly be influenced by elastification of control surfaces at high dynamic pressure. The static deformations there leads to reductions of control surface efficiencies and therefore to reductions of stability and damping of flight dynamic modes and to degrading effects on handling (for A/C with swept back wing).

In the case of an aircraft with flight control system only the rigid aircraft state is of interest to optimise handling criteria, any additional feedback of elastic contributions will lead to degrading effects through signal deterioration, structural coupling, vibration excitation and dynamic load amplification. Means are necessary here to avoid these coupling problems (optimisation of sensor location and signal filter technique like notch filtering for the elastic vibrations or optimisation of local stiffness for static deformations).

The problem area is different for an aircraft with gust alleviation system. There dynamic loads or vibrations should be minimised by the control system.

Often the consequence is that the low frequency vibration and the corresponding dynamic loads are reduced, however, the A/C handling characteristics are changed. Similar contradictory results occur when *elastic mode control* or *flutter suppression* systems are installed. The vibrations are alleviated, however, dynamic loads may increase.

Compatibility will be achieved only by looking into all aspects from the beginning of structural design.

The requirements which are important for the structural and control design are therefore mainly

- Small static elastic deformations for high control surface efficiency and low static signal contamination.
- Adequate sensor positioning to avoid structural coupling.
- Notch filter layout.

3. Examples of Active Control

3.1. *Gust Alleviation on a Transport Airplane*

A gust alleviation system with respect to ride comfort improvement can be installed to a commuter airplane. Such a system, which is based on the closed loop feedback of vertical fuselage acceleration to drive symmetrically the aileron installed in the wing, has been investigated by MBB and AIT [4]. The design criteria here is the so-called RIDE DISCOMFORT INDEX $D_i\,g$ which is defined in Fig. 2. The requirement for a flight phase duration of 0.5 to 1.5 hours is $D_i \leq 0.2g$.

D_i is calculated by the expression given in Fig. 2 where $T_{cs}(f)$ is the transfer function of the vertical acceleration at each passenger seat of the aircraft. Then the power spectral density of the vertical acceleration is built by the use of the

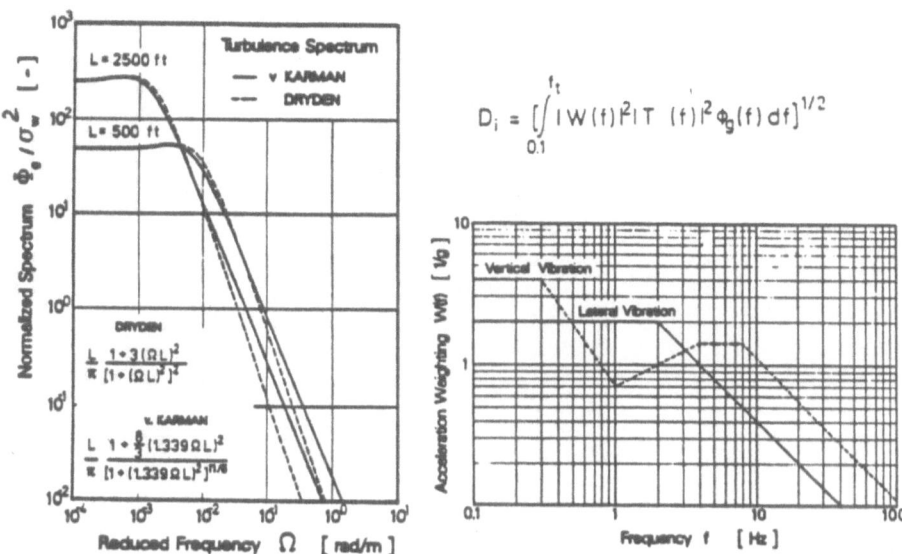

Figure 2: Ride comfort criteria

PSD of turbulent gust in the form of v. Karman or Dryden and multiplied by a weighting function, which takes into account human receptance. When the system is installed the stability of the aircraft is changed, the damping is strongly increased as demonstrated in Fig. 3a by the gain root locus diagram and the comfort is improved by about 50% as shown in Fig. 3b in the diagram of the ride discomfort index D_i which also shows the amount of aileron deflection (rms deg).

a) *First aspect* here is to fulfill the requirements of the aileron actuators in terms of maximum deflection under load and of maximum aileron rate (actuator power).

b) The *second aspect* is the aim to meet the handling requirements in terms of the short period mode frequency and damping. As illustrated in Fig. 3c, the PSD of C.G. acceleration is strongly reduced for a feedback of the vertical acceleration of $K_{n_z} = 0.3$. However, the frequency of the short period is reduced compared to the unaugmented result which indicated degradation of flying qualities, though the requirements are met.

c)   The third aspect is to consider global aircraft elastic vibration modes and their influence on the dynamic response and the passenger comfort. This is illustrated in Fig. 4 by the comparison of the results of vertical acceleration along the fuselage axis of the commuter aircraft for the rigid and elastic case.

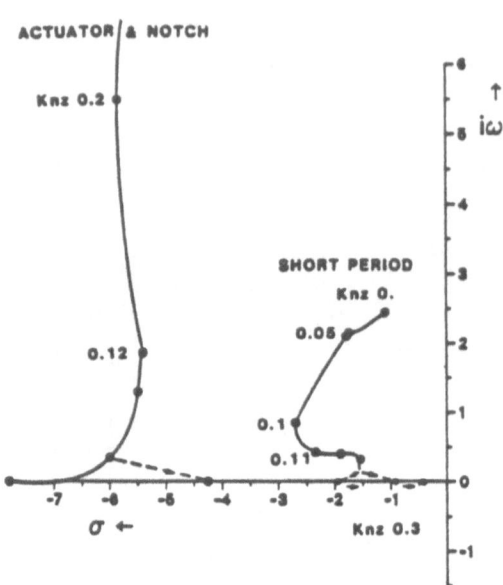

Figure 9a: *Transport airplane with ride improvement system: Gain root locus diagram*

d)   The fourth step which needs full attention is the prevention of structural coupling. In Fig. 4 it is demonstrated, that a control system without an adequate notch filter will lead to excitation or degradation of the comfort (here especially at the front fuselage) due to additional elastic mode feedback. The no notch example shown does include a low pass filter, therefore the effect is not as severe.

e)   The next important step in system layout is to demonstrate compatibility with design loads, in this case with design loads of the unaugmented aircraft. The design loads with control system should be equal or less compared to the loads without system. In Fig. 5 the distributed loads along wing span are demonstrated for the unaugmented airplane and the airplane with open and closed loop gust alleviation system. It is illustrated, that for both open and closed loop ride improvement systems with deflection and rate limitations for the aileron ± 8 deg and ± 30 deg/s the local shear and bending moment due to discrete gusts are decreased to a considerable amount. Thus the ride improvement system is compatible with the original structural design and can also serve as load alleviation system for design gusts. It is also demonstrated that the same system will locally increase the design loads in the local region of the ailerons, when the aileron activities are not limited.

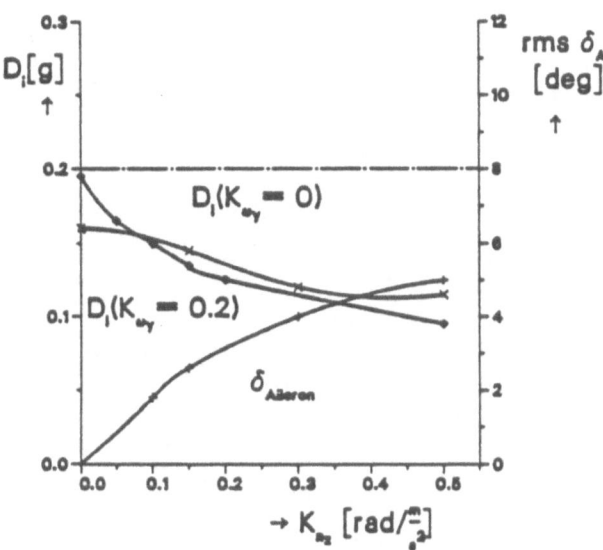

Figure 3b: Transport airplane with ride improvement system: Ride discomfort index

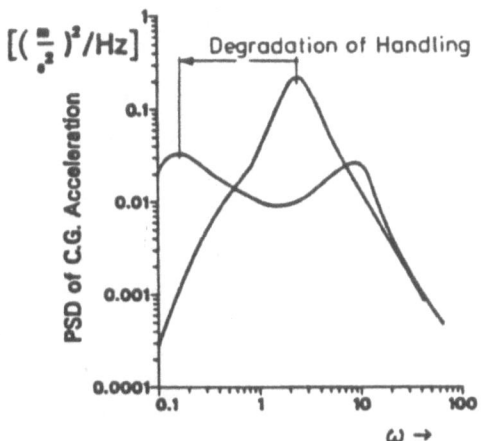

Figure 3c: Transport airplane with ride improvement system: PSD of C.G. acceleration

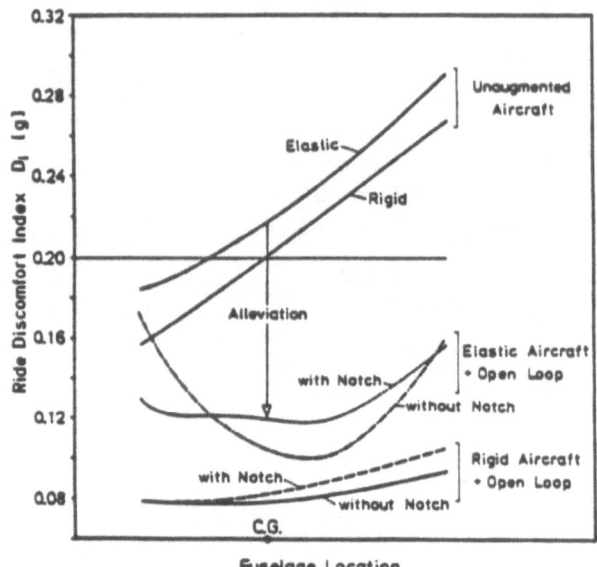

Figure 4: Ride discomfort index along fuselage

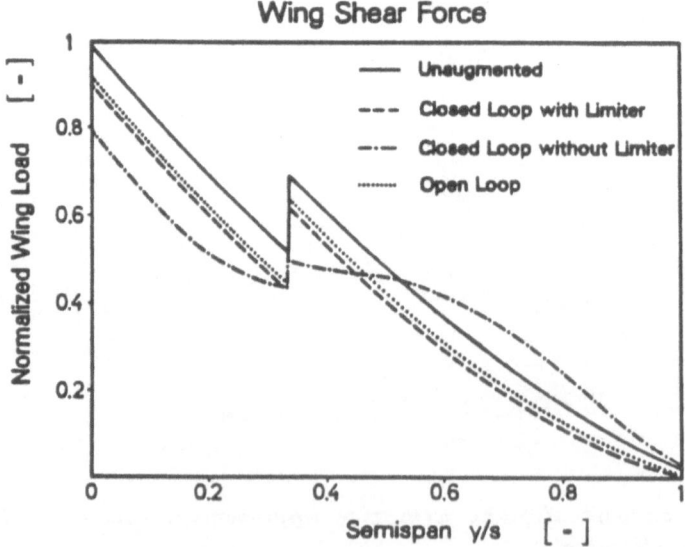

Figure 5a: Discrete gust analysis: Maximum local shear force

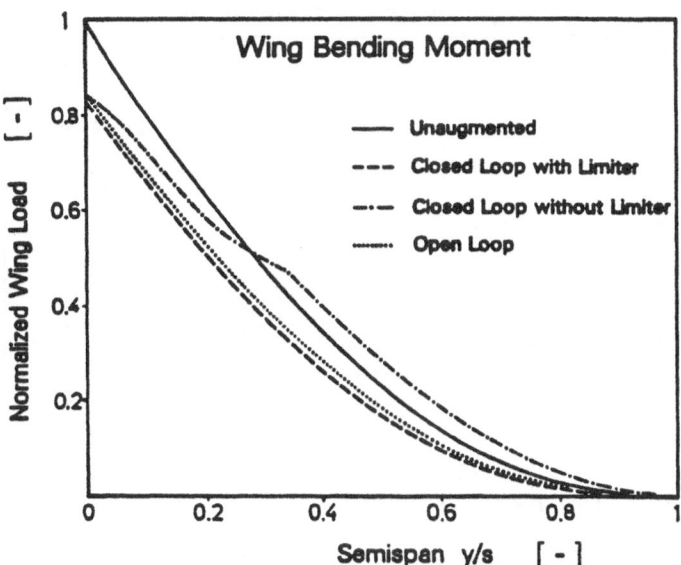

Figure 5b: Discrete gust analysis: Maximum local bending moment

f)   Finally the fatigue load aspect is considered.

The airplane response is investigated in turbulent gust environment. Both, the dynamic load distribution on the wing with and without ride improvement system is calculated. The excitation used in time domain computations is the Dryden spectrum defined in Fig. 2 with the gust intensity $\sigma_{w_j} = 1.4\,m/s$. As illustrated in Fig. 6 by the comparison of augmented and unaugmented results of the wing bending moment distribtuions, the control system generates an increase in rms wing bending moments, which would result in higher fatigue loads of the wing. The effect of limiters in the feedback loop would result in a reduction of fatigue loads as shown for the shear distribution on the wing in Fig. 6b. This result should be substantiated by total mission fatigue load evaluation.

The loads shown in the previous figures had been derived for rigid airplane response. It is of interest to note that the dynamic loads are influenced by the global airplane elastic mode response. Figures 7a and 7b illustrates the amount of structural dynamic loads by a comparison of the PSD of wing root shear and bending moment caused by the gust itself and the wing vibrations. Fatigue loads should be recalculated with the effects of the global elastic airplane.

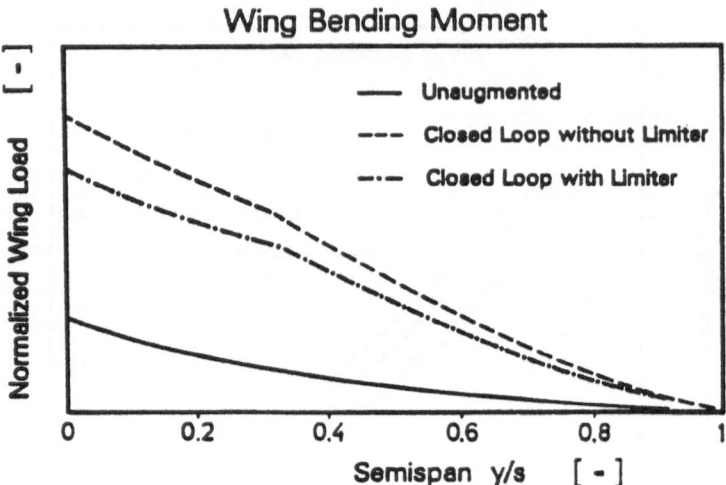

Figure 6a: Turbulent gust analysis: RMS local bending moment

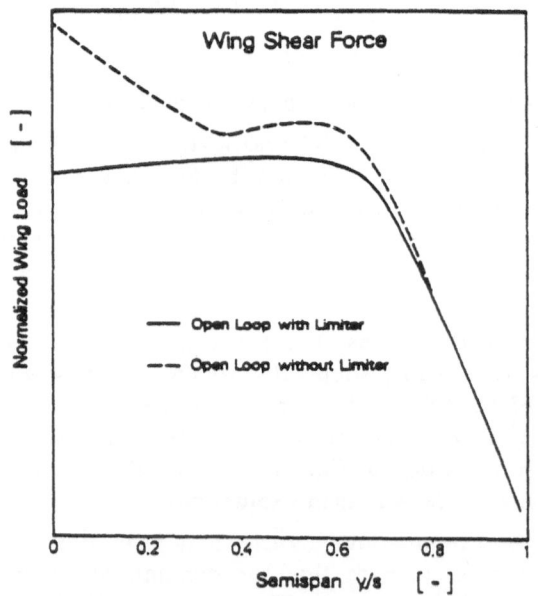

Figure 6b: Turbulent gust analysis: Maximum local shear force

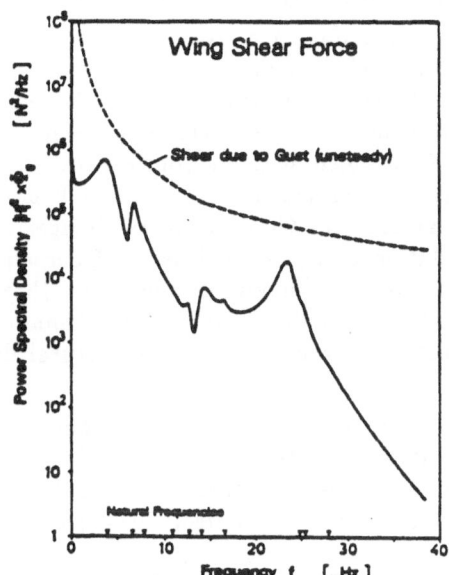

Figure 7a: PSD of wing root shear

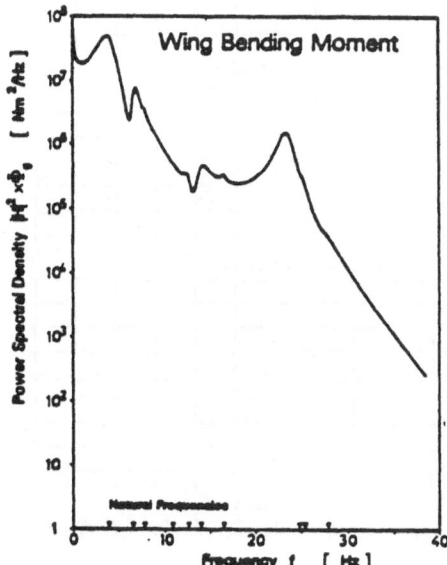

Figure 7b: PSD of wing root bending

3.2. Military aircraft with flight control system

This chapter deals with the impact of ACT on the dynamic response of an aircraft with introduced artificial stability (a control configured vehicle CCV) and the consequences on structural design mainly with respect to aeroelastic and structural dynamic influences. In the preliminary design of the FCS often rigid aircraft properties are assumed without consideration of reductions of control surface efficiencies due to aeroelastic effects at high dynamic pressures. These effects could result especially for CCV in severe changes of the flight mechanical behaviour and degradation of flying qualities and consequently in an unwanted increase of the global aircraft dynamic response and therefore in an increase of dynamic loads. To reach the requirements both for handling and for the loads the aeroelastic effects must be introduced in the structural design as well as in the FCS layout.

The example depicted in Fig. 8 shows the comparison of the incremental rigid and elastic aircraft response around a trimmed condition due to stick input. The FCS is designed for rigid conditions. The angle of attack oscillates with aeroelastic effects included, and the control surface deflections η and δ respond in the same manner.

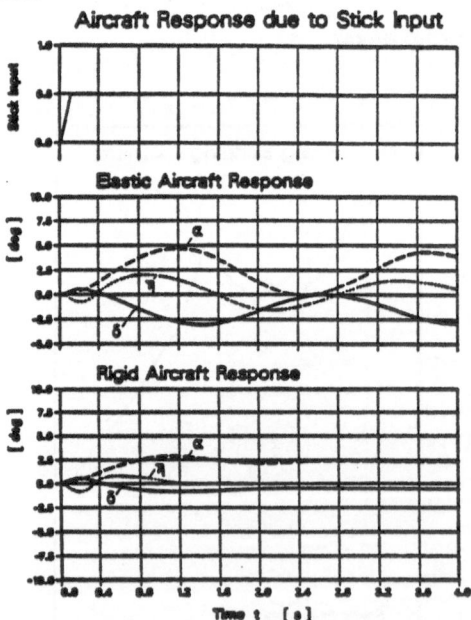

Figure 8: Aircraft response due to stick input (influence of control surface elastification)

If the structural design would be based on the rigid response behaviour the overswing of the angle of attack, which is due to loss of stability, would indicate an overshoot of design loads.

If the same aircraft is excited by turbulent gust as defined in Fig. 2 the power spectral density of the vertical acceleration at the pilot seat may be derived in order to evaluate the ride qualities. Figure 9 demonstrates a higher peak in the spectrum for the elastic aircraft which would correspond to unacceptable ride qualities. The reason here again is the reduction of control surface derivatives caused by aeroelastic effects.

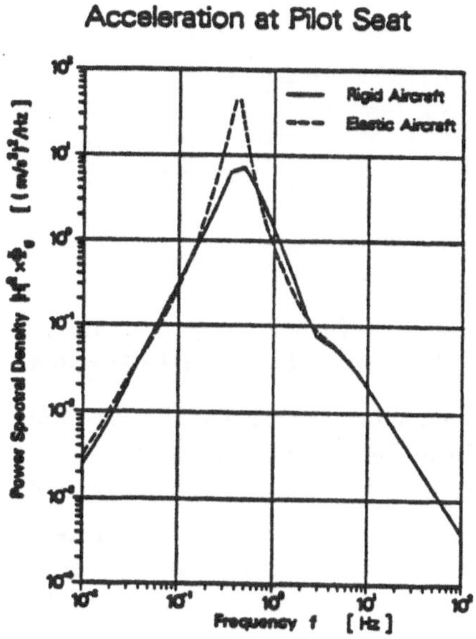

Figure 9: PSD of pilot seat acceleration

Structural coupling problems are of specific interest when CCV technology is applied. Gain and phase margins of the open feedback loop must meet the requirements for flight mechanical and elastic modes. Since the stability margins for the flight mechanical mode are difficult to fulfill at extreme flight conditions all additional means to minimize structural coupling problems, for instance low pass filtering and notch filtering will reduce the margins. Therefore the filter characteristics should be carefully predicted using a mathematical model of the global aircraft including coupled flight dynamics and structure dynamics. This model should be updated using results from ground resonance test and from flight tests when they are available. Figure 10 shows the comparison of the gain and phase margins in Bode plots for an unstable aircraft with FCS with and without notch filters.

BODE Diagram

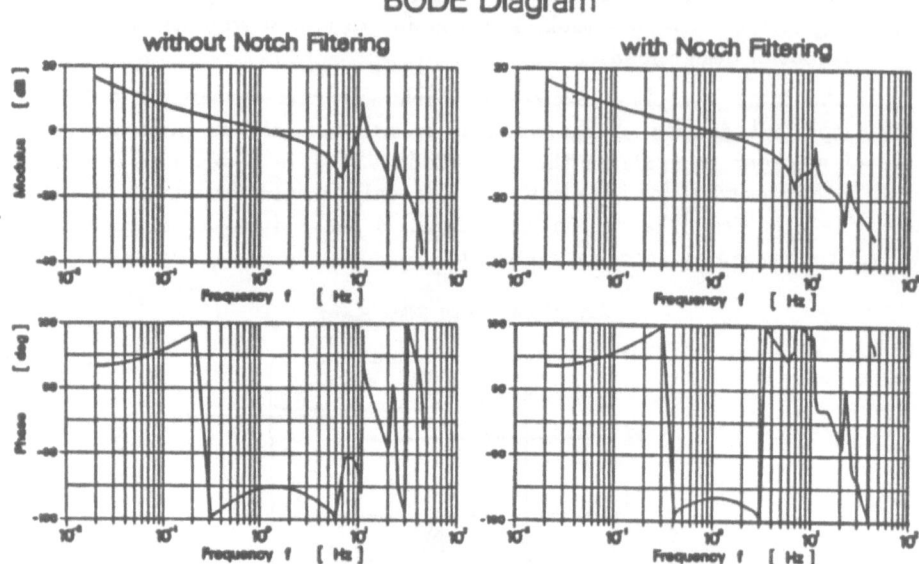

Figure 10: Open loop stability gain and phase margins. Comparison of BODE diagrams for the case with and without notch filter

With notch filters the elastic modes at 11 and 23 Hz have the proper gain margins, the phase margin of the flight mechanical mode (short period) reduces from 60 deg to 40 deg (at 0 dB) and also the gain margin (at -180 deg phase).

This demonstrates that the phase lag of notch filters at low frequencies (in flight dynamic region) should be minimised. Adequate sensor positioning together with structural design of the fuselage will reduce this problem area.

A discrete gust design load calculation which includes structural dynamic and unsteady aerodynamic effects is very seldom performed for military aircraft wing design purposes, since the construction for the high g maneuvers and corresponding static load cases are believed to cover also all dynamic effects. However, structural dynamic effects for wing design may play an important role.

In Figs. 11 and 12 the spanwise distribution of shear and bending moment on a military aircraft wing are demonstrated for the rigid and elastic aircraft for different gust lengths. At a gust length of 75 m the maximum wing root shear force and bending moment is reached. A smaller and higher gust length reduces the values at the root both for the rigid and elastic airplane. At the root the elastic influence due to wing vibrations can reach a 20% increase of the rigid bending moment at the root. The outer wing could experience however very high loads due to elastic structural vibrations at small gust lengths which could overrule static load requirements. This is shown for the outer wing at the gust length of 12.5 and 25 m.

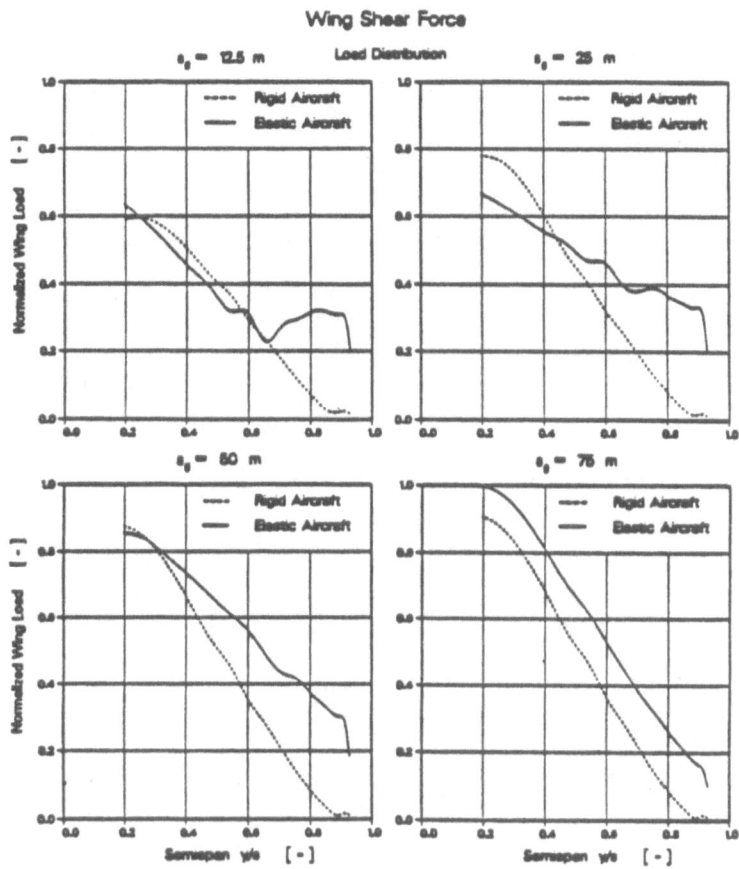

Figure 11: Design gust calculation: wing shear force

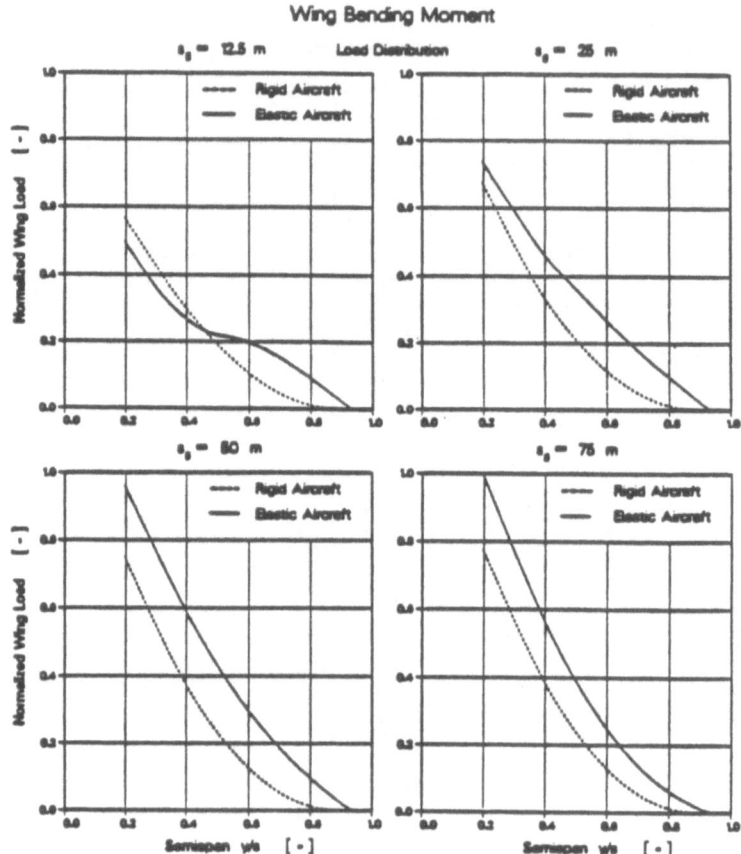

Figure 12: Design gust calculation: wing bending moment

4. Conclusions

Some aspects of contradictory requirements involved in structural design of aircraft structure had been described. These considerations are by no means complete and are mainly influenced by the work of the authors in the field of structural dynamic aircraft response.

References

[1] Collmann, K.D. and Sensburg, O., "Impact of a command and stability augmentation system on gust response of a combat aircraft", 44th Meeting of the SMP of AGARD, Lisboa, April, 1977.

[2] Zimmermann, H. and Sensburg, O., "Impact of active control technology on structures", Presentation at the 37th Annual Conference of the Society of Allied Weight Engineers (SAWE), SAWE-Paper No. 1242, Index Category No. 22, May, 1978.

[3] Sensburg, O., Becker, J. and Hönlinger, H., "Active control of flutter and vibration of an aircraft", *Proceedings of the International IUTAM Symposium*, June, 1979.

[4] Sensburg, O., Becker, J., Lusebrink, H. and Weiss, F., "Gust load alleviation on Airbus A300", Presentation at the 13th Congress of the International Council of the Aeronautical Sciences (ICAS), August, 1982.

[5] Becker, J., Weiss, F., Cavatorta, E. and Caldarelli, G., "Gust alleviation on a transport airplane", AGARD-Symposium on 'Unsteady Aerodynamics', Fundamentals and applications to aircraft dynamics, May, 1985.

STABILIZATION OF STRUCTURES
UNDER NON-CONSERVATIVE LOADS

Roman Bogacz
IFTR
Polish Academy of Sciences
Warsaw, Poland

1. Introduction

The study of the non-conservative systems stability is an important problem in engineering practice. The investigation of such systems as structures subjected to follower load or some lumped and continuous systems in relative motion, are similar. Both of them are described by a non-self-adjoint boundary value problem.

Assuming a steady state periodic solution of the problem in the form of standing or travelling waves one obtains a set of characteristic equations. The analysis of configuration of the characteristic curves in the parameters-plane provides the evaluation of critical states [1], [2]. The determination of the regions of stability makes possible a design of stable system for a given condition or it constitutes a basis for the active control and stabilization of such systems.

Two particular examples are presented. The first one deals with a column consisting of some elastic segments connected by elastic hinge-joints and subjected to a circulatory load. The second example in question is devoted to the stability analysis of relative motion of a single oscillator interacting with waves of a continuous system. A qualitatively new lay-out of the instability regions is obtained for the case of constant velocity of motion and two interaction points at least. A possibility of active control will be presented for the both examples.

2. Column under Circulatory Load

Various generalizations of Beck's problem have received considerable attention in recent years. Several investigators as Hauger [3], Wahed [4] have studied the influence of an elastic foundation and damping on the stability of a cantilever beam subjected to a follower force at its free end. In some cases, however, the critical load values were obtained by approximate methods without estimation of accuracy of the received results. Moreover, when supports or dampers are used as structural members acting at several distinct points on the column it is possible to obtain an exact solution [1], [5]. As shown in [1] the critical force for the case of single damper is independent of the damping coefficient, but depends mainly on the position of the damper. The first part of this paper is devoted to the generalization of above results to the case of Beck's column with some hinge-joints.

2.1. Formulation of the problem of column

In the following the structure principally shown in Fig. 1 will be considered. It consists of segments connected by some hinge-joints. In the general the joints are located at positions $x_1, x_2, ..., x_n$ and are characterized by the stiffness coefficients $k_1, k_2, ..., k_n$, respectively, or the dimensionless ones

$$K_j = \frac{k_j L^2}{EI} \quad , \; j=1,2,...,n \; .$$
(2.1)

The parameters L, EI refer to length and bending stiffness of the column. For the case of a simple structure, particularly when it is subjected to distributed follower load, good results can be obtained by use of Leipholz' generalization of the adjointness principle [6].

In the case of a more complicated structure with discontinuities of stiffness, cross-section or foundation at several distinct points the exact solution can be obtained by using transfer matrix technique [1], [5].

The simplest form of equation of motion of a uniform segment reads

$$EI \frac{\partial^4 y}{\partial x^4} + P \frac{\partial^2 y}{\partial x^2} + \rho A \frac{\partial^2 y}{\partial t} = 0 \quad ,$$
(2.2)

with

P longitudinal force

ρ density

A cross-sectional area .

The boundary conditions for the case of clamped end are

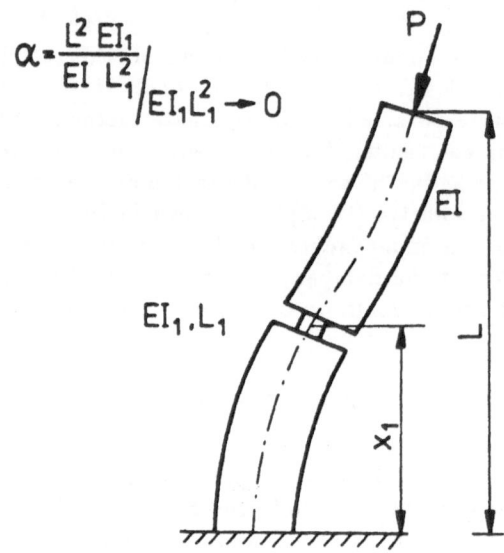

Figure 1: Column with Hinge-Joint

$$y=0 \quad , \quad \frac{\partial y}{\partial x}=0 \quad . \tag{2.3}$$

For the case of free end with tangential force we have

$$\frac{\partial^2 y}{\partial x^2}=0 \quad , \quad \frac{\partial}{\partial x}(EI \frac{\partial^2 y}{\partial x^2})=0 \quad , \tag{2.4}$$

and for the free end of Reut's case it follows

$$EI \frac{\partial^2 y}{\partial x^2}+Py=0 \quad , \quad \frac{\partial}{\partial x}(EI \frac{\partial^2 y}{\partial x^2}+Py)=0 \quad . \tag{2.5}$$

The exact solution for this segment of constant mass and stiffness distribution has the form

$$y(x,t)=e^{i\omega t}(A_1 sh \lambda_1 x + A_2 ch \lambda_1 x + A_3 \sin \lambda_2 x + A_4 \cos \lambda_2 x) \quad . \tag{2.6}$$

where

$$\lambda_{1/2} = \left[\frac{\pm P}{2EI} + \left[(\frac{P}{2EI})^2 + \frac{\rho A \omega^2}{EI} \right]^{1/2} \right]^{1/2} \quad . \tag{2.7}$$

Since all dependent variables y, ϕ, M, Q have a similar constitutive form (2.8), the state vector S and the partial transfer matrix T_i can be expressed as follows. The state vector S is given by

$$S=[y,\phi,M,Q]^T=[y,y',EIy'',-EIy''']^T,$$

$$S^o_{i+1}=T_iS^o_i \quad , \quad S^o_j=S_j(x_j=0) \quad . \tag{2.8}$$

The transfer matrix for the segment is defined in [7]. Non-zero elements of the transfer matrix for a support sensitive to deflection are

$$t_{ii}=1 \quad , \quad t_{41}=K_j \tag{2.9}$$

and in case of a support sensitive to rotation

$$t_{ii}=1 \quad , \quad t_{32}=K_j \quad . \tag{2.10}$$

For a hinge-joint the non-zero elements are

$$t_{ii}=1 \quad , \quad t_{23}=K_j \quad . \tag{2.11}$$

The transfer matrix for the whole structure can be expressed by

$$T=T_n T_{n-1} \cdots T_2 T_1 \quad . \tag{2.12}$$

Satisfying the boundary conditions we get a characteristic equation as relation between force and frequency,

$$\Phi(P.\omega)=0 \quad . \tag{2.13}$$

It is to be noted that in the case of a dissipative structure, the characteristic equation (2.13) is of a complex form. In order to get the critical values, one may use the Mikhaylov stability criterion or a similar one [3].

2.2. Results of numerical calculations

Let us now consider the behaviour of the column with an elastic hinge-joint with stiffness characterized by a parameter

$$\alpha=\frac{L^2 E_1 I_1}{EI L_1^2}\bigg|_{L_1 \to 0} \quad . \tag{2.14}$$

It is interesting that for the case of stiffness of an elastic hinge-joint smaller than the stiffness of the whole column ($\alpha<1$) the critical value is greater as in the case of uniform column without a hinge-joint.

The critical forces versus stiffness of a hinge-joint for various values of x_1 describing the position of the joint are shown in Fig. 2. It can be seen that for $x_1=0.5$ the critical value of load increases rapidly in the range $\alpha\in(0.1, 1.0)$ with decreasing stiffness taking for $\alpha=0.001$, a value about four times greater than for the case without joint ($\alpha\to\infty$).

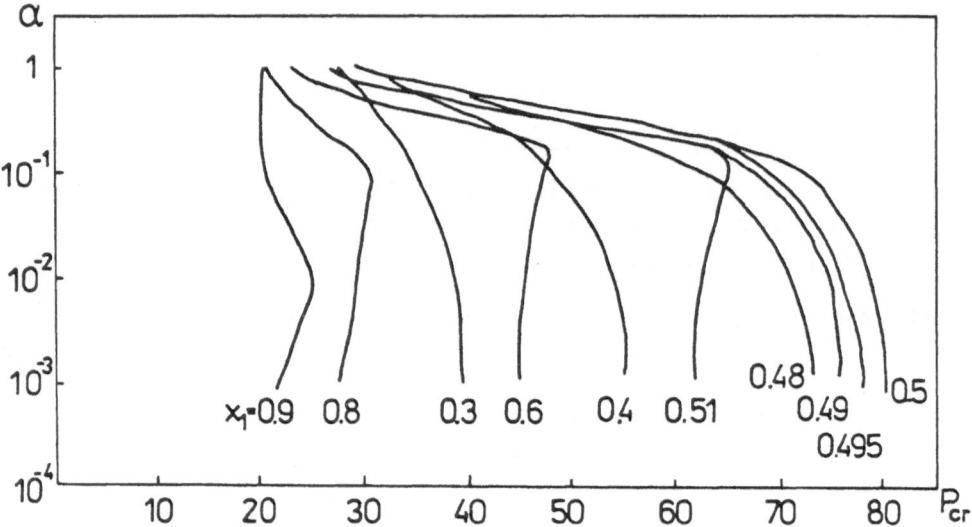

Figure 2: Critical Load Versus Hinge-Joint Stiffness

For the values of $x_1 \in (0.5, 1.0)$ one can observe that there exists an optimal stiffness of the elastic joint for which the critical force reaches a maximum. The configuration of the characteristic curves for various joint locations x_1 and a joint stiffness $\alpha = 0.001$ is shown in Fig. 3.

It is interesting that for the joint location $x_1 \in (0.5, 1.0)$ we have a classic shape of characteristic curves with an increasing critical value if the position x_1 tends from $x_1 = 1.0$ to $x_1 = 0.5$.

For $x_1 = 0.5$ we observe an intersection of the characteristic curves and a jump of the critical force from $P_{cr} \simeq 71$ to the value $P_{cr} \simeq 81$ with a discontinuous decreasing of critical frequency from $\omega \simeq 20$ to $\omega \simeq 4.0$. Further change of the joint position in the direction of the damped end causes decrease of critical force and critical frequency. For the case of the column with elastic joint of stiffness $\alpha = 0.001$ located at the position $x_1 = 0.001$ the critical load is smaller as in the case without joint. The critical load as a function of dimensionless joint location is shown in Fig. 4 for some values of α. Similar as in the case of viscoelastic support there is a significant influence of damping on the critical load in case of inelastic hinge-joint. This problem will be discussed in a separate paper. Considerable increase and decrease of the critical load depending on the parameters of hinge-joint and sensitivity of the system to small variations of the joint location make possible passive and active control of the structure under consideration.

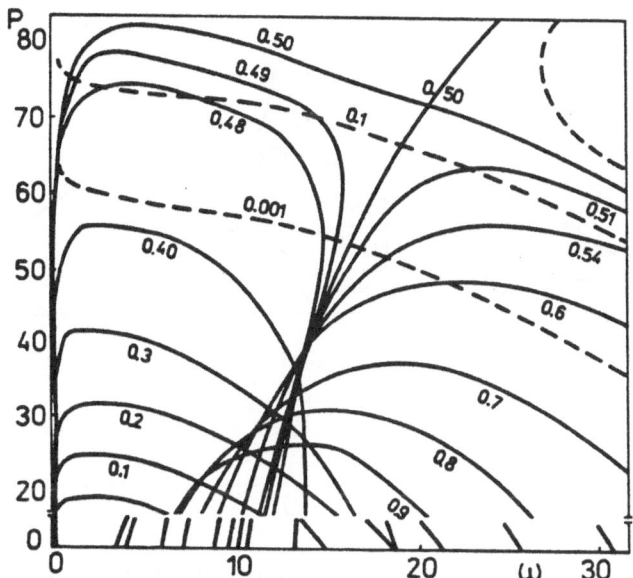

Figure 3: Force-Frequency Plane. Characteristic Curves for Various Joint Locations ($\alpha = 0.001$)

3. Formulation of the Problem of Interaction between Moving Lumped System and an Infinite Continuous System

We consider now similar as in [8] a system which consists of a one degree-of-freedom vehicle with mass m, spring constants c, distance between two contact points L, moving on an infinite beam supported by a linear elastic foundation or in the case $EI_1 = 0$, $T < 0$, a string under tension. The remaining parameters and the reference frames are specified analogously to ref. [8].

Now we analyze the case of a single periodic contact force $F(t)$ acting at the point $x_1 = 0$, cf. Mathews [8]. The corresponding equation of motion reads,

$$EI_1 \frac{\partial^4 w_1}{\partial x_1^4} + T_1 \frac{\partial^2 w_1}{\partial x_1^2} + \mu_1 \left(\frac{\partial^2 w_1}{\partial t^2} - 2V_o \frac{\partial^2 w_1}{\partial x_1 \partial t} + V_o^2 \frac{\partial^2 w_1}{\partial x_1^2} \right) + b_1 \left(\frac{\partial w_1}{\partial t} - \frac{\partial w_1}{\partial x} \right)$$

$$+ c\, w_1 - p(x_1, t) = 0 \ . \tag{3.1}$$

$$p(x_1, t) = F(t)\, \delta(x_1) = p\, e^{i\omega t} \delta(x_1) \ .$$

The solution can be composed of two parts,

$$w_1(x_1, t) = W_1(x_1, t)\, H(-x_1) + W_2(x_1, t)\, H(x_1) \ , \tag{3.2}$$

where $H(x)$ is the Heaviside unit function, i.e. $H(x) = 1$ if $x > 0$ and $H(x) = 0$ if $x < 0$. The functions W_1 and W_2 fulfill the following compatibility conditions at

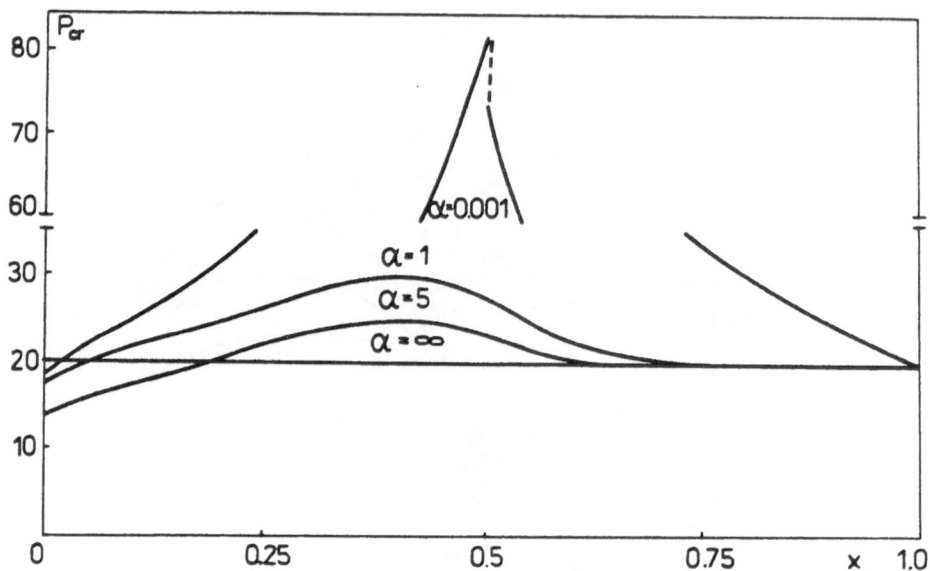

Figure 4: Critical Load Versus Joint Location for Various Joint Stiffnesses

$x_1 = 0$:

$$\frac{\partial^n W_1}{\partial x_1^n} = \frac{\partial^n W_2}{\partial x_1^n} \quad , \quad n = 0, 1, 2, \tag{3.3}$$

$$EI_1 \left(\frac{\partial^3 W_1}{\partial x_1^3} - \frac{\partial^3 W_2}{\partial x_1^3} \right) + p \ e^{-i\omega t} = 0 \quad . \tag{3.4}$$

Utilizing the condition of radiation and (3.2) ÷ (3.4) we obtain a relation between beam deflection $w_1(x_1, t)$, force $F(t)$, velocity V_o and frequency ω. In a steady-state it follows

$$\frac{w_1(x_1, t)}{F(t)} \bigg|_{t \to \infty} = \frac{w_{1\infty}(x_1)}{p} = G(x, \omega, V_o) \ . \tag{3.5}$$

For the free elastic waves the relation between wave velocity v and wave number k reads

$$v^2 = (EI_1 \ k^4 - T \ k^2 + c) \ (\mu \ k^2)^{-1} \ , \tag{3.6}$$

which leads to the critical velocity V_{cr},

$$\frac{\partial v}{\partial k} \bigg|_{v = V_{cr}} = 0 \quad \Rightarrow \quad V_{cr}^2 = \left[\frac{4c \ EI_1}{\mu_1^2} \right]^{1/2} - \frac{T_1}{\mu_i} . \tag{3.7}$$

Now we analyze the model shown in Fig. 5. Since symmetry is assumed, the equation of motion for the mass m takes the simple form

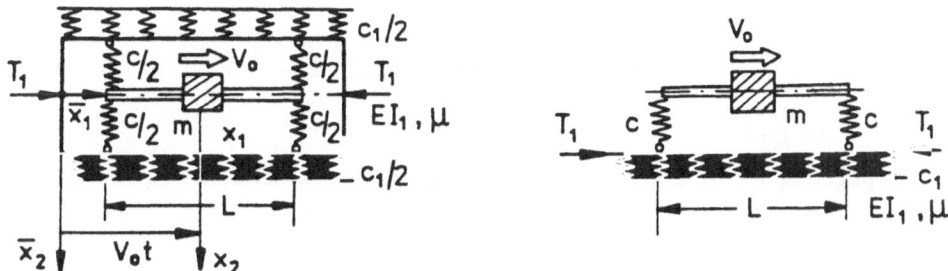

Figure 5

$$m \, \frac{d^2y}{dt^2} + f(t) = 0 \, , \qquad (3.8)$$

$$f(t) = c \left[2y(t) - w_1(\frac{-L}{2}, t) - w_1(\frac{L}{2}, t) + b \{ 2\frac{dy}{dt} - \frac{d \, w_1(-\frac{L}{2})}{dt} - \frac{d \, w_1(+\frac{L}{2})}{dt} \} \right] \qquad (3.9)$$

where y characterizes the mass displacement and $\frac{1}{2}f(t)$ denotes the contact force in x_2-direction. The equation of the continuous systems motion is given by (3.1), where the load reads

$$p(x_1,t) = \frac{1}{2} f(t) \delta(x_1 - \frac{L}{2}) + \frac{1}{2} f(t) \delta(x_1 + \frac{L}{2}) \, . \qquad (3.10)$$

For the steady-state motion from (3.8), (3.9) and (3.5), we obtain the characteristic equation in the form

$$G_o(\omega) + 2 G(0, \omega, V_o) + G(L, \omega, V_o) + G(-L, \omega, V_o) = 0 \, . \qquad (3.11)$$

For $V_o^2 < V_{cr}^2$ the function $G(x_1, \omega, V_o)$ reads

$$G(x_1, \omega, V_o) = [G_R(|x_1|, \omega, V_o) + i \, G_I(|x_1|, \omega, V_o) \, H(x_1) + G_R(|x_1|, \omega, V_o) +$$

$$-i \, G_I(|x_1|, \omega, V_o)] \, H(-x_1), \qquad (3.12)$$

where forms of the functions G_R and G_I for the case of beam are given in [7] and for the case of string eq. (3.11) takes a form:

$$[\frac{1}{2} a^3 - \beta_1 L \, \Phi(-\wedge, V_o)] \wedge^2 - [\kappa_1 \, \Phi(-\wedge, V_o) - \beta_1 L \, a^3] \wedge + \kappa_1 a^3 = 0 \, , \qquad (3.13)$$

where a denotes the velocity of elastic waves, $\beta_1 = b \, m^{-1}$, $\kappa_1 = c \, L^2 m^{-1}$ and $\wedge = (\epsilon + i\omega) L = \Sigma + i\Omega$.

For the case of the string under tension and $V_o \gtreqless a$ the function $\Phi(-\wedge, V_o)$ is given by the formula:

$$\Phi(-\wedge, V_o) = \frac{\exp\dfrac{-\wedge}{V_o - a} - \exp\dfrac{-\wedge}{V_o + a}}{2\,\gamma_o}\ (V_o^2 - a^2) \tag{3.14}$$

where $\gamma_o = 4\mu_1 L\,m^{-1}$.

Separating the real and imaginary part of the characteristic equation (3.13) we obtain:

$$[F_c(\Sigma,\Omega)\beta + \frac{a^3}{2}]\Sigma^2 + [F_c(\Sigma,\Omega)\beta + \frac{a^3}{2}]\Omega^2 + [F_c(\Sigma,\Omega)\,\kappa_1 - \beta\,a^3]\Sigma$$

$$-[2F_s(\Sigma,\Omega)\beta]\Sigma\Omega + [F_s(\Sigma,\Omega)\kappa_1]\,\Omega - \kappa_1 a^3 = 0\,, \tag{3.15}$$

$$i\{F_s(\Sigma,\Omega)\,\beta\Sigma^2 + F_s(\Sigma,\Omega)\,\beta\Omega^2 + F_s(\Sigma,\Omega)\kappa_1\Sigma - [2F_c(\Sigma,\Omega)\,\beta - a^3]\,\Sigma\Omega$$

$$-[F_c(\Sigma,\Omega)\,\kappa_1 - \beta\,a^3]\} = 0\,, \tag{3.16}$$

where

$$F_c(\Sigma,\Omega) = \frac{V_o^2 - a^2}{2\gamma_o}\exp\frac{-\Sigma}{V_o + a}\left(\exp\frac{2\Sigma}{V_o^2 - a^2}\cos\frac{\Omega}{V_o + a}\right.$$

$$\left. +\ \exp\frac{-2\Sigma}{V_o^2 - a^2}\cos\frac{\Omega}{V_o - a}\right), \tag{3.17}$$

$$F_s(\Sigma,\Omega) = \frac{V_o^2 - a^2}{2\,\gamma_o}\exp\frac{-\Sigma}{V_o + a}\left(\exp\frac{2\Sigma}{V_o^2 - a^2}\sin\frac{\Omega}{V + a}\right.$$

$$\left. +\exp\frac{-2\Sigma}{V_o^2 - a^2}\sin\frac{\Omega}{V_o - a}\right) \tag{3.18}$$

Now, using (3.15) $\div$ (3.18) and the instability conditions the boundaries of regions where instability is possible can be plotted in the V_o,Ω-plane. The results of estimations of instability in the case without damping were given in [9] (Fig. 8) and for the case of damping are shown in Fig. 6 and Fig. 7.

It is interesting to note that also in the case of two points of interaction the region of instability for the damping intensity equal zero are different as in the case of damping tending to zero. If the value of a generalized damping coefficient β increases the regions of instability change the form and for a very great value of β the configurations take the form shown in Fig. 7. Above estimations obtained from the characteristic equations (3.15), (3.16) give only a qualitative impression on the influence of damping on stability of a hybrid system in relative motion. It is easy to see that for the real case ($\beta > 0$) the regions of instability are separated in the V_o,Ω-plane and bounded as well in Ω as in V_o direction. This

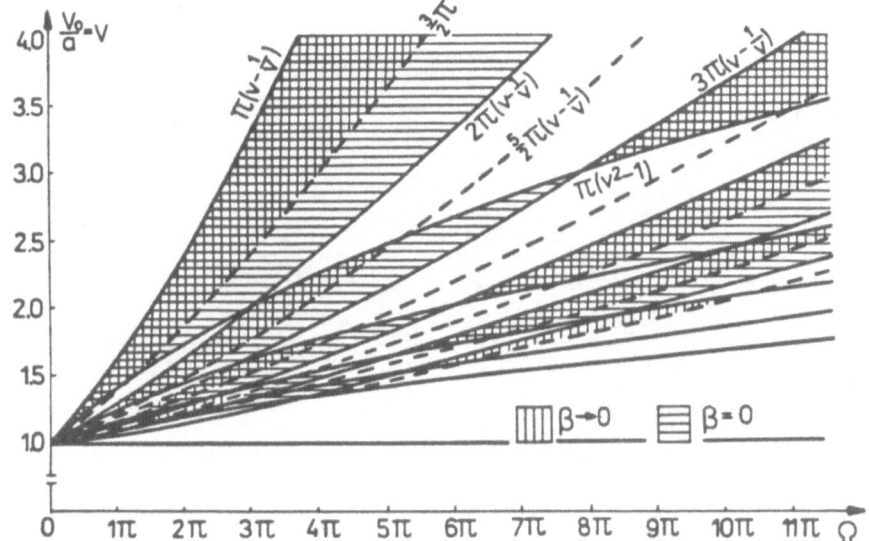

Figure 6

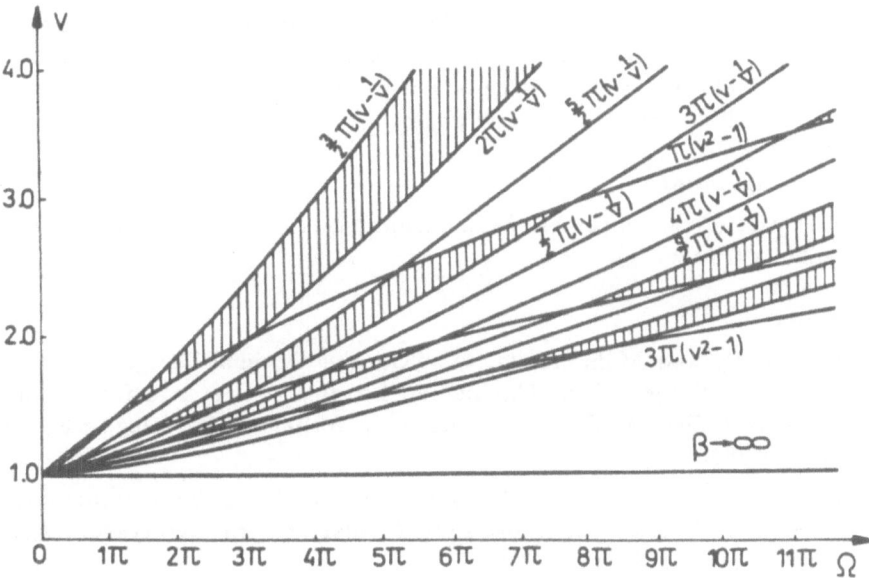

Figure 7

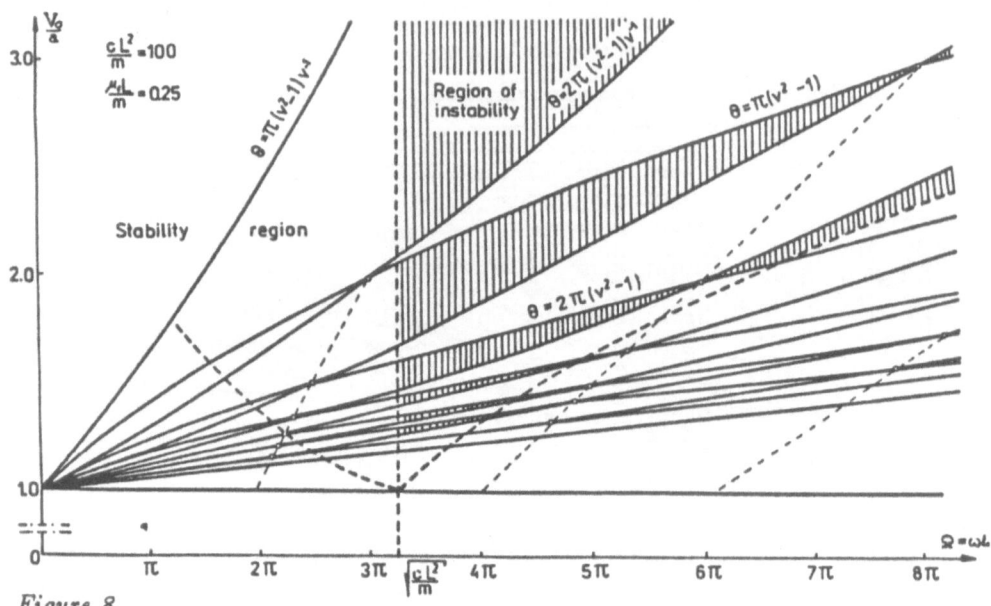

Figure 8

makes it possible to stabilize such a system by the control of a distance between both points of the interaction L, rigidity of springs c or damping intensity b. Furthermore, more precise estimations of instability regions show that also the first critical velocity of relative motion depends on the value of damping. This effect will be discussed in a separate paper.

Acknowledgements

The first part of this paper has been performed in cooperation with Professor O. Mahrenholtz (Hamburg).

References

[1] Bogacz, R. and Mahrenholtz, O., "On the Optimal Design of Viscoelastic Structures Subjected to Circulatory Loading", *Optimization Methods in Structural Design*, Ed. H. Eschenauer, Wissenschaftsverlag, Siegen, 1983, pp. 281-289.

[2] Bogacz, R. and Popp, K., "Dynamics and Stability of Train-Track-Systems", *Proc. 2nd Intern. Conf. on Recent Advances in Structural Dynamics*, Southampton, 1984, pp. 711-725.

[3] Hauger, W. and Vetter, K., "Influence of an Elastic Foundation on the Stability of Tangentially Loaded Column", *J. Sound and Vibration*, 47, 1976, pp. 296-299.

[4] Wahed, I.F.A., "The Instability of a Cantilever Beam on Elastic Foundation under Influence of Follower Force", *J. Mech. Engng. Sci.*, 1975, pp. 219-222.

[5] Bogacz, R. and Mahrenholtz, O., "Optimally Stable Structures Subjected to Follower Forces", *Structural Control*, Ed. H.H.E. Leipholz, North Holland Publ. Co., 1980, pp. 139-157.

[6] Leipholz, H.H.E., "On a Variational Principle for the Clamped-Free Rod Subjected to Tangential Follower Forces", *Mech. Res. Comm.*, 5, 1978, pp. 335-359.

[7] Bogacz, R. and Mahrenholtz, O., "Modal Analysis in Application to Design of Inelastic Structures Subjected to Circulatory Loading", *Proc. Euromech 174 on Inelastic Structures under Variable Load*, Ed. C. Polizzotto, Palermo, 1984, pp. 377-386.

[8] Mathews, P.M., "Vibrations of a Beam on Elastic Foundation", *Z. Angew. Math. uMech.*, 39, 1959, pp. 13-19.

[9] Bogacz, R., "On Self-Excitation of Moving Oscillator Interacting at Two Points with a Continuous System", *Non. Vibr. Probl.*, 19, 1979, pp. 240-250.

AN ACTIVE PROTECTION SYSTEM FOR WIND INDUCED VIBRATIONS OF PIPE-LINE SUSPENSION BRIDGES

A. Carotti
Polytechnic of Milan
Milan, Italy
M. De Miranda
de Miranda & Associates
Consulting Engineers
Milan, Italy
E. Turci
Polytechnic of Milan and
AERITALIA Space Systems Group
Turin, Italy

1. Introduction

1.1. Aims and limits of the study

In recent years some large pipe-line suspension bridges, managed by an Italian chemical manufacturer, presented excessive oscillation amplitudes due to wind. Although some passive type measures are available which have proved to be fairly efficient, the problem of protection of the serviceability limit state of these structures (or, in the worst cases, of the ultimate one) is still open to further research and alternative proposals. The particular working conditions of active control immediately dictate the following main characteristics of the preliminary design:

76

i)   the relative structural simplicity of the feedback scheme and its components,

ii) robustness of control, adoption of project and management strategies which permit efficient intervention even given unfavourable initial conditions and with relatively low costs,

iii) possibility of an autonomous energy source.

The present study which focuses attention on actual case (Fig. 1) of a tubular deck with a symmetricl cross section should be understood as a feasibility study of the active installation. The problem is immediately underlined as to how to project the instrumentation in view of the setting up of experiments on physical models and/or prototypes.

The first preliminary numerical simulations of various working conditions (resonance under Karman vortex discharge) and the evaluation of the orders of magnitude involved permit a first comparison of the costs and benefits of various solutions.

1.2. General aspects of the problem

Cable supported bridges, generally being light and deformable constructions, are rather sensitive to wind action which can induce, in particular conditions, oscillation amplitudes incompatible with the serviceability state of the bridge itself. The dynamic action of the wind reveals itself in the form of impulses induced by gusts or by buffeting due to upstream obstacle, in the effects of vortex shedding (Karman), or in self-excitation resulting from oscillations of the structure (flutter). The project criteria and measures adopted in order to limit the oscillations to acceptable levels have until now been the following:

*   flexural and torsional stiffening of the girder by the addition of a stay system or by inclining the hangers in suspension bridges,

*   mechanical damping devices,

*   aerodynamic devices,

*   aerodynamic profiling of the deck section.

However, these systems may sometimes be difficult to apply especially if the measures are taken after the structure has been completed, and anyway they impose serious limitations on the structure project itself. Structures destined to carry low working loads such as footbridges or road bridges with a long span must be adequately weighted or stiffened in order to withstand the dynamic effects of wind. This is also the case with pipeline suspension bridges, i.e., with structures where there is no real stiffening girder and where the weight is extremely low (i.e. gas carrying pipelines). For these types of structure the limitation of oscillations has until now been achieved by using aerodynamic devices (Fig. 2) or by stiffening the structure by means of, for example, an arrangement of lower stays connecting the frame to the bases of the pylons (Fig. 3). An active control system for the oscillations of this type of structure, and in general for

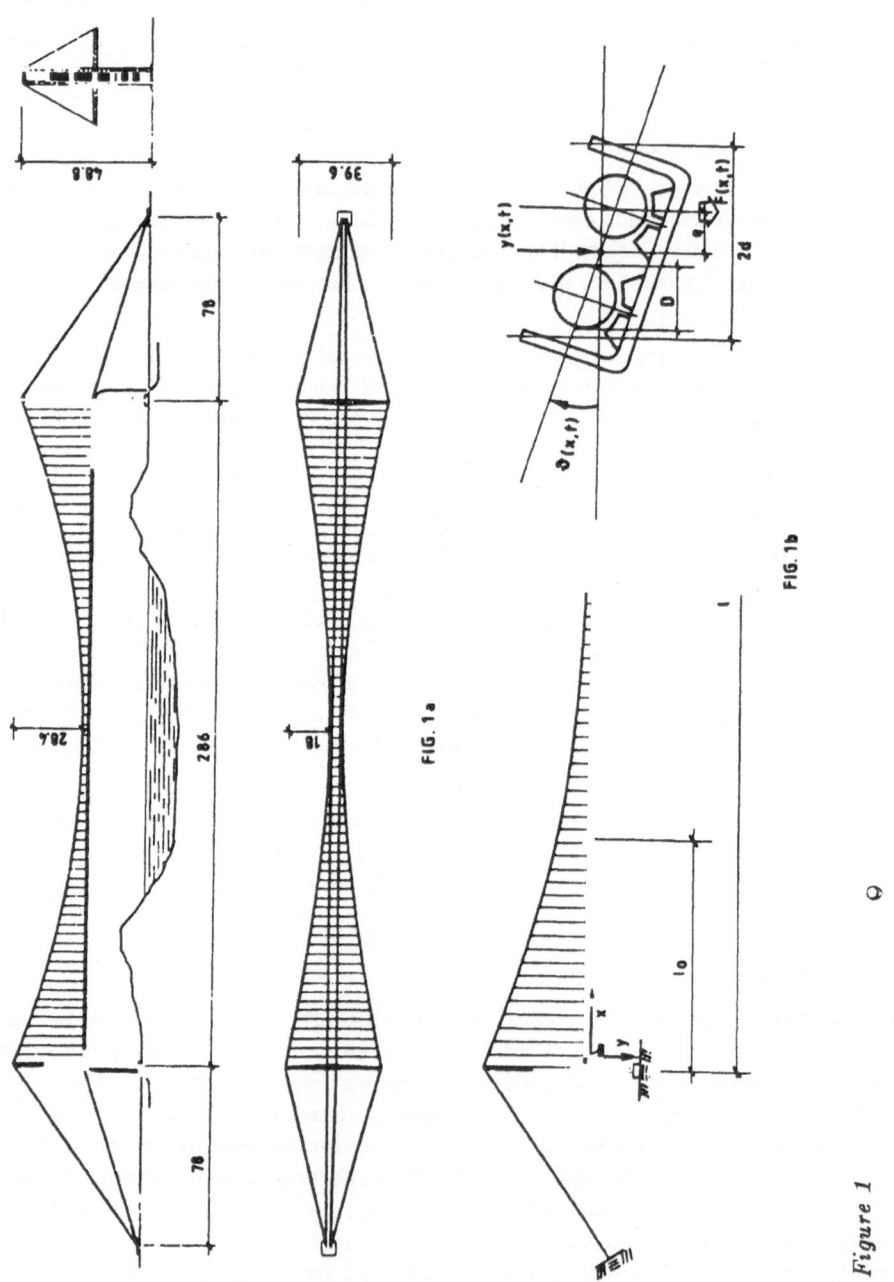

FIG. 1a

FIG. 1b

Figure 1

very slender and light bridge structures may have the following advantages:

- eliminating, or at least reducing, the project limitations imposed by the dynamic effect of wind,

- providing a general control system for oscillations induced by various causes: vortex shedding, buffeting, flutter, crowd action on pedestrian bridges, etc.,

- providing a control system applicable to existing structures with a minimum of interventions.

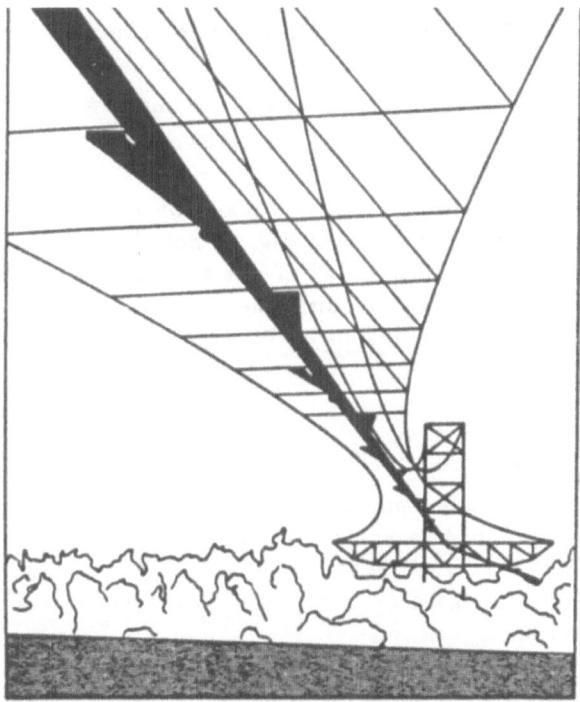

Figure 2: Field installation of aerodynamic damper system

The specific theme dealt with, although particular, can however be considered representative of an original approach to a general theme of active oscillation control in suspension bridges.

1.3. Configuration of the control system

For the practical application of the active control to bridge structures it is necessary to resolve first the problem of ensuring the functioning of the control devices (if possible independently from an energy network) and of minimizing the size of the force applied. The control system envisaged has therefore the following charactristics:

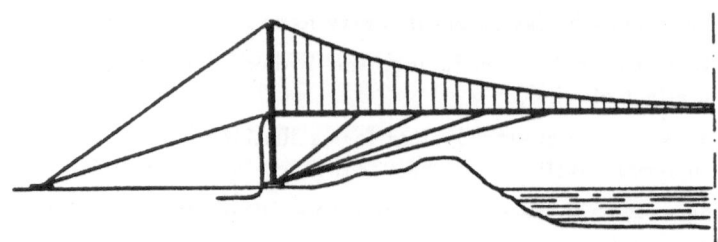

Figure 3

- the energy necessary for the application of the control forces is supplied by an Aeolian generator,

- the forces are applied to the structure by means of secondary cables loaded with forces acting orthogonally to the cables themselves, benefiting thereby of a favourable "lever effect".

The choice of the final configuration was made by critically examining various possible configurations (Figs. 4a,b,c and Fig. 5). Solution "c" seemed the simplest having the avantage over solution "a" of not encroaching on the zone beneath the deck. This solution, however, requires the pre-stressing of the control tendons with the aims of providing the structure with both positive and negative forces.

2. Math-Modelling

2.1. Equations for flexural-torsional bridge dynamics

Consider the two-dimensional pipeline suspension bridge described in Figs. 1a,b with the symbols listed in Appendix A. A portion $F^*(x,t)$ of the external disturbance density $F(x,t)$ is carried by the cable and the remaining $(F(x,t)-F^*(x,t))$ by the double tubular bridge deck. From the static equations of the cable and of the beam one obtains:

$$T_o y_c''(x) + \Delta T y_o''(x) = F^*(x) \qquad (2.1.1)$$

$$EJ y^{IV}(x) = F(x) - F^*(x) \qquad (2.1.1)'$$

and from $y_c = y$ (stiffness of the hangers) it follows that $F^*(x)$ can be expressed through y. The (2.1.1)', when inertia forces and structural flexural damping are also considered, leads to the equation governing vertical oscillations of the pipe-deck:

$$m y_{tt}(x,t) + c_y y_t(x,t) + (EJ\nabla_x^4 - T_o\nabla_x^2)\cdot y(x,t) = F(x,t) + \Delta T_y y_o''(x) \quad (2.1.2)$$

From the static torsional equation

$$-GC\theta''(x) = M(x)$$

considering the resisting torsional effect due to the distance 2d between the suspension cables, the torsional inertia $(J_p\theta_{tt}(x,t))$ and the structural damping,

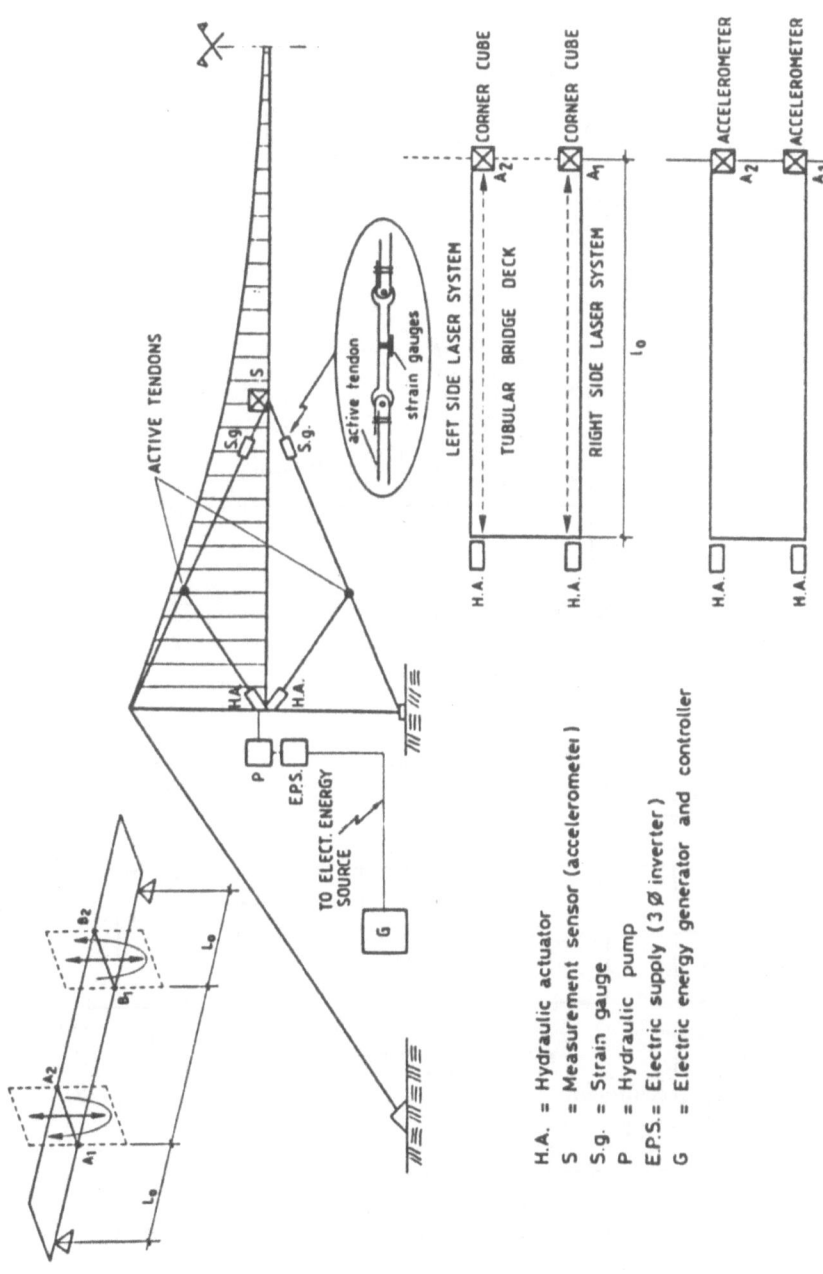

H.A. = Hydraulic actuator
S   = Measurement sensor (accelerometer)
S.g. = Strain gauge
P   = Hydraulic pump
E.P.S.= Electric supply (3 Ø inverter)
G   = Electric energy generator and controller

Figure 4a

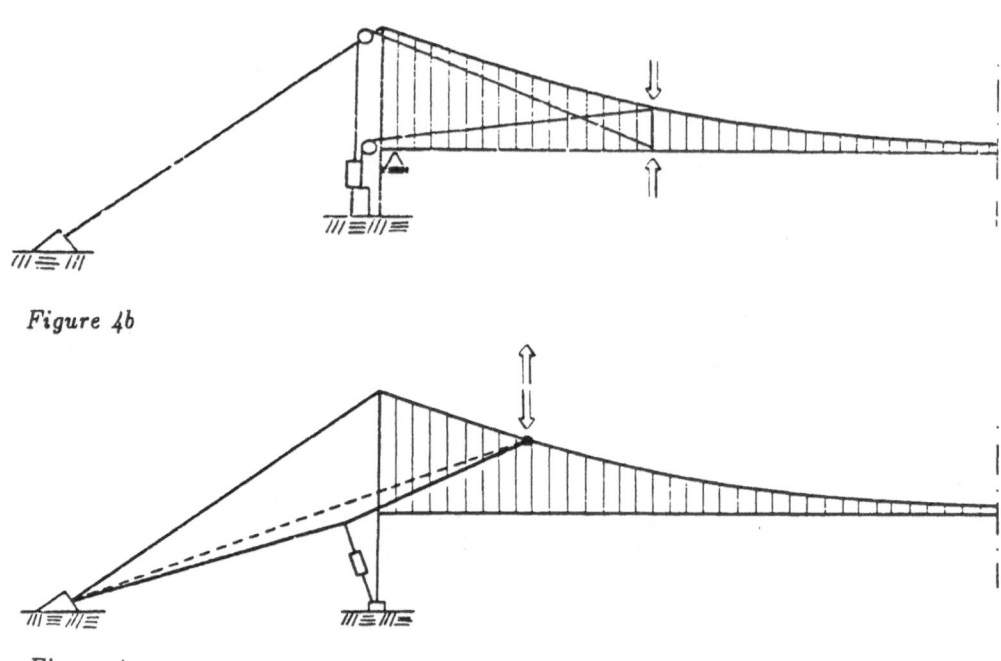

Figure 4b

Figure 4c

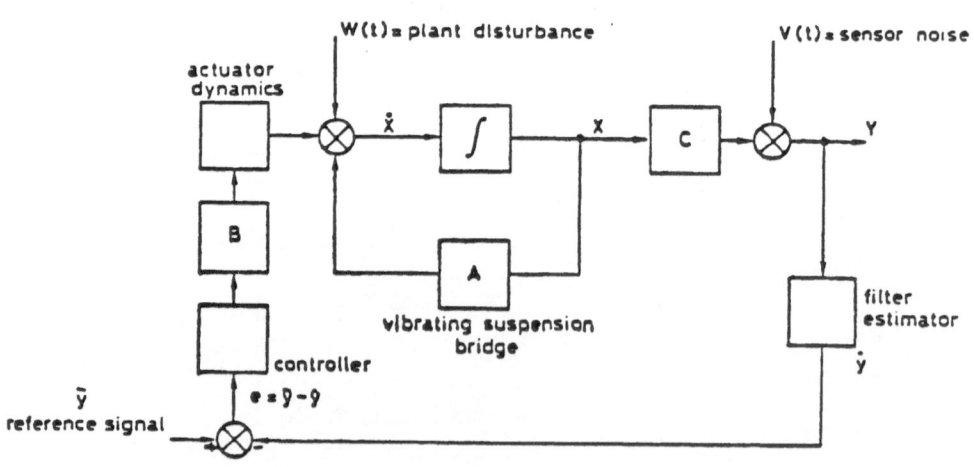

Figure 5: State-space normal form of the time invariant linear model

the following torsional equation of motion is obtained:

$$J_p \theta_{tt}(x,t) + c_\theta \theta_t(x,t) + d^2 \cdot [EJ\nabla_x^4 - (T_o + CG)\nabla_x^2] \cdot$$

$$\theta(x,t) = M(x,t) + \Delta T_\theta y_o''(x) \tag{2.1.3}$$

The external disturbance function $F(x,t)$ will be subsequently used to modelize the density of the pulsating lift force induced in the downstream pipeline by Karman vortex shedding along the span of the bridge; the function $M(x,t)$ will describe the corresponding density of torsional moment due to the eccentricity of $F(x,t)$. Horizontal motion due to drag is not considered in the present study. From this point on, formal development will mainly be referred to non-extensional vibrations (i.e. to the 1-st flexural and torsional asymmetric mode), and also to the 1-st symmetrical one:

$$Y_k(x) = \sin\frac{k\pi x}{l} \quad \text{frequency } \omega_k, \text{ damping ratio } \xi_k \quad (k=2) \tag{2.1.4}$$

$$\theta_k(x) = \sin\frac{k\pi x}{l} \quad \text{frequency } \widetilde{\omega}_k = \mu\omega_k, \text{ damping ratio } \widetilde{\xi}_k; \ \mu > 1 \ (k=2) \tag{2.1.4'}$$

The former is often the first to become dangerous since it is excitable without extension of the suspension cables. It follows that $\Delta T_y = \Delta T_\theta = 0$, and if the torsional stiffness is neglected, the partial differential vector equation of the 2-dimensional motion is:

$$\begin{bmatrix} m & 0 \\ 0 & J_p \end{bmatrix} \begin{Bmatrix} y_{tt}(x,t) \\ \theta_{tt}(x,t) \end{Bmatrix} + \begin{bmatrix} c_y & 0 \\ 0 & c_\theta \end{bmatrix} \begin{Bmatrix} y_t(x,t) \\ \theta_t(x,t) \end{Bmatrix}$$

$$+ \begin{bmatrix} \Sigma & 0 \\ 0 & d^2 \cdot \Sigma \end{bmatrix} \begin{Bmatrix} y(x,t) \\ \theta(x,t) \end{Bmatrix} = \begin{Bmatrix} F(x,t) \\ M(x,t) \end{Bmatrix} \tag{2.1.5}$$

with i.e.

$$\begin{Bmatrix} y(x,0) \\ \theta(x,0) \end{Bmatrix} = \begin{Bmatrix} \bar{y} \\ \bar{\theta} \end{Bmatrix}, \begin{Bmatrix} y_t(x,0) \\ \theta_t(x,0) \end{Bmatrix} = \begin{Bmatrix} \bar{y}_t \\ \bar{\theta}_t \end{Bmatrix}$$

Σ indicating the differential operator $\{EJ\nabla_x^2\nabla_x^2 - T_o\nabla_x^2\}$.

2.2. Wind distribution modelling

The main aerodynamic disturbance for the structures under examination are those connected with Karman's periodic vortex shedding. In the case of multi-tube bridges (side by side tubes) the resulting aerodynamic load induced by vortexes is generally eccentric with respect to the torsion axis thereby also inducing torsional effects. Furthermore, tubes close together or faired are capable of producing lift varying with the angle of incidence of the wind. As a consequence, the aerodynamic forces can induce the flexotorsional flutter phenomena, which can occur under conditions of rather low wind speed because the ratio between the flexural and torsional oscillations is in these cases very close to one. In the present study reference is essentially made to controlling the effect of Karman vortexes, while regarding the control of flutter only some formal preliminary developments are reported. The phenomena of Karman vortexes, as is known, occurs when wind encounters a non-streamlined body creating at the rear of the obstacle two vortex wakes which discharge themselves with a periodic alternating cadence. These vortexes induce variations of pressure in the cylinder and, there-fore, forces which are normal to the wind direction; they are periodic with a fre-quency equal to that of the vortex discharges. This frequency is expressed by: $n_s = V \cdot D / S_t$, where V is the wind speed, D is the diameter of the pipeline, S_t is the Strouhal number. When the oscillation frequency of the structure being examined becomes close or equal to n_s for a given wind velocity V_c, resonance phenomena occur and the oscillations increase up to a limit vaue which is a func-tion of the global damping of the system.

The pulsating transverse forces can be expressed by:

$$F_s = \frac{1}{2} \cdot \rho \cdot V_c^2 \cdot D \cdot C_L \cdot \sin(2\pi n \cdot t)$$

in which C_L, the lift coefficient, assumes values ranging from 0.2 to 0.8 according to the Reynolds number and to the amplitude in the oscillation of the structure, increasing with the latter. The increase of the lift coefficient with the amplitude characterizes this oscillation, therefore, as partially self-excited and no longer as simply forced (Fig. 6). Moreover, from experiments has been found that with an increase in the amplitude of the oscillation the frequency of discharge of the vortexes tends to coincide precisely with that of the structure even with a wind velocity higher than critical. This fact produces an apparent reduction in the Strouhal number. The aerodynamic force connected with Karman vortexes can, therefore, be expressed in the form:

$$F_s = \frac{1}{2} \cdot \rho \cdot \overline{V}_c^2 \cdot D \cdot \overline{C}_L \cdot \sin(2\pi n t) \quad \text{with } \overline{v} = K \cdot V_c \text{ and } \overline{C}_L = C_L(R_e, Y/D). \tag{2.2.1}$$

In the case of pipeline suspension bridges, the oscillation frequencies are low and such that they induce resonance with the vortexes with wind velocity of only a few meters per second and, therefore, with a much reduced turbulence and high degree of correlation between the forces fluctuating along the structure.

Furthermore, the damping parameters are also very low in that the deforma-tions take place mainly through variations in shape with small vibrations in ten-sion.

As a consequence, we often observe high flexural oscillation amplitudes. By applying the theory of an equivalent oscillator we obtain in resonance conditions:

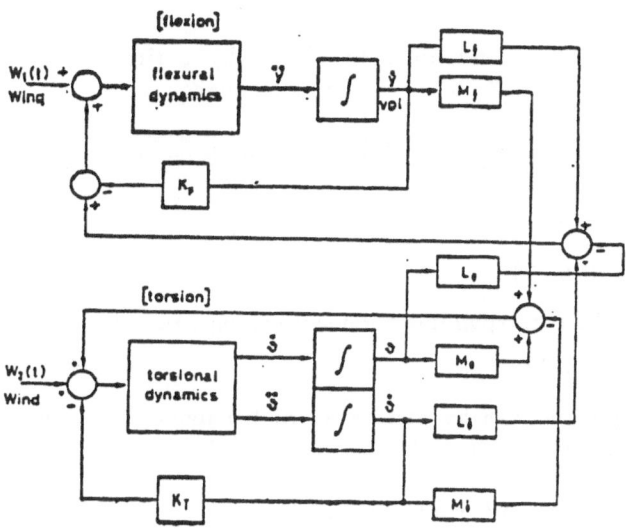

Figure 6: Closed-loop block diagram with forcing of 2-dimensional self-excited wind disturbance

$$F_c = \pi/\delta \cdot \rho/2 \cdot (D \cdot w/(2\pi S_T))^2 \cdot D \cdot C_L \ ,$$

and with:

$$K_e = m_e \omega^2 \ ; \quad m_e = \int m \cdot \mu_i^2(x) = m/2 \ \text{for} \ \mu_i(x) = \sin(\frac{\pi x}{l})$$

and hence:

$$Y_{max}/D = F_c/(K_e \cdot D) = C_L \cdot \rho \cdot D^2/(\delta \cdot M/2) \ .$$

If we introduce the data given in Appendix A together with the following typical values: D=0.50-1.00 m; m=200-300 Kg/m (for spans of 150-300 m), we obtain: Y_{max}/D=0.15-1.00, values of the same order of size as those actually observed in some real cases. For the numerical simulations we considered the flexural and torsional excitation of the 1-st asymmetric mode (see (2.1.4) and (2.1.4)') and of the 1-st symmetric one, given by (2.1.4), (2.1.4)' with K=1, and the 2-nd symmetric ((2.1.4), (2.1.4)' with K=4). The particular distribution of modal frequencies and the low damping ratios lead to separate excitations of the single modes, each time with the same "mechanism" of excitation and of active control (the choice of the points $x_A = l_o$ and $x_B = l - l_o$ for the application of control forces and couples (Fig. 4a) permits active work at least in the first three modes). Inside each flexural-torsional mode, the numerical simulations tested both the flexural and the torsional resonance (wind disturbance frequency $\overline{\omega}$ ranging from ω_k and $\mu\omega_k = \omega_k$, K=1,2,4,μ>1). Concentrating on the 1-st asymmetric mode and supposing that the oscillating deck immediately imposes its own phase on the vortexes which are discharging along its span, the density $F(x,t)$ of the pulsating lift force acting on the downstream pipeline can be described as a space square wave with period $T=l$ and amplitude given by a time square wave pulsating between $\pm F$, (see Eq. (2.2.1)) at resonating frequency $\overline{w}=\omega_k$ (2.1.4):

$$F(x,t)=F(x){\cdot}F(t)=\frac{8F_{\prime}}{\pi}\sum_{o}^{N}{}_k\frac{1}{a_k}\sin\frac{\pi a_k x}{l}{\cdot}\sum_{o}^{N}{}_k\frac{1}{a_k}\sin\frac{a_k \overline{w} t}{2} \qquad (2.2.2)$$

with

$$a_k = 2(2k+1) .$$

The density of the corresponding torsional moment is:

$$M(x,t)=F(x,t){\cdot}d . \qquad (2.2.3)$$

Analogous expressions can be written for higher modes. The flutter formalization will be briefly discussed later.

2.3. Discretization and modal control

The active control of the flexo-torsional freedoms of the bridge is effected by exerting 2 appropriate forces at points A_1 and A_2 of the section $x_A=l_o$ as in Fig. 4, and/or other 2 forces at points B_1 and B_2 of the section $x_B=(l-l_o)$. The possible configurations of the control actuation devices were introduced in Section 1.2. We now focus our attention on the pair of active tendons acting at point A. The two active forces at point A_1 and A_2 are equivalent to a resultant flexural control forces $U^a(l_o,t)$ and to a torsional control torque $W^a(l_o,t)$ (Fig. 4a):

$$U^a(l_o,t)=F_1^a(l_o,t)+F_2^a(l_o,t)$$

$$W^a(l_o,t)=d\{ F_2^a(l_o,t)-F_1^a(l_o,t)\} \qquad (2.3.1)$$

If $a_1(l_o,t)$ and $a_2(l_o,t)$ respectively indicate the displacement of the hydraulic actuator's ram which actuates the active tendons and α is the angle of the tendon with the vertical axis, the flexural-torsional displacement of the section due to the active control is given by

$$u(l_o,t)=\frac{a_2(l_o,t)+a_1(l_o,t)}{2}\cos\alpha$$

$$w(l_o,t)=\frac{a_2(l_o,t)-a_1(l_o,t)}{2d} . \qquad (2.3.2)$$

From (2.3.1), being:

$$F_i^a(l_o,t)=\frac{-EA_c}{l}a_i(l_o,t)\cos\alpha \qquad (i=1,2)$$

we have for $x_A=l_o$ (and similarly for $x_3=l-l_o$):

$$U^a(l_o,t)=\frac{-2EA}{l}u(l_o,t)$$

$$;t\in T_c$$

$$W^a(l_o,t)=\frac{-2EA}{l}d^2w(l_o,t) \qquad (2.3.1)'$$

T_c being the control time interval.

When control is acting, the 2-nd member of the vectorial differential equation of motion (2.1.5) must be completed with one or both of the following additional vectors:

$$\begin{Bmatrix} U(l_o,t)\cdot\delta(x-l_o) \\ W(l_o,t)\cdot\delta(x-l_o) \end{Bmatrix} ; \begin{Bmatrix} U(l-l_o,t)\cdot\delta(x-l+l_o) \\ W(l-l_o,t)\cdot\delta(x-l+l_o) \end{Bmatrix} \qquad (2.3.3)$$

which refer, respectively, to the control action in $x_A=l_o$ and in $x_B=l-l_o$ ($\delta(x-a)$) being the Dirac spatial distribution centered in $x=a$). Returning to the modal description which was introduced in Section 2.1, $r_k(t)$ and $p_k(t)$ indicate the amplitude of the k-th flexural mode and of the corresponding k-th torsional one, respectively. The lagrangian description of the k-th controlled modal motion is then obtained from (2.1.5), (2.3.1), (2.3.3), after product by $Y_k(x)$ (and $\Theta_k(x)$ respectively) and taking into account the modal orthogonality conditions. We then have:

$$\begin{Bmatrix} \ddot{r}_k \\ \ddot{p}_k \end{Bmatrix} + \begin{bmatrix} 2\xi_k\omega_k & 0 \\ 0 & 2\widetilde{\xi}_k\widetilde{\omega}_k \end{bmatrix}\begin{Bmatrix} \dot{r}_k \\ \dot{p}_k \end{Bmatrix} + \begin{bmatrix} \omega_k^2 & 0 \\ 0 & \widetilde{\omega}_k^2 \end{bmatrix}\begin{Bmatrix} r_k \\ p_k \end{Bmatrix}$$

$$= \begin{Bmatrix} \mathcal{F}_k(t)/M \\ \mathcal{M}_k(t)/I \end{Bmatrix} + \begin{Bmatrix} \mathcal{U}_k^a(t)/M \\ \mathcal{W}_k^a(t)/I \end{Bmatrix} \qquad (k=1,2,4).$$

For $k=2$ (1-st asymmetric mode), the torsional frequency is $\widetilde{\omega}_2=\mu\omega_2$ with $\mu=1.25$; M,I are the generalized mass and moment of inertia, $\mathcal{F}_k(t)$ and $\mathcal{U}_k(t)$ the generalized flexural and torsional Karman disturbances, and $\mathcal{U}_k^a(t)$ and $\mathcal{W}_k^a(t)$ the generalized flexural and torsional active control in $x_A=l_o$ and $x_B=(l-l_o)$:

$$\mathcal{U}_k^a(t)=U^a(l_o,t)Y_k(l_o)+U^a(l-l_o,t)Y_k(l-l_o)$$

$$\mathcal{W}_k^a(t)=W^a(l_o,t)\Theta_k(l_o)+W^a(l-l_o,t)\Theta_k(l-l_o) . \qquad (2.3.5)$$

The control addend in the 2-nd member of (2.3.4) can thus be written:

$$\begin{bmatrix} Y_k(l_o) & Y_k(l-l_o) & 0 & 0 \\ 0 & 0 & \Theta_k(l_o) & \Theta_k(l-l_o) \end{bmatrix} \left\{ U^a(l_o,t) \quad U^a(l-l_o,t) \quad W(l_o,t) \quad W(l-l_o,t) \right\}^T$$

$$(k=1,2,4) \,.(2.3.4)'$$

For flexural and torsional damping we assumed a compound logarithmic decrement which is the sum of a structural addend plus an aerodynamic one: $\delta = \delta_S + \delta_A$ with $\delta_S = .006$ well confirmed by experimental tests on pipeline-bridges and

$$\delta_A = \frac{1}{2}\rho V D C_D / (\frac{\omega}{2\pi} \cdot m)$$

We also took into consideration the results of experimental research which showed, at least on bridge sections, that mechanical damping does not appreciably change in the band of the first natural frequencies of the system. It was operated with the following modal damping $\xi_k = \tilde{\xi}_k = .001 \div .002$. When a passive control contribution can also be considered because of the pulsating stretching of the control tendons there is a shifting of the modal eigenvalues deeper into the left Gaussian half plane, i.e. there is a modified frequency matrix in (2.3.4) given by:

$$\begin{bmatrix} \omega_k^2 + \dfrac{GEA}{lM}\cos^2\alpha\sin^2\dfrac{k\pi l_o}{l} & 0 \\ 0 & \tilde{\omega}_k^2 + \dfrac{GEAd^2}{Il}\cos^2\alpha\sin^2\dfrac{k\pi l_o}{l} \end{bmatrix} . \qquad (2.3.4)''$$

As already mentioned in Section 2.2 the case of aeroelastic flutter phenomena set off at a critical wind speed $V_F = K_F V_C$ was finally formalized. We considered the following two formal structures (in a preliminary numerical simulation we have assumed the following values: $K_F = 15., C_{L,F} = 1.72, C_{M,F} = .36$):

a)

$$M\ddot{q} + 2C\dot{q} + Kq = Fq \qquad (2.3.5)$$

with $q = (y \quad \theta)^T$ and $F \in R^{2 \times 2}$, being for 2 pipelines side by side:

$$F_{12} = \frac{1}{2}\rho V_F^2 \cdot 2b \cdot C_{L,F}; \quad F_{22} = \frac{1}{2}\rho b^2 V_F^2 \cdot C_{M,F}; \quad F_{11} = F_{21} = 0 .$$

b)

$$M\ddot{q} + 2C\dot{q} + K q = H\dot{q} + Qq \qquad (2.3.6)$$

with $H, L \in R^{2 \times 2}$ (see Fig. 6):

$$H_{11} = L_{\dot{y}} = \frac{1}{2}\rho \cdot b \cdot V_F \cdot C_{L,F} \; ; \quad H_{12} = L_{\theta} = \frac{1}{2}\rho \cdot b^2 \cdot V_F \cdot C_{L,F}$$

Pipe-Line Suspension Bridges

89

$$H_{21}=M_{\dot{y}}=\frac{1}{4}\rho\cdot b\cdot V_F\cdot C_{M,F} \quad;\quad H_{22}=M_{\dot{\theta}}=\frac{1}{4}\rho\cdot b^3\cdot V_F\cdot C_{M,F}$$

$$Q_{12}=L_{\theta}=\frac{1}{2}\rho\cdot b\cdot V_F^2\cdot C_{L,F} \quad;\quad Q_{22}=M_{\theta}=\frac{1}{4}\rho\cdot b^2\cdot V_F\cdot C_{M,F} \quad;\quad Q_{11}=Q_{21}=0 \ .$$

In this latter case, the destabilizing effect of aeroelastic self-excitation leads to different damping and stiffness matrices C and K i.e. a relocation of system eigenvalues.

2.4. State-space description. Formal Properties

Introducing the following state vector for the K-th structural mode:

$$x_k(t)\underline{\triangleq}(r_k\dot{r}_k p_k\dot{p}_k)^T \tag{2.4.1}$$

when all the first 3 modes are considered an $x\in R^1$ state vector is obtained

$$x(t)\underline{\triangleq}(r_1\dot{r}_1 r_2\dot{r}_2 \cdots p_3\dot{p}_3)^T \ . \tag{2.4.1'}$$

From (2.4.1)' the following natural matrix is obtained for the K-th flexural mode and the corresponding K-th torsional one:

$$A=\begin{bmatrix}A^{fl} & 0 \\ 0 & A^{tor}\end{bmatrix}\in R^{4\times4} \quad;\quad A^{fl},A^{tor}\in R^{2\times2}$$

with flexural and torsional submatrices given by:

$$A^{fl}=\begin{bmatrix}0 & 1 \\ -\omega_k^2 & -2\xi\omega_k\end{bmatrix}; \quad A^{tor}=\begin{bmatrix}0 & 1 \\ -\omega_k^2 & -2\mu\tilde{\omega}_k\end{bmatrix}$$

$\hat{\omega}_k$ and $\tilde{\omega}_k$ being the flexural and torsional modal frequencies modified by passive control ("hat" notation). When the following control vector $u\in R^4$ and a control distribution matrix of the actuators' configuration $b\in R^{4\times4}$ are introduced:

$$u(t)=\begin{Bmatrix}U^a(l_o,t)\\U^a(l-l_o,t)\\W^a(l_o,t)\\W^a(l-l_o,t)\end{Bmatrix}\in R^4; \quad B=\begin{bmatrix}0 & 0 & 0 & 0\\ \dfrac{Y_k(l_o)}{M} & \dfrac{Y_k(l-l_o)}{M} & & \\ 0 & 0 & 0 & 0\\ & & \dfrac{\Theta_k(l_o)}{I} & \dfrac{\Theta_k(l-l_o)}{I}\end{bmatrix}$$

with the following R^4 Karman disturbance vector $d(t)$:

$$d(t)=(\ \mathcal{F}_k(t) \quad \mathcal{M}_k(t))^T\in R^2$$

then, the open-loop state-space description of the K-th mode of the vibrating bridge under wind disturbance and control can finally be written:

$$\dot{x} = Ax(t) + Bu(t) + Dd(t) \tag{2.4.2}$$

with

$$D = \begin{bmatrix} 0 & 0 \\ 1/M & 0 \\ 0 & 0 \\ 0 & 1/I \end{bmatrix} \in \mathbf{R}^{4 \times 4} \ , \ t \in T_c$$

being the matrix of plant disturbance distribution. With reference to control implementation given later in Section 3, the output transform (observation model) has the form

$$y = Cx \qquad y \in \mathbf{R}^2 \tag{2.4.2'}$$

with

$$C = \begin{bmatrix} 0 & Y_k(l_o) & 0 & 0 \\ 0 & Y_k(l-l_o) & 0 & 0 \\ 0 & 0 & 0 & \Theta_k(l_o) \\ 0 & 0 & 0 & \Theta_k(1-l_o) \end{bmatrix} \in \mathbf{R}^{4 \times 4}$$

with only velocity sensors. Formal properties of (2.4.2), (2.4.2)' system:

- asymptotical stability of A with respect to zero state,
- complete controllability of the pair (A,B),
- complete observability of the pair (A,C).

3. Feedback Control Design

Because of the intrinsic simplicity of our type of control, a computer simulation approach is preferable to the algorithm of the optimal control theory. The state measurement sensors are co-positioned with the forces and the controller has been designed taking into account a Karman time-dependent wind disturbance. In the presence of disturbance on measurements the complete state-space Hamiltonian model given in (2.4.2), (2.4.2)' becomes (Fig. 3):

$$\begin{cases} \dot{x}(t) = Ax(t) + Bu(t) + w(t) \\ y(t) = Cx(t) \end{cases} \tag{3.1}$$

where $w(t)$ represents the wind disturbance.

The selection problem of the actuators and sensors (optimized position and force values) involves the following major assumptions:

i)   the actuator dynamics has not been considered in the mathematical model
       because the chosen actuators have a large response frequency band com-
       pared to the closed loop frequencies and also because of the propagation
       delays in the cables,

ii)   the outputs of the measurement sensors (with their associated electronics
       handling) are assumed to be linear functions of the speed with high signal-
       to-noise ratios. This assumption is well represented by the actuators and sen-
       sors proposed in Chapter 4. When a half-mode control is considered, i.e.,

control forces only at points A_1 nd A_2 in $x_A = l_o$, the matrix B of the actuators
and the matrix C of the sensors respectively become (K-th mode):

$$B = \begin{bmatrix} 0 & 0 \\ Y_k(x_a)/M & 0 \\ 0 & 0 \\ 0 & \Theta_k(x_a)/I \end{bmatrix} \; ; \; C = \begin{bmatrix} 0 & Y_k(x_s) & 0 & 0 \\ 0 & 0 & 0 & \Theta_k(x_s) \end{bmatrix} \qquad (3.2)$$

x_a and x_s being the position of actuators and sensors respectively: $x_a = x_s = 1/3$.
The considered control vector is the following:

$$u = Gy = (u^{fl} u^{tor})^T \; , \; G \in \mathbf{R}^{2\times2} \qquad (3.3)$$

with

$$y = Cx = (\hat{\dot{y}}_k \quad \hat{\dot{\theta}}_k)^T \qquad (3.4)$$

$\hat{\dot{y}}_k$ and $\hat{\dot{\theta}}$ being the measured modal values of the flexural and torsional velocity,
and

$$G = \begin{bmatrix} -K_F & 0 \\ 0 & -K_T \end{bmatrix} \qquad (3.5)$$

the modal feedback gain matrix. From (3.1), according to (3.3), the state-space
bridge dynamics (without wind disturbance) for the K-th flexural and torsional
mode is:

$$\dot{x} = (A + BGC)x \qquad x \in \mathbf{R}^4 \qquad (3.6)$$

with

$$BCH = \begin{bmatrix} 0 & 0 & 0 & 0 \\ 0 & \dfrac{-K_F Y_k(l_o)}{M} & 0 & 0 \\ 0 & 0 & 0 & 0 \\ 0 & 0 & 0 & \dfrac{-K_T \Theta_k(l_o)}{I} \end{bmatrix} .$$

Feedback gains have been estimated in order to steer the natural open loop damping ratio $\xi_i^{0.l.} = .001$ to a prefixed closed loop value $\xi_i^{c.l.} = .6$.

When flutter is also considered the state space model is described in Fig. 6.

4. Sensors, Actuators and Electronics. Energy Sources

The concept of the active structural control described in Section 3 requires displacement rate measurements and high speed linear double acting actuators which provide relevant forces and strokes; both require an electric energy source in the proximity of the bridge. Many methods have been investigated for each subject but the most promising solutions are the following:

- Sensors: inertial or $AsGa$ laser system,

- actuators: hydraulic with servo-valve, automatic control,

- electronics: analogic and microprocessor,

- electric energy sources: aeolic, or sun, or conversion of the vibration of the structure.

These sources can be considered as alternatives to the electric power network where non-existant; and can be designed taking into account potential services to agricultural communities and/or factories.

4.1. Sensors

The choice of the sensor is influenced by the following main factors: very long design life, high conceptual reliability, minimum maintenance, all weather functioning and robust devices, low cost. Given these requirements it follows that the best candidate is a servo-accelerometer. The servo-linear accelerometer is located on the bridge and is hard mounted with the output vertical axis. The $1.g$ output signal is set at zero by a bias and any variation electronically integrated. The integration procedure is as follows: the signal is sampled by the microprocessor with a sampling period ΔT compatible with the measurement frequency band. In the time interval ΔT the signal is analogically integrated; the velocity variation is $\Delta v = \int_o^{\Delta T} a(t)dt$. The total speed at the instant "i" is calculated by: $v_i = v_{i-1} + \Delta v_i$, v_{i-1} being the previous measurement.

The integration process involves a measurement error which increases with time; this error can be periodically set to zero calculating the average values over a long period of time and updating the measurements as the effective mean values are zero. Unfortunately the angular displacements of the bridge due to torsion and flexion, generate considerable errors at the output of the accelerometers. These errors are $|\delta a| = g|1 - \cos\epsilon|$ where ϵ is the tilt angle with respect to the horizontal plane. Assuming: $|a|_{\min} = y_{\min}(\omega_2)^2$ where $y_{\min}$ is the minimum displacement to be controlled and $|a|_{\min}/|\delta a| = 2$. It follows that: $\epsilon \sim 2.4°$. Lateral rotations are greater than $2.4°$ because of the foreseen torsions and unpredictable structural instabilities, but the longitudinal rotations are of a lower order of magnitude. Due to this fact a correction of this systematic error is possible utilizing a 2-nd accelerometer with its output axis horizontal and transverse (as in Fig. 7). The signal outputs of the two accelerometers are: $a_y = g\cos\epsilon + \dot{v}$, $a_x = g\,sen\,\epsilon$, $\dot{v} = (a_y - g) + a_x^2/2g^2$, $\epsilon \simeq a_x/g$.

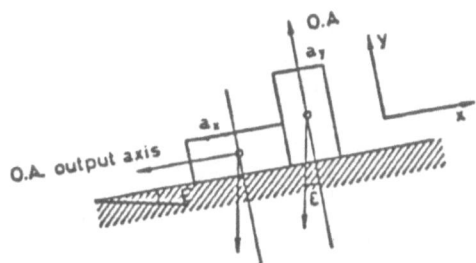

Figure 7: Hard mounted x, y servo-accelerometers

In the previous equations $\dot{v}_x$ have been assumed to be negligible, and assumed $\dot{v}_y \sim \dot{v}$. The processing of the equations can be developed by the electronic unit in a simple way with no penalty. A $10.g$ range, $10^4·g$ sensitivity fluid damped flexure suspended servo-accelerometer can be used (see Fig. 8).

A more attractive concept is the $AsGa$ laser system described in Fig. 9. The laser pulsed infrared energy ($\lambda = 0.9\mu m$) is reflected back by the corner cube located on the bridge at an appropriate distance. A silicon sensor system receives the reflected energy and measures the vertical displacements. Appropriate PRF (Pulse Repetition Frequencies) and measurements sampling rates, will provide accurate electronic derivate measurements by using the following algorithm: $\dot{y}_i = \dot{y}_{i-1} + \Delta y_i/\Delta T$ where Δy_i is the displacement measured in the ΔT sampling period at the time instant "i". It will be observed that the displacement measured by the receiver is $2·\Delta y$ were Δy is the displacement of the corner cube. At a distance of 100 m the measurement accuracy can be within a few millimeters. $AsGa$ laser system is a static device with no moving mechanisms, with an

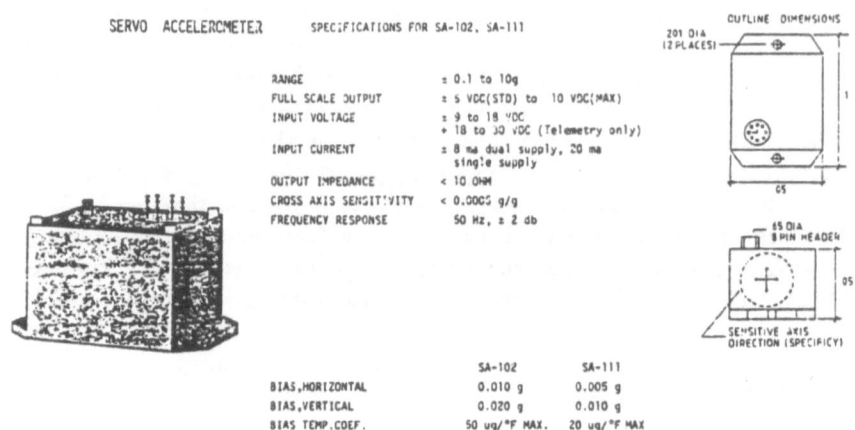

Figure 8: Servo-Accelerometer

intrinsic high reliability and very long operative life. Similar devices have been
used for all weather monitoring of dams.

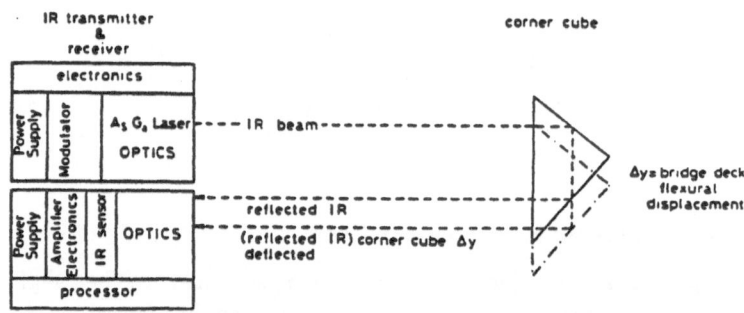

Figure 9: AsGa laser system: remote displacement measurements

4.2. Actuators

The hydraulic double-acting actuator is, given the state of the art, the appropriate choice where high forces (tons), large strokes (0,2 m or more) and a large frequency band (0-5 Hz) response is necessary. The actuator consists mainly of an
electric pump, an hydraulic network, a jack and a servo-valve controlling the
force and the stroke by means of an electric signal. In this project the electrical
command to the servo-valve utilizes a closed loop control, the feedback being a
strain gauge transducer located on the cable at the point of application of the
force. This is necessary because of unpredictable force losses on the pulleys,
structural deformations due to temperature changes and preloading variations.

The main characteristics of each of the actuators are indicated in Fig. 10 as a possible solution:

- force: 1500 kg (active); 3000 Kg (holding)
- stroke: 100-250 mm
- oil pressure: 270 bars (STD)
- rate of flow: 50 liters/min
- electric motor: asyncronous 3 ϕ 20 kw
- frequency band (sinusoidal): 2 Hz.

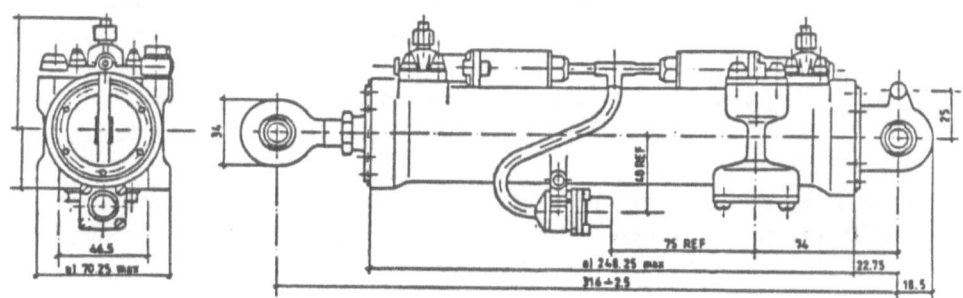

Figure 10

A leverage adapter is considered appropriate for the best utilization of the actuator. Thermal dilatations of the cables are of approximately 8 cm over the expected temperature range. This large expansion must be compensated by holding preloads fixed via closed loop strain gauge measurements. When not activated, the actuators hold load by means of an automatic locking device. The calculation of the energy required by the actuator system has been carried out as follows:

$$L_c = \int_o^{T_c} (F(t)\dot{y}(t) + M(t)\dot{\theta}(t))dt = \int_o^{T_c} (K_F \dot{y}^2 + K_T \dot{\theta}^2)dt$$

with T_c control interval dependent on K_F and K_T.

4.3. Electronics

The active control design described in this paper requires a microprocessor capable of performing the following main tasks:

- regulator algorithms for the closed loop control system;
- handling of the sensor and actuator outputs: sampling, integration, registers, algorithms;
- conversion from analog to digital signals and viceversa.

An analogic unit is also necessary:

1) for the standardization of all analogic signals (state measurements and strain gauges) so as to provide an appropriate interface with the microprocessor,

2) for the integrations in sampled intervals of time (where accelerometers are used),

3) for the servovalve power amplifiers of the hydraulic actuators.

Batteries can be used for the power supplies charged by external electric energy sources. The high electric power required by the pumps on the hydraulic actuators can be provided by a separate power supply unit (three phase static inverter). An electronic functional system diagram is presented in Fig. 11.

4.4. Energy sources

As the wind is responsible for the main external disturbances this could also be a possible energy source. Where no other electric energy is available we propose a wind generator as an alternative solution (see Figs. 12 and 13).

5. In-Service Control Strategies

5.1 With reference to the numerical data input given in Appendix A, from a simple statistical analysis of the return periods of critical wind (Karman or flutter) at the bridge site, relatively low average return frequencies have been found. The simulations carried out by us in these conditions indicate that the most advantageous control strategy is that which generally remains inactive (turned off) and comes into action only at the critical wind level, when this has already been continuously present for at least some tens of seconds (isolated disturbances of an impulsive type do not switch on the control). We, therefore, have adopted an "on-off" control strategy which is equivalent to a random sequence of control cycles on the time axis with a mean return period that coincides with that of the critical wind speed. The state of the vibrating structure is reduced to zero in one or more tens of seconds at the most. As previously mentioned, the control produces a simulated damping equal to, or less than, the critical value. The duration of the control time interval depends on the feedback gain adopted, but is, however, limited to within 1 to 3 times the natural period of the controlled mode.

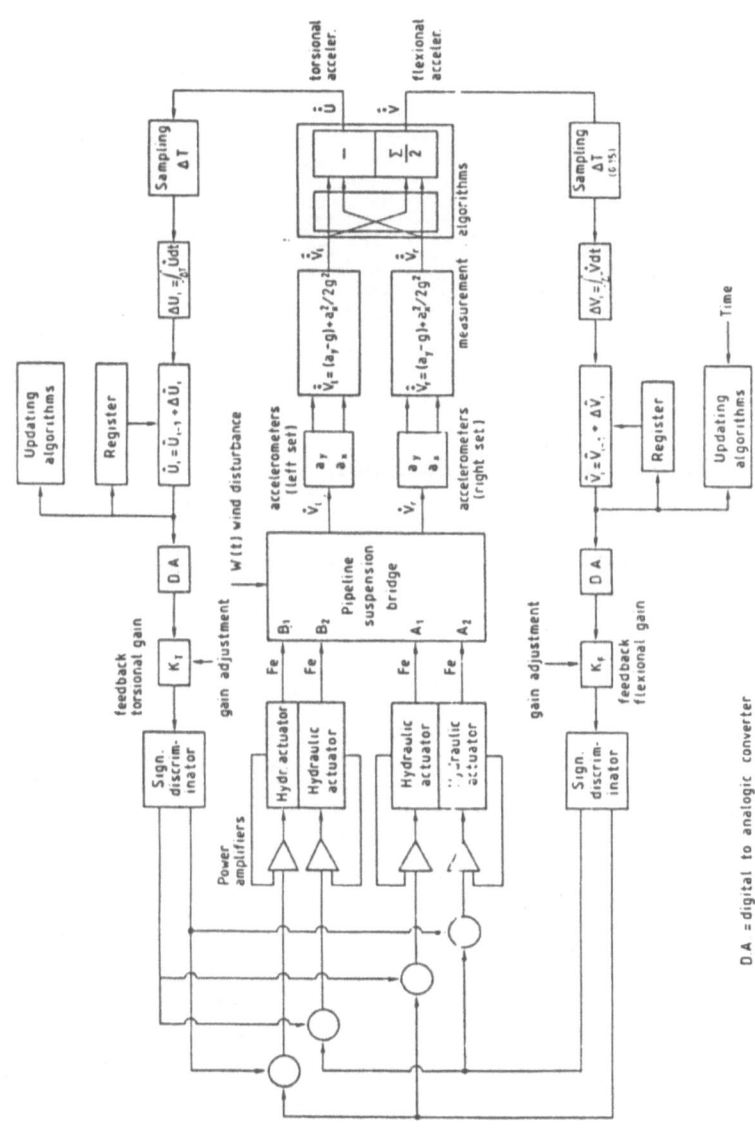

DA = digital to analogic converter

Figure 11: Block diagram of the feedback control - Electronic functional diagram

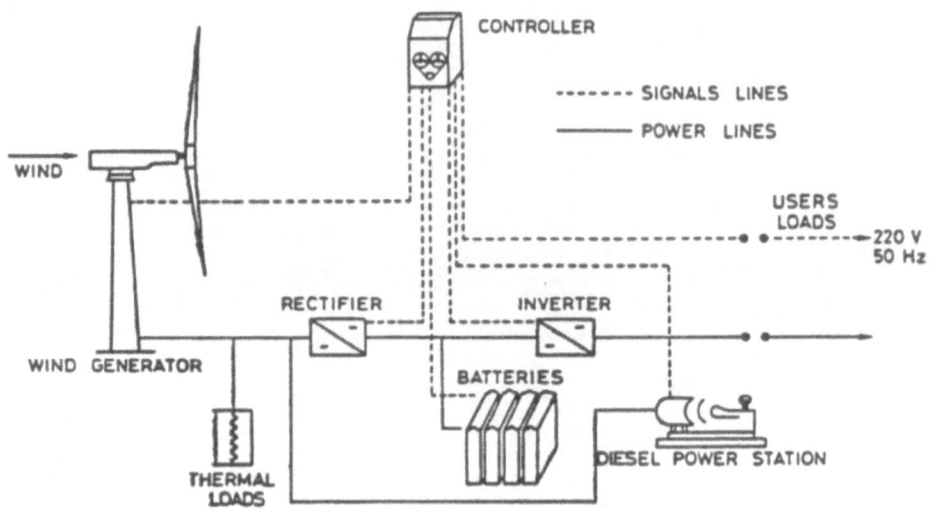

Figure 12:   Aeritalia eoloc generator (A.I.G.E.-03)

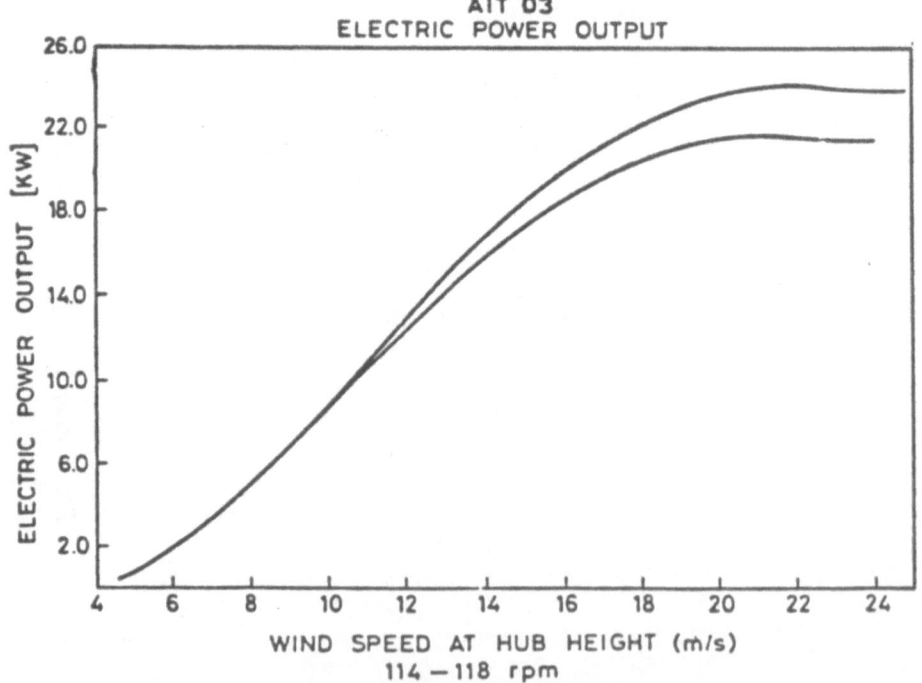

Figure 13:   Eolic generator electric power characteristics

5.2 In Figs. 14-15 the overall response of the 4 actuators is described together with that of the 1-st asymmetrical mode in flexion (with frequency ω_1, when the "on-off" control strategy is particularly unfavourable due to a very delayed onset; i.e. one considers a sustained Karman critical wind with a frequency of $\overline{\omega}$ for about 15 minutes. In this time interval the steady resonating response ($\overline{\omega}=\omega_1$) is reached with a maximum displacement of 0.54 meters; at this point control begins. A critical and a sub-critical feedback gain are shown in Figs. 14 and 15 respectively. In the latter case a control time interval of about 4.5 secs. was found and control outputs at every actuator of about 3000 daN. In the presence of sustained wind, the control cycle can be repeated. The instantaneous power absorbed is in the order of some tens of HP. In Figs. 16-17 two different strategies are given for torque control, and the corresponding 1-st asymmetrical torsional mode responses. The functioning of the controller according to the "on-off" strategy, may take place in much more favorable conditions than those considered here: in fact if one fixes the initial delay Δt, between the wind start-up and automatic turning-on of control at about 5 minutes, one obtains much better i.c. at the onset of control (maximum mode displacements of about .18-.25 m at steady state) with possibility of limiting the actuator output to a thousand daNs.

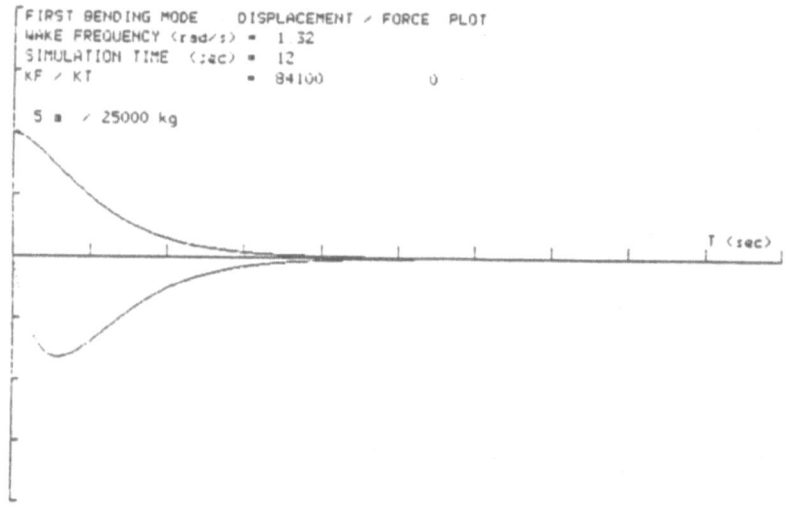

Figure 14
An alternative strategy which envisages the installation being continuously turned on even for very long time intervals starting from an instant of incipient displacement of the bridge deck (flexural or torsional), would permit an onset of control almost simultaneously with the beginning of the disturbance and reductions in the peak values of the control's strokes of an order of magnitude lower than those found for the "on-off" strategy.

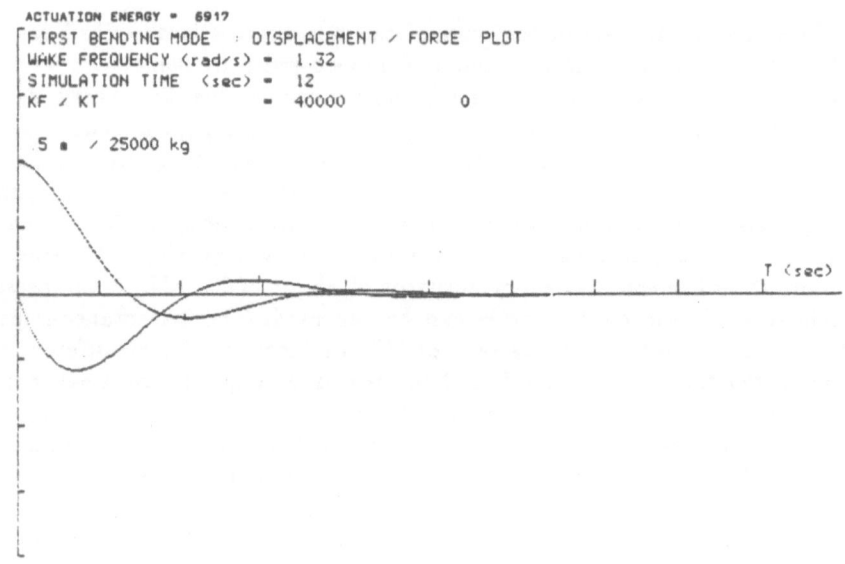

Figure 15

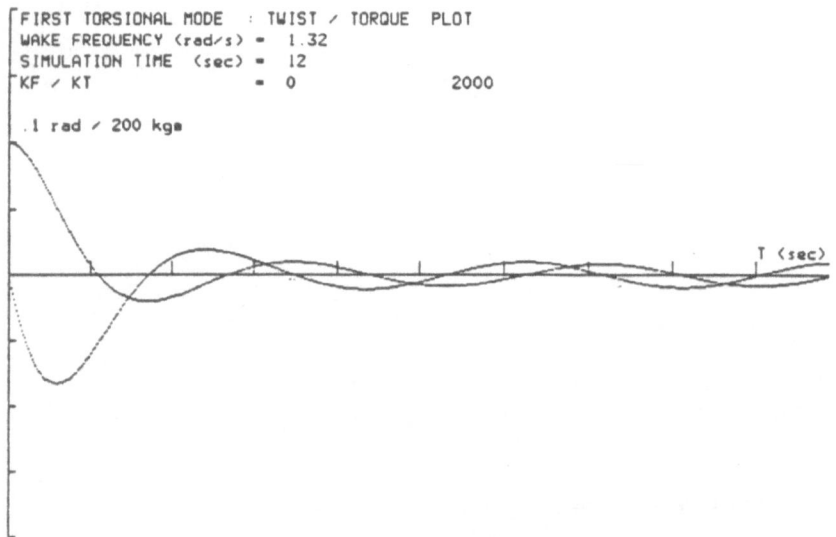

Figure 16

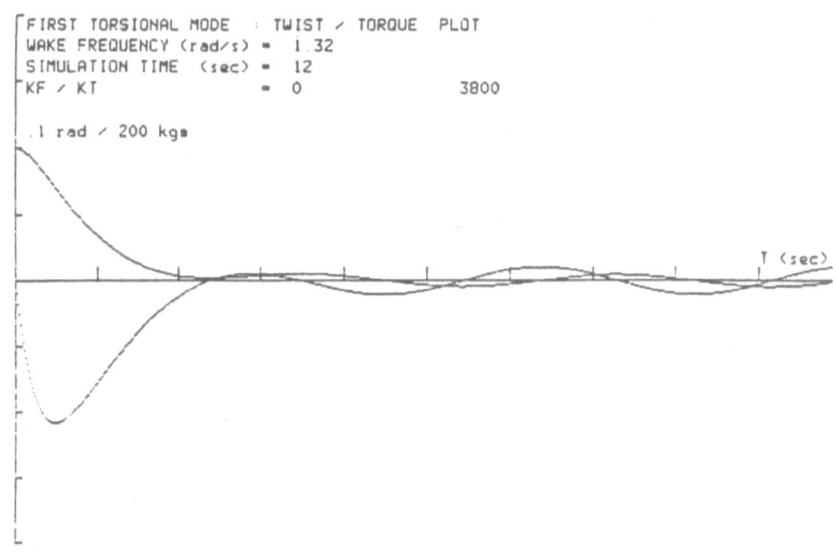

FIRST TORSIONAL MODE : TWIST ∕ TORQUE PLOT
WAKE FREQUENCY (rad∕s) = 1.32
SIMULATION TIME (sec) = 12
KF ∕ KT   = 0   3800

.1 rad ∕ 200 kgs

T (sec)

ACTUATION ENERGY = 17.73389

Figure 17

6. Conclusions

The study carried out permitted the critical evaluation of the feasibility of applying active control techniques in order to control oscillations induced by wind in
pipeline suspension bridges. This feasibility was confirmed by studying an actual
case and on the basis of this it is possible to forward the following considerations:

- the peak values of the forces of control can even be quite limited; in fact,
  they largely depend on the strategy adopted for the working of the installation dropping to values of only a few hundred daNs for very early control
  interventions,

- the energy required also depends on the time interval programmed to reduce
  the oscillations to the prescribed values.

Therefore, a "soft and early" strategy would permit the installation to be of
minimum size, and would reduce the additional forces induced in the structure by
the control action. On the other hand a lower limit is imposed on the speed of
intervention by the warm-up times of the hydraulic plant. It seems, therefore,
that on the whole it may be more advantageous to use a strategy of intermittent
functioning of the control installation with successive, isolated control cycles
appropriately distributed on the time axis.

The cost of an active control system, considering the cost of materials,
apparatus and assembly, but excluding the cost of the aeolic generator and the
possible energy accumulator, was found to be, after a first analysis, equal to
approximately 6-7% of the cost of the whole bridge. One is, therefore, dealing

with a higher cost, but still close to that of other types of passive controls, such as simple lower staying or the application of aerodynmaic devices. These, however, are not always easy to apply in all situations and are mainly controls which are not systematic and are set up case by case.

The application of active control on the other hand offers a general solution to a vast range of problems connected with oscillations in these structures, both in operations where one is starting from scratch and in renewal operations. Moreover, the structural behaviour allowing, among other things, useful information to be acquired for refining control techniques.

Acknowledgements

The financial support of the Italian industry "VIBROSTOP Italiana srl", Milan, in performing this work is greatfully acknowledged. The authors wish also to thank Professor Fabrizio de Miranda, for preliminary discussions and hints, and Jacek Marczyk, P.Eng. (Polytechnic of Milan) for computer simulations and participation in controller strategy design.

References

[1] Baird, R.C., "Wind Induced Vibration of a Pipe-Line Suspension Bridge and Its Cure", *Trans. ASME*, August, 1955, pp. 797-804.

[2] Carotti, A., "Actively controlled steel frames with a variational-type constraint on the control gradient", *Ingegneria*, 3-4, 1984, pp. 1-7.

[3] Carotti, A., "On the application of active bracings in the automatic control of steel frames vibrations", *Tecnica Italiana XLIX*, 1-2, 1984, pp. 41-54. (in Italian).

[4] E.C.C.S., European Conv. for Constr. Steelwork, "Recommendations for the calculation of wind effect on building and structures", 1978.

[5] E.S.D.U., Eng. Sc. Data Units, Item n° 78006, "Across-flow response due to vortex shedding".

[6] E.S.D.U., Item n° 70013, "Fluid forces acting on circular cylinders for application in general engineering".

[7] de Miranda, M., "Some remarks on the aerodynamc stability of pipe-line suspension bridges", *Atti Convegno C.T.A.*, 1981 (in Italian).

[8] Fradsen, A.G., "Wind stability of suspension bridges — Application of the theory of thin foils", *Symp. on Suspension Bridges*, Lisbon, 1966.

[9] Marczyk, J. and Terletti, D., "Contribution to the study of active modal control of large space structures", M.A.Sc. Thesis, Polytechnic of Milan, 1983. (in Italian).

Pipe-Line Suspension Bridges 103

[10] Turci, E., "Fundamentals on Robotics", Milan, 1985 (in Italian).

[11] Turci, E., "Systems for the automatic control and attitude of aircrafts and space vehicles", Milan, 1982 (in Italian)

Appendix A

Main Symbols & Numerical Data Input. Units used: m, s, Kg_m; daN.

l	====	286. m length of bridge span
f	====	28.4 m central sag
$2 \cdot d$	====	1.50 m bridge deck width
q	====	223 daN/m unitary weight
D	====	pipeline diameter
J	====	$1.25 \cdot 10^4$ flexural inertia
J_p	====	$8.07\ Kg_m\, m^2/m$ unit tors. inertia
E	====	$201 \cdot 10^{10} daN/m^2$ Young modulus
A	====	$.00165\ m^2$ cross sectional area of suspension cable
$y_o(x)$	====	cable static configuration
$y_c(x)$	====	cable deflection from $y_o(x)$
$y(x)$	====	static bridge deck deflection
T_o	====	horizontal tension in cable due to dead weight
ΔT_y, ΔT_θ	====	changes in tension due to bridge dynamics
$y(x,t)$, $\theta(x,t)$	====	flex. tors. freedom of the deck section
EJ, GC	====	bending, torsional stiffness of bridge deck
c_y, c_θ	====	flex. tors. viscous damping
$F(x)$, $F(x,t)$	====	static, dynamic external flexural disturbance
$M(x)$, $M(x,t)$	====	static, dynamic external tors. disturbance
ρ	====	$1.25 Kg_m/m^3$ = air density
V_c	====	critical wind speed (Karman)
C_L	====	lift coefficient
$\pm F_S$	====	amplitude of the pulsating lift (Karman)
H, Q	====	matrices of flutter coeff.
ω_K, ξ_K	====	natural pulsation and damping ratio of K-th flexural mode
$\widetilde{\omega}_K$, $\widetilde{\xi}_K$	====	natural pulsation and damping ratio of K-th

		torsional mode
μ	$=$	$\tilde{\omega}_K / \omega_K$
$Y_K(t), \Theta_k(t)$	$=$	K-th flexural, torsional mode
$r_K(t), p_K(t)$	$=$	amplitude of K-th flex., tors. mode
M, I	$=$	generalized mass and moment of inertia
$\mathcal{F}_K(t), \mathcal{M}_K(t)$	$=$	generalized external disturbance (flex. tors.)
$\mathcal{U}_K^a(t), \mathcal{W}_K^a(t)$	$=$	generalized flex. and torsional active control
$\mathcal{U}^a(\cdot, t), \mathcal{W}^a(\cdot, t)$	$=$	active control forces and torque
$A, x(t), u(t)$	$=$	system state matrix, state vector, control vector
B, D, C	$=$	control, disturbance, sensor distribution matrix

$$y_t(\cdot, t), y_{tt}(\cdot, t) : \frac{\partial y}{\partial t}, \frac{\partial^2 y}{\partial t^2}$$

$$\nabla_x^2 y(x, \cdot) : \frac{\partial^2 y}{\partial x^2}$$

$$\nabla_x^4 y(x, \cdot) : \frac{\partial^4 y}{\partial x^4}$$

THE PAST AND FUTURE
OF SEISMIC EFFECTIVENESS
OF TUNED MASS DAMPERS

A.H. Chowdhury, M.D. Iwuchukwu
Department of Civil Engineering
North Dakota State University
Fargo, ND 58105
J.J Garske
Jorgenson Engineering
Fargo, ND 58102

1. Introduction

The tuned mass dampers (TMDs) have been found effective to reduce the wind induced response of tall buildings [5, 6, 13, 14, 15, 16, 21, 22]. The effectiveness of tuned mass dampers under transient type ground excitations such as earthquakes has been studied by several investigators [2, 4, 8, 12, 19, 23, 24, 25]. Grandall and Mark [4] investigated the effect of single-degree-of-freedom tuned mass dampers mounted on single-degree-of-freedom primary structural systems subjected to stationary ideal white noise excitation. They found that as the frequency of the tuned mass damper approaches the frequency of the primary structure, the tuned mass damper is effective in reducing the displacement and acceleration response of the primary structure. Their results show that the acceleration has sharp tuning when the mass ratio μ is small and has broad tuning when μ is large. They have defined the mass ratio μ as the ratio of the mass of the TMD to the mass of the primary structure.

Chandrasekaran and Gupta [2] studied the effect of multiple linear elastic TMDs connected in parallel on the seismic response of a single-degree-of-freedom primary system. They subjected the primary system to the N-S component of

1940 El Centro earthquake. Their results indicated that the tuned mass dampers are effective for earthquake type excitations. Wirsching and Campbell [23] investigated the effectiveness of TMD in reducing the structural response for both single- and multi-degree-of-freedom linear primary systems having a white noise base excitation. Their study included the calculation of the TMD parameters which minimize the response variance of a single-degree-of-freedom system and the first mode response variance of a 5- and 10-story shear structure. Their results demonstrated that the TMD is quite effective in reducing the first mode response for 5- and 10-story shear structures as well as the response of the single-degree-of-freedom structure. Wirsching and Yao [24, 25] subjected multistory shear structures with a tuned mass damper to pseudo-random nonstationary earthquake motion. They found that the TMD is effective in reducing the linear first mode seismic response of multistory shear structures.

The discussion in the previous paragraphs reveals that the TMDs are useful aseismic control devices. But these findings have been contradicted by more recent studies [12, 19] indicating that the TMDs under seismic loads are less effective than previously thought. Kaynia, Veneziano and Biggs [12] investigated statistically the effect of tuned mass dampers by subjecting elastic and inelastic multistory buildings to a population of real earthquakes. They found some reduction in response of the primary structures but the statistical variability of this response reduction is large enough to cause uncertainty as to the effectiveness of tuned mass dampers for individual earthquakes. Sladek and Klinger [19] studied the effect of tuned mass dampers on the elastic and inelastic seismic response of a 25-story building. They designed the optimum TMD using available TMD design procedures [6, 22, 23] and subjected the building to the N-S component of 1940 El Centro earthquake. Their results show that the optimum TMD does not reduce the maximum seismic response of the structure. Gupta and Chandrasekaran [8] investigated the seismic effect of a group of elasto-plastic TMDs connected in parallel to a single-degree-of-freedom system. They concluded that TMDs have only minor influence on the peak response of the primary system.

The findings by Kaynia, et al., Sladek, et al., and Gupta, et al., discourage indiscriminate use of tuned mass dampers as aseismic control devices. However, the validity of most of the findings to date is limited to the studies where the response quantity of interest is calculated primarily by considering the first mode of vibration of the primary structures. As the first mode response contributes more than 80% to the total seismic response of tall buildings [26] the tuned mass damper is generally tuned to the fundamental frequency of the primary structure. The seismic effectiveness of such a tuned mass damper tuned to the fundamental frequency of the primary structure will vary from mode to mode. The objectives of the present study are: (i) Linear elastic parametric analysis on effectiveness of tuned mass dampers on seismic response of tall buildings considering higher modes, (ii) Case studies on seismic effectiveness of tuned mass dampers involving multi-mode inelastic behavior of a typical tall building, and (iii) Assessing the future of tuned mass dampers as useful aseismic control devices.

2. Linear Elastic Parameter Analysis Considering Higher Modes of Vibration of Tall Buildings

The effect of various characteristics of primary structures, tuned mass dampers and earthquakes in determining the effectiveness of TMDs is investigated by subjecting a primary structural system to a set of six historical earthquakes listed in Table 1 [20]. The primary structure considered in this study is idealized as a uniform tall building consisting of interacting shear walls and frames, with effectively planar loading and deformation (Fig. 1). This structure represents a combination of flexural and shear vertical cantilever beams, i.e., deforming either in bending or shear configurations, respectively. It is assumed that the floor slabs are rigid in plane and that the structure does not twist; thus the vertical components are constrained to translate identically at each floor. Consequently, shear and flexure components in separate parallel planes interact as though in the same plane. Heidebrecht and Stafford-Smith [9] have developed a single mathematical model (Fig. 2) for dynamic analysis of this type of structure consisting of uniform flexural and shear vertical cantilever beams which are constrained to have identical horizontal deflection throughout their height. In the present study, it is also assumed that the presence of tuned mass damper does not modify the frequencies and made shapes of the primary structure.

Table 1: Historical earthquakes used in analysis

Cal. Tech. Number	Earthquake	Component	Maximum Acceleration (G)	Strong Motion Duration (Sec.)	Approximate Source Distance (KM)
(1)	(2)	(3)	(4)	(5)	(6)
A001-1	El Centro	S00E	0.348	7.5	8
A001-2	El Centro	S90W	0.214	16.68	8
A004-4	Kern County	S69E	0.179	10.72	43
A008-2	Eureka	N79E	0.258	4.23	24
A009-2	Eurela	N46W	0.201	3.52	40
A018-2	Hollister	N89W	0.179	2.72	21

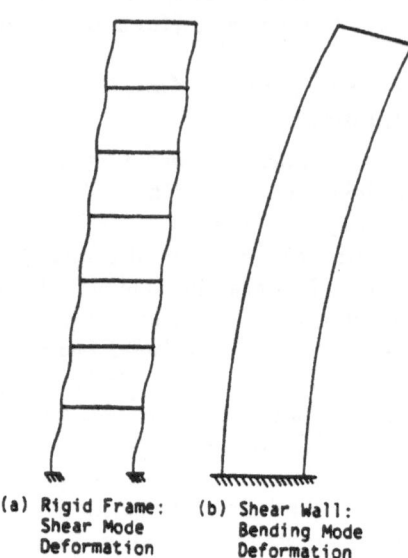

(a) Rigid Frame: (b) Shear Wall:
      Shear Mode   Bending Mode
      Deformation   Deformation

Figure 1: Shear-flexure building

3. Mathematical Model of Uniform Tall Building Without Tuned Mass Damper

The equation of motion of a linear elastic uniform structure (Fig. 2a) with viscous damping and subjected to seismic base acceleration $\ddot{y}_g$ is given by [9]:

$$\frac{\partial^4 y}{\partial x^4} - \alpha^2 \frac{\partial^2 y}{\partial x^2} + \frac{1}{a^2}\frac{\partial^2 y}{\partial t^2} + \frac{C}{EI}\frac{\partial y}{\partial t} = -\frac{1}{a^2}\ddot{y}_g, \tag{1}$$

in which

$$\alpha^2 = \frac{GA}{EI}, \tag{2}$$

$$a^2 = \frac{EI}{\rho A}, \tag{3}$$

and E and I = modulus of elasticity and the second moment of area, respectively, of the flexural beam; G and A = the shear modulus and cross-sectional area, respectively, of the shear beam; ρ = the mass per unit volume of the uniform structure; and C = viscous damping coefficient of the structure.

Equation (1) is the equation of motion for forced vibration of a damped uniform shear-flexure building subjected to seismic base motion. The equation of motion for the undamped free vibration of the building can be obtained by eliminating the two terms involving damping C and ground acceleration $\ddot{y}_g$ from Eq. (1) and is given by:

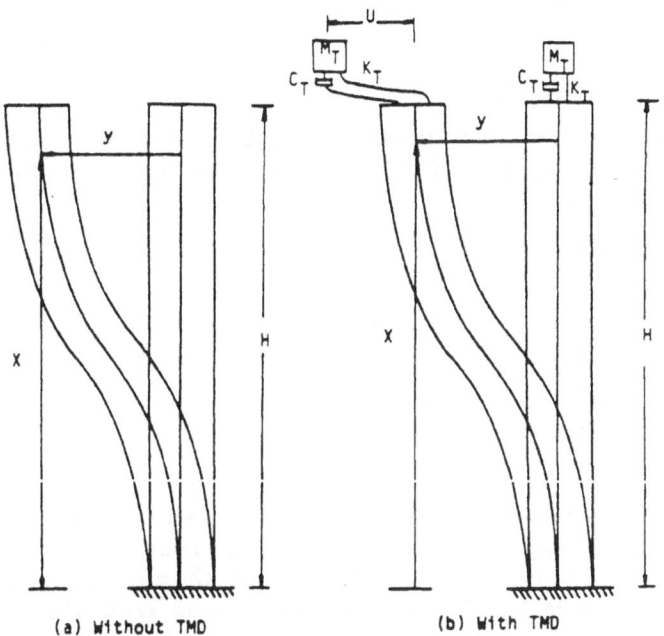

(a) Without TMD

(b) With TMD

Figure 2: Mathematical model of shear-flexure uniform building

$$\frac{\partial^4 y}{\partial x^4} - \alpha^2 \frac{\partial^2 y}{\partial x^2} = -\frac{1}{a^2}\frac{\partial^2 y}{\partial t^2}. \tag{4}$$

Assuming a product solution for Eq. (4) of the form:

$$y(x,t) = \psi(x)q(t), \tag{5}$$

in which $\psi(x)$ is the mode shape, and for free vibration $q(t)$ is the time dependent harmonic response function with frequency, ω, and substituting Eq. (5) into Eq. (4) and rearranging terms yield:

$$\frac{d^4\psi(x)}{dx^4} - \alpha^2\frac{d^2\psi(x)}{dx^2} - k^4\psi(x) = 0, \tag{6}$$

in which

$$k^4 = \frac{\omega^2}{a^2}. \tag{7}$$

The solution of Eq. (6) can be written in the form [9]:

$$\psi(x) = C_1 \mathrm{Cos}\lambda_1 x + C_2 \mathrm{Sin}\lambda_1 x + C_3 \mathrm{Cosh}\lambda_2 x + C_4 \mathrm{Sinh}\lambda_2 x, \tag{8}$$

in which

$$\lambda_1^2 = \sqrt{(\alpha^2/2)^2 + k^4} - \alpha^2/2, \tag{9}$$

and

$$\lambda_2^2 = \lambda_1^2 + \alpha^2. \tag{10}$$

A fixed supported shear-flexure tall building has zero rotation and zero deflection at the base, and is free at the top. The boundary conditions for this building are [3]:

Deflection: $\psi(0) = 0.$ (11)

Slope: $\dfrac{d\psi(0)}{dx} = 0.$ (12)

Moment: $\dfrac{d^2\psi(H)}{dx^2} = 0.$ (13)

Shear: $\alpha^2\dfrac{d\psi(H)}{dx} - \dfrac{d^3\psi(H)}{dx^3} = 0.$ (14)

Substituting boundary conditions (Eqs. (11)-(14)) into Eq. (8) and expressing in matrix notation yield Eq. (15) (Table 2). The nontrivial solution of Eq. (15), that is, the solution for not all C_1, C_2, C_3 and C_4 are zeros, requires that the determinant of the matrix factor of $\{C\}$ be equal to zero. For a given value of shear-flexure stiffness proportion, αH, the roots $\lambda_1 H$ and $\lambda_2 H$ of Eq. (15) for each mode of vibration of the tall building can be calculated using half-interval search technique. For each pair of their values of $\lambda_1 H$ and $\lambda_2 H$, we can solve Eq. (15) for C_1, C_2, C_3 and C_4. Substituting the values of C_1, C_2, C_3 and C_4 of a given mode into Eq. (8), the mode shape for that mode of vibration of the building is obtained.

Finally, from Eqs. (3), (7), and (9) we have

$$\omega_1 = \left[\frac{EI}{\rho A}\frac{(\lambda_{1i}H)^2\{(\lambda_{1i}H)^2 + (\alpha H)^2\}}{H^4}\right]^{1/2}, \tag{16}$$

in which the subscript, i, refers to the ith mode.

The orthogonality property of normal modes for uniform shear-flexure structure is given by [10]

$$\frac{1}{a^2}\int_0^H \psi_i(x)\psi_j(x)\,dx = 0, \quad \text{for } i \neq j, \tag{17}$$

and the normalization factor, f_i, for normalization of the ith mode shapes with respect to mass is given by [10]

$$f_i = \frac{1}{\int_0^H \psi_i^2(x)\,dx}. \tag{18}$$

Table 2: Equation 15

$$
\begin{bmatrix}
1 & 0 & 1 & 0 \\
0 & \lambda_1 H & 0 & \lambda_2 H \\
(-\lambda_1^2 H^2 \cos\lambda_1 H) & (-\lambda_1^2 H^2 \sin\lambda_1 H) & (\lambda_2^2 H^2 \cosh\lambda_2 H) & (\lambda_2^2 H^2 \sinh\lambda_2 H) \\
(-a^2 H^2 \lambda_1 \sin\lambda_1 H - \lambda_1^2 H^2 \sin\lambda_1 H) & (a^2 \lambda_1 H^2 \cos\lambda_1 H + \lambda_1^2 H^2 \cos\lambda_1 H) & (a^2 \lambda_2 H^2 \sinh\lambda_2 H - \lambda_2^3 H^2 \sinh\lambda_2 H) & (a^2 \lambda_2 H^2 \cosh\lambda_2 H - \lambda_2^2 H^2 \cosh\lambda_2 H)
\end{bmatrix}
\begin{bmatrix} c_1 \\ c_2 \\ c_3 \\ c_4 \end{bmatrix}
=
\begin{bmatrix} 0 \\ 0 \\ 0 \\ 0 \end{bmatrix}
$$

The modal seismic response of a uniform shear-flexure building without TMD can be obtained by assuming a product solution for Eq. (1) similar to Eq. (5) but of the form:

$$y(x,t) = \sum_{i=1}^{\alpha} \psi_i(x)q_i(t), \tag{19}$$

in which $\psi_i(x)$ is the ith mode shape and $q_i(t)$ is the generalized time dependent response function of the ith mode; and substituting Eq. (4) into Eq. (1), then multiplying by the jth mode shape $\psi_j(x)$ and integrating over the building height, H yield:

$$\sum_{i=1}^{\infty} \left[\left(\int_0^H \psi_j \frac{d^4\psi_i}{dx^4} dx - \alpha^2 \int_0^H \psi_j \frac{d^2\psi_i}{dx^2} dx \right) q_i(t) + \left(\int_0^H \psi_j \psi_i \ dx \right) \frac{\ddot{q}_i(t)}{\alpha^2} \right.$$

$$\left. + \left[\frac{C_i}{EI} \int_0^H \psi_j \psi_i \ dx \right] \dot{q}_i(t) \right] = -\frac{1}{\alpha^2} \ddot{y}_g \int_0^H \psi_j \ dx. \tag{20}$$

Substituting the orthogonality property of normal modes Eq. (17) into Eq. (20) and simplifying the terms yield the modal equation of motion of the uniform shear-flexure building relative to the base:

$$\ddot{q}_i(t) + 2\xi\omega_i \dot{q}_i(t) + \omega_i^2 q_i(t) = -p_i \ddot{y}_g, \tag{21}$$

in which

$$p_i = \frac{\int_0^H \psi_i \ dx}{\int_0^H \psi_i^2 \ dx} = modal \ participation \ factor \ of \ mode \ i. \tag{22}$$

Equation (21) is a set of infinite number of uncoupled differential equations $(i=1,2,...)$, each for one mode of vibration of the building.

If the mode shape is normalized with respect to mass, that is, if

$$\Phi_i(x) = \frac{\psi_i(x)}{\int_0^H \psi_i^2(x) \ dx}, \tag{23}$$

in which $\Phi_i(x)$ is the ith mode shape normalized with respect to mass, then

$$\int_0^H \Phi_i^2(x) \ dx = 1, \tag{24}$$

and the ith modal participation factor, p_i, is given by:

$$p_i = \int_0^H \Phi_i(x) \ dx. \tag{25}$$

4. Mathematical Model of Uniform Tall Building with Tuned Mass Damper

The primary structure with a linear elastic tuned mass damper is shown in Fig. 2b. When the tuned mass damper is added to the building, the equation of motion of the building becomes:

$$\frac{\partial^4 y}{\partial x^4} - \alpha^2 \frac{\partial^2 y}{\partial x^2} + \frac{1}{a^2}\frac{\partial^2 y}{\partial t^2} + \frac{C}{EI}\frac{\partial y}{\partial t} - \frac{F(H)}{EI} = -\frac{1}{a^2}\ddot{y}_g, \qquad (26)$$

in which $F(H)$ is the external force applied at height H of the primary structure by the TMD and is given by:

$$F(H) = C_T \dot{U} + K_T U, \qquad (27)$$

where C_T is the viscous damping coefficient of TMD and K_T is the stiffness of the TMD.

Operations similar to the ones preceeding Eq. (20) leads to the following equation:

$$\sum_{i=1}^{\infty}\left[\left[\int_0^H \psi_j \frac{d^4\psi_i}{dx^4}dx - \alpha^2\int_0^H \psi_j \frac{d^2\psi_i}{dx^2}dx\right]q_i(t) + (\int_0^H \psi_j\psi_i\,dx)\frac{\ddot{q}_i(t)}{a^2}\right.$$

$$\left. + \left[\frac{C_i}{EI}\int_0^H \psi_j\psi_i\,dx\right]\dot{q}_i(t) - \left[\frac{C_T\dot{U}_i + K_T U_i}{EI}\right]\psi_j(H)\right] = -\frac{1}{a^2}\ddot{y}_g\int_0^H \psi_j\,dx. \qquad (28)$$

With the assumption that the tuned mass damper does not modify the mode shapes and the frequencies of the primary structure, operations similar to those preceding Eq. (21) yields:

$$\ddot{q}_i(t)+2\xi_i\omega_i\dot{q}_i(t)+\omega_i^2 q_i(t)-\frac{M_T\psi_i(H)}{\rho A\int_0^H \psi_i^2 dx}(2\xi_T\omega_T\dot{U}_i + \omega_T^2 U_i)=-\frac{\int_0^H \psi_i\,dx}{\int_0^H \psi_i^2 dx}\ddot{y}_g. \qquad (29)$$

Alternately,

$$\ddot{q}_i(t) + 2\xi_i\omega_i\dot{q}(t) + \omega_i^2 q_i(t) - \mu_i(2\xi_T\omega_T\dot{U}_i + \omega_T^2 U_i) = -p_i\ddot{y}_g, \qquad (30)$$

in which

$$\mu_i = \text{Generalized ith modal mass ratio}$$

$$= \frac{M_T\psi_i(H)}{\rho A\int_0^H \psi_i^2\,dx}, \qquad (31)$$

$\xi_T = \%$ of critical damping of TMD, $\omega_T = $ uncoupled frequency of TMD.

The equation of motion of the linear elastic tuned mass damper (Fig. 2b) for the ith mode of vibration of the primary structure is given by:

$$M_T(\ddot{y}_i(H) + \ddot{U}_i + \ddot{y}_g) + C_T\dot{U}_i + k_TU_i = 0, \tag{32}$$

in which

$$y_i(H) = \psi_i(H)q_i(t). \tag{33}$$

Substitution of Eq. (33) into Eq. (32) and simplification yields:

$$\ddot{U}_i + 2\xi_T\omega_T\dot{U}_i + \omega_T^2 U_i + \psi_i(H)\ddot{q}_i(t) = -\ddot{y}_g. \tag{34}$$

Equations (30) and (34) are the modal equations of motion for uniform building with tuned mass damper. They are coupled equations in $q_i(t)$ and U_i and must be integrated directly for a deterministic analysis. In this study 'constant velocity method' [1] has been used for the numerical solution of Eqs. (30) and (34).

5. Numerical Evaluation for Linear Elastic Parametric Analysis

The parametric study was conducted considering a 25-story shear-flexure uniform building [17]. The building constants and the frequencies of the first eight modes of the building are given in Table 3. The peak response ratio is studied as a function of parameters such as frequency ratio, generalized mass ratio, damping of the building, and damping of the tuned mass damper; one parameter being varied at a time when all other parameters are kept fixed at a given value. The peak response ratio and frequency ratio are defined respectively as $y_{T\max}/y_{\max}$ and ω_T/ω, where $y_{T\max}$, $y_{\max}$ = maximum relative displacement at top floor of building with and without tuned mass damper, ω_T = natural frequency of tuned mass damper, and ω = natural frequency of the primary structure. The generalized mass ratio is defined as the ratio of the generalized modal mass of tuned mass damper to generalized modal mass of the primary structure. The response of the lowest eight modes of the primary structure is studied in this investigation. Only selected results are given here, a more complete result is given in Ref. [10].

Table 3: Building constants and frequencies of the primary structure

Building Constants		Modal Frequencies		
		Mode No.	Frequency; w (rad/sec)	Period, T (sec.)
Parameters	Numerical Values			
		1	2.952	2.129
		2	9.792	0.642
Height, H(ft.)	300	3	19.242	0.327
EI_y(k–ft^2)	2.18×10^9	4	32.301	0.196
GA_y (k)	1346.4×10^3	5	49.394	0.127
ρA(k–sec^2/ft^2)	5.8	6	70.668	0.089
αH	7.4555	7	96.169	0.065
a^2	375862000	8	125.915	0.050

The effect of frequency ratio on the sensitivity of peak response ratio when the primary structure is subjected to earthquakes A001-1, A001-2, and A008-2 is given in Figs. 3, 4, and 5, respectively. It is clear from these plots that for a given mode the maximum modal response reduction occurs when the frequency of the tuned mass damper is tuned close to the modal frequency of the structure. Since a tuned mass damper is generally tuned to the fundamental frequency of the primary structure, the modal frequency ratio for the higher modes of vibration of the primary structure will be far from unity and as can be seen in Figs. 3, 4, and 5, the higher mode response will in fact increase. A summary of peak modal response ratio reduction for all the six earthquakes is given in Table 4. The results of Table 4 show that the degree of reduction of peak modal response ratio vary from earthquake to earthquake. Figure 6 and Table 4 reveal the effect of earthquake characteristics on seismic effectiveness of tuned mass dampers. The TMD is a passive device which is to be excited by the motion of the primary structure whose peak response it is supposed to reduce. The effectiveness of the TMD to control the peak response of the primary structure depends upon the nature of the pre-peak motion of the primary structure. This pre-peak motion is dependent, in addition to the other factors, on the magnitude of the earhtquake accelerations and their distribution with respect to time. This study also shows (Fig. 7) that for all modes, the peak response ratio decreases as the generalized modal mass ratio increases but the increase of the damping of the primary structure increases the peak response ratio (Fig. 8). Finally, the increase of the TMD damping increases the peak response ratio slowly (Fig. 9) but decreases the TMD motion relative to the primary structure significantly. Figures 7, 8, and 9 also show that the effectiveness of TMD to influence the modal response of the primary structure varies from mode to mode.

Table 4: Reduction in peak modal response for different earthquakes

Cal. Tech. No.	Reduction in Peak Modal Response							
	1st Mode	2nd Mode	3rd Mode	4th Mode	5th Mode	6th Mode	7th Mode	8th Mode
A001-1	0.90	0.80	0.77	0.75	0.60	0.81	0.84	0.86
A001-2	0.50	0.87	0.62	0.67	1.00	0.88	0.81	1.01
A004-2	0.80	0.63	0.46	0.63	1.00	0.88	1.05	1.22
A008-2	0.78	0.60	0.71	0.60	1.05	1.01	1.03	1.02
A009-2	0.86	0.88	0.78	0.87	0.76	0.90	0.86	0.85
A018-2	0.74	0.65	0.77	0.83	0.92	1.06	1.24	1.22

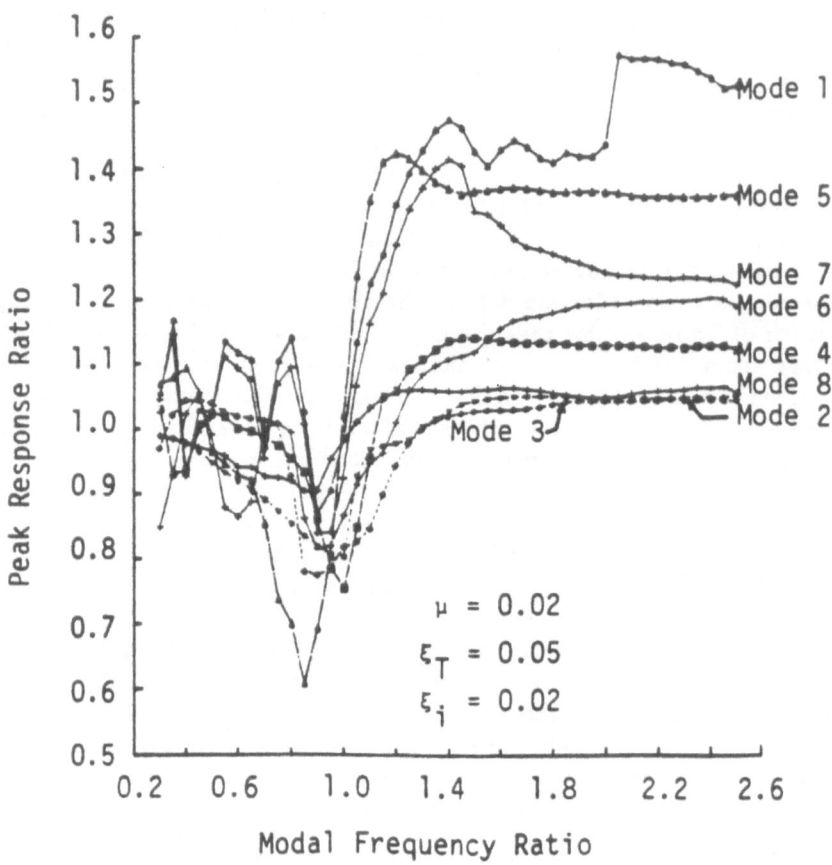

Figure 3: Sensitivity of peak response ratio to modal frequency ratio for earth-quake: A001-1

6. Inelastic Case Study Considering High Modes of Vibration of Tall Buildings

The second part of the present study evaluates the effectiveness of tuned mass dampers to reduce the inelastic seismic response of a 25-story building as described in Refs. [18] and [19]. The building is a steel frame-shear wall type structure, with the shear wall represented by two elevator cores. The steel frame is assumed to carry all the gravity loads, while primarily the shear wall is to resist the lateral forces. The building was modelled as an eight node lumped mass system with one translational and one rotational degree of freedom at each node, but at each node only translational inertia was considered for seismic analysis. The building's mass was assumed to be distributed uniformly over its height, but the flexural stiffness of the building was assumed to decrease from a maximum at the base to 55% of this value at the top of the structure as given in

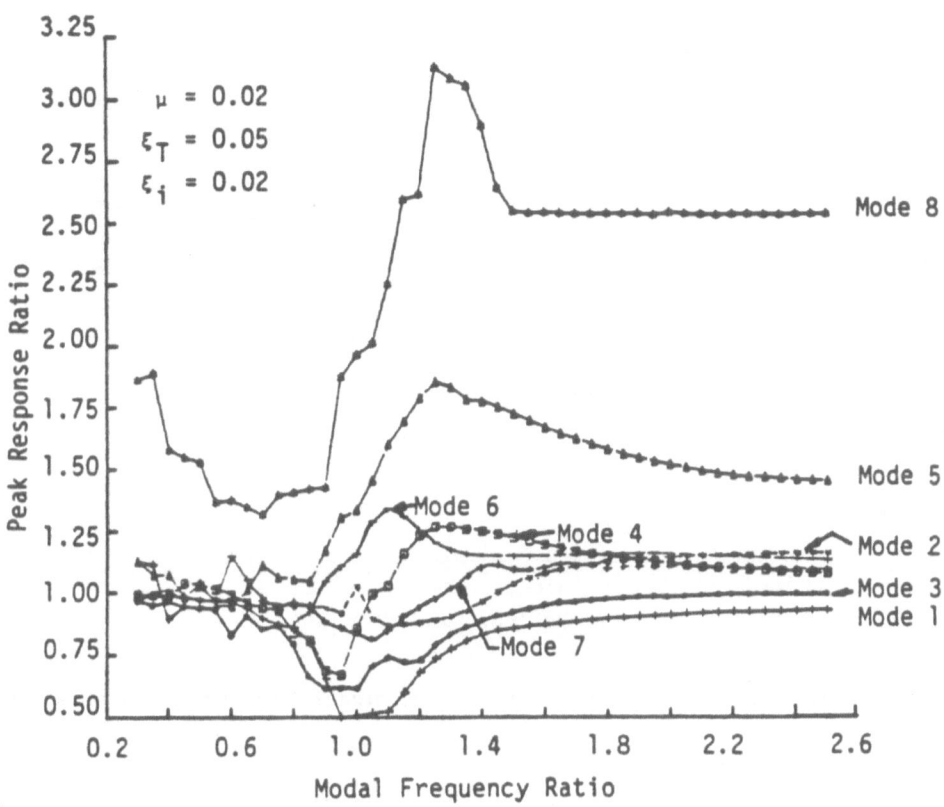

Figure 4: Sensitivity of peak response ratio to modal frequency ratio for earthquake: A001-2

Ref. [19].

The moment-curvature relationship of the building was used as the input structural property for the inelastic seismic analysis. The detailed derivation of the moment-curvature curve of the building with degrading stiffness is given in Ref. [18]. The elastic building stiffness is based on an empirical formula [18] which took into account the contributions of nonstructural elements, which may add significant stiffness to the building. But as the building is subjected to cyclic seismic motions which cause inelastic behavior, the contribution of some elements is reduced. This type of behavior was represented by the inelastic model with degrading stiffness.

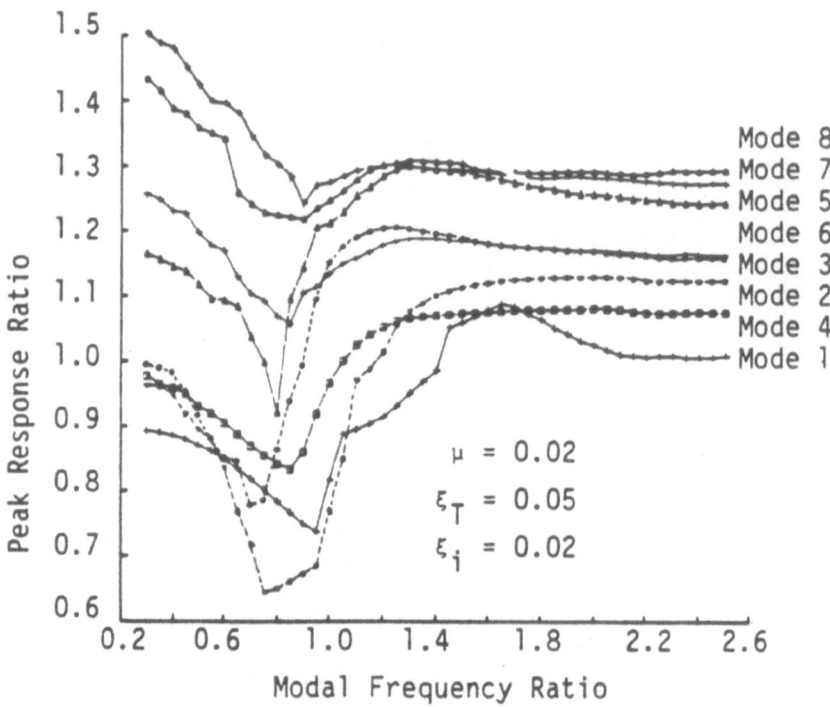

Figure 5: Sensitivity of peak response ratio to modal frequency ratio for earth-quake: A008-2

The building was subjected to a set of six historical earthquakes (Table 1) which have also been used for elastic parametric study in the first part of this study. Nine different TMDs were used, with mass ratios ranging from 0.125% to 4.5% of the total building mass. The design of the TMDs was based on a method developed by Wiesner [22]. This method is based on optimum values for damping and frequency ratio of a generalized mass model. The properties of TMDs are given in Table 5.

7. Equations of Motion

The equation of motion of the lumped mass modal without the tuned mass damper (Fig. 7) is given by:

$$[M]\{\ddot{y}\} + [C]\{\dot{y}\} + [K]\{y\} = -[M]\{\gamma\}\ddot{y}_g, \tag{35}$$

in which $[M]$, $[C]$ and $[K]$ = the mass, viscous damping, and stiffness matrices; $\{\gamma\}$ = vector of unit values; $\{y\}$ = the column vector of horizontal nodal displacements relative to the ground, and $\{\dot{y}\}$ and $\{\ddot{y}\}$ are the derivatives of $\{y\}$.

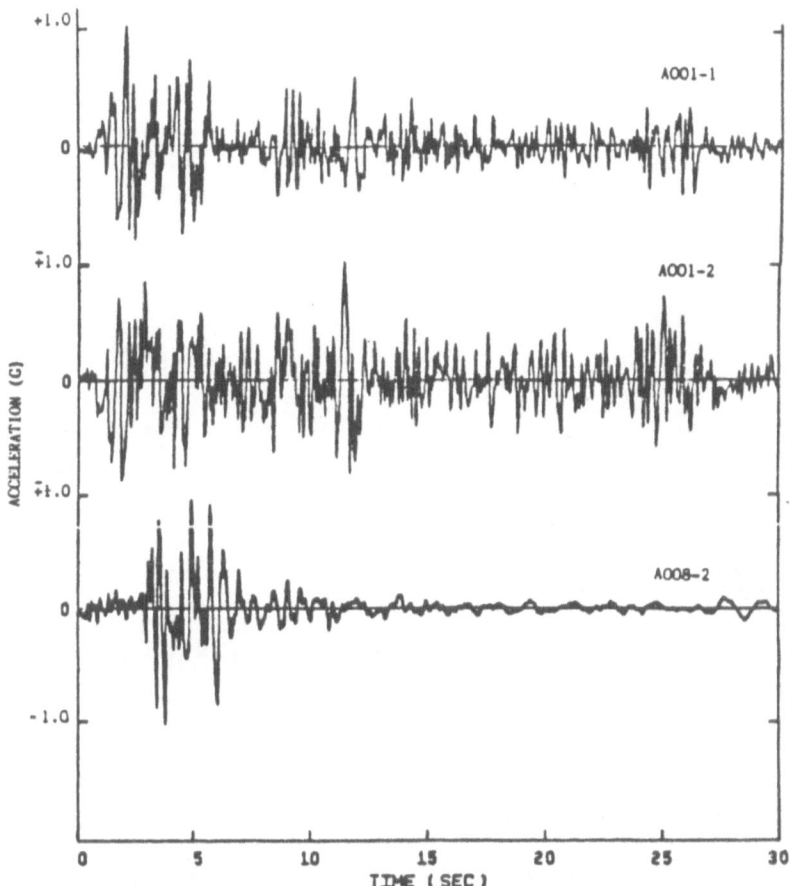

Figure 6: Time history of historical earthquake normalized to 1G

The equation of motion of the lumped mass model with the tuned mass damper is given by:

$$[M]\{\ddot{y}\} + [C]\{\dot{y}\} + [K]\{y\} - \{F\} = -[M]\{\gamma\}\ddot{y}_g, \tag{36}$$

in which

$$\{F\} = \left\{ \begin{array}{c} 0 \\ 0 \\ . \\ . \\ . \\ C_T\dot{U}_T + K_T U_T \end{array} \right\}, \tag{37}$$

where U_T, K_T and C_T are the displacement of the TMD relative to the top floor

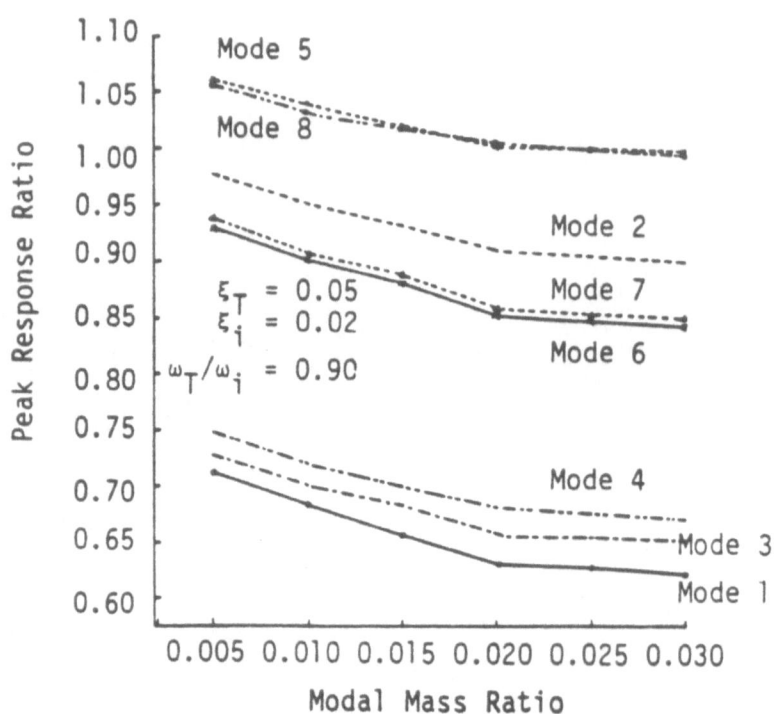

Figure 7: Sensitivity of peak response ratio to generalized modal mass ratio for earthquake: A001-2

of the building, the associated spring constant and the associated viscous damping coefficient. Equations (35) and (36) were solved by nonlinear direct integration technique using the constant acceleration method [1]. The analysis of the structure with and without a TMD was done with the use of two computer programs, SALBO [18] and DRAIN-2D [11].

8. Numerical Evaluation for Inelastic Case Study

The computer program SALBO was used to calculate the inelastic response to the 25-story building with and without a TMD. Computer program DRAIN-2D was used to check the results obtained from SALBO for the building without a TMD. Because of low magnitude ground accelerations of earthquakes A004-2 and A018-2, the building did not exhibit any inelastic behavior under these two seismic motions.

Table 6 shows the peak response ratio of the building for the seismic motions A001-1, A001-2, A008-2 and A009-2 for nine different TMDs. The results in Table 6 confirm the findings of the elastic parametric study that the effectiveness

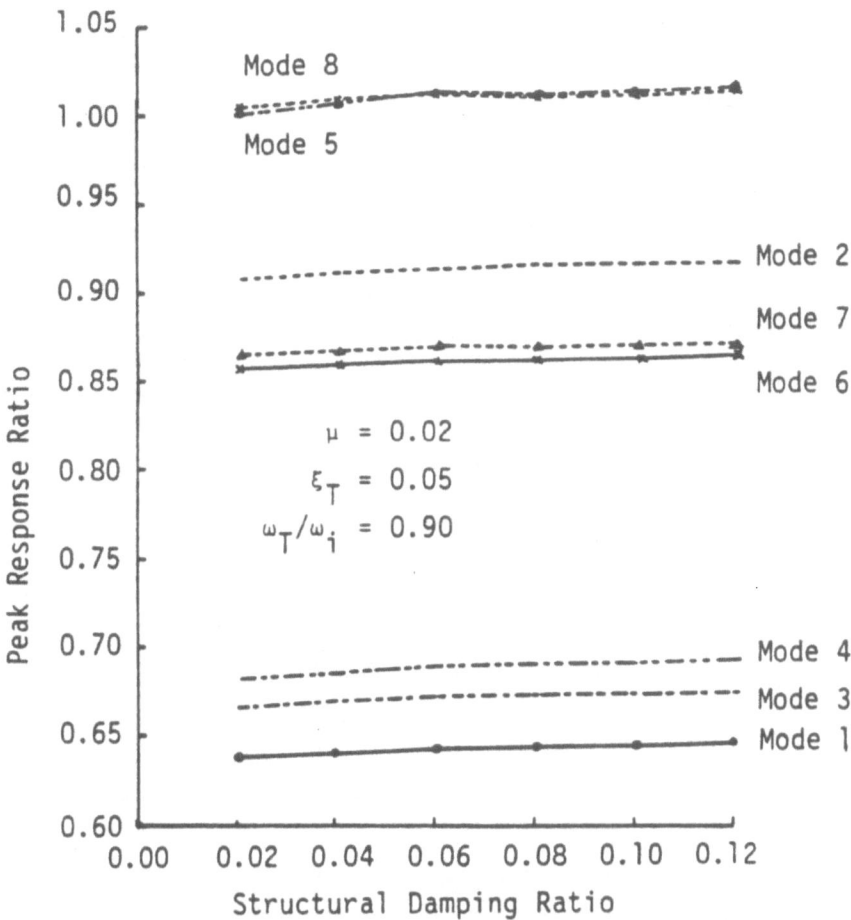

Figure 8: Sensitivity of peak response ratio to structural damping ratio for earthquake: A001-2

of the TMD to control the peak response of the primary structure depends on the nature of the pre-peak motion of the primary structure. It also shows that for effective TMDs the peak response ratio decreases as the mass ratio increases. Figure 10 illustrates that the TMD once excited by the motion of the primary structure, does reduce the response of the primary structure, though not necessarily the peak response. This reduction of inelastic response of the building will have an effect on the degrading stiffness caused by the earthquakes. A more complete result on the effectiveness of tuned mass dampers to reduce the inelastic seismic response of a tall building is given in Ref. [7].

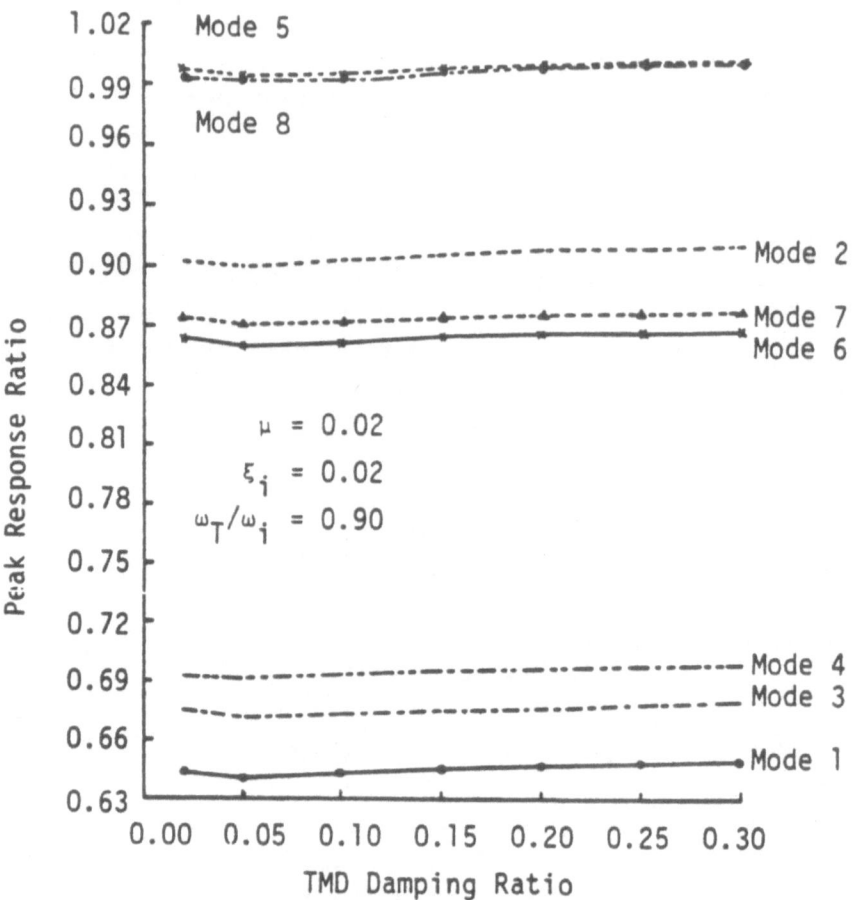

Figure 9: Sensitivity of peak response ratio to TMD damping ratio for earth-quake: A001-2

9. Assessment of Tuned Mass Dampers as Useful Aseismic Control Devices

The effectiveness of tuned mass dampers to reduce the seismic response of tall buildings is assessed in this section using the available literature and the present study by the authors. All the studies to date on the seismic effectiveness of TMD reveal that once it is set in motion, the tuned mass damper reduces the modal response of the primary structure for the mode to which the TMD has been tuned. However, the passive TMD is to be excited by the motion of the primary structure whose peak response it is supposed to reduce. If the peak response of the primary structure occurs without significant pre-peak motion, the passive TMD does not get enough excitation to be effective and is unable to prevent the

Table 5: Properties of tuned mass dampers

Generalized mass ratio	TMD frequency Rad./Sec.	TMD damping %	TMD mass K–Sec² per in	TMD damping coefficient K-Sec per in	TMD stiffness K/in.
.005	3.27	.04	.265	.066	2.834
.010	3.25	.05	.529	.151	5.591
.020	3.23	.07	1.060	.479	11.060
.030	3.21	.09	1.588	.897	16.363
.040	3.19	.10	2.118	1.324	21.553
.060	3.14	.12	3.177	2.394	31.324
.080	3.08	.13	4.236	3.523	40.184
.120	2.99	.16	6.354	6.155	56.805
.180	2.85	.19	9.530	10.593	77.407

Table 6: Peak response ratio for the top floor displacement

Earthquake Record No.	A009-2	A008-2	A001-2	A001-1
Generalized Mass Ratio	Peak Response Ratio			
.005	.98	.98	.98	1.00
.010	.97	.97	.92	.99
.020	.95	.94	.87	.99
.030	.94	.92	.87	1.02
.040	.92	.90	.88	1.03
.060	.89	.87	.88	1.05
.080	.86	.85	.87	1.05
.120	.81	.80	.83	1.05
.180	.74	.74	.84	1.00

primary structure from reaching maximum response level. Thus the effectiveness of the tuned mass dampers to control the peak response of the primary structure depends on the nature of the pre-peak motion of the primary structure which is influenced, in addition to other factors, by the magnitudes of seismic accelerations, the duration of strong motions and their distribution with respect to time. Since the earthquake characteristics vary from earthquake to earthquake, the peak-response ratios from individual earthquakes have large statistical dispersion. The statistical variability of the peak-response reduction is large enough to cause uncertainty as to the effect of adding a TMD to control seismic motion.

A tuned mass damper tuned to the fundamental frequency of the primary structure is even less effective to reduce the response of the structure at higher modes because the TMD is out of tune with those modes. Additionally, as the primary structure experiences elasto-plastic deformations, its frequencies decrease

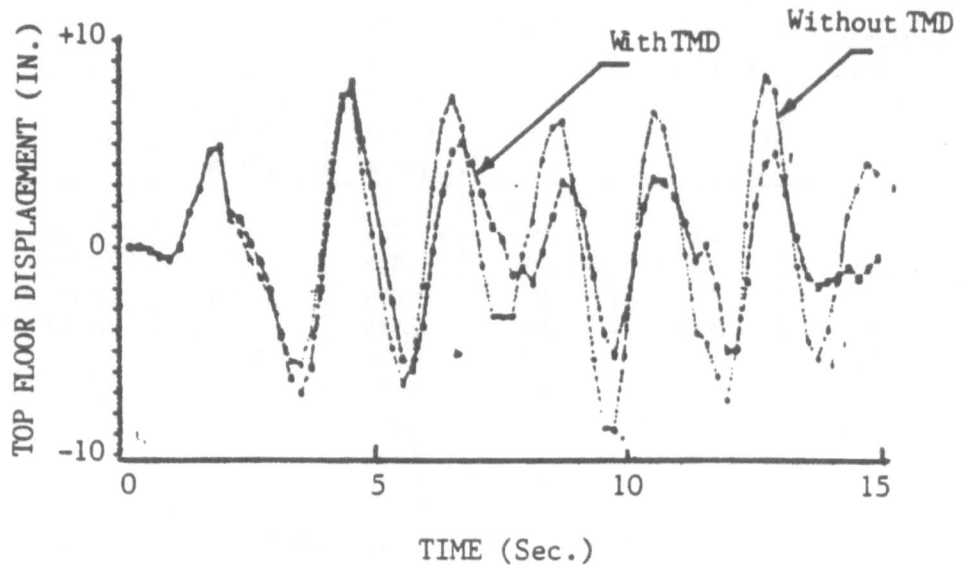

Figure 10: *Top floor displacement of inelastic model with degrading stiffness subjected to earthquake, A001-2, mass ratio, $\mu = 0.080$*

thereby making the TMD out of tune even with the fundamental mode of the damaged building [19]. Thus the uncertain seismic effectiveness of TMD on elastic systems becomes smaller on inelastic systems. The passive devices such as tuned mass dampers are not recommended for using as aseismic control devices.

10. Conclusions

This paper presents the effect of tuned mass dampers on the modal response of primary structure subjected to earthquake motions. This paper uses the results of the present study by the authors and the earlier studies by others to assess the tuned mass dampers as useful aseismic control devices. This paper shows that:

(i)   A tuned mass damper tuned to the fundamental frequency of the primary structure is less effective in reducing the response of the primary structure at higher modes, the higher mode response will in fact increase.

(ii) At all modes, the peak modal response ratio of the primary structure decreases as the generalized modal mass ratio increases.

(iii) An increase of the damping of the primary structure increases its peak modal response ratio at all modes.

(iv) An increase of the damping of the tuned mass damper increases the peak modal response ratio of the primary structure but decreases the motion of the tuned mass damper relative to the primary structure.

(v) A tuned mass damper tuned to the fundamental frequency of an elastic primary structure is even less effective on the inelastic response of the primary structure.

(vi) The passive tuned mass dampers cannot be used as reliable aseismic control devices.

References

[1] Biggs, J.M., *Introduction to Structural Dynamics*, McGraw-Hill Book Co., Inc., New York, N.Y., 1964.

[2] Chandrasekaran, A.R. and Gupta, Y.P., "Vibration Absorbers for Single Degree Freedom Systems," Proceedings of the Third Symposium on Earthquake Engineering, University of Roorkee, U.P., India, November 1966, pp. 23-32.

[3] Chowdhury, A.H., "Effect of Foundation Flexibility on Dynamic Response of Tall Buildings," Proceedings of the International Symposium on Dynamic Soil-Structure Interaction, Minneapolis, September 1984, pp. 159-168.

[4] Crandall, S.H. and Mark, W.D., *Random Vibration in Mechanical Systems*, Academic Press, New York, N.Y., 1973, pp. 80-101.

[5] "Dampers Blunt the Wind's Force on Tall Buildings," *Architectural Record*, September 1971, pp. 155-158.

[6] Dan Hartog, J.P., *Mechanical Vibrations*, McGraw-Hill Book Co., Inc., New York, N.Y., 1956.

[7] Garske, J.J., "Effectiveness of Tuned Mass Dampers on Seismic Behavior of a Tall Building," Master of Science Plan B Paper, Department of Civil Engineering, North Dakota State University, Fargo, 1985.

[8] Gupta, Y.P. and Chandrasekaran, A.R., "Absorber System for Earthquake Excitations," Proceeding of the Fourth World Conference on Earthquake Engineering, Santiago, Chile, January 1969, pp. 139-148.

[9] Heidebrecht, A.C. and Stafford-Smith, B., "Approximate Analysis of Tall Wall-Frame Buildings," *Journal of the Structural Division, ASCE*, Vol. 99, No. ST2, February 1973, pp. 199-221.

[10] Iwuchukwu, M.D., "Seismic Effectiveness of Tuned Mass Dampers Considering Higher Modes of Vibration," Master of Science Thesis, Department of Civil Engineering, North Dakota State University, Fargo, 1985.

[11] Kanaan, A.E. and Powell, G.H., "DRAIN-2D: A General Purpose Computer Program for Dynamic Analysis of Inelastic Plane Structures," Reports No. EERC73-6 and EERC73-22, Earthquake Engineering Research Center,

University of California, Berkely, California, April 1973 (revised 1973 and 1975).

[12] Kaynia, A.M., Veneziano, D., and Biggs, J.M., "Seismic Effectiveness of Tuned Mass Dampers," *Journal of the Structural Division, ASCE*, Vol. 107, No. ST8, Proc. Paper 16427, August 1981, pp. 1465-1484.

[13] Kwok, K.C.S., "Damping Increase in Building With Tuned-Mass Damper," Proceedings of the Fourth Engineering Mechanics Division Specialty Conference, Purdue University, West Layfayette, Indiana, May 23-25 1983, pp. 323-326.

[14] "Lead Hula Hoops Stabilize Antenna," *Engineering News-Record*, July 22 1976, pp. 10.

[15] McNamara, R.J., "Tuned Mass Dampers for Buildings," *Journal of the Structural Division, ASCE*, Vol. 103, No. ST9, September 1977, pp. 1985-1998.

[16] Ormondroyd, J. and Dan Hartog, J.P., "The Theory of the Dynamic Vibration Absorber," *Transactions, American Society of Mechanical Engineers*, APM-50-7, 1928, pp. 9-22.

[17] Rutenberg, A., Tso, W.K., and Heidebrecht, A.C., "Dynamic Properties of Asymmetric Wall-Frame Structures," *Earthquake Engineering and Structural Dynamics*, Vol. 5, 1977. pp. 41-51.

[18] Sladek, J.R., "Use of Vibration Absorbers to Reduce Seismic Response," Master of Science Thesis, University of Texas at Austin, Austin, Texas, 1979.

[19] Sladek, J.R. and Klinger, R.E., "Effect of Tuned-Mass Dampers on Seismic Response," *Journal of the Structural Division, ASCE*, Vol. 109, No. ST8, Proc. Paper 18136, August 1983, pp. 2004-2009.

[20] "Strong Motion Earthquake Accelerograms - Digitized and Plotted Data," Earhtquake Engineering Research Laboratory, California Institute of Technology, Pasadena, California, Vol. II, Part A, September 1971.

[21] "Tuned Mass Dampers Steady Sway of Skyscrappers in Wind," *Engineering News-Record*, August 18 1977, pp. 28-29.

[22] Wiesner, K.B., "Tuned Mass Dampers to Reduce Building Wind Motion," Preprint 3510, ASCE Convention and Exposition, Boston, Massachusetts, April 2-6 1979.

[23] Wirsching, P.H. and Campbell, G.W., "Minimal Structural Response Under Random Excitation Using the Vibration Absorber," *Earthquake Engineering and Structural Dynamics*, John Wiley & Sons, Inc., New York, N.Y., 1974, pp. 303-312.

[24] Wirsching, P.H. and Yao, J.T.P., "A Statistical Study of Some Design Concepts in Earthquake Engineering," Technical Report CE-21(70) NSF-065, *Bureau of Engineering Research*, The University of New Mexico,

Albequerque, New Mexico, May 1970.

[25] Wirsching, P.H. and Yao, J.T.P., "Safety Design Concepts for Seismic Structures," *Computers and Structures*, Vol. 3, Pergamon Press, Inc., New York, N.Y., 1973, pp. 809-826.

[26] Wirsching, P.H. and Yao, J.T.P., "Modal Response of Structures," *Journal of the Structural Division, ASCE*, Vol. 96, No. ST4, April 1970, pp. 879-883.

ACTIVE MINIMISATION OF VIBRATIONAL ENERGY IN PERIODICALLY EXCITED STRUCTURES

A.R.D. Curtis, P.A. Nelson, S.J. Elliott
Institute of Sound and Vibration Research
The University
Southampton, England

1. Introduction

In many practical cases the vibration of a structure can be considered as periodic or quasi-periodic in nature. Rotating and reciprocating machinery such as internal combustion engines, a.c. electric motors, gear boxes and aeronautical and nautical propellers are common sources of periodic vibration. Vibration from these machines is transmitted to surrounding structures through the machinery mountings and through acoustic excitation and can result in high levels of noise and vibration at some distance from the source.

A particular problem of interest is the reduction of internal noise levels in the cabins of propeller driven aircraft. The proposed fuel efficient high speed propeller driven passenger aircraft are expected to be subject to very high levels of acoustic loading of the structure at low frequency and of essentially periodic nature. Reduction of the vibration of the structure would impede the transmission of the sound and reduce cabin noise levels.

The theory of active control of distributed vibrating systems has been extensively studied [1]. Practical implementations [2] have involved impressive use of sophisticated system identification and adaptive control algorithms. However a considerable simplification of the control system is possible if use is made of the periodic nature of disturbances. Chaplin [6] showed how the periodicity of a noise could be used in a simple noise cancelling system and cited many possible applications of his technique.

This paper outlines an approach to the suppression of resonances in periodicaly excited structures. The technique was proposed for the active control of periodic sould fields by Nelson *et al* [3] and has been adapted here for use with structural vibration. The objective of the technique is to suppress resonances by using secondary active sources of vibration to minimize the strain energy in the structure. A practical approximation to this is achieved by minimizing the weighted sum of the squares of the signals from a number of vibration sensors, by adjusting the inputs to a number of actuators which are fed with a periodic signal related to the excitation. The basic periodic control signal is obtained directly and independently from the source for example by a tachometer. By careful choice of the position of the actuators and sensors their numbers can be kept small whilst still achieving resonance suppression.

The theoretical basis of this approach is presented in a simple formulation which shows the strain energy to be a quadratic function of the secondary force inputs. A simulation of a simple structure is used to illustrate the mechanisms of reduction and to demonstrate the suppression of multiple resonances by the careful positioning of a few sources and sensors.

This system has wide application in situations where the vibrational excitation is of low frequency and periodic. Applications include ship engine rafts, electrical transformer noise, diesel engines in trucks, plant isolation in buildings, helicopters and automobiles.

2. The Minimisation of Strain Energy

Consider a linear distributed vibrating system with negligible damping which can be described by the generalised equation [4]

$$M(\ddot{u}) + L(u) = 0$$

where $u(x,t)$ is the displacement, $\ddot{u}(x,t)$ the acceleration and M and L are linear differential operators involving only the space variable x. The potential or strain energy function V is [4]

$$V = 1/2 \int_{\Sigma} u L(u) d\sigma$$

where Σ is the domain of the system. The vibration is assumed to have a harmonic time dependence of frequency ω. The system is excited by a primary pressure distribution $p(x,\omega)e^{j\omega t}$ and N independent secondary forces

$$s_1(\omega)e^{j\omega t}, s_2(\omega)e^{j\omega t}, \ldots, s_N(\omega)e^{j\omega t}.$$

As the system is linear both u and $L(u)$ will be linear functions of each $s_i(\omega)$. The time averaged strain energy function V will thus be a quadratic function of each $s_i(\omega)$ and can be written in the complex quadratic vector form [3] given by

$$V = s^H A s + b^H s + s^H b + c \tag{1}$$

where s is the vector of $s_i(\omega)$'s and s^H denotes the complex conjugate transpose of the vector s. The matrix A, vector b and scalar c depend upon the system and the primary pressure distribution. V will have a unique global minimum calculated by setting the differential of V with respect to s to zero. This minimum is defined by a secondary force vector

$$s_o = -A^{-1} b \tag{2}$$

and a minimum value of

$$V_o = c - b^H A^{-1} b \ . \tag{3}$$

If V as a function of s is available, the minimum can be found using standard optimisation techniques [5] which do not require direct knowledge of A, b and c.

In practice the strain energy function V is not directly measurable and so cannot be used as an objective of minimisation or control. A practical method of reducing strain energy is found to be the minimization of the time averaged sum of the weighted squares of the signals from R vibration sensors. If $u_i(\omega)e^{j\omega t}$ is the displacement at the ith sensor then the minimization objective J is

$$J = 1/2 \sum_{i=1}^{R} u_i^*(\omega) w_i u_i(\omega)$$

or in vector form

$$J = 1/2 u^H W u \tag{4}$$

where u is the vector of $u_i(\omega)$'s and W is the diagonal matrix of weighting factors w_i. It can be seen that J is also a quadratic function of the secondary force vector s and will also have a unique global minimum. If J is available as a function of s it may also be minimized by standard techniques making special use of techniques for least squares optimisation.

3. A Simulation

The model chosen for the simulation is a beam of length L, Young's modulus E, 2nd moment of area I and mass per unit length m. The beam is simply supported at either end, and is shown in Fig. 1. The response at x to a point force input $Fe^{j\omega t}$ at ℓ is given by solving the wave equation with the appropriate boundary conditions. Thus

$$u(x,t)=\frac{Fe^{j\omega t}}{2\beta^3EI}\begin{cases}\dfrac{\sin\beta(L-\ell)\sin\beta x}{\sin\beta L}-\dfrac{\sinh\beta(L-\ell)\sinh\beta x}{\sinh\beta L} & 0\leq x\leq\ell \\[2ex] \dfrac{\sin\beta\ell\sin\beta(L-x)}{\sin\beta L}-\dfrac{\sinh\beta\ell\sinh\beta(L-x)}{\sinh\beta L} & \ell\leq x\leq L\end{cases}$$

where $\beta^4=\omega^2m/EI$.

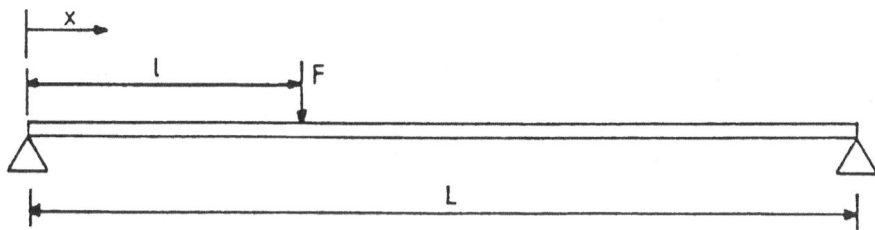

Figure 1: The simulated beam

The strain energy function for a beam in bending is [4]

$$V=1/2EI\int_o^L[\partial^2u/\partial x^2]^2dx \ .$$

The beam is excited by a point primary force located at ℓ_1 and a set of N point secondary forces located at $\ell_2,\ldots,\ell_{N+1}$. The displacement u and curvature $\partial^2u/\partial x^2$ are calculated for each set of source locations and strengths by superposition of the displacement and curvature for each source individually. The strain energy function is calculated by numerical integration of the curvature. The optimum set of secondary forces is calculated by numerical optimisation of the time averaged strain energy function. The analysis is simplified by neglecting damping as the generally complex secondary forces become purely real. For the vibration due to a single primary force the displacements are in phase with that force. Thus any cancelling secondary force must be purely in phase or out of phase with the primary force.

The function J is formed by calculating the time average of the sum of the squares of the displacements at R locations $r_1,r_2,\ldots,r_R$. The weights are chosen in this case to be unity. The set of secondary forces which minimise J is calculated by numerical optimisation. The strain energy function is calculated for these secondary forces in order to compare the resulting strain energy with the minimum strain energy.

4. Results

4.1. The minimisation of V

The primary source location was chosen to be $\ell_2 = 0.3L$ as the first seven resonances are excited as shown by Fig. 2, which shows the time averaged strain energy function against non-dimensional frequency α where $\alpha = \omega L^2 (m/EI)^{1/4}$. There are resonances when $\beta L = n\pi$, i.e. where $\alpha = (n\pi)^2$.

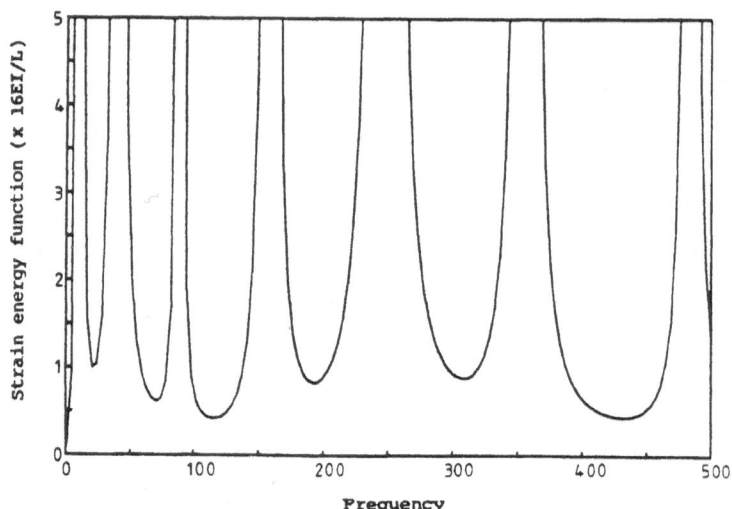

Figure 2: Time averaged strain energy function versus non-dimensional frquency for a primary source at 0.3L

If a secondary source is introduced at $\ell_2 = 0.5L$ and the strain energy function is minimised then it can be seen in Fig. 3 that the even resonances, $n = 2, 4, 6$ have been suppressed. The odd resonances, $n = 1, 2, 5, 7$ have nodes at $x = 0.5L$ and so the secondary source is unable to excite them and thus is unable to cancel them. A secondary source located at $\ell_2 = 0.1L$ is able to excite all the resonances and thus suppression of all resonances in the frequency band is possible as seen in Fig. 4. A secondary source at $\ell_2 = 0.28L$ which is close to the primary source at $\ell = 0.30L$ suppresses resonances well up to the 6th when it is near a node (at $x = 0.286L$) as seen in Fig. 5.

When two secondary sources are used the suppression achieved is greater than the suppression achieved by each source individually. In Fig. 6 can be seen the result of using single secondary sources $\ell_2 = 0.2L$ and $\ell_2 = 0.5L$ together with the result of using both secondary sources together.

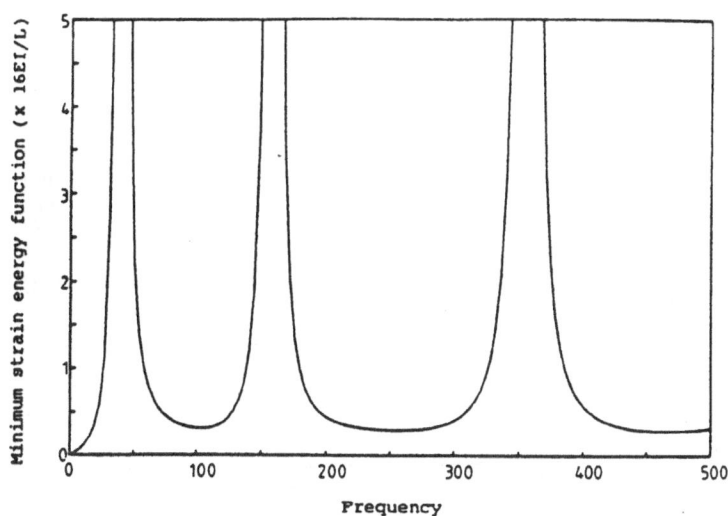

Figure 3: Minimum time averaged strain energy function versus non-dimensional frequency for a primary source at 0.3L and a secondary source at 0.5L

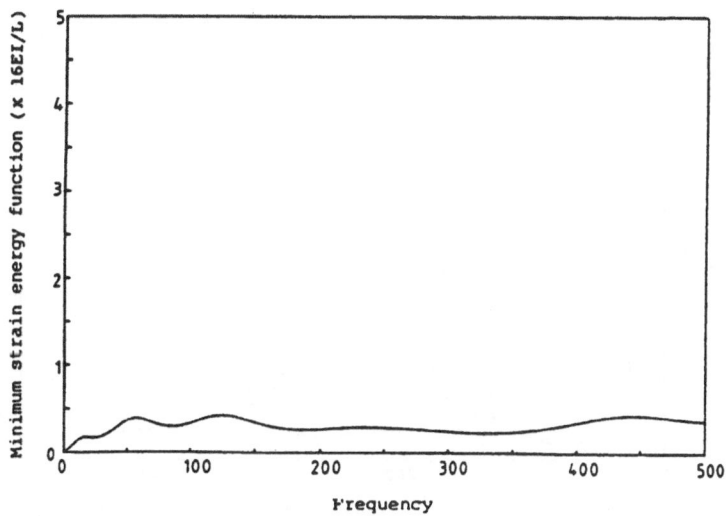

Figure 4: Minimum time averaged strain energy function versus non-dimensional frequency for a primary source at 0.3L and a secondary source at 0.1L

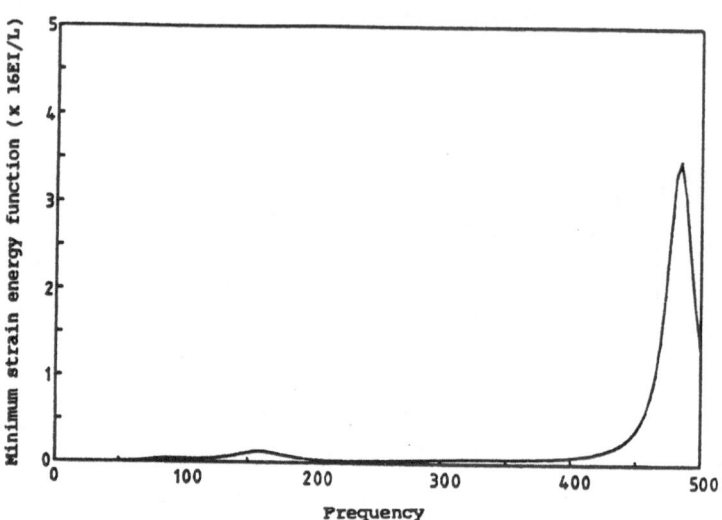

Figure 5: Minimum time averaged strain energy function versus non-dimensional frequency for a primary source at 0.3L and a secondary source at 0.28L

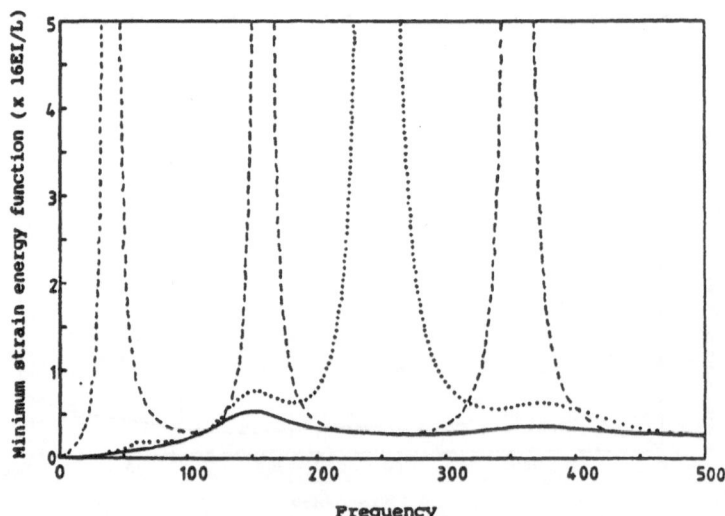

Figure 6: Minimum time averaged strain energy function versus non-dimensional frequency for a primary source at 0.3L and two secondary sources at 0.2L and 0.5L (—), and for a single secondary source at 0.2L (....), and at 0.5L (----)

5. The Minimisation of J

With a secondary source $\ell_2 = 0.1L$ and a single sensor location co-located at $r_1 = 0.1L$ the displacement there may be reduced completely, effectively pinning the beam. The resulting strain energy function seen in Fig. 7 shows the resonances of the new boundary conditions imposed on the beam. If a second sensor location is introduced at $r_2 = 0.3L$ co-located with the primary source and the function J is minimised then resonance suppression occurs as can be seen in Fig. 8. If a third sensor location is introduced at $r_3 = 0.5L$ then a strain energy function close to the minimum is achieved as can be seen in Fig. 9. Figure 10 shows the effect of locating three sensors at $r_1 = 0.25L$, $r_2 = 0.5L$, $r_3 = 0.75L$. The fourth resonance is not suppressed because the sensor locations are at nodes of the resonance and thus are unable to sense it.

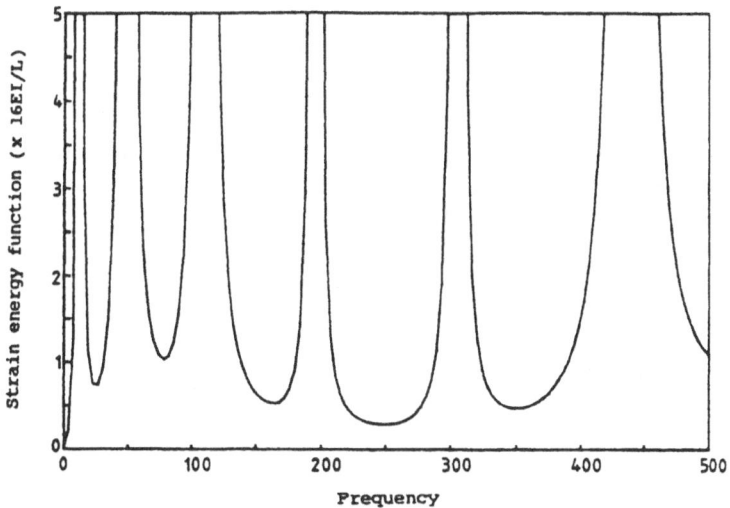

Figure 7: *Time averaged strain energy function versus non-dimensional frequency for a primary source at 0.3L, a second source at 0.1L and a single sensor at 0.1L*

6. Summary of Results

Vibrational resonances are suppressed by the minimisation of the strain energy function. A single secondary source is sufficient to suppress a single resonance if it is located away from any node of the resonance. A single secondary source is sufficient to suppress all the resonances over a band of frequencies if it is located away from any node of all the resonances in the band. Greater suppression is achieved within this limitation if the secondary source is located close to the primary source. When damping is negligible two secondary sources do not give twice the suppression of a single secondary source but at least the suppression

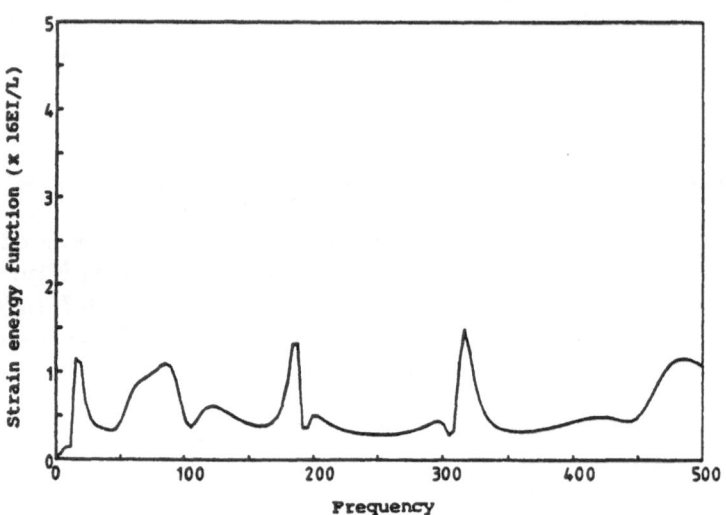

Figure 8: Time averaged strain energy function versus non-dimensional frequency for a primary source at 0.3L, a secondary source at 0.1L and two sensors at 0.1L and 0.3L

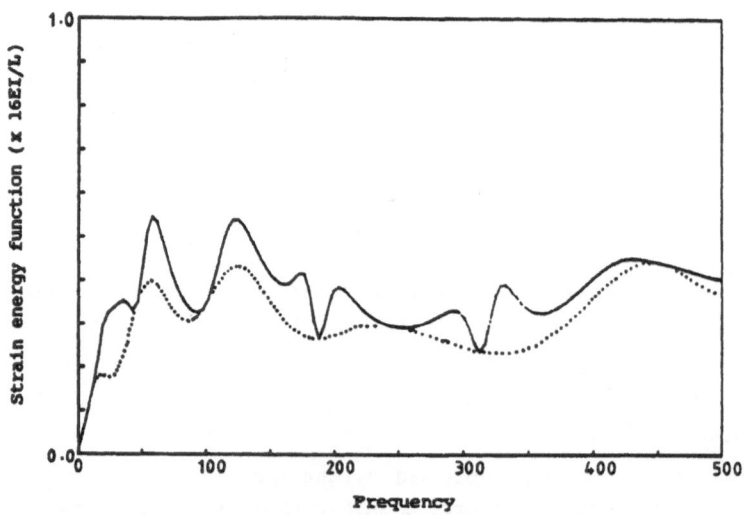

Figure 9: Time averaged strain energy function versus non-dimensional frequency for a primary source at 0.3L, a secondary source at 0.1L and three sensors at 0.1L, 0.3L and 0.5L (—), and the minimum time averaged strain energy function (.....)

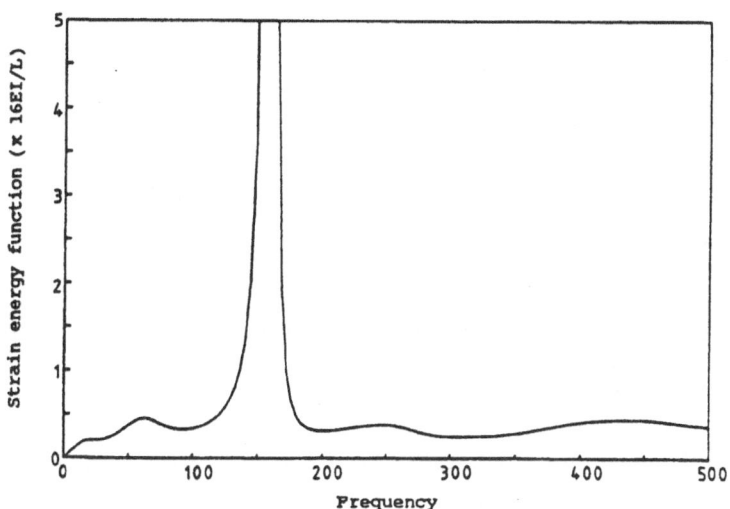

Figure 10: *Time averaged strain energy function versus non-dimensional fre-*
quency for a primary source at 0.3L, a secondary source at 0.1L
and three sensors at 0.25L, 0.5L and 0.75L

achievable by each secondary source alone.

Vibration resonances are suppressed by the minimisation of the time averaged sum of the squares of the displacements at several locations on the structure. The number of sensor locations should exceed the number of secondary
sources to avoid the creation of new resonances. The sensor locations should be
chosen so that at least one location is away from any node of a resonance to be
suppressed.

Further work is required to determine the effect of damping and closely
spaced resonances on the theoretical performance of the proposed system.

7. A Discussion of Results using Modal Theory

The control system envisaged will function independently of any model used in
the analysis. However a modal analysis is useful in discussing the performance of
the control system.

Consider a distributed vibrating system which is lightly damped and can be
described by a set of uncoupled mutually orthogonal modes. The displacement u
at a location specified by the position vector $\mathbf{x}$ is described in terms of a finite
number of modes having characteristic functions $\phi_i(\mathbf{x})$ and complex mode amplitudes $q_i(\omega)$. Thus

$$u(\mathbf{x},t)=\sum_{i=1}^{M}\phi_i(\mathbf{x})q_i(\omega)e^{j\omega t}$$

or in vector form

$$u(\mathbf{x},t)=\boldsymbol{\phi}^T(\mathbf{x})\mathbf{q}(\omega)e^{j\omega t} \qquad (5)$$

where

$$\boldsymbol{\phi}^T(x)=[\phi_1(\mathbf{x}),\phi_2(\mathbf{x}),\ldots,\phi_M(\mathbf{x})]$$

$$\mathbf{q}^T(\omega)=[q_1(\omega),q_2(\omega),\ldots,q_M(\omega)]$$

The time averaged strain energy function V can be shown to be [4]

$$V=1/4\sum_{i=1}^{M}k_{ii}q_i^*(\omega)q_i(\omega),$$

where k_{ii} is the generalised stiffness of the ith mode, or in vector form

$$V=1/4\mathbf{q}(\omega)^H\mathbf{K}\mathbf{q}(\omega) \qquad (6)$$

where $\mathbf{q}^H(\omega)$ is the complex conjugate transpose of the vector of complex modal amplitudes and $\mathbf{K}$ is the diagonal matrix of generalised stiffness.

The system is excited by a primary pressure distribution with corresponding generalised forces $p_i(\omega)$ and N point secondary forces $s_1(\omega),s_2(\omega),\ldots,s_N(\omega)$ at locations specified by the position vectors $\boldsymbol{\ell}_1,\boldsymbol{\ell}_2,\ldots,\boldsymbol{\ell}_N$. The equation of motion of the ith mode can be written, neglecting damping,

$$[k_{ii}-\omega^2 m_{ii}]q_i(\omega)=p_i(\omega)+\sum_{j=1}^{N}\phi_i[\boldsymbol{\ell}_j]s_j(\omega)$$

where m_{ii} is the generalised mass of the ith mode.

Thus in vector form

$$\mathbf{q}=\mathbf{p}+\boldsymbol{\phi}\mathbf{s} \qquad (7)$$

where

$$\mathbf{p}^T(\omega)=\left[\frac{p_1(\omega),}{k_{11}-\omega^2 m_{11}}\frac{p_2(\omega),}{k_{22}-\omega^2 m_{22}},\ldots,\frac{p_M(\omega)}{k_{MM}-\omega^2 m_{MM}}\right]$$

$$\mathbf{s}^T(\omega)=[s_1(\omega),s_2(\omega),\ldots,s_N(\omega)]$$

and

$$\phi(\omega)=\begin{bmatrix} \dfrac{\phi_1[\ell_1]}{k_{11}-\omega^2 m_{11}} & \cdots & \dfrac{\phi_1[\ell_N]}{k_{11}-\omega^2 m_{11}} \\ \cdots & \cdots & \cdots \\ \dfrac{\phi_M[\ell_1]}{k_{MM}-\omega^2 m_{MM}} & \cdots & \dfrac{\phi_M[\ell_n]}{k_{MM}-\omega^2 m_{MM}} \end{bmatrix}.$$

The strain energy function is thus a quadratic function of the secondary force vector s, from Equation (6) and Equation (7)

$$V=1/4[\mathbf{p}^H \mathbf{K}\mathbf{p}+\mathbf{p}^H \mathbf{K}\phi\mathbf{s}+\mathbf{s}^H \phi^H \mathbf{K}\mathbf{p}+\mathbf{s}^H \phi^H \mathbf{K}\phi\mathbf{s}] \tag{8}$$

and has a minimum defined from Equation (2) by

$$\mathbf{s}_o=-[\phi^H \mathbf{K}\phi]^{-1}\phi^H \mathbf{K}\mathbf{p}$$

and a minimum value from Equation (3).

$$V_o=1/4[\mathbf{p}^H \mathbf{K}\mathbf{p}-\mathbf{p}^H \mathbf{K}\phi[\phi^H \mathbf{K}\phi]^{-1}\phi^H \mathbf{K}\mathbf{p}]$$

The function J is the time average of the weighted sum of the squares of the displacements at the R locations $\mathbf{r}_1,\mathbf{r}_2,\ldots,\mathbf{r}_R$. From Equation (5)

$$\mathbf{u}=\mathbf{B}\mathbf{q}$$

where

$$\mathbf{B}=\begin{bmatrix} \phi^T(\mathbf{r}_1) \\ \phi^T(\mathbf{r}_2) \\ \cdots \\ \phi^T(\mathbf{r}_R) \end{bmatrix}$$

Thus substituting into Equation (4)

$$J=1/2\mathbf{q}^H \mathbf{B}^H \mathbf{W}\mathbf{B}\mathbf{q}$$

and writing $\mathbf{K}'=2\mathbf{B}^H \mathbf{W}\mathbf{B}$ and substituting for q from Equation (7))

$$J=1/4[\mathbf{p}^H \mathbf{K}'\mathbf{p}+\mathbf{p}^H \mathbf{K}'\phi\mathbf{s}+\mathbf{s}^H \phi^H \mathbf{K}'\mathbf{p}+\mathbf{s}^H \phi^H \mathbf{K}'\phi\mathbf{s}] \tag{9}$$

which is the right hand side of Equation (8) with K replaced by K'. If K' is in some way similar to K then minimisation of J will reduce V. K' is a constant matrix which depends only upon the sensor locations. When the system is excited at a resonance the $q_i(\omega)$ corresponding to that mode will be much larger than the rest. Both V and J will then be proportional to $|q_i|^2$ and so reducing J will ensure reduction of V. This explains why the minimisation of the time averaged weighted sum of the squares of the signals from several sensors reduces the strain energy at resonances.

8. Conclusions

The active minimisation of the vibrational strain energy of a periodically excited structure by the action of suitably located secondary source will suppress vibrational resonances. A practical means of achieving this is to minimise the time averaged weighted sum of the squares of the signals from a number of suitably located vibration sensors.

Acknowledgements

A.R.D. Curtis acknowledges the support of the U.K. Department of Trade and Industry. P.A. Nelson and S.J. Elliott also acknowledge the support of the U.K. Science and Engineering Research Council.

References

[1] Balas, M.J., "Toward a more practical control theory for distributed parameter systems", *Control and Dynamic Systems*, Vol. 18, 1982, pp. 361-421.

[2] Sundararajan, N., Montgomery, R.C. and Williams, J.P., "Adaptive identification and control of structural dynamic systems using recursive lattice filters", NASA Techanical Paper 2371, 1985.

[3] Nelson, P.A., Curtis, A.R.D. and Elliott, S.J., "Quardratic optimisation problems in the active control of free and enclosed sound fields", *Proceedings of the Institute of Acoustics*, Vol. 7, Part 2, 1985, pp. 45-54.

[4] Tong, K.N., *Theory of Mechanical Vibration*, John Wiley and Sons, New York, 1960.

[5] Gill, P.E., Murray, W. and Wright, M.H., *Practical Optimization*, Academic Press, London, 1981.

[6] Chaplin, G.B.B., "Anti-noise — the Essex breakthrough", *Chartered Mechanical Engineer*, January, 1983, pp. 41-47.

ON-LINE PARAMETER CONTROL OF NONLINEAR FLEXIBLE STRUCTURES

T.J. Dehghanyar, S.F. Masri, R.K. Miller
Department of Civil Engineering
University of Southern California
Los Angeles
T.K. Caughey
Division of Engineering and Applied Science
California Institute of Technology
Pasadena

1. Introduction

The use of passive auxiliary mass dampers [1], in which a relatively small auxiliary mass is attached to the primary system by a resilient element, to attenuate the response of oscillating systems has long been a standard method for vibration control. The dynamic vibration neutralizer (DVN), also known as the Frahm damper, is a leading member of the sub-class of linear dampers. Its principle of operation relies on the "tuning" of the auxiliary mass so that under steady-state conditions it will generate an opposing force capable of appreciably neutralizing the motion of the primary system. Another member of this group of devices is the Lanchester damper which relies exclusively on mechanical energy dissipation obtained by making the coupling resilient element consist solely of a viscous dash-pot.

In situations involving transient and/or wide-band excitations, where linear auxiliary mass dampers are inefficient, nonlinear devices have been used to overcome some of the limitations of their linear counterparts. One such highly non-linear device that has found applications in practice is the impact damper [2], also known as the "acceleration damper".

141

The impact damper consists of a solid mass that is constrained to oscillate unidirectionally in a container with a certain clearance. When attached to an oscillating primary system, this device (essentially a highly nonlinear DVN in which the coupling "spring" has deadpace characteristics) employs plastic deformation, Coulomb friction and momentum transfer between the two masses during collision, to reduce the vibrations of the primary system considerably.

Analytical and experimental studies [3] have shown that, even with a relatively small auxiliary mass ratio, properly designed impact dampers are superior to DVNs in attenuating the response of structures subjected to nonstationary wide-band random excitatins such as earthquake ground motion. The main reason for the effectiveness of the impact damper in limiting the vibrations of a structure eminatng from arbitrary dynamic environments is that the small damping forces generated by the impacting mass introduces chaos in the dynamic system response by disorganizing the orderly process of amplitude buildup, thus reducing the structural response drastically.

However, as in any passive device, even when the characteristics of a particular damper have been optimized for a given operating condition, its vibration damping efficiency is limited in handling wide-band excitations due to the inability of continuously adapting its governing characteristics to the situation at hand. This limitation of passive dampers is particularly pertinent in applications where not only the root-mean-square (rms) level of the response but also the peak levels of the primary structure response are of concern, as is the case in most structural applications.

Motivated by the previous discussion, the authors have developed and implemented two on-line active control algorithms (henceforth refered to as Methods 1 and 2) that utilize pulse generators to emulate the action of optimally designed impact dampers, to control the vibrations of linear as well as nonlinear multidegree-of-freedom (MDOF) flexible structures responding to arbitrary dynamic environments. The main limitation of these pulse-control techniques is that, unlike optimal control methods, they are incapable of completely eliminating the undesirable oscillations of vibrating structures. This drawback is more than compensated for by the fact that the pulse control methods have features that are appealing in realistic structural dynamics applications: (1) capability of controlling linear or nonlinear time-invariant or deteriorating systems with equal ease; (2) detailed knowledge of the system structure is not needed; only local measurements of the displacement and velocity are needed with Method 2 to determine the control force to be applied at a specific location in the nonlinear flexible structure; (3) wide latitude in the choice of hardware characteristics; and (4) simple on-line implementation, since a negligible amount of computational effort is needed to determine the instantaneous value of the control pulse. Details of some of the analytical and experimental studies concerned with the applications of these algorithms to a variety of mathematical and physical models are available in [4-6]. The characteristics of a recently-developed and tested cold-gas pulse generator suitable for applying the pulse-control methods under discussion to some civil structures are given in [7-8].

The above-mentioned on-line control procedures have been shown to be quite effective in greatly reducing the rms response as well as the peak response of vibrating structures even when the excitation is nonstationary wide-band random. This significant improvement in efficiency is achieved because the active control algorithms under discussion are designed to maximize the influence of the control actuators either by (1) optimizing the relative magnitude of the control pulses (Method 1) or by (2) choosing the optimum time for applying the control forces (Method 2). Both methods assume the availability of an external energy source to produce the control pulses on demand.

Since in many practical cases the amount of energy available for control purposes is limited, the present study explores an alternate pulse-control strategy that economizes the use of control energy. This is accomplished by devising an on-line control procedure that attempts to optimize the *parameters* of incorporated impact vibration dampers attached to different locations within the vibrating flexible structure. Instead of using mass-ejection techniques (or equivalent methods) to directly furnish the needed control forces, the internal mechanism of momentum transfer between the primary structure and the auxiliary masses is employed. It is shown that the tradeoff between vibration damping efficiency and control energy economy does not lead to major deterioration in the overall vibration reduction of the primary system as compared to what can be achieved with fully active pulse-control methods.

The proposed parameter control method is applicable to arbitrary distributed systems with dynamic excitations which are directly applied or supplied through base motion. However, for the sake of clarity in explaining the procedure and illustrating its applications, a single-degree-of-freedom (SDOF) system of the type shown in Fig. 1 will be used.

2. Formulation

The equations of motion for the system in Fig. 1 are

$$m_1 \ddot{x}_1 + f_1(x_1, \dot{x}_1) - f_2(z, \dot{z}) = F_1(t) ,$$
$$m_2 \ddot{x}_2 + f_2(z, \dot{z}) = 0 , \tag{1}$$

where

f_1 represents the resistance characteristics of the element connecting the primary system to ground,

f_2 represents the resistance characteristics of the element interposed between m_1 and the auxiliary mass m_2,

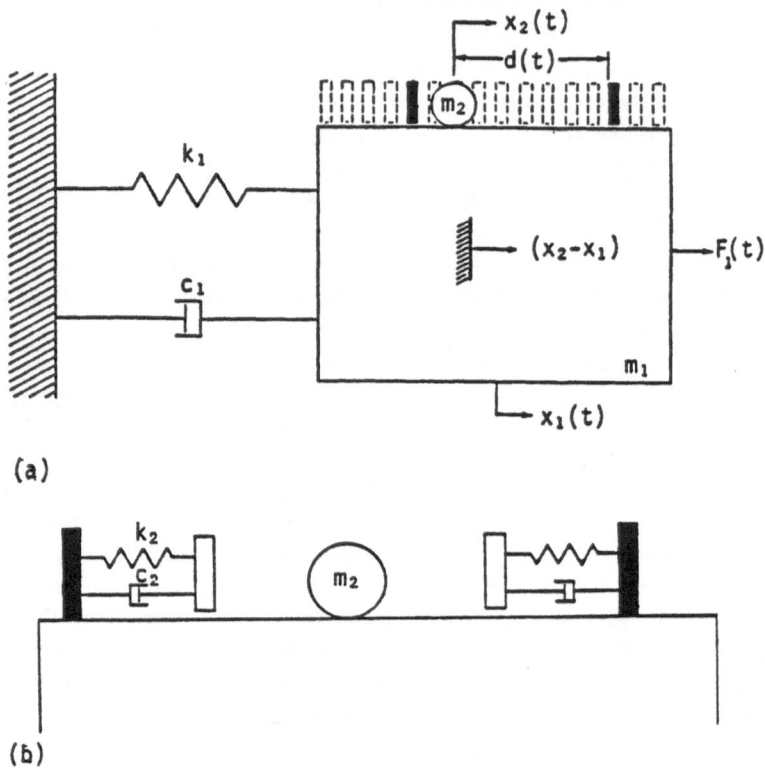

(a)

(b)

Figure 1: *Model of a linear single-degree-of-freedom system provided with a nonlinear auxiliary mass damper; (a) semi-active impact damper attached to a SDOF system; (b) model of impact dampr resilient stops.*

$F_1(t)$ represents the dynamic load acting on m_1,

$x_i(t)$ is the absolute displacement of mass i; $i = 1,2$,

$z(t)$ is the relative displacement between m_1 and m_2; $z = x_2 - x_1$.

In the case of a linear SDOF system provided with an impact damper having stiff, resilient stops, the equations of motion become

$$\ddot{x}_1 + 2\varsigma_1 \omega_1 \dot{x}_1 + \omega^2 x_1 - \mu q(z, \dot{z}, \mathbf{p}) = F(t)$$

$$\ddot{x}_2 + q(z, \dot{z}, \mathbf{p}) = 0 , \tag{2}$$

where

$$f_1(x_1,\dot{x}_1)=c_1\dot{x}_1+k_1x_1 \ ,$$

c_1 viscous damping coefficient associated with f_1,

k_1 linear spring constant associated with f_1,

$\omega_1=(k_1/m_1)^{1/2}$,

$\varsigma_1=0.5c_1(k_1m_1)^{-1/2}$,

$F(t)=F_1(t)/m_1$,

$\mu=m_2/m_1$,

$q(z,\dot{z},\mathbf{p})=2\varsigma_2\omega_2h(z,\dot{z},d)+\omega^2g(z,d)$,

$\omega_2=(k_2/m_2)^{1/2}$,

$\varsigma_2=0.5c_2(k_2m_2)^{-1/2}$,

$g(z,d)=[z-sgn(z)d]u(|z|-d)$,

$h(z,\dot{z},d)=\dot{z}u(|z|-d)$,

d impact damper clearance; equal to half of total gap size in the passive damper,

$sgn(.)$ indicates the algebraic sign of its argument,

$u(a)$ unit step function defined by

$u(a)=1$; if $a>0$,

$u(a)=0$; if $a<$ or $=0$,

k_2 stiffness of the resilient damper stops,

c_2 equivalent viscous damping parameter involved during impacts.

It is clear from Equation (2) that the term $q(z,\dot{z},\mathbf{p})$, which depends on the relative displacement z and the velocity $\dot{z}$ between the two oscillating masses, and on the damper parameters that constitute the components of vector $\mathbf{p}$, represents the cumulative effects of the normalized control force exerted by the auxiliary mass m_2 on m_1. The influence of this damping force on m_1 is directly proportional to the mass ratio $\mu=m_2/m_1$.

If the system in Fig. 1 is provided with a frictionless passive impact damper (PID) having infinitely stiff stops with a coefficient of restitution e, its steady-state motion with two symmetric impacts per cycle of the harmonic excitation $F_0\sin\Omega t$ is exactly described by the following simple expression:

$$x_1(t) = \exp(-\varsigma_1\omega_1t)(B_1\sin\omega_1\eta_1t+B_2\cos\omega_1\eta_1t)$$

$$+ A\sin(\Omega t+\tau) ; \quad 0<\Omega t<\pi \tag{3}$$

where

$\eta_1 = (1 - \varsigma_1^2)^{1/2}$,

r dimensionless excitation frequency ratio, Ω/ω_1,

$A = (F_0/k_1)\,|H(r,\varsigma_1)|$,

$|H(r,\varsigma)|$ dimensionless transfer function; $|H| = [(1-r^2)^2 + (2\varsigma r)^2]^{-1/2}$,

and B_1, B_2 and τ are functions of the impact damper parameters μ, e and d [9].

   An important feature of the exact solution in Equation (3) is that, for a PID of a particular size μ using a material with a particular e for its stops, the peak value of $x_1(t)$ under steady-state excitation can be optimized (minimized) by a proper choice of the damper clearance d. A graphical illustration of this feature is shown in Fig. 2.

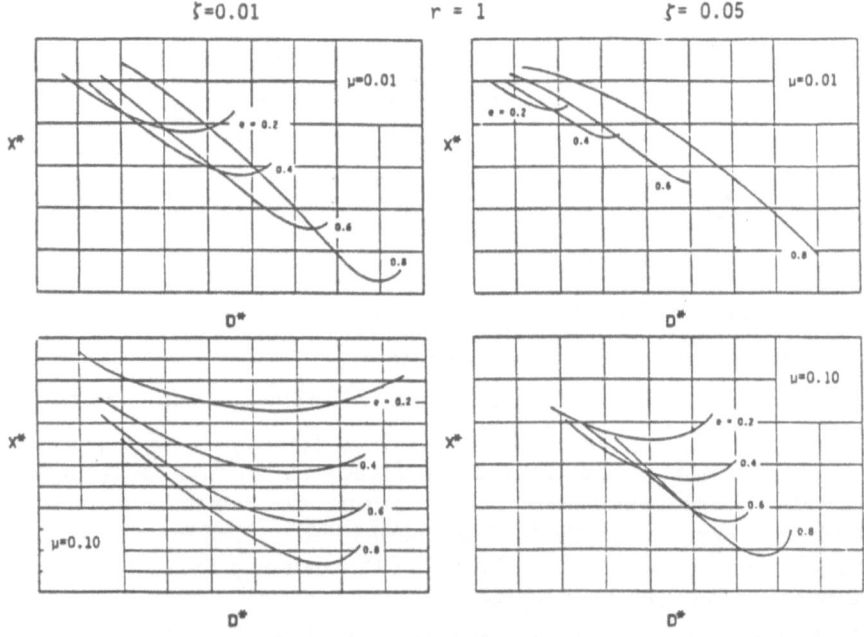

Figure 2: *Steady-state response characteristics of a harmonically excited SDOF system with ratio of critical damping ς_1 operating at resonance and provided with a passive impact damper of mass $\mu = m_2/m_1$ with a total clearance d and coefficient of restitution e.*

3. Semi-Active Control Strategy

The basic idea behind the proposed semi-active control strategy is to provide the primary vibrating structure with the most beneficial damping effects resulting from the collisions between the auxiliary and primary masses. This objective is achieved by actively adjusting the position of the resilient motion limiting stops, so that the generated impacts occur in optimum phase with respect to the response.

Mathematically, the problem can be formulated by considering a typical cost function J in the form of

$$J(d) = \alpha_1 x_{1_{max}} + \alpha_2 \int_{t_\ell}^{t_\ell + T_{opt}} x_1^2(t) dt \ , \qquad (4)$$

where

$x_{1_{max}}$ is the peak displacement of the primary structure during the time $t_i < t < t_i + T_{opt}$,

t_i is the time the last impact was completed,

T_{opt} is a time interval related to the time constant of the system,

α_1, α_2 are weighting constants.

The control approach can now be stated in terms of an optimization process, involving Equation (4) as the cost function and the clearance distance d as the critical parameter. In principle, standard optimization techniques can be used at this stage to determine the optimum value of the gap size d_{opt}. However, due to the severe nonlinearity associated with the vibroimpact nature of the system, standard analytical solution methods are not applicable for the development of accurate closed form solutions that predict the behavior of this nonlinear MDOF system under transient excitation. Alternative numerical approaches are also computationally involved and unsuitable for on-line implementation in situations involving nondeterministic excitations, in which the optimization process has to be repeated numerous times.

On the other hand, it is possible to obtain qualitative insights into the physics of the optimum impacts by studying the damper's off-line behavior. This examination may lead to useful observations, which will assist in the practical utilization of the damping device.

To pursue this task, a Direct Search Method (DSM) of optimization is employed. The goal of this approach is to explore the possibility of establishing a simple relationship between the cost function, $J(d)$, and an appropriate measure of the physical characteristics of the damping mechanism. The technique simply tests a set of prescribed values for $d(t)$ and determines that set which will minimize the cost function. This process is repeated for time segments of length T_{opt}, each starting from the time the last optimum impact was ended.

Before discussing the qualitative behavior of Equation (4), it is worth noting that the first term on the right-hand-side (RHS) of Equation (4) measures the contribution of the rms level of the transient response during the optimization interval. Depending on the relative magnitude of α_1 and α_2 varying amounts of emphasis can be placed on the minimization of the peaks or the rms level.

To illustrate this approach, consider a linear SDOF system with inherent damping $\varsigma_1=0.05$ that is provided with an impact damper having a mass ratio $\mu=0.1$ and rigid stops with a coefficient of restitution $e=0.75$. If the primary system m_1 is now subjected to a transient load consisting of swept-sine excitation of the form $F_i(t)=F_0 \sin[\Omega(t)t]$, where $\Omega(t)=at+b$, the auxiliary mass m_2 will sustain repetitive (generally chaotic) impacts on different sides of its container. The number, location, and intensity of these irregular impacts is a highly nonlinear function of the system characteristics and the nature of the excitation. If the time of occurance of one of these impacts is used to define the reference time t, appearing in Equation (4), then the variation of the peak and rms levels of $x_1(t)$ with the gap size d that governs the time of occurance of the succeeding impact, will be as indicated in Fig. 3.

Since the predominant mechanism that governs the interaction between m_1 and m_2 is momentum transfer, it is reasonable to expect a strong dependence of the criterion function $J(d)$ on the discontinuity in the velocity of $\dot{x}_1$ and/or $\dot{x}_2$ during the impact process. This expectation is borne out by the results depicted in Fig. 3 where the negative (to ease comparison) value of the momentum transfer is superimposed on the graph of the two constituants of $J(d)$. It is thus clear that, for the situation under consideration, optimizing $J(d)$ is practically identical to seeking an extremum value of the momentum transfer involved in the impact process. For the class of problems under discussion, this condition is equivalent to having an impacts occur when the primary system's velocity is at its peak value. Representative results are shown in Fig. 4 regarding the dependence of the peak steady-state response of the primary system on the static gap size (curve "a") and the simultaneous dependence of the primary system's velocity before impact on the static gap size (curve "b"). Clearly this result supports the conclusion that the most beneficial impacts occur when the primary structure is at its peak velocity.

4. On-Line Implementation of Semi-Active Impact Dampers

The preceeding discussion established the guidelines for a simple procedure to optimize the operation of semi-active impact dampers configured as mentioned above. To maximize the efficiency of an impact damper between two consecutive impacts, the gap size d should be adjusted so that the following conditions are satisfied:

1. An impact is made to occur when the velocity of the primary system has reached its peak value. This instant corresponds to the zero crossing of the system displacement.

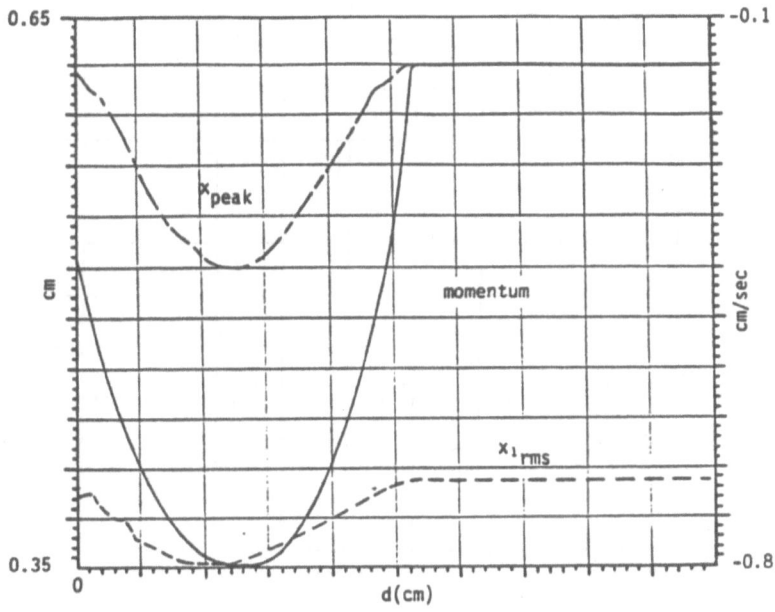

Figure 3: The plotted curves show the variation of the indicated quantity with the gap size d, all other parametrs remaining the saem. Curve (a): Momentum transfer during the impact at the end of the observation segment. Curve (b): Peak value of $x_1(t)$. Curve (c): RMS value of $x_1(t)$.

2. The velocities of the two colliding masses must be opposite to each other at time of impact. This condition insures that the impact process will stabilize the motion of the primary system.

On this basis, two slightly different control algorithms for on-line implementation of the damping device are proposed. The methods employ the same basic principle; selection of one method over the other mainly depends on the hardware design of the impact damper and the physical characteristics of the vibrating system.

4.1. Adaptive Control Strategy (ACS)

This control strategy consists of detecting the displacement (absolute or relative to a moving support) zero crossing of the oscillating structure and generating an impulsive control force by inducing a collision between the auxiliary mass and the structure. The essential features of this approach can be summarized as follows:

1. No global information regarding the dynamic system characteristics is needed.

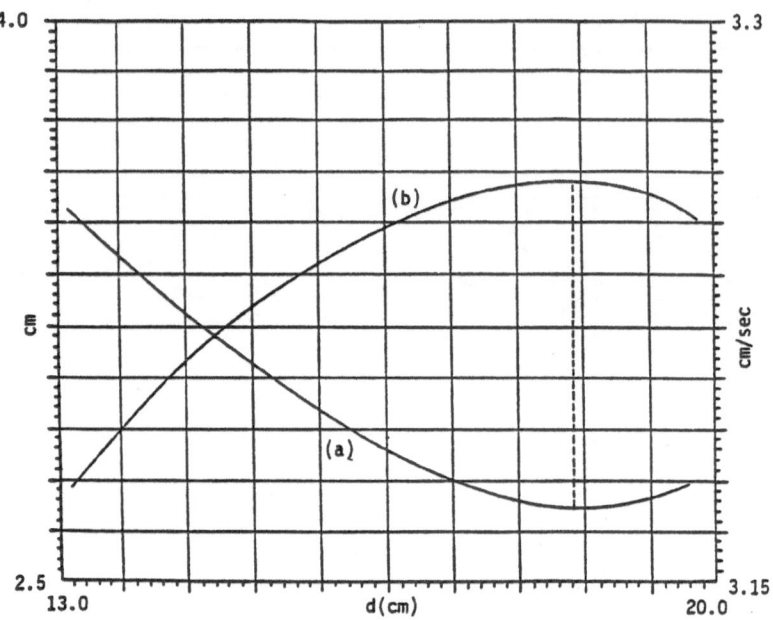

Figure 4: Curve (a): Dependence of the steady-state peak displacement of the primary system on the constant gap size d. Curve (b): Dependence of the velocity of the primary system before impact on the gap size d.

2.   Monitoring of only the system displacement is required.

3.   The on-line computation of the clearance distance d is reduced to a simple detection process.

To illustrate the application of this approach, a representative segtment of the motion being controlled by such a semi-active damper is shown in Fig. 5. These graphs represent the absolute and relative state variables of the system and the nonlinear conservative and nonconservative control functions, g and h. The time histories in this figure have been normalized to lie between -1.0 and +1.0. As seen from these results, suitable control impacts are applied twice every fundamental period of the system. The total control energy exerted on the structure during an impact is the sum of the areas under the g and h functions.

The only significant disadvantage of this technique is the lack of consideration for possible hardware delays in the activation of the impacting mechanism. Two possible provisions may be adopted to overcome this inadequacy:

1.   The first obvious choice is to design a high speed activation system with delays that are small when compared to the fundamental period of the structural system.

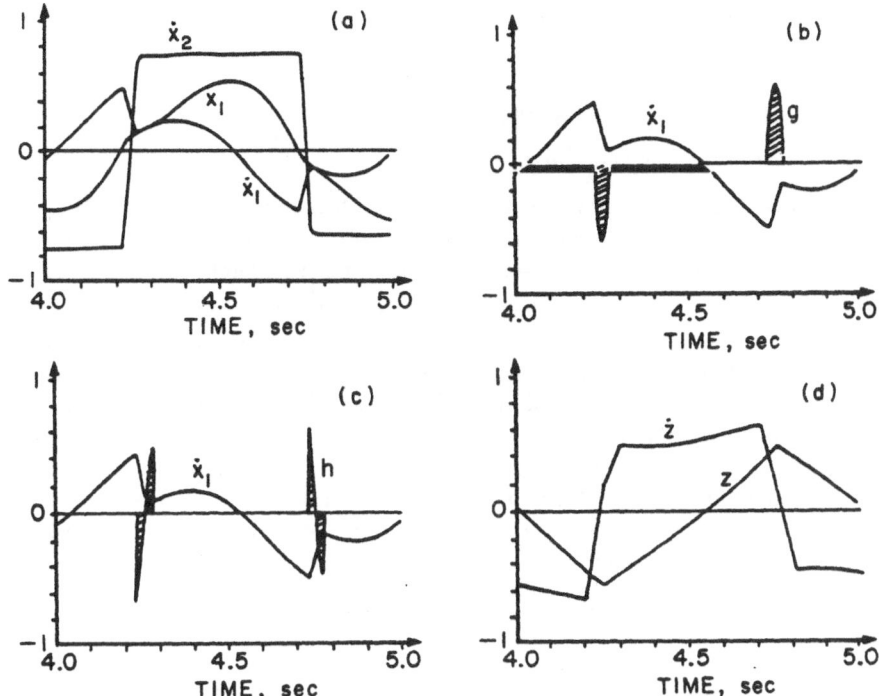

Figure 5: Detailed response of a segment of the steady-state motion of an SDOF system provided with a semi-active impact damper using an adaptive control strategy. The primary system has a ratio of critical damping $\varsigma_1 = 0.05$ and is excited at resonance. The impact damper mass ratio $\mu = 0.1$ and its adjustable-gap stops have a coefficient of restitution $e = 0.75$.

2. The alternative solution is to activate the impacting mechanism when the displacement of the structure has crossed a certain prescribed threshold level.

In the event that these remedies could not be accommodated, a modified version of the control algorithm will be presented next.

4.2. Predictive Control Stragtegy (PCS)

The alternative control strategy is based on the assumption that, when dealing with nonstationary dynamic loads, the excitation can be treated as a zero mean random process between two consecutive impacts. Therefore, during this time the expected value of $x_1(t)$ will not depend on $F_1(t)$ (or the base motion $s(t)$) and can be written as

$$E\,x_1(t) = u(t-t_.)x_1(t_.)+v(t-t_.)\dot{x}_1(t_.)\tag{5}$$

where

$$u(t)=\exp(-\varsigma_1\omega_1t)[(\varsigma_1/\eta_n)\sin\omega_1\eta_1t+\cos\omega_1\eta_1t]$$
$$v(t)=\exp(-\varsigma_1\omega_1t)[(\omega_1\eta_1)^{-1}\sin\omega_1\eta_1t]$$

and t denotes the state of the system at completion of the last impact. Hence the task of the control process is to use the results of Equation (5) to predict the time of the next displacement zero crossing and to synchronize the hardware activation time so as to insure an impact at the anticipated time.

It is evident from an on-line computation stand point that the new approach is more involved; however, it provides the hardware mechanism with a lead time (approximately half of the fundamental period) for proper adjustment. Illustrated in Fig. 6 are the various state variables of the system for a typical time segment. Graph (d) represents a comparison between the predicted and the actual displacement of the structure. It is worth noting that, unlike the adaptive control strategy, the predictive control strategy requires a priori knowledge of the physical characteristics of the system to be controlled.

5. Numerical Simulation

The efficiency of the proposed control strategies is demonstrated by presenting numerical simulation results of several single-degree-of-freedom models with diverse characteristics, subjected to deterministic and stochastic dynamic environments.

Example (1): Harmonic Excitation

The results shown in Fig. 7 correspond to a linear viscously damped SDOF system of mass m_1 having a ratio of critical damping $\varsigma_1=0.01$, and initially at rest, that is subjected to harmonic excitation, of magnitude $F_0=1$, whose constant frequency Ω equals the undamped natural frequency ω_1 of m_1 (thus $r=\Omega/\omega_1=1$). The transient displacement $x_1(t)$ and velocity $\dot{x}_1(t)$ response of this system during a time segment covering a span of 10 fundamental periods (i.e., $t/T_1=10$) is shown in Figs. 7(a) and (b), respectively. Had the forced response been allowed to reach its steady state (at $t/T_1\simeq25$), the peak displacement $x_{1_{max}}$ would have become equal to $50(F_0/k_1)$.

Suppose now that this SDOF system is provided with a passive impact damper whose mass $m_2=0.1m_1$, with stiff stops characterized by $\omega_2/\omega_1=10$ and energy dissipation equivalent to $\varsigma_2=0.1$ (corresponding to a coefficient of restitution $e=(0.75)$. By setting the gap size d equal to a value d_{opt} chosen to yield a minimum peak displacement response during the test period $0<t<T_{max}$, the results shown in Figs. 7(c) and (d) are obtained. Due to the nonlinear nature of the impact-damped system, numerical techniques have to be used to determine d_{opt} for this transient phase of the motion. Had the value of d been kept equal to

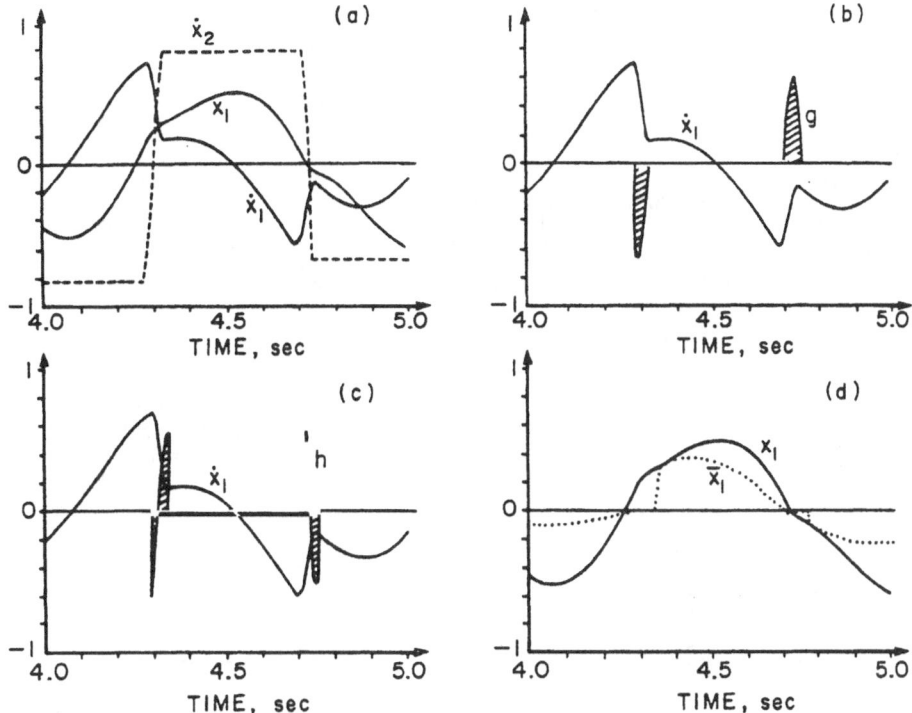

*Figure 6: Detailed response of a segment of the steady-state motion of a har-
monically excited SDOF system provided with a semi-active impact
damper using a predictive control strategy.*

d_{opt} but the excitation time T_{max} increased from $10T_1$ to $25T_1$, the peak response
of m_1 would have increased substantially. This fact illustrates a typical limitation
of the passive impact damper: while it can be designed to be a very effective
damper over a relatively narrow excitation frquency range, its operation over a
relatively wide frequency and/or amplitude range, compared to the conditions
under which its gap was optimized, may result in unacceptable performance
characteristics (possibly, even amplifying rather than attenuating the motion of
m_1). In other words, had the value of d been set equal to the optimum d when
the system is undergoing steady-state motion (i.e., $t/T_1 \gg 1$), the response of m_1
during the transient phase of the motion covering the time span $0 < t/T_1 < 10$
would have been much larger than the level shown in Figs. 7(c) and (d). The
value of d used here is 1.23 while the optimum value under steady state conditions is 1.25.

If the SDOF system is now provided with a semi-active impact damper
(SAID) whose gap size is repetitively adjusted according to the adaptive control
strategy discussed above, the resulting response of the primary system during the
test period is as shown in Figs. 7(e) and (f). Unlike what happens with the PID,

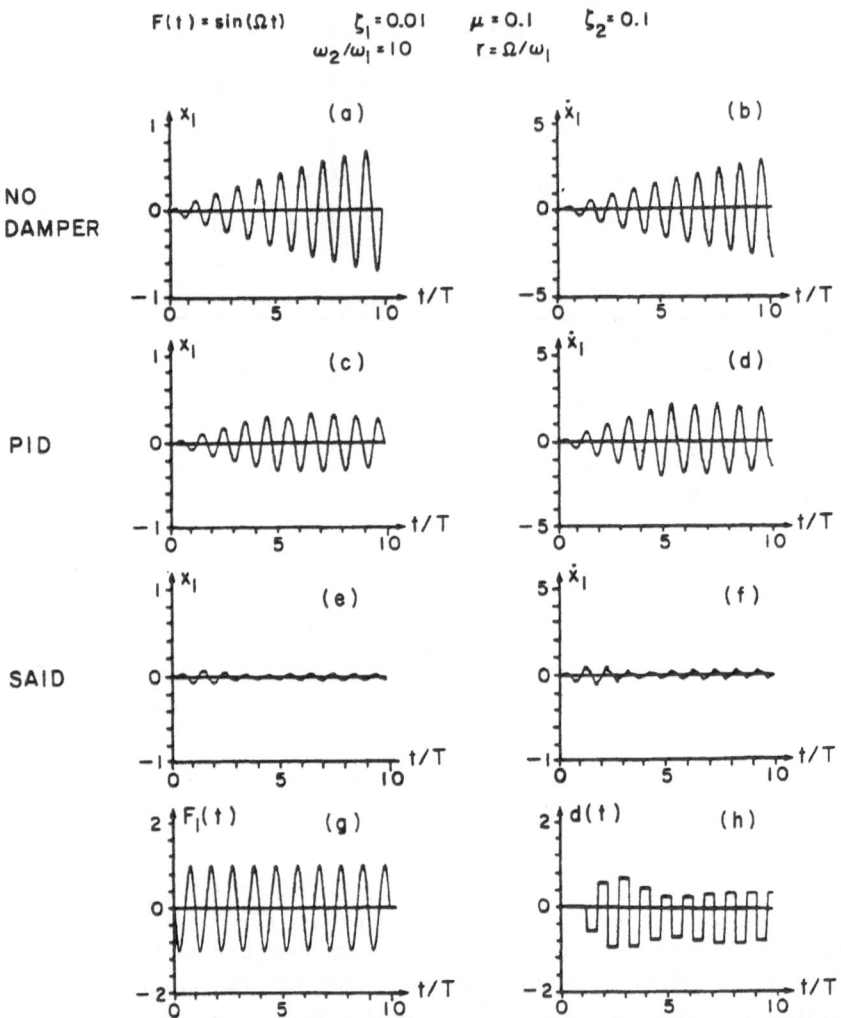

Figure 7: *Figs. (a) and (b) show the transient displacement and velocity of m_1 in the absence of an auxiliary mass damper. Figs. (c) and (d) show the response of m_1 when an optimized passive impact damper (PID) is used. Figs. (e) and (f) show the response of m_1 when a semi-active impact (SAID) is used. Fig. (g) shows the applied harmonic excitation, and Fig. (h) shows the evolution of the damper stops when the SAID is used.*

the level of attenuation evident in this case during the transient test period shown is indefinitely sustained. The corresponding values of $d(t)$ for the test period are shown in Fig. 7(h). As might be expected, the limits of $d(t)$ gradually approach a steady-state value. It is worth noting that when the system reaches steady-state conditions at $t/T_1 \gg 1$, the value of $d(t)$, as determined by the control algorithm under discussion, becomes *identical* to the value of d_{opt} based on minimizing the exact analytical solution (Eq. (3)) corresponding to steady-state motion of the impact-damped system.

Example (2): Impulsive Excitation

If the same primary system under discussion in the previous example is now subjected to an impulsive excitation, its free vibrations over a period of $\simeq 40$ natural periods is as shown in Fig. 8(a). When provided with a SAID having the same characteristics as those of the one used in the preceding example, the controlled motion is as shown in Fig. 8(b) and the evolution of the gap size is illustrated in Fig. 8(c).

It is worth noting that the peak response occurs during the first excursion of the motion before the impact damper (assumed to be initially at rest in the middle of its container) can be mobilized to exchange momentum with the primary system. Note also that after about 5 natural periods, the damper ceases operation (no more impacts occur) since the primary system's response falls below a prescribed threshold below which the damping mechanism is assumed deactivated (as would likely occur in practical cases).

Inspite of the only slight reduction in the peak response of the primary system, it is clear that the SAID dramatically increases (by about an order of magnitude in this case) the effective amount of damping that controls the rate of mechanial energy dissipation in the system (from $\varsigma_1 = 0.01$ to $\varsigma_{eq} \simeq 0.10$).

Example (3): Random Excitation

The previous two cases involved disturbances and/or responses characterized by smoothly changing oscillations. By contrast, the present example deals with a zero-mean wide-band Gaussian random excitation of which a representative time history segment is shown in Fig. 9(e). The portion of the displacement and velocity response history shown in Figs. 9(a) and (b) span a time segment of 40 natural periods T_1. Due to the slight amount of inherent damping ($\varsigma_1 = 0.01$) in the primary system, the response of m_1 exhibits the characteristics of a narrow-band process.

When the primary system is provided with the same SAID used in the previous examples, the resulting controlled motion becomes as shown in Figs. 9(c) and (d). The level of attenuation in the peak of the displacement is $(\hat{x}_1/x_1) = 1.41/3.65 \simeq 0.39$ and the corresponding reduction in the peak velocity is $(\hat{\dot{x}}_1/\dot{x}_1) = 14.2/48.1 \simeq 0.29$. Even better results are obtained for the rms reduction level.

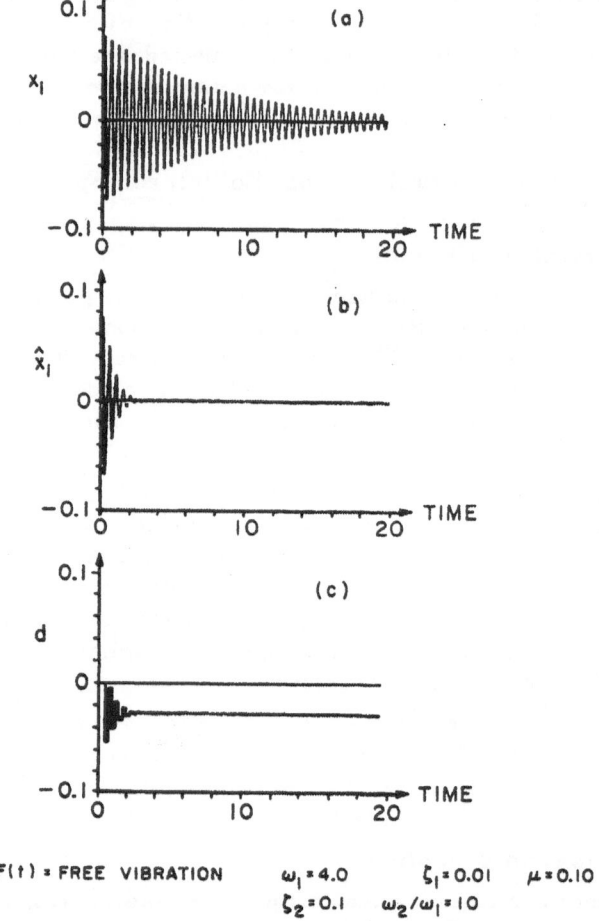

F(t) = FREE VIBRATION $\omega_1 = 4.0$ $\zeta_1 = 0.01$ $\mu = 0.10$
$\zeta_2 = 0.1$ $\omega_2/\omega_1 = 10$

*Figure 8: Response of the SDOF system in Fig. 7 under impulsive excitatin.
(a) Response of m_1 without an auxiliary mass damper. (b) Response
of m_1 when provided with a SAID. (c) Evolution of the damper
stops when the SAID is used.*

The time history of the adaptive gap size is shown in Fig. 9(e). The lack of
any discernible pattern in the evolution of the optimum gap size reflects the
nature of the random disturbance. The complex changes of $d(t)$ between impacts
clearly illustrate the handicaps passive dampers have to cope with since their initial (fixed) gap size cannot change with time to accommodate quiescent or active
episodes of the random response.

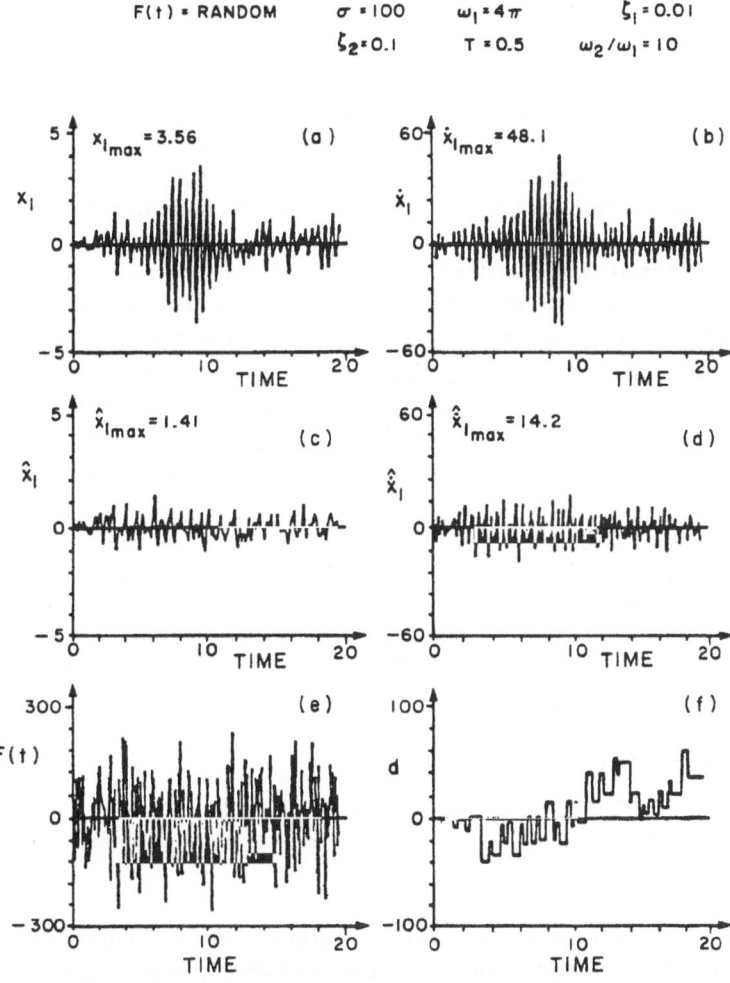

Figure 9: Response of the SDOF system in Fig. 7 under random excitation. Figs. (a) and (b) show the response of m_1 without an auxiliary mass damper. Figs. (c) and (d) show the response of m_1 when provided with a SAID. Fig. (e) shows the random excitation, and Fig. (f) shows the evolution of the damper stops when the SAID is used.

6. Summary and Conclusions

A simple, yet efficient method is presented for the on-line *parameter* control of linear as well as nonlinear, multi-degree-of-freedom systems provided with adjustable-gap impact dampers responding to arbitrary dynamic loads. The on-line control algorithm is suitable for situations in which detailed knowledge of the system structure is not available; only *local* measurements in the vicinity of each of the attached impact dampers are needed with this adaptive control method to determine the evolution of each impact damper clearance so as to optimize the vibration attenuation efficiency of the individual dampers.

In order to accommodate time delays associated with available actuator hardware, a second on-line control method is presented in which partial knowledge of the primary system's properties is used to anticipate the evolution of the dynamic system trajectory thereby allowing for incorporation of a certain lead time into the control process in order to compensate for the hardware limitations.

A stability analysis, simulation studies, and experimental tests with a mechanical model have demonstrated the feasibility, reliability, and robustness of the proposed semi-active on-line control method [10]. Currently under investigation are the effect on the efficiency of the damper of a finite range of possible values for d, and the effectiveness of the damper for MDOF and continuous systems.

Acknowledgement

This study was supported in part by a grant from the U.S. National Science Foundation.

References

[1] Reed, F.E., "Dynamic Vibration Absorbers and Auxiliary Mass Dampers", Chapter 6 in *Shock and Vibration Handbook*, Harris and Crede (Eds.), McGraw-Hill, New York, 1961.

[2] Paget, A., "The Acceleration Damper", *Engineering*, 1934, p. 557.

[3] Masri, S.F. and Yang, L., "Earthquake Response Spectra of Systems Provided with Nonlinear Auxiliary Mass Dampers", *Proc. Fifth World Conference on Earthquake Engineering*, Rome, Italy, 1973.

[4] Dehghanyar, T.J., Masri, S.F., Miller, R.K., Bekey, G.A. and Caughey, T.K., "An Analytical and Experimental Study into the Stability and Control of Nonlinear Flexible Structures", *Proc. Fourth VPI and SU/AIAA Symposium on Dynamics and Control of Large Structures*, Balcksburg, VA, 6-8 June, 1983.

[5] Dehghanyar, T.J., Masri, S.F., Miller, R.K., Bekey, G.A. and Caughey, T.K., "Sub-Optimal Control of Nonlinear Flexible Space Structures", *Proc. NASA/JPL Workshop on Identification and Control of Flexible Space Structures*, San Diego, CA., 4-6 June, 1984.

[6] Miller, R.K., Masri, S.F., Dehghanyar, T.J. and Caughey, T.K., "Active Vibration Control of Large Civil Structures", Paper No. AIAA-85-0681-CP, *Proc. AIAA 26th SDM Conference*, Orlando, Florida, 15-17 April, 1985.

[7] Agbabian Associates, "Validation of Pulse Techniques for the Simulation of Earthquake Motions in Civil Structures", AA Report No. R-7824-5489, Agbabian Associates, El Segondo, CA, May, 1983.

[8] Agababian Associates, "Induced Earthquake Motions in Civil Structures by Pulse Methods", AA Report No. R-8428-5764, Agbabian Associates, El Segondo, CA, August, 1984.

[9] Masri, S.F. and Caughey, T.K., "On the Stability of the Impact Damper", *Journal of Applied Mechanics*, Vol. 33, *Trans. ASME*, Vol. 88, 1966, pp. 586-592.

[10] Dehghanyar, T.J., Masri, S.F., Miller, R.K. and Caughey, T.K., "Semi-Active Control of Nonlinear Flexible Structures", *Proc. XVIth International Congress of Theoretical and Applied Mechanics*, Tech. Univ. of Denmark, Lyngby, Denmark, 19-25 August 1984.

PASSIVE STRUCTURAL CONTROL BY MEANS OF CONNECTIONS WITH ADAPTABLE PARAMETERS

N. Dimitrov and A. Pocanschi
Department of Structures and Building Design
University of Stuttgart
Stuttgart, West Germany

1. Introduction

In order to limit structural response under environmental actions, such as earthquake, within acceptable ranges, two main control ways can be considered:

• the active structural control, which assumed a "fight" of the structures against earthquake attack by neutralizing the impact motion through the generation of internal energy opposite to the induced energy at exactly the right time and in the right manner, and

• the structural "self defence", i.e. passive control, where the dissipation of motion energy is the key factor.

From a practical standpoint, since active control requires the ability to generate and apply instantaneously large controlled forces to the structures by means of mechanical devices, its implementation is usually only effective for slender and flexible structures, such as high rise buildings, which undergo large deflections under dynamic actions, which could lead to discomfort of the tenants. On the other hand, passive control techniques use simple energy absorbing devices, such as shock isolators and dampers, located at the base of the structure or at some structural areas in such a way, that during the motion dissipative mechanisms are developed [1],[2],[3].

The main idea of our contribution refers to a control technique of structural behavior under seismic actions using connections whose characteristics are changeable and hence, adaptable to different working conditions. The control objective is to effect direct changes of specific dynamic modes and stiffnesses of the structures by adapting their response parameters to the action characteristics. Conceptually, it will be assumed that the structural deformability is conditioned mainly by connections with complex functionality which are alone able to control the partial and overall structural stiffness, the force transmission and the energy dissipation process.

The paper presents two types of connections for controlling the rigid-body response of buildings as developed by the German Comp. GUMBA, München as well as some types of complex connections to control the response of shearwall structures.

2. General Considerations

2.1 Generally, the first step in structural control represents the adoption of a basic mechanism that fits best to the building form and structural concept and whose deformability can be controlled in the simplest manner.

When a building is conceived as a solid, for example low and medium rise buildings with three dimensional structures, the mechanism can be obtained simply by partitioning the whole building into one or more rigid-body subsystems and afterwards connecting them as desired (Figure 1).

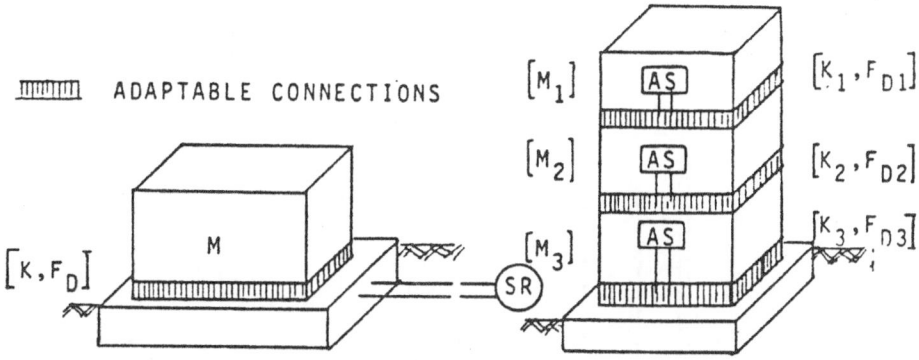

Figure 1: Control of rigid-body system motion

The very rigid systems of reinforced concrete shearwalls can be converted into controllable mechanisms by dividing them with one or more vertical and horizontal joints extending the full height and width of the wall and then coupling the resulting panels together with controllable connections. Some possible arrangements of partitioned shearwalls are shown in Table 1.

Table 1:   *Arrangements of shearwall mechanisms*

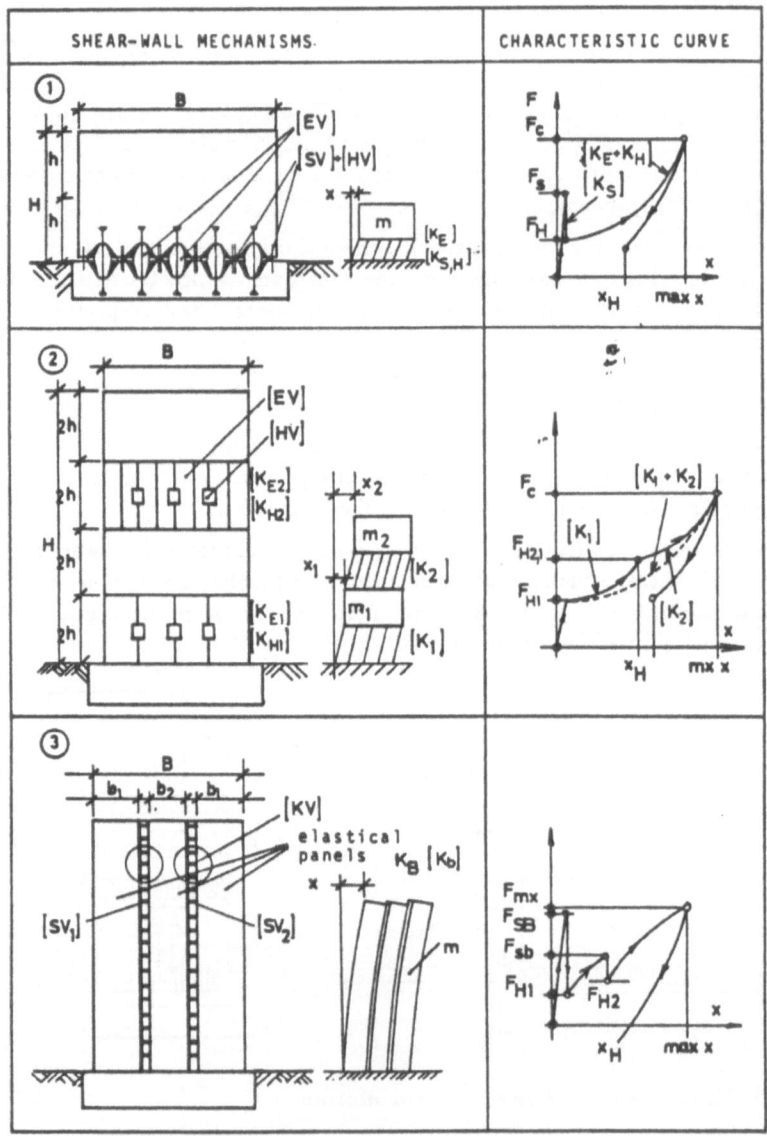

In fact, a partitioned structure with controllable connections can be looked at as a dynamic adaptable system which represents in a whole two systems: a primary rigid one for transfering of normal horizontal and vertical loads and a secondary kinematical one, which will be activated under dynamic actions. As an example, the primary systems of the buildings in Figure 1 and the shearwalls in Table 1 can be assumed as more or less rigid cantilever beams while the secondary systems represent dynamic models with one or more degree of freedom, depending on the chosen mechanism.

2.2 Essentially one can classify the connections to couple the subsystems into two main groups: permanent connections and temporary — or destructive-connections. The temporary connections consist mainly of rigid connectors, such as special bearings, brittle elements or of hysteretic devices with small yield capacity They are responsible for the elastic and strength properties of the primary systems. Permanent connections, in order to be used, have to be made of elastic connectors that are coupled with energy absorbing devices such as viscous, frictional or hysteretic dampers. The dynamic characteristics of the secondary systems are determined essentially by the nondestructive connections.

A connection with adaptable parameters in the sense of this paper comprises in a compact unit both temporary and simple nondestructive connections.

2.3 Viewed in terms of dynamic behavior, the structures with adaptable connections are to be treated as non-stationary dynamic models whose response can only be investigated by means of nonlinear dynamic analysis, that assumes in generally form the step by step integration of the matrix differential equation:

$$\{M\}\{\ddot{x}\}+\{K(x)\}\{x\}+\{F_D\}=-\{M\}\{\ddot{u}\}$$

where:

$\{M\}$ inertial matrix

$\{x\}$ vector of relative displacements

$\{K(x)\}$ lateral stiffness matrix of the secondary system

$\{F_D\}$ matrix of damping forces

$\{\ddot{u}\}$ vector of earthquake accelerations.

Considerations on the estimation of stiffness and of damping forces for some types of connections follow next.

2.4 As mentioned, to transfer the horizontal wind forces to the ground and to assure stability against overturning, rigid connectors are required which must vanish in the earliest phase of the earthquake. It is also necessary to consider a control technique for the transition mechanism between the primary and secondary system.

Conceptually, there are two possibilities to release the rigid connectors:

- the "self control" way, which assumes that the connection possesses adequate stiffness to exhibit essentially linear elastic behavior under maximum wind loading but vanishes when subjected to inertial forces slightly greater than these. Thus, the connection itself sets an upper limit to the force that can be induced in the structure, and

- the active control way, which involves a mechanical releasing of the rigid connectors by means of sensors and electrical impulses as shown in Figure 1. One can use acceleration sensors (AS) placed inside the subsystems, which release the rigid connections when critical acceleration values are at hand as well as seismic receiver (SR) placed at certain distances around groups of building, which are triggered by the attack of the very first signals of an earthquake (minor tremor waves).

3. Rigid-Body Mechanism Connections

Adaptable connections for rigid-body deformation can be conceived to control the space motion of three dimensional systems and also as connections to control in-plane motion, the latter ones being suited for planar structures, such as low height shearwalls. Practically, the in-space acting connections are complex bearings which satisfy in a compact technique requirements of rigidity, damping, elasticity and stability in such a manner that during the time between two earthquakes they behave as three dimensional bearings able to transfer external actions according to strict conditions of deformation, and are then able to change their characteristics when an earthquake occurs.

3.1 An active adaptable connection with rigid transmission in the primary system is shown in Figure 2.

The vanishing rigid connector consists of a certain number of detachable steel wedges placed between two cylindrical steel casings and held in vertical position by prestressed circular steel rings. In the gaps between these wedges there are a number of prismatic supports made of lead, conceived to act as temporary hysteretic connections during the transition phase, after the releasing of rigid connector.

The elastic permanent element consists of a steel laminated ring rubber bearing with central stabilizer which also provides the active element of a viscous damper placed at the top of the upper casing. To release the rigid connection an electric detonator coupled to the sensors is provided in the closing ring.

The following features of this connection unit are to be pointed out:

- high degree of reliability in maintaining vertical supports for the entire lifetime of the building,

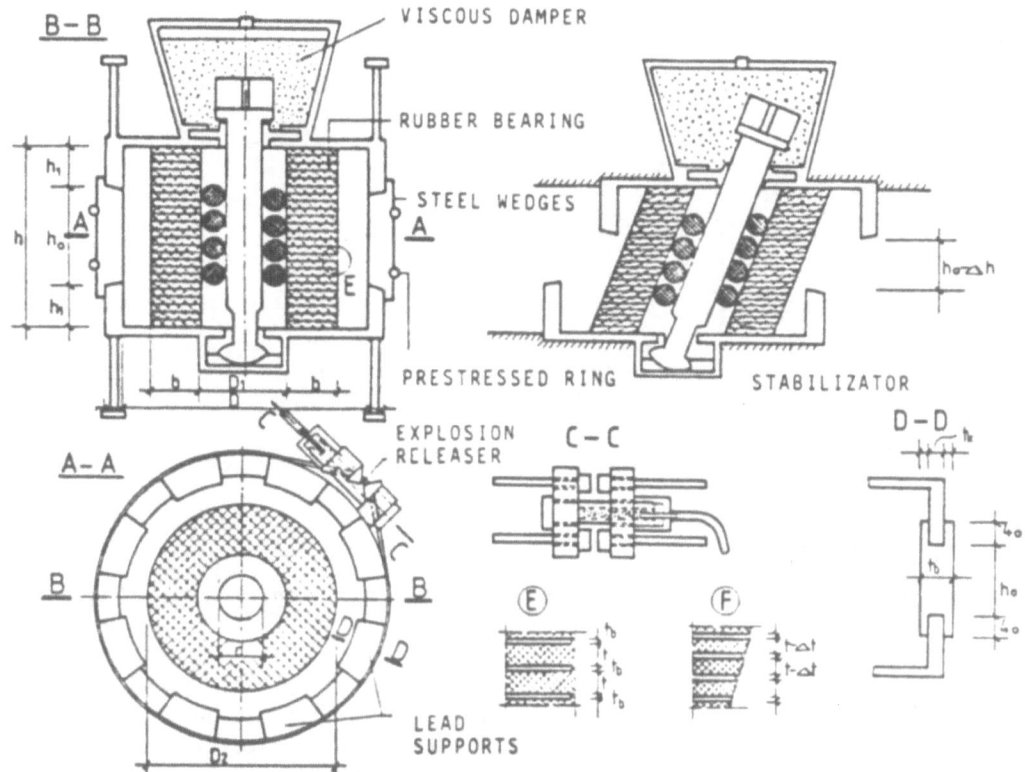

Figure 2: Active adaptable connection with rigid transmission in primary system

- the rubber bearing is only loaded during the earthquake and for a short time afterwards, hence his life will be substantially prolonged,

- prevents the lift up tendency of the superstructure, and

- can be brought back to its original function through simple operations.

3.2 A drawing of a simpler, passive adaptable connection unit with elastic transmission in primary phase is shown in Figure 3.

The elastic connector consists of a ball bearing between two elastomeric pads. In the primary system the excessive deformation of the bearing is prevented by steel wedges placed around the bearings circumference. These are held together by a prestressed closing ring provided with a mechanical fuse. The permanent hysteretic component consists of mild steel bars welded onto the cylindrical walls of the upper and lower casing, placed among the wedges. The advantageous features of this connection-unit are: the elastic behavior in the primary system and due to its form and kinematical qualities, high safety even

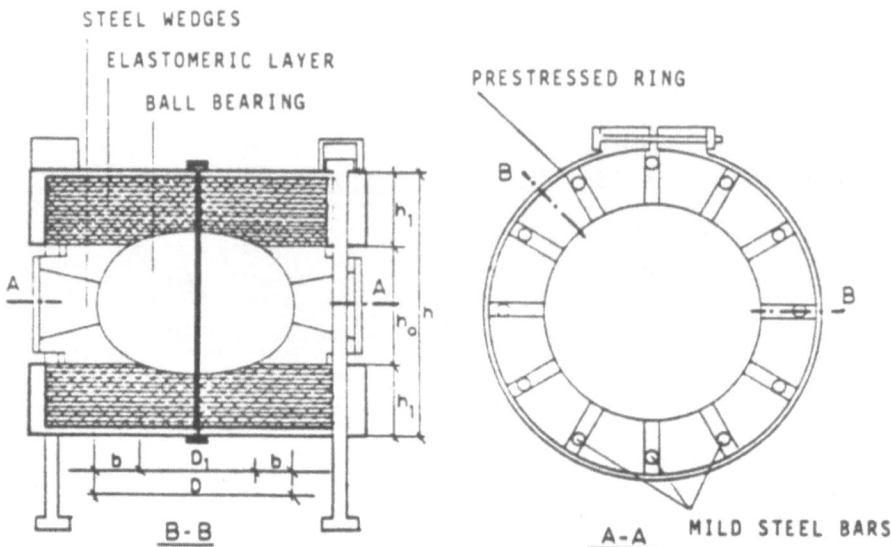

Figure 3   Passive adaptable connection with elastic transmission in primary system

under strongest load conditions.

3.3 To transform a low rise shearwall into a rigid-body mechanism with controlled motion one can make use of the following connectors which are to be introduced into a cut along the base line, as shown in Table 1:

- pendular connectors (EV), consisting of a series of short rhombiodal R.C. columns hinged on the top at the shear wall and on the bottom at the base and laterally embeded in an elastomeric mass as elastic connections,

- hysteretic connectors (HV), built up from vertical steel bars anchored at the top and bottom in the concrete mass and working in flexure in inelastic range during the motion, and

- temporary rigid shear connections (SV), consisting of a concrete fuse placed in the joint between two pendular units which restrain the parallel relative motion between the wall and foundation.

The main characteristics of the pendular connector are described in Table 2. Due to its force-displacement diagram (elastic nonlinear rigid type), the connection is capable of controlling the lateral sways of the superstructure. By increasing restoring force it becomes practically rigid and hence, stability of the motion can be assured without large plastic reserve of the hysteretic connectors being necessary.

Table 2:   Main characteristics of a pendular connector

PENDULAR CONNECTION	STIFFNESS	TRANSFERRED FORCE
$\Delta_o \cong 0.3\, a_o$ $\Delta_c \cong 0.8\, a_o$	LINEAR $K_o \doteq \dfrac{E_o h_o^3 b}{12\, H^2 a_o}$ NONLINEAR $K = K_o \left[1 + \left(\dfrac{F_{NL}}{2 K_o \Delta_c}\right)^2\right]$	$\Delta \cong \Delta_o$ $F_o = K_o x$ $\Delta_o < \Delta < \Delta_c$ $F_{NL} = \dfrac{2 K_o}{\pi} \Delta_c \tan \dfrac{\pi \Delta}{2 \Delta_c}$ $\Delta = \Delta_c$ K of rigid wall

Available relations to evaluate the stiffness and force transmission characteristics of the hysteretic connections are given in Table 4b. In these relations Q_{pl} denotes the yielding shear force, σ_{pl} the yielding bending stress and L_{dp} the energy dissipated during each cycle.

2. Shear Mechanism Connections

Shear mechanism connections are thought to be used for controlling medium and high rise shearwall structures where rigid body mechanisms with one degree of freedom, due to their relative reduced energy absorbtion capacity, are no longer effective.

On the other hand, adopting one mass rigid motion mechanisms to control high rise structures may encounter insurmountable problems. Because the elastic centre of the system does not coincide with the centroid of the rigid-body superstructure there will be a strong tendency for pendulum oscillations to occur. This can cause uplift and great stress concentrations in the perimetral connection elements.

There are many alternatives to obtain mechanisms from shearwalls. Two of them are shown in the examples 2 and 3 in Table 1. According to the first example, the monolithical wall is sectionalized with horizontal joints at every story level and then the story panels will be again alternatively divided along common vertical slip lines.

To couple the panels first of all elastic elements consisting of elastomeric strips with different thicknesses (t_o and t_u, see Table 3) have to be inserted in the joints. In this way every second story-height panel becomes a shear-wall mechanism whose elastic properties can be controlled by choosing adequate dimension rates for the panel and elastic strips.

Table 3:   One story inserted panel: main characteristics

INSERTED SHEAR-WALL	STIFFNESS
$J = \dfrac{cb^3}{12}$ $A_l = 2\,ch$ n= Number of panels	a. Wide panels $b > h$ $K = \dfrac{n E_n\,J}{4 h^2\,t_0} + \dfrac{(n-1)}{4}\dfrac{G_0 A_l}{t_v}\left(\dfrac{b}{h}\right)^2$ b. Slender panels $b < h$ $K = \dfrac{3n E_b \cdot J}{4(h - t_0)^2\left[h + t_0(3n_b - 1)\right]} + \dfrac{(n-1)\,G_0 A_l}{4\,t_v}\left(\dfrac{b}{h}\right)^2$ $n_b = \dfrac{E_b}{E_n}$

Practically, inserted shearwalls can be achieved by assembling R.C. panels together with elastomeric strips and prestressing them by means of horizontal steel bars.

Inside a shearwall structure a one story inserted panel represents a unit of elastic connection, the number of which determines the number of degrees of freedom of the secondary system. The expression to evaluate the stiffness of inserted wall units are listed in Table 3. In these relations E_n and E_b are the modulus of elasticity of the elastomeric respectively of the concrete and G_o is the shear modulus of the vertical elastomeric strips.

Since the axial stiffness of the elastomeric mass is a non constant value that depends on the load intensity, it will be assumed for the modulus of elasticity E_n a tangential function as follows:

$$E_n = E_o \left[1 + (\frac{1}{\pi}\tan g \frac{\pi \Delta v}{2 \Delta c})^2 \right]$$

where

E_o modulus of elasticity in non-loaded state

Δv vertical bending compression deformation

Δc characteristic height of the elastomeric strip.

A specific shear connection of the hysteretic type is shown in Table 4a. It consists of shear-blocks made of R.C. or, as the example in Table 4 shows, of lead which couple two adjacent panels and yield when slips occur.

For increased safety active controlled shear connections can be considered.

Table 4:   Characteristics of hysteretic connections

HYSTERETIC CONNECTIONS	CHARACTERISTICS
a. 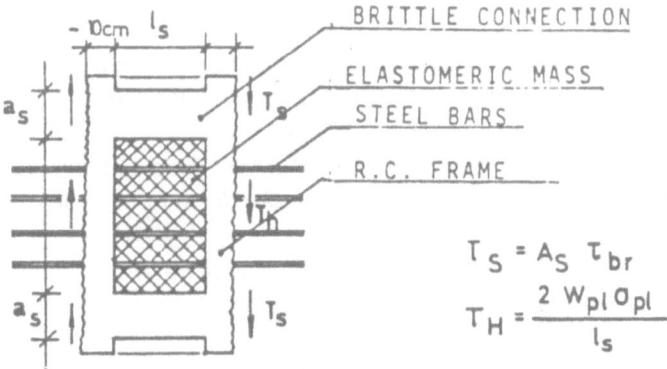	$Q_{pl} = h_b \cdot c_b \cdot \tau_{pl}$; $\quad \tau_{pl} = 1200 \ N/cm^2$ $\quad G = 55000 \ N/cm^2$ $\quad Y < Y_{br} = 0.35$ $L_{dp} = 4 \ Q_{pl} \ ymx \cdot t_v$ (per Periode)
b.	$Q_{pl} = \dfrac{2 \ W_{pl} \ \sigma_{pl}}{l_s}$ $\quad W_{pl} = 1.7 \ W_e = \dfrac{D^3}{6}$ $L_{dp} = 4 \ Q_{pl} \ \Delta$ (per Period)

The latter example in Table 1 originates from a mechanism proposed in [1]. The wall is sectioned only with vertical joints extending from the foundation to the full height of the building, so that more elastic cantilevered R.C. strips are obtained. These are then to be coupled together with connection units as shown in Figure 4. One recognizes the fact that this element brings together into a compact form some of the already described permanent and temporary connection types. These elements can be easily mounted in the forms and embedded into the concrete mass during the erection of the structures.

- 0cm l_s

BRITTLE CONNECTION

ELASTOMERIC MASS

STEEL BARS

R.C. FRAME

$T_S = A_S \ \tau_{br}$

$T_H = \dfrac{2 \ W_{pl} \ \sigma_{pl}}{l_s}$

Figure 4: Compact unit of shear adaptable connection

5. Concluding Remarks

The intention of the presented paper is to suggest some original passive control techniques for structures most usually constructed: low rise and medium rise buildings subjected to seismic loads.

The use of connections with adaptable parameters is, in fact, a simple way to obtain dynamic adaptable structures, i.e. to bring their seismic response under control. The described technique assumes principally that a structures consists of two systems:

• a primary system, which optimally transfers external horizontal and vertical loads, and

• a secondary system which responds in the best manner to a large spectrum of earthquake actions.

The transition from the first to the second system can be both actively or passively controlled.

Approach techniques to design and predimensioning of the connections using global energy balance criteria were examined in earlier papers [6],[7].

References

[1] Avtar, S., Pall and Marsh, C., "Friction — Damped Concrete Shearwalls", *ACI Journal*, May-June 1981, pp. 187-193.

[2] Pillai, S.U. and Kirk, D.W., "Ductile Beam Column Connections in Precast Concrete", *ACI Journal*, November-Decembe, 1981, pp. 480-487.

[3] Paulay, T., "Earthquake — Resisting Shearwall New Zealand Design Trends", *ACI Journal*, May-June, 1980, pp. 144-152.

[4] Abdel-Rohman and Leipholz, H.H.E., "A General Approach to Active Structural Control", in *Structural Control*, H.H.E. Leipholz (Ed.), Comp. IUTAM 1980, North Holland Publishers, pp. 1-28.

[5] Skinner, J., Beck, J. and Bycroft, G., "A Practical System for Isolating Structures from Earthquake Attack", *Earth. Eng. and Str. Dyn.*, 3, 1975, pp. 297-309.

[6] Dimitrov, N. and Pocanschi, A., "Wandscheiben mit dynamischer Anpassungsfähighkeit für Bauten in Erdbebengebieten", *Bauingenieur*, No. 60, 1985, pp. 91-98.

[7] Pocanschi, A., "Dynamic Adaptable Bearings: A Compact Technique for Isolating Buildings from Earthquake Attack", Topical Technical Information GUMBA GmbH München, 1984.

[8] Pocanschi, A., "Kontrolle des Tragverhaltens von erdbebengefährdeten Bauwerken durch Dämpfungsmaßnahmen", *Forschungsbericht*, Nr. 14, Inst. für Tragkonstruktionen, Univ. Stuttgart, 1982.

[9] Bahtti, M., Pister, K.S. and Polak, E., "Optimization of Control Devices in Base Isolation Systems for Aseismic Design", *Structural Control*, H.H.E. Leipholz (Ed.), North-Holland Publishers Comp. IUTAM, 1980, pp. 128-136.

[10] Olariu, I and Pocanschi, A., "On the Earthquake Isolation of the Buildings", VII-th European Conference on Earthquake Eng., Athens, September, 1982.

[11] Kelly, J.M. and Skinner, M.S., "The Design of Steel Engery Absorbing Restrainers and their Incorporation into Nuclear Power Plants for Enhanced Safety", Vol. IV, Rep. UCB/EERC-79/10, University of California, Berkeley, 1979.

INCREASE IN EFFICIENCY AND RELIABILITY OF LOAD BEARING MEMBERS BY ACTIVE DEFORMATION CONTROL

SURVEY, ADVANTAGES AND RESTRAINTS OF ADC IN CIVIL ENGINEERING

H. Domke
Civil Engineering
Technical University
Aachen, Germany

1. Definition of Active Deformation Control (ADC)

The basic idea of Active Deformation Control requires that every beginning deformation of a structural member be compensated instantaneously by counter-forces. The active elements producing these counterforces are supported by independent components of arbitrary but limited yield. Simultaneously, all stresses in relation to deformations subject to control will disappear.

The principle of ADC may be used to:

1) Adjust inadmissible movements of foundation by actively controlled shifting of building supports in vertical and horizontal direction.

    1.1 compensate slow movements due to arbitrary causes

    1.2 absorb horizontal earthquake shockwaves

    1.3 reduce strain caused by impeded temperature dilation.

2) Adjust differing deformations of members belonging to the same structure

    2.1 subdivision of a structure into subassemblies free of deflection and supporting members

    2.2 support of vibrating members against freeswinging inert masses.

2. Application of ADC to Girder Structures

A structure equipped by ADC consists of a girder with a preferably ⊓ shaped cross section, which is coupled to movable supporting cables by means of actively adjustable elements. These cables are anchored in the neutral axis at the girder's ends. The active elements are operated automatically in such a way that the girder remains free of deflection and hence free of bending or buckling stress. The cables react to the distribution of varying load by altering their geometrical form only. Deflections of the girders axis exceeding 1/10 mm are continuously measured against an independent reference plane. These data of measurement are transformed into subtle control impulses which trigger the required counterforces generated by pneumatic or hydraulic gear.

The described function requires that the cables alone must sustain dead weight and live load of the whole structure. Their load bearing capacity depends on their design working stress, their curvature and the girder's resistance capacity in compression. Continuously arranged coupling elements effectuate that all parts of the structure are exposed to tension or pressure only, thus allowing ultimate exploitation of the materials' strength.

The independent reference plane can be realized by wires spanned between the structures' abutments. These wires are protected against exposure to weather and physical damages by tubes fastened to the structure. Any deviation between wire and structure can be measured by sensors installed in these tubes without actual contact. To cover wider spans a reference plane created by laser beams may be more reliable.

The theoretically most effective conception of a structure equipped by ADC is shown in the following sketch:

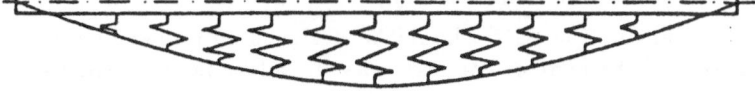 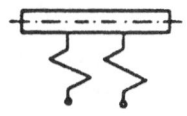

However, this system is considered unsuitable for practical application in views of the following arguments:

1.   The structure turns instable, if the ADC equipment fails.

2.   A continuously working coupling gear is very expensive.

3.   Cables and ADC gear are unprotected.

4.   The height of the structure varies with the actual load.

These disadvantages can be avoided by supplying the girder with a greater structural stiffness. Girders with the neutral axis in 2/3 of their height seem to be most appropriate.

Even with the ADC out of action the structure retains a load bearing capacity corresponsing to that of a passively reacting cable supported girder.

The number of coupling elements may be reduced if minor additional stresses by bending and shearing forces are accepted.

3. Requirements on Control Equipment

The variation of load usually does not exceed frequencies of 1,5 Hz. For these low frequencies the forces of inertia are neglible. The control apparatus may be considered as operating between successive static conditions and may thus be denoted as quasi-static.

At the same time a moving load on a rough girder surface may produce shock forces resulting in vibrations with frequencies up to 10 Hz. But, due to the damping effect of the now effective forces of inertia the amplitudes of these vibrations decrease with growing frequencies. The counterforces required to suppress the amplitudes of such vibrations less than one tenth of the quasi-static. Higher counterforces are only necessary, if the resonance frequencies of the structure correspond to the generated vibrations.

From the view point of failure consequences, the control of quasi-static alterations must be associated with a very high degree of reliability since failure may lead to a collapse of the structure. In contrast, a complete suppression of vibrations is usually not vital, as long as resonances can be avoided. Often active control of vibrations will prove to be unnecessary.

As a principle requirement the ADC should be designed to prevent unintentional oscillations caused by the control apparatus itself. Delays in the response of the pneumatic or hydraulic gear to control impulses should be compensated by accelerating the growth of counterforces, such that they attain their peak value at the right instant.

As a result of our recent tests it is recommended to strictly separate control of quasi-static varying loads and vibrations with regard to all control devices.

4. Attainable Advantages

4.1. Eliminated deflection

Evading deflection allows thorough exploitation of the strength of the material employed. A low elasticity module need not be compensated by a high momentum of inertia. More slender structures and the extended use of plastics may be persued.

4.2. Augmented load bearing capacity

A conventional cable supported girder has to sustain bending forces in addition to the compression resulting from the anchored cable forces. In comparison the same structure with ADC equipment permits the use of stronger cables, because bending forces in the girder are eliminated. This results in a higher load bearing capacity or wider span.

This could be effected by increasing the initial tension of the cables with growing load using the normal prestress gear, which would become an additional part of the ADC equipment.

The same effect can be brought about much simpler by the coupling members only, if the structure in its passive state allows to use a higher prestress linked to a smaller curvature of the cables. Increasing load will heighten the curvature and as a consequence strain the cables to maximum stress.

The theoretical relations are demonstrated for the example of a prestressed concrete girder. For simplicity reference is still made to the conventional permissible stress format.

A	section area
$A \cdot h \cdot c_B = W_{\min}$	section modulus
h	height of structure
ℓ	span
$A \cdot \gamma = g$	dead weight
γ	volumetric weight
h/ℓ	slenderness
σ/γ	tension length
σ_v	centric prestress
$\sigma = \sigma_{\max}$	safe stress

bending stress due to:

σ_{v_m}	prestress

$c_A = 2/3$ specified constraint

σ_g dead weight

σ_p live load

$c_A \cdot h = f$ max sag in active state

$\bar{c}_A \cdot h = f$ max sag in passive state

Boundary conditions for calculating this prestressed girder are as follows:

upper fiber (loaded structure)

$$-\sigma_v + \frac{\sigma_{v_m}}{2} - \frac{\sigma_g}{2} - \frac{\sigma_p}{2} = -\sigma$$

lower fiber (dead weight only)

$$-\sigma_v - \sigma_{v_m} + \sigma_g = \sigma$$

Restricted to a range of $\sigma_p = \sigma$:

$$-\sigma_{v_m} = (c_A - 1)\sigma - \frac{\gamma \cdot \ell^2}{8h \cdot c_B} = -\frac{M_v}{W_{min}} = -\frac{c_A}{c_B}\bar{c}_A \cdot \sigma$$

or

$$\bar{c}_A = \frac{c_B}{c_A}\left(\frac{\gamma}{\sigma} \cdot \frac{\ell}{h} \cdot \frac{1}{8 \cdot c_B} \cdot \ell + 1 \right) - c_B$$

solved to ℓ :

$$\ell = 8 \left(\frac{c_A}{c_B} \cdot \bar{c}_A + c_A - 1 \right) \frac{\sigma}{\gamma} \cdot \frac{h}{\ell} \cdot c_B \; .$$

When using the simple method an increase in performance can only be achieved within the range of $\bar{c}_A < c_A$. The widest length obtainable results from $\bar{c}_A = c_A$

$$l_k = 8 \left(\frac{c_A^2}{c_B} + c_A - 1 \right) \frac{\sigma}{\gamma} \cdot \frac{h}{\ell} \cdot c_B \; .$$

Beyond the critical length an increase of live load is possible only by activating additional cables using permanently installed prestress gear.

Figure 1 shows the rate of activity to passively carried load in relation to the critical length.

Figure 2 shows the possible reduction of the girder's dead weight when only the peak load is carried by means of ADC.

The largest saving in material could be obtained if the passive load bearing capacity of the girder is just sufficient for its dead weight. Any live load would then be carried by ADC.

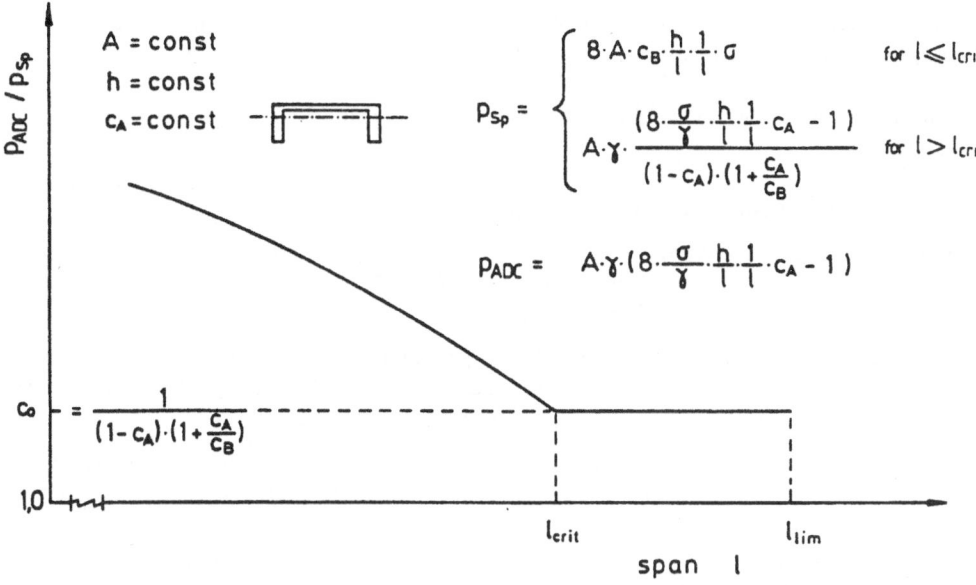

Figure 1: Increase in load by ADC

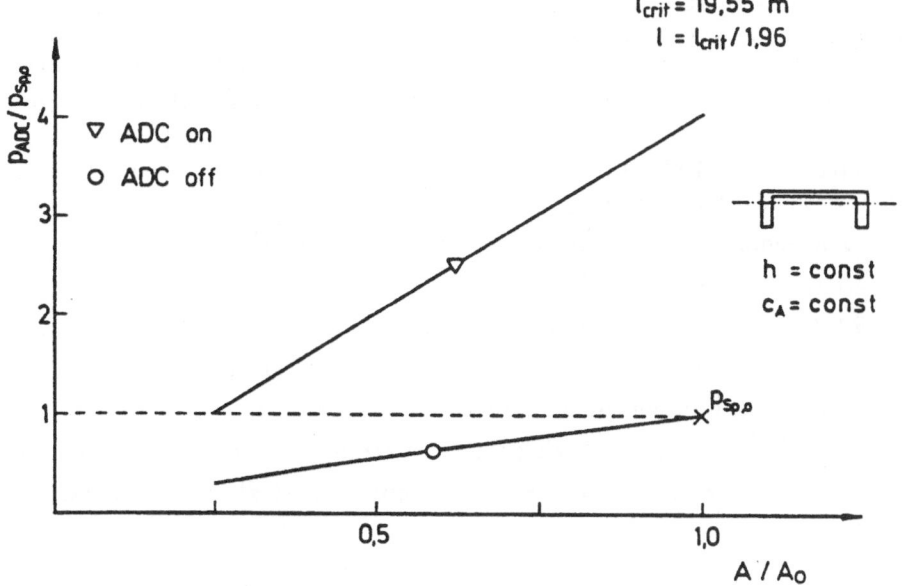

Figure 2: Decrease of dead weight by ADC

Economic and technical deliberations, however, recommend that the load supported in the structure's passive state should comprise at least 95% of all possibilities. This load is defined as basic load. Loads exceeding this basic load initiale the ADC. The rate of basic to maximum sustained top-load varies from 1:1,5 to 1:4.

5. Automatic Supervision of Performance

The fact that the model behaviour of a cable under specified variable load is practically identical to that of an actual cable of any size results in a very close correspondence between calculated and measureable cable reactions.

Differences identified by comparing precalculated and measured response of the structure to a given load can be used to analyse the kind and location of the cause of difference by computers. (Examples see Appendix)

If defined values of tolerance are surpassed, a distant technical control service may be alarmed. Subsequent damage to the structure may be avoided by timely counteraction.

In addition, it may be noted that before a really dangerous situation develops the ADC equipment will automatically operate to suppress increasing deformations regardless of the actual load.

These precautions against unexpected failure substantially increase the reliability of the structure. Hence, a reduction of safety factors may be considered. Such a reduction in structural engineering amounts to a rise of the design working stress and consequently results in a smaller cross section of the girder. The section decreases more than proportional in view of the reduced dead weight.

The following deviations from the structure's nominal shape may be corrected automatically by activating the ADC. At the same time the cause of the deviations are recorded.

5.1 to be eliminated by countermovements of the supports.

  5.1.1   differing subsidences

  5.1.2   impeded temperature dilation

  5.1.3   change of support forces as a result of creep

  5.1.4   alterations of the structure's dead weight

5.2 to be eliminated by modifying the counterforces and the cable's geometry

  5.2.1   decrease of support due to rupture of single wires in a cable

  5.2.2   creep

  5.2.3   growing shock stress as a result of surface damages

  5.2.4   weakening of load bearing capacity caused by cracks in the girder.

The procedure to analyse discrepancies in performance resembles the methods described in the following section.

6. Checking of Design Data Against the Measured Data

At present most of the design data can be estimated only. As a consequence rather adverse conditions are generally assumed. Checking the real data at the structure itself usually fails because of insufficient accuracy attainable.

The actual load bearing capacity of a structure in most cases does not correspond to the calculated prediction, because the calculation model is based on various idealizing assumptions. Moreover, deviations from the intended shape and presumed material homogenity are practically unavoidable.

The consequences thereof may be detected and quantified in the following cases by the method already described.

6.1. Imperfections

Concerted operations of the ADC will deliver the following results:

6.1.1   If the measured counterforces differ individually from the calculated ones but if their sums are equal to the girder's neutral axis pendulats around the pressure axis. The differences between calculated and measured counterforces result in bending stresses of the girder and match those produced by the compressive force and its excentricity from the neutral axis.

6.1.2   The neutral axis is deflected downward if the sum of the counterforces is higher than calculated and lower if deflected upward. In both cases a constant amount of additional bending stress will exist in the girder when ADC is in operation.

6.2. Calibrating the ADC equipment

Exact knowledge is required about magnitude and distribution of dead weight and load, so that the activated counterforces are identical to the calculated ones. Noted deviations can be eliminated by adjusting the readings in the controlling computer to the measured distances to the references plane. The pressure axis then coincides with the neutral axis in all coupling points.

6.3. Recording the counterforces alone renders the following information:

6.3.1   live load

6.3.2   vertical shock forces

Alteration of counterforces and cable goemetry allow to analyse:

6.3.3   modulus of elsticity

6.3.4   creep module

6.3.5   temperature strain coefficient

6.3.6   bending stress resulting from non-uniform distribution of temperature in its cross section

The ADC equipment regardless of operation status renders:

6.3.7   The natural frequency of the unloaded and loaded structure.

6.3.8   Registration of anomalies in the vibration process compared to the theoretically expected course.

6.3.9   Bending creep produced during the passive state of the structure and the differences in the necessary counterforces and their effect on the structure after starting the ADC.

Examples of the actual application to the quoted problems including the calculation programme are given in the Appendix.

7. Measures for Improving the Reliability of the ADC Equipment

The chance of realising the described advantages of an ADC equipment depends on the evidence, that the probability of failure of an ADC equipped structure is equal to or smaller than the failure probability of a passively reacting structure. Among others the following steps and their contribution to an improvement of reliability may be considered.

7.1 A survey of possible events, which might cause the failure of vital elements.

7.2 Assessment of failure rates of such elements and measures to improve their life time.

7.3 Elimination of failure caused by structural alterations.

7.4 Structural measures to reduce the consequences of a possible failure.

7.5 Self-control of a function to detect impending failure.

7.6 Redundant elements to compensate defects of vital parts.

7.7 Emergency control in case of a breakdown of the whole control system.

The measures 7.5 to 7.7 will be efficient only, if additional software and hardware will reliably identify malfunction and failure in the control system. Even the activation of redundant parts may fail and thus demands additional supervision. On the other hand, increasing complexity of the control system results in an increasing number of potential failure causes.

One optimal solution to the problem presented will be difficult to find. We try to simplify the task by subdividing the entire system into 4 independent feedback control systems of which 3 are active and one is passive.

They are coordinated by a master control system. The active systems comprehend the scanning, the converting of scanning data into control impulses and the gear to produce the counterforces. The passive system is the structure itself. Each subsystem is prepared to manage the functions 7.5 to 7.7 autonomously.

Defective parts must be recognized as such by surpassing tolerance values. The act of switching them off must automatically activate redundant elements or processes.

Basically it is possible to plan a sufficient sequence of redundancies to attain a very low rate of failure of the subcontrol systems and so the necessary probability of failure for the whole control equipment. A scheme of how this may be done is given in Figure 3.

The realisation of this design may prove too expensive. As an alternative the basic control function could be reduced to its bare necessities. Additional control systems may improve the regulation quality with regard to the demanded standard. In case of disturbances in these additional systems they may be switched off without impeding the function of the basic system. The reliability of the basic control consisting of few and simple parts could easily be improved by redundancies (Figure 4).

8. Quantification of Failures

Lacking experience it is obvious that no verified statement can be made as to the statistical probability of failure concerning a complete ADC equipment.

It is, however, possible to obtain such information for individual elements, which have been in longer use for other purposes. They should be discerned into those whose possible failure becomes apparent some time before and those which fail without warning.

8.1. Predictable Failure

All parts of mechanical gear are subjected to wear and aging. This is valid notably for the gear to produce counterforces. Methods to recognize possible damage at an early state have successfully been developed for pneumatic or hydraulic compounds.

8.2. Limited Predictable Failure

Failure of non-mechanical structural parts is most frequent within the first hours of operation and in a progressed state of aging. Between these extremes a very low rate of failure is predominant.

A test run of sufficient duration will eliminate defective parts. Most of the remaining components will function reliably for a long term. With the beginning of aging the rate of failure will rise significantly. Now at the latest all

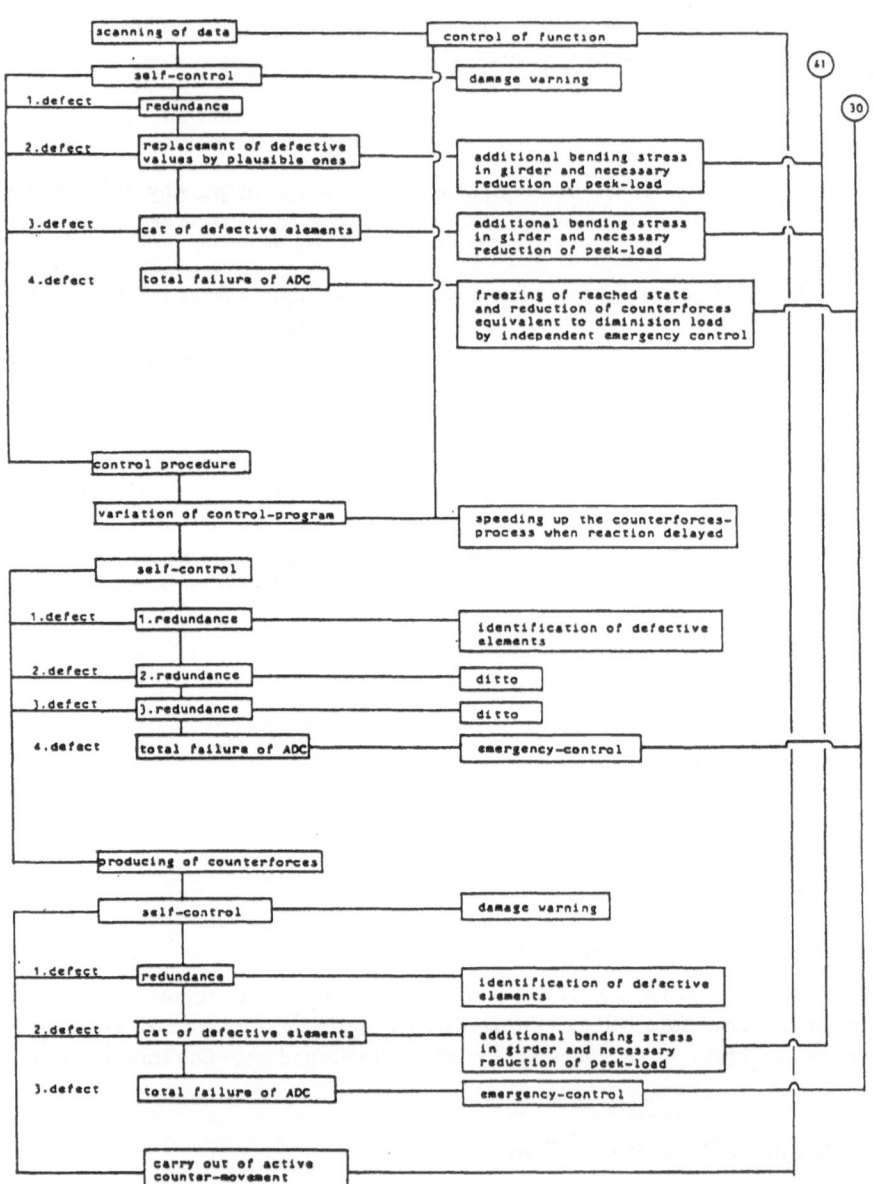

Figure 3: Improvement of reliability I

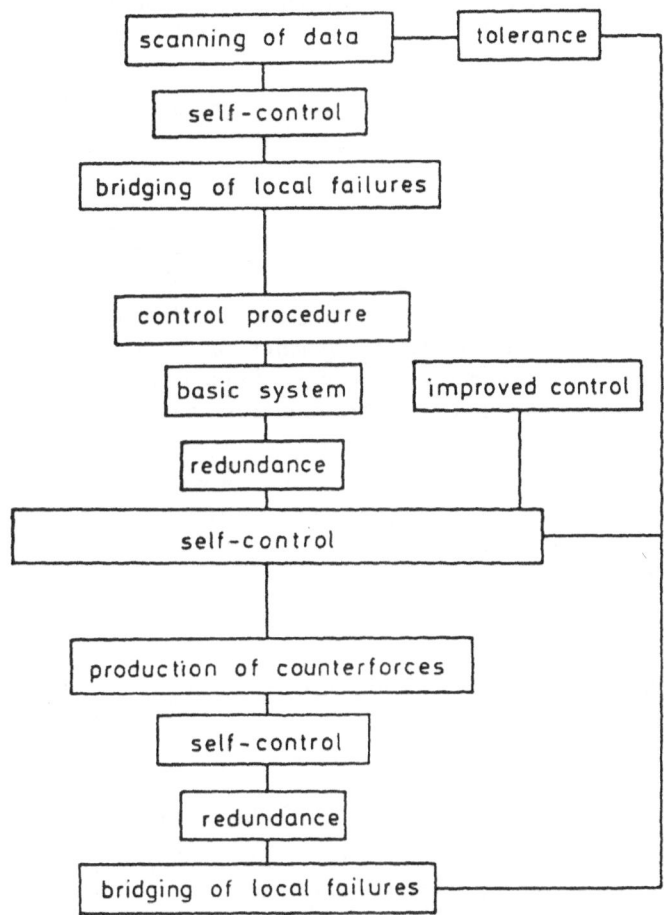

Figure 4: Improvement of reliability II

components concerned should be replaced.

9. Verification by an Experimental Girder

All statements in this paper have been tested at a concrete girder of 10 m length, 1.0 m breadth and 0.3 m height with ⊓ shaped section of 0.1 m². The girder is supported by 3 prestressed cables with a tension force varying between 660 and 1000 kN. Inflatable cushions serve as coupling elements. They are supplied by compressed air of 8 bar developing counterforces of 30 kN at a lift of 10 cm and are regulated by simple electrically controlled valves. The reference plane is realized by two wires spanned beside the cables and anchored to the girder's ends. The distance between girder and wire are scanned by optical sensors. As a check the distance to the floor is measured independently by inductive gadgets. All data thus gained are processed by computers into control impulses for the regulation of the counterforces.

After the eliminations of obvious errors in measurement the practical results with test loads certified the theoretical predictions.

10. Energy Requirements

The energy consumed by the counterforces gear roughly corresponds to the deformation energy achieved by the structure when ADC is out of action.

Whilst it is always possible to calculate the dimensions of the necessary counterforces gear, the power demand depends on the frequency and duration of the ADC being in operation. High energy demand in sporadic cases may be secured by storing compressed air in accumulators, which would keep the running energy supply low.

Counterforces can be produced by cylinders with a working pressure up to 350 bar or cushions with only 10 bar. The overall height of these components varies from starting position to full lift at a rate of 1:1.8 for the plain cylinder and of 1:5 for cushions.

References

[1] Domke, H., "Sicherungsmaßnahmen gegen Bergschäden und Erdbeben sowie ihre Auswirkungen auf neuere konstruktive Entwicklungen im Bauwesen", Westdeutscher Verlag, 1979.

[2] Roorda, J., "Experiments in feedback control of structures", in *Structural Control*, H.H.E. Leipholz (Ed.), North Holland Publ., 1980, pp. 629-661.

[3] Domke, H., Backé, W., Meyr, H., Hirsch, G. and Goffin, H., "Aktive Verformungskontrolle von Bauwerken", *Bauingenieur*, 56, Springer Verlag, Berlin, 1981, pp. 405-412.

[4] Domke, H., Backé, W., Theissen, H., Meyr, H., Bouten, H., Zach, B., Witte, B., Busch, W., Goffin, H., "Leistungssteigerung von Biegetragwerken durch Aktive Verfformungskontrolle", *Bauingenieur*, 59, Springer Verlag, Berlin, 1984, pp. 1-8.

[5] Domke, H., "Neue Möglichkeiten in der konstruktiven Gestaltung von Bauwerken", Westdeutscher Verlag, 1984.

APPENDIX

H. Domke and D. Streck
Civil Engineering
Technical University
Aachen, Germany

A1. A Comparison Between the ADC-Girder and a Cable Supported Girder

To show in principle the possibility of increased efficiency by using Active Deformation Control comparisons were made between a cable supported girder with a fixed cable and a girder with ADC-equipment.

As is customary for cable supported girders, prestressing is calculated to the furthest extent, and not only for the compensation of dead weight.

Used labels

l_s	length of cable
l_o	unstretched cable
σ_{St}	cable stress
A_S	section area of cable
E_S	elasticity module of cables
$\sigma_{St,s}$	safe cable stress
S	cable force
σ_D	safe compressive stress
D	pressure force on girder
E	elasticity module of girder

185

The following theoretical carrying capacity of girders results are on the condition that live load is equally distributed.

1. Girder with fixed cable on condition of:

(dead weight only)

upper fiber: $\qquad -\sigma_v + \sigma_{vm} - \sigma_g \leq 0 \qquad$ 1

lower fiber: $\qquad -\sigma_v - \sigma_{vm} + \sigma_g \geq -\sigma_D \qquad$ 2

(loaded structure)

upper fiber: $\qquad -\sigma_v + \sigma_{vm} - \sigma_g - \sigma_p \geq -\sigma_D \qquad$ 3

lower fiber: $\qquad -\sigma_v - \sigma_{vm} + \sigma_g + \sigma_p \leq 0 \qquad$ 4

Below the critical length it is possible to take a constant centric prestress of:

$$\sigma_v = c_A \cdot \sigma_D \ .$$

In general the critical length can be calculated with $\bar{c}_A = c_A$ and 2 and $h/\ell \neq$ const

$$\ell_K = \left[\frac{8 \cdot h \cdot c_B}{\gamma} \cdot \sigma_D \cdot \left(\frac{c_A^2}{c_B} + c_A - 1 \right) \right]^{1/2}$$

Below the critical length one gets in the lower fiber: 4 - 2

$$\sigma_p \leq \sigma_D$$

$$p = \frac{p \cdot \ell^2}{8 \cdot W_u} = \sigma_D \ .$$

The bearing capacity can thus be calculated as:

$$p = 8 \cdot h \cdot A \cdot c_B \cdot \frac{1}{\ell^2} \cdot \sigma_D \ .$$

Beyond the critical length the centric prestress must be increased (whereby the lower fiber is a decisive factor)

$$\text{where} \quad \sigma_{vm} = \frac{c_A}{c_B} \cdot \sigma_v \quad \text{and} \quad 2$$

results in

$$\sigma_v = \frac{\sigma_D + \sigma_g}{1 + \dfrac{c_A}{c_B}}$$

lower fiber:

$$\sigma_p = \sigma_D - \sigma_v + \frac{(\sigma_D - \sigma_v) \cdot c_A}{1 - c_A}$$

and

$$\sigma_p = \frac{p \cdot \ell^2}{8 \cdot W_u} \ .$$

Then the load bearing capacity results as:

$$p = A \cdot \gamma \cdot \left[\frac{\sigma_D}{\gamma} \cdot 8 \cdot \frac{h}{\ell} \cdot c_A \cdot \frac{1}{\ell} - 1 \right] \cdot \frac{1}{(1 - c_A)\left(1 + \dfrac{c_A}{c_B}\right)}$$

2. ADC-girder:

$$\frac{g + p}{8} \cdot \ell^2 = A \cdot c_A \cdot h \cdot \sigma_D$$

$$p = A \cdot \gamma \cdot \left[\frac{\sigma_D}{\gamma} \cdot 8 \cdot \frac{h}{\ell} \cdot c_A \cdot \frac{1}{\ell} - 1 \right] \ .$$

The results of the comparison are shown in Figure 1 of *Survey, Advantages and Restraints of ADC in Civil Engineering*, by H. Domke.

Another comparison was made to show the possible reduction of the girder's dead weight by using the ADC.

By reduction of the cross section A, on the condition that height and width of the structure and the neutral axis will be constant, the possible carrying capacity is calculated.

In the state when ADC is on: (top load: p_{on})

$$\sigma_{St} = \sigma_{St,z}$$

compressive force

$$D = A \cdot \sigma_v \cdot S$$

required section area of cable

$$A_{St} = \frac{A \cdot \sigma_v}{\sigma_{St}}$$

The cable's geometry approximated by a parabolic curve, is:

$$\ell_{\bullet}=\ell+\ell\cdot\frac{8}{3}\cdot\left(\frac{c_A\cdot h}{\ell}\right)^2 .$$

It is possible to calculate the unstretched length of the cable as:

$$\sigma_{St,ADC}=\epsilon\cdot E_S=\frac{\ell_{\bullet}-\ell_o}{\ell_o}\cdot E_{St}$$

$$\ell_o=\frac{\ell_{\bullet}\cdot E_{St}}{\sigma_{St,ADC}+E_{St}} .$$

By switching off the ADC-equipment, the sag should be reduced to $\bar{c}_A\cdot h$. (basic load: p_{off})

Therefore, the stress is reduced to

$$\sigma_{St,off}=E_{St}\cdot\frac{\ell+\frac{8}{3}\cdot\ell\cdot(\frac{\bar{c}_A\cdot h}{\ell})^2-\ell_o}{\ell_o}$$

tension force:

$$S_{off}=\sigma_{St,off}\cdot A_{St}$$

centric prestress:

$$\sigma_{v,off}=\frac{S_{off}}{A}=\frac{A_{St}\cdot\sigma_{St,off}}{A} .$$

When ADC-equipment is off, the conditions 1 to 4 must be realised to enable an iterative calculation of the value of $\bar{c}_A$, where in passive state the carrying capacity is at a maximum.

The results of the comparison are shown in Figure 2 of *Survey, Advantages and Restraints of ADC in Civil Engineering*, by H. Domke.

In addition to this comparison Figure A1 shows to what extent the carrying capacity can be increased by using ADC with the preselected values $\bar{c}_A/c_A$.

In this case an increase in efficiency is obtained by enlarging the sag and stretching the cable to maximum stress.

A2.   **Gain of Design Data (Examples) by ADC-Equipment**

1) Imperfection

    Requirements: Counterforces acting continuously. Dead weight and live load known in distribution and size.

    Model and actual

    Structure: Imperfections become apparant by differences between calculated and actual counterforces.

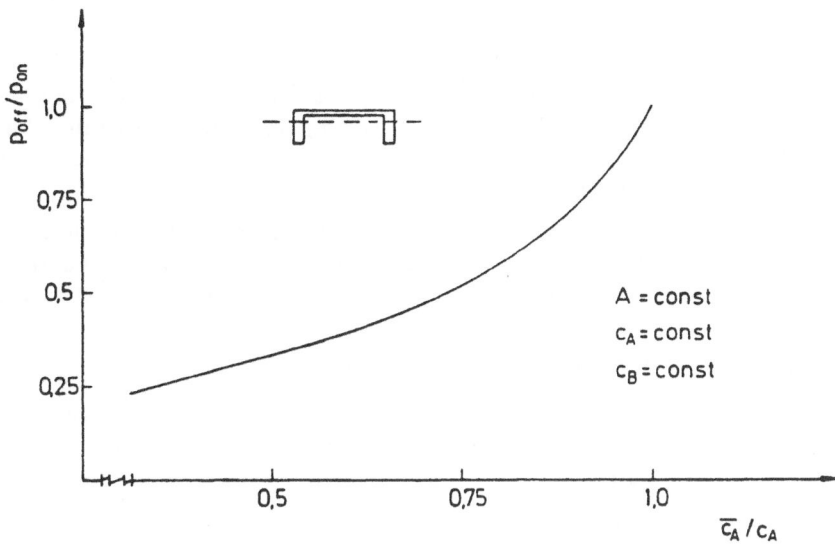

Figure A1: Design data $\bar{c}_A/c_A$ for p_{off}/p_{on} required

Cause:　　　　　Neutral axis of girder deviates from pressure axis, owing to geometrical deviations or inhomogenity of material.

Extent of

imperfection:　　The momentum produced by the difference of calculated to measured counterforces corresponds to the momentum caused by the pressure force and its excentricity to the neutral axis.

2) Calibration of structure

Requirements as before:

Calibration:　　Calculated and actual counterforces are made identical by inducing the control apparatus to transform the measured distances between girder and reference plane until those conditions are satisfied.

Result:　　　　The distribution of momentum produced by the altered counterforces correspond mirror symmetrical to that of the given imperfections and so neutralize each other. Any additional live load corresponds directly to the measurable size of the counterforces.

Precision:　　　The amount of the necessary adjustments varies slightly with the altering cable force. The amount of inaccuracy is however negligible, when the calibration is effected under low load.

3) Determination of Elasticity-modulus

Requirements: All measurable data of the structure under varying load can be gained accurately by the ADC alone. Bending forces in the cable will be prevented by restricted variations of its curvature.

Mathematic: Model and realized structure correspond to each other with such accuracy that deviations between calculated and measured deformations originate in different boundary conditions only.

Procedure: 1st step:

length of girder, cable sag and counterforces under low load are identical in the calculated model and the actual structure.

2nd step:

The alteration of the cable's geometry under higher load effecting higher counterforces is calculated under the assumption that the girder is absolutely rigid.

The actual change in the cable's geometry under the same counterforces will differ from the calculated one because here the girder is not rigid.

In a first approximation the length of the cable may in both cases be considered identical. The compression of the girder is then easily determined by the difference in the cable sag. In a second approximation the influence of the decrease of cable force as the result of its greater sag can be stated.

The horizontal share of the cable's force and the girders cross section, known the actual E-modulus, results from $E = \sigma/\epsilon$.

measured: U_i counterforces

f_i sag at position i

approximation $S_h = $ const $=$ horizontal component of cable force.

1. Calculation:

$$\ell_i = f(\ell, f_{i,1})$$
$$S_{h,1} \cdot f_{i,1} = M_{i,1} = f(U_{i,1}, \ell)$$

$$S_1 = f(Sh, 1, f_{i,1})$$

2. Calculation:

$$\iota_{\bullet} = f(\Delta \ell, \ell, f_{i,1})$$

$$S_{h,2} \cdot f_{i,2} = M_{i,2} = f(U_{i,2}, \ell, \Delta \ell)$$

$$S_2 = f(S_{h,2}, f_{i,2}) \qquad 1$$

$$S_2 = S_1 + \Delta S$$

with

$$S = f(\epsilon, E_{\bullet})$$

$$S_2 = S_1 + f(\epsilon, E_S) \qquad 2$$

$\Delta \ell$ can be calculated on the condition, that 1 and 2 must result in the same value of S_2.

The elasticity modulus can then be calculated as:

$$E = \frac{S_{h,2} - S_{h,1}}{A} \cdot \frac{\ell}{\Delta \ell} \quad .$$

REFERENCE MEASURING SYSTEMS AS A CONSTRUCTIONAL ELEMENT OF LOAD-BEARING ASSEMBLIES WITH ADC

W. Busch
Institute of Mine Surveying
B. Witte
Geodetic Institute
Rheinisch-Westfälische Technische Hochschule
Aachen, West Germany

1. Concept of the Measuring System

Load-bearing assemblies with Active Deformation Control (ADC) need measuring devices which continuously detect the vertical deflections of the girder axis. Exacting requirements on the reliability of these devices must be set because they give the basic information for the adjustment. Furthermore, the security of the construction demands a certain degree of redundancy. Therefore, it is proposed to apply two different measuring systems which are connected together in a suitable way [1].

Both systems are based on the idea to measure the deflections directly as changes in vertical distance of the girder beam relative to a stable reference which is connected with the abutments. Dependent upon the realization of the references their duties and specifications can be summarized as in Figure 1.

The measuring System I consists of two tightened wires (Figure 2) which are anchored to the abutments, and several new developed electro-optical displacement transducers which are connected with the girder beam at those places where the counterforces are generated. These transducers measure the vertical beam's deflection without touching the wires. Hereby, vibrations of the beam cannot affect the stability of the reference wires. Because of the application of two parallel wires it is possible to regulate not only vertical deflections of the girder axis but also deflections transverse of the axis. Moreover, the redundancy of the system is increased.

	System I	System II
measuring method	measurement of the changes in vertical distance relative to a solid reference which is anchored to the abutments	measurement of the vertical movements of the complete object relative to one abutment or a fixed point out of the load-bearing assembly
realization of the reference by a	tightened wire	fluid (hydrostatic pressure)
measuring transducer	electro-optical displacement transducer	differential pressure transducer
measuring range	< 50 mm	> 100 mm
accuracy	0.1 mm	0.1 mm
measuring frequency	50 Hz	1 Hz
duties	detection of the girder beam's deflections, measurement results for the control loop	detection of long-term changes in elevation of the complete construction including the abutments, supervision of system I

Figure 1: Measuring systems for load-bearing assemblies with ADC

To guarantee the high accuracy of 0.1 mm the reference wires must have a very good stabilty. Detailed investigations [5] show that the magnitude of the changes in the sag of the wire depend on the used wire material. Today it is possible to use different high-strength, high-modulus, low-density filaments [6] which have such good properties that changes in the sag, caused by different influences, will be smaller than 0.1 mm (relative to a length of the tightened wire of 100 meter and with an initial stress of 1 000 Newton). Furthermore, the adaption of such wires to the abutments and their handling is quite easy. Only a mechanical protection against wind and mechanical destruction is necessary.

In contrast to the System I, that produces the values for regulating the girder beam, there is the duty of measuring System II to detect the long-term changes in elevation of the abutments and to supervise System I. Therefore, it's measuring points are located near those of System I and upon the abutments (Figure 2). This system based on the measuring principle of hydrostatic leveling instruments which has proved itself time and again for the continuous long-term

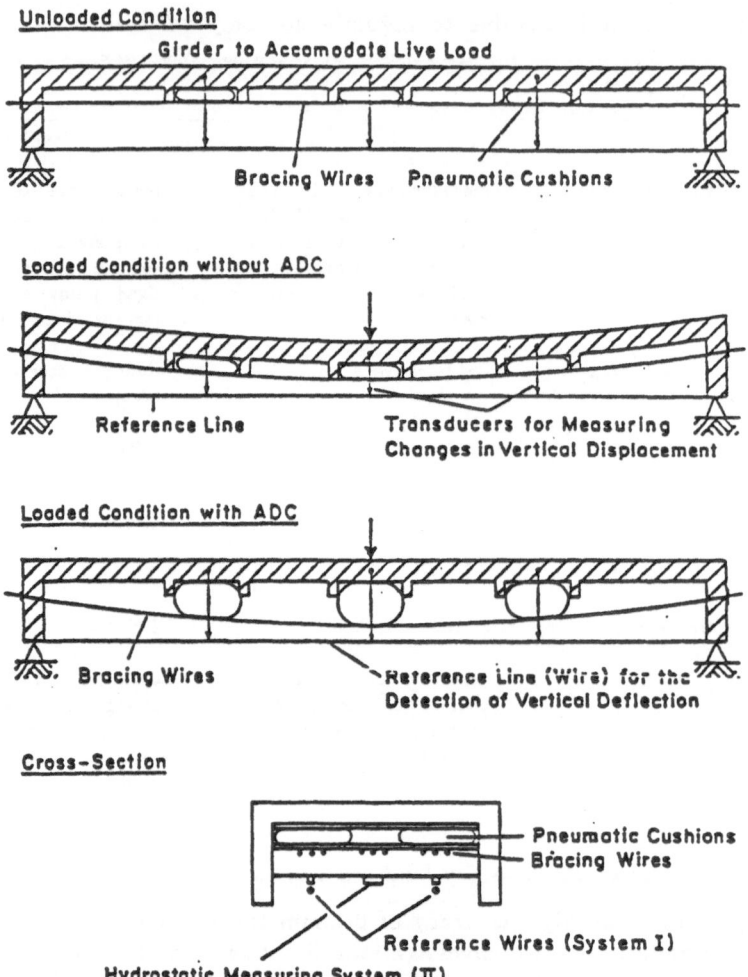

*Figure 2: Principle of the detection and regulated compensation of vertical
deflections of the girder beam caused by varying loads*

monitoring of large objects [1]. For detection of faster changes in elevation there
is developed a new hydrostatic leveling system based on differential pressure
transducers [2]. It consists of several pressure transducers which are connected
with each other by a tube or a pipe filled with distilled water. A change in elevation of a transducer produces a change in hydrostatic pressure which is measured
with electrical pressure transducers.

All sensors of System I and II produce digital or analogous signals. After
A/D conversion these digital signals (12 bit) are transmitted corresponding to a
modified version of the EIA standard RS-422 A. Each transducer is connected

with the central self-developed data scanner (with temporary storage) across a separate data cable (Figure 3). This star-like structure of the data transmission system increases the security and reliability of the measuring and the regulating system because the failure of one data cable does not affect the data transmission from the other sensors.

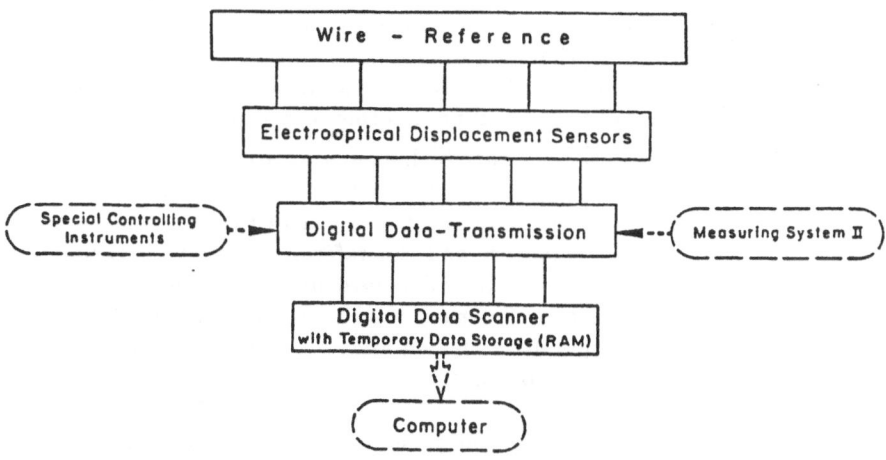

Figure 3: Parts of the complete measuring system

2. Possibilities to Increase the Reliability and Redundancy

The adapted measuring systems have to master a lot of duties when they are integrated in a supporting structure equipped with ADC. They shall give measured values

1. to the control loop for the compensation of vertical deflections of a structural member under load,

2. to calculate and supervise special data of the construction (for instance dead weight, modulus of elasticity, eigenfrequency),

3. to observe the deflections of the girder beam caused by different environmental influences on the abutments (for instance as a result of changes in temperature, tilt, settlement),

4. to check the algorithm and data of the control and the function of the pneumatic or hydraulic elements.

Therefore, it is necessary that the two measuring systems can guarantee

- a high durability
- a constant accuracy over a long time
- negligible systematic errors without long-term changes.

Because there is nearly no measuring system that can guarantee this under rough environments and with low costs, we think that it is necessary to use systems which have possibilities and facilities to supervise their own behaviour. Then it is possible to calculate new corresponding parameters and to exchange or to repair a "sick" element of the measurement system before the security of the construction is endangered. Beside this, the system must be of a high degree of redundancy because of sudden and unpredictable interruptions.

In general we have to differentiate between a stochastic and deterministic behaviour or event. Accidental errors, like outliers in the measured data, we can eliminate by using special mathematical tests and algorithms or a marginal checking. Systematic errors of the sensors and the references can be determinted with different methods. They can be subdivided into methods using additional measurements (of temperature or the tension of the wire for instance) and methods using a complex mathematical model which allows a good estimation of the deterministic changes by interpretation of the on-line gained sensor values [4],[8]. The application of those methods shows a further advantage of such measuring systems with a reference. Because of the common interaction of all sensors by the reference it is possible to:

- prove the sensors by moving the reference,
- calculate changes in the position of the reference,
- identify defective sensor data by comparison with a mathematical model calculated with the measured data of the other sources.

Although such measuring systems have enough possibilities of their own to increase their redundancy there might remain always the question "What will happen....if?" because some unexpected events can cause a failure of the measuring system or of the complete system [7]. Such failures and their consequences are collected and compiled in Figure 4.

Finally we can summarize that the proposed measuring system consisting of two independent systems with different references and sensors has a high degree of reliability and redundancy. Furthermore, the implantation of such systems in an ADC-girder structure allows a continuous supervision of the vertical deflections of the complete structure. Therefore, such measuring systems are not only necessary for an ADC-supporting structure, but also increase the reliability and security of such load-bearing assemblies.

Failure of	Effect	Compensation by	Restriction	Safety Precaution
one reference of system I	values of the sensors of that reference are not utilizable	values of the sensors of the second reference (system I) and completion through system II	under certain conditions lateral tilt is not determinable with the required accuracy	temporary not necessary
both references of system I	values of all sensors of system I are not utilizable	system II	compensation of fast varying load is not possible	reduction of the permissible peak-load and maximum speed
system II complete	long-term supervision of system I and of the abutments is not possible		temporary none	none
some sensors of one reference of system I	values of the sensors are missing for the control and compensation	measured values of adequate sensors of the second reference (I) and values calculated by interpolation between measured values of neighbouring sensors	under certain conditions lateral tilt is not determinable with the required accuracy	mostly not necessary

Figure 4: Redundancy of measuring system — failures and their consequences

References

[1] Busch, W., Eigenschaften stationärer hydrostatischer Präzisions-Höhenmeßsysteme für kontinuierliche Langzeitbeobachtungen, untersucht an einem neu entwickelten Schlauchwaagensystem. Veröffentlichung des Geodätischen Instituts der RWTH Aachen, Nr. 31, Aachen, 1982.

[2] Busch, W., Hydrostatische Vielstellen-Höhenmeßsysteme mit elektrischen Druckaufnehmern. IV. Intern. Symposium über Deformationsmessungen mit geodätischen Methoden, Katowice 9. - 16. Juni, 1985.

[3] Busch, W. and Witte, B., "Die Aktive Verformungskontrolle als neues Konstruktionsprinzip für Tragwerke — Grundlagen und Anforderungen an die Meßtechnik", in *Ingenieurvermessung*, 84, Band 2, Rinner, Schelling, Brandstätter (Eds.), Dümmler Verlag, Bonn, 1984, pp. E1-E13.

[4] Frank, P.M., "Detektion von Sensorausfällen: Per Software-Redundanz",
     Indsutrie - elektrik + elektronik 30, 1985, 4, pp. 42-50.

[5] Jakobs, M., Untersuchung statischer und dynamischer Durchhangänderungen
     eines gespannten Drahtes im Hinblick auf seine meßtechnische Anwendung
     als Referenzlinie zur kontinuierlichen Erfassung vertikaler Bauwerksverfor-
     mungen. Unveröffentl. Diplomarbeit, Institut für Markscheidewesen,
     Bergschadenkunde und Geophysik im Bergbau der RWTH Aachen, 1984.

[6] Kaufmann, S. and Lauck, L., "Einsatz von hochleistungsfaserverstärktem
     Plast zur Minimierung der Fehler beim Messen großer Längen",
     Feingerätetechnik, 32, 1983, 10, pp. 460-464.

[7] Schrüfer, E., *Zuverlässigkeit von Meß- und Automatisierungseinri-*
     chtungen, Carl Hanser Verlag, München, Wien, 1984.

[8] Young, P., *Recursive Estimation and Time-Series Analysis*, Springer-
     Verlag, Berlin, Heidelberg, New York, Tokyo, 1984.

CONTROL DESIGN FOR ADC-GIRDER

H. Bouten and H. Meyr
Lehrstuhl für Elektrische Regelungstechnik
RWTH Aachen, West Germany

1. Introduction

In civil engineering the most effective load bearing constructions are cables. For any load there is only tension stress. On the other hand the geometry of a cable depends on position and weight of the loads. Because varying geometry is not desired in civil engineering, cables are mainly used to bear static loads. For varying loads inflexible materials such as concrete are used.

In a prestressed concrete girder the concrete is subjected to the compressive force with results from the cable tension and the momentum caused by bending stress. Because of the momentum a homogeneous utilization of the cross-section of the concrete is not possible. Cable forces necessary for maximum load would snap the girder if the load was removed. This changes if the supplement forces of the cables are continuously adapted to the prevailing load. This is made possible through placing active coupling links between cable and girder. Bending is constantly measured and results in changing supplement forces until the bending stress is eliminated. Without this bending stress we get an even compressive stress in the concrete, whereas the cables change their geometry as if they are directly bearing the load. This principle of control is called Active Deformation Control (ADC).

Without changing the cross-section of the girder an increase of up to 4 times the present live load can be achieved and where the load remains the same the present span can be raised by up to 1.5 times, and it is also possible to reduce dead load to 70% for the same load and span. Arguments for this principle and its importance in civil engineering were presented at this conference by Domke et al [1] in a survey paper, which also contains information on possible drives.

Our paper deals with the control design of an ADC-girder.

2. Description of Experimental Girder

Theoretical results were tested on a 10 m experimental girder (Figure 1). This girder functions as a normal prestressed concrete girder when coupling links are inactive. The high cable force allows only a small sag.

Figure 1: General view of the experimental girder

When subjected to increased loads the coupling links, which are pneumatic cushions in this case, push the steel cables down until the resulting counterforces are equal to the sum of dead weight and life load (Figure 2). An electrical control system measures the fluctuation in the distance between girder and a constant reference (here: floor) and computes the appropriate orders to set the pneumatic valves. This directly shows the necessity of a regulator for this ADC-girder. The cushion pressure, i.e. the volume flow into the cushions is to be controlled in such a manner that, independent of varying load, the girder stays free of deflection without deviating from the above mentioned reference.

Ideally only the cables bear the load, while the concrete merely returns the compressive force, but only axial compressive stress.

To facilitate the fast and easy implementation of several different control algorithm a microprocessor is used as controller.

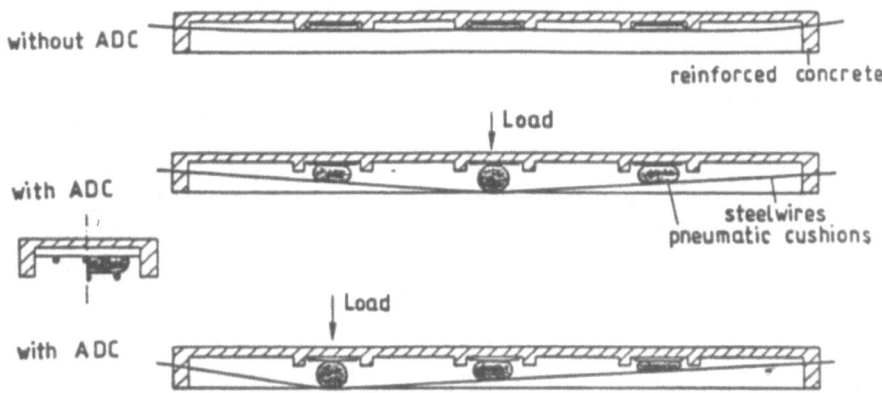

Figure 2: Function of experimental girder

The following block diagram is obtained:

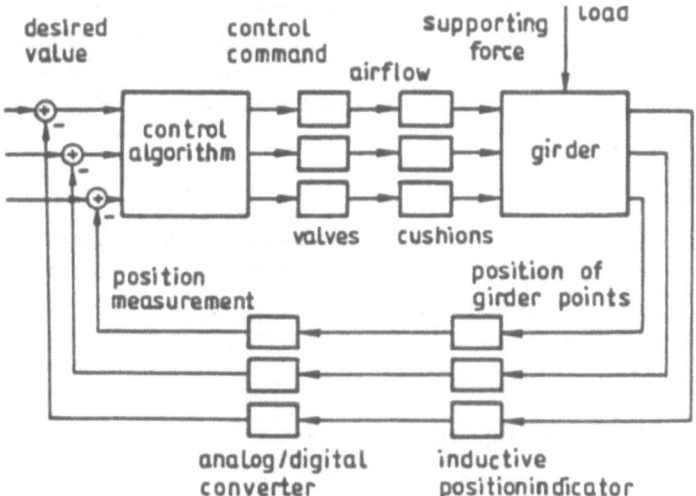

Figure 3: Block diagram of closed loop

3. Modelling of Controlled System

The first requirement of control design is to model an automatic control system to obtain a quantitative description of relations between input (here: airflow into the cushions), disturbance (here: varying load) and output quantities (here: deflection of girder).

Modelling is done in several steps. First the system is divided into components which are described by their equations of motion. This leads to the structure of the model.

The next step is to determine the parameter of the model through appropriate experiments and measurements.

Finally to test the model, behaviour of the total system is simulated on a digital computer and compared to the real system.

Because there are only three points with supplement forces (by cushions), deflection is measured only at these points, with the assumption that the mass of the girder is concentrated at these three points and live loads are put on the girder only at these points. Later the model is extended to take loads at various points.

At these three particular points (mid of grider and both quarterpoints) which have mass m_i, the girder is exerted to by forces of the cushions k_i and live loads which have a mass of m_{ai} and a force of k_{ai}. The description of girder movement (function of time and place) is reduced to a description of three particular points (discrete description of a system with distributed parameters). The following subsystems arise:

3.1. Concrete beam

By taking twice the integral over the momentums, the deflection of a massless flexible beam at the points y_1, y_2, y_3 can be computed as a function of the forces k_{g1}, k_{g2}, k_{g3}. This yields (Figure 4):

$$Y = C_B^{-1} \cdot K_g \tag{1}$$

with:

Y : vector of deflection of girder y_i at point i

K_g: vector of forces k_{gi} on the girder at point i

C_B: spring rate matrix of beam (3×3).

The force k_{gi} at point i is the sum of the live load k_{ai}, the force from cushion to beam k_i, and the dead weight at this point.

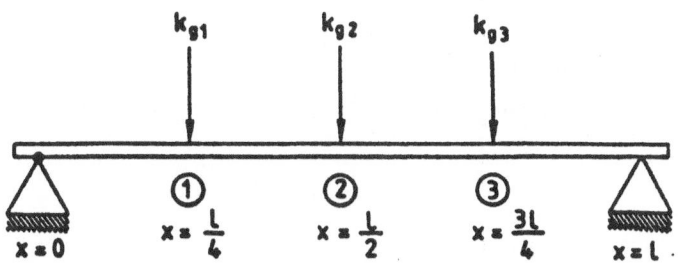

Figure 4: Beam with system of forces

3.2. Span cable

Changing the cable geometry through cushions causes changes in the cable stress of 660 kN at minimum sag and up to 1000 kN at maximum sag. Small changes in the length of the beam resulting from changing pressure can be ignored. Figure 5 shows the geometry. The relation between three forces k_i and the amplitude $(X+Y)$ of the wire can be represented analytically.

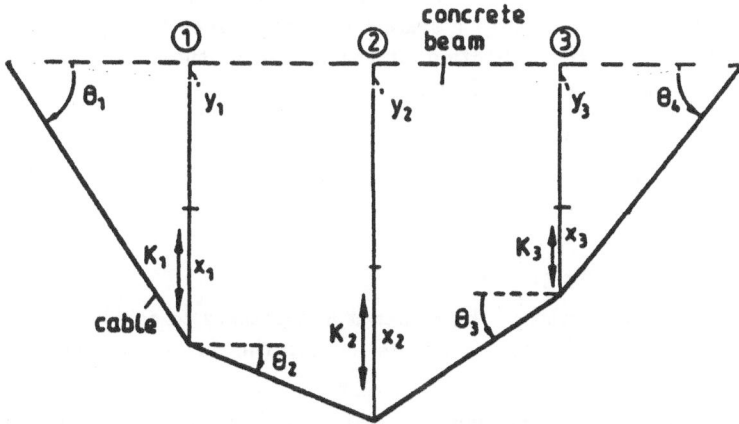

Figure 5: Geometry of the cable

The nonlinearities, because of the trigonometrical functions, can be linearized very well in the area of interest:

$$K = C_s \cdot (X+Y) \tag{2}$$

with:

K : vector of forces k_i between span cable and beam

C_s: spring rate matrix of wires (3×3)

X : vector of cushion height.

    Thus a block diagram of the girder for the static case without mass forces is obtained (Figure 6).

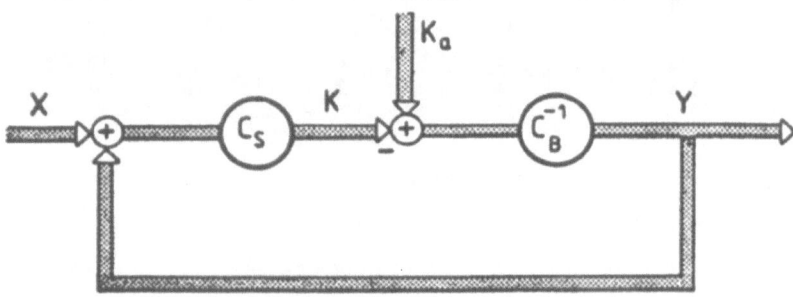

Figure 6: Block diagram of the static linear system (K_a: vector of load)

3.3. Force transformation

The restriction of forces to three points is eliminated by a transformation. A single force k_e is transformed into a system of three forces k_{a1}, k_{a2}, k_{a3} at the three fixed points, so that the deflection of the girder at the three points equals the deflection caused by the single force:

$$Y = f(k_e, a) \overset{!}{=} C_B^{-1} \cdot K_a \ . \tag{3}$$

This yields the unknown force system. A similar transformation is possible for a square load.

3.4. Dynamics of the girder

In describing the dynamics of the girder it is necessary to consider mass forces. In relation to the mass of the concrete beam, the mass of the cables can be neglected.

    By comparing the system with an elastic chain it can be shown that the mass matrix M is diagonal. This is not true however for the damping matrix and the spring rate matrix, which are full and describe the coupling of points by damping and spring forces.

    With acceleration forces we get the differential equation:

$$M\ddot{Y} + D\dot{Y} + (C_B + C_s)Y = K_a - C_s X \tag{4}$$

with:

M: mass matrix (3×3)

D: damping matrix (3×3)

which describes the following block diagram (Figure 7):

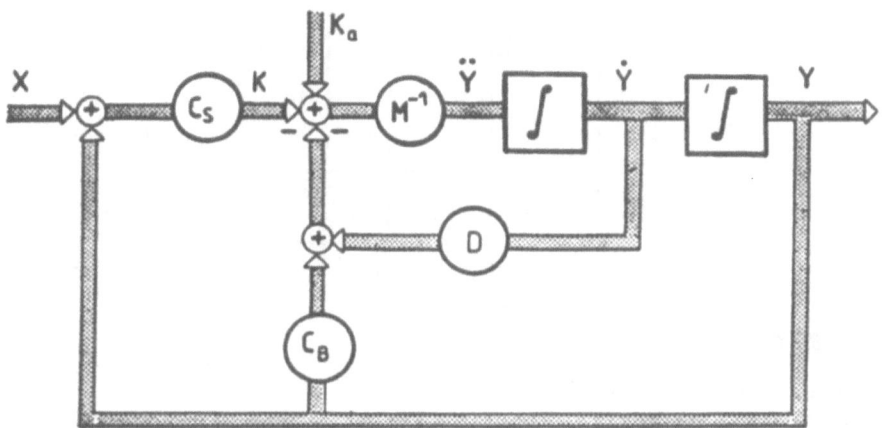

Figure 7: Block diagram of linear dynamic model of control system

3.5. The cushions as actuator

The regulator sets the valves in three possible positions:

a) filling the cushions (=upward),

b) stop airflow, pressure remains constant (=neutral),

c) emptying of cushions (=downward).

Figure 8 shows a possible curve of the cushion height with the positions of the corresponding valves. The analytical description of the cushion height as a function of the valve position and force onto the cushions is very complex and therefore not useful here.

For two major reasons a simplified description can be given:

a) The force between cables and beam (elevating force) has almost no influence on the velocity of upward/downward movement. The velocity depends only on the cross section area of the valves.

b) Neither switching the valves nor oscillating the girder yields oscillations between cables and beam, confirming the assumption that the mass of cables is negligible in comparison with the mass of beam.

Thus the height of the cusion as a function of the valve position can be described sectionally with lines, parabolas and exponential functions, the cushion pressure having no influence on the cushion height.

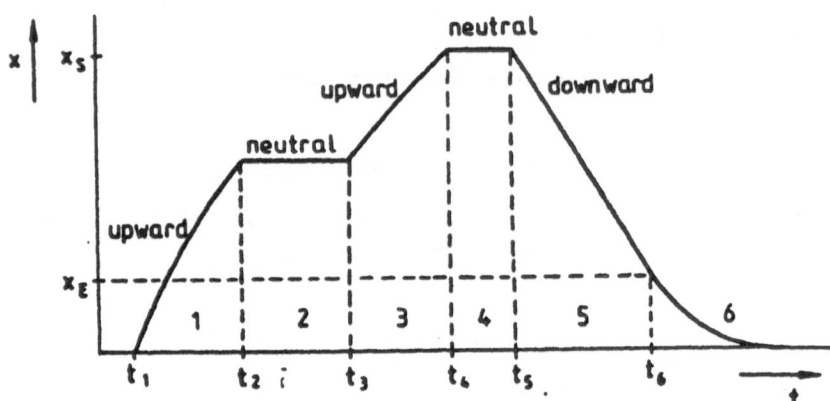

Figure 8: Possible curve of cushion height

3.6. Determinatin of model parameters

Starting with the differential equation of beam and cables (4), the parameters can be derived from appropriate measurements. Without changing the cushion height (X=const.) the matrix $C_s + C_B$ can be determined by changing load K_a and measuring the deflection Y. Then changing X until $Y=0$ yields the matrix C_s.

The measured matrices C_s and C_B match the computed values quite well. Differences result from unsymmetrical assembly of the girder.

A comparison between measurement and simulation with assumed parameters gives the mass matrix M. Because the measurement points are symmetrical it is assumed that the mass m in the three points is equal. This yields a system with three almost equal natural frequencies. The natural frequency of the girder is compared to a simulation with an assumed mass $\hat{m}$. Then mass m of the girder can be computed as follows:

$$m = \hat{m} \left(\frac{f_{\text{measured}}}{f_{\text{simulated}}} \right)^2 . \tag{5}$$

The mass of a load on the girder is added to the mass in the three points. In the absence of load at the three particular points, the load is transformed as described above for forces.

Damping forces are inner forces as are spring forces. Both result from material properties. Therefore it is assumed that the damping matrix D is proportional to spring rate matrix:

$$D = const. \cdot (C_s + C_B) .$$

The proportional factor can be determined by a comparison between the curves for the decay in simulation and measurement.

To test the model and the accuracy of the parameter a digital simulation is made of the behavior of the complete girder.

Two tests are presented:

In the first test a load of 1650 kg passes over the girder, the deflection of the three measurement points is shown against time (Figures 9 and 10). Because of unsymmetrical clearance at the ends of the girder, deflection at the beginnng and the end of test is not identical to the symmetrical simulation. The second test shows the dynamic behavior of the model. A man on the girder excites the girder (loaded with 1500 kg) by jumping.

Figures 11 and 12 show the decay of oscillation.

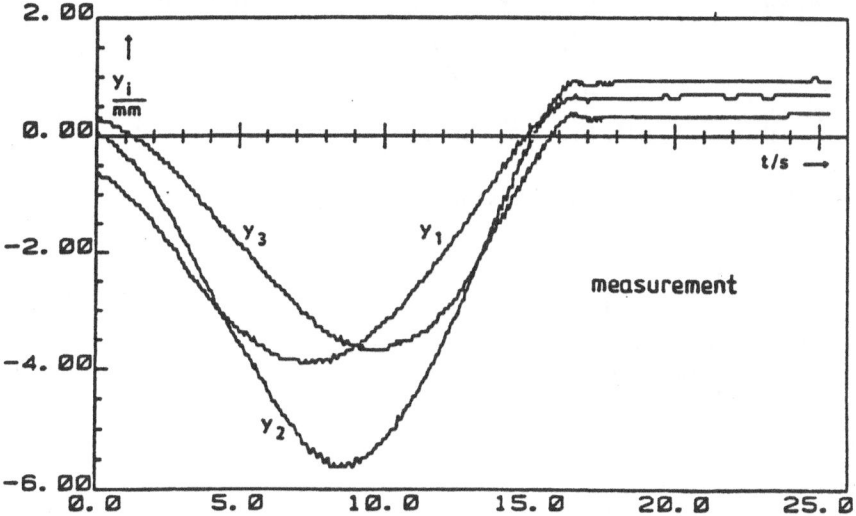

Figure 9: Measurement: load passes over the girder

4. Control Design

Because a result of several nonlinearities in the control system and the actuator (valves, changes in oscillating mass by load) the control design cannot be done by a linear theory. Therefore a control concept is developed through simulating the regulator and control system.

The measurement system gives three voltages which are proportional to the difference between girder and the floor. These voltages are sampled and quantized. The differences between desired values (horizontal girder) and the measured values are used to set the control algorithm, which decides where the girder is to be lifted or to be lowered. The appropriate voltage is transmitted to the switching coil. The appropriate valve switches on the pressure supply to the cushion (valve to rise) or empties the cushion (valve to lower).

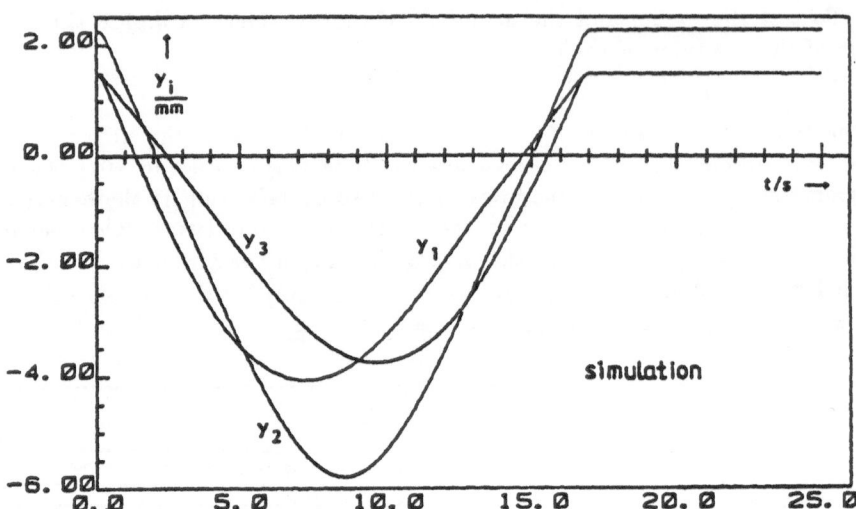

Figure 10: Simulation: load passes over the girder

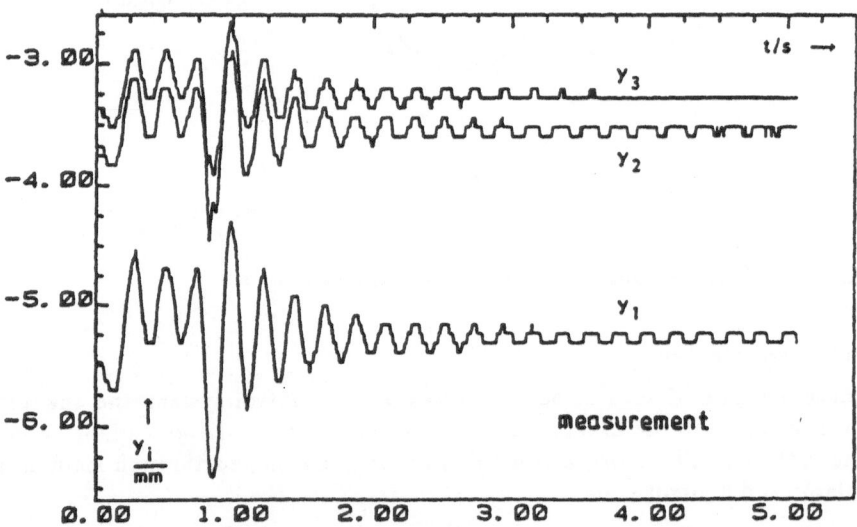

Figure 11: Measurement: decay of oscillations

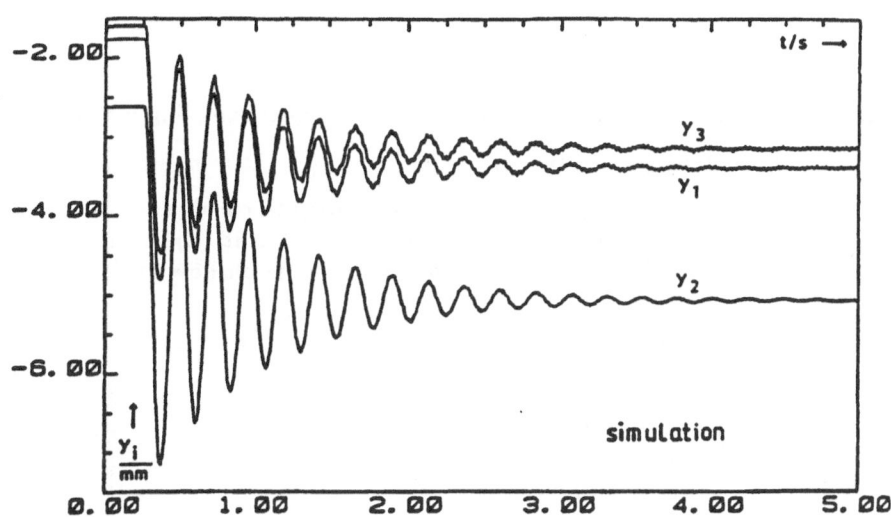

Figure 12: Simulation: decay of oscillations

Because of their limited power the cushions are not able to damp the girder in the range of its natural frequency. But there is sufficient power to rapidly and effectively prevent deformations of the girder caused by live load.

Movement of the girder can be divided into oscillations in the range of natural frequency (dependent on live load 6.15 Hz to 3 Hz) and quasistatic deformation by live load (below 1 Hz). Therefore control strategy is divided into the appropriate parts.

The pneumatic cushions have to prevent large deformations caused by varying loads. (The next section deals with the damping of oscillations). Measured data are filtered to get a quasistatic signal for the deformation control. Several low-pass filters are investigated. If the filter is too slow, oscillations are removed efficiently, but the accuracy of the regulator is reduced and hunting problems can arise. If filtering is poor, a number of switching operations occur without altering deflection. This causes further loss of power and reduces the life time of the valves. The filter can only be designed corresponding to the regulator. From simulations it can be seen that the load estimating regulator is the best.

Starting with the linear model (Figure 7), the force applied to the girder at the sampling time n can be computed from the deflection and cushion height:

$$\hat{K}_a(n) = C_s \cdot X(n) + (C_s + C_B)Y(n) . \tag{6}$$

By linear extrapolation, the load at a later sampling time can be estimated as:

$$\hat{K}_a(n+1) = \hat{K}_a(n) + exf \cdot (\hat{K}_a(n) - \hat{K}_a(n-1)) . \tag{7}$$

Factor exf determines, to what extent linear extrapolation is carried out. With the objective

$$Y(n+1)=0 \tag{8}$$

we get from (6) the estimated cushion height at the next sampling instant $(n+1)$:

$$\hat{X}(n+1)=C_s^{-1} \cdot \hat{K}_a(n+1) . \tag{9}$$

Inserting (6) into (7) into (9) yields an estimate for the cushion height in such a manner, that the deflection $Y(n+1)$ becomes insignificant. By comparison with the actual cushion height $X(n)$ a decision can be made on how cushion height is to change:

$$\hat{X}(n+1)-X(n)=\Delta X(n) \tag{10}$$

with:

$\Delta X>$ upper bound: upward
$\Delta X<$ lower bound: downward
                    else: nothing

Figure 13 shows the resulting block diagram. The low-pass filter, which should reduce only oscillations, introduces a delay in force estimation too. Therefore factor $exf \gg 1$ has to compensate for this.

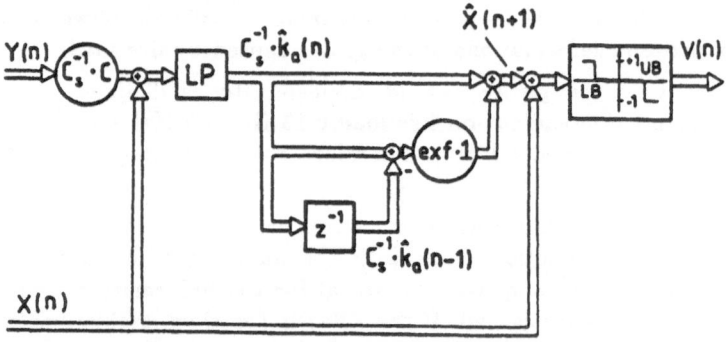

Figure 13: Block diagram of the regulator

By computer simulation optimal parameters of filter and regulator can be obtained.

The developed regulator and filter are used on the ADC-girder. Figure 14 shows the test of Figure 10 with active regulator. The maximum girder deflection of 6 mm is reduced to less than 0.5 mm by controlling. Further reduction seems to be nearly impossible through the use of these actuators and may not be economical.

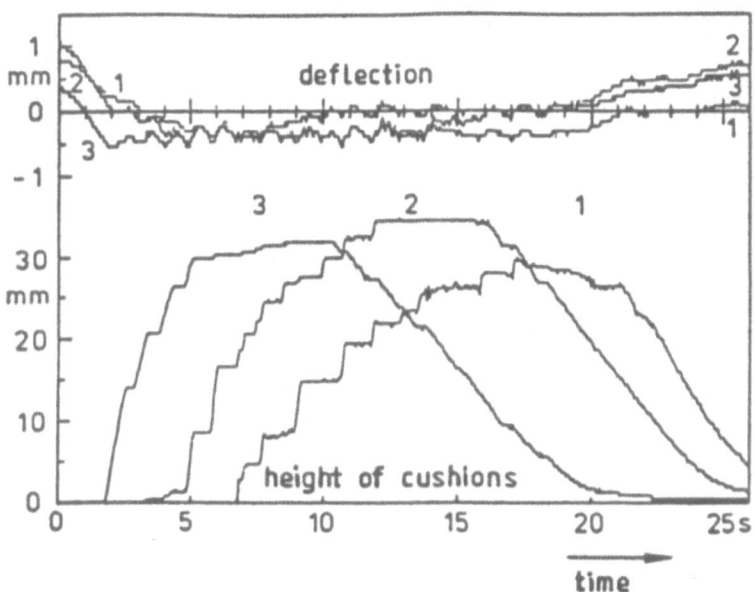

Figure 14: Cushion height and girder deflection with regulator, as load of 1650 kg passes over the girder

5. Damping of Oscillations

The roughness of the road surface results in oscillations of the girder through passing loads. Figure 15 shows an enlarged section of a passing load test like Figure 9.

As is shown above, damping through active forces does not need to be feasible. Using a passive damper tuned to a fixed frequency is senseless, because the oscillation frequency varies within a wide range depending on oscillation mass (mass of girder and mass of load).

Therefore we plan to tune a passive damper online to the actual oscillation frequency of the system. The frequency is determined and used to tune the damping system.

Frequency measurement derived from deflection Y and cushion height X gives impractical results. Using the dynamic model (Eq. (4)) computation is very extensive, and using the static model (Figure 6) the computed frequency is not exact enough. In both cases inaccurate parameters (inexact measurements, changes while operating, nonlinearities of girder) strongly falsify the result. Therefore frequency is measured directly from the alternating component of the girder deflection.

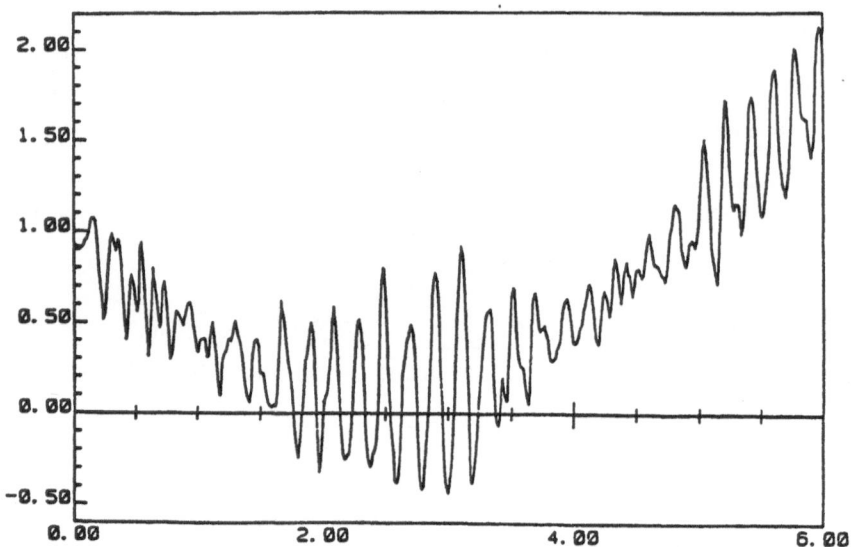

Figure 15: Enlarged section of a passing load test with vibrations

With changing loads the frequency measurement has to be fast e: ough to follow the actual frequency. At the experimental girder maximum chang: in frequency while a load is passing is about 0.75 Hz/sec. So only measurem-nt from an interval of less than one second can be used, if a resolution limit of).5 Hz is to be achieved.

This may be done by following the argument of Maximum Entr py with filtering as shown by van den Bos [8] and the recursive "least mean squa:e adaption algorithm" outlined by Widrow [9] and Griffiths [6],[5]. Figure 16 s.iows the load passing test from Figure 10 and 15 with frequency estimation. If tl :re is no oscillation of the girder, the frequency estimation is stopped at the last detected frequency and is held. These sections can be recognized by their even lin s.

If only one load with a known mass is on the girder, deflection siows the position and thereby the frequency. Therefore mid-girder deflection as a measure for the frequency is shown too.

Filtering of data to smooth the curve is not done in order to show :he capacity and limits of the method.

The next step is to design a tuneable passive damper to test the interaction with the measurement system. This is being carried out at present.

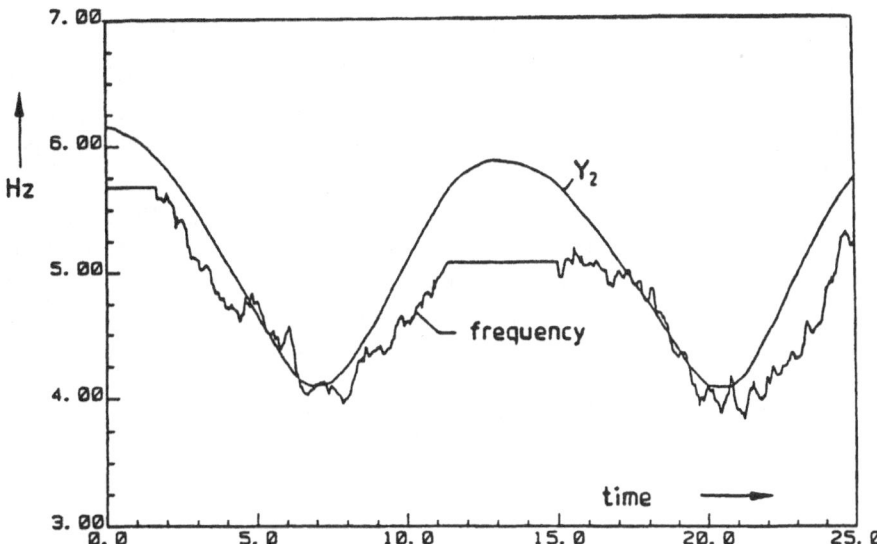

Figure 16:Frequency estimation for a passing load test

6. Conclusion

The principle of Active Deformation Control has been successfully tested on a 10 m experimental girder.

This paper describes the control design. Analysis of subsystems yields the mathematical description of the system. By computer simulation a control concept is developed, which exploits the power of control system and actuator optimaly. The regulator can be successfully used on the ADC-girder. Dividing the regulator into a quasistatic part and an oscillation damping part makes it economical to use.

The further development of the ADC-concept is therefore most promising.

References

[1] Domke, H., "Increase in Efficiency and Reliability of Load Bearing Members by Active Deformation Control", presented at The Second International Symposium on Structural Control, Waterloo, Ontario, Canada, 1985.

[2] Domke,H., "Leistungsteigerung von Biegetragwerken durch Aktive Verformungskontrolle", *Bauingineur*, 59, Springer-Verlag, Berlin, 1984.

[3] Bouten, H., "Demonstrationsmodell Hängebrücke", Internal Report No. 765/0, Lehrstuhl für Elektrische Regelungstechnik, RWTH Aachen, West Germany.

[4] Burg, J.P., "Maximum Entropy Spectral Analysis", presented at the Society of Explor. Geophys., Oklahoma City, Oklahoma, October, 1976.

[5] Griffiths, L.J., "A Simple Adaptive Algorithm for Real-Time-Processing in Antenna Arrays", *Proceedings of the IEEE*, Vol. 57, No. 10, October, 1976.

[6] Grifiths, L.J., "Rapid Measurement of Digital Instantaneous Frequency", *IEEE Transactions on Acoustics, Speech and Signal Processing*, Vol. ASSP-23, No. 2, 1975, pp. 207-222.

[7] Haykin, S. and Kesler, S., "Prediction-Error-Filtering and Maximum Entropie Spectral Estimation", in *Nonlinear Methods of Spectral Analysis*, S. Haykin (Ed.), Springer-Verlag, Berlin, 1979.

[8] Van den Bos, A., "Alternative Interpretation of Maximum Entropie Spectral Analysis", *IEEE Transactions on Information Theory*, 17, 1971, pp. 493-494.

[9] Widrow, B., "Adaptive Filter", in *Aspects of Network and System Theory*, R.E. Kalman and N. De Claris (Eds.), Holt, Rinehart and Winston, New York, 1979.

[10] T.J. Ulrych and T.N. Bishop, "Maximum Entropy Spectral Analysis and Autoregressive Decomposition," *Reviews of Geophysics and Space Physics*, 13, 1975, pp. 183-200.

[11] Ulrych, T.J. and Ooe, M., "Autoregressive- and Mixed Autoregresive-Moving-Average-Models and Spectra", in *Nonlinear Methods of Spektral Analysis*, S. Haykin (Ed.), Springer-Verlag, Berling, 1979.

FLUID-TECHNICAL UNIT FOR ACTIVE
CONTROL OF DEFORMATIONS

W. Backé and I. Forster
Fluid Engineering
Technical University
Aachen, Germany

Fluid-technical units are used to correct deformations of bending structures, because of the necessary high forces. Plants for these purposes consist of three sections, the generating section, the conductive section and the motoric section (Figure 1).

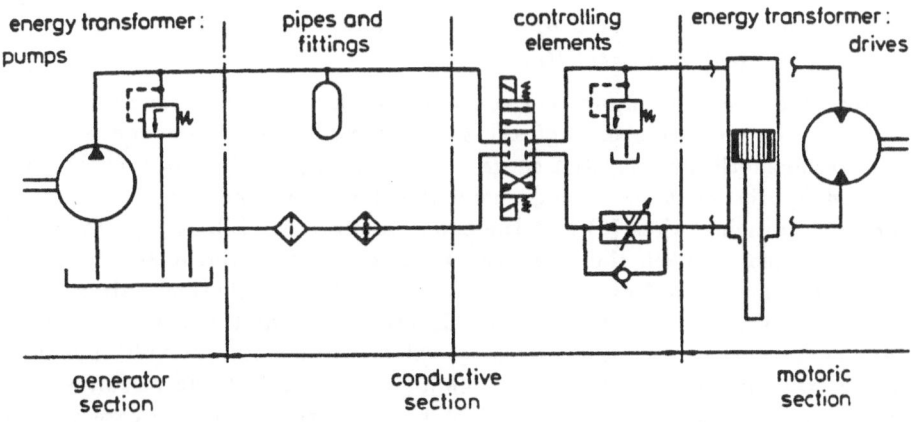

Figure 1: Principle structure of fluid-technical units

The mechanical energy taken from the output shaft of an electric motor or an internal combustion engine is transformed into fluidal energy in the generating section. The conductive section comprises the piping system, including fittings and the controlling elements, which determine and govern the amount and direction of the fluidal energy. The hydraulic or pneumatic energy is transformed into mechanical energy in the motoric section of the unit.

The technical efficiency of the principle of deformation control (ADC) was demonstrated with an experimental structure (ten meters long) equipped with a pneumatic unit. Figure 2 shows the circuit diagram of the structure.

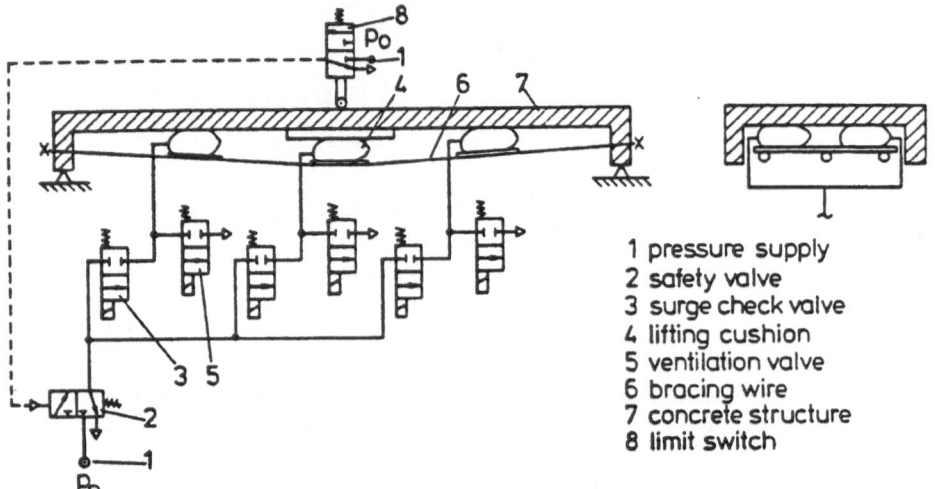

1 pressure supply
2 safety valve
3 surge check valve
4 lifting cushion
5 ventilation valve
6 bracing wire
7 concrete structure
8 limit switch

Figure 2: Pneumatic circuitry of the experimental structure for active deformation control tests [1]

From the pressure supply (1) consisting of a pressure tank charged by a compressor, the air passes through the safety valve (2) and subsequently through the electromagnetically operated surge check valves (3). The surge check valves are gated by the control. Lifting cushions (4) are arranged at three positions between the bracing wires (6) and the concrete structure (7). They are supplied with compressed air from the surge check valves. If the given tolerance limits of the actual value of deformation are exceeded, the ventilation valves (5) are activated by the control system and air escapes from the cushions into the environment. The combination of safety valve (2) (depicted in "off"-position in Figure 2) and pneumatic push-button limit switch is a fail-safe circuitry, which ensures that the concrete structure is not being overloaded in the case of failure of the electric control system.

The controlling elements of the experimental structure for active deformation control are pilot valves with a nominal flow rate of $Q_n = 300$ ℓ/min and switching times of $t_{on} \approx t_{off} = 15$ ms. The positioning accuracy obtainable with these very basic electropneumatic transformers is ± 0,5 mm. It can be achieved without triggering continuous controlling action. This tolerance band can be maintained at loads of 2,5 tonnes moving across the structures at the speed of a pedestrian. The natural oscillation of the structure, however, cannot be stabilized with this basic set up; this oscillation depends on the load, it varies between 3 and 6,5 Hz. Electrohydraulic units appear to be more suitable if active damping of oscillations and high regulating power are required.

Whether to utilize a hydraulic or, as it was done in this case, a pneumatic fluid-technical unit is a question occuring only at low power requirements. If the power rating is low, the special suitability of fluid-technical systems for use in active deformation controlled plants has to be discussed. Among others the following factors should be taken into account:

- reliability,
- controllability,
- purchase price.

Components of pneumatic units are relatively moderate in price, due to the low level of pressure (maximum of approx. 15 bar) at which they are operating. The pressuremedium air can be taken from the environment, it can also be dismissed without having to deal with waste disposal problems. In case of leakage due to pipe fracture or leakiness, the environment is not being contaminated. Secondary damages, for example, such as accidents due to oil leakage from structures reaching over roads, do not occur either. Hence precautionary measures are not necessary.

Lifting cushions can be used as positioning elements. The cushions ensure an areal force application to the structure without further measures. This type of positioning elements is characterized in that the effective surface of force application reduces with the height of lift. Figure 3 shows that the applied force at constant internal pressure of the cushion decreases with increasing height of lift.

If off-center loads are applied, adjoining cushions arranged along the transverse axis of the structure may be wired parallely as a consequence. The force-lift correlation then has a stabilizing effect. It may thus be possible to reduce the number of controlling elements if conventional positioning elements are used. In any case a separate control has to be provided.

Due to the large surface and low pressure level of these elements large valve opening areas are necessary in order to realize high variable displacement rates. This is particularly true for the lowering action.

The economically favourable operating pressures of oil hydraulics are between 100 and 400 bar. Hence only cylinders with accordingly small piston areas can be used as positioning elements. In the case of differential pistons, it is additionally possible to apply pressure to the ring area of the piston in order to accelerate the lowering action.

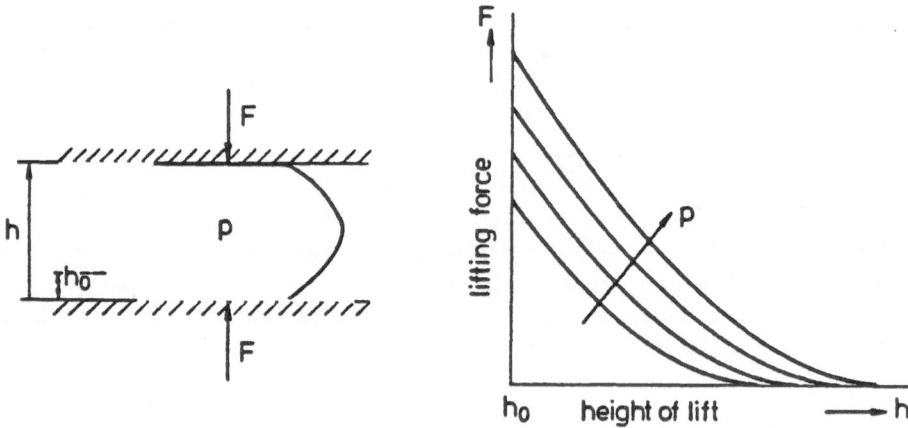

*Figure 3: Correlation between lifting force F and height of lift h in the case of
cushions. The internal pressure of the cushion p is the parameter of
the function*

It is a disadvantage of hydraulic cylinders, that the area of force application
is almost reduced to a point. Special constructive measures are necessary to avoid
intolerable load concentration at certain points of the concrete structure and to
avoid the occurence of transversal forces at the cylinder. In the case of an
extracted piston rod the unfavourable rate of installation height to lift may result
in that the cylinder block extends over the bottom edge of the concrete support,
if hydraulic cylinders are used in slim buildings. In addition it should be noted
that slide ring packings are always subject to wear. Thus, contaminations can
effect the hydraulic circuit. The occuring leakages as well as wear and the conse-
quent contamination result in increased maintenance expenses.

In principle, intermittently as well as continuously operating valves may be
used as controlling elements.

Switching valves have the advantage to be sturdy and of simple design. The
simple construction of these valves ensures low susceptibility to damage and
defects. In addition these valves are not very critical to contamination, because
the valve-piston opens the entire opening section at each switching action. No
leakage currents will occur, if valves with seats are used and if they are not in
operating position.

The disadvantage of these two-position (on/off) valves, however, is the
resulting poor controlling behaviour. The switching times result in restricted
accuracy of control. Moreover, each switching of the discontinuously operating
valves initiates shock pressures. Particularly in the case of incompressible pressure media, these shock pressures can imply additional dynamic stresses for the
building.

Continuous action valves (proportional action valves and servo valves) do not exhibit the above mentioned disadvantages. They were specially designed for the use in control systems. The opening section of these continuously operating valves, determined by the spool, is proportional to the electrical input signal. With this valve design, gating of the electromagnetical transducer can only be accomplished with an expensive interface. Particularly with micro-computer controlled systems an additional D/A transducer is generally necessary. Servo valves are especially expensive and susceptible to disturbances, due to the relatively expensive electronics and close tolerances at the production. Proportional action valves are favourable with respect to their lower susceptibility.

Taking all advantages and disadvantages of the respective system into account, pneumatic units seem to be suitable for use in active deformation controlled plants. Due to the fact that appropriate continuous action valves are not yet available to pneumatic systems, it will be undertaken to improve the performance (improved displacement rates and positioning accuracy) of systems with switching valves. One possible solution could be, for example, to use parallel wired valves with improved switching times. This measure will result in improved redundancy as well. In the case of power ratings which cannot be met by pneumatic systems electrohydraulic units have to be utilized. They can be operated with either servo valves or the more economic proportional valves, according to the respective requirements on the dynamics of the system.

The specific task of active deformation control makes high operating reliability and safety of the fluid-technical systems essential. If valve controlled hydraulic drives are used, systems designed for aircraft hydraulics, for example, may be taken into consideration. In the case of pneumatic units, correct operation of the switching valves may be checked, for example, by monitoring the lifting end-positions of the spool. The control signal obtained by this measure, then has to be compared to the input signal of the valve. A set-up consisting of a calibrated orifice and a differential pressure switch, for example, may be utilized for leakage checks. The use of electric control devices has the advantage of allowing simple processing of the control signals. The electronics necessary for this have to perform monitoring tasks as well as controlling functions. This is particularly significant with redundantly designed system-components, whenever it is necessary to change over from defective elements into intact elements. This kind of redundancy is called passive redundancy or stand-by redundancy. In comparison, active redundancy signifies that all redundant elements are operated simultaneously and parallel. In addition to the redundancy for the improvement of safety, the early diagnosis of faults and damages is of essential importance. Often components do not fail all of a sudden, but gradual changes of the operating behaviour indicate damages ahead of time. Quantities for the supervision of the unit condition could be, for example, the leakage flow rate or the sound signal in the case of positive displacement units. The methods for early diagnosis of damages make it possible to schedule the exchange of faulty components a certain period ahead of time [2].

References

[1] Autorenkollektiv, "Leistungssteigerung von Biegetragwerken durch Aktive Verformungskontrolle", *Bauingenieur*, 59, 1984.

[2] Backé, W., Stand und Entwicklung elektronisch gesteuerter hydraulischer Anlagen zur aktiven Verformungskontrolle, Möglichkeiten und Grenzen ihrer derzeitigen Leistungsfähigkeit, Fachbeitrag zum 1. Kolloquium *Aktive Verformungskontrolle von Bauwerken,* vom 4. und 5., Februar 1982, Aachen.

Acknowledgement

The authors of these five presented papers spontaneously formed a research group at the Technische Hochschule Aachen, which hence was supported by the building industry and governmental means (Ministerium für Wissenschaft und Forschung des Landes Nordrhein-Westfalen).

MODELING AND CONTROL OF AN INDUSTRIAL ROBOT WITH ELASTIC COMPONENTS

Bernd Gebler
Lehrstuhl B für Mechanik
Technische Universität München
München, West Germany

1. Introduction

Due to the rapidly increasing number of industrial robots the question of economical efficiency on one hand and the question of enlargement of the applicability on the other hand will become more and more important. For both questions consideration of elasticity and deformation of the elements of the manipulator is of great importance.

The payload/manipulator weight ratio of todays industrial robots is small and operating as well as production costs are relatively high. A lightweight construction would improve economical efficiency and allow higher operating speeds. A wide field of additional applications could be opened by raising effector position accuracy. To match the two objectives precision and cost, detailed modeling and suitable control of the robot can be very helpful.

2. Manipulator Dynamic Model

In this study an industrial robot with three revolute joints is considered (Figure 1). The arms of the manipulator are modeled as Euler-Bernoulli-beams with two-fold bending and distortion. They are connected with rigid bodies representing joints and payload. Torques applied at the joints of the robotic system are computed from a simplified joint model consisting of a direct-current motor (moment of inertia) and a drive unit (torsion spring, gear ratio) Figure 2.

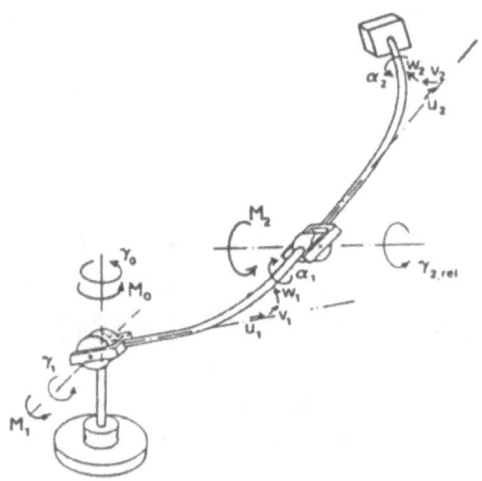

Figure 1: Three-axis manipulator with deformable links

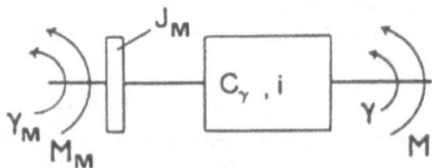

Figure 2: Mechanical model of the joints

The position of the mechanical system is described by the motor coordinates γ_{M0}, γ_{M1}, $\gamma_{M2,rel}$, the joint angles γ_0, γ_1, $\gamma_{2,rel}$ and the distributed coordinates v_i, w_i, α_i, $i=1,2$.

However, a two joint model with in-plane-motions only will mainly enter further calculations in order to discuss physical effects and numerical results (Figure 3).

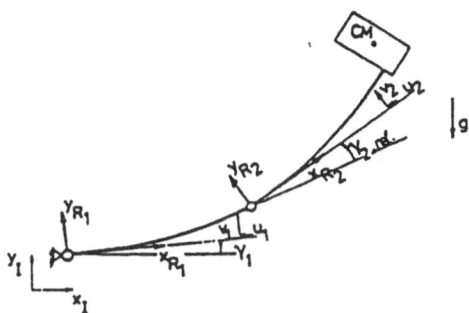

Figure 3: Two-dimensional model of the industrial robot

3. Mathematical Description of the System

3.1. Deformation geometry of the beams

The elastic deformations of the arms are assumed to be small. The position of an arbitrary beam element is in general described by six coordinates, quite often three displacements $u(x,t)$, $v(x,t)$, $w(x,t)$ and three rotations $\alpha(x,t)$, $\beta(x,t)$, $\delta(x,t)$ around specified axes (Figure 4). The assumption that the beam elements remain perpendicular to the elastic axis and that the elastic axis is unextensionable leads to the fact that there are geometric relations between these six coordinates which have to be taken into account. To get correct equations of motion all kinematic quantities have to be formulated up to second order. The rotational matrix S_{RE} and the position vector r_{dm} are functions which depend on the generalized coordinates.

The position vector r_{dm} reads:

$$r_{dm} = \begin{bmatrix} x+u \\ v \\ w \end{bmatrix}$$

where $u=-\int_o^x (v'^2+w'^2)/2dx$ since the neutral axis is assumed to have constant length.

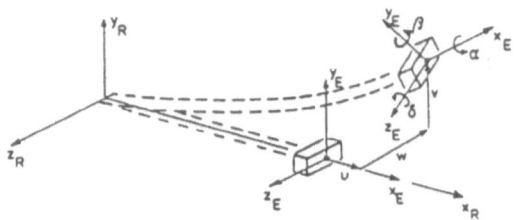

Figure 4: Geometry of the deformed beam

The rotational matrix can be determined using the unit tangent vector to the neutral axis:

$$r'_{dm} = S_{RE}\begin{bmatrix} 1 \\ 0 \\ 0 \end{bmatrix}, \qquad (\ )' = \frac{\partial}{\partial s}(\ ).$$

After some transformations the matrix S_{RE} results to be be [1,2]:

$$S_{RE} = \begin{bmatrix} 1 - v'^2/2 - w'^2/2 & -v' - w'\alpha & -w' + v'\alpha \\ v' & 1 - v'^2/2 - \alpha^2/2 & -\alpha - v'w' - \int_o^z w'v''d\xi \\ w' & \alpha - \int_o^z w'v''d\xi & 1 - w'^2/2 - \alpha^2/2 \end{bmatrix}.$$

The quadratic formulation of these kinematic quantities is necessary in order to retain all linear terms in the equation of motion which have significant influence on the analysis [3].

For this purpose, an elastic beam bearing a rigid body at one end is considered (Figure 5). This can be demonstrated by two examples:

1) Acceleration a_o of the system in x_R-direction

The eigenvalues of the system depend on the magnitude of a_o. If a_o exceeds a certain value the system becomes unstable. In comparison, if the quadratic terms of the position vector r_{dm} would be neglected, the eigenvalues would not depend on the acceleration, yielding wrong results (Figure 6).

2) Rotation of the beam about the z_R-axis with the angular velocity ω_o.

Here as well only the quadratic formulation of kinematics leads to correct results. Taking into account only the linear terms of S_{RE} and r_{dm} leads to the physically unreasonable result that centrifugal forces would destabilize the system (Figure 7).

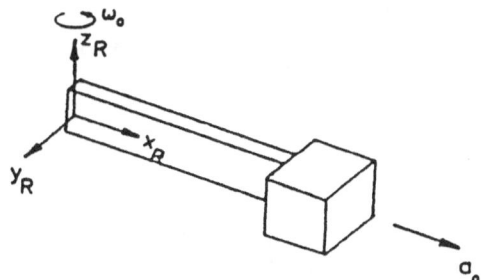

Figure 5: Elastic beam

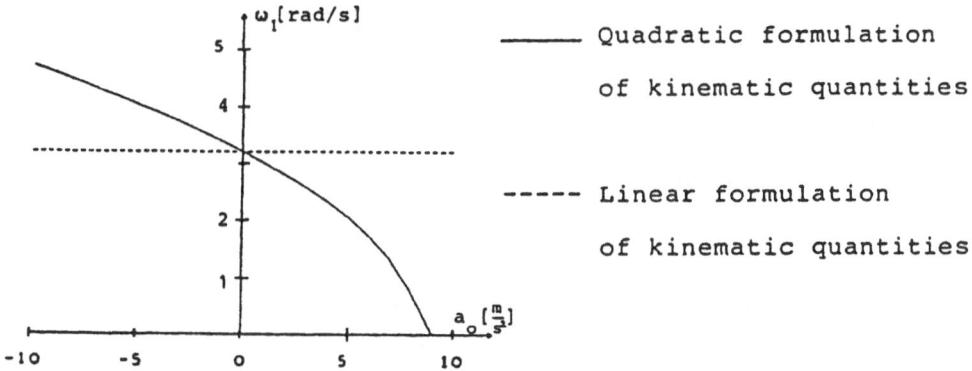

Figure 6: First eigenvalue of an elastic beam depending on the acceleration a_0

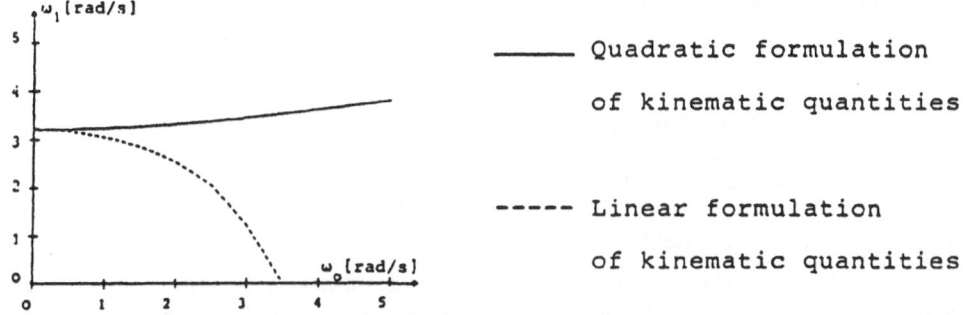

Figure 7: First eigenvalue of the elastic beam depending on the angular velocity ω_0

3.2. The equations of motion

To establish the differential equations of the robotic system d'Alembert's principle in the Lagrangian formulation is used.

$$\delta z^T \sum_{i=1}^{n} \int \left(\frac{\partial r_{IK}}{\partial z} \right)^T (\ddot{r}_{IK} - f) dm = 0 \tag{1}$$

n : number of bodies .

Mathematical evaluation of this equation yields:

$$\sum_{\text{elastic bodies } i} \int J_{Ti}^T (a^{Si} dm_i - dF_i^e) + (\frac{\partial \Pi_i}{\partial z})^T$$

$$+ \sum_{\text{rigid bodies } j} \left[(J_{Tj}^T (a^{Sj} m_j - F_{j,N}^e) + J_{Rj}^T (J^{Sj} \dot{\omega}_j + \omega_j \times J^{Sj} \omega_j \right.$$

$$\left. - r^{Sj,Fj} \times F_{j,N}^e - M_{j,N}^e) + (\frac{\partial V_j}{\partial z})^T \right] = 0 \tag{2}$$

z	: coordinate vector
J^T	: jacobian matrix of translation
J^R	: jacobian matrix of rotation
$\Pi(z,t)$	: potential function of the elastic beam
$V(x,t)$	: potential function of applied forces
$F_{i,N}^e$	: applied non-potential forces
$M_{i,N}^e$	: applied non-potential torques

In these equations for the distributed coordinates a Bernoulli-Ritz-approximation was introduced:

$$v(x,t) = \bar{\bar{v}}^T(x)\bar{v}(t)$$

$$w(x,t) = \bar{\bar{w}}^T(x)\bar{w}(t)$$

$$\alpha(x,t) = \bar{\bar{\alpha}}^T(x)\bar{\alpha}(t)$$

where $\bar{\bar{v}}$, $\bar{\bar{w}}$, $\bar{\bar{\alpha}}$ are admissible shapefunctions.

Equation (2) is evaluated analytically and numerically. The computations lead to a set of nonlinear ordinary differential equations of the form:

$$M(z)\ddot{z} + g(z,\dot{z}) = BM_M + h(z,\dot{z}) \tag{3}$$

Linearization with regard to a nominal rigid body motion, which is reasonable for many manipulator tasks, and separation of the motor torques into nominal torques M_{Mo} and additional torques ΔM_M:

$$z = z_o + y \tag{4}$$

$$M_M = M_{Mo} + \Delta M_M$$

results in a set of linear timevariant differential equations to describe the motion of the system.

$$M(z_o)\ddot{y} + P(z_o,\dot{z}_o)\dot{y} + Q(z_o,\dot{z}_o,\ddot{z}_o)y = h(z_o,\dot{z}_o,\ddot{z}_o) + B\Delta M_M \tag{5}$$

4. Shapefunctions

4.1. Selection of shapefunctions

For the numerical analysis of the system the selection of shapefunctions is of great importance. Cubic B-splines yield very good results for many hybrid multi-body-systems even when only a small number of shapefunctions is used, offering the advantage of easy differentiation or integration. For further reduction of the necessary number of shapefunctions, the property that the considered robotic system is characterized by a very low ratio of armweight vs. payload or jointweight is efficiently used. As far as forces or torques originating from robot motion or gravitation are concerned, those forces acting on the rigid bodies (joints, payload) dominate the forces acting on the mass elements of the beams. Therefore the static deflection bending lines of a cantilever beam with only one concentrated force (torque, resp.) acting on the outer end are selected as shapefunctions (Figure 8).

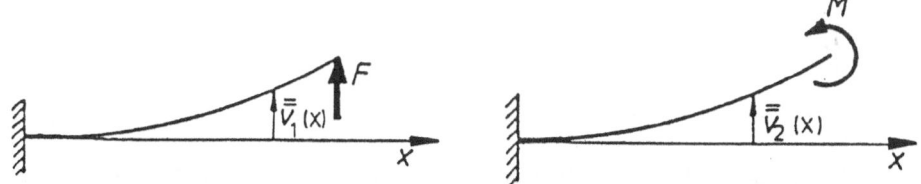

Figure 8: Static deflection bending lines

The first shapefunction $\bar{v}_1$ is characterized by linear increasing, the second shapefunction $\bar{v}_2$ by constant curvature. This leads to:

$$\bar{v}_2 = (x/1)^2$$

$$\bar{v}_1 = 1.5(x/1)^2 - 0.5(x/1)^3 .$$

4.2. Numerical results

To test what accuracy can be reached using these two shapefunctions only, the natural frequencies of the system for two different arm positions computed with two shapefunctions and computed with eight shapefunctions were compared (Table 1).

Table 1: Comparison of natural frequencies computed with 2 and with 8 shapefunctions for each beam

	Position 1		Position 2	
	Number of Shapefunctions		Number of Shapefunctions	
	$N = 2$	$N = 8$	$N = 2$	$N = 8$
ω_1	1.39	1.39	1.46	1.46
ω_2	1.79	1.79	2.57	2.57
ω_3	21.18	21.18	19.91	19.91
ω_4	52.23	52.23	40.26	40.26
ω_5	307.1	287.9	301.3	280.8
ω_6	576.3	529.8	576.4	528.9
ω_7	972.5	829.5	972.6	819.7
ω_8	3340	981.9	3270	980.7
ω_9	--	1250	--	1250
ω_{10}	--	1760	--	1760
ω_{11}	--	2560	--	2550
ω_{12}	--	3230	--	3220
ω_{13}	--	4830	--	4820
ω_{14}	--	5270	--	5260

Up to the 4. eigenvalue there is evidently no difference. Even for the 6. eigenvalue the inaccuracy is less than nine percent. Moreover, Table 1 shows, that for the different positions 1 and 2 the higher frequencies are equal. This result can be confirmed by further computations and by measurements and means, that the higher normal modes of the system are nearly the normal modes of beams with adequate boundary conditions due to increasing inertia effects of joint and payload.

Numerical simulation of the robot motion along a given trajectory yields the same result. The differences between the tip positions of the manipulator modeled with two or modeled with six shapefunctions are less than 0.01 mm (Figure 9). This is of course negligible. However, the time ratio of computation is 1/20.

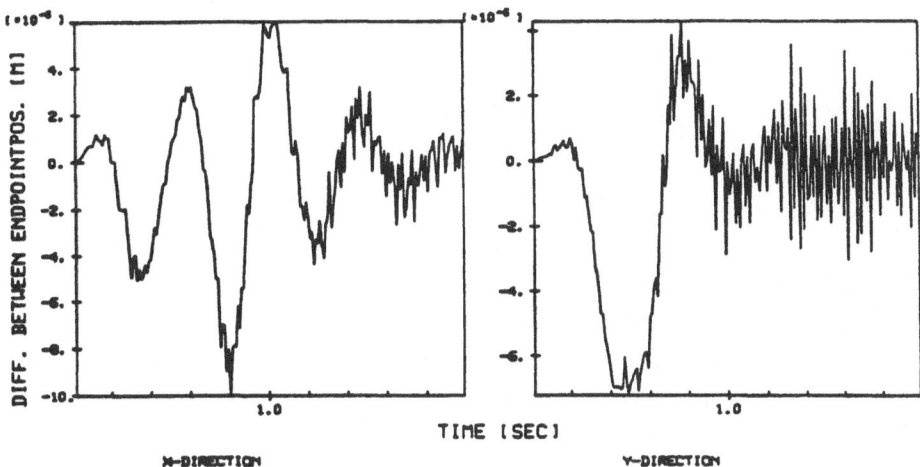

Figure 9: Difference in x and y direction of end point position for simulations performed with 2 and with 6 shapefunctions

The description of the robotic system with two selected shapefunctions yields, therefore, with minimum effort, very good results. The higher normal frequencies which cannot be represented by the selected shapefunctions are of less importance to the technical analysis of the system, an effect which is confirmed by measurements.

1.1. Measurability of elastic body coordinates

If only two shapefunctions are used for each beam the direct measurability of the elastic body coordinates with only two strain gauges will facilitate control realization. Using strain gauges at the end of each beam the curvatures can be measured (Figure 10). With the used shapefunctions $\bar{v}_1''(x)$ as a linear and $\bar{v}_2''(x)$ as a constant function, the coordinates $\bar{v}_1(t)$ and $\bar{v}_2(t)$ follow to be:

$$\bar{v}_2(t) = K_S u_E(t) / \bar{v}_2''(L_E)$$

$$\bar{v}_1(t) = K_S (u_B(t) - u_E(t)) / \bar{v}_1''(L_B)$$

where $u_E(t)$ and $u_B(t)$ are the measured voltages.

With this simple and cheap measurements total information about deformation of the arms is obtained, which can be efficiently used in a control algorithm.

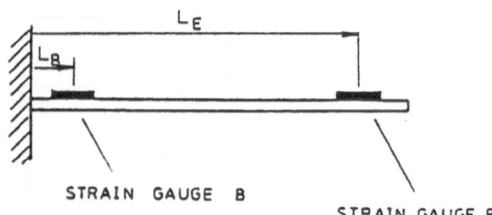

Figure 10:   Measurement of elastic body coordinates

5. Control of the Robotic System

Examination of the linear time-variant differential equation (5) shows, that of course following a prescribed trajectory γ_{00}, γ_{10}, γ_{20} will induce elastic vibrations. This leads to oscillations of the endpoint of the robot around the desired path which are nearly undamped when using a control concept which is based on a rigid body model of the robot, neglecting elastic deformations (Figure 11).

The distance Δr of the end point from the nominal position can be computed by means of the jacobian-matrix F. (Figure 12)

$$\Delta r = F y$$

The intent of any control is $\Delta r = 0$.

In the following a method is developed how a given trajectory can be followed with $\Delta r = 0$ by modifying the control torques. The vector y of generalized coordinates is split up into three parts:

$$y^T = (y_M^T y_S^T y_E^T)$$

y_M : coordinates of the direct-current motors

y_S : coordinates of the rigid body motion

y_E : elastic body coordinates

The differential equation can be written in the form (for the model according to Figure 3):

$$
\begin{vmatrix} M_{11} & M_{12} & M_{13} \\ M_{12}^T & M_{22} & M_{23} \\ M_{13}^T & M_{23}^T & M_{33} \end{vmatrix}
\begin{vmatrix} y_M \\ y_S \\ y_E \end{vmatrix}
+
\begin{vmatrix} 0 & 0 & 0 \\ 0 & P_{22} & P_{23} \\ 0 & P_{32} & P_{33} \end{vmatrix}
\begin{vmatrix} y_M \\ y_S \\ y_E \end{vmatrix}
$$

$$
+
\begin{vmatrix} Q_{11} & Q_{12} & 0 \\ Q_{21} & Q_{22} & Q_{23} \\ 0 & Q_{33} & Q_{33} \end{vmatrix}
\begin{vmatrix} y_M \\ y_S \\ y_E \end{vmatrix}
=
\begin{vmatrix} \Delta M_M \\ 0 \\ 0 \end{vmatrix}
+
\begin{vmatrix} h_M \\ h_S \\ h_E \end{vmatrix}
\tag{6}
$$

In addition the equation

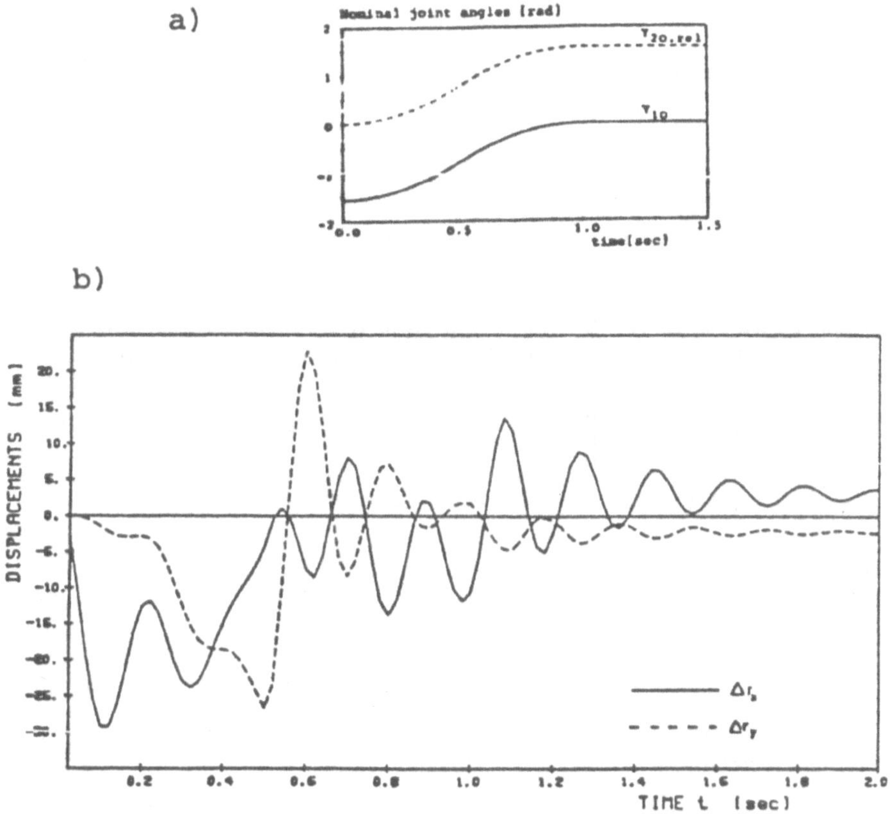

a)

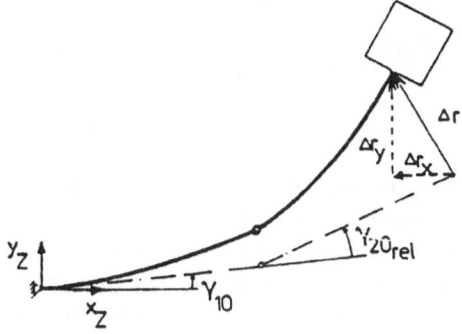

b)

Figure 11: a) Nominal trajectory (joint angels); b) Displacement of the endpoint of the robot from the nominal trajectory

Figure 12: Displacement Δr

$$\Delta r = |0\,F_2\,F_3| \begin{vmatrix} y_M \\ y_S \\ y_E \end{vmatrix} = 0 \tag{7}$$

shall hold.

From the first row of matrix equation (6) the additional control torques ΔM_M can be determined if the coordinate vector y is known:

$$\Delta M_M = M_{11}\ddot{y}_M + M_{12}\ddot{y}_S + M_{13}\ddot{y}_E + Q_{11}y_M + Q_{12}y_S - h_1 \ . \tag{8}$$

The coordinates of y_S can be eliminated from the second and third row of matrix equation (6) using condition (7) yielding a coordinate transformation;

$$y = \begin{vmatrix} y_M \\ y_S \\ y_E \end{vmatrix} = T\hat{y} = T \begin{vmatrix} y_M \\ y_E \end{vmatrix} \tag{9}$$

with

$$T = \begin{vmatrix} E & 0 \\ 0 & -F_2^{-1}F_3 \\ 0 & E \end{vmatrix} \quad E\text{: Unit Matrix} \ .$$

Calculating $\dot{y}$ and $\ddot{y}$ by differentiating equation (9) and inserting into equation (6) yields the reduced and transformed system:

$$\overline{M}\ddot{y}_E + \overline{P}\dot{y}_E + \overline{Q}y_E = \overline{h} \tag{11}$$

where

$$\overline{M} = -M_{23}T_{22} + M_{33}$$
$$\overline{P} = P_{32}T_{22} + P_{33} + M_{23}T_{22}$$
$$\overline{Q} = Q_{32}T_{22} + Q_{33} + P_{32}T_{22} + M_{23}^T T_{22}$$
$$\overline{h} = h_E$$

The motor coordinates y_M do not appear anymore in equation (11) as they can be determined by algebraic equations. The solution for the vector y_E of the elastic body coordinates can be obtained by numerical integration of differential equation (11) and subsequently y_M can be determined from the algebraic equations and y_S from condition (7). Therefore the torques ΔM_M are given by equation (8), and in addition with the torques M_{Mo} determined by the pure rigid body motion, control of the system including the elastic deformations is achieved. However, to guarantee asymptotic stability an additional regulation is necessary.

6. Experimental Results

A rather complex mathematical model was used to analyze the dynamics of a laboratory industrial robot with very light and slender arms. Computational results and measurements were compared to test the capability of the model. In Table 2 measured and computed eigenvalues are shown for two chosen robot configurations.

Table 2:   *Comparison of computed and measured normal frequencies*

	Position 1			Position 2		
	measured	computed	error	measured	computed	error
ω_1	24.6	24.5	-0.4%	29.4	29.4	-
ω_2	75.4	75.8	+0.5%	56.3	57.4	+1.9%
ω_3	436	402	-7.8%	430	386	-10.2%
ω_4	704	577	-18.0%	660	577	-12.6%
ω_5	1380	1520	+10.1%	1340	1490	+11.2%
ω_6	1420	1570	+10.6%	-	1580	-
ω_7	-	2850	-	-	2850	-
ω_8	3550	3560	+0.3%	-	3350	-
ω_9	4950	4490	-9.3%	4960	4490	-9.5%

The two first frequencies are matched with an error less than 2% and the higher modes are at least given approximately with a mismatch of less than 10% (one exception). As higher frequencies do have little influence on the system motion as long as they are not excited by additional external forces the model will be efficient for many applications.

7. Conclusions

Adequate modeling of industrial robots, comprising the influence of elasticity is necessary for analyization and control design of light weight constructions. It is shown, that with only two shapefunctions for each arm good results can be obtained.

Furthermore, a concept was developed to determine control torques for a given trajectory, taking into account the elastic properties of the system, so that following this trajectory will not lead to oscillations of the endpoint of the robot.

References

[1] Hodges, D.H. and Ormiston, R.A., "On the Nonlinear Deformation Geometry of the Euler-Bernoulli Beams", NASA Technical Paper 1566, 1980.

[2] Gebler, B., "Mechanisches Ersatzmodell und Bewegungsgleichungen für einen Industrieroboter mit elastischen Komponenten", *ZAMM*, 65, 1985, 4, T53-55.

[3] Johanni, R., *Automatisches Aufstellen der Bewegungsgleichungen von baumstrukturierten Mehrkörpersystemen mit elastischen Körpern*. Diplomarbeit, Techn. Universität München, Lehrstuhl B für Mechanik, 1985.

STRUCTURAL CONTROL
OF AIRCRAFT

Sabry F. Girgis
Aeronautical Engineering
Al-Fateh University
Tripoli, Libya

1. Introduction

Structural control is needed for aircraft in view of weight and aerodynamic limitations. Some control measures should be tried on the prototype at the design stage such as controlling the wing resonant mode shapes. The aim is to avoid aeroelastic problems. Other modifications were found necessary to be introduced at service life, such as that needed to avoid the resulting acoustic fatigue failure of aircraft components in the vicinity of the jet efflux. Both methods were presented by the Author at the Colloquium (DFVLR) for aeroelasticity, September 1980, Göttingen, West Germany, and the 106th meeting of the Acoustical Society of America at San Diego, California at 11 November 1983, respectively. This paper presents a method of structural control required to increase the load carrying capacity of wings of aircraft used for agricultural and military purposes within engine thrust limitations.

Normally, the design data is kept confidential at aircraft factories. Static tests are made on the wing to find the shear centres distribution along the span. During the tests, the wing is mounted on a test rig in a similar manner to that of the aircraft, usually by three bolts. Warping cannot be prevented totally due to fuselage flexibility. The upper skin is removed at some wing stations along the span. The distribution of webs, stiffners, skin and the number of cells at each station are recorded. The positions of the centre of gravity and shear centre are calculated at each station, along the wing span. It is recommended to transfer the shear centre location to the effective location of the effective inertia load of the external loads. The locus of the shear centres along the span must be a straight line, so that the external load that produces no twist at one station will not produce twist at another wing station. Opening or closing cells at any station of the wing section consisting of multicell structure was found effective in transferring the shear centre location along the chord of the wing section. The theoretical results were in good agreement with the results of the static tests. Structural control of the wing for that purpose was fulfilled. The increased external loads will not create a torque that may increase the twisting angle beyond the angle of attack that makes the wing stall or result in the loss of aircraft controlability. The method adopted in this paper can be applied to bridges that are subjected to wind forces which increase with the twisting angle.

Figure 1a: Static Test to Find Shear Centre Locations

2. Formulation of the Problem

Aircraft wings are complex, thin-walled shell structures, supported by longitudinal stringers and lateral ribs, see Figure 1b. The wing components should resist the bending moment, shear forces and torsional loads. These external loads result from aerodynamic and inertia loads. A typical external loading for one flight loading condition of Cessna - 150, is shown in Figure 2.

Figure 1b: The Cessna - 150 Wing with skin removed

Ribs provide the aerofoil shape of the wings. They resist the aerodynamic pressure loads together with the skins. They increase the column-buckling stress of longitudinal stringers and increase the plate-buckling stress of the skin panels [1,5], see Figure 3a, 3b. Spar webs and skin resist shear and torsional loads, while, spar flanges and stringers carry the bending loads, see Figure 4.

Wing distortion due to its flexibility results in significant changes in the aerodynamic forces that has been calculated for the rigid wing [2], see Figure 5.

Moreover, the wing flexibility results in an increase in wing incidence due to the twisting moment about the centre of twist. The increased wing incidence produces a further increase in lift forces and so on. At speeds below what is called divergence speed (U_D) the torsional moment of the aerodynamic forces about the centre of twist is balanced by the torsional rigidity of the wing. But, the aerodynamic forces increase with the square of the aircraft velocity while, the

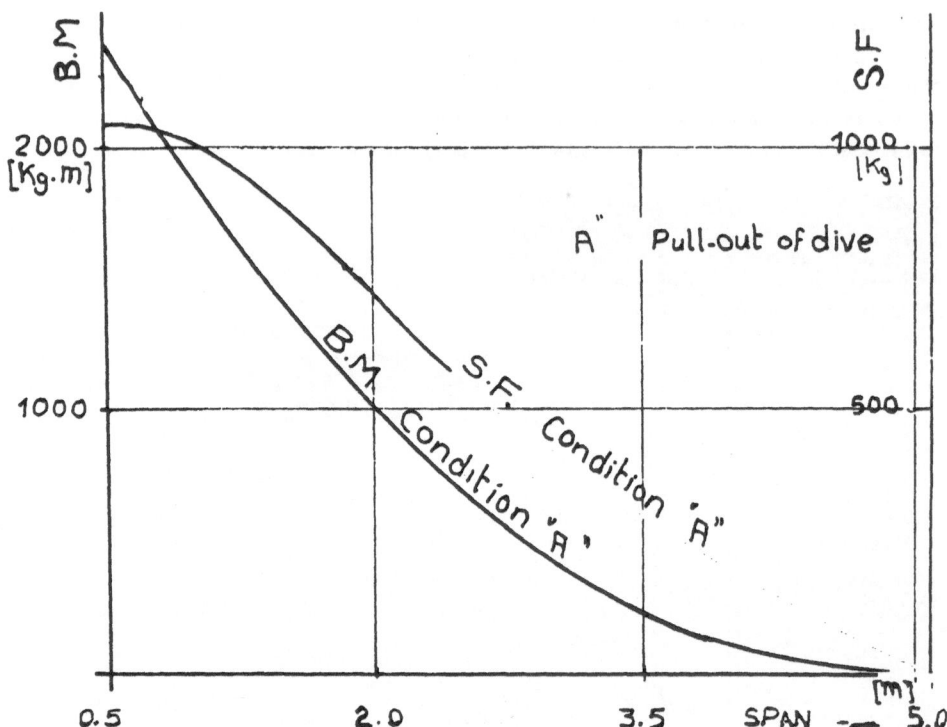

Figure 2: Typical wing external loads

torsional rigidity of the wing (GJ) is constant, practically. Static instability is expected, if the speed exceeds the divergence speed (U_D). The design parameter affecting this instability is the offset distance (e) between the elastic centre (E.S.) and the centre of pressure (C.P.). Bisplinghoff [2] presents this relation for a wing of span 1 and mean chord C:

$$U_D = \frac{\Pi}{2l} \left[GJ \Big/ (cea_o\rho/2) \right]^{1/2} ,$$

where ρ is the air density and a_o is the 2-dimensional liftslope curve.

Also, at flutter speed (U_F) of a wing in which the flexural and torsional modes are coupled, is an important example of instability. The relative position of the centre of gravity (C.G.) location and the elastic centre (E.S.) is responsible for this coupling [3], see Figure 6.

Due to the above mentioned importance of the elastic centre (E.S.) locations along wing span, with respect to centre of gravity and centre of pressure locations, they should be known.

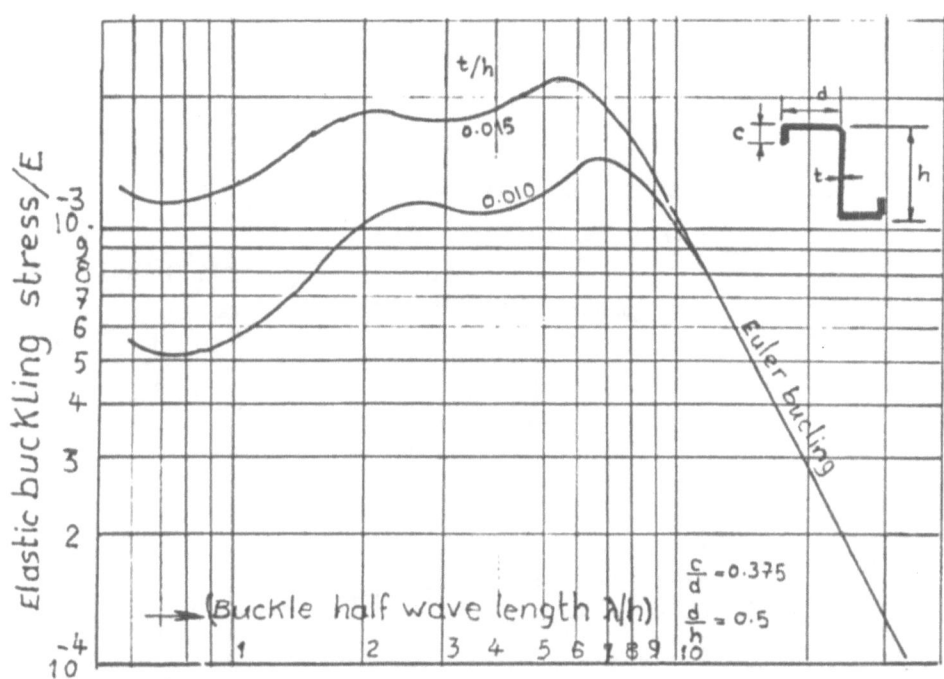

Figure 3a: *The buckling stress of stringers increases with decreasing rib spacing* (λ)*[5]*

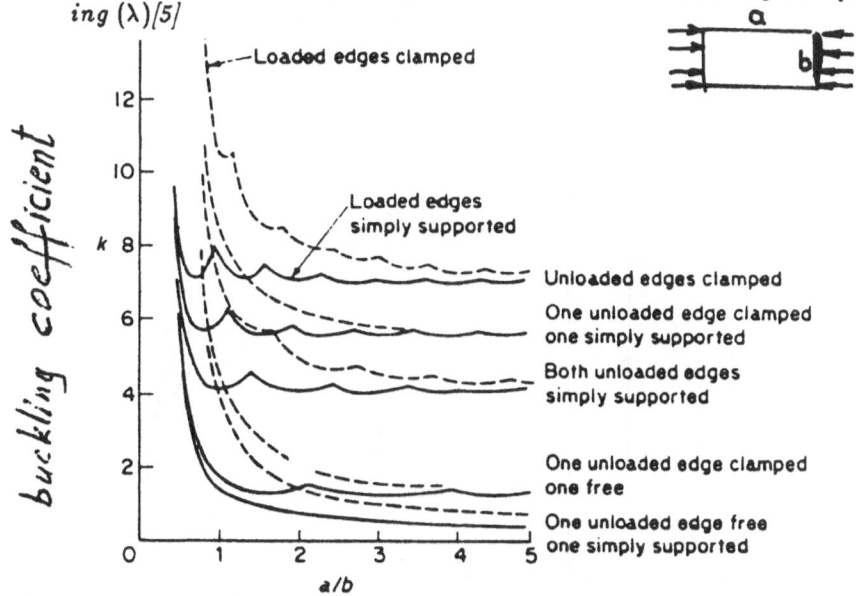

Figure 3b: Plate buckling[1,4]

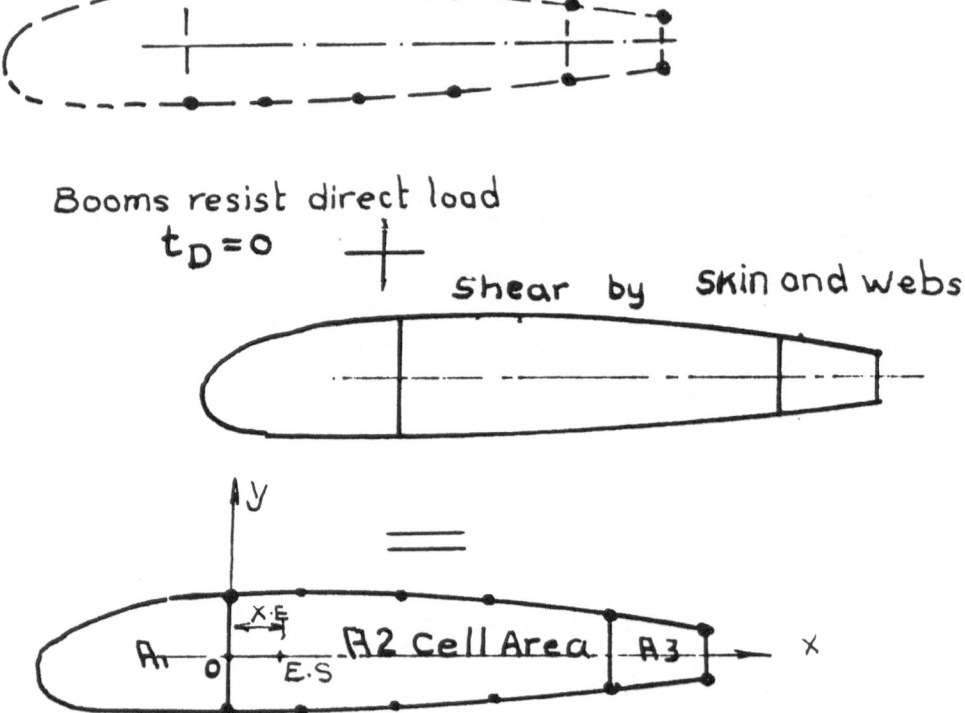

Figure 4: *Wing skin is ineffective to carry direct stress "σ"*

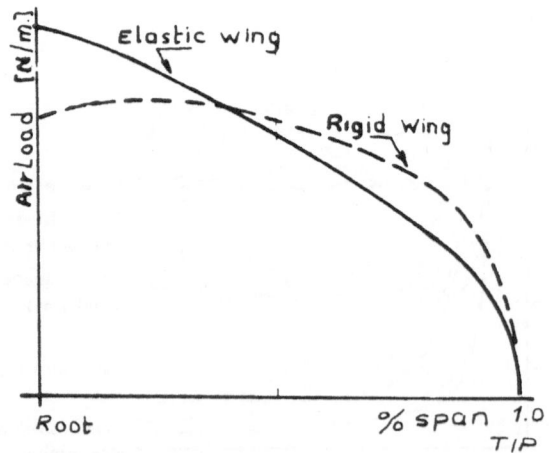

Figure 5: *Airload distribution swept wing*

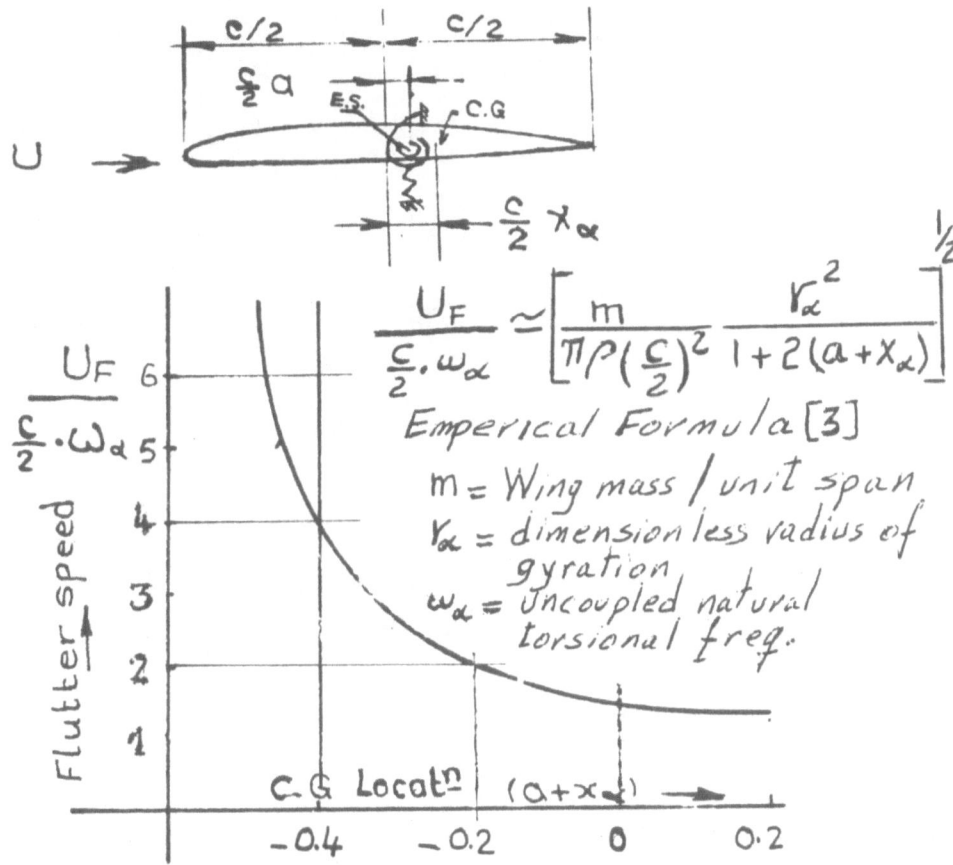

Figure 6: *The dimensionless flutter speed plotted against center of gravity location*

The addition of external stores under the wing should be close to the elastic centre. A slight modification in few cells is needed to transfer the elastic centre along the chord of the wing section, so that the locus of the elastic centres along the span is to be a straight line. This modification insures that the external stores that produced no twist at one station will not produce twist at another wing station.

The modification was simply by opening the wing cells at some stations along the span. This was accomplished by removing part of the web of the rear spar and leaving its flanges to pass across the ribs, see Figure 7.

Theoretical calculations were made for the opened cells to find the elastic centre distributions along the span. It was found in good agreement with the test results on the actual modified wing, see Figure 8.

Figure 7: The wing with opened cell

3. Theoretical Work

An idealized model was needed for this complex wing structure. The condition was, that, the model behaves under the given loading conditions in the same manner as the actual structures. To evaluate the internal geometrical properties of that wing, the upper skin was removed from one crashed Aircraft. The structural details were reported, see Tables 1,2. Accordingly, the models were proposed, see Figure 9.

The idealized model of each wing station was of the 3-celled type (R=3) and the booms (B_r) were symmetric about one axis. To find the elastic centre (E.S.) [1,4], a shear force S_y was assumed passing through (E.S.) of that wing section. No twist (θ) would result. For simplification, the top and bottom sheet covering and the nose skin together with the two vertical webs will be considered ineffective in carrying direct stress (t_D) resulting from bending loads.

Wing stations are selected to be away from the wing root. This condition allowed sections to warp. The same wing fuselage fitting was used which fulfilled the above condition, even to some extent at sections close to the root. The shear flow $[q_s]$ at any point in each cell is unknown. Therefore, to make the shear flow statically determinate, a cut in the skin is made, and a value of shear flow in each cell will be assumed: $[q_{so,1}, q_{so,2}, q_{so,3}]$. The basic shear flow $[q_b]$ for the

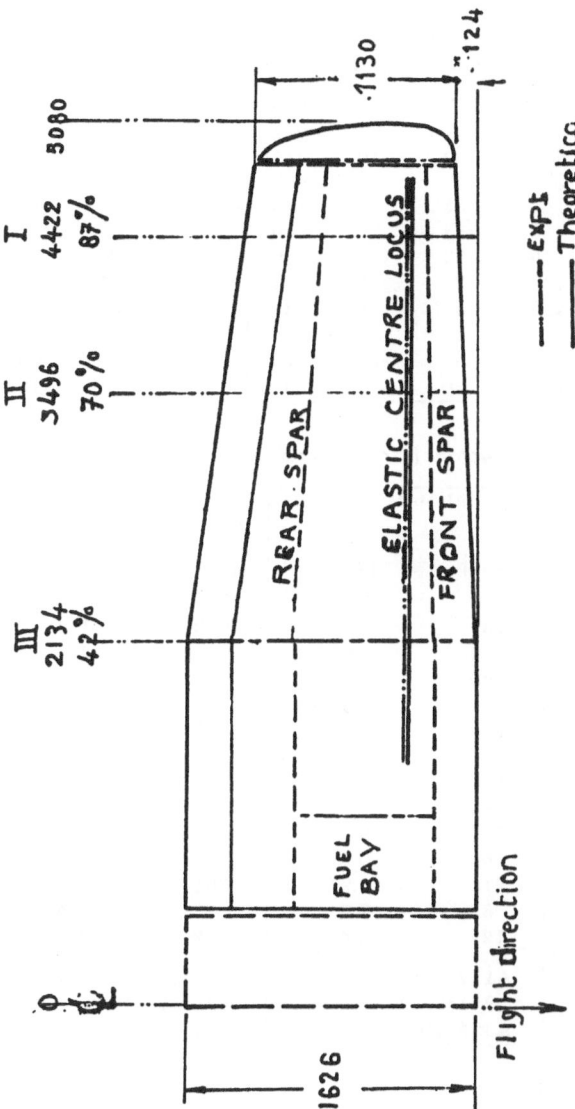

Figure 8: The elastic centres distribution along the Cessna-150 Wing performed experimentally and realized theoretically, for the modified wing.

Table 1: Cessna - 150, wing-station geometric properties

Boom No.	station I, 87%			station II, 70%			station III, 42%		
	Boom Area [mm²]	y [mm]	x [mm]	B_r [mm²]	y [mm]	x [mm]	B_r [mm²]	y [mm]	x [mm]
1	39	42	722	40	43	888	64	39	1105
2	35	45	579	35	63	679	94	69	801
3	35	56	434	35	70	514	35	84	599
4	35	64	279	35	78	335	35	89	396
5	35	68	140	35	80	155	35	97	187
6	122	60	0	353	70	0	358	84	0
7	122	-60	0	353	-78	0	358	-84	0
8	35	-68	140	35	-80	155	35	-97	187
9	35	-64	279	35	-78	335	35	-89	396
10	35	-56	434	35	-70	514	35	-84	599
11	35	-45	579	35	-63	679	94	-69	801
12	39	-42	721	40	-43	888	64	-39	1105

Table 2: Cessna - 150 wing stations geometric properties

Wall	station I, 87%		station II, 70%		station III, 42%	
	length s [mm]	thick t [mm]	s [mm]	t [mm]	s [mm]	t [mm]
1/2	155	0.97	210	0.97	305	1.3
2/3	145	0.97	165	0.97	203	1.3
3/4	155	0.97	180	0.97	203	1.3
4/5	155	0.97	180	0.97	208	1.5
5/6	145	0.97	155	0.97	188	2.0
$6/7_i$	120	3.94	140	0.97	168	5.4
$6/7_o$	370	0.97	420	2.87	528	0.97
7/8	145	0.97	155	0.97	188	2.0
8/9	155	0.97	180	0.97	208	1.5
9/10	155	0.97	180	0.97	203	1.3
10/11	145	0.97	165	0.97	203	1.3
11/12	155	0.97	210	0.97	304	1.3
12/1	84	3.0	85	2.9	76	1.3
2/11	90	3.4	125	3.4	142	3.4
	A1: 15086 mm² A2: 71328 A3: 12395		A1: 11896 A2: 119391 A3: 21109		A1: 29670 A2: 129000 A3: 32250	

opened cells will then be calculated.

Also, the two cells will not rotate ($d\theta/dZ = 0$), due to the fact that (S_y) passes through (E.S.),

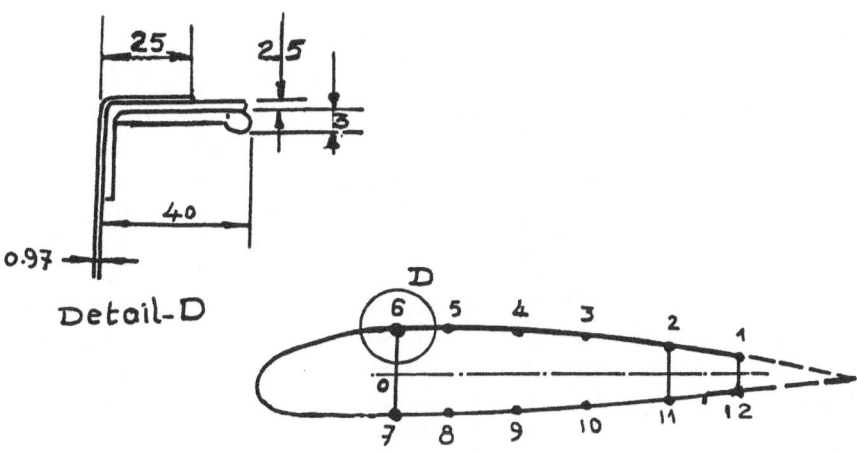

Figure 9: The idealized wing section II

$$q_b = -\frac{\bar{S}_y}{I_{zz}}\left[\int_0^s t_D\, y\; ds + \sum_{r=1}^{n} B_r\, y_r\right] - \frac{\bar{S}_z}{I_{yy}}\left[\int_0^s t_D\, x\; ds + \sum_{r=1}^{n} B_r\, x_r\right].$$

Since the boom areas (B_r) are symmetrical with respect to $x-x$: $I_{zy} = 0$, and

$$\bar{S}_y = \frac{S_y - S_z\, I_{zy}/I_{yy}}{1 - I_{zy}^2/(I_{zz}\, I_{yy})}\;,\qquad \bar{S}_y = S_y\;.$$

Also, the elastic centre lies on $x-x$, hence, no need to assume s_z.

The skins are assumed ineffective in carrying direct load. Therefore $t_D = 0$,

$$q_b = -\frac{S_y}{I_{zz}}\sum_{r=1}^{n} B_r\, y_r\;,$$

$$\left[\frac{d\theta}{dz}\right]_1 = \left[\frac{d\theta}{dz}\right]_2\;,$$

$$\frac{d\theta}{dz} = \frac{1}{2A_R G}\oint_R q_s\frac{ds}{t} = \frac{1}{2A_R G}\oint_R \left(q_b + q_{so,R}\right)\frac{ds}{t}\;,$$

r is the boom No. . See Tables 3,4 for the shear flow distribution (q_s).

The (E.S) location is derived from the moment of the external and internal shear forces about any point (o) in the plane of the cross section that must equal zero.

Table 3: *The shear flow distribution for wing stations before introducing the modification*

wall	station I, 87%	station II, 79%	station III, 42%
1/2	$0.003 \times 10^{-3} \, S_y$	$0.003 \times 10^{-3} \, S_y$	$0.117 \times 10^{-3} \, S_y$
2/3	$0.114 \times 10^{-3} \, S_y$	$0.267 \times 10^{-3} \, S_y$	$0.58 \times 10^{-3} \, S_y$
3/4	$-0.706 \times 10^{-3} \, S_y$	$-0.214 \times 10^{-3} \, S_y$	$-0.200 \times 10^{-3} \, S_y$
4/5	$-1.646 \times 10^{-3} \, S_y$	$-0.744 \times 10^{-3} \, S_y$	$2.024 \times 10^{-3} \, S_y$
5/6	$-2.636 \times 10^{-3} \, S_y$	$-1.294 \times 10^{-3} \, S_y$	$0.64 \times 10^{-3} \, S_y$
$6/7_i$	$-5.726 \times 10^{-3} \, S_y$	$-2.746 \times 10^{-3} \, S_y$	$4.229 \times 10^{-3} \, S_y$
$6/7_o$	$-0.424 \times 10^{-3} \, S_y$	$-3.398 \times 10^{-3} \, S_y$	$-2.409 \times 10^{-3} \, S_y$
7/8	$-2.636 \times 10^{-3} \, S_y$	$-1.294 \times 10^{-3} \, S_y$	$0.64 \times 10^{-3} \, S_y$
8/9	$-1.646 \times 10^{-3} \, S_y$	$-0.744 \times 10^{-3} \, S_y$	$0.202 \times 10^{-3} \, S_y$
9/10	$-0.706 \times 10^{-3} \, S_y$	$-0.214 \times 10^{-3} \, S_y$	$-0.200 \times 10^{-3} \, S_y$
10/11	$0.114 \times 10^{-3} \, S_y$	$0.266 \times 10^{-3} \, S_y$	$0.58 \times 10^{-3} \, S_y$
11/12	$0.003 \times 10^{-3} \, S_y$	$0.033 \times 10^{-3} \, S_y$	$0.117 \times 10^{-3} \, S_y$
12/1	$0.693 \times 10^{-3} \, S_y$	$0.963 \times 10^{-3} \, S_y$	$0.204 \times 10^{-3} \, S_y$
11/2	$0.771 \times 10^{-3} \, S_y$	$0.663 \times 10^{-3} \, S_y$	$-1.52 \times 10^{-3} \, S_y$

Table 4: *The shear flow distribution for the modified win stations I & II*

wall	station I, 87%	station II, 70%
1/2	$1.65 \times 10^{-3} \, S_y$	$1.096 \times 10^{-3} \, S_y$
2/3	$0.987 \times 10^{-3} \, S_y'$	$-0.667 \times 10^{-3} \, S_y$
3/4	$-0.17 \times 10^{-3} \, S_y$	$-0.185 \times 10^{-3} \, S_y$
4/5	$0.76 \times 10^{-3} \, S_y$	$0.344 \times 10^{-3} \, S_y$
5/6	$1.75 \times 10^{-3} \, S_y$	$0.894 \times 10^{-3} \, S_y$
$6/7_i$	$4.49 \times 10^{-3} \, S_y$	$2.897 \times 10^{-3} \, S_y$
$6/7_o$	$-0.36 \times 10^{-3} \, S_y$	$-2.857 \times 10^{-3} \, S_y$
7/8	$1.75 \times 10^{-3} \, S_y$	$0.894 \times 10^{-3} \, S_y$
8/9	$0.76 \times 10^{-3} \, S_y$	$0.344 \times 10^{-3} \, S_y$
9/10	$-0.17 \times 10^{-3} \, S_y$	$-0.185 \times 10^{-3} \, S_y$
10/11	$-0.987 \times 10^{-3} \, S_y$	$-0.667 \times 10^{-3} \, S_y$
11/12	$1.65 \times 10^{-3} \, S_y$	$1.096 \times 10^{-3} \, S_y$
12/1	$0.23 \times 10^{-3} \, S_y$	$1.43 \times 10^{-3} \, S_y$

$$S_y x_{E.S} = \sum_{R=1}^{N} \oint_R q_b \, p \, ds + \sum_{R=1}^{N} 2 A_R q_{so,R} \; ,$$

where R: Cell No., A: Cell Area. Figure 10 shows the elastic centres distribution along the wing span for the original wing span for the original wing and the modified wing derived theoretically.

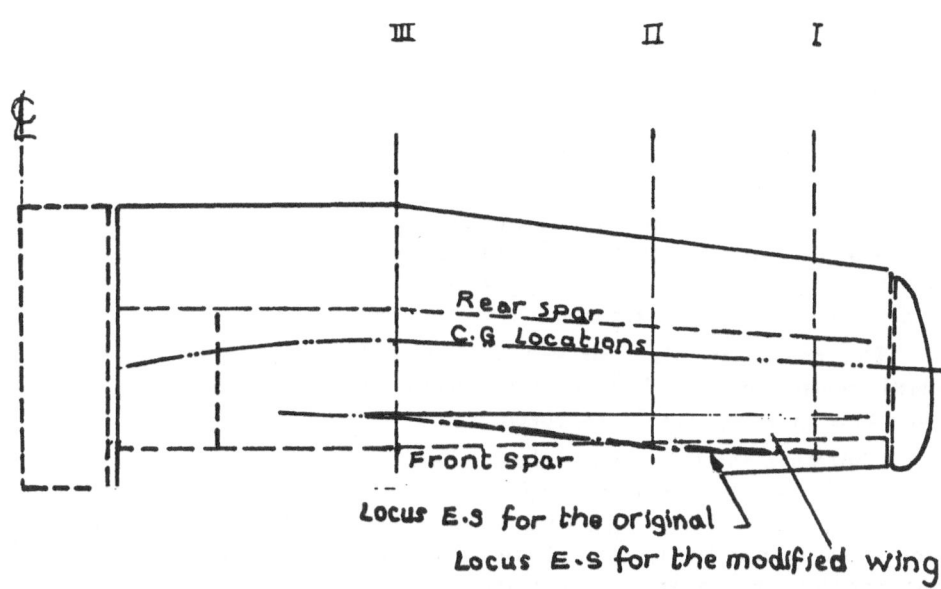

Figure 10: The improved wing after modification

4. Experimental Work

Static tests were made on a Cessna-150 wing mounted on the wall of the laboratory. A similar wing-fuselage fitting was used as that used in the original aircraft. A torque was applied at each predetermined wing station. The nearest
wing-section was 42% of the span, to be away from the fixation points, and to
allow for warping. This condition was the basic assumption upon which the
theoretical derivation was made. A torque of 42.5 kg.m was applied gradually on
each wing station through a system of wooden blocks. A dial indicator showed
the points of zero vertical deflections along the chrod of each wing station. The
locus of these points is the elastic centre axis.

The skin of the wing was removed and the lightening hole was enlarged at
the web of the rear spar, in the places of large deviation of (E.S). The flanges
were left to pass through the ribs. The skin was attached again, and the previous test procedure was done again to find the E.S distribution on the modified
wing, see Figure 1a.

2. Conclusion and Discussion

The experimental results were compared with the theoretical work. Great coincidence was found. This coincidence was expected due to the realization of the boundary conditions for the theoretical calculations. Using the same wing-fuselage fitting of a crashed Aircraft provided close similarity. Selecting the nearest wing section to be 42% of the span overcame the effect of the end-constraint, thus, the farthest wing sections are allowed to warp.

Also, the skin and web thicknesses are of an order of 1.5 mm., which is much less than the minimum cross-section dimension that is of order 100 mm... . This fact supported the assumption of $(t_D = 0)$, which means that skin and web were ineffective in carrying direct load.

This paper presents a method to control the structure by transferring the elastic centre more closely to both the centre of gravity and the centre of pressure, thus improving the flutter speed.

Also, when the locus of the elastic centres along the wing span is a straight line, the increased external loads will not introduce extra twisting moments about the elastic centre. The external loads can be located at properly selected suspension points around the locus of elastic centres in tandem arrangement by placing the rear stores in the wake of the front stores.

Generally, aircraft design data are kept confidential in aircraft factories. Static tests followed by skin removal is required to indentify the geometric properties of the wing which are essential for deciding the modifications required to be introduced. Upon this information, the theoretical calculation can be made for the original wing. The effect of opening or closing some cells will be evaluated theoretically. The final result will be applied to the real wing. Attach the skin and conduct the static test to justify the result.

The method adopted in this paper by closing or opening some cells to transfer the elastic centre location with respect to the centre of gravity and centre of pressure can be applid to tall buildings and large bridges subjected to high wind speed.

Acknowledgement

The author wishes to thank Professor Roger W. fife, the Field Supervisor of Cessna Aircraft Company - Pawnee Division, Wichita, Kansas for his assistance. Thanks are due to Professor Megson of the University of Leeds, U.K. for his sincere help and encouragement to run the aircraft structures course. I wish to thank Al-Fateh University, Faculty of Engineering, where this work was performed. Thanks are due also to Professor Dr. H. Försching of DFVLR, W. germany, Aeroelastic Institute, for his never ceasing help.

References

[1] Bruhn, E. F., "Analysis and Design of Flight Vehicle Structures," June 1973.

[2] Bisplinghoff, Ashley, Halfman, "Aeroelasticity" 1957.

[3] Theordorsen, T. and Garrick, I. E., "Mechanism of Flutter, a Theoretical and Experimenal Investigation of the flutter problem." N.A.C.A. Report 685, 1940.

[4] Megson, T. H. G., "Aircraft structures for Engineering students," 1972.

[5] ESDU 77030, "Buckling of struts lipped and unlipped Z-sections," 1977.

THE EFFECT OF SMALL STRUCTURAL CHANGES ON OPTIMIZED ACTIVE CONTROL SYSTEMS

Raphael T. Haftka, Zoran N. Martinovic,
William L. Hallauer Jr., George Schamel
Department of Aerospace and Ocean Engineering
Virginia Polytechnic Institute and State University
Blacksburg, Virginia, U.S.A.

1. Introduction

The structural model used for designing an active control system is rarely accurate. Even if the model is accurate when the structure is originally built, small changes during the lifetime of the structure can destroy that accuracy. Because of this problem one major concern in the design of such a system is to minimize its sensitivity to variations in the structural model (e.g. [1]). Design for insensitivity will usually compromise the performance of the control system. Therefore, it is important to find the structural parameters that the control system is most sensitive to. Once these parameters are identified the designer of the control system has three options at his disposal. He can try to obtain more accurate values of the most troublesome parameters, he can redesign the control system to reduce the sensitivity or he can seek minor changes in the structure to reduce the sensitivity.

A procedure for calculating the sensitivity of an optimized control system to small changes in the structure was developed in [2]. That procedure was used to identify minor changes in the structure that can substantially improve the performance of the control system. In the present work the same procedure is used to identify small changes in the structural model that have a large effect on the design of the control system. The procedure is applied to a laboratory structure

250

consisting of a cruciform beam suspended by cables. Two types of structural changes are investigated. The first is a change in the thickness of the beam that may represent modelling errors. The second is the addition of a small concentrated mass which may represent the effect of a small appendage that may be added to a structure during its lifetime. Some of the analytical predictions are tested by a laboratory experiment.

2. Analysis, Optimization and Sensitivity

The structure-control system equations of motion for a structure with n_n degrees of freedom (DOF) and the rate feedback control produced by pairs of colocated velocity sensors and actuators are

$$[M]\{\ddot{u}\}+[C]\{\dot{u}\}+[K]\{u\}=\{0\} . \tag{1}$$

Neglecting the inherent damping, the damping matrix $[C]$ has nonzero entries equal to feedback gains c_i $i=1,2,..,n_c$, at the diagonal positions corresponding to the controlled DOFs.

The equations of motion are reduced to a system of $2n_n$ first order equations, whose solution may be written as

$$\{q(t)\}=e^{\mu t}\{q_o\} \tag{2}$$

where

$$\{q(t)\}^T=[\{\dot{u}\}^T\{u\}^T] \tag{3}$$

The stability of the system is determined by the n_n possible pairs of complex conjugate eigenvalues μ_r, $r=1,2,...,n_n$

$$\mu_r=\sigma_r\pm j\omega_r , \quad j=(-1)^{1/2} . \tag{4}$$

In particular, in the present investigation the modal damping ratio ς_r is used as a measure of stability

$$\varsigma_r=-\sigma_r/(\sigma_r^2+\omega_r^2)^{1/2} \tag{5}$$

Typically, the proposed procedure consists of two steps. First, the control system is designed for a given structure by solving an optimization problem of the following type:

find $\{x\}$

to minimize a performance index $f(\{x\})$

subject to some constraints

$g_j(\{x\})\geq0 , \quad j=1,2,...,n_g.$

Here, $\{x\}$ is a vector of control system design variables, which in the present study are the gains c_i of the actuators. The performance index is defined here as the sum of all control gains. The $g_j(\{x\})$ represent lower limits on the modal damping ratios ς_r of the system of the form,

$$g_j(\{x\}) = \varsigma_r - \varsigma_{rL} \tag{6}$$

and side constraints specifying upper and lower limits on the design variables. The optimization problem was solved by the NEWSUMT optimization program [3] which employs an extended interior penalty function formulation.

Second, the sensitivity of the optimized control system to changes in structural parameters is calculated in order to estimate whether such changes would have significant effects on control system performance or not. A vector of structural parametrs $\{p\}$ is selected which represents possible structural changes. These parameters are kept fixed during the control system optimization. It can be shown [4] that the change in optimized performance index $f_{opt} = f(\{x\}_{opt})$ due to a change in parameter p_k is

$$df_{opt}/dp_k = -\sum_{j=1}^{n_a} \lambda_j(\partial g_j/\partial p_k) \tag{7}$$

where λ_j are Lagrange multipliers obtained at the optimum design of the control system. The derivatives of the constraints with respect to the structural parameters (typically, derivatives of damping ratios) are calculated analytically (see [5]).

It is also necessary to estimate the changes in the optimum design. These changes may be estimated from the sensitivity of optimized control system design variables $\{x\}$ with respect to a design parameter p_k. For the case when the number of active constraints is equal to the number of design variables [4]

$$\partial \{x\}_{opt}/\partial p_k = -([N]^T)^{-1}\partial \{g_a\}/\partial p_k \tag{8}$$

where the matrix $[N]$ has components $n_{ij} = \partial g_{aj}/\partial x_i$. The vector $\{g_a\}$ consists of constraints g_{aj} which are active at the optimum design.

3. Cruciform Beam Model

The procedure described in the previous section has been applied to a small laboratory structure. The structure consists of a vertical beam and an attached horizontal crosspiece, with the vertical beam suspended by cables in tension at its top and bottom, Figure 1. The crosspiece was designed so that the structure would have third and fourth vibration modes with relatively close natural frequencies. The vertical beam is a uniform steel beam 80 inches long with a rectangular cross section $2 \times 1/8$ inches. The crosspiece is an aluminum beam 32 inches long with a rectangular cross section $2 \times 1/8$ inches. Small masses consisting of two ceramic magnets are located at both ends of the crosspiece, which is secured to the vertical beam by a clamp. The vertical beam is attached to floor and ceiling by four 0.09 inch diameter steel cables, as shown on Figure 1.

Eight 10-inch beam finite elements with translations and rotations as structural DOFs are used to model the vertical beam. The crosspiece is symmetrical relative to the vertical beam, and only symmetrical out-of-plane motion of the crosspiece is represented in modelling. The flexible portion of the crosspiece is

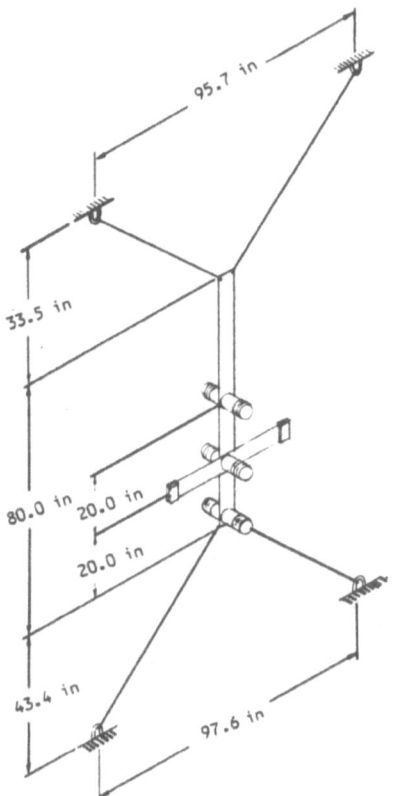

Figure 1: Drawing of the laboratory structure

modelled as a two-DOF spring-mass system, as shown in Figure 2. A string-in-tension finite element represents each cable. The model includes geometric stiffness matrices accounting for tension in the beam elements. Small lumped masses representing the control system coils, cable clamps, and the crosspiece clamp are added to the model. Complete details of the modelling are given in [6].

The six lowest modes of the cruciform beam are of the primary interest. Their calculated and measured natural frequencies and calculated mode shapes are shown in Figure 2.

Inherent damping is not included in the analysis because it is small in comparison with the active damping imposed by the controller. Active damping is effected by rate feedback control involving three force actuators, each colocated with a velocity sensor. The instantaneous control force at each actuator location is directly proportional, but opposite in sign, to the instantaneous velocity. This is equivalent to the attachment of a viscous dashpot to the structural DOF of the sensor, with the ratio of controlling force to sensed velocity being the viscous damping coefficient c. Therefore, the sum of the c's is defined to be the measure of control strength. The three control sensor-actuator pairs are located at grid

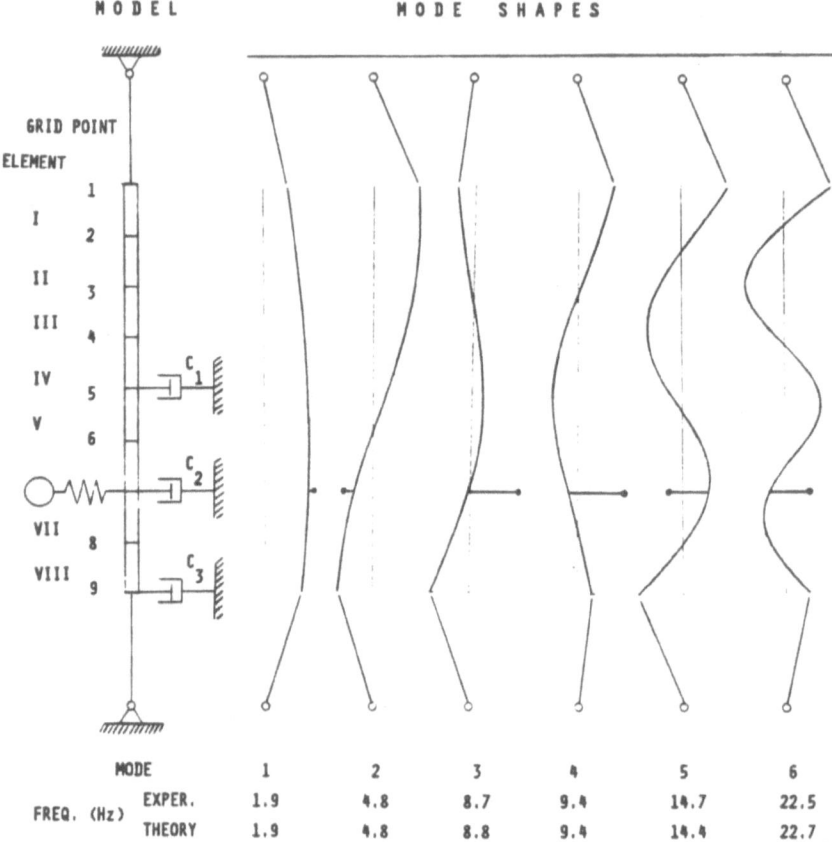

Figure 2: Finite element model and mode shapes

points 5, 7, and 9 as shown in Figure 2.

4. Analytical Results and Discussion

The control system was designed for the cruciform beam so as to minimze the sum of the c_i supplied by the three actuators. The design variables were the individual c_i. The requirements imposed on the control system were $0.9 \geq \varsigma_r \geq 0.03$ for the first six modes, $r = 1,...,6$. The control optimization procedure produced the baseline design shown in Table 1.

The fourth, fifth and sixth damping ratios were at the lower limit of 0.03 while the first three damping ratios were above this value.

Next, structural parameters were selected for sensitivity calculations. First, the thickness of the beam was considered. Local variation in the thickness may reflect modelling errors or changes as the structure ages. The sensitivity derivatives of the performance index with respect to the thickness t_k of each beam

Table 1: Controller gains c_i (lb-sec/in) and performance index $(c_1+c_2+c_3)$ for
baseline and added-mass designs

Grain	Baseline Design	Added Mass Design (0.86%)	
		from Eq. (8)	by reoptimization
c_1	0.06289	0.05735	0.05779
c_2	0.01165	0.04992	0.05185
c_3	0.02870	0.01663	0.01595
Σc_i	0.10324	0.12390	0.12559

finite element were calculated from Equation (7) as shown in Table 2. It is seen
from Table 2 that the thickness of the top element had the largest effect on the
objective function. However, even that derivative is small in that one percent
change in the thickness of the first element produces only about one percent
change in the performance index.

Table 2: Sensitivity derivatives of the sum of control gains with respect to
the beam element thicknesses for the baseline design

Element k	$d\Sigma c_i/dt_k$ (lb-sec/in)
I	0.869
II	0.059
III	-0.065
IV	0.218
V	0.098
VI	0.404
VII	0.393
VIII	0.009

Next, the effect of adding small concentrated masses m_k to the nine grid
points of the finite element model was investigated. Such an added mass may
simulate a small appendage added to the original design of the structure. The
sensitivity of the baseline performance index with respect to each added mass
was calculated from Equation (7) and is given in Table 3. It is seen from Table 3
that the largest derivative is for the first grid point. Adding a small mass at that
point of 0.0739 lb or 0.86% of the mass of the structure results in a drastic
change in the optimum control and about 20% deterioration in the performance
index as is shown in Table 1. The results in Table 1 also show good agreement
between the prediction of the sensitivity analysis, Equation (8), and the exact

results obtained by reoptimizing the control system for the added mass structure. The results indicate that small changes in the structure can indeed have large effects on the design of the control system and its performance.

Table 3:   Sensitivity derivatives of the sum of control gains with respect to added masses for the baseline design

Grid Point k	$d\Sigma c_i/dm_k$ (sec^{-1})
1	107.7
2	33.5
3	5.6
4	26.3
5	60.6
6	51.3
7	-0.5
8	-18.1
9	12.8

5. Experimental Apparatus and Procedure

An experiment was conducted to check the predicted deleterious effect of a small mass located at the first grid point. The large effect of the mass on the control system design was found to be due to a large reduction of the damping ratio of the fourth mode for the baseline control system. This necessitated large changes in the control system to restore the damping ratio to its prescribed level. Therefore, the experiment was intended to check whether such reduction occurs in reality.

The basic experimental apparatus and procedures are described in detail in references [2], [5] and [6]. A summary of the apparatus used and the procedure relevant to this paper is provided here. Figure 3 is a photograph of the lower portion of the structure showing the three sensor-actuator pairs with the supporting framework.

For the added mass design, the concentrated mass consisted of a small ceramic magnet and other small pieces of magnetic metal so that the collective weight was 1.31% of the total structural weight. The magnet and other metallic pieces were held firmly to the steel beam by the magnetic field. The added mass was placed on one side of the beam at grid point 1.

Three colocated sensor-actuator pairs along with an analog feedback gain circuit and a controlled-current power amplifier for each sensor-actuator pair made up the control system. Noncontacting velocity sensors produced voltage proportional to the velocity, and noncontacting force actuators generated force proportional to the control feedback signal. Each device consisted of a conducting coil attached to the beam and a magnetic field structure attached to the

Figure 3: Photograph of the lower portion of the structure

supporting framework (Figure 4). A data acquisition and analysis system was used to generate an excitation signal, acquire all the response and excitation signals, and perform all data analysis. The excitation signal was added to the control feedback signal for grid point 9 so that one actuator coil served the dual purpose of control actuator and exciter. The relationship of the control system parameters to the theoretical viscous damping constant is developed in reference [5].

Translation-to-force frequency response functions (FRFs) were measured at four locations on the vertical beam. The translation sensors used were noncontacting inductive type proximity probes. Random excitation was used. To achieve a good signal-to-noise ratio, the general excitation level was set as high as possible consistent with maintaining linear behavior of the proximity probes, velocity sensors, and force actuators. Fast Fourier Transforms of the response and excitation signals were calculated, and the former were divided by the latter to

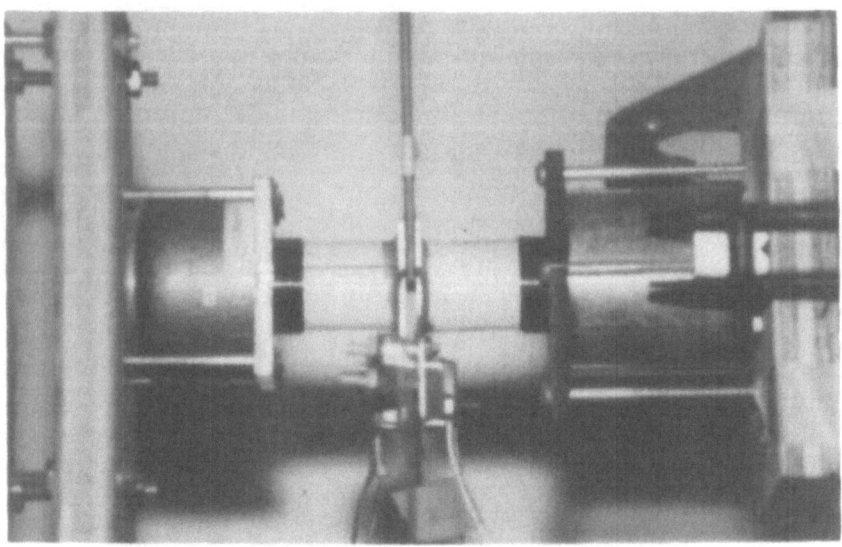

Figure 4: Closeup photograph of a sensor (at right) and actuator (at left) at grid point 9

produce FRFs. The frequency resolution was 0.0781 Hz. The repeatability was verified by performing each experiment three times and comparing the results.

6. Experimental Results and Comparison with Theory

A representative FRF is plotted in Figure 5. The solid curve is experimentally measured data for the baseline design and the dashed curve is a curve fit to the experimental data. The curve fitting was based on a six-mode theoretical model where the frequencies and damping ratios were calculated for each of the four FRFs and global averages were calculated. The curve fits are clearly very good, with no significant deviations from the experimental data.

Two cases were tested, both having the baseline control system (Table 1). One was the original structure and the other, the same structure with the additional mass at the location corresponding the the first grid point. The experimental modal damping factors for the two cases tested are given in Table 4 along with the corresponding theoretical predictions. There are some substantial differences between predictions and measurements, but in general the agreement is good. In particular it is clear that the additional mass severely reduces the damping ratio for the fourth mode as predicted by the analysis. These results validate the theoretical prediction of the large detrimental effect of a small mass on the performance of the control system.

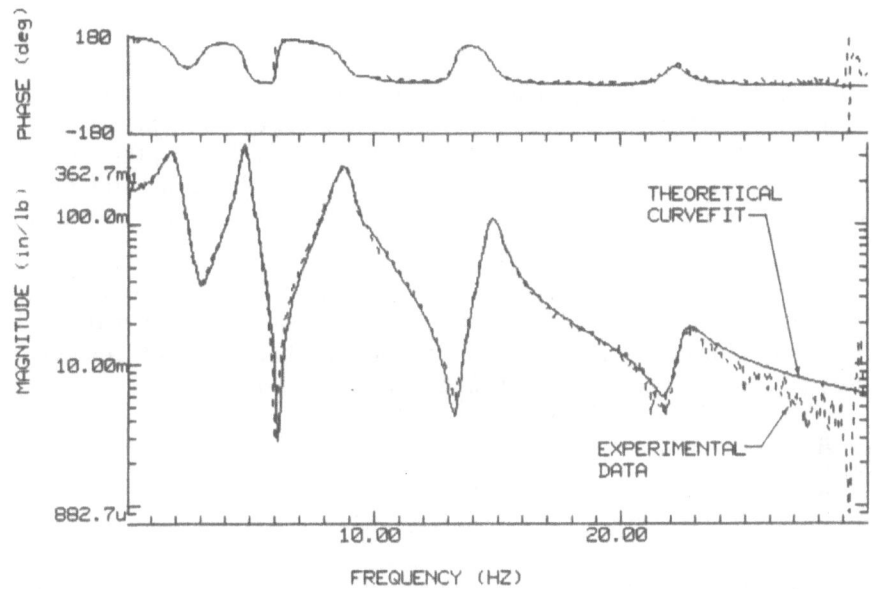

Figure 5: Translation-to-force FRF of response 2 inches above grid point 9 due
to excitation at grid point 9, for the baseline control system

Table 4 :   Theoretical and experimental modal damping ratios

Mode	Baseline Design		Added Mass Case (1.31%)	
	Theory	Experiment	Theory	Experiment
1	0.219	0.200	0.218	0.187
2	0.046	0.040	0.043	0.038
3	0.071	0.050	0.085	0.059
4	0.030	0.036	0.017	0.020
5	0.030	0.024	0.029	0.021
6	0.030	0.029	0.032	0.031

7. Concluding Remarks

A procedure for assessing the sensitivity of optimized control systems to errors or change in the structural model was described. The procedure was applied to a cruciform beam supported by cables and controlled by several rate-feedback colocated sensor-actuator pairs. It was found that the control system was very sensitive to the addition of the mass at the top of the beam. In fact, the addition of a concentrated mass equal in magnitude to 0.86% of the structural mass completely changed the design of the control system and degraded its performance by over 20%.

The drastic effect of the additional mass was traced to its effect on one of the modal damping ratios. This effect was verified by an experiment in which a 1.31% mass addition without any corresponding change in the control system resulted in a 44% reduction of that modal damping ratio.

Acknowledgement

The research reported in this paper was supported by NASA Grant NAG-1-224.

References

[1] Skelton, R.E. and Wagie, D.A., "Minimal Root Sensitivity in Linear Systems", *AIAA J. Guidance & Control*, Vol. 7, No. 5, 1983, pp. 570-574.

[2] Haftka, R.T., Martinovic, Z.N., Hallauer, W.L. Jr. and Schamel, G., "Sensitivity of Optimized Control Systems to Minor Structural Modification", AIAA Paper No. 85-0801-CP, presented at the AIAA/ASME/ASCE/AHS 26th Structures, Structural Dynamics and Materials Conference, Orlando, Florida, April, 1985.

[3] Miura, H. and Schmit, L.A., "NEWSUMT — A Fortran Program for Inequality Constrained Function Minimization — Users Guide", NASA Contractor Report 159070, June, 1979.

[4] Haftka, R.T. and Kamat, M.P., *Elements of Structural Optimization*, Martinus Nijhoff, The Netherlands, 1985, pp. 181-183.

[5] Haftka, R.T., Martinovic, Z.N. and Hallauer, W.L. Jr., "Enhanced Vibration Controlability by Minor Structural Modification", *AIAA Journal*, Vol. 23, No. 8, 1985, pp. 1260-1266.

[6] Schamel, G.C., "Active Damping of a Structure with Low Frequency and Closely Spaced Modes: Experiments and Theory", M.Sc. Thesis, Virginia Polytechnic Institute and State University, March, 1985.

ON A NEW CONCEPT OF
ACTIVE VIBRATION DAMPING OF
ELASTIC STRUCTURES

P. Hagedorn
Institut für Mechanik
TH Darmstadt, Germany

1. Introduction

In this paper we deal with flexible, continuous systems, which for the sake of simplicity are assumed to be undamped; this does not however represent a limiting restraint. The active vibration damping of such distributed parameter systems, i.e., the attenuation of vibration by means of active control devices, was first studied in the last decade mainly with a view toward potential applications in aeronautics and astronautics. A typical application in aeronautical engineering is the case of active flutter control.

Structures in space are of course extremely flexible, since they are designed for minimum mass; and some of them would even collapse under their own weight on the earth's surface. Their costs being so high anyhow, the inclusion here of sophisticated active damping devices seems quite practicable. Also in the more traditional branches of engineering, such as mechanical and civil engineering, the realization of active damping control recently is becoming more likely.

In order to illustrate the main concepts we refer to an extremely simple structure: the prestressed string of Figure 1a. If the tensile force in the string is T, its mass per unit length is ρA and, if the string is subjected to the distributed transverse force $q(x,t)$, then its transverse vibrations are described by the partial differential equation

$$\rho A\ddot{w}(x,t) = Tw''(x,t) + q(x,t) , \tag{1}$$

where the dots indicate differentiation with respect to the time t and the prime differentiation with respect to x. In addition to the differential equation (1) the boundary conditions

$$w(0,t)=0 \quad , \quad w(l,t)=0 \tag{2}$$

are prescribed. With $q(x,t) \equiv 0$ the string oscillates freely, its motion being given by

$$w(x,t) = \sum_{i=1}^{\infty} A_i \sin\frac{i\pi x}{l}\cos(\omega_i t + \alpha_i) \tag{3}$$

with the eigenfrequencies $\omega_i = i\pi/l\sqrt{T/(\rho A)}$, $i=1,2,...$, and the constant amplitude and phase coefficients A_i and α_i, $i=1,2,...$, respectively. Control forces act upon the string in such a way as to attenuate the vibrations or to eliminate them completely, even in finite time if possible. The control force may be distributed in x as shown in Figure 1a or may consist of a number of concentrated forces $f_1(t), f_2(t),...,f_k(t)$ as in Figure 1b.

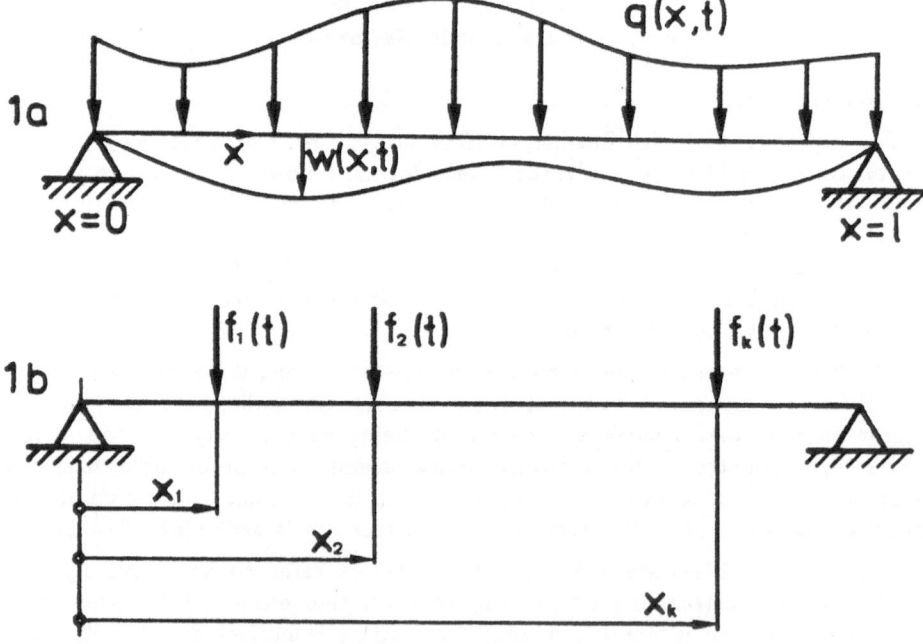

Figure 1: *String with control forces a) distributed control forces; b) concentrated (localized) control forces*

One of the first considerations is then whether it is at all possible to eliminate the vibrations completely by choosing appropriate control forces, i.e., whether the system is "completely controllable" [1]. This is of course the case with the distributed control force $q(x,t)$ acting upon all points of the string according to Figure 1a. But even if $q(x,t)$ is assumed to vanish identically except for values of x belonging to some arbitrarily small interval $[a-\epsilon,a+\epsilon]$, $0<a<l$, $\epsilon>0$, the system is still completely controllable.

Distributed control forces $q(x,t)$, as in Figure 1a, are certainly of theoretical interest and, while they may lead to fascinating mathematical problems, they do not however seem to be of great practical importance in vibration problems, since they are almost impossible to produce in a real structure. Concentrated control forces, as in Figure 1b, are more realistic. It turns out that even a single concentrated control force $f_1(t)$ acting at x_1 gives complete controllability of the distributed system, provided x_1/l is an irrational number! This is, of course, again not a very practical form of controllability, since for modes of sufficiently high order the point $x=x_1$ will be arbitrarily close to a vibration node, and the corresponding mode becomes very difficult to control. Due to this reason, it may be more convenient to use several localized control forces acting at distinct points, even if mathematically complete controllability would seem to be ensured with a conveniently located single control force. It should be observed that the question of complete controllability is intimately related to the question of the pervasiveness of damping. In fact, a single dashpot attached to the string at $x=x_1$, damps out all vibrations if x_1/l is irrational, although it will not work well for some of the high modes. The relation between controllability and pervasiveness of damping is discussed for discrete systems in [2].

Besides this classification into distributed and localized control forces another division is important: control forces may be external or internal forces with respect to the system under consideration. Internal control forces act between different parts of one and the same system, all the other control forces are termed external; in the example of Figure 1 all the control forces are external. In space structures, external forces have to be produced by means of reaction jets usually, which require fuel, i.e., additional mass to be carried into orbit. Internal forces on the other hand can be produced in a different manner, e.g., by means of electric motors, requiring energy only, which may be obtained from solar radiation, with no extra mass. This is of course of little consequence in other engineering applications. Also, in structures floating freely in space the internal control forces will not change the net momentum and moment of momentum, which may be advantageous, since in this manner the problem of station keeping is decoupled from the vibration control problem.

If a structure is completely controllable by means of appropriate control forces, then active vibration damping can be achieved in many different ways, i.e., control laws are far from being uniquely determined. The techniques developed for discrete control problems are then applied to formulate problems which yield good engineering solutions, the main approaches being pole placement and optimal control. In the latter case the arbitrariness often lays in the choice of a cost function. The optimal control problems usually are either minimum time problems — if there are bounds on the control forces — or minimum cost with respect to some quadratic cost function, if the control forces are considered as unbounded.

For discrete systems, powerful tools have been fully developed for the solution of linear-quadratic optimal control problems and therefore modal analysis or some other form of discretization is generally used in designing the optimal control of distributed systems. For the example of Figure 1b, described by

$$\rho A\ddot{w}(x,t)=Tw''(x,t)+\sum_{s=1}^{k}\delta(x-x_s)f_s(t) \ , \tag{4}$$

modal analysis consists of writing $w(x,t)$ as

$$w(x,t)=\sum_{i=1}^{m}p_i(t)\sin\frac{i\pi x}{l} \ . \tag{5}$$

Using the orthogonality conditions this leads to the system of ordinary differential equations

$$\ddot{p}_i+\omega_i^2 p_i=\frac{2}{\rho Al}\sum_{s=1}^{k}\sin\frac{\pi i x_s}{l}f_s(t) \ , \quad i=1,2,...,m \ . \tag{6}$$

The number m of modes taken into account in modelling the string vibrations is of course finite for all practial purposes. In designing the control of the system, the conceptually simplest approach is that of controlling the m modelled modes independently, as has been extensively discussed by Meirovitch, et al. [3,4]. In this case, the number k of actuators has to be equal to m and (6) can then be written as

$$\ddot{p}_i+\omega_i^2 p_i=u_i(t) \ , \tag{7}$$

provided

$$\frac{2}{\rho Al}\sum_{s=1}^{m}\sin\frac{\pi i x_s}{l}f_s(t)=u_i(t) \ , \quad i=1,2,...,m \tag{8}$$

has a unique solution in $f_1,f_2,...,f_m$, which is the case if

$$\det(c_{is})\neq 0 \ , \quad c_{is}:=\sin\frac{\pi i x_s}{l} \ . \tag{9}$$

The modal control functions u_i, $i=1,2,...,m$ can then be taken as independent and the control forces in Figure 1b are given by relations of the type

$$f_s(t) = \sum_{i=1}^{m} b_{si} u_i(t) \; . \tag{10}$$

If in the optimal control problem the cost function is such that the different modes are not coupled, then a number of independent and extremely simple problems have to be solved, making the independent modal control very appealing from a formal point of view.

An important disadvantage of independent modal control is that the number of actuators must be equal to the number of modes to be controlled, and thus will not be feasable in many practical applications. As mentioned before, the taut string is completely controllable with a single force only, and in this case (6) reduces to

$$\ddot{p}_i + \omega_i^2 p_i = \frac{2}{\rho A l} \sin \frac{\pi i x_1}{l} f_1(t) \; , \quad i = 1,2,...,m \; . \tag{11}$$

The control force $f_1(t)$ however now couples all the modes, so that one has less insight into the control problem and the design of the optimal control may become more laborious.

In any case, in discretizing a continuous structure with modal analysis or via some other method, the number of degrees of freedom taken into account in the first stage of the design of an active vibration control is usually maintained small in order to keep the mathematical problem simple. The number of modes used in designing the control is almost always smaller than the number of modes which play a role in the technical vibration problem. On the other hand, the higher, nonmodelled modes are in general affected by the controls designed for the reduced order system and this fact is known as control spill-over (see [5,6]). Spill-over may lead to instabilities, since a control designed to damp out vibrations in the low modes in some optimal fashion may very well pump energy into some of the higher modes. This is a basic problem which may be difficult to circumvent in practical applications.

An additional difficulty is that sometimes the modes are not well known, since in technical problems boundary conditions are often not well defined. A clamped beam for example, will not usually be clamped rigidly but in some other way, which possibly can be modelled via an elasto-plastic support. This lack of precise information on the system further aggravates the problem of control spill-over.

In designing the control complete knowledge of the state of the system at each instant is usually assumed. In practice, part of the state is measured and the complete state, i.e., the functions $w(x,t)$, $\dot{w}(x,t)$, $0 \le x \le l$ in the case of the string (Figure 2), are then reconstructed in some manner. Again, in this observation problem the system is discretized and an observation spill-over analogous to the control spill-over mentioned earlier does occur (see [7]).

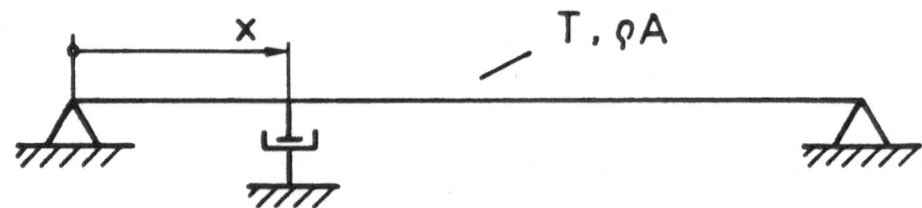

Figure 2: String with damper

It is well known that in some vibration problems a solution in terms of traveling waves has certain advantages over the modal approach. This is of course particularly true for the systems modelled by the one- or more-dimensional wave equation, but it also holds true for certain problems in somewhat more complex structural elements such as beams, plates, etc. The concept has for example been used successfully in the analysis of overhead transmission lines with dampers (see [8,9]), and some of its aspects have also been discussed recently in [10-12]. The traveling wave concept provides a particularly convenient approach to the study of the problem of passive and active vibration control at the boundary of a simple structural element.

In what follows, the design of active damping control via this traveling wave concept is discussed, first for controls at the boundary and later for controls displaced from the boundary. It is shown that in particular the problem of control spill-over can be thus completely avoided.

2. Vibration Control at the Boundary Using Traveling Wave Approach

For simple structural elements, namely for unidimensional systems like strings, rods and beams it is in principle easy to damp out vibrations either actively or passively by controlling the system at a boundary point. In fact, considering the string of Figure 3, with the single control force $f(t)$ at the point $x=0$, it is not difficult to show that this system is completely controllable.

A convenient choice for the vibration control of this structure is found if we keep in mind that for $0<x<l$ the system is described by the homogeneous equation (1), i.e., with $q(x,t)=0$, the general solution of which can be written as

$$w(x,t)=g_1(x-ct)+g_2(x+ct) \tag{12}$$

with $c=\sqrt{T/(\rho A)}$ and where $g_1(.)$ and $g_2(.)$ are arbitrary functions, which physically correspond to waves traveling to the right and to the left respectively, with speed c. Any wave arriving at $x=0$ and coming from the right will be completely reflected back, if the left end of the string is held fixed. On the other hand, a control can be devised such as to minimize the reflection of the incoming energy. For the string the solution is simple: if a damper with damping coefficient d is attached to the end of the string and if d is chosen as $d=\sqrt{T\rho A}$, then all the

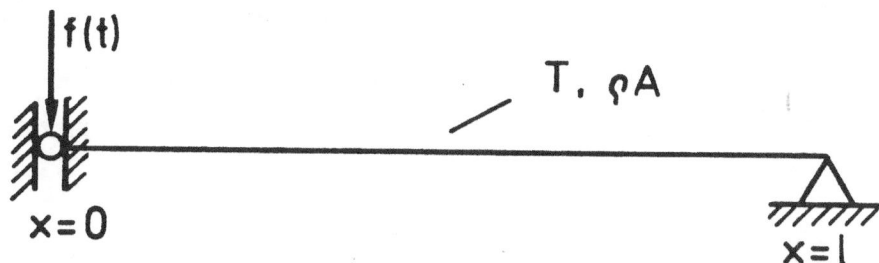

Figure 3: String with end control

waves arriving at the end are completely absorbed and no reflection occurs. This follows immediately from the boundary condition

$$f(t) = -Tw'(0,t) \tag{13}$$

if one sets

$$f(t) = d\dot{w}(0,t) , \tag{14}$$

observing that

$$\dot{w}(x,t) = -cg_1'(x-ct) + cg_2'(x+ct) \tag{15}$$

and calculating the reflected wave $g_1(x,t)$ for a given arbitrary incoming wave $g_2(x,t)$. In other words, the damper at $x=0$ in Figure 4a with $d = \sqrt{T\rho A}$ behaves as if the string at $x=0$ would continue towards $x \rightarrow -\infty$ (no reflection !) and the impedance of the semi-infinite string is exactly that of the damper. Note that no use was made of the boundary condition at $x=l$ and the same argument holds for the string elastically supported at $x=1$ by a spring of arbitrary stiffness, as in Figure 4b.

In any case, if there is some vibration energy in the string at $t=0$ and the damper is perfectly tuned, the string will be at rest for $t > 2l/c$, which is exactly the time needed for a wave to travel from $x=0$ to $x=l$ and back. It may also be useful to look at the eigenvalue problem for the string of Figure 4a. Since the system is damped, the eigenfunctions (and eigenvalues) are complex in general and we charactrize complex quantities by **boldface** symbols. Looking for solutions of

$$\rho A\ddot{\mathbf{w}}(x,t) = T\mathbf{w}''(x,t) , \quad \mathbf{w}(l,t) = 0 , \quad T\mathbf{w}'(0,t) = -d\dot{\mathbf{w}}(0,t) \tag{16}$$

we set

$$\mathbf{w}(x,t) = \mathbf{W}(x)e^{\mathbf{s}t} , \tag{17}$$

which leads to

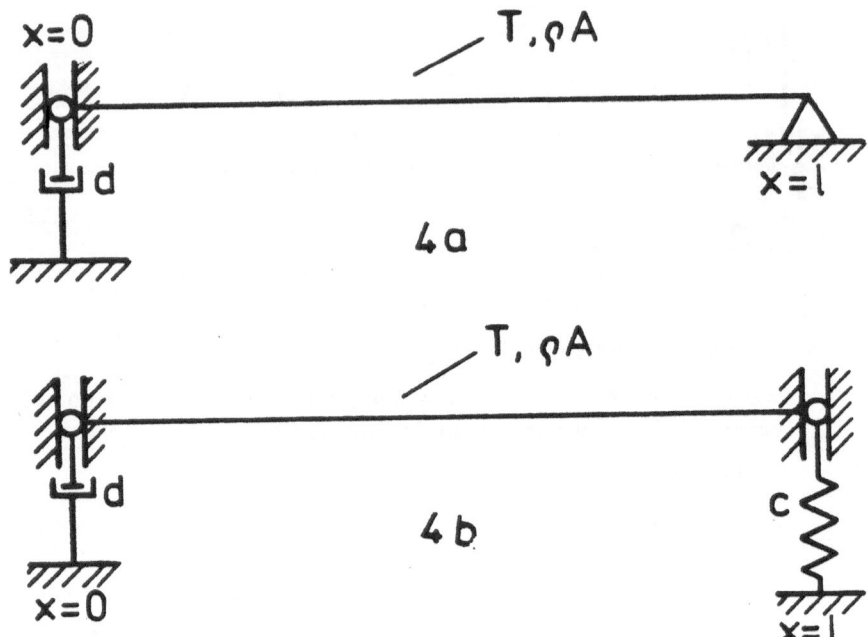

Figure 4: String with end damper a) right hand end fixed; b) right hand end elastically supported.

$$W''(x)-\frac{s^2}{c^2}W(x)=0 ,\qquad(18)$$

finally giving

$$w(x,t)=C_1 e^{s(ct+x)/c}+C_2 e^{s(ct-x)/c} .\qquad(19)$$

The boundary condition at $x=0$ with $d=T/c$ gives $C_1=0$, so that the remaining boundary condition at $x=l$ leads to

$$C_2 e^{sl/c}= 0 .\qquad(20)$$

Clearly the problem has no eigenvalues at all! This does however not come as a surprise, since we know that there are no solutions decaying exponentially in time, all solutions vanishing identically for $t>2l/c$.

For the wave equation, it is therefore possible to damp vibrations very efficiently at the boundary by using a perfectly tuned damper, i.e., a simple passive element, and this can be easily understood by using traveling waves. The analogous problem for the beam is far more involved since bending waves are dispersive, i.e., the speed of propagation is a function of the wave length. In fact, the equation of motion of the beam with constant cross-section

$$\rho A \ddot{w}(x,t) + EI w^{IV}(x,t) = 0 \tag{21}$$

has the solution (boundary conditions omitted!)

$$w(x,t) = \int_o^\infty [C_-(\lambda)\cos k(x - c_B t) + S_-(\lambda)\sin k(x - c_B t)$$
$$+ C_+(\lambda)\cos k(x + c_B t) + S_+(\lambda)\sin k(x + c_B t)] d\lambda , \tag{22}$$

where the parameter λ is the wave length, $k = 2\pi/\lambda$ the wave number and

$$c_B(\lambda) = \sqrt{E/\rho}\sqrt{I/A}\, 2\pi/\lambda \tag{23}$$

the speed of propagation of bending waves as a function of the wave length. The functions $C_-(\lambda)$, $S_-(\lambda)$ and $C_+(\lambda)$, $S_+(\lambda)$ characterize respectively waves traveling to the right and to the left.

Suppose now, that at the left end of a beam there is an active vibration damping device which absorbs all the vibration energy arriving from the right towards this end (Figure 5) by applying moments $m(t)$ and forces $f(t)$. It follows from [9] that the controls defined by

$$m(t) = EI \int_o^\infty k^2 [C_+(\lambda)\cos k c_B t + S_+(\lambda)\sin k c_B t] d\lambda , \tag{24}$$

$$f(t) = -\sqrt{EI\rho A} \int_o^\infty k^2 c_B [-C_+(\lambda)\sin k c_B t + S_+(\lambda)\cos k c_B t] d\lambda \tag{25}$$

do precisely that; the expressions (24), (25) are easily determined if one keeps in mind that the control device should simply simulate a semi-infinite beam, so that no reflection takes place at $x = 0$. Again, in the case of the beam it follows easily that the complete vibration energy is absorbed after a time interval of length $2l/c_B(l)$, regardless of the boundary conditions at $x = l$, the beam being at rest after that.

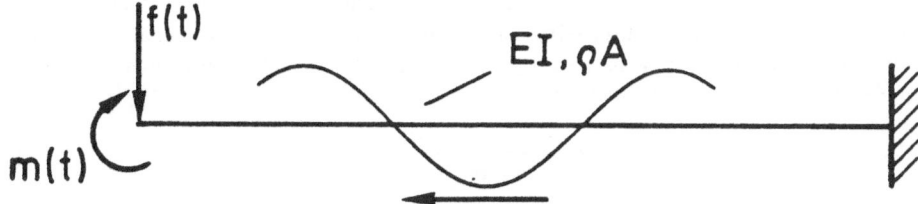

Figure 5: Beam with end control

What makes the implementation of this vibration control device at the end of the beam far more complicated than that of the string, is that now the control forces and moments are not given by simple formulae as functions of the local displacement $w(0,t)$ and velocity $\dot{w}(0,t)$. In any practical application the expressions on the right hand sides of (24), (25) will have to be evaluated at each instant and the vibration control can no longer be enforced by elementary passive components. To this end, the functions $C_+(\lambda)$, $S_+(\lambda)$ have to be known, which is of course the case if the complete state of the system in $0 \leq x \leq l$ is given. These functions can be obtained from measurements carried out at discrete points $z_1, z_2, ..., z_v$, i.e., they can be calculated if the functions $w(z_1, t)$, $w(z_2, t), ..., w(z_v, t)$, are given. The right hand sides of (24), (25) can be also expressed by means of the derivatives of $w(x, t)$ with respect to x, which then have to be calculated from measurements at several points in the immediate neighborhood of x.

The absoprtion of the vibration energy at $x = 0$ can also be achieved by controls different from $m(t)$, $f(t)$, for example with $m(t) \equiv 0$ and a suitable chosen $f(t)$ as shown in [9]. In this case however the will not be a perfect absorption but a local distortion of the vibrations in the neighborhood of the end. Similarly, the vibration control can also be carried out with $f(t) \equiv 0$ and $m(t)$ adequately selected.

In many applications one wants to absorb vibration energy in a string or in a beam by placing a vibration control device not at the end point but at a different location. We deal with this general case in the following section, but consider a special case presently. If the control device is located not too far from the end, then a simple solution can again be given, at least for the string. Consider the string of Figure 6a with the control device at $x = l_1$ applying a control force to the string. Let $\bar{f}$ be the highest frequency of interest in the problem, i.e., the highest frequency still to be absorbed. The corresponding wavelength is then $\lambda = c / \bar{f}$ and one has to choose $l_1 < \lambda/2$ in order to assure that the damping device does not act at a node of any vibration mode in the frequency range of interest. Typically one would choose $\lambda/4 \leq l_1 < \lambda/3$.

In this case, one can design the damping device in such a way that all the vibration energy arriving from the right towards the end $x = 0$ in Figure 6b in the form of traveling waves with intensity I_I (intensity of incident wave) is completely absorbed, and the intensity I_R of the reflected waves is zero. It is not difficult to show that this is exactly the case if the damping device has the impedance

$$ Z = \frac{d}{\sin k l_1} e^{-j(k l_1 - \pi/2)} \qquad (26) $$

with $d = T/c = \sqrt{T \rho A}$ which depends upon the frequency through the wave number $k = 2\pi/\lambda = 2\pi f /c$ and where j is the imaginary unit (see [8]). In practical applications such an impedance will have to be approximated by means of a number of simple linear active and/or passive elements. If in the string example of Figures 4a and 4b, the left end had been free in the transverse direction, instead of fixed, a different impedance would have been obtained instead of (26). This impedance is also a function of l_1 and it tends towards $d = T/c$ for $l_1 \rightarrow 0$.

The analogous problem for the beam will not be treated here. It is again somewhat more complicated than that of the string due to the dispersiveness of the bending waves. In both cases however, with the beam and the string, it was

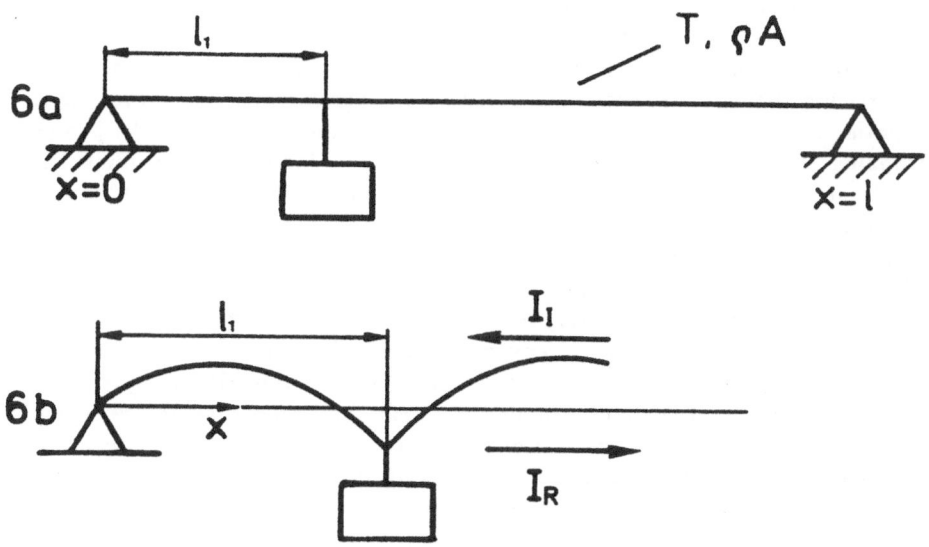

Figure 6: String with control device near end a) boundary value problem; b) end section, incident and reflected

possible to devise a vibration control with the traveling wave concept, without using the boundary conditions at the end $x=l$, and not even the length of the string or beam was of any importance in this analysis. In the formulae (24), (25), it is however clear that the incoming wave has to be measured and this may not be easy in a real system.

Different mathematical aspects of vibration control at the boundary, not only of one dimensional systems but also of plates have recently been studied by Eichenauer without using the traveling wave aspect. It seems however that in many engineering applications this latter approach considerably simplifies the problem. On the other hand, there are many systems in which control at the boundary is not feasable, for example if the systems are very large and/or the boundry conditions are not well defined.

3. Vibration Control Away from the Boundaries

We now consider a section of a beam away from the boundary, at which k control forces $f_1(t)$, $f_2(t),...,f_k(t)$ and moments $m_1(t)$, $m_2(t),...,m_k(t)$ are acting. The distances between the points of action of the control forces and moments can be arbitrarily for now. Two of these points may for example be almost adjacent and the corresponding control moments may be opposite to each other, so that the case of internal control moments is included. The system of Figure 7 may also be considered as a section of a string, and in this case the control moments are of course not present. In the usual approach, the control problem is studied by discretizing the structure, i.e., the underlying boundary value problem, thus reducing the problem to a classical control problem with ordinary differential equations. In what follows, we suggest a different simplification: instead of discretizing the boundary value problem, we suggest maintaining the partial differential equations describing the beam vibrations and to simplify the problem by dropping the boundary conditions. In other words, instead of designing the control in such a way as to damp out the vibrations of the discretized structure in all its modes, we think in terms of traveling waves and we choose the controls so that as much as possible of the energy of the incident waves arriving from the left and right is absorbed in the section. If one succeeds in solving such a problem for the non-discretized structure, the control spill-over mentioned in the introduction is completely avoided. Moreover, the controls being designed so that they only absorb vibration energy, they will damp out the vibrations, regardless of the particular type of boundary conditions. Also, in a large structure containing many beams and other structural elements, the active damping devices could be matched to some of the beams, without taking into account the rest of the structure at least in the initial stage of the project. In a second step, the whole structure could be discretized and the performance of the previously designed active vibration controls checked and improvements made if necessary. Again, control spill-over would be avoided completely, because the discretized model would only be used to test the performance of the control while the detailed design would be done in the continuous model.

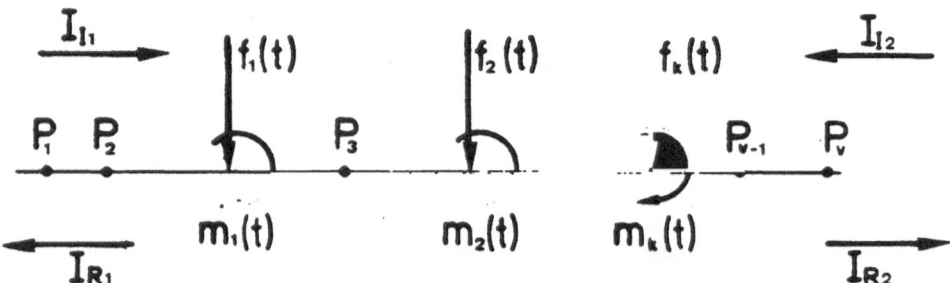

Figure 7: Section of beam with control forces and moments and points $P_1, P_2, ..., P_r$ of measurement.

According to the type and number of actuators, different mathematical problems of varying degrees of sophistication can be formulated and some of them are presently being studied. Some related problems are discussed in [14] but many questions remain unanswered so far. In order to gain some insight into the matter, consider the simple problem of a string section with two control forces only at the points $x=-l$ and $x=+l$ respectively (see Figure 8). The distance between these two points is $2l$ and it has to be smaller than $\lambda/2$. In the three regions, the solutions to the wave equations are respectively

$$w(x,t)=e_1(x-ct)+a_1(x+ct) \quad \text{for } x\leq l \ , \tag{27}$$

$$w(x,t)=b(x-ct)+g(x+ct) \quad \text{for } l\geq x\geq -l \ , \tag{28}$$

$$w(x,t)=e_2(x+ct) \quad \text{for } x\geq l \tag{29}$$

(see Figure 8). At the points of the control forces, the conditions

$$f_1(t)=T(w'(-l^-,t)-w'(-l^+,t)) \ , \tag{30}$$

$$f_2(t)=T(w'(l^-,t)-w'(l^+,t)) \tag{31}$$

as well as continuity hold, and this leads to

$$f_1(t)/T=e_1'(-l-ct)+a_1'(-l+ct)-b'(-l-ct)-g'(-l+ct) \ , \tag{32}$$

$$f_2(t)/T=b'(l-ct)+g'(l+ct)-e_2'(l+ct)-a_2'(l-ct) \ , \tag{33}$$

$$e_1(-l-ct)+a_1(-l+ct)=b(-l-ct)+g(-l+ct) \ , \tag{34}$$

$$e_2(l+ct)+a_2(l+ct)= b(l-ct)+g(l+ct) \ . \tag{35}$$

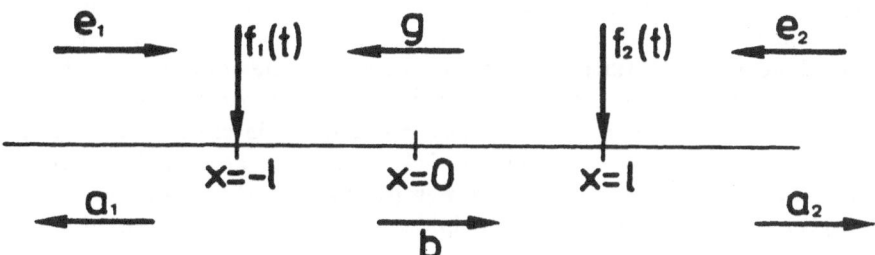

Figure 8: Section of string with two control forces

For given incident waves $e_1(x-ct)$, $e_2(x+ct)$ the control forces $f_1(t)$, $f_2(t)$ are now to be chosen in such a way that the energy contained in the reflected waves $a_1(x+ct)$, $a_2(x-ct)$ is minimized, while the oscillations in the section $-l\leq x\leq l$, described by $b(x-ct)$, $g(x+ct)$, are of little direct interest. They are however important, since the dynamics of the problem is of the type described by delay equations, i.e., given $e_1(-l+ct)$, $e_2(l-ct)$, $f_1(t)$ and $f_2(t)$ for $t\geq 0$ this does not suffice to compute $a_1(-l+ct)$, $a_2(l-ct)$ for $t\geq 0$, but the functions $b(l-ct)$ and $g(-l+ct)$ do in addition have to be known in the complete past interval $-2l/c\leq t\leq 0$.

The problem becomes more manageable if harmonic waves only are considered and steady state solutions are sought. In this case, the control can be described as a black box with the two inputs $e_1(-l,t)$, $e_2(l,t)$ and the two outputs $a_1(-l,t)$, $a_2(l,t)$, its dynamical behavior being described by 2×2 transfer matrix, to which also a 2×2 mechanical impedance matrix can be associated. This impedance matrix can then be chosen in such a way that the reflected energy is

minimized. A somewhat different approach is to represent the dynamics of each
of the control forces by a single scalar impedance and then to optimize these
(equal) impedances for maximum energy absorption. The calculations will not be
carried out here but they present no difficulty.

From what was said above, it becomes clear that different formulations of
the control problem are possible, depending also upon the knowledge available on
the traveling waves. It is generally necessary for active vibration damping using
the traveling wave concept, to measure not only local vibration levels, i.e., the
function $w(l,t)$, but to determine the energy flows or intensities, i.e., to obtain
the two components traveling in opposite directions which give rise to $w(l,t)$. It is
clear that this is impossible by measuring the function $w(x,t)$ at a single point
only, in general one will have to measure these functions at two neighboring
points at least. Until very recently such intensity or energy flow measurements
were not possible. During the last two decades however, methods have been
developed to measure intensities of sound waves. These methods require two
microphones located close to each other, but at a distance much smaller than the
wavelengths of interest, and from the two corresponding signals the intensity, i.e.,
the energy flow, in the direction of the two microphones can be obtained [15].
This technique is having a tremendous impact on acoustic measurements, since
with their help it is possible for example to measure the noise irradiated by a certain machine in a factory complex, without having to install it in an anechoic or
to a reverberant chamber. With the new method the intensities are measured at
many points along the normals of some closed surface around the machine, and
an integration then gives the total power irradiated; other acoustic sources outside this closed surface should in principle not affect this net measured power.

Analogous intensity measurements would of course also be extremely useful
in other areas. If one could measure intensities of bending waves, for example in
plates, it would then be possible not only to know the local vibration levels but
also to study the energy flows. For example in the walls of a gear box one could
possibly identify the particular bearing as the source of mechanical vibrations
which would not have to be located at a point of very high local vibration levels.
Similar applications exist in civil engineering. Intensity measurements in bending
waves are however more complicated than in the case of the wave equation, due
to the dispersiveness of the bending waves. Techniques of measuring bending
waves in beams and plates are however being developed presently in several
laboratories and should become quite common in the near future.

The development of these new intensity measurement techniques together
with the systematic study of active vibration control via the traveling wave concept should provide a powerful tool for the design of active damping devices of
large flexible structures.

4. Final Remarks

It has been shown in the previous sections that the traveling wave concept may be quite useful in designing active vibration control for flexible structures. Using this concept, the control problem is simplified by dropping the boundary conditions in the first stage of the design, instead of discretizing the boundary value problem, and in this manner control spill-over is completely avoided. With the controls designed in this way, the next step of the project usually will require the structure to be discretized in order to test the performance of the control in the discretized model. If only a few modes of a simple structure with well defined boundary conditions are to be damped, the discretized model of course can be used from the outset.

In the practical implementation of active damping control, there are understandably some difficulties which are only gradually being overcome. Because of this only a relatively small number of papers dealing with such experiments is available [16-18]. The behavior of electromechanical actuators for example, which is quite important in these implementations, has been discussed in [17]. In the traveling wave approach to the active vibration control problem, it is essential that traveling waves and not only local vibration levels be measured. These techniques are currently under development in several laboratories and should be available in the near future.

The present knowledge of active vibration control design via the traveling wave approach looks so promising that further research in this area will be intensified. It is felt that this may lead to an important and practical tool in the control of the vibrations of large flexible systems.

The author thanks the Volkswagenwerk Foundation (Stiftung Volkswagenwerk) for their support of this work.

References

[1] Ogata, K., *State Space Analysis of Control System*, Prentice-Hall, Englewood Cliffs, New Jersey, 1967.

[2] Müller, P.C., *Stabilität und Matrizen*, Springer-Verlag, Berlin, 1977.

[3] Meirovitch, L. and Rajaram, S., "On the Uniqueness of the Independent Modal-Space Control Method, Dynamics and Control of Large Flexible Spacecraft," *Proceedings of the Third VPI & SU/AIAA Symposium*, Blacksburg, Va., 1981.

[4] Baruh, H. and Meirovitch, L., "Implementation of the IMSC Method by Means of a Varying Number of Actuators, Dynamics and Control of Large Structures", *Proceedings of the Third VPI & SU/AIAA Symposium*, Blacksburg, Va., 1983.

[5] Balas, M.C. and Meisner, T.L., "Spillover and Model Error Bounding Techniques for Large Scale Systems, Dynamics and Control of Large Flexible Spacecraft", *Proceedings of the Third VPI & SU/AIAA Symposium*, Blacksburg, Va., 1981.

[6] Calico, R.A. and Janiszewski, A.M., "Control of a Flexible Satellite Via Elimination of Observation Spillover, Dynamics and Control of Large Flexible Spacecraft", *Proceedings of the Third VPI & SU/AIAA Symposium*, Blacksburg, Va., 1981.

[7] Meirovitch, L. and Baruh, H., "On the Problem of Observation Spillover in Distributed-Parameter Systems, Dynamics and Control of Large Flexible Spacecraft", *Proceedings of the Third VIP & SU/AIAA Symposium*, Blacksburg, Va., 1981.

[8] Hagedorn, P., "Ein einfaches Rechenmodell zur Berechnung winderregter Schwingungen an Hochspannungsleitungen mit Dämpfern", *Ingenieur-Archiv*, 49, 1980, pp. 161-177.

[9] Hagedorn, P., "On the Computation of Damped Wind-Excited Vibrations of Overhead Transmission Lines", *J. Sound & Vibrations*, 83, 1982, pp. 253-271.

[10] Flotow, A.H. von, "A Traveling Wave Approach to the Dynamic Analysis of Large Space Structures", Paper 83-0964 presented at the 24th AIAA Structures, Structures Dynamics and Materials Conference, Lake Tahoe, May 2-4, 1983.

[11] Flotow, A.H. von, "Low-Authority Control Synthesis for Large Spacecraft Structures, Using Disturbance Propagation Concepts", paper presented at the 26th AIAA Structures, Structural Dynamics and Materials Conference, Orlando, Florida, April 15-17, 1985.

[12] Flotow, A.H. von, "Disturbance Propagation in Structural Networks, Control of Large Space Structures", Ph.D. Thesis, Stanford University, Stanford, California, 1984.

[13] Eichenauer, W., "Über trigonometrische Momentenprobleme und deren Anwendung auf gewisse Schwingungskontrollprobleme", Doctoral Thesis, TH Darmstadt, 1982.

[14] Scheuren, J., "Energieverhältnisse bei aktiver Körperschallminderung, Fortschritte der Akustik", *Proceedings of the DAGA Conference*, Darmstadt, 1984, pp. 315-318.

[15] Rasmussen, G. et al., *Intensity Measurements*, Brüel & Kjaer, Denmark, 1984.

[16] Montgomery, R.C., Horner, G.C. and Cole, S.R., "Experimental Research on Structural Dynamics and Control, Dynamics and Control of Large Structures", *Proceedings of the Third VPI & SU/AIAA Symposium*, Blacksburg,

Va., 1981.

[17] McClamroch, H.N., "Modelling and Control of Large Flexible Structures Using Electromechanical Actuators, Dynamics and Control of Large Structures", *Proceedings of the Fourth VPI & SU/AIAA Symposium*, Blacksburg, Va., 1983.

[18] Schäfer, B. and Holzach, H., "Experimental Research on Flexible Beam Modal Control," paper presented at the 25th AIAA Structurs, Structural Dynamics and Materials Conference, Palm Springs, Cal., May 14-16, 1984.

PRACTICAL EXPERIENCES IN PASSIVE VIBRATION CONTROL OF CHIMNEYS (CONCLUSIONS FROM WIND TUNNEL AND FULL SCALE TESTS)

G. Hirsch
Institut für Leichtbau
Technische Hochschule Aachen
Aachen, Germany

1. Introduction

Cylindrical chimney structures with low structural damping are very susceptible to wind induced oscillations. It is well known that chimneys under wind loading respond dynamically in the across-wind as well as in the along-wind direction. For such structures the dynamic response in the across-wind direction is often greater than along-wind.

Significant amplitudes usually only occur when the vortex shedding frequency (which increases as the wind-speed arises) coincides with the structural frequency of the fundamental mode. The vortex shedding frequency may be determined by $f_s = SV/D$, where S = the Strouhal-Number, which exhibits over a large range of Reynolds-Numbers a nearly constant value of 0.2. The Strouhal Number decreases with decreasing distance (a) of nearby chimneys in a row arrangement. For a/D the Strouhal-Number is $S = 0.10 + 0.085 \, Log \, a/D$. In the above mentioned formula V = the flow velocity and D = the diameter of the chimney shell (averaged over the top third).

This dynamic response during vortex shedding resonance depends on the movement of the chimney. Large movements will be close to harmonic oscillations and small movements ($<.03D$) will be random. In this region the response is nearly Gaussian.

A treatment of the detailed approach to the evaluation of the across-wind response of chimneys has been given by Vickery and Basu [1]. The variation of RMS amplitude with the damping parameter $Ks = Sc/4\pi$, where Sc = the Scruton-Number $(4 \cdot \pi \cdot m \cdot \varsigma/\rho \cdot D^2)$ and m = the mass per unit length over the top third, ς = the damping ratio, ρ = the density of air, shown in Figure 1.

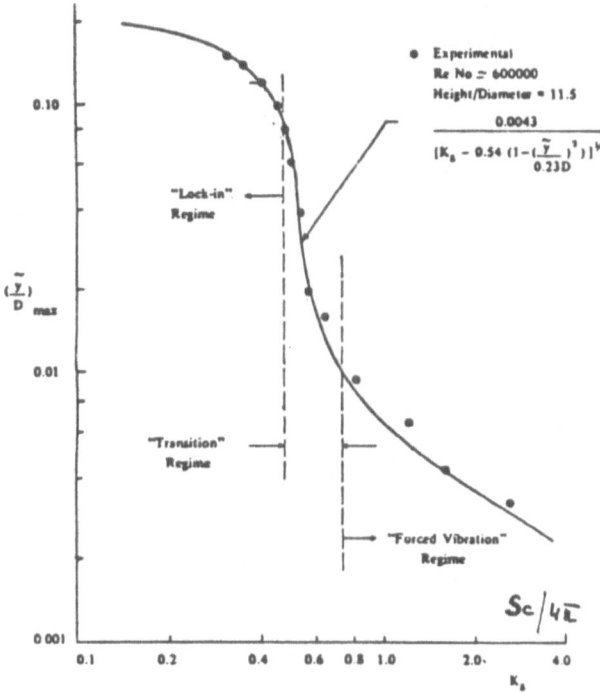

Figure 1: Variation of RMS amplitude with Scruton-Number

A large amplitude or "lock-in" region corresponds to low values of Ks (<0.45) and a small amplitude region corresponds to great values of Ks (>0.7). Full scale tests of self-supporting steel chimneys show that the structural damping often is rather well below to be a safe figure ($Ks > 2$ or $Sc > 25$).

In general the problem is the interaction and feedback between a vibrating structure and an oscillating flow field, which requires the knowledge of the aero-elastically excited time dependent forces upon the structure, which actually are nonlinear. For practical purposes more thought should be given to the reduction or avoidance of such vibrations. Increasing the damping, which decreases the resonant amplitude of the vibration, may be achieved by using external dampers. A survey on vibration damping has been given Rogers [2]. A variety of damping devices have been suggested from Walsh and Wooton [3]. Other devices for energy dissipation were used such as draping chains over tubes or permitting scraping and the incorporation of energy absorbing materials such as rubber, plastics, wood, etc.

Moreover, across-wind vibrations can be reduced by aerodynamic stabilizers. The useful effect of three helical vanes has been proven on many steel chimneys. The radial width of the vanes must be 10% of the diameter. The pitch of the vanes must be about $5D$. The vanes must be fitted over the upper $1/3$ of the height. The disadvantage is the increasing of the drag. The vanes will not reduce the oscillations of nearby chimneys. Moreover they lose the strakes of their damping effect if the Scruton-Number leads to small values, as Figure 2 shows.

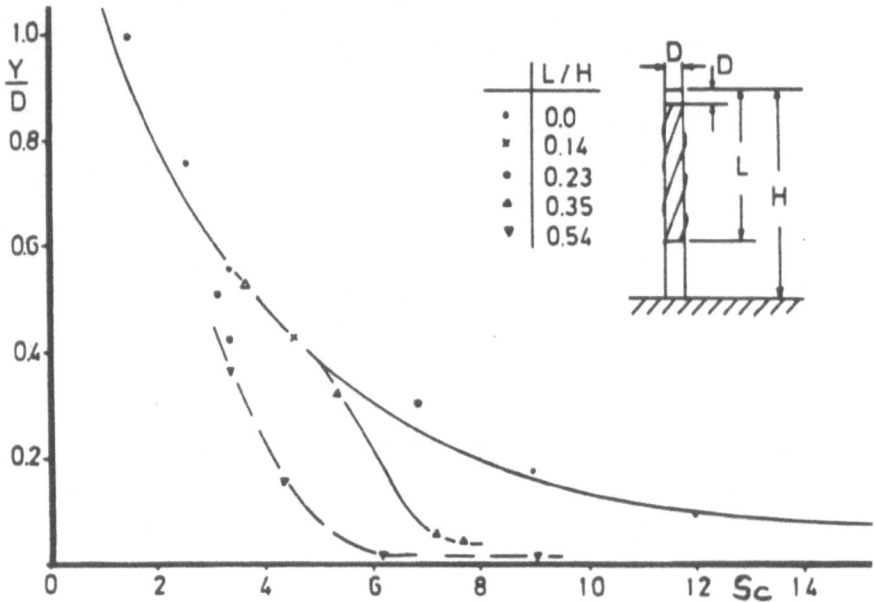

Figure 2: Variation of maximum across-wind amplitude versus Scruton-Number

The present paper is concerned with a particular type of added damping treatment, characterized by low maintenance display. The techniques and the results of full scale tests will be given in the presentation respecting chimneys up to a height of 120 m.

2. Passive Vibration Control by Means of Auxiliary Damping Systems

The control of wind-induced vibrations due to vortex shedding, buffeting and galloping can be realized by means of the installation of dynamic vibration absorbers [4]. The optimum absorber has to minimize the dynamic response of the chimney, which is excited by an harmonic force due to vortex shedding resonance.

A detailed theoretical treatment of the response of a one degree of freedom system fitted with an auxiliary mass damper to a sinusoidal excitation can be found in [5]. Design charts concerning the required characteristics of damper systems were presented from Vickery, Isyumov and Davenport [6]. Figure 3 shows such a chart relating a mass ratio $\mu = M_2/M_1 = 0.05$, where M_1 is the generalized mass of the fundamental vibration mode of the chimney and $M_2 =$ the absorber mass. From the practical point of view the mass ratio 0.05 is an optimum value concerning economical aspects.

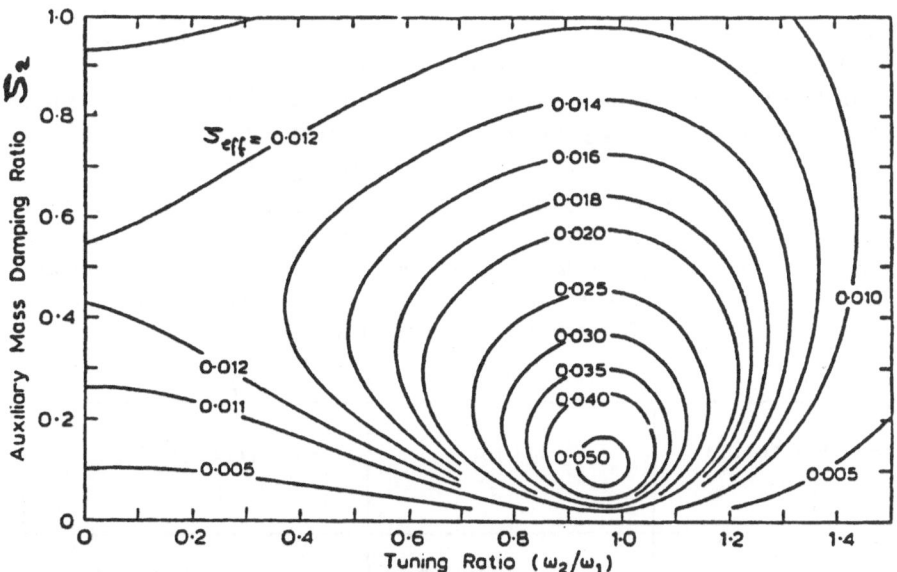

Figure 3: Effective damping versus tuning ratio and mass damping ratio

The magnitude of improvement in vibration response Yd/Y as a result of using a damped absorber can be approximated by the expression

$$Yd/Y = 2 \cdot \varsigma_1 \cdot (1 + 2M_1/M_2)^{1/2} \tag{2}$$

where $Yd =$ the peak amplitude at the top of the chimney (together with the damping device) and $Y =$ the amplitude without external damping. Figure 4 shows Yd/d versus μ for different damping values ς_1, where $\varsigma_1 =$ the actual damping ratio of the chimney structure without the damper. Equation (2) can be used as a basis for selecting a practical value of mass ratio.

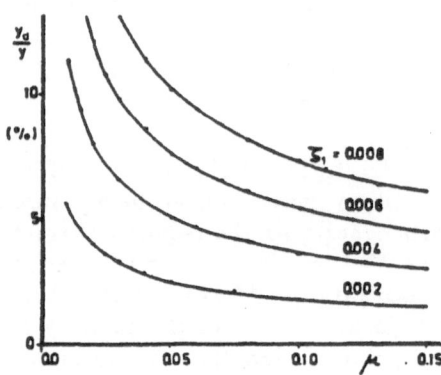

Figure 4: Amplitude ratio versus mass ratio

It also must be recognized that the relative movement between the main system and the mass of the damping device depends on the mass ratio μ. For a μ value of 0.05 the relative movement is 36 times the size of the static amplitude under influence of the exciting forces (i.e. without dynamic magnification).

Finally, a remark with regard to the tuning (natural frequency and damping) of auxiliary systems should be given. In cases of greater μ values it must be remembered that the excitation due to vortex shedding is proportional to frequency-squared. To illustrate these aspects the optimum tuning values are summarized in Figure 5. Sauer and Garland [7] studied the optimum tuning of absorbers concerning the frequency-squared excitation.

	$\mu \lesssim 0.05$	$\mu > 0.05$	
frequency-tuning ω_2/ω_1	$1/1+\mu$	$(1/1+\mu)^{1/2}$	(3)
damping ratio ς_2	$(3\mu/8(1+\mu)^3)^{1/2}$	$(3\mu/4(1+\mu)(2+\mu))^{1/2}$	(4)

Figure 5: Optimum tuning values

Figure 6 shows the dynamic response of chimneys (Y/Y_{st}) versus the frequency ratio η without (1) and with (2) consideration of the frequency-squared excitation ($\mu=0.05$). Y_{st} is the static deflection of the chimney (top value) at critical wind speed ($V_{cr}=5 \cdot f_1 \cdot D$) without dynamic magnification.

The dynamic response of chimneys including a damper system to a random excitation can be obtained by examining its response to white noise input. Vickery and Davenport [8] showed that the effective damping provided by auxiliary mass damper becomes

$$\varsigma_{eff} = \pi/4 \int_o^\infty (H_{Y1}(\eta))^2 d\eta \qquad (5)$$

where H_{Y1} is the dynamic response of the main system (chimney) versus the frequency ratio η in case of harmonic excitation. As calculations in the range of μ values from 0.012 to 0.05 showed, the effective damping is approximately 40% greater for harmonic forcing than for white noise excitation.

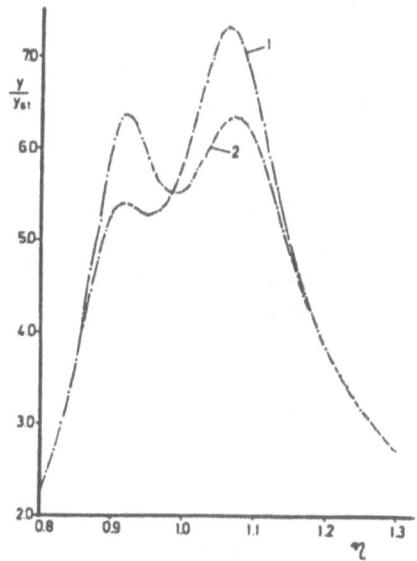

Figure 6: Dynamic response Y/Y_{st} versus frequency ratio η

3. Critical Comparison between Passive and Active Vibration Control

In recent years, the application of active control devices to reduce the response of civil engineering structures to dynamic excitation has been widely studied. The first summary of the state of the art was given on the IUTAM Symposium on Structural Control [9], held at the University of Waterloo in 1979.

Although impressive examples of applications of control systems are already known, it became obvious that a number of questions are still open. To date, many papers have been published concernng the application of active and/or passive control devices in civil engineering structures. However, few of these studies dealt with the reliability aspects of structures with control systems. Basharka and Yao [10] studied the reliability aspects of structural control.

Concerning active vibration control of chimneys prestressing tendons have the potential of realization of this technique. Freyssinet and Zetlin [15] studied the use of active control by means of tendons.

Carotti [11] discussed the concept of active vibration control of chimneys. Unfortunately no papers have been published to date concerning practical examples of active vibration control realization.

With regard to the effort and to avoid maintenance the author [12] investigated an new type passive control system in cooperation with the KABE-Company, Oberursel, Germany (DP 28.06.757).

The paper describes practical experiences concerning these passive systems, based on wind tunnel and full scale tests.

4. Characteristics of a New Type of Auxiliary Passive System

In general the greatest disadvantage is that the auxiliary systems require maintenance (e.g. in case of hydraulic shock absorbers). Moreover, if rubber elements are used for stiffening and damping the additional system, the elements will change their dynamic properties during the life time of the main structure.

The damping elements of the new type of absorber system are stranded wire rope helicals held between rugged metal retainers. The helicals have specific dynamic properties which are determined by cable diameter, number of strands per cable, cable length, cable twist or lay, and number of cables per absorber system.

The attachment of the pendulum-type vibration absorber is quite close at the top of the chimney, as Figure 7 shows. The length of the pendulum is approximately 30% more than the length of the mathematical pendulum with regard to the (uncoupled) natural frequency of the absorber system.

Design details of the absorber are shown in Figure 8. The ring mass of the damper system and the chimney shell to avoid falls of the damper mass against the chimney shell due to gust loads.

5. Practical Experiences with the New Type of Damper System

To date, approximately 90 damper units of the new type are installed and in operation and tests on these have proven them to be satisfactory. The damper weights extend from 100 to 10 000 kg for steel chimneys and 1 500 kg (Telecommunication tower Nürnberg, Germany) as soon as 18 000 kg (EOLE vertical-axis wind turbine, Cape Chat on the Gaspé Peninsula, Canada, installation 1986).

Fifteen absorber systems of this type must be installed, because vortex shedding induced vibrations of extreme intensity occurred after erection of steel chimneys.

Figure 7: Dynamic absorber quite close at the top of chimney (KABE-type damper)

5.1. Prototype

Figure 8a shows the 1978 installed prototype of the new damper system. The ringbeam mass of the damper system is 1.7% (827 kg) of the weight of the 70 m high × 1.8 m diameter self supported chimney. In this case six sets of damping elements with four 175 × 3 mm diameter windings are used.

It was possible to adjust the damper system natural frequency (uncoupled), and damping before installation in the manufacturer's workshop. The dynamic properties have been estimated by the records of the decay curve of the oscillations after switching off the prestressing. The natural frequency of the damper system was + 4.5% and the damping -40% from the optimum value. Nevertheless the effective damping of the coupled system (chimney with absorber system) is satisfactory. A 7.8% damping ratio has been calculated by the damper system tests. Full scale tests were carried out after the erection of the chimney. The natural frequency of the fundamental bending mode of the chimney was found out by the decay rate to be 0.4 Hz. The test results confirmed the calculation of the natural frequency. It is significant that by the Rayleigh's method, calculated natural frequencies of steel chimneys could be confirmed by the results of full scale tests. The discrepancies were not more than approximately 5% in a series of examples.

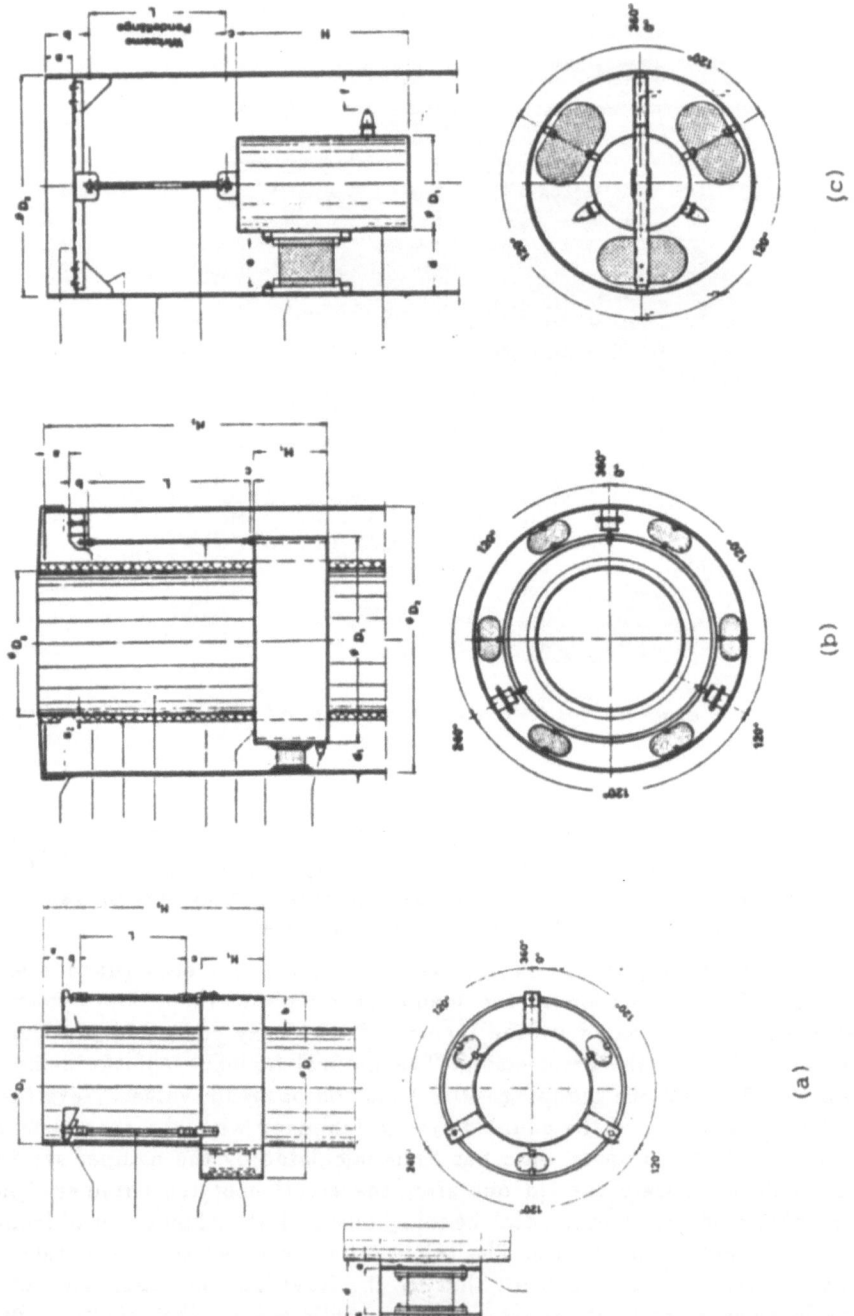

Figure 8: Design details of dynamic absorbers (KABE-type)

The effective damping of the chimney with attached absorber system was found out by full scale testing to be approximately 6% of critical. The absorber increased the structural damping of the chimney ($M_1 = 8\ 270$ kg, $\varsigma_1 = 0.0035$, $\mu = 0.1$) by a factor of approximately fifteen. The prototype absorber could be considered to be adequate to limit wind-induced vibrations of the steel chimney to tolerable levels.

The absorber operates satisfactorily for seven years at present. Not any changing in the dynamic properties and/or injuries of the absorber damping elements have occured as shown by the yearly inspections. Finally, the successful operation of the passive control system explains the increasing of damping reduces the oscillation amplitudes and therefore also the amplitude depending exciting forces.

Figure 9 shows records of the oscillation decay. The dynamic properties of the tested absorber are $M_2 = 13\ 500$ kg, $f_2 = 0.46$ Hz, $\varsigma_2 = 0.203$ (absorber with added helical springs and damping elements). The decay curve is approximately of viscous character.

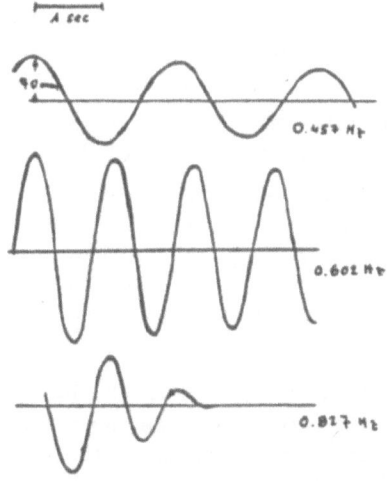

Figure 9: Decay curves of (uncoupled) absorber system

5.2. Heavyweight absorbrs

Heavyweight absorbers of the above mentioned type have been installed to avoid wind-induced vibrations of steel chimneys in a row arrangement. Chimneys in a row or in a grouped arrangement are excited not only by their own vortex separation but also by buffeting due to the surrounding chimneys. Helical strakes fixed to single chimneys usually suppress the vortex-shedding vibrations, but because of their sensitivity to buffeting their damping effect may be decreased for multiple chimneys. Ruscheweyh [13], Beaumont and Walshe [14] investigated by wind tunnel tests and full scale tests these effects and confirmed earlier investigations of Vickery [16]. Furthermore, a locking-in effect may also occur for a straked design in case of low mass-damping-parameter (Scruton-Number).

Five 120 m high × 8.0 m diameter self supporting steel chimneys, each of which is fitted with an absorber system, were erected in 1982 in Saudi-Arabia. The distance between the stacks is 60 m. The chimneys are of identical DIN 4133 (German Code for Praxis for Steel Chimneys) standard and are 4.9 m ϕ at the top.

Each chimney is insulated with 100 mm thickness of mineral wool, covered with 1.5 mm aluminium cladding. The total mass of each chimney is 300 tons and the mass per unit length over the top third is 2 113 kg/m. The natural frequency of the fundamental bending vibration mode was calculated to be 0.46 Hz. The influence of the elasticity of the foundation has been considered in the calculation.

Before the damper systems have been installed the downstream chimneys suffered severe oscillations due to vortex shedding and buffeting when the critical wind ($V = 11.27$ m/sec) was approximately in line with the chimney row.

A dynamic model of the stacks was used for the wind tunnel tests at the Institute of Light Weight Structures, carried out by Ruscheweyh [13]. The mass of the auxiliary systems was chosen to be 2% of the total chimney mass.

The tests showed the high effectiveness of the absorber system. The maximum recorded response Yd/D was only 0.02. That means an increase of the effective Scruton-Number by a factor of 30. The effectiveness of the absorber systems depends not only on the increasing of the damping but also on the shortening of the correlation length of the vortex separation due to decreasing the amplitude of oscillations.

The damper system used was tuned to the fundamental frequency of each chimney. The system consists of a ringbeam mass with opening for the ladder, concentric with the chimney and suspended at three points, as Figure 10 shows. Twelve damping elements with eight windings each are installed between the ring mass and the chimney shell. The damper mass is 10% of the generalized mass and the damper weight is 7 tons (2.3% of the total mass of the chimney).

The tuning of this heavyweight type damper system has been carried out at the factory. The tests showed that the systems have approximately the optimum dynamical properties (0.413 Hz natural frequency and 28% damping). The deviations of the optimum values are neglegible.

However, also at the optimum absorber parameters it is necessary to check that the resulting maximum amplitude is considerably larger than the amplitude of the structure at the attachment point. In this case the relative movement is 11

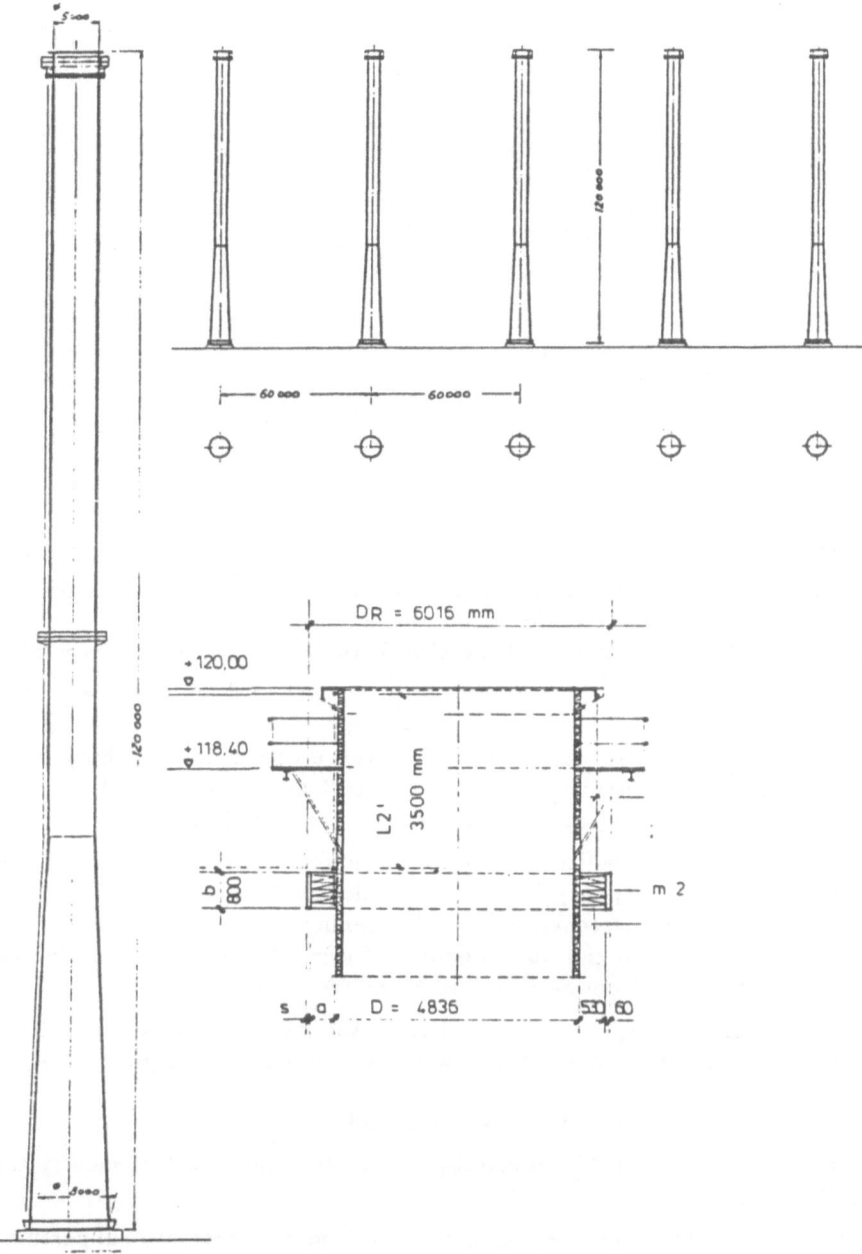

Figure 10: Chimney row arrangement and damper system

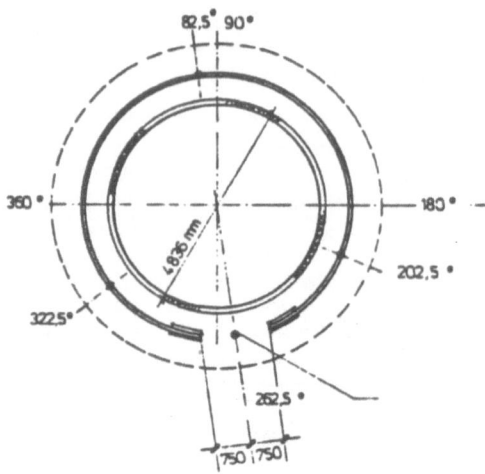

Figure 10: (continued)

times the steady state deflection of the chimney without dynamic magnification (i.e. 144 mm). The clearance between the damper ring mass and the chimney shell is approximately 340 mm. Shock absorbers of the Continental type 58499 of 65 Shore are incorporated to avoid impact loads due to extremely high wind gust loads.

Full scale tests were carried out after the erection of two chimneys. Each stack was set into an oscillatory motion due to transient load by the prestressed absorber and the decay of the oscillations was recorded electronically.

The theoretically predicted results were in good agreement with those measured in situ. The discrepancy between the calculated (Rayleigh method) and the measured fundamental frequency of each chimney was found to be 1.6%. The effective damping was found out to be 5% of critical. That means an increase of the effective damping from $Sc = 3.2$ to $Sc = 50$.

The critical wind speed occured often in the past, but no remarkable oscillations could be observed. Maintenance was not necessary in the past three years.

5.3. Full scale tests at critical wind speed

In some cases it was possible to investigate the damping effect of such systems at critical wind speed.

In the first case three 86 m high × 3.7 m diameter self-supporting steel chimneys were erected at a power plant in North-Germany. Each chimney has two flues. The distance between the stacks is 22.8 m. The total mass of each chimney is 173 tons and the mass per unit length at top third is 2 000 kg/m.

The fundamental frequency was calculated to be 0.52 Hz. Across-wind oscillations could be started at 9.6 m/sec. To avoid vortex shedding oscillations and buffeting effects absorbers have been fitted, as Figure 8b shows (12 damping elements $\times$ 9 windings). The mass ratio concerning the damper weight of 4 500 kg is 0.11.

The damper system has been tuned (0.48 Hz and 13% damping) and full scale tests showed the coupled fundamental frequency of the chimney-absorber system (0.42 Hz). Fortunately it was possible to have test results at the critical wind speed. The maximum deflection due to across-wind oscillations have been found out to be $\pm$ 50 mm ($\pm$ 80 mm relative movement). It was to expect that in case of critical wind direction (approximately in line) the dynamic deflections will be two times the above mentioned values.

The chimney project was finished in 1983 and significantly vibrations could not be observed.

In a second case a group of three steel chimneys of 72 and 80 m height and 4.75 (4.4) m diameter were erected in Saudi, Arabia. After the erection of the third chimney critical across-wind oscillations have been observed at the critical wind speed (13 m/sec). The distance between chimneys is 33 m. The dynamic properties of the third chimney are $f_1 = 0.5$ Hz and $Sc = 30$.

In February 1984, movements have been observed which were of varying intensity depending on direction and power of wind. This phenomenon prompted measurements which have been carried out by the site management.

Although evaluation of these movements by experts has been different (maximum amplitude ± 110 mm), the management decided to install a vibration absorber. In December 1984 a site inspection was carried out. A strong and approximately critical wind prevailed during these days which, moreover, blew from the most unfavourable direction. Figure 11 shows clearly the direction of the attacking wind. The observations made enable drawing the conclusion that the movements of the critical excited chimney were within a range of a few millimeters.

The working principle of the absorber could be clearly observed. It made pronounced movements transverse to the direction of the attacking wind. In accordance with the fact that in connection with previously observed amplitudes the chimney had performed movements of random nature and caused by turbulences the mode of operation of the absorber was the oscillating type, too. Its reactions practically corresponded to the response of the stack provided no absorber had been installed. As expected, the absorber is a vibration measuring instrument.

The frequency of movement clearly corresponded to the coupled stack-absorber natural frequency (0.55 Hz), which furnishes proof of absorbing systems being of optimum design. The relative movement corresponds very exactly with the pre-calculated value ($36.56/1.5 = \pm 24$ mm).

Figure 11: Group arrangement of steel chimneys

Concerning the relative movement the amplitude of the chimney top will be ± 8 mm. This maximum value is approximately 7% of the maximum value as observed at the chimney without the absorber.

According to the draft for EUROCODE 3, fatigue, the alternating stress at critical wind speed has to be considered harmless ($\Delta\sigma = 6.6$ N/mm^2).

6. Concluding Remarks

The purpose of this paper was to give informations concerning practical experiences in passive vibration control of chimneys. At the present, steel chimneys are due to low structural damping very susceptible to oscillations. However, the concrete chimney will be growing lighter and in consequence more sensitive against dynamic wind forces.

In conclusion we are able to state the following. A passive damper system without necessity of maintenance is a very effective means of damping wind-induced vibrations of steel chimneys and it seems promising to continue investigations in view of the growing importance of the structural control techniques.

Finally an extension of the damper system will be shown in the Figurs 12 and 13 with regard to the damping of the vertical axis wind turbine EOLE, Canada.

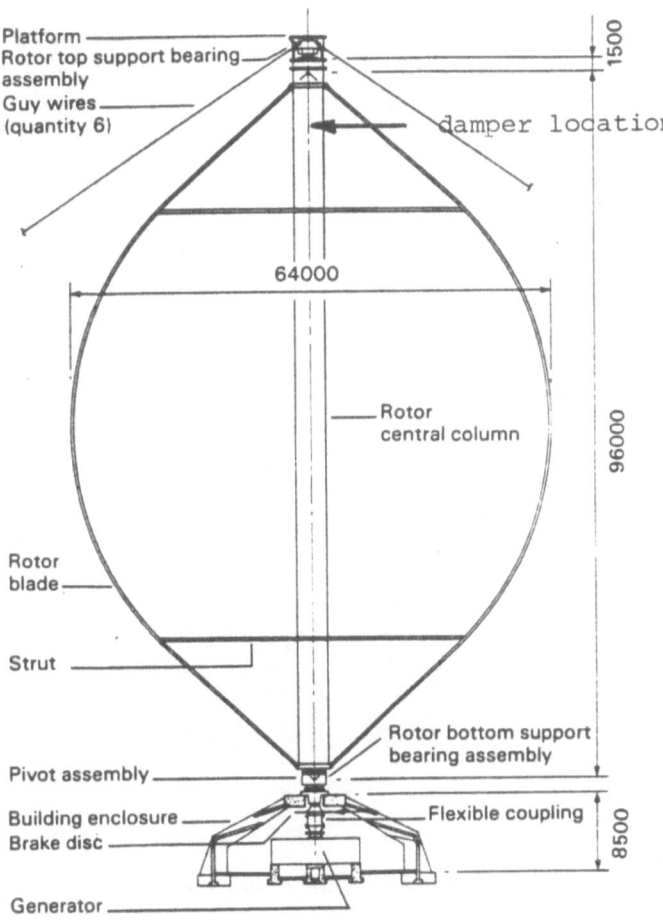

Figure 12: Vertical-axis wind turbine EOLE (4 MW) - Canada

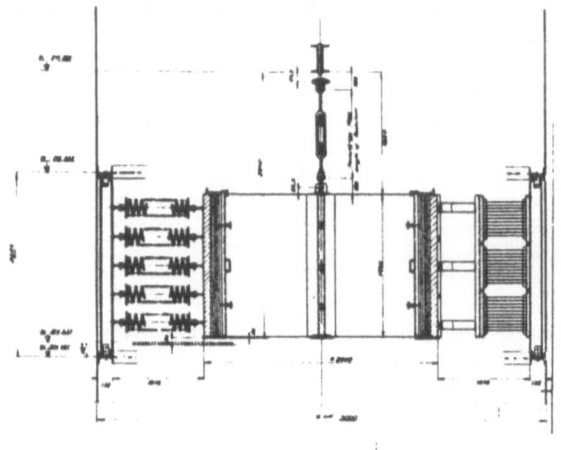

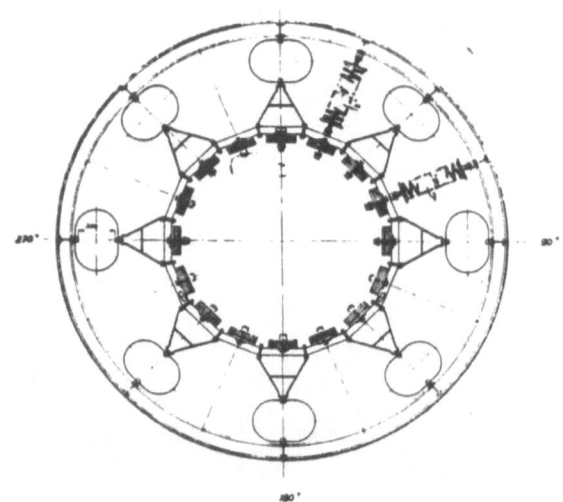

Figure 13: Design details of the EOLE damper system

References

[1]   Vickery, B.J. and Basu, R., "Simplified approaches to the evaluation of the across-wind response of chimneys", *Journal of Wind Engineering and Industrial Aerodynamics*, 14, 1983, pp. 153-166.

[2]   Rogers, L., *Vibration Damping 1984 Workshop Proceedings*, AFWAL-TR-84-3064, Flight Dynamics Laboratory, Air Forces Wright Aeronautical Laboratories, Wright-Patterson Air Force Base, Ohio.

[3]   Walsh, D.E. and Wooton, L.R., "Preventing wind induced oscillations of structures of circular section", *Proc. Inst. Civ. Engrs.*, London, 1970.

[4]   Hirsch, G.H., "Passive control of steel chimney vibrations originated by wind and earthquake", *Proc. 4th Symp. on Ind. Chimneys*, The Hague, 1981, pp. 111-126.

[5]   Den Hartog, J.P., *Mechanical Vibrations*, McGraw-Hill Book Co., 1956.

[6]   Vickery, B.J., Isyumov, N. and Davenport, A.G., "The role of damping, mass and stiffness in the reduction of wind effects on structures", *Proc. of the 5th Coll. on Ind. Aerodynamics*, Aachen, 1982, pp. 63-73 (Annex).

[7]   Sauer, F.M. and Garland, C.F., "Performance of the viscously damped vibration absorber applied to systems having frequency squared excitation", *Journal of Appl Mech.*, 1949, pp. 109-117.

[8]   Isyumov, N., Holmes, J.D., Surry, D. and Davenport, A.G., "A study of wind effects for the first National City Corporation project", BLWT-SS1-75, Fac. of Eng. Sc., University of Western Ontario, Canada.

[9]   Leipholz, H.H.E., "Structural Control", *Proc. of the Int. IUTAM Symp. of Structural Control*, held at the University of Waterloo, Canada, June 4-7, 1979, North-Holland, 1980.

[10] Basharkhah, M.A. and Yao, J.T.B., "Reliability aspects of structural control", *Civ. Engng. Syst.*, Vol. 1, June, 1984, pp. 224-229.

[11] Carotti, A., "Automatic control of drift vibrations in steel stacks subject to Bernard-Karman vortex discharges", *Proc. of the 5th Int. Chimney Congress*, CICIND, Paper 27, Essen, October, 1984.

[12] Hirsch, G.H., "Critical comparison between active and passive control of wind-induced vibrations of structures by means of mechanical devices", *Proc. of the Int. IUTAM Symp. on Structural Control*, University of Waterloo, 1979, pp. 313-339.

[13] Ruscheweyh, H.P., "Straked in-line steel chimneys with low mass-damping parameter", *Journ. of Wind-Eng. and Ind. Aerd.*, 8, 1981, pp. 203-210.

[14] Beaumont, M. and Walshe, D.E., "Investigation into wind-induced oscillation on three 76 m high self-supporting steel chimneys", *Proc. 4th Int. Symp. on Ind. Chimneys*, The Hauge, 1981, pp. 127-135. Zuk, W., "The past and future of active structural control systems", in [9], pp. 779-794.

[15] Zuk, W., "The past and future of active structural control systems", in [9], pp. 779-794.

[16] Vickery, B. J. and Watkins, R. D., "Flow induced vibrations of cylindrical structures", Proc. of the 1st Australian Conf. on Hydr. and Fluid Mech., 1962, Pergamon-Press, London, 1963.

ACTIVE CONTROL OF STRESSES AND DEFLECTIONS OF ELASTIC STRUCTURES BY MEANS OF IMPOSED DISTORTIONS

J. Holnicki-Szulc and Z. Mróz
Institute of Fundamental Technological Research
Warsaw, Poland

1. Introduction

Active control of structures is usually performed by application of additional external forces, (tendon or support action, actuator action, [2,5,6]) which vary according to a specified rule depending on the external load variation. In this paper, the concept of active structure adaptation to varying loads will be discussed by assuming that initial distortion fields $\epsilon^o(x,t)$ within a structure can be generated and varied. Such distortion fields could, for instance, be generated by temperature fields associated with proper heating or cooling of structure elements or by introducing mechanical dislocations and disclinations in truss or beam structures. For instance, shortening or elongating of bars OC and OD of a hyperstatic truss shown in Fig. 4 induces self-equilibrated stress states σ^r and the associated initial strain states ϵ^r. These states can be used to control the total stress and strain states $\sigma = \sigma^r + \sigma^\ell$ and $\epsilon = \epsilon^r + \epsilon^i$ of a structure where σ^ℓ, and ϵ^i are the stress and strain states due to external loading. The concept of active control of civil engineering or space structures by means of distortion fields provides an additional feasible option especially in cases where quasistatic control is required, for instance in counteracting the environmental temperature changes and their effect on structure stress and displacement fields.

297

In Section 2, the problem formulation for stress control and optimality conditions will be presented whereas in Section 3 some illustrative examples will be discussed. In Section 4, the problem of active control of structure deflections will be presented with simple illustrative examples.

2. Active Control of Structure Stresses

Consider an elastic body (in particular, truss or beam structure) subjected to an external loading $L(t)$ specified on its boundary surface and varying in time. Assume that besides this loading, there is also a distortion field $\epsilon^o(t,x)$ introduced into a structure in order to reduce the excessive stress level. The resulting governing equations are as follows

$$R[\sigma(t),L(t)]=0$$
$$C[\epsilon(t)]=0 \tag{1}$$
$$\sigma(t)=D\cdot[\epsilon(t)-\epsilon^o(t)]$$

where

$$\sigma=\sigma^r+\sigma^\ell$$
$$\epsilon=\epsilon^r+\epsilon^\ell \tag{2}$$

In (1), R denotes the equilibrium operator (including internal equilibrium equations and boundary conditions), C is the compatibility operator and D denotes the elasticity (stiffness) matrix. When only initial distortions ϵ^o are introduced into the structure, the constitutive relation takes the form

$$\sigma^r=D\cdot(\epsilon^r-\epsilon^o)$$
$$\epsilon^r=\epsilon^o+\epsilon_r^o=\epsilon^o+C\cdot\sigma^r \ , \ \ C=D^{-1} \tag{3}$$

where C denotes the elastic compliance matrix and $\epsilon_r^o=C\cdot\sigma^r$ is the elastic strain tensor associated by Hooke's law to the residual stress σ^r, (cf. Fig. 1 illustrating decomposition of strain and stress). Note that the field ϵ^r satisfies compatibility conditions and is associated with the displacement field u^r satisfying the kinematic boundary conditions, whereas the stress field σ^r is self-equilibrated, so that

$$\int \sigma^r\cdot\epsilon^r\,dV=\int T\cdot u^b\,dS_u \tag{4}$$

where u^b are the specified displacements on the boundary portion S_u and T are the surface tractions. In the following, we shall discuss the cases when $u^b=0$ on S_u, that is rigidly supported structures, loaded on the boundary portion S_T.

The aim of introducing active control of distortions is to set bounds on local stresses, for instance, by imposing the inequality

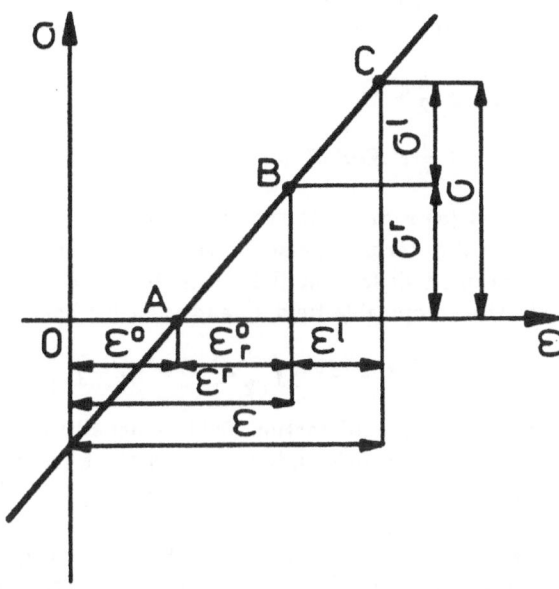

Figure 1: Strain and stress decomposition with initial prestrain

$$F[\boldsymbol{\sigma}(t)] \leq \sigma_o^2 \tag{5}$$

where F is a scalar function of stress tensor and σ_o is the admissible stress level. The optimal distribution of initial distortions satisfying (5) depends on the form of objective (or cost) function assumed in the problem. The first formulation is based on an energy measure of initial distortions, thus

$$\min I_{\epsilon^o(t)} = \min \int \frac{1}{2} \, \epsilon^o(t) \cdot \mathbf{D} \cdot \epsilon^o(t) dV \quad . \tag{6}$$

The other formulation could be based on the energy of distortion rates during the loading process, thus

$$\min J_{\dot{\epsilon}^o(t)} = \min \int \frac{1}{2} \, \dot{\epsilon}^o(t) \cdot \mathbf{D} \cdot \dot{\epsilon}^o(t) dV \tag{7}$$

where $\dot{\epsilon} = d\epsilon^o/dt$ denotes the instantaneous distortion rate. Whereas the cost function (6) is associated with the magnitudes of distortions, the cost function (7) is associated with their rate of variation. Both functions have their practical significance. For instance, when thermally induced distortions are considered, the objective function (6) is more proper since the continuing heat input is required to maintain distortion field at the specified values. On the other hand, for distortions induced by mechanical devices, the cost is associated with their variation requiring operation of such devices.

Using the constitutive equations (3) and the equality (4) with $u^b=0$ the objective function (6) can be presented in the alternative form

$$I_{\epsilon^o(t)}=\int \frac{1}{2}\epsilon^r\cdot\mathbf{D}\cdot\epsilon^r\,dV+\int \frac{1}{2}\epsilon^o_r\cdot\sigma^r\,dV$$

$$=\int \frac{1}{2}\epsilon^r\cdot\mathbf{D}\cdot\epsilon^r\,dV+\int \frac{1}{2}\epsilon^o_r\cdot\mathbf{D}\cdot\epsilon^o_r\,dV \tag{8}$$

The first term of (8) represents the energy norm of the compatible strain field ϵ^r whereas the second term represents the elastic energy of residual stresses σ^r. When unconstrained distortion fields can be introduced into the structure, one can eliminate the compatible field ϵ^r and then (8) becomes

$$H_{\epsilon^o(t)}=\int \frac{1}{2}\epsilon^o_r\cdot\sigma^r\,dV=\int \frac{1}{2}\sigma^r\cdot\mathbf{C}\cdot\sigma^r\,dV=\int \frac{1}{2}\epsilon^o_r\cdot\mathbf{D}\cdot\epsilon^o_r\,dV \tag{9}$$

However, the introduced distortion field is usually constrained, for instance, by governing equations of a field inducing distortions (such as heat conduction equation in the case of thermal distortions) or by technological constraints. It is therefore not always possible to generate a field satisfying the condition $\epsilon^r=0$, $\epsilon^o=-\epsilon^o_r$. Nevertheless, the residual stress energy measure (9) can be used in many cases as the objective function or cost of prestressing the structure also for non-vanishing compatible field ϵ^r. A more general measure would not be related to elastic energy and takes a form

$$G=\int \Phi(\sigma^r)dV=\int \Psi(\epsilon^r-\epsilon^o)dV \tag{10}$$

where Φ and Ψ are arbitrary, positive definite functions of the residual stress σ^r, or strains ϵ^r and ϵ^o.

It can be proved that the assumption (9) corresponds to the simplest formulation for which the general optimality conditions provide the analogy with the deformation theory of plasticity. Consider the following optimization problem

$$\min H_{\epsilon^o(t)}=\min \int \frac{1}{2}\sigma^r\cdot\mathbf{C}\cdot\sigma^r\,dV \tag{11}$$

subject to

$$\mathrm{div}\ \sigma^r=0\ ,\quad \sigma^r\cdot n=0\ \mathrm{on}\ S_T$$

$$F(\sigma)=F(\sigma^r+\sigma^t)\le\sigma^2_o \tag{12}$$

Introducing the augmented functional

$$H'(\sigma^r,\lambda,\Psi)=\int \{\frac{1}{2}\sigma^r\cdot\mathbf{C}\cdot\sigma^r\,dV+\lambda\ \mathrm{div}\ \sigma^r+\Psi[F(\sigma^r+\sigma^{'})-\sigma^2_o]\}dV+\lambda\int \sigma^r\cdot ndS_T \tag{13}$$

where λ and Ψ are vector and scalar Lagrange multipliers, the stationarity H' with respect to σ^r, λ and Ψ provide the optimality conditions and constraint equations, namely (cf. Appendix)

$$\text{grad}^s \lambda = \epsilon_r^o + \Psi \frac{\partial F}{\partial \sigma} \ , \quad \lambda = u$$

$$\text{div } \sigma' = 0 \ , \quad \sigma' \cdot n = 0 \text{ on } S_T \tag{14}$$

$$F(\sigma) - \sigma_o^2 \leq 0 \ , \quad \Psi \cdot (F - \sigma_o^2) = 0 \ , \quad \Psi \geq 0$$

where grad^s denotes the symmetric part of the gradient. As λ can be identified with the displacement field u, the first equation (14) provides the relation

$$\epsilon^o = \begin{cases} \Psi \dfrac{\partial F}{\partial \sigma} & F - \sigma_o^2 = 0 \\[2mm] 0 & F - \sigma_o^2 < 0 \end{cases} \tag{15}$$

that is the initial distortion field is generated by the gradient rule associated with the constraint surface $F - \sigma_o^2 = 0$ where Ψ is a scalar multiplier in such rule. The relations (15) resemble the constitutive equations for the deformation theory of plasticity associated with the yield function $F - \sigma_o^2 = 0$ or the constitutive equations for a perfectly soft elastic material [9], cf. Fig. 2.

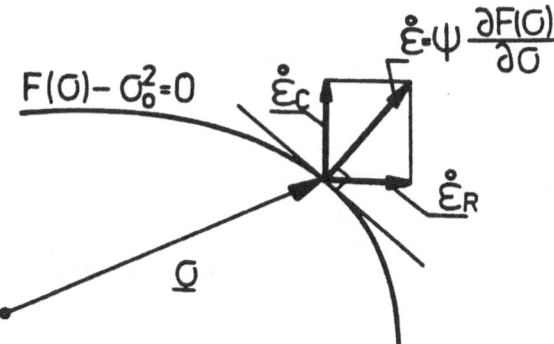

Figure 2: Normality rule for optimal distortion field

An alternative proof of optimality condition can be provided by applying the virtual work principle. The elastic stress energy of the body $\Pi(\sigma)$ equals

$$\Pi(\sigma) = \int \frac{1}{2} \sigma \cdot C \cdot \sigma \, dV = \int W(\sigma^r) dV + \int W(\sigma^i) dV$$

$$W(\sigma^r) = \frac{1}{2} \sigma^r \cdot C \cdot \sigma^r \ , \quad W(\sigma^i) = \frac{1}{2} \sigma^i \cdot C \cdot \sigma^i \tag{16}$$

For a rigidly supported structure $\Pi(\sigma)$ is equal to the complementary energy of the structure. Since the stress and strain fields σ^i and ϵ^i, due to external loading are not subject to variation, $\delta \sigma^i = \delta \epsilon^i = 0$, the variation of $\Pi(\sigma)$ equals

$$\delta \, \Pi_\sigma = \int \frac{\partial W}{\partial \sigma^r} \cdot \delta \, \sigma^r \, dV = \int \epsilon_r^0 \cdot \delta \, \sigma^r \, dV = \int (\epsilon^r - \epsilon^0) \cdot \delta \, \sigma^r \, dV$$

$$= - \int \epsilon^0 \cdot \delta \, \sigma^r \, dV = - \int \epsilon^0 \cdot \delta \, \sigma \, dV$$

Assume now that ϵ^0 is normal to the constraint surface $F(\sigma) - \sigma_o^2 = 0$. This implies that

$$\epsilon^0 \cdot \delta \, \sigma \leq 0 \tag{17}$$

and in view of (17), there is

$$\delta \, \Pi_\sigma \geq 0 \tag{18}$$

that is a local minimum of $\Pi(\sigma)$ occurs within the class of variations of σ^r satisfying equilibrium conditions and constraint condition (12). It is seen that the minimum property of Π_σ is equivalent to well known Haar-von Kármán principle formulated for elastic-perfectly plastic materials obeying the constitutive law (15). In other words, the residual stresses developed during progressive plastic deformation, are optimal from the point of view of present formulation (that is minimizing residual stress energy and satisfying the constraint (12)). When the compatible strain field ϵ^r can be eliminated by the unconstrained distortion field, there is $\epsilon^0 = -\epsilon_r^0$ and in view of (3) we have

$$\sigma^r = -\mathbf{D} \cdot \frac{\partial F}{\partial \sigma} \cdot \Psi \tag{19}$$

Consider now the case when a set of point distortions are introduced within the structure. Denote by $\alpha^{(i)} \epsilon_{(i)}^0$ $((i) = 1, 2, \dots n)$ the values of localized distortions and by $\sigma_{(i)}^r(\mathbf{x})$ the associated residual stress fields corresponding to $\alpha^{(i)} = 1$, where $\alpha^{(i)}$ denote the scalar intensity factors of particular distortions. The total stress field now equals

$$\sigma = \sigma^i + \sum_{i=1}^n \alpha_{(i)} \sigma_{(i)}^r \tag{20}$$

and the optimization problem is formulated as follows

$$\min I = \min \frac{1}{2} \sum_{i=1}^n \alpha_{(i)} \epsilon_{(i)}^0 \cdot \mathbf{D} \cdot \alpha_{(i)} \epsilon_{(i)}^0$$

subject to $\quad F(\sigma^\ell + \sum_{i=1}^n \alpha_{(i)} \sigma_{(i)}^r) - \sigma_o^2 \leq 0 \tag{21}$

and equilibrium conditions. The augmented functional

$$I'(\alpha_{(i)}, \Psi) = \sum \frac{1}{2} \alpha_{(i)} \epsilon_{(i)}^0 \cdot \mathbf{D} \cdot \alpha_{(i)} \epsilon_{(i)}^0 + \int \Psi[F(\sigma^i + \sum \alpha_{(i)} \sigma_{(i)}^r) - \sigma_o^2] dV \tag{22}$$

now depends on a set of scalar parameters $\alpha_{(i)}$ and the multiplier Ψ. The stationarity conditions now are

$$\alpha_{(i)} \cdot \mathbf{D} \cdot \epsilon_{(i)}^{o} + \int \Psi \frac{\partial F}{\partial \sigma} \cdot \sigma_{(i)}^{r} dV = 0$$

$$F(\sigma) - \sigma_o^2 \leq 0 , \quad \Psi(F(\sigma) - \sigma_o^2) = 0, \quad \Psi \geq 0 \tag{23}$$

and constitute a set of equations from which localized distortion tensors can be specified, namely

$$\alpha_{(i)} \epsilon_{(i)}^{o} = \mathbf{C} \cdot \int \Psi \cdot \frac{\partial F}{\partial \sigma} \cdot \sigma_{(i)}^{r} dV \quad . \tag{24}$$

In the case of truss or beam structures, the formulation (21) can be reduced to a quadratic programming problem.

Similar relations are obtained when the distortion field depends on a set of parameters $a_i (i=1,2,...,n)$ thus $\epsilon^o = \epsilon^o(\underline{x}, a_i)$. The optimality conditions associated with the objective function (6) now take the form

$$\int \epsilon^o \cdot \mathbf{D} \frac{\partial \epsilon^o}{\partial a_i} dV = \int \Psi \cdot \frac{\partial F}{\partial \sigma} \cdot \frac{\partial F}{\partial a_i} dV , \quad i=1,2,...n \tag{25}$$

More general optimality conditions associated with the stress intensity measure (10) not coinciding with the energy measure are discussed in the Appendix.

The active structure adaptation to varying in time, quasistatic loading can be executed by an automatic control system whose scheme is shown in Fig. 3a. In the case of active control based on the rate norm (7) the idea of a respective control system is depicted in Fig. 3b.

3. Examples of Active Control of Stresses

3.1. Active control of a truss cantilever

Let us discuss an example of truss cantilever loaded by the force $P = $ const (Fig. 4) rotating around the joint 0. Assuming that distortions could be generated only in rods OC and OD the optimal active control problem (21) can be expressed in the following form:

$$\min(\overset{o}{\epsilon}{}_1^2 + \overset{o}{\epsilon}{}_2^2) \tag{26}$$

under constraints:

$$-aP \leq F_1 \leq aP \qquad -aP \leq X_1 \leq aP$$

$$-aP \leq F_2 \leq aP \qquad -aP \leq X_2 \leq aP \tag{27}$$

where

$\overset{o}{\epsilon}{}_1, \overset{o}{\epsilon}{}_2$ - the distortions generated independently in rods OC and OD, respectively

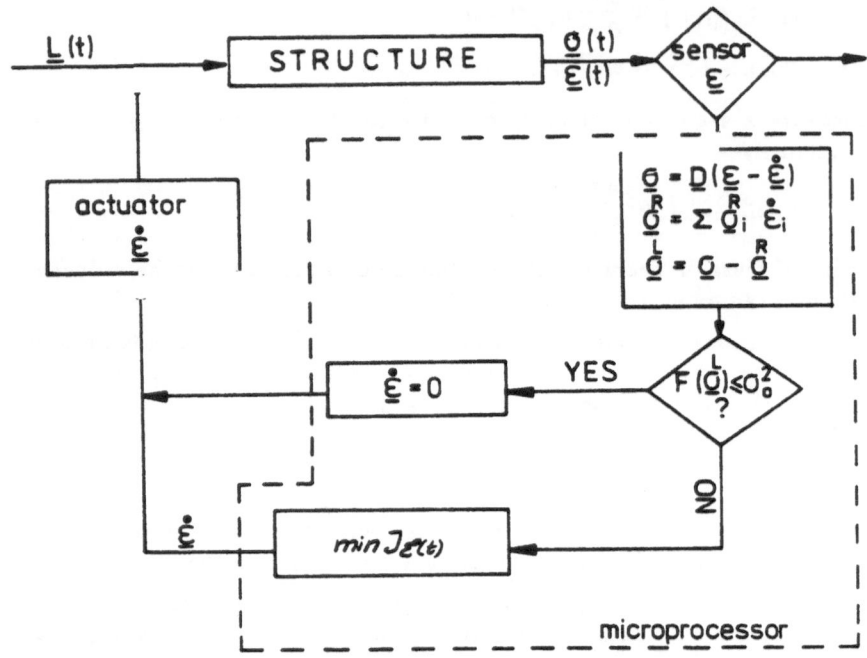

Figure 3a: Schematic diagram of active control system

F_1, F_2, X_1, X_2 - the internal forces in rods OA, OB, OC, OD respectively

aP - the limit value of internal forces (in extension and compression). The same cross-sectional area C for all elements is assumed.

Any self-equilibrated state of internal forces can be caused in the structure by some distortions $\mathring{\epsilon}_1, \mathring{\epsilon}_2$, because the degree of statical indeterminancy of the discussed truss is two.

Let us assume $a=0.4842$. It is easy to check that without prestressing $(\mathring{\epsilon}_1 = \mathring{\epsilon}_2 = 0)$ restrictions (27) are violated. For $\psi = 11/6$ e.g. the force $X_1 = 0.6023 P$ exceeds the limit value aP by 24.4%.

Determining the internal forces in the rods of the structure due to external load P and distortions $\mathring{\epsilon}_1, \mathring{\epsilon}_2$ as well, one can obtain:

$$X_1 = P\left[\frac{3}{3\sqrt{3}+1}\cos\psi + \frac{\sqrt{3}}{\sqrt{3}+3}\sin\psi\right] - \frac{EC}{2}\left[\left(\frac{1}{3\sqrt{3}+1} + \frac{3}{\sqrt{3}+3}\right)\mathring{\epsilon}_1 + \left(\frac{1}{3\sqrt{3}+1} - \frac{3}{\sqrt{3}+3}\right)\mathring{\epsilon}_2\right]$$

$$X_2 = P\left[\frac{3}{3\sqrt{3}+1}\cos\psi - \frac{\sqrt{3}}{\sqrt{3}-3}\sin\psi\right] - \frac{EC}{2}\left[\left(\frac{1}{3\sqrt{3}-1} - \frac{3}{\sqrt{3}+3}\right)\mathring{\epsilon}_1 + \left(\frac{1}{3\sqrt{3}+1} + \frac{3}{\sqrt{3}+3}\right)\mathring{\epsilon}_2\right]$$

$$F_1 = P\left[\left(1 - \frac{3\sqrt{3}}{3\sqrt{3}+1}\right)\cos\psi + \left(\frac{1}{\sqrt{3}} - \frac{1}{\sqrt{3}+3}\right)\sin\psi\right] + \frac{EC\sqrt{3}}{2}\left[\left(\frac{1}{3\sqrt{3}+1} + \frac{1}{\sqrt{3}+3}\right)\mathring{\epsilon}_1 + \right. \tag{28}$$

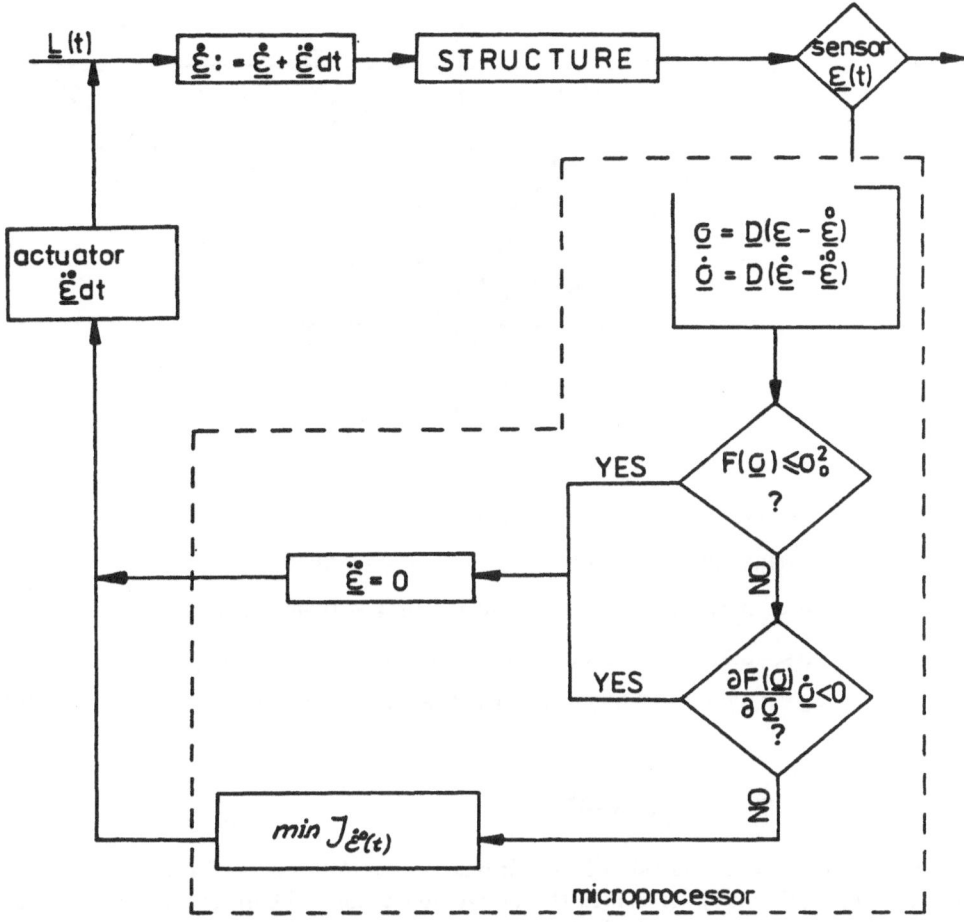

Figure 9b: Schematic diagram of active control system with rate dependent objective function

$$F_2 = P[1 - \frac{3\sqrt{3}}{3\sqrt{3}+1}]\cos\psi - (\frac{1}{\sqrt{3}} - \frac{1}{\sqrt{3}+3})\sin\psi] + \frac{EC\sqrt{3}}{2}[(\frac{1}{3\sqrt{3}+1} - \frac{1}{\sqrt{3}+3})\dot{\epsilon}_1 +$$

$$+ (\frac{1}{3\sqrt{3}+1} - \frac{1}{\sqrt{3}+3})\dot{\epsilon}_2]$$

$$+ (\frac{1}{3\sqrt{3}+1} + \frac{1}{\sqrt{3}+3})\dot{\epsilon}_2]$$

where: E - the Young's modulus.

Substituting formulae (28) to constraints (27) the set of inequalities describing the area of admissible solutions take the following form:

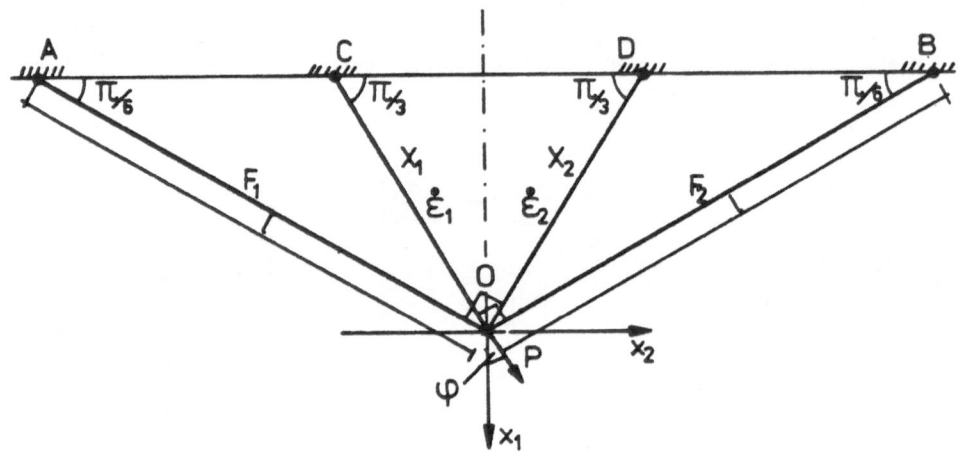

Figure 4: Truss loaded at O by rotating force P

$$-a \leq 0.3228\,\widetilde{\mathring{\epsilon}}_1 - 0.0432\,\widetilde{\mathring{\epsilon}}_2 + (0.1616\cos\psi + 0.3660\sin\psi) \leq a$$

$$-a \leq -0.0432\,\widetilde{\mathring{\epsilon}}_1 + 0.3228\,\widetilde{\mathring{\epsilon}}_2 + (0.1616\cos\psi - 0.3660\sin\psi) \leq a$$

$$-a \leq -0.3977\,\widetilde{\mathring{\epsilon}}_1 + 0.2363\,\widetilde{\mathring{\epsilon}}_2 + (0.4842\cos\psi + 03660\sin\psi) \leq a$$

$$-a \leq 0.2363\,\widetilde{\mathring{\epsilon}}_1 - 0.3977\,\widetilde{\mathring{\epsilon}}_2 + (0.4842\cos\psi - 0.3660\sin\psi) \leq a \qquad (29)$$

where

$$\widetilde{\mathring{\epsilon}}_1 = \frac{EC}{P}\mathring{\epsilon}_1, \quad \widetilde{\mathring{\epsilon}}_2 = \frac{EC}{P}\mathring{\epsilon}_2$$

From the numerical analysis of conditions (29) it follows that for $\Psi \in <0, \Pi/2>$ the left inequality $(29)^3$ describes the active constraint. Therefore the active control problem in the range $<0, \Pi/2>$ one can express in the form:

$$\min(\widetilde{\mathring{\epsilon}}_1^2 + \widetilde{\mathring{\epsilon}}_2^2) \qquad (30)$$

subject to the constraint

$$0.3977\,\widetilde{\mathring{\epsilon}}_1 - 02368\,\widetilde{\mathring{\epsilon}}_2 \leq a + 0.4842\cos\psi - 0.3660\sin\psi \qquad (31)$$

which leads to the following optimal solution:

$$\widetilde{\mathring{\epsilon}}_1 = \frac{0.3977}{0.3977^2 + 0.2363^2}(0.4842\cos\psi + 0.3660\sin\psi - a)$$

$$\widetilde{\mathring{\epsilon}}_2 = \frac{-0.2363}{0.3977^2 + 0.2363^2}(0.4842\cos\psi + 0.3660\sin\psi - a) \qquad (32)$$

Making use of periodicity of the solution (with respect to ψ), the result (32) is easily extended for the range $<0, \pi>$ (Fig. 5). The actual state of distortions depends only on the actual state of external load (that is the angle ψ). From the criterion (6) it follows that prestressing is not necessary and distortions vanish for $\psi=0$ ($\psi=\pi$) and in the vicinity of $\psi=(\pi/2)$ ($\psi=\dfrac{3}{2}\pi$).

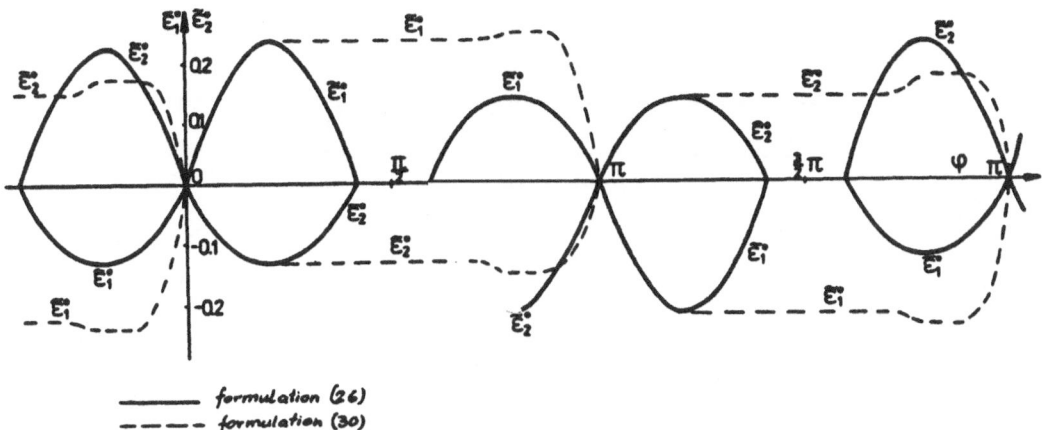

Figure 5: Optimal active prestrain of truss for two objective functions /26/ and /30/

Taking into account the incremental formulation (7) the active control problem leads in the case considered to minimization of the following expression

$$\min\left[(\frac{d\overset{\circ}{\epsilon}_1}{dt})^2+(\frac{d\overset{\circ}{\epsilon}_2}{dt})^2\right] \tag{33}$$

subject to constraints (29). The resultant distortions depend in this case on the whole history of external load. Knowing the values of the unknown functions $\overset{\circ}{\epsilon}_1(t)$, $\overset{\circ}{\epsilon}_2(t)$ it is necessary to determine their increments $\dfrac{d\overset{\circ}{\epsilon}_1}{dt}\Delta t$, $\dfrac{d\overset{\circ}{\epsilon}_2}{dt}\Delta t$ for successive instants. Substituting the values: $\overset{\circ}{\epsilon}_1(t)$, $\overset{\circ}{\epsilon}_2(t)$, $\psi(t)$, $\dfrac{d\psi}{dt}\Delta t$ into the set of conditions (29) one can obtain the inequalities constraining the increments $\dfrac{d\overset{\circ}{\epsilon}_1}{dt}\Delta t$, $\dfrac{d\overset{\circ}{\epsilon}_2}{dt}\Delta t$ for the time t. The latest constraints together with conditions (33) describe the optimal increment of distortions. A numerical procedure determines the resultant distortions $\overset{\circ}{\epsilon}_1(\psi)$, $\overset{\circ}{\epsilon}_2(\psi)$ (c.f. Fig. 5) using the starting point $\psi=0$ (with vanishing prestressing).

3.2. Active control of a three span beam

Let us discuss an example of a three span beam loaded by the constant force P
moving along the beam (Fig. 6).

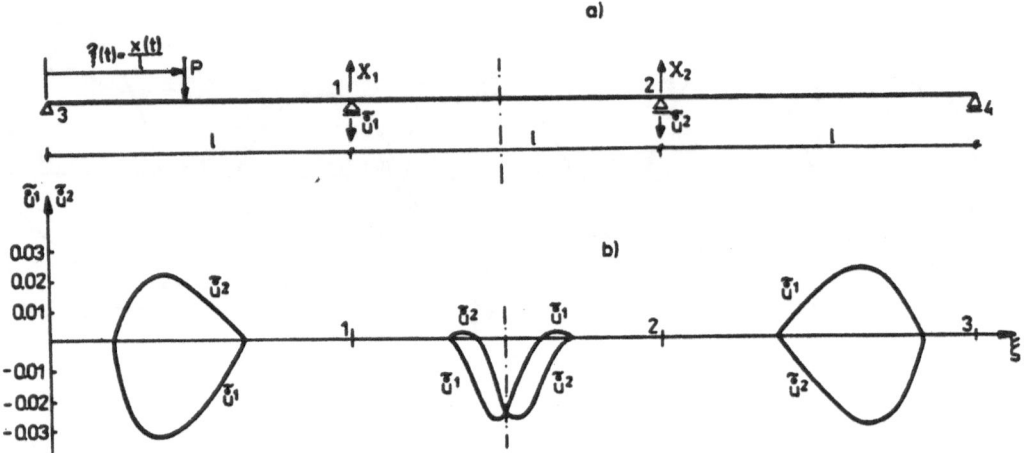

Figure 6: *a. Three-span beam loaded by a moving force P*
b. Optimal active prestrain by support displacements

The constraints (5) take the form of limit conditions for bending moments

$$-aP\ell \leq M \leq aP\ell \tag{34}$$

Assuming $a=0.15$, the non-prestressed structure is unable to carry the external
load. It is easy to check that the limit value $aP\ell$ is exceeded by 50% in this
case.

Let us assume that the initial states of stresses can be generated in the struc-
ture by shifting the supports 1 and 2 by the displacement $\overset{\circ}{u}_1$ and $\overset{\circ}{u}_2$. Any self-
equilibrated state of moments can be caused in the structure by the vertical
movements $\overset{\circ}{u}_1, \overset{\circ}{u}_2$ because the degree of statical indeterminancy of the discussed
beam is two.

The active control problem (21) can be expressed as follows

$$\min[(\overset{\circ}{u}_1)^2 + (\overset{\circ}{u}_2)^2] \tag{35}$$

subject to constraints (34).

The support reactions X_1, X_2 due to initial displacements $\overset{\circ}{u}_1, \overset{\circ}{u}_2$ and the external
load moving in the first span $X \in <0, \ell>$ take the form

$$X_1 = \frac{-6EJ}{5\ell^3}(8\mathring{u}_1 - 7\mathring{u}_2) + \frac{Px}{5\ell^3}(8\ell^2 - 3x^2)$$

$$X_2 = \frac{-6EJ}{5\ell^3}(-7\mathring{u}_1 + 8\mathring{u}_2) + \frac{2Px}{5\ell^3}(x^2 - \ell^2) \qquad (36)$$

while for the load P moving in the second span $x \in \langle \ell, 2\ell \rangle$ take another form

$$X_1 = \frac{-6EJ}{5\ell^3}(8\mathring{u}_1 - 7\mathring{u}_2) + \frac{P}{15\ell^3}\{8(3\ell - x)[(6\ell - x)x - \ell^2] - 7x(8\ell^2 - x^2)\}$$

$$X_2 = -\frac{6EJ}{5\ell^3}(-7\mathring{u}_1 + 8\mathring{u}_2) + \frac{P}{15\ell^3}\{-7(3\ell - x)[(6\ell - x)x - \ell^2] + 8x(8\ell^2 - x^2)\} \qquad (37)$$

The problem is symmetrical and our considerations can be restricted to the range $\langle 0, \frac{3}{2}\ell \rangle$. Since the function M of bending moments is piecewise-linear, it is sufficient to satisfy the constraints (34) only at the cross-sections under the load P ($y=x$) and above the supports 1 and 2 ($y=\ell$ and $y=2\ell$). Therefore the following set of constraints has to be taken into account

$$-aP\ell \leq M_{|y=x} = \frac{Px}{3\ell}(3\ell - x) - \frac{2}{3}X_1 x - \frac{1}{3}X_2 x \leq aP\ell$$

$$-aP\ell \leq M_{|y=\ell} = \frac{2Px}{3} - \frac{2}{3}X_1\ell - \frac{1}{3}X_2\ell \leq aP\ell \qquad (38)$$

$$-aP\ell \leq M_{|y=2\ell} = \frac{Px}{3\ell}\ell - \frac{1}{3}X_1\ell - \frac{2}{3}X_2\ell \leq aP\ell$$

for $x \in \langle 0, \ell \rangle$ and

$$-aP\ell \leq M_{|y=\ell} = \frac{P(3\ell - x)}{3\ell} - \frac{2}{3}X_1\ell - \frac{1}{3}X_2\ell \leq aP\ell$$

$$-aP\ell \leq M_{|y=x} = \frac{Px(3\ell - x)}{3\ell} - \frac{3\ell - x}{3}X_1 - \frac{1}{3}X_2 x \leq aP\ell \qquad (39)$$

$$-aP\ell \leq M_{|y=2\ell} = \frac{Px}{3\ell}\ell - \frac{1}{3}X_1\ell - \frac{2}{3}X_2\ell \leq aP\ell$$

for $x \in \langle \ell, 2\ell \rangle$

Substituting (36) to (38) and (37) to (39) the description of the admissible solution domain is obtained: for $x \in \langle 0, \ell \rangle$

$$-a - t(\xi) \leq (3\widetilde{\mathring{u}}_1 - 2\widetilde{\mathring{u}}_2)\xi \leq a - t(\xi)$$

$$-a - s(\xi) \leq (3\widetilde{\mathring{u}}_1 - 2\widetilde{\mathring{u}}_2) \leq a - s(\xi) \qquad (40)$$

$$-a - u(\xi) \leq (-2\widetilde{\mathring{u}}_1 + 3\widetilde{\mathring{u}}_2) \leq a - u(\xi)$$

where

$$\widetilde{\overset{\circ}{u}}_1 = \frac{6EJ}{5P\ell^3}\overset{\circ}{u}_1, \quad \widetilde{\overset{\circ}{u}}_2 = \frac{GEJ}{5P\ell^3}\overset{\circ}{u}_2, \quad \xi = \frac{x}{\ell}$$

$$t(\xi) = \xi[1 - \frac{\xi}{3} - \frac{12}{5}\xi(7 - 2\xi^2)]$$

$$s(\xi) = \xi[\frac{2}{3} - \frac{2}{15}(7 - 2\xi^2)]$$

$$u(\xi) = \xi[\frac{1}{3} - \frac{1}{15}(4 + \xi^2)]$$

and for $x \in <\ell, 2\ell>$ there is

$$-a - s(\xi) \le (3\widetilde{\overset{\circ}{u}}_1 - 2\widetilde{\overset{\circ}{u}}_2) \le a - s(\xi)$$

$$-a - t(\xi) \le (8 - 5\xi)\widetilde{\overset{\circ}{u}}_1 - (7 - 5\xi)\widetilde{\overset{\circ}{u}}_2 \le a - t(\xi) \tag{41}$$

$$-a - u(\xi) \le (-2\widetilde{\overset{\circ}{u}}_1 + 3\widetilde{\overset{\circ}{u}}_2) \le a - u(\xi)$$

where

$$s(\xi) = 1 - \frac{\xi}{3} - \frac{1}{15}(3d_\xi - 2b_\xi)$$

$$t(\xi) = \frac{\xi}{3}(3 - \xi) - \frac{1}{15}[(8 - 5\xi)d_\xi - (7 - 5\xi)b_\xi]$$

$$u(\xi) = \frac{\xi}{3} - \frac{1}{15}[-2d_\xi + 3b_\xi]$$

$$d_\xi = (3 - \xi)[(6 - \xi)\xi - 1]$$

$$b_\xi = \xi(8 - \xi^2)$$

From the numerical analysis of the functions $t(\xi)$, $s(\xi)$, $u(\xi)$ it follows that the active constraint in the set (40) is

$$(3\widetilde{\overset{\circ}{u}}_1 - 2\widetilde{\overset{\circ}{u}}_2) \le a - t(\xi) \quad \text{for} \quad \xi \in <0,1> \tag{42}$$

while in the set (41) there is

$$(8 - 5\xi)\widetilde{\overset{\circ}{u}}_1 - (7 - 5\xi)\widetilde{\overset{\circ}{u}}_2 \le a - t(\xi) \quad \text{for} \quad \xi \in <1,2> \tag{43}$$

Finally, the optimal active control problem (35) subject to constraints (42), (43) gives the result (cf. Fig. 6)

$$\left. \begin{aligned} \widetilde{\overset{\circ}{u}}_1 &= \frac{6EJ}{5P\ell^3}\overset{\circ}{u}_1 = \frac{3}{13}\frac{a - t(\xi)}{\xi} \\ \widetilde{\overset{\circ}{u}}_2 &= \frac{6EJ}{5P\ell^3}\overset{\circ}{u}_2 = \frac{-2}{12}\frac{a - t(\xi)}{\xi} \end{aligned} \right\} \quad \text{for} \quad \xi \in <0,1>$$

$$\left.\begin{aligned}
\widetilde{u}_1 &= \frac{6EJ}{5P\ell^3}\overset{\circ}{u}_1 = \frac{(8-5\xi)(a-t(\xi))}{(8-5\xi)^2+(7-5\xi)^2} \\
\widetilde{u}_2 &= \frac{6EJ}{5P\ell^3}\overset{\circ}{u}_2 = \frac{-(7-5\xi)(a-t(\xi))}{(8-5\xi)^2+(7-5\xi)^2}
\end{aligned}\right\} \quad \text{for } \xi \in <1,2> \tag{44}$$

The prestressing is necessary when the load is moving near the midpoint of the span and is vanishing when the load is moving near the support points.

4. Active Control of Structure Deflections

In this section, we shall discuss the case of active control in order to minimize local structure deflections. Consider a structure whose displacement at a point D along the direction d is to be minimized. Considering an adjoint structure with the same support conditions and loaded by a unit force $\mathbf{P}^a=1$ at D along d, we can write

$$\mathbf{P}^a \cdot u_D = \int \sigma^a \cdot \epsilon(t)dV = \int \sigma^a \cdot \epsilon'(t)dV + \int \sigma^a \cdot \epsilon^r(t)dV \tag{45}$$

where σ^a denotes the stress state in the adjoint structure loaded by the concentrated force $\mathbf{P}^a=1$ at D, $\epsilon'(t)$ is the strain field in the primary structure due to external loading $\mathbf{L}(t)$ and $\epsilon^r(t)$ is the compatible strain field due to initial distortion field $\epsilon^o(t)$. The problem of optimal active distortion control can be formulated as follows

$$\min |u_D| = \min \left| \int \sigma^a \cdot \epsilon'(t)dV + \int \sigma^a \cdot \epsilon^r(t)dV \right| \tag{46}$$

with the global or local constraint set on distortions, for instance

$$\int \Psi(\epsilon^o)dv \leq A \tag{47}$$

or

$$H(\epsilon^o) \leq K \tag{48}$$

where Ψ and H are scalar functions of distortions and A, K are admissible upper bounds on function values.

In the case of truss or beam structures with point distortions $\epsilon^o_{(i)}(i=1,2,...\ell)$, the initial strain field is specified as follows

$$\epsilon^r(t) = \sum_{i=1}^{l} \epsilon^r_{(i)} \cdot \epsilon^o_{(i)}(t) \tag{49}$$

where $\epsilon^r_{(i)}$ denotes the initial strain field due to unit distortion $\epsilon^o_{(i)}=1$. The optimal prestress is now reduced to a linear programming problem

$$\min \left| \sum_{i=1}^{\ell} \epsilon_{(i)}^{o} \int \epsilon_{(i)}^{r} \cdot \sigma^{a} dV + \int \epsilon^{r}(t) \cdot \sigma^{a} dV \right| \tag{50}$$

subject to $|\epsilon_{(i)}^{o}(t)| \leq \bar{\epsilon}$ $i = 1, 2, \dots \ell$.

The following example illustrates the applicability of this formulation to structural problems.

4.1. Example of active control of truss cantilever

Let us discuss the previously considered example of truss cantilever (Fig. 4) with minimized displacements of the joint 0 due to the active prestressing (cf. formulation (50))

$$\min(|u_1| + |u_2|) \tag{51}$$

subject to constraints (48). The trivial result is obtained

$$u_1 = u_2 = 0 \tag{52}$$

in the case of nonactive constraints (48). The displacements u_1, u_2 can be expressed in terms of forces F_1, F_2 and then, making use of $(28)^{3,4}$ in terms of distortions $\mathring{\epsilon}_1, \mathring{\epsilon}_2$:

$$u_1 = \frac{\ell}{2EC\cos\Pi/3}(F_1 + F_2) = \frac{\ell}{2EC\cos\Pi/3}[2P\cos\psi(1 - \frac{3\sqrt{3}}{3\sqrt{3}+1}$$
$$+ \frac{EC\sqrt{3}}{3\sqrt{3}+1}\mathring{\epsilon}_1 + \frac{EC\sqrt{3}}{3\sqrt{3}+1}\mathring{\epsilon}_2]$$

$$u_2 = \frac{\ell}{2EC\cos\Pi/6}(F_1 - F_2) = \frac{\ell}{2EC\cos\Pi/6}[2P\sin\psi(\frac{1}{\sqrt{3}} - \frac{1}{\sqrt{3}+3}) \tag{53}$$
$$+ \frac{EC\sqrt{3}}{\sqrt{3}+3}\mathring{\epsilon}_1 - \frac{EC\sqrt{3}}{\sqrt{3}+3}\mathring{\epsilon}_2]$$

Substituting (53) to (52) one can obtain

$$\mathring{\epsilon}_1 + \mathring{\epsilon}_2 = -\frac{2P\cos\psi}{\sqrt{3}EC}$$

$$\mathring{\epsilon}_1 - \mathring{\epsilon}_2 = -\frac{2P\sin\psi}{EC} \tag{54}$$

and finally, the solution of distortions (cf. Fig. 7)

$$\mathring{\epsilon}_1 = -\frac{P}{EC}(\frac{\cos\psi}{\sqrt{3}} + \sin\psi)$$

$$\mathring{\epsilon}_2 = \frac{P}{EC}(\sin\psi - \frac{\cos\psi}{\sqrt{3}}) \tag{55}$$

In the case of active constraints (48), for example $\bar{\epsilon}/\frac{P}{EC} = 1$, the problem (50) can be expressed as follows

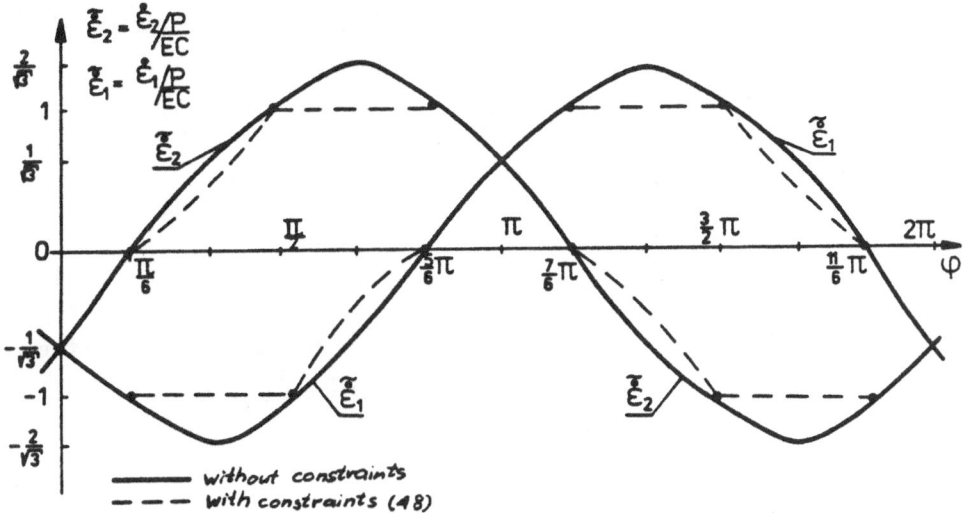

Figure 7: Optimal truss prestrain for minimum displacement

$$\min(|\overset{\circ}{\epsilon}_1 \int_V \epsilon_1^r \cdot \sigma_1'' dV + \overset{\circ}{\epsilon}_2 \int_V \epsilon_2^r \cdot \sigma_1'' dV + \int_V \epsilon^i(\psi) \cdot \sigma_1'' dV| +$$

$$|\overset{\circ}{\epsilon}_1 \int_V \epsilon_1^r \cdot \sigma_2'' dV + \epsilon_2^o \int_V \epsilon_2^r \cdot \sigma_2'' dV + \int_V \epsilon^i(\psi) \cdot \sigma_2'' dV|) \tag{56}$$

$$|\overset{\circ}{\epsilon}_1| \leq \frac{P}{EC}, \quad |\overset{\circ}{\epsilon}_2| \leq \frac{P}{EC}$$

where:

σ_1'' - the state of stresses in the structure caused by the vertical force $P=1$ $(\psi=0)$

σ_2'' - the state of stresses in the structure caused by the horizontal force $P=1$ $(\psi=\frac{\Pi}{2})$. Calculating values of the integrals, the formulation (56) can be written in a more simple way

$$\min \frac{\ell P^2}{EC}(\frac{3}{\sqrt{3}+3}|\tilde{\epsilon}_1 + \tilde{\epsilon}_2 + \frac{2}{\sqrt{3}}\cos\psi| + \frac{1}{\sqrt{3}+3}|\tilde{\epsilon}_1 - \tilde{\epsilon}_2 + 2\sin\psi|) \tag{57}$$

$$|\tilde{\epsilon}_1| \leq 1, \quad |\tilde{\epsilon}_2| \leq 1$$

where: $\tilde{\epsilon}_1 = \frac{\overset{\circ}{\epsilon}_1 EC}{P}, \quad \tilde{\epsilon}_2 = \frac{\overset{\circ}{\epsilon}_2 EC}{P}$.

In the following ranges $<0,\Pi/6>$, $<\frac{5}{6}\Pi, \frac{7}{6}\Pi>$, $<\frac{11}{6}\Pi, 2\Pi>$ the solution of this problem is identical to the solution (55) for the problem with nonactive constraints (48). In the range $<\Pi/6, \Pi/2>$ the distortion $\overset{\circ}{\epsilon}_1$ takes its limit value $\overset{\circ}{\epsilon}_1 = -1$, while the second distortion $\overset{\circ}{\epsilon}_2$ can be determined from the problem (57)[1] without constraints. One can check, that minimization of the expression $(57)^1$ leads to vanishing of the vertical part of displacement

$$u_1 = \overset{\circ}{\widetilde{\epsilon}}_1 + \overset{\circ}{\widetilde{\epsilon}}_2 + \frac{2}{\sqrt{3}} \cos \psi = 0 \tag{58}$$

Finally, the following distortions describe the active prestressing

$$\overset{\circ}{\widetilde{\epsilon}}_1 = -1$$

$$\overset{\circ}{\widetilde{\epsilon}}_2 = 1 - \frac{2}{\sqrt{3}} \cos \psi \tag{59}$$

Analogously, in the range $<\Pi/2, \frac{5}{6}\Pi>$, the solution is described by the condition of horizontal displacement vanishing while the distortion $\overset{\circ}{\widetilde{\epsilon}}_2$ takes its limit value $\overset{\circ}{\widetilde{\epsilon}}_2 = 1$:

$$\overset{\circ}{\widetilde{\epsilon}}_1 = -\left(1 + \frac{2}{\sqrt{3}} \cos \psi\right)$$

$$\overset{\circ}{\widetilde{\epsilon}}_2 = 1 \tag{60}$$

In the range $<\Pi, 2\Pi>$ the optimal solution is symmetrical (cf. Fig. 7). When the objective function (51) is taken in the quadratic form $J = (u_1)^2 + (u_2)^2$ it leads to a different solution within the ranges $<\frac{\Pi}{6}, \frac{5}{6}\Pi> <\frac{7}{6}\Pi, \frac{11}{6}\Pi>$. Vanishing of particular components of displacement does not occur in this case.

5. Concluding Remarks

The present paper provided a discussion of a relatively novel concept of active control of structures through varying distortion fields. Such distortion fields could be introduced by thermal gradients or mechanical devices changing lengths of structural members. Only quasistatic problems were considered though dynamic applications could be elaborated as well. The potential applications can be related to civil engineering structures (control and excessive displacements of tall buildings, counteraction to large thermal stresses and deflections) and to space structures or astronomical telescope mirrors. Usually the control of surface configuration was attained in such structures by application of concentrated forces [3 - 5]. However in a recent paper [10] the shape control by thermal distortions was analysed and demonstrated to be an effective method. It seems that the active control by distortion field offers a new option worthy of theoretical study and engineering application.

Appendix

Let us discuss a generalized case of optimal prestress by means of imposed distortions (cf. § 2) when the cost function takes the form

$$J(\sigma')=\int \Phi'(\sigma')dV \tag{A.1}$$

Separating the part describing the stress potential

$$J(\sigma')=\int \frac{1}{2}\sigma' \cdot C \cdot \sigma' \, dV + \int \Phi(\sigma')dV$$

where $\quad \int \Phi(\sigma')dV \overset{df}{=} \int \Phi'(\sigma') - \int \frac{1}{2}\sigma' \cdot C \cdot \sigma' \, dV \tag{A.2}$

the problem of minimization of the functional (1) subject to the following constraints

$$\text{div } \sigma' = 0 \ , \quad \sigma' \cdot n = 0 \ \text{ on } S_T$$

$$F(\sigma' + \sigma') \le \sigma_o^2 \tag{A.3}$$

can be equivalently described as the minimization problem of the substitute functional

$$J(\sigma',\lambda,\Psi)=\int\limits_V \{\sigma' \cdot C \cdot \sigma' + \Phi(\sigma') + \lambda \ \text{div } \sigma' +$$

$$+ \Psi[F(\sigma' + \sigma') - \sigma_o^2]\}dV + \int\limits_{S_T} \lambda \cdot \sigma' \cdot n \, dS \tag{A.4}$$

where

λ - the vector Lagrange's coefficient,

Ψ - the scalar Lagrange's coefficient.

Substituting the relation:

$$\lambda \cdot \text{div } \sigma' = \text{div } (\lambda \cdot \sigma') - (\text{grad}' \lambda) \cdot \sigma' \tag{A.5}$$

and making use of the Ostrogradsky's theorem the functional J can be expressed as follows

$$J(\sigma',\lambda,\Psi)=\int\limits_V \{\frac{1}{2}\sigma' \cdot C \cdot \sigma' + \Phi(\sigma') - (\text{grad}' \lambda)\sigma' +$$

$$+ \Psi[F(\sigma' + \sigma') - \sigma_o^2]\}dV + \int\limits_S \lambda \cdot \sigma' \cdot n \, dS + \int\limits_{S_T} \lambda \cdot \sigma' \cdot n \, dS \overset{df}{=} \tag{A.6}$$

$$\overset{df}{=} \int\limits_V F_1(\sigma',\lambda,\Psi)dV + \int\limits_{S_a} F_2(\sigma',\lambda)dS + \int\limits_{S_T} 2.F_2(\sigma',\lambda)dS$$

Now one can write the variation of the functional:

$$\delta J = \int_V [F_{,\sigma'} \cdot \delta \sigma' - \text{div } F_{1,\text{grad}'\lambda} \cdot \delta \lambda + F_{1,\Psi} \delta \Psi]dV +$$

$$+\int_{S_u}[(F_{1,\text{grad}'\lambda}\cdot\mathbf{n}+F_{2,\lambda}\cdot\delta\lambda+F_{2,\sigma'}\cdot\delta\sigma']dS$$

$$+\int_{S_T}[(F_{1,\text{grad}'\lambda}\cdot\mathbf{n}+2F_{2,\lambda})\cdot\delta\lambda+2F_{2,\sigma'}\cdot\delta\sigma']dS \qquad (A.7)$$

The stationarity condition $\delta J=0$ is assured by the following relations within the domain V:

$$F_{1,\sigma'}=C\cdot\sigma'+\frac{\partial\Phi}{\partial\sigma'}-\text{grad}'\lambda+\Psi\frac{\partial F}{\partial\sigma}=0$$

$$\text{div } F_{1,\text{grad}'\lambda}=\text{div } \sigma'=0$$

$$F_{1,\Psi}=F(\sigma'+\sigma^{\ell})-\sigma_o^2\leq0 \qquad\qquad (A.8)$$

$$\Psi F_{1,\Psi}=\Psi[F(\sigma'+\sigma^{\ell})-\sigma_o^2]=0 \ , \quad \Psi\geq0$$

and the equation for the boundary integrals:

$$\int_{S_u}\lambda\cdot\mathbf{n}\cdot\delta\sigma' dS+\int_{S_T}(\sigma'\cdot\mathbf{n}\cdot\delta\lambda+2\lambda\cdot\mathbf{n}\cdot\delta\sigma')dS=0. \qquad (A.9)$$

The last equation leads for any variations $\delta\lambda$ and $\delta\sigma'$ (self-equilibrated) to the boundary conditions:

$$\sigma'\cdot\mathbf{n}=0 \text{ on } S_T \text{ and } \lambda=0 \text{ on } S_u. \qquad (A.10)$$

Interpreting the stationarity conditions (8), (10) one can define the optimal distortion field by the following non-associated constitutive law (Fig. A.8a):

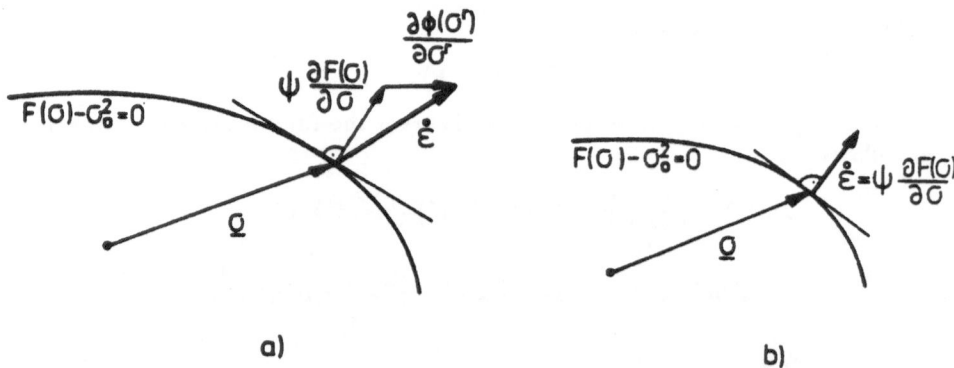

a)   b)

Figure A.8: *Non-associated and associated rules specifying optimal distortion fields associated with stress objective function /10/ and energy function /9/.*

$$\overset{\circ}{\varepsilon} = \text{grad}'\lambda - C\cdot\sigma' = \Psi\frac{\partial K(\sigma)}{\partial\sigma}+\frac{\partial\Phi(\sigma')}{\partial\sigma'} \qquad (A.11)$$

The component $\text{grad}^S\lambda$ describes the compatible part of distortions. If the function $\Phi(\sigma^r)$ vanishes the final soluion (A.11) describes distortions by the associated consitutive law (Fig. A.8b).

References

[1] Yao, J.T.P., "Concept of Structural Conrol" presented at the April 19-23, 1971 ASCE National Structural Enginering Meeting-Baltimore.

[2] Roorda, J., "Tendon Control in Tall Structures", *J. Struc. Div. ASCE ST3*, March 1975.

[3] Howell, W.E. and Creedon, J.F., "A Technique for Designing Active Control Systems for Astronomical Telescope Mirrors", NASA TN D-7090, 1973.

[4] Crawford, F.S., Schweimin, A.J., Smits, R.G., Muller, R.A., and Buffington, A., "Active Image Restoration with a Flexible Mirror", Berkeley Lawrence Radiation Lab. Report No. CONF-750645-2, June 1975.

[5] Bushnell, D., "Control of Surface Configuration by Application of Concentrated Load", *AIAA, Vol. 17, No. 1, January 1979.*

[6] Bushnell, D., "Control of Surface Configuration of Nonuniformly Heated Shells", *AIAA,* Vol. 17, NO. 1, January 1979.

[7] Gawecki, Z. and Mróz, Z., "Optimal Design Supports of Elastic Structures Subjected to Loads and Initial Distortions", *J. Structural Mech.*

[8] Holnicki-Szulc, J., "Initial Distortion Problems in Mechanics" and "Minisystem of Optimal Structural Adaptation", *Communicaciones tecnicas IIMAS-UNAM* No. 261, 274 Universidad de Mexico, Mexico, 1981.

[9] Mróz, Z., *Mathematical Models of Inelastic Material Behaviour,* University of Waterloo Press, 1983.

[10] Haftka, R.I. and Adelman, H.M., "An Analytical Investigation of Shape Control of Large Space Structures by Applied Temperatures", *AIAA J.,* Vol. 23, No. 3, 450-457, 1985.

ACTIVE CONTROL PERFORMANCE ENHANCEMENT FOR REDUCED ORDER MODELS OF LARGE SCALE SYSTEMS†

O. Ibidapo-Obe
University of Lagos
Akoka, Lagos
Nigeria

1. Introduction

Active structural control offers an exciting alternative solution to the dual problems of safety and comfort prevalent in design studies associated with tall buildings, long bridges, large spinning spacecrafts and other flexible engineering structures. This has led to an upsurge of interest by structural engineers exploring the various means by which control concepts may be utilized to advantage in design. A key problem in active control of structures is the problem of spillover which results when design procedures based on reduced order models (ROM) are applied to the actual large scale structures (LSS). A singular value decomposition (SVD) scheme for model reduction of LSS with a direct and simple procedure that provides a methodology for performance enhancement and spillover compensation is presented in this paper.

---

† See also Journal of Mathematical Analysis and Applications, Vol. 117, 1986.

Let the "full-order" discretized system (FOS) be

$$S_1: \dot{x} = Ax + Bu$$

$$y = Cx \tag{1}$$

where x is the n-dimensional state vector; A is the $n \times n$ system matrix, B is an $n \times p$ control matrix, u is the pth control vector; y is a q-dimensional output vector and C is a $q \times n$ observation matrix. It is assumed that the system is observed through m $(m < n)$ vector related to x in the following manner:

$$z = Dx \tag{2}$$

where D is the $(m \times n)$ aggregation matrix, which, from the geometric point of view, is the projection operator from R^n of x variables to the space R^m of z-variables.

The model reduction problem is to find a ROM

$$S_2: \dot{z} = Fz + Gu$$

$$w = Ez \tag{3}$$

where the matrix F is of reduced order m, G is an $(m \times p)$ matrix and E is an $(r \times m)$ matrix; under certain invariance constraints such that the system S_2 approximates S_1 optimally in sense of minimum cost of computation for system control, since this varies directly with the square of system order. It is often required that the ROM contains the most dominant modes, as well as modes determined by the disturbance environment and performance objectives, which, in most cases, for civil engineering structures, are the first few low fundamental frequency modes that are more likely to excite the system than the higher modes.

The approach is to differentiate equation (2) and substitute equation (1) into the result to obtain

$$\dot{z} = DAx + DBu . \tag{4}$$

If equation (2) is put in equation (3) we obtain

$$\dot{z} = FDx + Gu . \tag{5}$$

Now,

$$FD = DA \tag{6}$$

and

$$G = DB ; \quad C = ED . \tag{7}$$

Equations (6) and (7) indicats that the reduction process for the ROM is a direct function of the aggregation matrix D.

The matrices D and B commute, therefore the evaluation of G reduces to a simple matrix multiplication problem; the matrices F and E cannot, however, be obtained easily and hence their approximations are usually required. Current approximation techniques (Davison [4], Anderson [1], Skelton [8]) are tedious and become increasingly inefficient if the order of the reduced system is not reasonably small.

2. Singular Value Decompostion Approach

The search for an efficient and effective method for the evaluation of the matrices F and E in this paper follows the ad hoc scheme proposed by Aoki [2] and the statistical estimation procedure for the associated matching error by Soong [9]. This technique is based on the method of matrix singular value decomposition (SVD) (Golub [5], Ibidapo-Obe [6]) that has found useful applications in several linear least squares problems.

The SVD concept gives the inverse D^+ of the matrix D as

$$D^+ = V \Lambda U^T \tag{8}$$

where U, V are unitary matrices whose columns are the eigenvectors of matrices DD^T and D^TD respectively. D^T is the transpose of matrix D and

$$\Lambda = \begin{bmatrix} \sigma_1^{-1} & & 0 \\ & \sigma_2^{-1} & \\ 0 & & \sigma_n^{-1} \\ & & & 0 \end{bmatrix}_n \times n$$

where, $\sigma_1 \geq \sigma_2 \geq \cdots \geq \sigma_m \geq 0$, called singular values, are the non-negative square roots of the eigenvalues of D^TD. For the utilization of the SVD algorithm, zero and close-to-zero singular values are discarded.

The SVD process may be performed in two stages for large-scale systems:

(i) the matrix D is reduced to upper bidiagonal form by a sequence of Givens/Householder transformations,

(ii) an iterative QR algorithm computes the decomposition of the bidiagonal form and transforms it back to equation (8).

The matrices F and E may now be obtained from equation (6) as

$$F = DAD^+ \text{ and } E = CD^+ \tag{9}$$

so that substituting equation (8), we get

$$F = DA(V \Lambda U^T) \text{ and } E = C(V \Lambda U^T) . \tag{10}$$

The matrix D can also be written in the form

$$D = U\Sigma V^T \tag{11}$$

where

$$\Sigma = \begin{bmatrix} \sigma_1 & & & 0 \\ & \sigma_2 & & \\ 0 & & \sigma_m & \\ & & & 0 \end{bmatrix}_n \times n$$

so that

$$F = U\Sigma V^T A V\Lambda U^T . \tag{12}$$

This is an improvement, for systems with large aggregation matrices, over previous models where the estimate $\hat{F}$ of F is given as

$$\hat{F} = DAD^T(DD^T)^{-1} . \tag{13}$$

The need for matrix inversion is completely eliminated.

(i) For example, given that

$$A = \begin{bmatrix} -1 & -1 & 6 & -2 \\ 0 & -2 & 2 & 0 \\ 0 & 0 & -3 & 1 \\ 0 & 0 & 0 & -4 \end{bmatrix} \text{ with } \mathbf{b} = \begin{bmatrix} 1 \\ 0 \\ 0 \\ 1 \end{bmatrix} \text{ then } \mathbf{F} = \begin{bmatrix} -1 & -1 \\ 0 & -2 \end{bmatrix} \text{ and } \mathbf{g} = \begin{bmatrix} 2 \\ 1 \end{bmatrix}$$

where b and g are vectors rather than matrices B and G, using

$$D = \begin{bmatrix} 1 & 0 & 4 & 1 \\ 0 & 1 & 2 & 1 \end{bmatrix} .$$

The error associated with the estimate $\hat{F}$ of F is given as

$$cov(vec\, F) = \sigma^2[(DD^T)^{-1} \otimes I_m] \tag{14}$$

where the Kronecker product $\otimes$ and the $vec(\ )$ operator is defined as follows:

(i) $P \otimes Q = [p_{ij}Q]$,

(ii) $vec(P) = [P_1 P_2 \cdots P_n]^T$,

where P, Q are matrices and P_k is the kth column of the matrix P; $vec(P)$ is an $(n^2 \times x)$ vector.

For the example above,

$$cov\,(vec\,F)=\frac{\sigma^2}{9}\begin{bmatrix} 2 & 0 & -3 & 0 \\ 0 & 2 & 0 & -3 \\ -3 & 0 & 6 & 0 \\ -0 & -3 & 0 & 6 \end{bmatrix};$$

$||cov(vec\,F)||=0.394\sigma^2$.

The norms of the error matrices are used as a measure of the system performance.

The SVD approach eliminates the need for the computation of the inverse; it also presents a significant result for performance enhancement.

Now,

$$(DD^T)^{-1} = (D^T)^+ D^+$$
$$= (U\Lambda V^T)(V\Lambda U^T)$$
$$(DD^T)^{-1} = UI_m\Lambda U^T$$
$$= U\Lambda^2 U^T$$

hence the covariance matrix of $\hat{F}$

$$cov(vec\,\hat{F})=\sigma^2[U\Lambda^2 U^T\otimes I_m] \ . \tag{15}$$

Equation (15) is now of a simpler structure than equation (14).

The above result is given in a theorem below that provides for the improvement of the aggregation accuracy as a function of the system dimension.

Theorem 1: In the model of multivariate linear regression

$$DA=FD+E \tag{16}$$

where the matrx DA of dependent variables is $m\times n$, the matrix D of explanatory variables is $m\times n$, and the elements of E are statistically independent with zero means and identical variances σ^2, the SVD estimate is

$$\hat{F}=U\Sigma V^T AV\Lambda U^T \tag{17}$$

and the linear, unbiased, minimum-variance covariance of

$$cov(vec\,\hat{F})=\sigma^2[U\Lambda^2 U^T\otimes I_m] \ . \tag{18}$$

The problem of performance improvement/degradation when an additional row or column d (rank 1 addition) is augumented to the aggregation matrix D is estimated through the use of Bordering Method/Partitioning Method for block decomposition [10]. This result is stated in a second theorem below:

Theorem 2: Let the aggregation matrix $D=D_{n-1}$; that is the matrix of explanatory variables after the $(n-1)$th update. If d is the additional row vector required, then

$$D_n = (D_{n-1} | \alpha)^T , \quad [D_{n-1} D_{n-1}^T]^{-1} = U \Lambda^2 U^T$$

such that

$$D_n D_n^T = \begin{bmatrix} D_{n-1} D_{n-1}^T & D_{n-1} d^T \\ d D_{n-1}^T & d D_{n-1}^T \end{bmatrix} \tag{19}$$

and

$$(D_n D_n^T)^{-1} = \begin{bmatrix} P_{n-1} & r_n \\ q_n & \alpha_n^{-1} \end{bmatrix} \tag{20}$$

where

$$\alpha_n = dd^T - (d D_{n-1}^T)[D_{n-1} D_{n-1}^T]^{-1} D_{n-1} d^T$$
$$q_n = -\alpha_n^{-1}(d D_{n-1}^T)[D_{n-1} D_{n-1}^T]^{-1}$$
$$r_n = -[D_{n-1} D_{n-1}^T]^{-1}(D_{n-1} d^T)\alpha_n - 1$$
$$P_{n-1} = [D_{n-1} D_{n-1}^T]^{-1} - [D_{n-1} D_{n-1}^T]^{-1}(D_{n-1} d^T)q_n$$

The estimate of F and its minimum-variance covariance can be similarly expressed.

(ii) If

$$d = [0 \ 0 \ 1 \ 1]$$

D can be updated to

$$D_n = \begin{bmatrix} 1 & 0 & 4 & 1 \\ 0 & 1 & 2 & 1 \\ 0 & 0 & 1 & 1 \end{bmatrix}.$$

In that case, the new estimate of $\hat{F}$ is

$$\hat{F} = \begin{bmatrix} -1 & -1 & 0 \\ 0 & -2 & 0 \\ 0 & 0 & -3 \end{bmatrix};$$

and the corresponding error of estimate is

$$cov(vec\hat{F})=\frac{\sigma_2}{9}\begin{pmatrix} 2.25 & 0 & 0 & -2.25 & 0 & 0 & -2.25 & 0 & 0 \\ 0 & 2.25 & 0 & 0 & -2.25 & 0 & 0 & -2.25 & 0 \\ 0 & 0 & 2.25 & 0 & 0 & -2.25 & 0 & 0 & -2.25 \\ -2.25 & 0 & 0 & 8.25 & 0 & 0 & -6.75 & 0 & 0 \\ 0 & -2.25 & 0 & 0 & 8.25 & 0 & 0 & -6.75 & 0 \\ 0 & 0 & -2.25 & 0 & 0 & 8.25 & 0 & 0 & -6.75 \\ -2.25 & 0 & 0 & -6.75 & 0 & 0 & 20.25 & 0 & 0 \\ 0 & -2.25 & 0 & 0 & -6.75 & 0 & 0 & 20.25 & 0 \\ 0 & 0 & -2.25 & 0 & 0 & -6.75 & 0 & 0 & 20.25 \end{pmatrix}$$

The norm of the above matrix is $0.881\sigma^2$; this is a 3.55 magnification of the original ROM.

(iii) Similarly if $d=[1,1,1,1]$;

$$\hat{F}=\begin{pmatrix} -2.14 & 2.43 & -2.57 \\ -0.86 & 0.57 & 2.57 \\ 2 & -2 & 8 \end{pmatrix}$$

and the corresponding error of the estimate is

$$cov(vec\hat{F})=\frac{\sigma^2}{9}\begin{pmatrix} -2 & 0 & 0 & 1.67 & 0 & 0 & 2 & 0 & 0 \\ 0 & -2 & 0 & 0 & 1.67 & 0 & 0 & 2 & 0 \\ 0 & 0 & -2 & 0 & 0 & 1.67 & 0 & 0 & 2 \\ 3 & 0 & 0 & -4 & 0 & 0 & -3 & 0 & 0 \\ 0 & 3 & 0 & 0 & -4 & 0 & 0 & -3 & 0 \\ 0 & 0 & 3 & 0 & 0 & -4 & 0 & 0 & -3 \\ 0 & 0 & 0 & 12 & 0 & 0 & -9 & 0 & 0 \\ 0 & 0 & 0 & 0 & 12 & 0 & 0 & -9 & 0 \\ 0 & 0 & 0 & 0 & 0 & 12 & 0 & 0 & -9 \end{pmatrix}$$

The norm of this matrix is $0.338\sigma^2$ which is about 38% of that of example (ii).

Comparing $cov(vec\,\hat{F})$ before and after updating it thus appears that the errors in the third and fourth components of the system are considerably amplified, whilst the errors in the first and second components are slightly reduced. The Euclidean norm of the matrices may be used as a measure of the performance after the update.

3. Spillover Compensation

For the efficient realisation of the ROM; spillovers due to the application of ROM controllers applied to LSS, have to be minimzed. A method proposed by Martin and Soong [7] for the state selection in situations where a maximum of m measurements can only be made is given below:

The reduced model given by equation (3) can be put in the form

$$\dot{z}=F+g_1u_1+g_2u_2+ \cdots +g_pu_p \tag{21}$$

where the matrix

$$G=\begin{bmatrix} g_1^1 & g_2^1 & \cdots & g_p^1 \\ g_1^2 & g_1^2 & \cdots & g_p^2 \\ g_1^m & g_2^m & \cdots & g_p^m \end{bmatrix}$$

such that g_k, $k=1,2,...,p$ are column vectors of G and u_ℓ, $\ell=1,2,...,p$ are unit step inputs to the system.

Let the measurements be

$$\xi=R^Tz \tag{22}$$

where

$$R^T=\begin{bmatrix} f_1^1 & f_2^1 & \cdots & f_m^1 \\ f_1^2 & f_2^2 & \cdots & f_m^2 \\ f_1^m & f_2^m & \cdots & f_m^m \end{bmatrix}.$$

The columns of R are the eigenvectors of F^T; so that the component ξ_j

$$\xi_j=f_1^j z_1+f_2^j z_2+ \cdots +f_m^j z_m \; ; \quad j=1,2,...,k \tag{23}$$

gives a measure of the contribution of the jth mode to the system. It is possible therefore to eliminate those terms of ξ_j with smaller values without significantly affecting its value. The design procedure is as follows:

(i)   Obtain the normalized average steady state z'' of system (21) when the input forcing functions are applied as unit step functions:

$$z''=-\frac{1}{m}F^{-1}\sum_{i=1}^m g_i /(g_i^T g_i)^{1/2} \tag{24}$$

(ii) Let $\Gamma=diag\,(z_1'',z_2'',...,z_m'')$ and form

$$\Gamma f^1,\Gamma f^2, \ldots ,\Gamma f^k$$

such that

$$\Gamma f^j = [\bar{f}_1^j, \bar{f}_2^j, \ldots, \bar{f}_m^j], \quad j = 1, 2, \ldots, k .$$

(iii) Eliminate negligible elements which are common to all vectors $\Gamma f^1, \Gamma f^2, \ldots, \Gamma f^k$. If, for example, z_r, z_s and z_t are the elements, then an approximation for ξ_j is

$$\xi_j = f_1^j z_1 + \cdots + f_{r-1}^j z_{r-1} + f_{r+1}^j z_{r+1} + \cdots$$
$$+ f_{s-1}^j z_{s-1} + f_{s+1}^j z_{s+1} + \cdots + f_m^j z_m . \qquad (25)$$

This scheme will minimize observation spillover. For purposes of illustration, the above scheme was applied to the problems below:

(i) Given that

$$A = \begin{bmatrix} -1 & -1 & 6 & -2 \\ 0 & -2 & 2 & 0 \\ 0 & 0 & -3 & 1 \\ 0 & 0 & 0 & -4 \end{bmatrix}; \quad B = \begin{bmatrix} 1 \\ 0 \\ 0 \\ 1 \end{bmatrix}$$

The reduced system (before updating) is

$$\hat{F} = \begin{bmatrix} -1 & -1 & -8 \\ 0 & -2 & -6 \\ -1/3 & -1 & -6 \end{bmatrix}; \quad q = \begin{bmatrix} 2 \\ 1 \\ 1 \end{bmatrix}$$

with the redesigned observer as

$$\begin{bmatrix} \xi_2 \\ \xi_2 \end{bmatrix} = \begin{bmatrix} 0.24 & -0.079 \\ 0.34 & 0 \end{bmatrix} \begin{bmatrix} z_1 \\ z_2 \end{bmatrix} .$$

In this particular case, z_2 is not significant.

After updating however,

$$\hat{F} = \begin{bmatrix} -1 & 1- & -8 \\ 0 & -2 & -6 \\ -1/3 & -1 & -6 \end{bmatrix}; \quad g = \begin{bmatrix} 2 \\ 1 \\ 1 \end{bmatrix}$$

and the redesigned observer is

$$\begin{bmatrix} \xi_1 \\ \xi_2 \\ \xi_3 \end{bmatrix} = \begin{bmatrix} 0.010 & -0.013 & 0.133 \\ 0.100 & -0.057 & 0.030 \\ 0.059 & -0.039 & 0.105 \end{bmatrix} \begin{bmatrix} z_1 \\ z_2 \\ z_3 \end{bmatrix} .$$

Clearly, all the three modes are significant; therefore, updating the aggregation matrix will reduce the modal spillovers.

(ii) Given that

$$A = \begin{bmatrix} -0.1 & 1 & 1 \\ 1 & -20 & 0 \\ 1 & 1 & -30 \end{bmatrix} : B = \begin{bmatrix} 1 \\ 1 \\ 1 \end{bmatrix} ; C = \begin{bmatrix} 1 & 0 & 0 \\ 0 & 1/2 & 1/2 \end{bmatrix}$$

then the reduced system is

$$F = \begin{bmatrix} -0.1 & \sqrt{2} \\ 1/2 & -15\sqrt{2} \end{bmatrix} ; G = \begin{bmatrix} 1 \\ 1 \end{bmatrix}$$

and the redesigned observer is

$$\begin{bmatrix} \xi_1 \\ \xi_2 \end{bmatrix} = \begin{bmatrix} 0.2 & 0 \\ -114.72 & 0 \end{bmatrix} \begin{bmatrix} z_1 \\ z_2 \end{bmatrix}$$

Only the parameter z_1 is significant.

(iii) Given that

$$A = \begin{bmatrix} -1 & 0 \\ 0 & -2 \end{bmatrix} ; B = \begin{bmatrix} 1 & 0 \\ 0 & 1 \end{bmatrix} ; C = \begin{bmatrix} 1 & 1 \\ 1 & 0 \end{bmatrix}$$

then a new system (not of reduced dimension) is

$$F = \begin{bmatrix} -9.145 & 5.653 \\ 5.653 & -3.494 \end{bmatrix} ; G = \begin{bmatrix} 1 & 1 \\ 1 & 0 \end{bmatrix}$$

and the redesigned observer is

$$\begin{bmatrix} \xi_1 \\ \xi_2 \end{bmatrix} = \begin{bmatrix} 4350.41 & 0 \\ -11387.78 & 0 \end{bmatrix} \begin{bmatrix} z_1 \\ z_2 \end{bmatrix}$$

Only parameter z_1 is significant.

(iv) It is given that if

$$A = \begin{bmatrix} -4 & -1 & -3 \\ 3 & 1 & 1 \\ 5 & 1 & 3 \end{bmatrix} ; B = \begin{bmatrix} 1 & 1 \\ -1 & 0 \\ -1 & 1 \end{bmatrix} ; C = \begin{bmatrix} 4 & -4 & 4 \\ 0 & 0 & 0 \end{bmatrix}$$

A reduced system cannot be obtained as the (FOS) system is not observable.

4. Conclusions

This paper discusses a scheme based on singular value decomposition technique for the model reduction of large scale systems; the paper completely circumvents the use of inverses and also provides a measure, through the minimum variance covariance, of performance.

It is further opined that the performance of the reduced order system may be enhanced by incrementing the order of the reduced system by rank one.

Observation spillovers are minimized by selecting the observer system through an eigenvalue retention approach. This is a plausible alternative to that proposed in Balas [3].

Acknowledgements

The assistance of Professor T.T. Soong (SUNY at Buffalo, New York) and the reviewers of this paper is thankfully acknowledged.

References

[1] Anderson, J.H., "Geometrical approach to the reduction of dynamical systems", *Proc. IEEE*, 114, 1967, pp. 1914-1981.

[2] Aoki, M., "Control of large-scale dynamic systems by aggregation", *IEEE Trans. Ant. Control*, AC-13, 1968, pp. 246-253.

[3] Balas, M., "Modal control of certain flexible dynamic systems", *SIAM J. Control and Opt.*, Vol. 16, 1978, pp. 450-462.

[4] Davidson, E.J., "A method for simplifying linear dynamic systems", *IEEE Trans. Ant. Control*, AC-11, 1966, pp. 93-101.

[5] Golub, G.H. and Reinsch, C., "Singular value decomposition and least squares solutions", *Numer. Math., 1970, pp. 403-420.*

[6] Ibidapo-Obe, O., "A note on model reduction of large-scale systems", *J. Math. Analy. Applics.*, 90(2), 1982, pp. 480-483.

[7] Martin, C.R. and Soong, T.T., "Modal control of multistorey structures", *ASCE J. of the Eng. Mech. Division*, 104(EM.4), 1976, pp. 613-623.

[8] Skelton, R.E., "Cost-Decomposition of linear systems with application to model reduction", *International J. of Control*, December, 1980.

[9] Soong, T.T., "On model reduction of large-scale systems", *J. Math. Analy. Applics.*, 60(2), 1977, pp. 477-482.

[10] Westlake, J.R., *A Handbook of Numerical Matrix Inversion and Solution of Linear Equations*, John Wiley, New York, 1968.

SYMMETRIZABLE STRUCTURES AND MODAL CONTROL

Daniel J. Inman
Department of Mechanical and Aerospace Engineering
State University of New York at Buffalo
Buffalo, New York

1. Introduction

Recently [1] it was shown that certain asymmetric lumped parameter systems have dynamical properties very similar to the properties of related symmetric systems. In fact, for certain systems there exists a similarity transformation that transforms the asymmetric system into an equivalent symmetric one. The work here intends to extend this idea to non-self-adjoint distributed parameter systems and to apply modal control theory to the results. The key to making this extension and the inspiration for this work has been provided by Leipholz [2] and Walker [3].

The work presented here examines distributed parameter systems with velocity dependent forces. In particular the question of the existence of eigenfunction expansions for non-self-adjoint systems is addressed. This has significance because of the recent interest in modal control methods for large flexible space structures [4] and because of popularity of modal testing and design in structural dynamics research.

2. Background

The existence of modes for a given structure is very important if one is interested in performing modal control. The results presented here consider distributed mass systems described by:

$$u_{tt}(x,t)+L_1 u_t(x,t)+L_2 u(x,t)=f(x,t) \text{ on } \Omega \tag{1}$$

plus boundary conditions of the form $Bu=0$ alone $\partial\Omega$ where:

$x=x(x_1,x_2,x_3)$ denotes the position in Ω a bounded open region in R^3 with boundary $\partial\Omega$,

$u(x,t)$ denotes the deflection in Ω,

superscript t denotes partial differentiation with respect to the time

$f(x,t)$ denotes external forces

L_1 and L_2 denote linear spatial differential operators of order n_1 and n_2, respectively,

B denotes a linear spatial operator reflecting the boundary conditions.

Let $\ell^2(\Omega)$ denote the Hilbert space of all square integrable functions in the Legesque sense defined by the inner product and norm denoted by:

$$(u,v)=\int_\Omega u(x)v(x)dx \tag{2}$$

$$||u(x)||=(u,u)^{1/2} \tag{3}$$

L_1 and L_2 are defined on a domain $D(L)$ consisting of the set of functions $u(\cdot,t)$ in $\ell^2(\Omega)$ such that all partial derivatives with respect to x of order up to and including k and in $\ell^2(\Omega)$, where $k=2\max(n_1,n_2)$ and such that $Bu=0$ along $\partial\Omega$ for all $t>0$. The above formulation can be used as a generic description of most vibrating systems including those with odd boundary conditions and physical configurations.

In the control problem, the function $f(x,t)$ becomes the applied control force. In the case under study here we are interested in position and velocity feedback control of a system subject to an impulsive disturbance (i.e., non-zero initial conditions). For the distributed actuator case $f(x,t)$ becomes

$$f(x,t)=-L_3 u_t(x,t)-L_4 u(x,t) \tag{4}$$

where L_3 and L_4 are operators having the same properties as L_1 and L_2. In other cases, $f(x,t)$ may represent a bounded finite dimensional (compact) feedback law representing a small number of actuators placed at various points in Ω.

3. Results

Following the results obtained for multidegree of freedom systems given in [1], a theory for the vibration and control of a subclass of non-self-adjoint distributed parameter systems can be developed. The partial differential equations and boundary conditions used to describe such systems require the use of unbounded operators making the mathematics a little more difficult and restrictive. In this section operator properties will be considered and used to transform the original non-self-adjoint problem into an equivalent self-adjoint problem. Then, results regarding the modal control of such systems will be considered.

4. Operator Products

The theory of unbounded operators is not as well developed as the theory for matrices is. However, several results for the factoring of an unbounded operator can be stated based on theories developed for matrices and bounded operators. First, consider an unbounded non-self-adjoint operator L defined on a domain $D(L)$ dense in $L^2(\Omega)$. Furthermore, assume that there exists a positive definite self-adjoint operator T such that its inverse is compact. A differential operator with a Green's function inverse is an example of such an operator T. Let the operator T be defined such that the product operator $P=TL$ is self-adjoint and the operator $T^{1/2}P^{-1}T^{1/2}$ is compact, then it can be shown that the operator L has real eigenvalues and a real set of eigenfunctions. The eigenfunctions of $T^{1/2}P^{-1}T^{1/2}$ are orthogonal and can be used to provide a modal expansion for $u(x,t)$. To see this, note that a positive definite operator T has a unique square root, denoted $T^{1/2}$, defined such that $T^{1/2}T^{1/2}=T$ (see Kato [5], for instance). Furthermore, the operator $T^{1/2}$ is positive definite and also as a compact inverse, $T^{-1/2}$. Assuming that the domains and ranges of the operators forming the following products are such that the products are in fact defined (see Figure 1), then we can write $P=TL$, by which we mean

$$Pu = TLu \tag{5}$$

for all u in the domain $D(P)$ defined by the set of all functions u such that u is in $D(L)$ and Lu is in $D(T)$. This is denoted by

$$D(P)=\{u \,|u \text{ in } D(L) \text{ and } Lu \text{ in } D(T)\}.$$

Note, that by assumption, P is self-adjoint with compact resolvent. Also note that since $T^{-1/2}$ is compact (hence bounded) it makes sense to let $u=T^{-1/2}w$ in (5) and operate on the result with $T^{-1/2}$ to get

$$T^{-1/2}PT^{-1/2}w = T^{1/2}LT^{-1/2}w \tag{6}$$

where w lies in $D(T^{-1/2}PT^{-1/2})$ defined by

$$D(T^{-1/2}PT^{-1/2})=\{u \,|u=T^{-1/2}w \text{ in } D(P)\} \tag{7}$$

Since the operator P is self-adjoint,

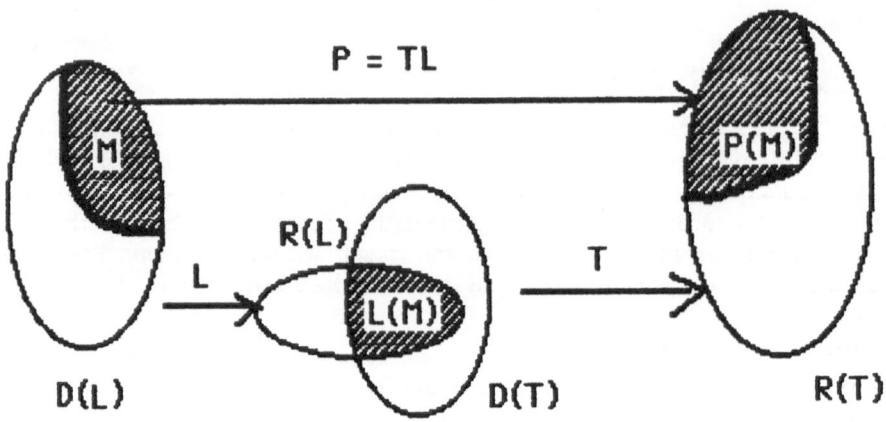

Figure 1

$$(Pu,u)=(u,Pu) \tag{8}$$

for all u in $D(P)=D(P^{*})$ where the superscript * denotes the adjoint operator. However, $u=T^{-1/2}w$, which is in $D(P)$, so that (8) becomes

$$(PT^{-1/2}w,T^{-1/2}w)=(T^{-1/2}w,PT^{-1/2}w) \tag{9}$$

The operator $T^{-1/2}$ is self-adjoint (because T is) so this becomes

$$(T^{-1/2}PT^{-1/2}w,w)=(w,T^{-1/2}PT^{-1/2}w) \tag{10}$$

so that the operator $T^{-1/2}PT^{-1/2}$ is indeed symmetric. Also, $D(P)=D(P^{*})$ and $D(T^{-1/2})=D(T^{-1/2^{*}})$, so that

$$
\begin{aligned}
D(T^{-1/2}PT^{-1/2}) &= \{u \mid T^{-1/2}u \text{ is in } D(P)\} \\
&= \{u \mid T^{-1/2}u \text{ is in } D(P^{*})\} \\
&= D((T^{-1/2}PT^{-1/2})^{*})
\end{aligned}
\tag{11}
$$

and hence the operator $T^{-1/2}PT^{-1/2}$ is actually self-adjoint. Thus, from Equation (6), the product operator $T^{1/2}LT^{-1/2}$ is also self-adjoint and hence has real eigenvalues (see for instance Kato [5]).

Since the operator $T^{-1/2}PT^{-1/2}$ has a compact inverse and is self-adjoint there exists a complete set of orthonormal eigenfunctions $\{w_{n}\}$ with corresponding eigenvalues, λ_{n}, associated with it. To see that this also yields a set of eigenfunctions for the non-self-adjoint operator L, note that the eigenvalue problem for $T^{-1/2}PT^{-1/2}$ is

$$T^{-1/2}PT^{-1/2}w_n = \lambda_n w_n \ . \tag{12}$$

This can also be written as

$$PT^{-1/2}w_n = \lambda_n T^{1/2}w_n \ . \tag{13}$$

Since $T^{1/2}$ is non-singular, define $u_n = T^{-1/2}w_n$

$$Pu_n = \lambda_n Tu_n$$

which becomes

$$Lu_n = T^{-1}Pu_n = \lambda_n u_n \ . \tag{14}$$

Hence, u_n is an eigenvector of L with associated eigenvalue λ_n. The relationship between the eigenfunctions, u_n, of L and those of $T^{-1/2}PT^{-1/2}$ is given by $u_n = T^{-1/2}w_n$. Note that the set $\{w_n\}$ is a complete orthonormal set while the set $\{u_n\}$ will not be. The next section comments on the effect of this result on the nature of the solution of (1) for the non-dissipative case, $L_1 = 0$.

5. Symmetrizable Systems

Consider Equation (1) with $L_1 = 0$, and L_2 such that the conditions of the preceding section prevail. That is, assume there exist a self-adjoint operator T such that $P = TL_2$ is self-adjoint and $T^{1/2}P^{-1}T^{1/2}$ is compact. Then let $u = T^{-1/2}w$ in Equation (1) to get:

$$T^{-1/2}w_{tt} + LT^{-1/2}w = f \ \text{in} \ \Omega \tag{15}$$

subject to the boundary conditions $BT^{-1/2}w = 0$ alone $\partial\Omega$. Assuming that f is in $D(T^{1/2})$ and premultiplying by $T^{1/2}$ yields the equivalent system,

$$w_{tt} + T^{1/2}LT^{-1/2}w = F \tag{16}$$

where $F = T^{1/2}f$. Because of (6), the system described by (16) is self-adjoint and has a complete set of eigenfunctions associated with it. Thus, $w(x,t)$ can be expanded as

$$w(x,t) = \sum_n^\infty a_n(t)w_n(t) \ . \tag{17}$$

Furthermore, $F(x,t)$ can also be expanded in terms of $w_n(x)$ to yield

$$F(x,t) = \sum_{n=1}^\infty b_n(t)w_n(x) \ . \tag{18}$$

Substitution of (17) and (18) into (16) multiplying by $w_n(t)$ and integrating over Ω yields that $a_n(T)$ must satisfy:

$$\ddot{a}_n(t) + \lambda_n a_n(t) = b_n(t) \quad n = 1,2 \ \cdots \tag{19}$$

and the appropriate initial conditions.

Note that the λ_n are real numbers and in fact will be positive real numbers if P is positive definite. Systems such that $L=T^{-1}P$ are called symmetrizable after the notion of symmetrizable (bounded) transformations [7] and are also called conservative systems of the second kind after Leipholz [2].

Equations (17), (18) and (19) constitute a theoretical modal analysis [8] of a particular subclass of non-self-adjoint systems. The eigenfunctions $w_n(x)$ are the mode shapes, λ_n are the square of the system's natural frequencies and the temporal coefficients $a_n(t)$ are referred to as the modal participation factors.

To see the connection between the solution $w(x,t)$ of the related self-adjoint problem (16) and the solution of the original non-self-adjoint system, note that for $u(x,t)=T^{-1/2}w(x,t)$ then

$$T^{-1/2}w(x,t)=\Sigma a_n(t)T^{-1/2}w_n(x)$$

converges since $T^{-1/2}$ is a compact (and hence continuous) operator and maps convergent sequences into convergent sequences. That is, if $u(x,t)$ is an arbitrary element in the intersection of $D(L)$ and $D(T^{1/2})$ then

$$||u(x,t)-\Sigma a_i u_i||=||T^{-1/2}w(x,t)-\Sigma a_i T^{-1/2}w_i||$$
$$\leq ||T^{-1/2}|| \, ||w(x,t)-\Sigma a_i w_i||$$
$$\leq M\epsilon$$

where ϵ is an arbitrary small number. Hence, the series $\{a_i u_i\}$ converges [6] to a solution of Equation (1).

6. Systems with Damping

For systems with dissipative viscous damping and gyroscopic forces the operator L_1 in Equation (1) is not zero and is also potentially non-self-adjoint. This section examines the subclass of non-self-adjoint systems with velocity dependent forces which have properties similar to self-adjoint systems.

The first obvious result is that if L_1 is proportional, and L_2 is symmetrizable, then the system described by Equation (1) will also be symmetrizable and there exists a complete set of eigenfunctions that can be used to expand the solution of the related symmetric form of (1) in.

To this end consider operators L_1 of the form

$$L_1=\alpha I+\beta L_2 \tag{20}$$

where I is the identity operator. Note, that since L_2 is symmetrizable,

$$TL_1u=\alpha Tu+\beta Pu \tag{21}$$

for all u in the proper domain and hence the operator L_1 is also symmetrizable.

The above yields a damped non-self-adjoint system with self-adjoint properties. Substitution of $u(x,t)=T^{-1/2}w(x,t)$ in (1) yields

$$T^{-1/2}w_{tt}+L_1T^{-1/2}w_t+L_2T^{-1/2}w=f \text{ in } \Omega \tag{22}$$

subject to boundary conditions $BT^{-1/2}w=0$ along $\partial\Omega$. As before assuming that
f is in $D(T^{1/2})$, premultiplying by $T^{1/2}$ yields

$$w_{tt}+T^{1/2}L_1T^{-1/2}w_t+T^{1/2}L_2T^{-1/2}w=F \text{ in } \Omega \tag{23}$$

where $F=T^{1/2}f$. Again because of (6), the coefficient of $w(x,t)$ is self-adjoint with
a complete set of orthonormal eigenfunctions $w_n(x)$ and has real eigenvalues.
Furthermore:

$$T^{1/2}L_1T^{-1/2}w_t=T^{1/2}(\alpha I+\beta L_2)T^{-1/2}w_t$$
$$=(\alpha I+\beta T^{-1/2}PT^{-1/2})w_t \tag{24}$$

so that the coefficient of $w_t(x,t)$ is also self-adjoint and has the same eigenfunctions as the coefficient of $w(x,t)$.

Substitution of the modal expansion (17) and (18) into (23), pre-multiplying
by $w_n(x)$, and integrating over Ω then yields the following formula for the temporal coefficients $a_n(t)$:

$$\ddot{a}_n(t)+(\alpha+\beta\lambda_n)\dot{a}_n(t)+\lambda_n a_n(t)=b_n(t) . \tag{25}$$

The solution, $a_n(t)$, of this expression represents the modal analysis of (1) under
the assumptions that L_1 and L_2 are non-self-adjoint, but that there exists a self-
adjoint positive definite operator T with compact inverse such that the product
operator $TL=P$ is defined and self-adjoint, the product operator $T^{1/2}P^{-1}T^{1/2}$ is
compact, and such that $L_1=\alpha I+\beta L_2$.

This last restriction can be somewhat relaxed, making the result more general, by making the additional assumption that L_1 and L_2 commute. Consider
the self-adjoint equation (23). It is known [9] that such systems have an expansion of the form given in Equation (17) where the complete orthonormal set of
functions $\{w_n(x)\}$ are the eigenfunctions of the operator $T^{1/2}L_2T^{-1/2}$, if and only
if the coefficients of w_t and w commute on the appropriate domain.
Define $\mathbf{P}_1=T^{1/2}L_1T^{-1/2}$ and $\mathbf{P}_2=T^{1/2}L_2T^{-1/2}$ and consider

$$\mathbf{P}_1\mathbf{P}_2=T^{1/2}L_1L_2T^{-1/2} \tag{26}$$
$$=T^{1/2}L_2L_1T^{-1/2} \tag{27}$$

This last expression is true since L_1 and L_2 are assumed to commute. Thus

$$\mathbf{P}_1\mathbf{P}_2=T^{1/2}L_2T^{-1/2}T^{1/2}L_1T^{-1/2} \tag{28}$$
$$=\mathbf{P}_2\mathbf{P}_1 . \tag{29}$$

Thus the coefficients of w_t and w commute, and (17) is a solution of (23).

Substitution of (17) into (23) and taking the inner product with $w_n(x)$ yields

$$\ddot{a}_n(t)+\beta_n \dot{a}_n(t)+\lambda_n a_n(t)=0 \tag{30}$$

where β_n are the eigenvalues of L_1 and λ_n are the eigenvalues of L_2. This follows from Equations (6) and (15). Note that the eigenfunctions are still those of $T^{1/2}L_2 T^{-1/2}$.

Equation (30) and (17) constitute the modal expansion and solution for the non-self-adjoint case with commuting symmetrizable coefficient operators. In the terms of languages used in [9], the assumption of commuting coefficient operators yields a "normal mode system" and the real eigenfunctions $\{w_n\}$ are referred' to as classical normal modes. Equation (30) also constitutes a theoretical modal analysis for the damped non-self-adjoint case.

7. Modal Control

In [10-12], theories for modal control have been presented for the self-adjoint version of Equation (1). These theories are extended here to include a certain subclass of non-self-adjoint systems. If the symmetrizable operators L_3 and L_4 also commute with the operator L_2 then Equation (30) becomes

$$\ddot{a}_n(t)+\beta_n \dot{a}_n(t)+\lambda_n a_n(t)=-\beta_n' \dot{a}_n(t)-\lambda_n' a_n(t)$$

or

$$\ddot{a}_n(t)+(\beta_n+\beta_n')\dot{a}_n(t)+(\lambda_n+\lambda_n')a_n(t)=0 \ . \tag{31}$$

Here the numbers β_n' and λ_n' are the eigenvalues (real) of the operators L_3 and L_4 respectively. In this case, the control problem is said to be both internally and externally decoupled [4,10] and the numbers β_n' and λ_n' can be chosen to shape each modal response $a_n(t)$ as desired. One could perform control on a system of this form by any method — root locus, optimal control, pole placement, etc. — and in each case select the appropriate β_n' and λ_n'. This is the crux of the argument presented in [10].

The major flaw with this approach is that an inverse problem must be solved to construct the operators L_3 and L_4 which have the spectrum β_n' and λ_n'. Once constructed, it may be difficult to build distributed actuators to yield L_3 and L_4. Consult reference [10] for details.

Another approach is to use point actuators which feedback velocity and position at a finite number of points in the domain Ω. When the function $f(x,t)$ is taken to be a control force of the form.

$$f(x,t)=\sum_{i=1}^{p} k_i \delta(x-x_i)u(x,t)+\sum_{i=1}^{p} c_i \delta(x-x_i)u_t(x,t) \tag{32}$$

usually stated for the control of flexible structures by using position and velocity feedback at p points on the structure. Here δ indicates the Dirac delta function. The control problem is to choose the constants c_i and k_i so that the response of

(1) has certain specified properties.

The common approach to solving the problem as first indicated by Berkman and Karnopp [13], is to truncate the series solution (1) and change the problem into a finite dimensional case. The general philosophy here being that for the free structure, the system can always be truncated at a large enough value of N (say the number of modes kept in a modal expansion of the solution) so that the response can be modeled as accurately as desired. With control applied to the structure, the truncated model is then split into two parts. The first part consists of a certain number of modes which define the controlled system. This is the model of the structure used in solving for the values of the control gains; c_i and k_i. The remaining portion of the truncated model, referred to as residual modes, are not considered in the calculation of the control law. However, since the action of the control also adds energy to the residual modes and hence the total response, the response may be effected in an undesirable way. This has been discussed by a number of authors (see [11] for instance). This action is sometimes referred to as control spillover into the residual modes.

Recently, Sakawa [12] has formulated the problem as a second order evolution equation for the special case of (1) with $L_1 = 2\alpha L_2$. Sakawa was able to show under certain assumptions on α that

$$||w(t)|| \leq Ke^{-\alpha t}||w(0)|| \tag{33}$$

where K is a constant dependent on truncation number and $w(t)$ is the state vector of modeled modes. This represents another approach to the spillover problem. The bound given in (33) is derived under the assumptions that

$$\alpha^2\lambda_i^2 - \lambda_i \neq 0 \qquad i = 1,2,3,... \tag{34}$$

and

$$\alpha^2\lambda_1^2 - \lambda_1 < 0 . \tag{35}$$

Here as before, λ_i are the eigenvalues of the operator L_2 and α is the damping coefficient.

Another approach to examining the effects of using a finite dimensional control law on a distributed parameter (flexible) system is to examine the convergence of the truncated control system and structure to a control law that is valid (optimal) for the full distributed system. Gibson [11] has examined this problem and found that if enough damping is modeled in the system that one can expect the finite dimensional control law to be valid for the full distributed system and that the free response of the system will have a uniform decay rate. These results again make an assumption about the damping in the structure. Namely that the damping is such that there exists a real number λ such that

$$||L_1u|| \leq \lambda^2||L_2u|| . \tag{36}$$

Both of these previously addressed problems, i.e., convergence and spillover, address the important question of the validity of a finite dimensional model of a distributed parameter system subject to finite dimensional control. In the following it will be shown that underdamped distributed parameter systems, as defined in the next section, represent a class of systems which are well behaved with respect to convergence, spillover and stability.

Both the approach of Sakawa and that of Gibson can be extended immediately to the subclass of non-self-adjoint structures discussed above. Symmetrizable systems combined with the assumptions given in [14] yield a class of problems which do not suffer from spillover or convergence difficulties.

First, we recall from [14] that the definition of an underdamped system is one in which *each* of the temporal functions $a_n(t)$ is an underdamped function of time. Motivated by the discriminant of Equation (30) we consider the operator $4TL_2 - TL_1^2$ which is self-adjoint. Using the results of Sakawa and Gibson it is shown that if a symmetrizable system is such that the self-adjoint operator $TL_2 - TL_1^2$ is positive definite, then the solution of the optimal control problem for the truncated optimal control solution will converge to a control law optimal for the full distributed system.

To see this, note first that $4TL_2 - TL_1^2$ is in fact self-adjoint. This is true since TL_2 is assumed to be self-adjoint and (with $u = T^{-1/2}w$, $w = T^{1/2}u$)

$$
\begin{aligned}
(TL_1^2 u, u) &= (TL_1 T^{1/2} T^{-1/2} L_1 u, u) \\
&= (TL_1 T^{-1/2} T^{1/2} L_1 T^{-1/2} w, T^{-1/2} w) \\
&= (T^{1/2} L_1 T^{-1/2} T^{1/2} L_1 T^{-1/2} w, w) \ .
\end{aligned}
$$

It has been shown previously that $T^{1/2} L_1 T^{-1/2}$ is self-adjoint so that

$$
\begin{aligned}
(TL_1^2 u, u) &= (T^{1/2} L_1^2 T^{-1/2} w, w) \\
&= (w, T^{1/2} L_1 T^{-1/2} T^{1/2} L_1 T^{-1/2} w) \\
&= (u, TL_1^2 u) \ .
\end{aligned}
$$

Next it will be shown that if a non-self-adjoint structure satisfies the symmetrizable condition and if $T(4L_2 - L_1^2)$ is positive definite then the modal control problem associated with (1) will not exhibit spillover (as discussed by Sakawa and Gibson).

8. Main Result

Before proceding with a formal proof, an overview is presented. First, since the operator $T^{1/2}L_2T^{-1/2}$ is assumed to have a compact inverse and is self-adjoint, it has an associated complete set of orthnormal eigenfunctions $\{w_n(x)\}$ which can be used to expand the solution of (23) in a uniformly convergent infinite series given by (17). The system is then defined to be *underdamped* if each $a_n(t)$ has the form

$$a_n(t)=A_n e^{-\varsigma_n \omega_n t}\sin(\omega_{dn} t+\phi_n)\tag{37}$$

where, ς_n, ω_n and ω_{dn} are the modal damping ratio, natural frequency and damped natural frequency respectively and where A_n and ϕ_n are constants determined by the initial conditions. It is first shown that if the operator $T(4T_1^2-L_1^2)$ is positive definite, then each $a_n(t)$ will have the form of (37). Then it is shown that the underdamping condition (37) yields the convergence of (17) to a solution of (23) and hence to a solution of (1). With the additional assumption that $T^{1/2}L_1T^{-1/2}$ is coercive it can be shown that the definiteness condition yields Gibson's convergence criteria *and* satisfies Sakawa's condition.

The first task is to verify that each $a_n(t)$ is of the form given in (37). To that end, consider the function $a_n(t)w_n(x)$. Substitution into (23) yields,

$$\ddot{a}_n(t)w_n(x)+\dot{a}_n(t)T^{1/2}L_1^{-1/2}w_n(x)+a_n(t)T^{1/2}L_2T^{-1/2}w_n(x)=0\tag{38}$$

Premultiplying by $w_n(x)$ and integrating over Ω yields that the characteristic roots associated with the initial value problem for $a_n(t)$ are determined by the discriminant (details can be found in [14]) given by

$$d(w_n)=(w_n,T^{1/2}L_1T^{-1/2}w_n)^2-4(w_n,L_2w_n) \ .\tag{39}$$

By applying the Cauchy inequality to the term $(w_n,T^{1/2}L_1^{-1/2}w_n)^2$ and comparing it with the assumed definiteness statement

$$(w_n,T(4L_2-L_1^2)w_n)>0\tag{40}$$

yields that $d(w_n)<0$ for *any* choice of $w_n(x)$. Hence, each of the temporal functions $a_n(t)$ must have the form given in (37).

Next consider the function formed by

$$w(x,t)=\sum_{n=1}^{\infty} a_n(t)w_n(x)\tag{41}$$

where each term satisfies Equation (38). This series will be a solution of (23) if (41) in fact converges. However, since

$$\sum_{n=1}^{\infty} |a_n(t)|^2<\infty, \quad \text{for all } t>0\tag{42}$$

the Riesz-Fischer theorem yields the desired convergence. By a similar argument w_t and w_{tt} also converge. In addition, since $T^{1/2}L_2T^{-1/2}$ has a compact inverse,

$T^{1/2}L_2T^{-1/2}w(x,t)$ can also be expanded (see page 261 of [5]) as

$$T^{1/2}L_2T^{-1/2}w(x,t)=\Sigma \lambda_n a_n(t)w_n(x) \tag{43}$$

$$=\Sigma \lambda_n(w,w_n)w_n(x) \tag{44}$$

Since $T^{1/2}L_1^{-1/2}$ has the same eigenfunctions we also get

$$T^{1/2}L_1T^{-1/2}w_t(x,t)=\Sigma \beta_n \dot{a}_n(t)w_n(x) \tag{45}$$

Again, for $u(x,t)=T^{-1/2}w(x,t)$, we have shown that the solution of (1) under the assumption stated above is

$$u(x,t)=\sum_{n=1}^{\infty} a_n(t)u_n(x) \tag{46}$$

where each temporal coefficient is underdamped.

It remains to show that Sakawa's and Gibson's results follow from the assumption of underdamping and hence that underdamped systems do not experience spillover. Calculation of the eigenvalues of $T(4L_2-L_1^2)$ yields

$$4\lambda_i(1-\alpha^2\lambda_i) \tag{47}$$

where L_1 is of the form $L_1=2\alpha L_2$ (Sakawa's assumption). Expression (27) must be positive for each value of i, since a self-adjoint operator is positive definite if each of its eigenvalues are positive. In particular, the first eigenvalue must be positive or

$$\alpha^2\lambda_1^2-\lambda_1<0 \; . \tag{48}$$

Hence, Sakawa's approach is valid for underdamped, non-self-adjoint systems (i.e., no spillover will occur in this case).

To see that Gibson's convergence criteria is satisfied by underdamped systems note that the Cauchy inequality is applied to $(w,T^{1/2}L_2T^{-1/2}w)$ and compared with (40) then

$$||w|| \, ||T^{1/2}L_2T^{-1/2}w|| \geq 1/4 \, ||T^{1/2}L_1T^{-1/2}w|| \tag{49}$$

Since L_1 is coercive, $||T^{1/2}L_1T^{-1/2}w|| \geq p \, ||w||$ and

$$||T^{1/2}L_1T^{-1/2}w||^2 \geq p \, ||T^{1/2}L_1T^{-1/2}w|| \, ||w|| \tag{50}$$

Comparing these two expressions yields

$$||w|| \, ||T^{1/2}L_2T^{-1/2}w|| \geq (p/4) \, ||T^{1/2}L_1T^{-1/2}w|| \, ||w|| \tag{51}$$

and thus there exist a positive constant λ^2 such that

$$\lambda^2 ||T^{1/2}L_2T^{-1/2}w|| \geq ||T^{1/2}L_1T^{-1/2}w|| \tag{52}$$

for all functions w. Thus the underdamping condition is that which is required to insure convergence and stability as delineated in [11].

9. Example

The standard example of the class of problems considered here is a Pflüger problem. A Pflüger problem is a model of a beam under an arbitrary tangential loading. Here we might consider the loading, denoted $f(x)$, as an applied distributed control. Walker [3] discussed the stability of this class of problems with

$$L_1 = 2\beta$$

$$L_2 = \partial^4$$

$$L_3 = 0$$

$$L_4 = -f(x)\partial^2$$

$$\Omega = (0,1)$$

where β is a damping coefficient and ∂^n denotes the nth partial derivative with respect to x. The $D(L_2 + L_4)$ is taken to be $D = \{u \mid u = \partial^2 u = 0 \text{ at } x = 0 \text{ and } x = 1\}$ which is dense in the underlying Hilbert space. The operator T in this case is defined by

$$T^{-1} = -\partial^2, D(T^{-1}) = \{u \mid u = 0 \text{ at } x = 0 \text{ and } 1\}$$

so that

$$TL = -(\partial^2 + f(x))$$

which is self-adjoint on the same domain. Hence, the operator

$$T^{1/2}(L_2 + L_4)T^{-1/2}$$

is self-adjoint and the above modal control theories can be applied.

10. Summary

A subclass of non-self-adjoint distributed parameter systems has been defined which is essentially self-adjoint in its behavior. Once classified, several results about the existence of modes in various cases were presented. Modal expansions for the related self-adjoint problem are then used to prove certain results for the control of symmetrizable distributed parameter systems in such a way as to avoid spillover.

References

[1]   Inman, D.J., "Dynamics of Asymmetric Nonconservative Systems", *ASME Journal of Applied Mechanics*, Vol. 50, 1983, pp. 199-203.

[2]   Leipholz, H.H.E., "On Conservative Elastic Systems of the First and Second Kind", *Ingenieur-Archiv.*, Vol. 43, 1974, pp. 255-271.

[3] Walker, J.A., "Liapunov Analysis of the Generalized Pflüger Problem", *ASME Journal of Applied Mechanics*, 1972, pp. 935-938.

[4] Inman, D.J., "Modal Decoupling Conditions for Distributed Control of Flexible Structure", *AIAA Journal of Guidance, Control and Dynamics*, Vol. 7, No. 6, 1985, pp. 750-752.

[5] Kato, T. *Perturbation Theory for Linear Operators*, Springer-Verlag, New York, 1966.

[6] Stakgold, I., *Boundary Value Problems of Mathematical Physics*, Vol. 1, MacMillan, New York, 1967.

[7] Lax, P.D., "Symmetrizable Linear Transformations", *Communications on Pure and Applied Mathematics*, Vol. VII, 1954, pp. 633-647.

[8] Meirovitch, L., *Analytical Methods in Vibration*, MacMillan, New York, 1965.

[9] Caughey, T.K. and O'Kelly, M.E.J., "Classical Normal Modes in Damped Linear Dynamic Systems," *ASME Journal of Applied Mechanics*, Vol. 32, 1965, pp. 583-588.

[10] Meirovitch, L. and Baruh, H., "Control of Self-Adjoint Distributed Parameter Systems", *Journal of Guidance, Control and Dynamics*, Vol. 5, Jan.-Feb., 1982, pp. 60-66.

[11] Gibson, J.S., "An Analysis of Optimal Modal Regulation: Convergence and Stability", *SIAM Journal of Control and Optimization*, Vol. 19, 1981, pp. 686-706.

[12] Sakawa, V., "Feedback Control of Second Order Evolution Equations with Damping", *SIAM Journal of Control and Optmization*, Vol. 23, 1984, pp. 343-361.

[13] Berkman and Karnopp, D., "Complete Response of Distributed Systems Controlled by a Finite Number of Linear Feedback Loops", *ASME Journal of Engineering for Industry*, Vol. 91, 1969, pp. 1063-1068.

[14] Inman D.J. and Andry, A.N., Jr., "The Nature of the Temporal Solutions of Damped Distributed Systems with Classical Normal Modes", *ASME Journal of Applied Mechanics*, Vol. 49, 1982, pp. 867-870.

APPLICATION OF LIMITING PERFORMANCE CONCEPTS TO STRUCTURAL CONTROL PROBLEMS

L. Kitis
Department of Mechanical Engineering
and Aerospace Sciences
University of Central Florida
Orlando, Florida
W.D. Pilkey and B.P. Robertson
Department of Mechanical and Aerospace Engineering
University of Virginia
Charlottesville, Virginia

1. Introduction

The absolute optimal response of a system for a given performance index or objective function is referred to as limiting performance. In computing the limiting performance, portions of a system being designed are replaced by active control forces. Since no restrictions are placed on the control forces when finding the limiting performance, this response is optimum over all possible design configurations. Knowledge of the limiting performance or control of a system is of great importance to the designer since it provides a convenient means for evaluating a current design and aids in the selection of optimal configurations and parameters.

To calculate the limiting performance, control forces are determined so that a performance index is optimized while typically satisfying a given set of constraints in response variables. If the performance index is defined to be a linear combination of response variables, control forces and external forces, then the limiting performance may be found using linear programming as outlined for transient systems by Sevin and Pilkey [1]. The linear programming formulation

343

requires that the response variables be expressed as functions of control forces. This has been accomplished using structural dynamic analysis programs which generate impulse responses at control force attachment points [2]. An alternative procedure set forth here is a modal approach. This approach is computationally convenient because it simplifies the steps necessary before linear programmng is initiated. Furthermore, modal truncation can be easily built in to handle large finite element models. Finally, the practicality of determining limiting performance characteristics is enhanced since modal properties may be experimentally available.

This paper summarizes work done to date in solution strategies for transient systems with specific attention given to large structural systems such as those occurring in civil engineering problems. In the limiting performance approach, the size of the linear program is primarily dependent on the number of discrete time intervals taken and not on the order of the structural control problem. This is an advantage when considering that large order finite element models routinely appear in civil engineering problems; however, what is an advantage from one point of view is a disadvantage from another. Specifically, a good solution usually requires a fine time discretization which increases the size of the linear program.

2. Formulation of the Linear Programming Problem

Many physical systems with n degrees of freedom subjected to excitation forces $\mathbf{f}$ and J control forces $\mathbf{u}$ may be described by the equations

$$\mathbf{M\ddot{x}} + \mathbf{C\dot{x}} + \mathbf{Kx} + \mathbf{Vu} = \mathbf{f} \tag{1}$$

where $\mathbf{x}$ is the displacement vector, $\mathbf{M}$, $\mathbf{C}$, and $\mathbf{K}$ are the $n \times n$ mass, damping and stiffness matrices, respectively, and $\mathbf{V}$ is the $n \times J$ control matrix. Control forces may or may not have replaced a portion of the system being designed, and can act as connections between substructures. The formulation of $\mathbf{V}$ may be deduced from the kinematics of the system. The i-th column of $\mathbf{V}$ corresponding to the i-th control force u_i can be determined by finding the component that a unit control force would have in the direction of the displacement coordinate for degrees of freedom associated with u_i.

In general, the measure of performance may be a linear combination of displacements, velocities, accelerations, control forces, and external forces. To define such a performance index let

$$\mathbf{h} = \mathbf{P}_1\mathbf{\ddot{x}} + \mathbf{P}_2\mathbf{\dot{x}} + \mathbf{P}_3\mathbf{X} + \mathbf{P}_4\mathbf{u} + \mathbf{P}_5\mathbf{f} \tag{2}$$

where $\mathbf{P}_k$ are prescribed coefficient matrices. For any system governed by (1), the system responses are linear functions of the control forces $\mathbf{u}$ to be determined and excitation forces $\mathbf{f}$ known. Thus, it is clear that the vector $\mathbf{h}$ may be considered to be explicitly dependent on time t and control forces $\mathbf{u}$,

$$h = h(t, u) \ .$$ (3)

The performance index ψ is then written in terms of h as

$$\psi = \max_i \ \max_t |h_i(t, u)|$$ (4)

where i varies over the elements of h and t over the time interval of concern. With these definitions, the optimization problem becomes a minimization of ψ with respect to the control forces u.

It is convenient at this point to discretize u in the time domain. If the time interval of interest is $0 \leq t \leq T_f$ and this interval is represented by L discrete instances of time, then each element of u takes on $(L-1)$ discrete values. This piecewise constant discretized function is denoted $\bar{u}$. Then, the discretized form of (2) becomes

$$\bar{h} = W\bar{u} + g(t)$$ (5)

where $g(t)$ is an explicit, known function of time and W is obtained by finding the response variables in terms of u and discretizing the results. The passage from (2) to (5) is an important step in developing the linear programming formulation.

To put the optimization procedure in standard linear programming form, define

$$\bar{u} = \begin{bmatrix} u(1) \\ u(2) \\ . \\ u(L-1) \end{bmatrix}$$ (6)

where $u(k)$ indicates $u(t)$ evaluated at $t = (k-1)T$.

Next, let

$$s = \begin{bmatrix} \psi \\ \bar{u} \end{bmatrix}$$ (7)

and

$$c^T = [1 \ 0 \ \cdots \ 0] \ .$$ (8)

Then the linear programming problem is to minimize

$$\psi = c^T s$$ (9)

subject to the constrants

$$\mathbf{Hz} > \mathbf{b} \ . \tag{10}$$

These constraints (10) always include the conditions

$$\psi \geq |h_i(t,\mathbf{u})| \tag{11}$$

in order to ensure that a minimum ψ is the least upper bound of the set $\{|h_i(t,\mathbf{u})|\}$ over time t and control forces $\mathbf{u}$. In addition to the constraints (11), (10) may also contain bounding constraints on response variables or control force magnitudes.

3. Modal Formulation for Undamped Systems

For an undamped system with $\mathbf{C}=0$, mode shapes ϕ_k and mode frequencies ω_k may be found from the eigenvalue problem

$$(-\omega_k^2 \mathbf{M} + \mathbf{K})\phi_k = 0 \ . \tag{12}$$

If mode shapes are normalized such that

$$\phi^T \mathbf{M} \phi = \mathbf{I} \tag{13}$$

$$\phi^T \mathbf{K} \phi = \Omega^2 = diag\,(\omega_1^2, \omega_2^2, \ldots, \omega_n^2) \tag{14}$$

where

$$\phi = [\phi_1 \phi_2 \cdots \phi_n] \tag{15}$$

then the solution to (1) with no rigid body modes is given by [3]

$$\mathbf{x}(t) = \sum_{k=1}^{n} C_k \phi_k e^{i\omega_k t} + \sum_{k=1}^{n} \phi_k \frac{1}{\omega_k}[\mathbf{f}(\tau) - \mathbf{V}\mathbf{u}(\tau)]\sin\omega_k(t-\tau)d\tau \ . \tag{16}$$

This may be written, for convenience, in the form

$$\mathbf{x}(t) = \mathbf{r}(t) \sum_{k=1}^{n} \phi_k \frac{1}{\omega_k} \int_0^t \phi_k^T \mathbf{V}\mathbf{u}(\tau)\sin\omega_k(t-\tau)d\tau \tag{17}$$

where $\mathbf{r}(t)$ represents contributions of the known external excitation $\mathbf{f}$ and the initial conditions.

Now if the control force vector $\mathbf{u}$ is discretized in piecewise constant fashion with uniform time increment T

$$T = t_{n+1} - t_n \tag{18}$$

then the discretized response at time $t = mT$ is obtained from (14)

$$\mathbf{x}(mT) = \mathbf{r}(mT) - \sum_{k=1}^{n} \phi_k \frac{1}{\omega_k^2} \sum_{i=0}^{m-1} \phi_k^T \mathbf{V}\mathbf{u}(iT)[\cos(m-i)\omega_k T] \tag{19}$$

To arrrive at (5) from (2), expressions for discretized velocity and acceleration are also needed. These may be obtained by differentiating (14) using Leibnitz's rule and discretizing the resulting equations. The final expressions are

$$x(mT)=r(mT)-\sum_{k=1}^{n}\phi_k\sum_{i=0}^{m=1}\frac{1}{\omega_k}\phi_k^T Vu(iT)[sin(m-i)\omega_k T-sin(m-i-1)\omega_k T] \qquad (20)$$

and

$$\ddot{x}(mT)=\ddot{r}(mT)-\sum_{k=l}^{n}\phi_k\sum_{i=0}^{m-1}\phi_k^T Vu(iT)[cos(m-i)\omega_k^T-cos(m-i-l)\omega_k T] \qquad (21)$$

$$-\sum_{k=1}^{n}\phi_k\phi_k^T Vu(mT) . \qquad (21)$$

In matrix notation, (19)-(21) have the form

$$x(mT)=r(mT)+Q\bar{u} \qquad (22)$$

$$\dot{x}(mT)=\dot{r}(mT)+R\bar{u} \qquad (23)$$

$$\ddot{x}(mT)=\ddot{r}(mT)+S\bar{u} \qquad (24)$$

where the matrices Q, R and S depend on modal properties V, and time. The discretized Equation (5) is obtained from (2) by using (22)-(24). Explicitly, the discretized vector $\bar{h}$ is given by

$$\bar{h}=(P_1 s+P_2 R+P_3 Q+P_4)\bar{u}+P_5\bar{f}+P_1\ddot{r}(mT)+P_2\dot{r}(mT)+P_3 r(mT)$$

Thus, the matrix W in (5) is identified as

$$W=P_1 S+P_2 R+P_3 Q+P_4 \qquad (26)$$

and the known function of time $g(t)$ is seen to be

$$g(t)=P_5\bar{f}+P_1\ddot{r}+P_2\dot{r}+P_3 r . \qquad (27)$$

The passage from (2) to the discretized form of (5) is now complete, so that for each time instant the inequalities (11) can be written as a linear combination of the entries of the vector z. This step essentially gives the linear programming problem in the standard form as described by equations (9) and (10).

3.1. Modal formulation for damped systems

When viscous damping is present, the eigenvalue problem

$$(\lambda_k^2 M+\lambda_k C+K)\psi_k=0 \quad 1\leq k\leq 2n \qquad (28)$$

is solved to determine the mode shapes ψ_k and the eigenvalues λ_k. The displacement response is then given in terms of these modal properties by a formula similar to (17). Specifically, if there are $2s$ real eigenvalues $\lambda_1,\lambda_2,\ldots,\lambda_{2S}$ and if the roots $\lambda_{n+s+1},\ldots,\lambda_{2n}$ are the complex conjugates of $\lambda_{2s+1},\ldots,\lambda_{n+s}$, then [3]

$$\mathbf{x}(t)=\mathbf{z}(t)-\sum_{k=1}^{2s}\boldsymbol{\psi}_k e^{\lambda_k t}\int_0^t e^{-\lambda_k \tau}\boldsymbol{\psi}_k^T\mathbf{V}\mathbf{u}(\tau)d\tau$$

$$-2\sum_{k=2s+1}^{n+s}\int_0^t\mathrm{Re}\{e^{\lambda_k(t-\tau)}\boldsymbol{\psi}_k\boldsymbol{\psi}_k^T\}\mathbf{V}\mathbf{u}(\tau)d\tau\ .\tag{29}$$

In (28) it is assumed that the eigenvectors have been defined so that, for each k, $1\le k\le 2n$

$$\boldsymbol{\psi}_k^T(2\lambda_k\mathbf{M}+\mathbf{C})\boldsymbol{\psi}_k=1$$

and $\mathbf{z}(t)$ denotes the part of the response that is independent of the control force u.

The steps necessary to arrive at the standard linear programming formulation are similar to those followed for the undamped case. Velocity and acceleration response are obtained by differentiating (29). Then, discretization in time and substitution into (2) leads to an expression for the discretized vector h. Additional inequality constraints, assumed to be expressible as linear functions of response variables, are included in discretized form using (29) and its derivatives. Thus, the matrix $\mathbf{H}$ and the vector $\mathbf{b}$ in (10) are obtained, which completes the formulation of the optimization method in standard linear programming form.

3.2. Example

As an example, the method presented above will be applied to a cantilever beam with two isolators, as shown in Figure 1. The elastic properties defining this uniform beam are given in Table 1. The beam is divided into ten elements each with four degrees of freedom. With the two degrees of freedom contributed by the two concentrated masses, the total number of active degrees of freedom for the system is 22. Traditional vibration isolation studies of such systems [4] start with a fixed isolator configuration and arrive at an optimum design for that particular configuration. The isolator systems would usually consist of discrete masses, springs, and dampers arranged in a predetermined pattern. In contrast, limiting performance studies do not specify the actual elements that make up the isolators. Figure 1 shows two control forces acting on the cantilever beam which may represent any isolator, active, passive, nonlinear, or even physically unrealizable. The optimum design achieved is limited only by the requirement that the equations of motion adequately describe the dynamics of the system and by the constraints imposed on response variables by the designer.

For the system in figure 1, the 22×2 matrix $\mathbf{V}$ has all zero elements except

$$V_{1,1}=1\quad V_{2,1}=-1\quad V_{12,2}=-1\quad V_{14,2}=1\ .\tag{31}$$

The forcing function f is chosen to be a rectangular pulse of magnitude p_0 and duration T_0 applied at the center of the beam at degree of freedom 12. Thus, the only nonzero element of the force vector f is

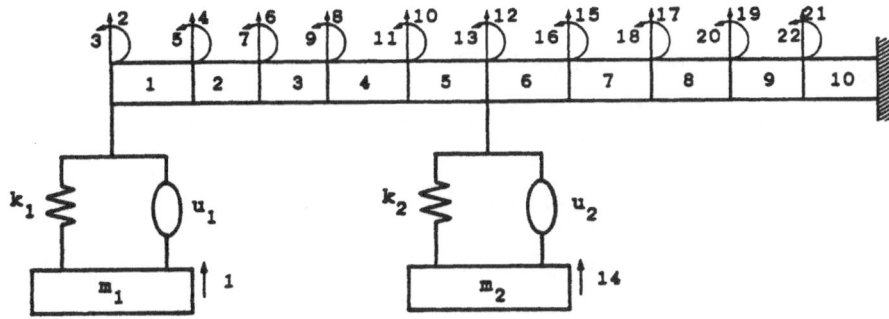

Figure 1: Cantilever beam with two control forces u_1 and u_2

Table 1:   Properties of the cantilever beam of Figure 1

E	29×10^6 psi
m_o	2.6 lb-sec^2/in^2
l_B	150 in
I	215.8 in^4

$$f_{12}(t) = \begin{cases} p_0 & 0 < t < T_0 \\ 0 & t > T_0 \end{cases} \tag{32}$$

With the forcing function and control force application points, specified, the design optimization problem formulation is completed by defining a performance index and adding constraints on response variables. For this example, the performance index is chosen to be the peak acceleration of the free end of the cantilever beam

$$\psi = \max_{0 \le t \le T_f} |\ddot{x}_2(t)| \, . \tag{33}$$

For the system considered, it is physically meaningful to limit the relative motion of the concentrated masses with respect to their points of attachment to the beam. Therefore, the following two rattle space constraints are needed

$$\max_{0 \le t \le T_f} |x_1 - x_1| \le d_1 \tag{34}$$

$$\max_{0 \le t \le T_f} |x_{14} - x_{12}| \le d_2 \, . \tag{35}$$

The convenience of the modal formulation becomes transparent from the explicit expression for the tip acceleration

$$\ddot{x}_2(mT)=\ddot{r}_2(mT)-\sum_{k=1}^{n}\phi_{2,k}[(\phi_{1,k}-\phi_{2,k})u_1(mT)$$

$$+(\phi_{14,k}-\phi_{12,k})u_2(mT)]$$

$$-\sum_{k=1}^{n}\sum_{i=0}^{m-1}\{[(\phi_{1,k}-\phi_{2,k})u_1(iT)$$

$$+(\phi_{14,k}-\phi_{12,k})u_2(iT)]\phi_{2,k}$$

$$.[cos(m-i)\omega_k T-\cos(m-i-1)\omega_k t]\}\tag{36}$$

in which

$$\ddot{r}_2(mT)=\begin{cases}\sum_{k=1}^{n}\phi_{2,k}\phi_{12,k}P_0\cos m\omega_k T & mT\le T_0\\[2ex]\sum_{k=1}^{n}\phi_{2,k}\phi_{12,k}p_0[\cos m\omega_k T-\cos\omega_k(mT-T_0)] & mT>T_0\end{cases}\tag{37}$$

where $\phi_{i,k}$ indicats the i-th element of the k-th eigenvector ϕ_k and all initial conditions are assumed to be zero. The constraint equations (34) jand (35) are also needed in discretized form. For instance, (34) takes the form

$$x_2-x_1(mT)=r_2(mT)-r_1(mT)$$

$$+\sum_{k=1}^{n}\sum_{i=0}^{m-1}\frac{1}{\omega_k^2}(\phi_{1,k}-\phi_{2,k})[(\phi_{1,k}-\phi_{2,k})u_1(iT)$$

$$+(\phi_{14,k}-\phi_{12,k})u_2(iT)][\cos(m-i-1)\omega_k T-\cos(m-i)\omega_k T]\tag{38}$$

where

$$r_2(mT)-r_1(mT)=$$

$$\begin{cases}\sum_{k=1}^{n}\frac{p_0}{\omega_k^2}\phi_{12,k}(\phi_{2,k}-\phi_{1,k})(1-\cos m\omega_k T) & mT\le T_0\\[2ex]\sum_{k=1}^{n}\frac{p_0}{\omega_k^2}\phi_{12,k}(\phi_{2,k}-\phi_{1,k})[\cos\omega_k(mT-T_{0)-cos}\omega_k mT] & mT>T_0\end{cases}\tag{39}$$

Following the general formulation of the linear programming problem given above, let L be the number of discrete time instances. Then the dimension of the unknown vector z in (7) is $2L-1$ since there are $2(L-1)$ control forces in (6) and ψ is added as one more variable. Most standard linear programming algorithms impose nonnegativity constraints on the unknowns, however, so that it is usually necessary to define [5]

$$u(k)=u'(k)-u''(k) \qquad (40)$$

where the primed vectors are nonnegative. This substitution increases the number of unknowns to $4L-3$.

The number of inequality constraints corresponding to the conditions

$$-\psi \leq \ddot{x}_2(t) \leq \psi \qquad (41)$$

is twice the number of time intervals, i.e., $2(L-1)$. Each of the constraints (34) and (35) contributes $2(L-1)$ inequalities, but since all displacements have been assumed zero at $t=0$, two of these are trivial, so that the total contribution of the rattlespace constraints is $4L-8$. Hence, the total number of inequalities, which is the row dimension of the matrix H in (10), is $6L-10$. The column dimension of H is equal to the total number of unknowns, $4L-3$. As a numerical example, this problem was solved with $T_0 = 0.1s$, $T_f = 0.1s$, $T = 0.008s$, $p_0 = 6000$ lb for a few different values of d_1 and d_2. The values for the spring constants were taken arbitrarily to be 1000 lb/in, and the concentrated masses were each one-tenth of the beam mass.

Table 2 shows the limiting performance values for tip acceleration for several rattle space constraints. The corresponding control for $u_1(t)$ was a rapidly varying discontinuous function and $u_2(t)$, the control force applied at the center, was zero, indicating that the constraint (35) was inactive for the values of d_2 chosen. These tip acceleration values may be compared to the values that would be obtained when the control is passive, i.e., when the control force is supplied by a spring-damper combination. For this particular example, the tip acceleration is insensitive to variations in spring and damper constant values. The maximum acceleration of the free end of the beam with two spring-damper isolators is about 242 in/sec^2 and occurs at $t=0.188s$.

Table 2:   Limiting performance charactristics for cantilever beam system

Rattle space constraints d_1, d_2 (in)	Limiting performance ψ (in/sec^2)
0.01	120
0.1	94
1.0	91
2.0	87

4. Structural Synthesis

The advantages of the modal formulation are especially apparent when control forces act as connections between substructures, as shown in Figure 2. The effect of the control force connections is to uncouple the system equations, so that system mass **M**, damping **C**, and stiffness **K** matrices may be obtained by arranging m substructure matrices such that

$$M = \begin{bmatrix} M_1 & & & \\ & M_2 & & \\ & & \cdot & \\ & & & M_m \end{bmatrix} \qquad (42)$$

$$C = \begin{bmatrix} C_1 & & & \\ & C_2 & & \\ & & \cdot & \\ & & & C_m \end{bmatrix} \qquad (43)$$

$$K = \begin{bmatrix} K_1 & & & \\ & K_2 & & \\ & & \cdot & \\ & & & K_m \end{bmatrix}. \qquad (44)$$

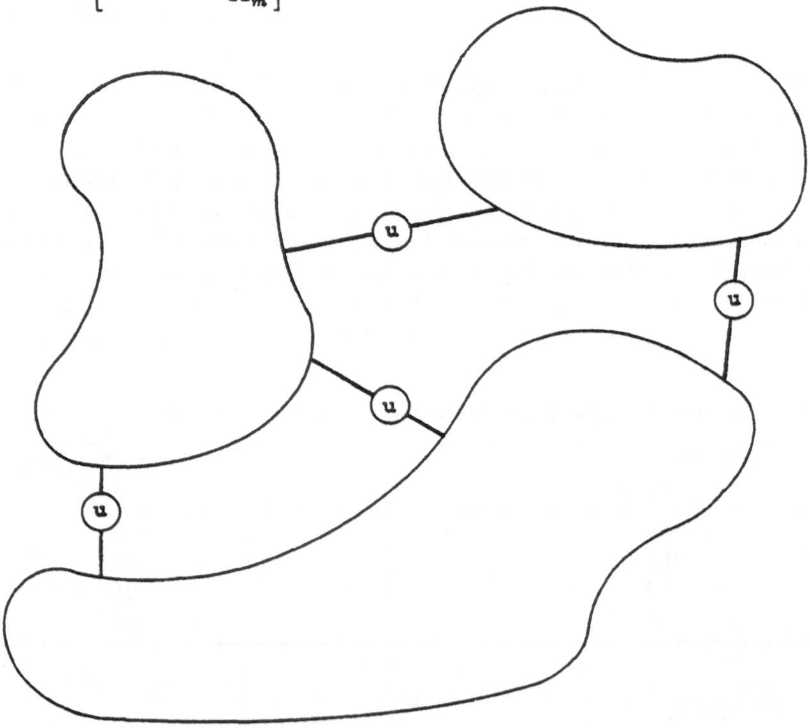

Figure 2: Substructures connected by control forces

In similar fashion, the modal matrix ϕ whose columns are the system modeshapes is given by

$$\phi = \begin{bmatrix} \phi_1 & & & \\ & \phi_2 & & \\ & & \cdot & \\ & & & \phi_m \end{bmatrix} = \begin{bmatrix} \phi_1 \phi_2 \cdots \phi_m \end{bmatrix}. \qquad (45)$$

It thus follows that complete system solutions given by (17) (no damping present) or (29) (damping present) may be evaluated using modeshapes ϕ_k obtained from (45) and mode frequencies ω_k which are the substructure natural frequencies.

5. Conclusion

Knowledge of the limiting control or performance of a system is of great importance to the designer since it aids in the selection of optimal configurations and parameters. It was demonstrated that a model analysis of systems undergoing transient loading leads to a computationally convenient form of linear programming which may be used to compute the limiting performance. It is emphasized that the linear program size is independent of the order of the structural control problem. The limiting performance formulation may be applied to systems with control forces imbedded within, or to systems with control forces acting as connections between substructures.

Acknowledgements

This work was supported by the Office of Naval Research, Arlington, Virginia.

References

[1] Sevin, E. and Pilkey, W.D. "Optimum Shock and Vibration Isolation," Shock and Vibration Information Center, Washington, D.C., 1971.

[2] Pilkey, W.D., Chen, Y.H. and Kalinowski, A.J., "The Use of General Purpose Computer Programs to Derive Equations of Motion for Optimal Isolation Studies", *The Shock and Vibration Bulletin*, August, 1976, pp. 269-279.

[3] Lancaster, P., *Lambda-Matrices and Vibrating Systems*, Pergamon Press, 1985.

[4] Snowdon, J.C., *Vibration and Shock in Damped Mechanical Systems*, John Wiley and Sons, Inc., 1968.

[5] Simmons, D.M., *Linear Programming for Operations Research*, Holden Day, Inc., San Franisco, 1972.

SEMI-ACTIVE CONTROL OF WIND INDUCED OSCILLATIONS IN STRUCTURES

R.E. Klein
Mechanical Engineering
University of Illinois at Urbana-Champaign
Urbana, Illinois
M.D. Healey
McDonnell Aircraft Company
St. Louis, Missouri

1. Introduction

During the past two decaded, there have been numerous accounts of wind-induced oscillations in tall buildings, bridges, and other civil engineering structures [1,2]. In an effort to reduce building costs, modern structures have become lighter, less stiff and, therefore, more vulnerable to wind excitation. This motion presents a serious problem because it jeopardizes the comfort and safety of the occupants as well as the structural integrity of the building. In general, a structure's inherent dampng is very low; thus, once oscillations have started, they will continue on for considerable periods with very little additional energy inputs. Also, in certain cases, motion of a structure can induce additional aerodynamic forces and a phenomenon with positive feedback (or flutter) occurs.

There are a variety of attacks that have been used or proposed in the literature to control unwanted vibrations in structures. Each method used to dissipate translational or torsional vibrations can be classified as either active, semi-active, passive with power assist, or purely passive. An active system is one that requires energy and information inputs to function, whereas semi-active systems require only information inputs. A purely passive system, on the other hand, requires no

354

energy or information inputs.

In this paper, a control system will be investigated that operates on pure information (sensors and logic), and minimal energy (power) inputs, and thus it can be classified as semi-active. The existing vibrational energy will be dissipated through the use of variable devices (ratchets and releases) and unidirectional forces (such as from cables). A cable system will assure unidirectional forces because cables, in the usual sense, are capable of developing only tensile forces. The variation in the nominal length of the cable will remove kinetic energy from the oscillating structure and transfer it to the cable (or cables) where it will be manifested, temporarily, in the form of potential energy.

Based on state feedback, a control algorithm is capable of dictating when slack should be taken up or let out in the cable, as well as holding the cable ends in a fixed configuration. A control mechanism will be discussed that requires small energy inputs to release the cable, hold the end(s) fixed, or to take in slack in the cable. It is assumed, *a priori*, that "taking up" will occur only during periods of available slack. Lastly, a recommendation will be made to improve control of all three vibrational directions in structures.

2. The Control Algorithm

Depending on the particular value of the displacement and velocity, the control mechanism will perform one of three commands: 1) take up slack in the cable, 2) let out the cable even while under load, and 3) hold the cable ends fixed.

For a single-degree-of-freedom oscillator (Figure 1), a phase plane portrait can be used to assist in determining which of the three commands should be given. In what follows, the small amount of inherent damping in the structure is omitted for simplicity. A hypothetical trajectory is shown in Figure 2, where its direction is clockwise. From the initial position to point a, the cable will be slack. Figure 3 illustrates the same system but with the added cable. Referring to the trajectory in Figure 2, at point a, the control mechanism will begin taking up slack in the cable, such that the cable will be pulled up tight at point b. Point b represents the maximum displacement in the positive x-direction. Between points b and c, the cable ends will be held tight; thus, the system in Figure 1 will become the system in Figure 3. At point c, the control mechanism will release the cable, and the energy that had been stored in the cable will be released via hydraulic drag and, thus, converted to heat. As a consequence, the structure's vibrational energy is transferred into the cable in the form of potential energy, and then it is converted to heat by use of an additional device, such as a hydraulic drag, not shown. Also, it is well known that state estimation techniques, using Kalman filters [3], for example, allow implementation of a control logic system to permit the release at point c to occur at the correct instant in time. Consequently, the state of the art in control systems theory is well developed which means that optimal determination in real time of the maximum stretch point is a problem for which a solution already exists. Now, the cycle of dissipating energy will continue until the bulk of the energy has been removed from the structure, and the oscillating body is brought to a steady position, relatively speaking.

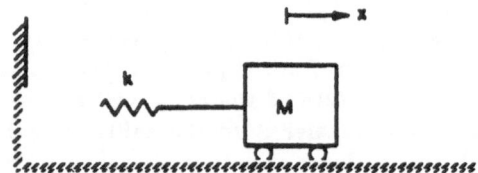

Figure 1: Single-degree-of-freedom system

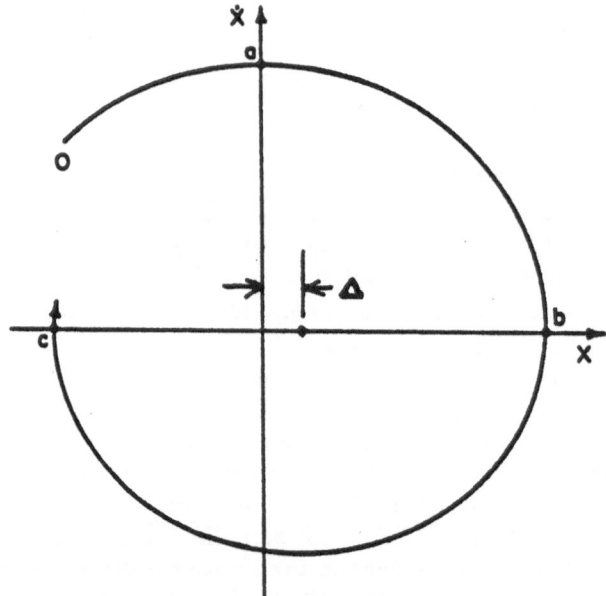

Figure 2: Hypothetical Trajectory of Motion

To examine the effectiveness of this strategy, the energy level in the system can be traced on the phase plane. The trajectory of motion will follow the ellipse of System I in the top half plane, see Figure 2. When $x = x_{max}$ (point b), the solution trajectory for System I will switch to System II by virtue of the available slack having been taken up, and the center of the ellipse will translate to the right by the amount Δ, where $\Delta = (k_c/(k + k_c))x_{max}$. It should be noted that Δ for each cycle will vary with x_{max}. When $x = x_{min}$ (point c), System II will switch back to System I and the center will translate back to the origin. Thus, the trajectory

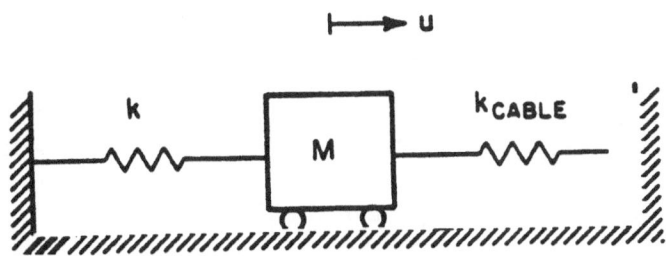

Figure 3: System II

is converging and will ultimately converge to the origin.

For every period of vibration, the maximum amplitude of vibration will decrease by an amount proportional to 2Δ. The new amplitude can be expressed as:

$$x_{\max}(n+1)=[1-2(\frac{k_c}{k+k_c})]x_{\max}(n) \ . \tag{2.1}$$

The natural logarithm of the ratio of the two successive amplitudes is called the logarithmic decrement, where $\delta = \ln[x(n)/x(n+1)]$. The logarithmic decrement can also be expressed in terms of the damping ratio, ξ_c, such that $\delta = [2\pi\xi_c]/[1-\xi_c^2]^{1/2}$. The damping ratio is the ratio of the system's damping to critical damping. Solving for ξ_c yields

$$\epsilon_c = ln(\frac{k+k_c}{k-k_c}) \ / \ \left\{[ln(\frac{k+k_c}{k-k_c})]^2 + [2\pi]^2\right\}^{1/2} \ . \tag{2.2}$$

This expression represents the damping ratio of the system and gives an indication of what effect the cable has on the cable system's damping of the oscillation of the mass. The value of ξ_c thus depends upon the values of k and k_c. It should be noted that while a building is physically large in size, this does not necesarily imply that k is "large", as the building actually consists of many elastic members in series, which tends to reduce the actual magnitude of k. On the other hand, the designer is free to achieve sizeable values of k_c with a cable and/or gas springs.

The same approach that was used for controlling one oscillating body can be applied to the problem of two adjacent oscillating bodies as shown in Figure 4. In this case, the acceleration of each mass may be assumed to be measured and the respective velocities and displacements can be obtained through integration. A cable will connect the two bodies and, based on state feedback, slack will be let out, taken up, or the cable will simply be held fixed at each end.

Figure 4: Two single-degree-of-freedom oscillators

Assuming both bodies are oscillating, the control algorithm will be slightly different than it was for the single oscillating body. A new variable will be formed $x_{rel} = x_2 - x_1$, where x_1 and x_2 are the respective positions of the masses. This variable, x_{rel}, will indicate whether the two masses have moved away from one another or closer. To take full advantage of the state variables $(x_1, \dot{x}_1, x_2, \dot{x}_2)$, another variable can be formed as $\dot{x}_{rel} = \dot{x}_2 - \dot{x}_1$ where $\dot{x}_1$ and $\dot{x}_2$ are the respective velocities of the two masses. A phase portrait of x_{rel} and $\dot{x}_{rel}$ can be formed with a hypothetical trajectory, as in Figure 5. Inspection of x_{rel} and $\dot{x}_{rel}$ will give an indication when the cable should be taken up or released by the control mechanism. All slack should be wound up at each minimum of x_{rel}, then the cable will stretch where $\omega_1 = (k_1/m_1)^{1/2}$, $\omega_2 = (k_2/m_2)^{1/2}$. Now that the governing equations have been cast into state space form, a controllability matrix can be formed which is defined as:

$$[S] = [B \ AB \ A^2B \ \cdots \ A^{n-1}B] \tag{2.6}$$

where, n equals the dimension of $[A]$.

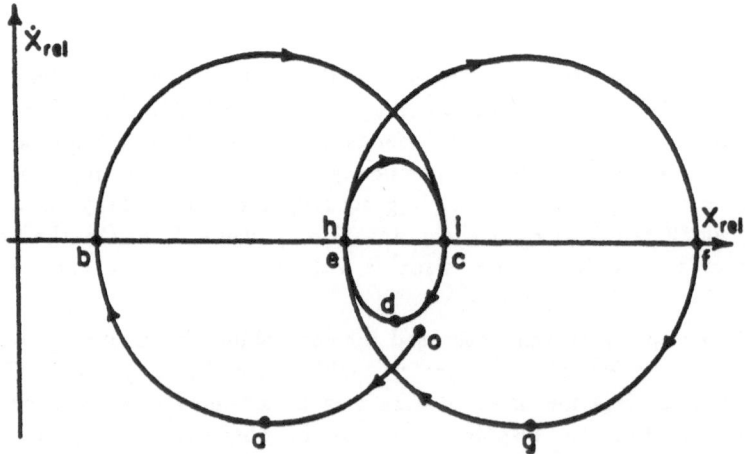

Figure 5: A hypothetical time history of x_{rel}

Therefore, the system described by (2.5) is completely controllable, if the matrix $[S]$ has a rank of n [4]. In other words, the system is controllable if the matrix $[S]$ is nonsingular. And the matrix, as defined previously, will be:

$$[S] = \begin{bmatrix} 0 & \dfrac{1}{m_1} & 0 & \dfrac{-(\omega_1^2)}{m_1} \\[2ex] \dfrac{1}{m_1} & 0 & \dfrac{-1(\omega_1^2)}{m_1} & 0 \\[2ex] 0 & -\dfrac{1}{m_2} & 0 & \dfrac{\omega_2^2}{m_2} \\[2ex] -\dfrac{1}{m_2} & 0 & \dfrac{\omega_2^2}{m_2} & 0 \end{bmatrix} \tag{2.7}$$

To find the determinant, algebra yields

$$|S| = \frac{1}{m_1^2 m_2^2} \{\omega_2^2 - \omega_1^2\}^2 . \tag{2.8}$$

Therefore, the determinant of the matrix will not equal zero when ω_2 and ω_1 are not equal and controllability of the system is assured. Hence, ω_1 not equal to ω_2 is a necessary and sufficient condition for controllability with a single force link (the cable) between the two masses. Saperstone and Yorke [5] establish that unidirectional forces, such as from cables, suffice for controllability of dynamic systems that have complex eigenvalues exclusively, which is the case in the civil engineering systems under consideration.

3. Physical Applications

Any structure that experiences wind induced oscillations could potentially lend itself to this system. In particular, several types of structures will be considered here: 1) adjacent sky scrapers, 2) single sky scrapers, and 3) bridges. The M.S. Thesis of the second author [6] provides discussion and preliminary design layout of specific hardware embellishments.

The case of adjacent sky scrapers is directly analogous to the two oscillating masses as presented previously. The analogy is based on the assumption that both sky scrapers experience wind induced vibrations, and that the natural frequencies of the two buildings are not equal, to assure controllability. It should be noted that a sky scrapper is a spatially distributive system and it is not modeled correctly by a discrete mass and spring. However, the algorithm that was developed for the two discrete masses is still valid here because it advocates putting the maximum stretch on the cable which will remove the maximum amount of kinetic energy from the structural system. This is valid for a multidegree-of-freedom system as well as the single-degree-of-freedom system.

The positioning of the control system is a straightforward problem. To have the greatest impact, the cables should be positioned usually on one of the top stories because the largest deflections occur there and because all vibrational modes have a non-zero amplitude due to the presence of a free boundary condition. Furthermore, the cable may best be installed to be off-center with respect to the building's center of twist. This would effectively permit reduction in torsional vibrations which the sky scraper may experience. As a result, the control system will dissipate vibrations in the one principal direction (north-south, for example) and the torsional direction, but it will not have a direct effect on vibrations in the transverse direction. The problem of transverse vibrational dissipation will be addressed subsequently.

In the event an adjacent sky scraper is unavailable, there is the possibility of tying the cable down. A possible alternative would be to tie the cable to the ground which is analogous to the practice with tall radio towers. The drawback with this method is that only the horizontal component of the cable's tensile force will exert a translational restoring force on the building, which will reduce the cable's effect. Also, the cable will exert a downward vertical component on the structure which will add to the vertical static load. While the increased vertical loading is a factor to be considered, it needs to be recalled that an alternative device, such as the Citicorp building's tuned mass vibration absorber, also adds a significant static vertical component as the mass employed is typically 2 percent of the structure's equivalent excited mass. Tendon control, proposed in the literature, also shares this drawback. The placing of such a loading as a concentrated vertical force requires probable strengthening and increased cost in building designs.

The last case, cable-stayed bridges, and suspension bridges, for example, employ the control algorithm for a single-degree-of-freedom oscillator. Again, it must be stated that a bridge is a spatially distributive system, but the algorithm developed for the discrete case is still valid. For this system, one will usually be concerned with two vibrational directions, the vertical and torsional directions, as opposed to three directions with the sky scraper. Therefore, the cables (two will be assumed to be in use although this is not necessary) shall be afixed to the bridge in such a manner that vibrations in both directions will be arrested.

For torsional vibrations, cables may be attached to each side of the bridge and extending downward to some support point, possibly underwater. This arrangement is capable of producing a moment that counteracts the torsional oscillation of the bridge. In effect, the cables will remove kinetic energy in the rotational (or roll) direction as well as translational vertical heave. Furthermore, the location of the cables along the span of the bridge should not coincide with the nodal points of a higher mode of vibration. The first inclination might be to afix the cable at the center of the span. However, in the event the second mode were excited, the point of attachment would experience no displacement and the opportunity to remove energy from the system for that mode would be obviated.

Therefore, the attachment should be displaced somewhat to the left or right of thé center point to assure the control of all excited transverse modes. See Cannon and Klein [7] for a related discussion of optimum sensor/actuator location.

4. Control Via Coupling

It has been shown that the semi-active control system is feasible in dissipating kinetic energy in the principal translational direction. However, it has little or no effect on the oscillations in the orthogonal direction, which is a problem for sky scrapers. However, if the three vibrational directions (x, y, θ) were dynamically coupled in some manner, then the dissipation of energy by the control system in one direction would, quite naturally, lead to the dissipation of energy in the other two. Such coupling would exist, for example, if the structural cross-section of the building were nonsymmetric.

When a cross-section is nonsymmetric, an eccentricity exists between the center of mass G, and the flexural center, C, as shown in simplistic terms, in Figure 6. It should be mentioned that it is only necessary for the structural members to be nonsymmetric but not necessarily the outer cladding of the building. A dynamic wind load not acting through the flexural or elastic center will result in torsional vibrations, as well as orthogonal transverse vibrations. For adjacent sky scrapers, in what follows it will be shown that all three vibrational directions of each sky scraper are simultaneously controllable by dissipating energy in one direction, such as between the two buildings, assuming that dynamic coupling in each structure is of sufficient magnitude.

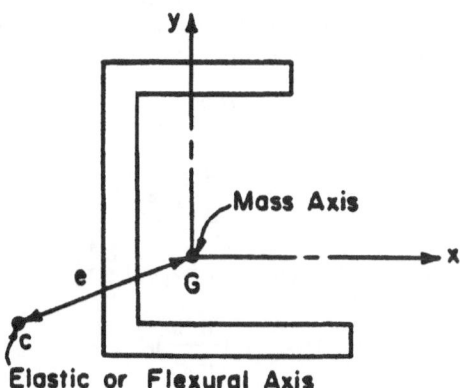

Figure 6: Cantilever beam with an unsymmetric cross-section

In order to prove controllability, the coupled equations of motion may be derived for a cantilever beam with an nonsymmetric cross-section. An extension of the work done by Garland [8] will assist in formulating the equations of motion. Applying the principle of motion to the beam's center of mass, the kinetic energy due to the translation of the center of mass and that due to the rotation about the flexural center may be derived. Therefore, the expression for the total kinetic energy becomes:

$$T = \frac{1}{2} \int_o^L [\rho A (\dot{u}^2 + \dot{v}^2) + J_c \dot{\theta}^2] dz \qquad (4.1)$$

where ρ is the mass density of the material of which the beam is made; A is the cross-sectional area of the beam; u is the x-axis displacement of the center of mass G; v is the y-axis displacement of G; $\dot{u}$ and $\dot{v}$ are the respective components of velocity of the center mass; $\dot{\theta}$ is the instantaneous angular velocity of rotation of the section; z is the axial (vertical) coordinate; and J_c is the polar moment of inertia with respect to the flexural center. The potential energy of the beam may be expressed as the sum of the energies stored in the beam due to the displacements u_c, v_c, and θ, respectively, or:

$$V = \frac{1}{2} \int_o^L [EI_x \left| \frac{\partial^2 u_c}{\partial z^2} \right|^2 + EI_y \left| \frac{\partial^2 v_c}{\partial z^2} \right|^2] dz + \frac{1}{2} \int_o^L GJ (\frac{\partial \theta}{\partial z})^2 dz \qquad (4.2)$$

where E is Young's modulus; I_x is the rectangular moment of inertia of the beam section relative to the x-axis; I_y is the rectangular moment of inertia relative to the y-axis; GJ is the torsional rigidity of the beam; u_c and v_c are the respective x and y translational displacements of the flexural center; and θ is the rotation of the cross-section with respect to the flexural center. Hamilton's principle is applicable to this system and the following equations of motion are obtained:

$$\rho A (\frac{\partial^2 u}{\partial t^2}) + EI_x (\frac{\partial^4 u}{\partial z^4}) = EI_x (\frac{\partial^4 \theta}{\partial z^4}) e_y$$

$$\rho A (\frac{\partial^2 v}{\partial t^2}) + EI_y (\frac{\partial^4 v}{\partial z^4}) = -EI_y (\frac{\partial^4 \theta}{\partial z^4}) e_x \qquad (4.3)$$

$$J_c (\frac{\partial^2 \theta}{\partial t^2}) - GJ (\frac{\partial^2 \theta}{\partial z^2}) + (EI_x e_y^2 + EI_y e_x^2) \frac{\partial^4 \theta}{\partial z^4}$$

$$= EI_y (\frac{\partial^4 u}{\partial z^4}) e_y - EI_x (\frac{\partial^4 v}{\partial z^4}) e_x .$$

The partial differential equations can be transformed into ordinary differential equations by separating the variables into space and time coordinates. Admissible functions can be selected for the space coordinates that satisfy the geometric boundary conditions yielding

$$v(z,t) = V(t)[1 - \cos\frac{\pi z}{2L}]$$

$$u(z,t) = U(t)[1 - \cos\frac{\pi z}{2L}]$$

$$\theta(z,t) = \theta(t)[\sin\frac{\pi z}{2L}] \ .$$

(4.4)

Substitution of these expressions into the equations of motion will yield three ordinary differential equations, which are an approximation of the equations of motion:

$$\ddot{U} - aU = b\theta$$

(4.5)

where

$$a = \frac{EI_z}{\rho a} \frac{\cos\frac{\pi z}{2L}(\frac{\pi}{2L})^4}{(1 - \cos\pi z/2L)} \ , \ b = \frac{EI_z}{\rho A} \frac{\sin\frac{\pi z}{2L}(\frac{\pi}{2L})^4}{(1 - \cos\pi z/2L)} e_y$$

$$\ddot{V} - cV = d\theta$$

(4.6)

where

$$c = \frac{EI_y}{\rho A} \frac{\cos\frac{\pi z}{2L}(\frac{\pi}{2L})^4}{(1 - \cos\pi z/2L)} \ , \ d = \frac{EI_z}{\rho A} \frac{\sin\frac{\pi z}{2L}(\frac{\pi}{2L})^4}{(1 - \cos\pi z/2L)} e_y$$

$$\ddot{\theta} + f\theta = -gU + hV$$

(4.7)

where

$$g = \frac{EI_y}{J_c} \frac{\cos\frac{\pi z}{2L}(\frac{\pi}{2L})^4}{\sin\pi z/2L} \ , \ h = \frac{EI_z}{J_c} \frac{\cos\frac{\pi z}{2L}(\frac{\pi}{2L})^4}{\sin\pi z/2L}$$

and

$$f = \frac{GJ(\frac{\pi}{2L})^2 + (EI_z e_y^2 + EI_y e_z^2)(\frac{\pi}{2L^4})^4}{J_c}$$

Equations (4.5) to (4.7) can also be cast into matrix form:

$$[M]\ddot{\alpha}+[K]\alpha=\begin{Bmatrix}1\\0\\0\end{Bmatrix} \qquad (4.8)$$

where the mass matrix is diagonal, and the stiffness matrix is nondiagonal because the equations of motion are elastically coupled. The vector $\ddot{\alpha}$ is the second time derivative of α. The vector

$$\begin{Bmatrix}1\\0\\0\end{Bmatrix}$$

represents a control input force of unity in the x-direction.

Assuming the dynamical matrix, $[[K]-\omega^2[M]]$, yields three distinct eigenvalues, Equation (4.8) can be simplified further:

$$\ddot{q}+[\omega^2]q=p \qquad (4.9)$$

where, $[Q]^T[M][Q]=[1]$, $[Q]^T[K][Q]=[\omega^2]$, $\alpha=[Q]q$, $[Q]^T\underline{1\ 0\ 0}^T =\underline{P_1\ P_2\ P_3}^T$, and $[Q]$ is the modal matrix. As an observation, Equation (4.9) is controllable if and only if p_1, p_2 and p_3 are all nonzero.

Equation (4.9) can be expressed in state space form by making the following assignments: $\dot{q}=v$, $\dot{v}=-[\omega^2]q+p$, and $z=q^T|v^{T\,T}$, which yields:

$$\dot{z}=\left[\begin{array}{ccc|ccc} & & & 1 & 0 & 0 \\ & 0 & & 0 & 1 & 0 \\ & & & 0 & 0 & 1 \\ \hline -\omega_1^2 & 0 & 0 & & & \\ 0 & -\omega_2^2 & 0 & & 0 & \\ 0 & 0 & -\omega_3^2 & & & \end{array}\right]\left(-\frac{q}{v}\right)+\begin{Bmatrix}0\\0\\0\\---\\P_1\\P_2\\P_3\end{Bmatrix}$$

The same principal can be extended to two cantilevers. This yields the state equation $\dot{W}=[A]W+(b)u$ where

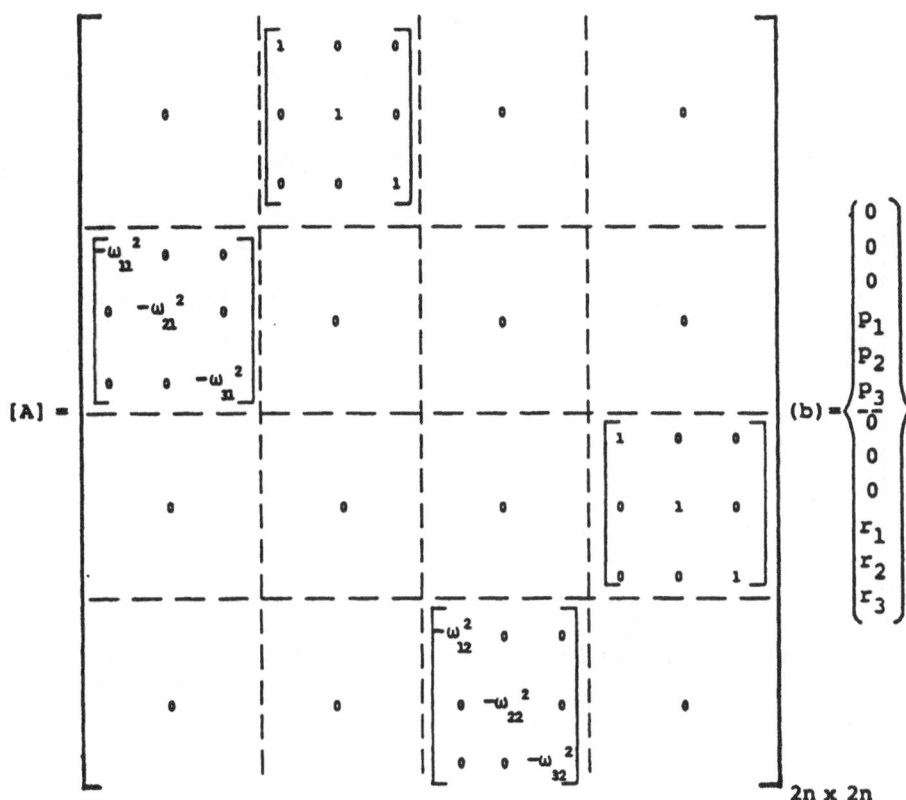

And as mentioned previously, a controllability matrix can be formed which is defined as $[S]=[B\,AB\,A^2B\,\cdots\,A^{n-1}B]$, where n equals the dimension of $[A]$. Therefore, the three directions of oscillation for each cantilever can be simultaneously controlled by a force input, acting between the structures in one direction if the matrix $[S]$ has a rank of n, which is twelve in this case. In other words, the system is controllable if the columns of $[S]$ are linerly independent and the controllability matrix would be nonsingular. Filling in the columns yields the following matrix:

$$[S] = \begin{bmatrix}
0 & p_1 & 0 & -p_1\omega_{11}^2 & 0 & p_1\omega_{11}^4 & 0 & -p_1\omega_{11}^6 & 0 & p_1\omega_{11}^8 & 0 & -p_1\omega_{11}^{10} \\
0 & p_2 & 0 & -p_2\omega_{21}^2 & 0 & p_2\omega_{21}^4 & 0 & -p_2\omega_{21}^6 & 0 & p_2\omega_{21}^8 & 0 & -p_2\omega_{21}^{10} \\
0 & p_3 & 0 & -p_3\omega_{31}^2 & 0 & p_3\omega_{32}^4 & 0 & -p_3\omega_{31}^6 & 0 & p_3\omega_{31}^8 & 0 & -p_3\omega_{31}^{10} \\
p_1 & 0 & -p_1\omega_{11}^2 & 0 & p_1\omega_{11}^4 & 0 & -p_1\omega_{11}^6 & 0 & p_1\omega_{11}^8 & 0 & -p_1\omega_{11}^{10} & 0 \\
p_2 & 0 & -p_2\omega_{21}^2 & 0 & p_2\omega_{21}^4 & 0 & -p_2\omega_{21}^6 & 0 & p_2\omega_{21}^8 & 0 & -p_2\omega_{21}^{10} & 0 \\
p_3 & 0 & -p_3\omega_{31}^2 & 0 & p_3\omega_{31}^4 & 0 & -p_3\omega_{31}^6 & 0 & p_3\omega_{31}^8 & 0 & -p_3\omega_{31}^{10} & 0 \\
0 & r_1 & 0 & -r_1\omega_{12}^2 & 0 & r_1\omega_{12}^4 & 0 & -r_1\omega_{12}^6 & 0 & r_1\omega_{12}^8 & 0 & -r_1\omega_{12}^{10} \\
0 & r_2 & 0 & -r_2\omega_{22}^2 & 0 & r_2\omega_{22}^4 & 0 & -r_2\omega_{22}^6 & 0 & r_2\omega_{22}^8 & 0 & -r_2\omega_{22}^{10} \\
0 & r_3 & 0 & -r_3\omega_{32}^2 & 0 & r_3\omega_{32}^4 & 0 & -r_3\omega_{32}^6 & 0 & r_3\omega_{32}^8 & 0 & -r_3\omega_{32}^{10} \\
r_1 & 0 & -r_1\omega_{12}^2 & 0 & r_1\omega_{12}^4 & 0 & -r_1\omega_{12}^6 & 0 & r_1\omega_{12}^8 & 0 & -r_1\omega_{12}^{10} & 0 \\
r_2 & 0 & -r_2\omega_{22}^2 & 0 & r_2\omega_{22}^4 & 0 & -r_2\omega_{22}^6 & 0 & r_2\omega_{22}^8 & 0 & -r_2\omega_{22}^{10} & 0 \\
r_3 & 0 & -r_3\omega_{32}^2 & 0 & r_3\omega_{32}^4 & 0 & -r_3\omega_{32}^6 & 0 & r_3\omega_{32}^8 & 0 & -r_3\omega_{32}^{10} & 0
\end{bmatrix}$$

Space limitations in this paper prohibit a proof, however, is suffices to state that the controllability matrix would be nonsingular, as a sufficient condition, if all six natural frequencies were distinct and each structure had sufficient dynamic coupling. As a note, in general, the torsional frequency is approximately twice the frequency of the first bending mode in many structures [9]. Furthermore, the probability that two separate structures would have identical natural frequencies is small, but it remains for a designer to verify this in each case. Also, weaker necessary conditions exist which would permit control in certain cases even if some natural frequencies were equal, for example, in the transverse modes of each structure, respectively. Other weaker special cases also exist. Obviously, building a structure with a nonsymmetric structural cross-section would be useful for control purposes because all three vibrational directions could be controlled most readily due to the strong coupling in each structure. On the other hand, a configuration of two identical towers, such as each built as symmetric rectangles or squares adjacent to each other, presents the worst of all worlds due to duplication of natural frequencies and lack of significant modal coupling.

5. Discussion and Conclusions

The proposed semi-active control system has enhanced the viability of using control theory to reduce wind-induced vibrations in structures. Currently, the employment of control systems in civil engineering types of structures is essentially nonexistent, whereas other industries, such as shipping and aerospace, have put them to good use. The semi-active control system, as proposed here, has several major advantages: an acceptable cost level, minimal space requirements, retrofit capability, and sufficiently low energy requirements so as to make continuous use a feasibility. In addition, the control system would be quite effective in removing unwanted kinetic energy from the structure in a reasonable period of time. Traditionally, steps have been taken to stiffen the structure once windinduced oscillations were detected. These corrections have proven to be costly and, in many instances, uneffective as well. It is the authors' opinion that the installation of the proposed control system would be a contender for adoption compared to other classical solutions. Also, it is quite clear that there would be a payoff to building a structure with an nonsymmetric structural cross-section. The resulting coupling between the modes of vibration would allow for energy dissipation from the three principal vibrations modes by applying a control input in just one direction, be it by cable devices or aerodynamic appendages [10]. It should be mentioned that eccentricities would exist in a structure that is perfectly symmetric. However, the eccentricity would be small as would the coupling between the vibrational directions. Also, horizontal deformation of a symmetric vertical structure, such as due to steady wind loading also induces certain types of coupling, but the magnitude is generally small. Therefore, the nonsymmetric crosssection would still be a valid recommendation.

The advantage of dynamic coupling between principal modes of vibration, those typically being the two translational directions, and then also the torsional mode, is suggested by implying that all of a set of coupled modes are controllable with a single force input or energy dissipation mechanism provided that adequate coupling is present within the structure. The method under investigation is designed such that the semi-active control system momentarily stores potential energy in its stretched cable, or cables, and this energy is then next dissipated as heat at the command of a logic device such that the energy is prevented from being converted back into kinetic or vibratory energy. Given the ability to dissipate energy from a single mode, the device also acts to dissipate energy from the entire structure due to the deliberate design of the structure to incorporate coupled modes. The device and methodology are also applicable to two adjacent structures and in this situation it is usually capable of dissipating vibrational energy from all of the principal modes in each of the two structures, for example. In this situation, the energy dissipation is quite akin to the farmer's trick of haltering an ill behaved cow to be trained, closely to a donkey. The both pull against each other and each eventually come to a new understanding about the world. The cow's education, in particular, is magnified as the "eigenvalue" of a donkey implies rigidity, stubborness, and swift kicks. Each animal acts as the energy sink or restraint for the other because of their differing "eigenvalues". The structures, or the two "halves" of a composite structure, do a similar thing except that the connecting link, or tether, the cable in this case, converts energy to heat while requiring no energy source in order to operate, but it does require information and a logic device. An additional possible design configuration would be to construct two "structures" in a shared or close proximity so as to act functionally like a single structure, and yet be dynamically two separate structures. Thus each "half" of a "building" could be the dynamic reaction mass for the other half. As an extension of this thinking, there is no reason why collections of structures, such as in a city, can't be interconnected with a network of energy dissipating cable mechanisms. City planning could include proper spacing of eigenvalues of the respective structures.

It is the position of the authors that the problem of structural control is much more of a philosophical problem than one of equations, mathematical models, optimal control theory, and the like. The central problem in a wide array of "structural control" problems is that of expending, defeating, or overcoming unwanted energy in a structure which necessarily manifests itself as vibratory motion. A suitable philosophical base with which to attack such a problem is characteristially close in viewpoint to certain eastern philosophies related to the harmony that naturally exists between heaven and earth, as stated by Lao-Tse in his bool *Tao Te Ching*. For example, "The Wu Wei approach to conflict-solving can be seen in the practice of the Taoist martial art T'ai Chi Ch'uan, the basic idea of which is to wear the opponent out either by sending his energy back at him or by deflecting it away, in order to weaken his power, balance, and position-for-defense. Never is a force opposed with force; instead it is overcome with yielding" [11]. A plausible Taoist strategy in structural control is to arrange affairs such that opponent looses energy by his very act of fighting. In a similar manner, the hooked fish on a fisherman's line wears itself out by moving about. As the fish moves away, the fish works against a drag mechanism in the reel, and yet if the fish approaches, the fisherman takes up all available slack with a simple ratchet. As a control strategy, the fisherman always takes up available slack. In recognition that control systems, fishing, animal training, and structural control all have strong analogies to the martial arts, it is all a matter of observing the opponent, having available a finger to place on him and knowing when and where to place that finger so as to cause the foe to defeat himself. Brute force controllers with high level energy inputs are often quite unnecessary and definitely anti-Taoist.

References

[1]   Enrich, T. and Perlman, J., "Dizzy Heights", *The Wall Street Journal*, January 10, 1973, pg. 1.

[2] *Engineering News Report*, "Tuned Mass Dampers Steady Sway of Skys-crapers in Wind", August 18, 1977, pp. 28-29.

[3] *IEEE Trans. on Automatic Control*, "Special Issue on the Linear-Quadratic-Gaussian Estimation and Control Problem", Vol. AC-16, No. 6, December, 1971.

[4] Ogata, K., *State Space Analytis of Control Systems*, Prentice-Hall, Engle-wood Cliffs, NJ, 1967.

[5] Saperstone, S.H. and Yorke, J.A., "Controllability of Linear Oscillatory Sys-tems Using Positive Controls," *SIAM Journal on Control*, Vol. 9, No. 2, May 1971, pp. 253-262.

[6] Healey, M.D., "Semi-Active Control of Wind Induced Oscillations in Struc-tures," M.S. Thesis, Department of Mechanical Engineering, University of Illinois at Urbana-Champaign, Urbana, IL, 1983.

[7] Cannon, J.R. and Klein, R.E., "Optimal Selection of Measurement Locations in a Conductor fo Approximate Determination of Temperature Distrbu-tions", *Trans. ASME, J. of Dynamic Systems, Measurement and Control*, Vol. 93, 1971, pp. 193-199.

[8] Garland, C., "The Normal Modes of Vibration of Beams Having Noncolliner Elastic and Mass Axes," *J. of Applied Mechanics*, Vol. A, 1940, pp. 90-105.

[9] Newmark, N. and Rosenblueth, E., *Fundamentals of Earthquake Engineer-ing*, Prentice-Hall, Englewood Cliffs, NJ, 1971.

[10] Klein, R.E., Cusano, C. and Stukel, J., "Investigation of a Method to Stabil-ize Wind Induced Oscillations in Large Structures," Paper No. 73-WA/AUT-11, presented at 1972 ASME Winter Annual Meeting.

[11] Hoff, B., *The Tao of Pooh*, E.P. Dutton Inc., New York, 1982, p. 87.

METHODS OF OPTIMAL CONTROL
APPLIED TO A PROBLEM OF
NONLINEAR SYSTEMS IDENTIFICATION

O. Klingmüller
Bilfinger & Berger Bauaktiengesellschaft
Carl-Resiss-Platz 1-5
68 Mannheim, Germany FRG
M. Lawo
Universität Essen - Gesamthochschule
Universitätsstrasse 13-17, 43 Essen
Germany FRG

1. Introduction

The measurement of structural behaviour is a tool used in structural control as well as in ordinary performance control of structures. Performance control is executed on completed structures during operation or on structural elements prior to construction during planning and design. In performance control two categories of measuring the structural behaviour can be distinguished:

1. Direct measurement, where the required quantity is the measured quantity,

2. Indirect measurement, where the required quantity has to be derived by interpreting the measured quantities.

In indirect measurement the use of a mathematical structural model for the interpretation of the measurements is essential. This structural model or system has to be determined or identified by use of the measurements [1]. This identification is done by altering model parameters in such a way that computer model behaviour coincides with measured behaviour under the same action.

Some problems of system identification allow closed form solutions of curve fitting, others demand iterative solution. For such a problem the advantage of an adequately applied optimal control strategy compared to currently used trial and error strategies is demonstrated.

The presentation will first give a detailed consideration of the problems of systems identification and then describe the special nonlinear problem to be solved. For the derived systematic formulation of the problem, a solution strategy is proposed and results are shown.

2. Systems Identification

A given action on a given structural system results in a definite behaviour given by quantities that in structural mechanics can be computed and expressed by the so-called state variables (e.g. stresses and displacements). This problem of 'system analysis' can be looked at as to be solved in a statistical sense, that is, according to the percentage of structures that are performing successfully. When together with the action the behaviour is given by measurements, to determine the system that exhibits this behaviour is a fairly new problem called 'systems identification' (Fig. 1).

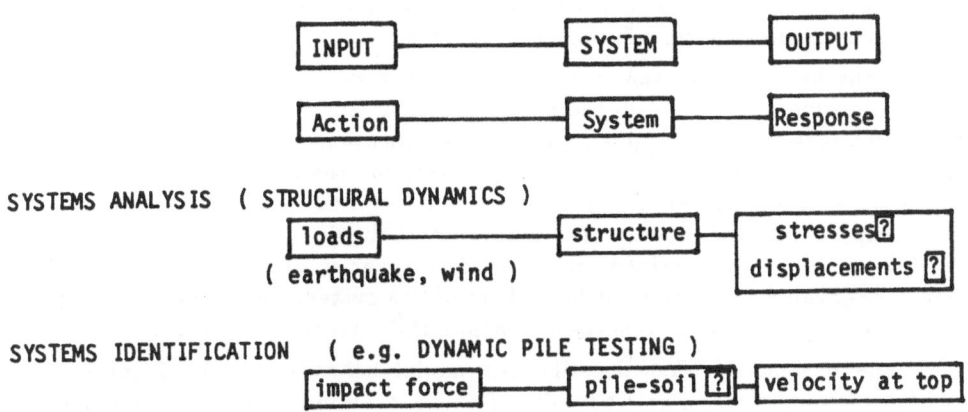

Figure 1: Systems Analysis and Systems Identification

In structural dynamics a mechanical model is usually described by the mathematical formulation of the equations of equilibrium

$$\underline{M}\,\ddot{r} + \underline{C}\,\dot{r} - \underline{K}r = \underline{R}(t) \tag{1}$$

where $\ddot{r}$, $\dot{r}$ and r : vectors of time dependent nodal accelerations, nodal velocities and nodal displacements respectively

$\underline{M}$: mass matrix,
$\underline{C}$: damping matrix,
$\underline{K}$: global stiffness matrix, and
$\underline{R}$: vector of dynamic forces acting at the nodal points.

Systems analysis is the computation of the elements of r for given $\underline{R}$. Computational procedures depend on the nature of the matrices $\underline{M}$, $\underline{C}$ and $\underline{K}$.

If the elements of the matrices are constants, the problem (1) is said to be linear and can be solve by modal analysis or by numerical integration. Explicit or implicit integration schemes are used [2]. Basically in systems identification the action $\underline{R}$ together with the response $\ddot{r}$, $\dot{r}$ and r is given and the elements of the matrices $\underline{M}$, $\underline{C}$ and $\underline{K}$ have to be determined. When matrices are assumed linear, solutions in the frequency domain have been derived and successfully applied. Of course, a reduction of unknowns, coupling of matrix elements, is essential for the solution [1].

For a reduced problem, i.e. unknown Rayleigh damping factors for given $\underline{K}$ and $\underline{M}$, even closed form solutions are possible.

If the behaviour of the system cannot be assumed to be linear the solution has to involve numerical integration of the basic equation (1) and only trial and error strategies can be applied for the identification problem. A system model given by values for the parameters that determine the matrices $\underline{M}$, $\underline{C}$ and $\underline{K}$ has to be chosen, for a given action $\underline{R}$ the response r has to be computed and has to be compared to the known (i.e. measured) behaviour r'. Alterations for the matrices have to be derived from the discrepancies of computed behaviour to given behaviour. If computed behaviour of the system model and known (measured) behaviour of the real system coincide, the system is identified.

The parameters that describe the model behaviour have of course to lie within physically reasonable bounds. If an adaptation of measured and computed behaviour leads to parameter values that do not lie inside these bounds, the chosen model is not able to represent the real physical behaviour [3].

For example, if the model equation (1) is considered, other than viscous damping cannot be accurately described. Whereas real structures are always continuous mechanic-mathematical modelling leads to structural systems with a finite number of degrees of freedom. Thus model degrees of freedom must be carefully chosen to describe the measured behaviour of the real system [3].

Deficiencies in the model may not only lead to physically unacceptable results that can be excluded but may even lead to wrong results that can not be ruled out unless existence and uniqueness of the true solution can be proved.

3. Determination of Bearing Capacity of Piles by Dynamic Measurements

Testing of piles for the determination of their vertical bearing capacity demands the application of loads with a certain safety margin compared to the design loads of serviceability or ultimate limit state. For hammer driven piles the design load lies in the range of 1 MN whereas for large bored piles the design load is about ten times as high and the testload must exceed 10 MN.

A static test with this load is very costly because the loading jacks have to be supported by some auxilliary construction. In dynamic testing the impulse of a falling mass or the hammer blow is transferred to the top of the pile in a few milliseconds and thus a momentary impact force exceeding the necessary testload can be easily produced. Masses of up to 20 tons have been used.

Compared to a static test the dynamic test reduces the financial effort to about 1/10. Whereas in static load testing the required static load-settlement behaviour can be determined by measuring the pressure in the jacks together with the achieved settlement at the pile top, in dynamic testing the static load-settlement behaviour has to be determined indirectly by the interpretation of the time histories of the impact force and the corresponding pile top velocity. This interpretation has to make use of a mechanic-mathematical model. Therefore the dynamic testing method is a problem of systems identification (see chap. 2). Because of the behaviour of the soil the problem is nonlinear.

In Figure 2 the set-up for dynamic testing of bored piles is given. The weight has to be dropped in free fall from a suitable lifting device. During the impact, strains and accelerations are recorded by the pile driving analyzer (PDA [4]). The PDA is a combined amplifier and filter with the ability to numerically analyze the signal. Additionally the PDA integrates the measured accelerations and scales the measured strains by use of the pile cross sectional area and the elastic modulus of the pile material to give the time history of the pile top force together with the time history of the pile top velocity.

Time histories are produced on the scope by adequate triggering and recorded on a magnetic analog tape. Analog storage gives the free choice of the digitizing rate for later evaluations.

Figure 3 shows two typical time histories of pile top force and velocity. During the beginning of the impact, pile top force and velocity are proportional to each other. The differences between force and velocity in the later part of the record reveal the characteristic features of the tested soil-pile system.

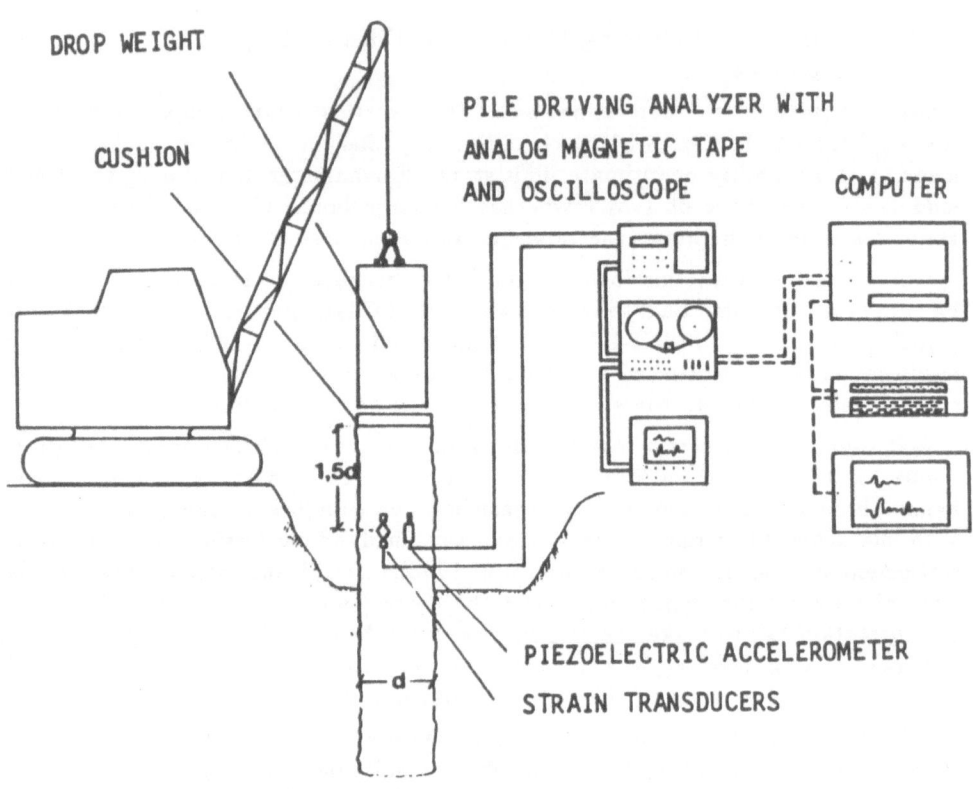

DROP WEIGHT

CUSHION

PILE DRIVING ANALYZER WITH
ANALOG MAGNETIC TAPE
AND OSCILLOSCOPE

COMPUTER

1,5d

d

PIEZOELECTRIC ACCELEROMETER
STRAIN TRANSDUCERS

Figure 2: Equipment for dynamic pile testing

Basically for the interpretation of the measurements the theory of one-dimensional wave propagation in an elastic medium has to be used. For a prismatic bar the propagation of waves has to be described by the well-known differential equation [5]:

$$\ddot{u} - c^2 u'' = 0 \tag{2}$$

where $u = u(x,t)$: displacement function,
       $\ddot{u}$: acceleration, second time derivative,
       u'' : differential strain, second space derivative
       c : wave velocity (material constant).

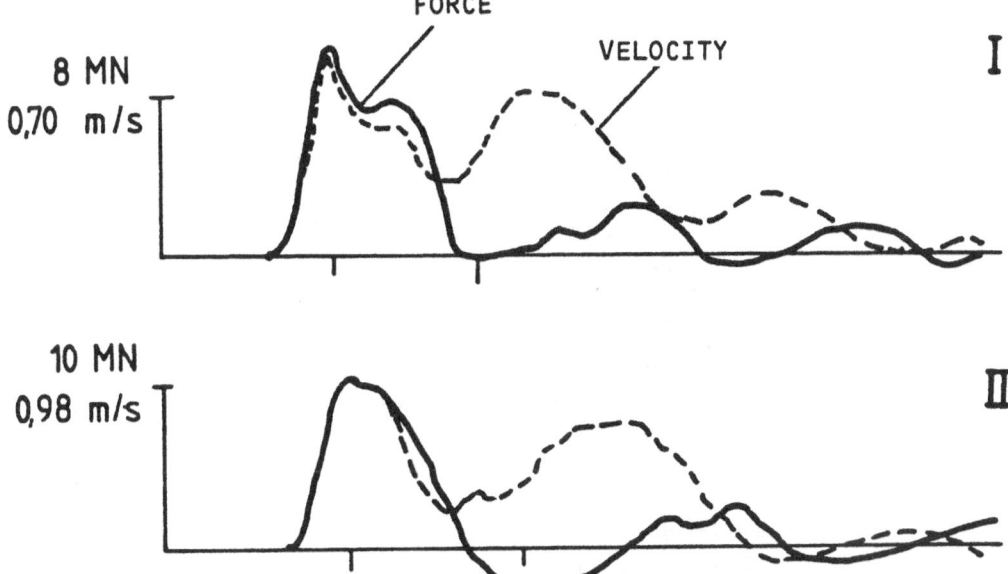

Figure 3: Time histories of recorded force and velocity

There exist two solutions for this differential equation. Bernoullis solution is suited to describe the longitudinal vibrations of the prismatic bar, whereas d'Alemberts solution is used in the investigation of wave propagation phenomena:

$$u(x,t) = u_1(x-ct) + u_2(x + ct) \tag{3}$$

As long as the impact wave is not reflected the wave proceeds only in one direction and equation (3) yields proportionality between velocity and strain. The proportionality is of course also valid for force and velocity. The respective factor EA/c, including Youngs modulus E, cross sectional area A, and material wave velocity c is called *impedance*. In dynamic testing of piles this relationship is used for an immediate interpretation of the recorded signals in such a way, that any deviations from porportionality that are not caused by reflections from the free pile tip must be caused by soil reactive forces. Difference soil layer properties and effects of damping can only be roughly approximated. The respective formulas can be evaluated by the pile driving analyzer immediately after recording (cf. Fig. 2). This so-called *CASE-Method* [4] is primarily applied for driven piles and in some countries has already been codified.

For piles with nonuniform cross-section, for layered soils that exhibit different mantle resistance in the layers, for very long piles and in case of unknown damping behaviour, the global approximation of soil resistance of the CASE-Method is not exact enough. Refined modelling makes use of a differential equation that includes side friction and damping explicitly:

$$m\ddot{u} - EA\,u'' - D\dot{u}'' + k_B u + D_B \dot{u} = 0 \tag{4}$$

In this equation $k_B u$ stands for the side friction, $D_B \dot{u}$ represents soil damping and $D\dot{u}''$ internal damping of the pile material. The impact force serves as a boundary condition.

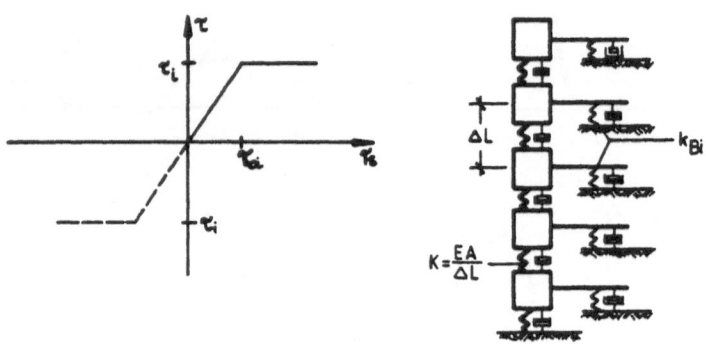

for soil frictional forces:

$k_{Bi} = (\tau_i / r_{oi}) \cdot A_{Mi}$, where A_{Mi} is the skin surface of the pile element

for tip resistance:

$k_{Bi} = (\tau_i / r_{oi}) \cdot A_i$, where A_i is the cross sectional area at the bottom

Figure 4: Nonlinear soil behaviour and pile-soil model

For side friction (mantle resistance) nonlinear behaviour has to be assumed, which is approximated by an elastic-ideally plastic material law (cf. Fig. 4). Values τ_i, r_{0i} in Figure 4 define the limits of the elastic domain.

As closed form analytical solution for equation (4) can only be given for special cases, a space discretization commonly used in the finite element method is performed ([5], [6]), symbolically visualized by the mass-spring-damper model (cf. Fig. 4).

After the discretization, the differential equation is replaced by the coupled system of ordinary differential equations

$$\underline{M}\ddot{\underline{r}} + (\underline{C} + \underline{C}_B)\dot{\underline{r}} + (\underline{K} + \underline{K}_B)\underline{r} = \underline{R}(t) \tag{5}$$

where $\underline{C} = \alpha\underline{M} + \beta\underline{K}$, α and β are material dependent parameters of Rayleigh damping,

$\underline{C}_B$: matrix of viscous soil damping coefficients,

$\underline{K}_B$: matrix of soil resistance at skin and tip,

for other notations see (1).

The element matrices for the soil behaviour are given by

$$\underline{k}_{Bi} = 1/6\, k_{Bi} \begin{bmatrix} 2 & 1 \\ 1 & 2 \end{bmatrix} \quad \text{and} \quad \underline{c}_{Bi} = \delta \; \underline{k}_{Bi}.$$

If the soil resistance reaches the specified plastic limit τ_1 or deformations reach the respective limit r_0, the corresponding elements in the vector $\underline{K}_B \underline{r}$ are replaced by the constant value $\tau_i A_i$. For a given impact $\underline{R}(t)$, the solution of equation (5) will be unique. For a measured time history $\underline{R}^*$, the computed behaviour $\underline{r}$ coincides with the measured behaviour $\underline{r}^*$ only if the mechanic-mathematical model is able to represent the real pile-soil behaviour.

In practical application of the method, the free model parameters (soil resistance limit R_0, elastic deformation limit r_0 and soil damping coefficient for each element) are to be varied according to the discrepancies of the measured and computed curves until both coincide.

Figure 5 shows several steps of an iteration. It is to be seen that tip reflections have an influence only in the final part of the record. This interdependency of time and space domain has to be observed in performing the iteration steps for adaptation. Model alterations should always start at the top of the pile.

Because of the complexity of the problem, model alterations have to be done in an interactive process using computer screen and plotter. On the plotter the discrepancy of the measured and computed curves after each iteration step is to be seen out of which the conclusions for the next step have to be drawn.

The procedure is terminated by demonstrating the completeness of fit of the last model.

By means of the known soil constants R_0 and r_0, a theoretical load settlement curve can be drawn as a final result. For a detailed discussion of this so-called CAPWAP-Method, see ref. [4], [6] and [7].

4. Mathematical Programming and Systems Identification

To answer the question whether the problem of systems identification is solvable and whether a solution will be unique, a mathematical formulation of the problem has to be found.

The verbal term *coincide* has to be replaced by an error function, the difference between measured and computed values.

Given are the m digitized values of the velocity at the pile top in the vector $\underline{\dot{r}}_1^*$. The solution of equation (5) for given parameters $\underline{z} \in \mathbb{R}^n$ at each time step are the m values of the velocities at the pile top

$$\underline{\dot{r}}_1 = \underline{\dot{r}}_1(z) \tag{6}$$

Instead of all the degrees of freedom of the finite element model only the pile top velocity at each time step is the state variable of the problem. It is presumed

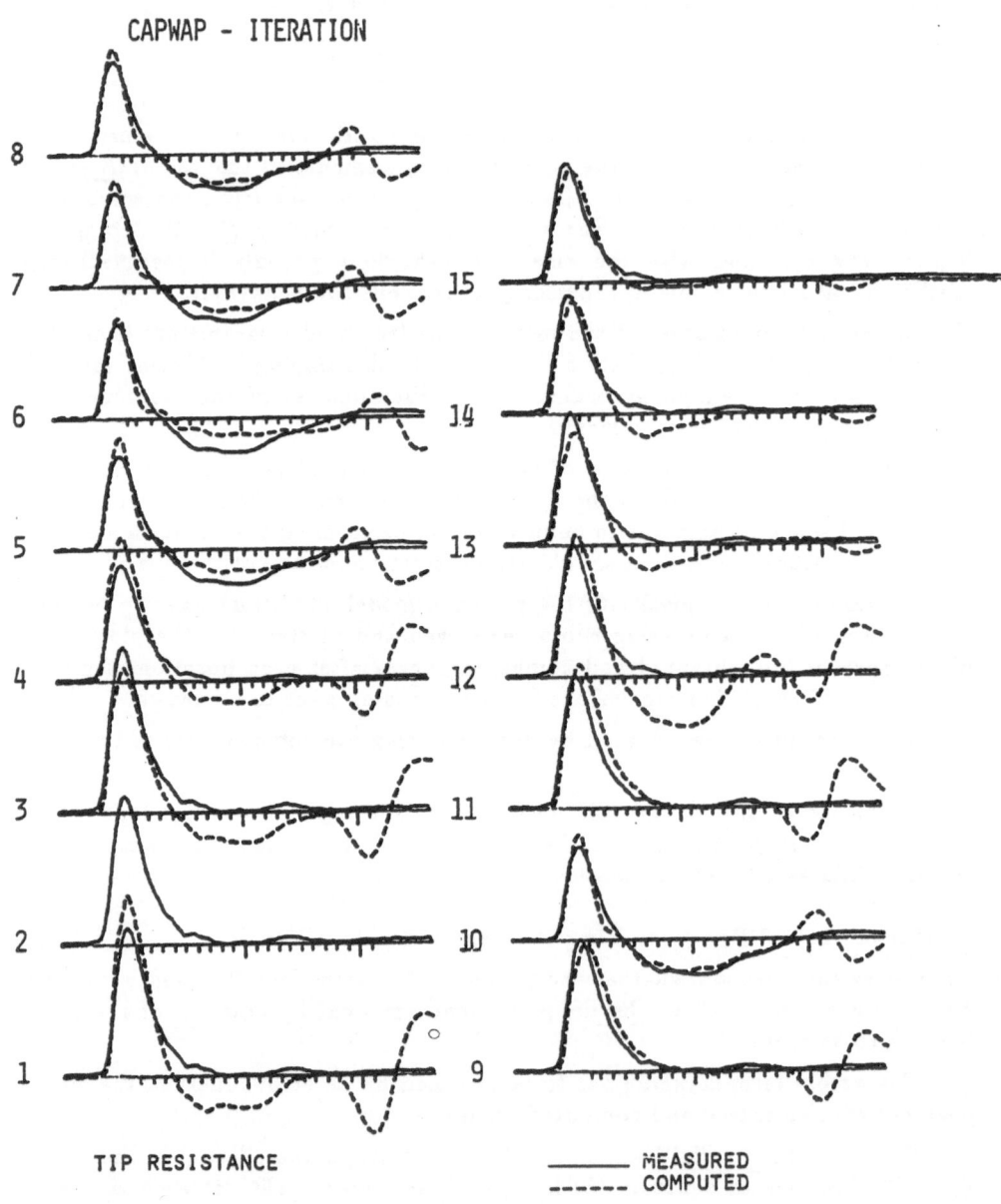

Figure 5: Iteration steps of dynamic pile testing

that the finite element idealization is fixed with regard to the wave propagation.

The variables of a mathematical programming problem are the parameters z (material properties of soil and pile) of which only a feasible domain Ω but not the exact values are known.

The problem is now to determine within the feasible domain the parameters z such that the difference between the measured velocities $\dot{r}_1{}^*$ and the computed velocities $\dot{r}_1(z)$ becomes a minimum:

$$\min_{z} \left\{ f(z,\dot{r}_1{}^*,\dot{r}_1) = \sum_{i=1}^{m} |\dot{r}_{1i}{}^* - \dot{r}_{1i}(z)| / m \right\} \qquad (7a)$$

subject to

$$\underline{M}\ddot{r}_i + \left[\underline{C}(z)+\underline{C}_B(z)\right]\dot{r}_i + \left[\underline{K}(z)+\underline{K}_B(z)\right]r_i = \underline{R}(t_i,\dot{r}_{1i}{}^*) \quad \text{for} \quad i = 1,\ldots,m, \qquad (7b)$$

$$z \in \Omega. \qquad (7c)$$

The coupled system of ordinary differential equations has to be solved for each time step i. The matrices of pile and soil damping and of pile and soil stiffness depend on the parameters z. The loading $\underline{R}$ of the pile is the impact load at the pile top.

A problem of mathematical programming has a solution, if the feasible domain for the variables described by problem (7) is bounded. The solution is unique if the domain is convex (cf. [8], [(]).

Both properties of the domain cannot easily be verified by strict mathematical proof. If the mechanical background of (7) is considered, it is clearly to be recognized that the domain must be bounded, i.e. there must be finite solution for the velocities for given finite (because measured) exciting forces for a reasonable mechanical model. So it can be concluded that there is at least one solution, if the mathematical and mechanical models are correct.

If the number of degrees of freedom of the pile-soil model is included in the vector of model parameters z the mathematical programming problem (7) will, of course, be not convex. Best matches may be found for different models, a unique solution cannot be given. On the other hand, the mechanical solution, i.e. the static pile capacity should be insensitive to model degrees of freedom if the error function itself is to be less than some specified error value.

With such a description of the formulated mathematical and the underlying mechanical model it must be concluded that methods of mathematical programming could be successfully applied.

As the number of model parameters z is about ten, the NLPQL - algorithm of Schittkowski [10] is used to solve the mathematical programming problem of equation (7).

NLPQL is a sequential quadratic programming algorithm. In each iteration, a quadratic programming subproblem is defined by approximating the corresponding Lagrange function quadratically using first- and second-order terms of the Taylor series expansion.

It should be pointed out that only explicit constraints (equation (7c)) are present. Equation (7b) has an influence only on the objective function. Different steps of approximation were carried out, within which different parameters were variable (Figure 6).

-   For step 1, it is presumed that the wave had not spread over zone 4 and 3 and only the damping of the soil δ and the yield stress of the soil τ_3 were variable. In zone 4 the soil-pile interaction is neglected in case of non-compact soil. This zone is problem dependent and may be absent.

-   In step 2, τ_3 from step 1 was fixed and τ_2 and τ_1 were varied in addition to δ. The time domain is extended to the pile tip.

-   In step 3, τ_3, τ_2 and τ_1 were fixed from step 1 and 2 respectively. The parameters δ and β (pile stiffness proportion coefficient of Rayleigh damping) were varied in the complete measured time interval.

In steps 1, 2 and 3, the usual procedure of trial and error is simulated. Investigations showed, however, that this procedure is not appropriate. Results are not better than good trial and error results. Therefore the complete set of parameters (δ, τ_1, τ_2, τ_3 and β) is chosen.

Furthermore, the parameters E (modulus of elasticity of the concrete pile) and r_0 (limit elastic deformation of soil) are considered. In this way improvements of more than 50% relative to be *trial and error* results can be achieved.

5. Example

For the pile in Figure 7, two impact loadings are examined. The measured pile top velocities are shown in Figure 8.

The finite element idealisation is done with 13 degrees of freedom (Figure 7). The feasible domain is described by

$$
\begin{array}{rcccl}
0.001 & \leq & \delta & \leq & 0.1 \\
0.1 & \leq & \tau_1 & \leq & 10.0 \ \mathrm{MN/m^2} \\
0.01 & \leq & \tau_2 & \leq & 0.5 \ \mathrm{MN/m^2} \\
0.01 & \leq & \tau_3 & \leq & 0.5 \ \mathrm{MN/m^2} \\
0.00001 & \leq & \beta & \leq & 0.005 \\
10000 & \leq & E & \leq & 50000 \ \mathrm{MN/m^2} \\
0.001 & \leq & r_0 & \leq & 0.02 \ \mathrm{m}
\end{array}
$$

In Figure 9 the results of steps 1, 2 and 3 for impact load no. 10 are given. The

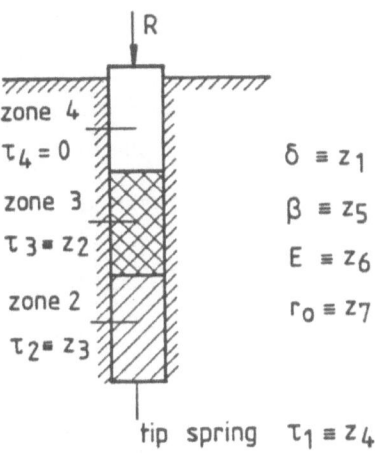

Figure 6: Pile with model parameters $\underline{z}$.

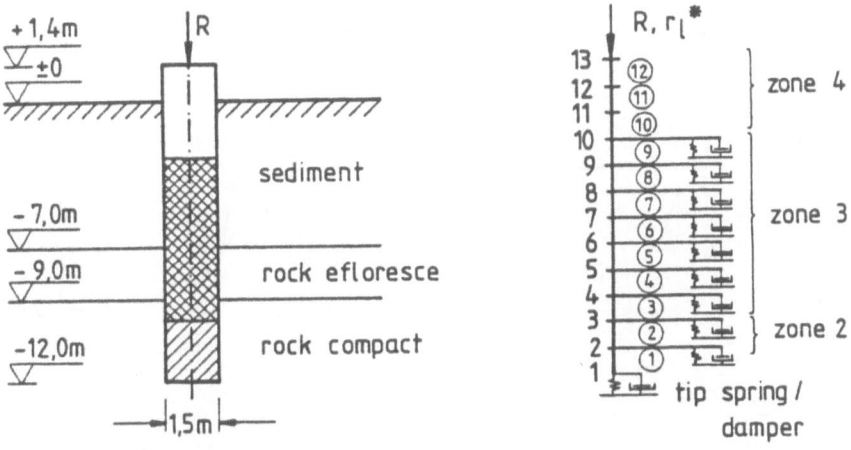

Figure 7: Pile II of Schwarzbachtal-bridge with finite element idealization

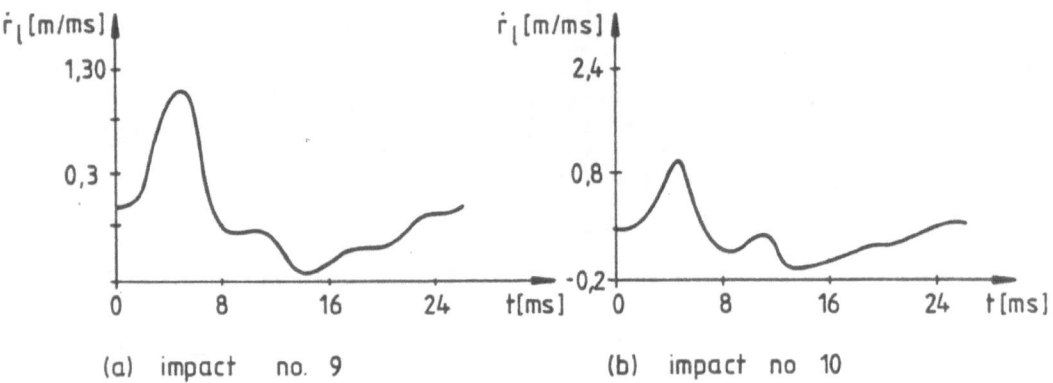

Figure 8: Measured top velocities of impact no. 9 and no. 10 of pile II.

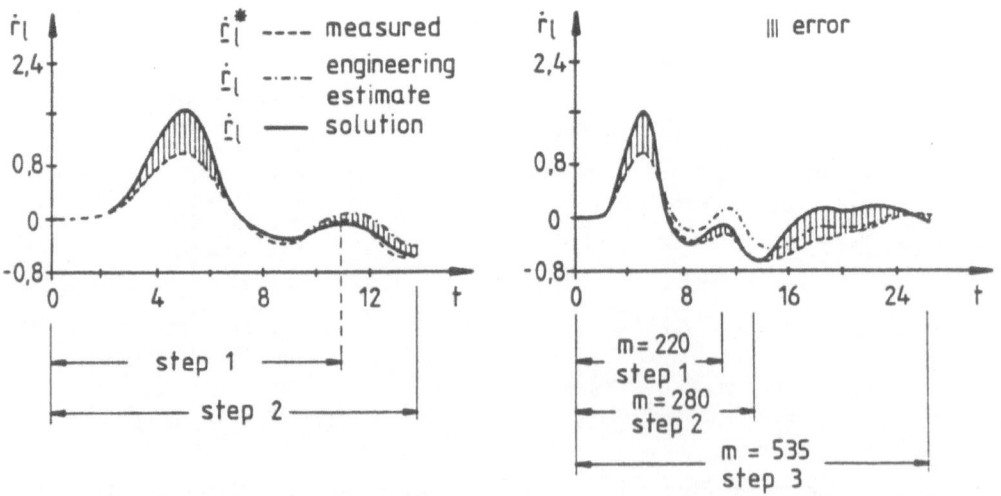

Figure 9: Results of steps 1, 2 and 3 for impact no. 10.

starting point was a good trial and error result. It can be seen that the results are even worse than the starting point. Using $\delta, \tau_1, \tau_2, \tau_3$ and β at the same time as free parameters gives not much better results (Figure 10a).

The results become really satisfactory when considering also E and r_0 (Figure 10b). In Table 1 the results are given for comparison. The improvement of the *best engineering estimate* is 58%.

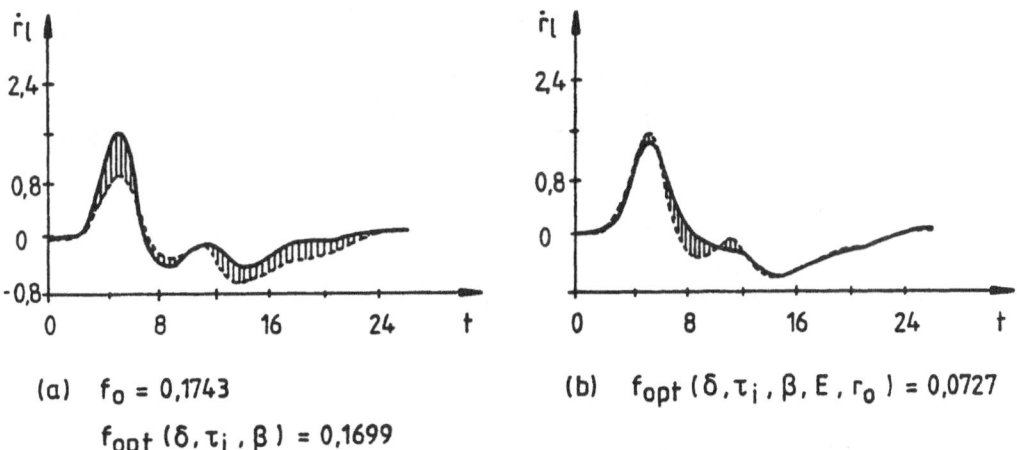

(a) $f_0 = 0,1743$

$f_{opt}(\delta, \tau_i, \beta) = 0,1699$

(b) $f_{opt}(\delta, \tau_i, \beta, E, r_0) = 0,0727$

Figure 10:   Results of error minimization for impact no. 10 including the different parameters.

Impact no. 9 was investigated in the same way. In Figure 11 the results for impact no. 9 are given besides the error functions for the starting and final point. Table 2 shows a comparison of the results for both impact loads. The results should be identical as pile and soil are the same. The only remarkable difference occurs with τ_3.

For the numerical solution, the trial and error code was adjusted by only the adaptation for the call of NLPQL. A scaling of the parameters to their upper bound was necessary because of the great difference in dimensions (e.g. E and β) and the computation of the gradients by difference quotients with a step-width of 10^{-5}.

The algorithm needed for an accuracy of 10^{-4} between 5 and 15 iterations and 10 and 50 complete solutions of the problem using the numerical integration scheme by Smith described in [4] page 47f.

Table 1: *Results of error minimization for impact no. 10 including the different parameters.*

parameters z	result of trial and error	step 1 δ, r_3	step2 δ, r_2, r_1	step 3 $\delta, 3$	δ, r_1, β	δ, r_1, β $E . r_0$
$\delta /10^{-2}$	0.295	0.243	0.154	0.154	0.280	0.604
r_1	1.762	→	1.828 →			1.772	4.860
r_2	0.11	→	0.427 →		0.135	0.130
r_3	0.11	0.237 →	→		0.175	0.033
$\beta /10^{-4}$	0.1	→	→	0.1	0.85	0.1
$E /10^{4}$	3.5	→	→	→	→	1.76
$r_0 /10^{-2}$	0.2	→	→	→	→	0.382
m	535	220	280	535 →	→	
f	0.1743	0.1417	0.1253	0.2104	0.1699	0.0727

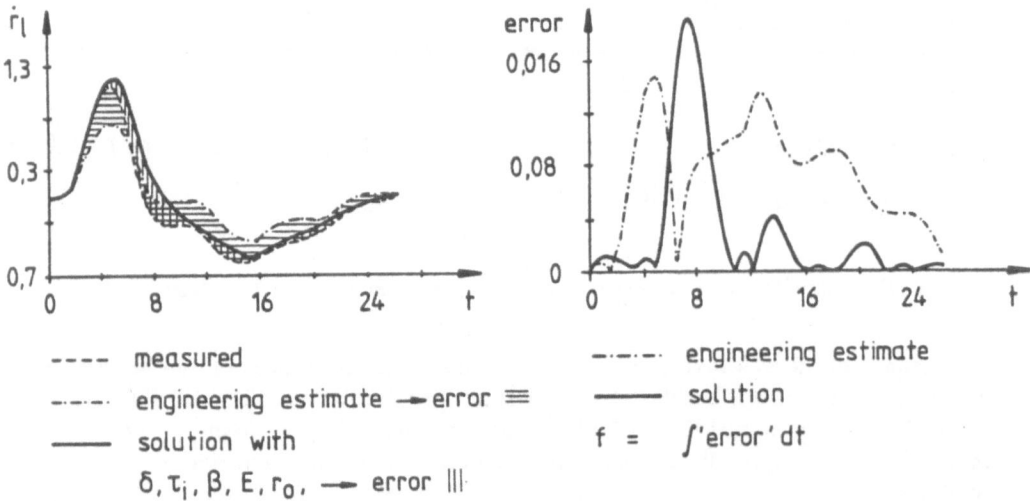

measured

engineering estimate → error ≡

solution with

$\delta, \tau_i, \beta, E, r_0$, → error |||

---- engineering estimate

—— solution

$f = \int' \text{error}' \, dt$

Figure 11: Results of error minimization for impact no. 9

Table 2: Comparison of results.

parameters $\underline{z}$	impact no. 9	impact no. 10	difference [%]
$\delta/10^{-2}$	0.594	0.604	1.7
τ_1	4.777	4.860	1.7
τ_2	0.130	0.130	0
τ_3	0.053	0.033	61
$\beta/10^{-4}$	0.1*	0.1*	0
$E/10^4$	1.75	1.76	0.6
$r_0/10^{-2}$	0.403	0.382	5.5
f	0.0744	0.0727	2.3

* lower bound

4. Conclusions

The present paper shows that a problem of nonlinear systems identification can be solved satisfactorily by a method of mathematical programming. In comparison to the trial and error CAPWAP method, remarkable improvements can be achieved.

The choice of the free parameters of the problem is crucial. Simulating the engineering process of trial and error by solving a sequency of mathematical programming problems for different time domains brings good results if the variable model parameters are correctly chosen.

Whereas some numerical values of the variables in the presented optimal solution have to be checked for consistency with structural and soil mechanic behaviour, the solution shows that the mechanical and mathematical models in general are appropriate.

References

[1] Natke, H. G.: Die systematische Anpassung von Rechenmodellen an Versuchswerte als Verfahren zum Nachweis dynamischen Systemverhaltens, Der Bauingenieur 57 (1982), 287-292, Springer-Verlag, Berlin.

[2] Bathe, K. J., Wilson, E. L.: Numerical methods in finite element analysis, Prentice-Hall Inc., Englewood Cliffs, N.J., 1976.

[3] Stanton, J. F., McNiven, H. D.: Towards an optimum model for the response of reinforced concrete beams to cyclic loads, Earthquake Engineering and Structural Dynamics, Vol. II, 299-312, 1983.

[4] Goble, G. G., Rausche, F., Likins, G.: The analysis of pile driving - A state-of-the-art, in 'Application of Stress-Wave Theory on Piles', ed. H. Bredenberg, Balkema, Rotterdam 1981.

[5] Clough, J. R., Penzien, P. H.: Dynamics of structures, McGraw Hill - Kokagusha, New York/Tokyo, 1976.

[6] Klingmüller, O.: Computational tools for dynamic pile testing, Sec. Int. Conf. on the Appl. of Stress Wave Theory on Piles, Stockholm 1984.

[7] Seitz, M., Klingmüller, O.: Dynamische Prüfung von Grossbohrpfählen, Baugrundtagung 1984, DGEG, Essen, 1984.

[8] Klingmüller, O.: Dynamische Pfahlprüfung als nichtlineare Systemidentifikation, in 'Dynamische Probleme - Modellierung und Wirklichkeit', Curt-Risch-Institut, Hannover 1984.

[9] Jakoby, S. L. S., Kowalik, J. S., Pizzo, J. T.: Iterative Methods for Nonlinear Optimization Problems, Prentice Hall Inc., Englewood Cliffs, New Jersey, 1972.

[10] Schittkowski, K.: On the convergence of a sequential quadratic programming method with an augmented Lagrangian line search function; Math. Operationsforschung und Statistik, Ser. Optimization, Vol. 14, No. 2, 1983, pp. 197-216.

BOUNDED STATE CONTROL
OF
LINEAR STRUCTURES

S. K. Lee
Agency for Defense Development
Korea
F. Kozin
Polytechnic Institute of New York
Route 110
Farmingdale, New York 11735

1. Introduction

In general, the main purpose of structural control is to maintain the controlled variables within an allowable region determined by the needs of structural safety and human comfort. This is a natural extension of the optimal control problem.

The performance concept of keeping system states within a prespecified region is not new. It was mentioned, for example in [1], as practical stability. The design of suboptimal controllers to maintain the lateral acceleration levels within comfortable bounds on high-speed rail vehicles was discussed in [2].

In the past decade the concept of optimal control of civil structures, developed and studied by Yao [3], Yang [4], Soong [5], Leipholz [6] and their co-workers has generated a significant interest in this field. Even more recently, the renewed interest in advanced structural design and control due to the NASA space station program has generated even greater activity in this field.

In the civil engineering structural control literature, linear quadratic cost functions, pole placement methods, modal control, as well as other techniques, have been applied. However, when bounded state constraints are incorporated, the results become highly complex and in general cannot be easily satisfied.

387

Indeed, the control of linear systems in the presence of set constrained external disturbances has been investigated maintaining the states in a prespecified
region by estimating the maximum reachable set [7], [8], [9]. Application of this
approach to structural control requires the solution of several non-linear optimization problems. Unfortunately, the existence of a solution for these non-linear
problems has not been clearly established at this time.

Our present approach takes a somewhat different and simpler direction. We
shall essentially construct an extension of the Lyapunov function method for
design of stabilizing controls.

In particular for the general linear structure

$$\dot{x} = Ax + Bu + Ef, \tag{1.1}$$

where $x(0) \in X_0$, $x \in R^{2n}$,
we wish to determine a controller u which will maintain the displacement and
velocity variables within a practically allowable region X_1 based upon the following assumptions:

(1) $f(t) \in \Omega_f$, a closed bounded region, for all $t \geq 0$

(2) $X_0 = \{x; x'P_0 x \leq c_0\}$

(3) $X_1 = \{x; x'P_1 x \leq c_0\}$,

where P_0, P_1 are positive definite matrices and $c_0 > 0$.

We will study linear feedback laws

$$u = -\Pi x \tag{1.3}$$

The problem is to determine Π, so that the invariant set$^{\perp}$, $X(\Pi)$, generated by Π
satisfies,

$$X_0 \subset X(\Pi) \subset X_1 \tag{1.4}$$

In section 2, we shall state and prove our main results and in Section 3 we
present an example.

2. Main Result

As stated earlier, the exact computation of a maximum reachable set is a very
difficult problem. Hence, we attempt to find a control u which generates an elipsoidal set X, which we call the invariant set, such that

$$X_0 \subseteq X \subseteq X_1, \quad t \geq t_0 \tag{2.1}$$

Let us first introduce several preliminary concepts. Consider a general linear
time-invariant system as

$^{\perp}$ Defined in Section 2.

$$\dot{x} = Ax \tag{2.2}$$

(A) Lyapunov functions: A Lyapunov function $V(x)$ of system (2.2) is defined as follows:

(i) $V(x)$ is smooth in a certain open set Ω around origin.

(ii) $V(x)>0$ for all x in Ω except for $x = 0$, i.e. the origin is an isolated minimum of $V(x)$.

(iii) Along any trajectory, $\dot{V}(x)\leq0$.

(B) Definition of invariant set.

Def. 2.1: For a fixed x_0 and a fixed t_0, the set of points $x(t,x_0,t_0)$ for $t\geq t_0$, is called the motion determined by x_0 and t_0.

Def. 2.2: Invariant set X: An invariant set $X\subseteq R^n$ is characterized by the dynamic property that if a point x_0 is in X, i.e., $x_0 \in X$, then the whole trajectory starting at x_0 lies in X, that is

$$x(t_0,x_0,t_0) \in X \Rightarrow x(t,x_0,t_0) \in X \ \forall \ t>t_0$$

Def. 2.3: The distance of a point y from X is defined as

$$d(y,X) = \text{Inf } ||y-x||, \text{ for } x \in X.$$

In view of Def. 2.1., 2.2 and 2.3, we further define

Def. 2.4: Stable invariant set: An invariant set X is called stable if

$$d(x(t,y,t_0),X)\leq\eta(d(y,X);t_0),$$

where $\eta(\cdot)$ is continuous, and strictly increasing with $\eta(0,t_0) = 0$.

Def. 2.5: An invariant set is called attractive if

$$d(x(t,y,t_0),X)\leq\epsilon(t - t_0;y,t_0)$$

provided $d(y,X) \leq \varsigma<\infty$, where $\epsilon(\cdot)$ is a continuous and strictly decreasing function for $t\geq\tau>t_0$.

Def. 2.6: If an invariant set is stable and attractive, then it is said to be an asymptotically stable invariant set.

It is possible to establish for an asymptotically stable invariant set that

$$d(x(t,y,t_0),X) \leq \eta(d(y,X);t_0)\epsilon(t-t_0;y,t_0)$$

The stability concept for invariant sets considers the dynamic behaviour of a subset of the state space, not an equilibrium point. Indeed, the stability concept of invariant set is an extension of Lyapunov stability.

For instance, consider the forced oscillator

$$\ddot{x} = -2\varsigma\omega\,\dot{x} - \omega^2 x + f(t),$$

where the external disturbance $f(t)$ is assumed to be in a closed-bounded set Ω_f.

Assume that there exists a unique solution P of the linear matrix equation

$$P\begin{bmatrix} 0 & 1 \\ -2\varsigma\omega & -\omega^2 \end{bmatrix} + \begin{bmatrix} 0 & -2\varsigma\omega \\ 1 & -\omega^2 \end{bmatrix}P = -1\,,\ P = P',$$

then, by the Lyapunov's stability theorem, the nominal system is asymptotically stable. However, asymptotic stability does not guarantee that all the solutions of the system starting in X_0 remains within an allowable region X_1 for all $f \in \Omega_f$ and for all $t \geq 0$. The above property of the system has been defined as practical stability [1]. Practical stability can be easily studied through the concept of the invariant set; the system is said to be practically stable, if the invariant set X satisfies the condition $X_0 \subseteq X \subseteq X_1$ for all $f(t) \in \Omega_f$: similarly, it is said that the system has strong practical stability, if the set X which satisfies that $X_0 \subseteq X \subseteq X_1$, is asymptotically stable (see [10], for the definition of strong practical stability).

Consider an n-dimensional linear, time-invariant system defined by

$$\dot{x} = Ax + Bu \tag{2.3}$$

It turns out that if there exists a linear map

$$\Pi : \mathcal{X} \to \mathcal{U} \tag{2.4}$$

in which $\mathcal{X}$ and $\mathcal{U}$ denote the state space and admissible control space, respectively, so that a real scalar function $V(x) \equiv x'Px$ can be a Lyapunov function of the closed loop system

$$\dot{x} = (A + B\Pi)x, \tag{2.5}$$

then an ellipsoidal set X, defined by

$$X \equiv \{x, x'Px \leq c, c \geq 0\} \tag{2.6}$$

is an asymptotically stable invariant set.

The invariant property of the set X defined by (2.6) is characterized in the following lemma.

Lemma 2.1: If $V(x) = x'Px$ has negative time derivative for the closed loop system (2.5); then the ellipsoidal set X defined by (2.6) is an asymptotically stable invariant set for all values of $c \geq 0$.

Proof: Suppose that $\dot{V}(x)$ be negative for the system (2.5). The time derivative of $V(x)$ can be represented as

$$\dot{V}(x) = (\dot{x}, \nabla V) = |\dot{x}| \, |\nabla V| \cos\theta ,$$

that is the projection of $\dot{x}$ on ∇V. On the other hand, since $V(x)$ is an increasing function of x, ∇V is an outward normal to the surface of $V(x) = c$. Furthermore, by assumption, since $\dot{V}(x) < 0$, θ must lie in an open interval, $90° < \theta < 270°$. Therefore, the tangential vector of x on the surface $V(x) = c$ must be inward (see Fig. 1).

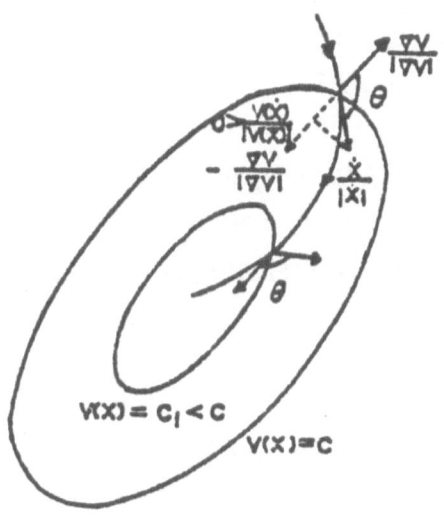

Figure 1

Hence, a trajectory starting at any point on $V(x) = c$ must move into a smaller surface than $V(x) = c$, i.e. from outside to inside, as time increases. Therefore, the trajectory starting at any point inside or outside X, remains in X, indeed it is attracted to the origin. Hence X is apparently an asymptotically stable invariant set for $c \geq 0$.

To characterize the control law, consider an elliposoidal set X defined as (2.6) with an arbitrary positive definite matrix P and $c > 0$. In order for X to be an asymptotically stable invariant set of the solutions of (2.5), the time derivative of $V(x) \equiv x'Px$ for the system (2.5) must satisfy the condition,

$$x'\{P(A + B\Pi) + (A + B\Pi)'P\}x < 0, \quad x \in R^{2n} \tag{2.7}$$

which follows from condition (iii) for Lyapunov functions.

The Π satisfying the condition (2.7) can be characterized in the following theorem.

Theorem 2.1: There exists a feedback control, for the system (2.3)

$$u \equiv \Pi x = -\beta B'Px \tag{2.8}$$

with arbitrary positive definite P and sufficiently large $\beta > 0$ (so that X is an asymptotically stable invariant set) if and only if

$$T'(PA + A'P)T < 0, \text{ provided rank } B < \text{rank } P, \tag{2.9}$$

in which $\mathrm{Span}(T)$ constructs the nullspace of $PBB'P$, i.e., $\mathrm{Span}(T) \equiv N(PBB'P)$.

Proof: First, suppose that there exists a control $u = -\beta B'Px$ which ensures the set X is asymptotically stable invariant. Then from (2.7), we know that

$$\dot{V}(x) = \dot{x}'Px + x'P\dot{x}$$
$$= x'(PA + A'P - 2\beta PBB'P)x < 0, \quad x \neq 0 \tag{2.10}$$

Since $x'PB = 0$ for all $x \in N(PBB'P)$, it follows from (2.10) that

$$x'(PA + A'P)x < 0 \quad \forall x \in N(PBB'P).$$

Furthermore, since $\mathrm{Span}(T) = N(PBB'P)$, i.e., $x = T\alpha$ for $x \in N(PBB'P)$, α any vector, we know that

$$T'(PA + A'P)T < 0.$$

In the other direction if $T'(PA + A'P)T < 0$, then

$$\dot{V}(x) = x'(PA + A'P)x < 0 \quad \forall x \in N(PBB'P).$$

On the other hand, let $\overline{N}(\cdot)$ be the orthogonal complement of $N(\cdot)$, then for all $x \in \overline{N}(PBB'P)$ since $x'PB \neq 0$, we know that $x'PBB'Px > 0$, that is, the symmetric matrix $PBB'P$ is positive semi-definite. For

$$x \in \overline{N}(PBB'P),$$

let $Q \equiv PA + A'P$ and $R \equiv PBB'P$, then we can write

$$\dot{V}(x) = x'Qx - 2\beta x'Rx. \tag{2.11}$$

Since R is an Hermitian matrix, its eigenvectors form a complete orthonormal basis. Let $Z = \{Z_i; i \in (1,2,...,n)\}$ denote the orthonormal basis of R and $dim(N(R)) = n-m$, then a vector $x \in \overline{N}(R)$ is given in the form,

$$x = \sum_{i=1}^{m} \varsigma_i Z_i = Z^m \varsigma, Z^m \equiv \{Z_i\}, i \in (1,2,...,m).$$ (2.12)

Therefore, rewriting (2.11), we have

$$\dot{V}(x) = \varsigma'(Z^m)'Q(Z^m)\varsigma - 2\beta \varsigma'(Z^m)'R(Z^m)\varsigma$$

$$= \varsigma'(Z^m)'Q(Z^m)\varsigma - 2\beta \sum_{i=1}^{m} \mu_i |\varsigma_i|^2,$$ (2.13)

in which μ_i is the i-th nonzero eigenvalue of R corresponding to Z_i. Furthermore, since $(Z^m)'Q(Z^m)$ is also Hermitian, there exists an orthonormal basis $Y = \{Y_i ; i \in (1,...,m)\}$, in which Y_i is an m-dimensional vector, such that $\varsigma = Ya$. Let $Z^m Y = S^m$, then we know that

$$x = S^m a = \sum_{i=1}^{m} a_i S_i$$ (2.14a)

and so,

$$||x||^2 = \sum_{i=1}^{m} |a_i|^2 = \sum_{i=1}^{m} |\varsigma_i|^2.$$ (2.14b)

Thus, substituting (2.14a) into (2.13), we get

$$\dot{V}(x) = \left[\sum_{i=1}^{m} a_i S_i\right]' Q \left[\sum_{i=1}^{m} a_i S_i\right] - 2\beta \sum_{i=1}^{m} \mu_i |\varsigma_i|^2$$

$$= \sum_{i=1}^{m} \lambda_i |a_i|^2 - 2\beta \sum_{i=1}^{m} \mu_i |\varsigma_i|^2$$

$$\le \lambda_{max}[Q] \sum_{i=1}^{m} |a_i|^2 - 2\beta \mu_{min}[R] \sum_{i=1}^{m} |\varsigma_1|^2$$

$$= ||x||^2 (\lambda_{max}[Q] - 2\beta \mu_{min}), \quad \forall x \in N(PBB'P),$$ (2.15)

in which $\lambda_{max}[Q]$ and $\mu_{min}[R]$ denote the largest eigenvalue of Q and the smallest nonzero eigenvalue of R, respectively. It is clear from (2.15) that there exists a sufficiently large β given by

$$\beta \ge \frac{\lambda_{max}[Q]}{2\mu_{min}[R]}$$

so that x can be an asymptotically stable invariant set with $\mu = -\beta B'Px$ for all $x \in \overline{N}(R)$.

Now let us define $J \equiv (Z^m)(Z^m)'$, then since $J^2 = J$, i.e., J is an idempotent matrix, we know that J is a projection of x onto $\overline{N}(R)$. Hence we can decompose x by $x = x^1 + x^2$, where $x^1 = Jx$ and $x^2 = (I - J)x$ and so $X^1 \in \overline{N}(R)$ and $x^2 \in N(R)$. That is the spaces $\overline{N}(R)$ and $N(R)$ generated by x^1 and x^2, respectively are the orthogonal complements of each other. On the other hand, x^1 and

x^2 can be also defined as follows:

$$x^1 = \sum_{i=1}^{m} \varsigma_i Z_i = Z^m \varsigma = \sum_{i=1}^{m} a_i S_i = S^m a, \qquad (2.16)$$

$$x^2 = \sum_{i=m+1}^{n} \varsigma_i Z_i = Z^{n-m} \varsigma = \sum_{i=m+1}^{n} a_i S_i = S^{n-m} a. \qquad (2.17)$$

Thus for $x \in \chi = \overline{N}(R) \oplus N(R)$, one can write

$$\dot{V}(x) = (x^{1'} + x^2)Q(x^1 + x^2) - 2\beta(x^{1'} + x^2)R(x^1 + x^2)$$

$$= (x^1)'Q(x^1) + (x^1)'Q(x^2) + (x^2)'Q(x^1) + (x^2)'Q(x^2) - 2(x^1)'R(x^1). \qquad (2.18)$$

Substituting (2.16) and (2.17) into (2.18), we obtain

$$\dot{V}(x) = \sum_{i=1}^{m} \lambda_i |a_i|^2 + (x^1)'Q(x^2) + (x^2)'Q(x^1) + \sum_{j=m+1}^{n} \lambda_j |a_j|^2 - 2\beta \sum_{i=1}^{m} \mu_1 |\varsigma_1|^2.$$

From the hypothesis, since $T'(PA + A'P)T = (Z^{n-m})'Q(z^{n-m}) < 0$, we see that $\lambda_j < 0$ for $j = m+1, \ldots, n$. Hence,

$$\dot{V}(x) \leq \max[\lambda_i] \sum_{i=1}^{m} |a_i|^2 + 2 \|Q\| \|x^1\| \|x^2\| - \min[|\lambda_j|] \sum_{j=m+1}^{n} |a_j|^2$$

$$- 2\beta \min[\mu_i] \sum_{i=1}^{m} |\varsigma_i|^2$$

$$= (\|x^1\|, \|x^2\|) \begin{bmatrix} \max[\lambda_i] - 2\beta\min[\mu_i] & \|Q\| \\ \|Q\| & -\min[|\lambda_j|] \end{bmatrix} \begin{bmatrix} \|x^1\| \\ \|x^2\| \end{bmatrix} \qquad (2.19)$$

in which Q is defined as a spectral norm. Thus, if Q is definite, then

$$\|Q\| = \max_{\{i\}} |\lambda_i(Q)|.$$

Therefore, it is very clear that there exists a sufficiently large β, which satisfies

$$\beta > \beta_1 = \frac{\max[\lambda_i]}{2\min[\mu_i]}, \qquad (2.20)$$

$$\beta > \beta_0 = \frac{\|Q\|^2 + \min[|\lambda_j|]\max[\lambda_i]}{2\min[\mu_i]\min[|\lambda_i|]}, \qquad (2.21)$$

so that $\dot{V}(x) < 0 \quad \forall x \in \chi \in R^n$. Hence the condition (2.9) is necessary and sufficient for the existence of control $u = -\beta B'Px$ which makes X asymptotically invariant for all x in X. Q.E.D.

Since the matrices A and B are determined at the design stage of the structure, the value of β which satisfies the invariant property of X depends on the matrix P and state x. The following table shows the required value of β for each different case.

Table 1

max $[\lambda_i]$	$x \in N(R)$	$x \in \overline{N}(R)$	$x \notin N(R)$
≤ 0	$\beta > 0$	$\beta > 0$	$\beta > \beta_2$
> 0	$\beta > 0$	$\beta > \beta_1$	$\beta > \beta_0$

where,

$$\beta_1 = \frac{\max[\lambda_i]}{2\min[\mu_i]}, \quad \beta_2 = \frac{\|Q\|^2 + \min[\,|\lambda_j\,|\,]\max[\lambda_i]}{2\min[\mu_i]\min[\,|\lambda_j\,|\,]}.$$

Corollary 2.1 The condition (2.9) can be equivalently written as

$$\Gamma'(P^{-1}A' + AP^{-1})\Gamma < 0, \quad \text{since } P^{-1}\text{exists,} \tag{2.22}$$

where $\text{Span}(\Gamma) \equiv \text{Null}(BB')$.

Proof: Suppose $x \in \text{Null}(PBB'P)$. Let $Z = Px$, then $x = P^{-1}Z$. Hence $x'PBB'Px = 0 = Z'BBZ$ for $x \in \text{Null}(PBB'P)$. It implies that $Z \in \text{Null}(BB')$ for all $x \in \text{Null}(PBB'P)$. Furthermore since $\text{Span}(T) = N(PBB'P)$ and P is positive definite, one can write

$$T = P^{-1}\Gamma \tag{2.23}$$

Hence,

$$T'(PA + A'P)T = (P^{-1}\Gamma)'(PA + A'P)(P^{-1}\Gamma)$$
$$= \Gamma'(P^1A' + AP^{-1})\Gamma < 0.$$

$$\text{Q.E.D.}$$

The significant feature of (2.1) is that it allows us to choose any positive definite matrix P, which satisfies $X_0 \subseteq X \subseteq X_1$, to construct a control $U = -\beta B'PX$. The control u with β determined by (2.21) guarantees that all the solutions of the system with respect to X_0 remain within X. This result also appears to be new in comparison with results which use the usual Lyapunov method.

In order to study the existence of the control u which guarantees the set constrained motion of a dynamic structure, it is necessary and sufficient to apply the condition (2.9) or (2.22). However, still there exists a question. That is, we have to consider the case of $N(PBB'P) = 0$, i.e., rank $B =$ rank P.

Theorem 2.2. For any positive definite P, there exists a sufficiently large positive $\beta > 0$ so that the time derivative of $V(x)$ for system (2.5) becomes negative if rank $B =$ rank P, i.e., $PBB'P \equiv R$ is a nonsingular.

Proof: From (2.12), it follows immediately that

$$\dot{V}(x) = x'(Q - \beta R_0)x,$$

where $R_0 \equiv 2R$. Since R_0 is a real symmetric positive definite matrix, it is known that $R_0 = FF$, where F is symmetric and positive definite. Hence, one can write

$$Q - \beta R_0 = F(F^{-1}QF^{-1} - \beta I)F. \tag{2.24}$$

Since $F > 0$, the investigation of the negative definiteness of $\dot{V}(x)$ is reduced to the evaluation of the eigenvalues of $(F^{-1}QF^{-1} - \beta I)$. Consider the characteristic equation given by

$$|F^{-1}QF^{-1} - \beta I - \lambda I| = 0. \tag{2.25}$$

Since $(\beta + \lambda)$ is the eigenvalue of $(F^{-1}QF^{-1})$, one obtains

$$\beta + \lambda_i = \mu_i(F^{-1}QF^{-1}), \tag{2.26}$$

where $\mu_i(\cdot)$ denotes the i-th eigenvalue of $(F^{-1}QF^{-1})$. Therefore,

$$\lambda_i = \mu_i(F^{-1}QF^{-1}) - \beta. \tag{2.27}$$

Hence, if $\beta > \mu_i(F^{-1}QF^{-1})$ for all i, that is

$$\beta > \mu_{max}(F^{-1}QF^{-1}), \tag{2.28}$$

in which $\mu_{max}(\cdot)$ denotes the largest eigenvalue, then $\dot{V}(x) < 0$ for all $x \neq 0$. Q.E.D.

Hence if X_0 is bounded, then there exists a positive definite matrix P such that $X_0 \subseteq X$. Therefore, the design of a feedback control law is reduced to choosing a proper P which ensures $X_0 \subseteq X \subseteq X_1$ for all $t \geq 0$ and satisfies (2.9) or (2.28).

The method by which we can find a matrix P satisfying condition (2.1) can be deduced directly from the following theorem, stated without proof [11].

Theorem 2.3

Let

$$X_0 = \{x \in R^n ; x'P_0x \leq c_0\},$$

$$X = \{x \in R^n ; x'Px \leq c\},$$

$$X_1 = \{x \in R^n ; x'P_1x \leq c_0\},$$

in which P_0, P_1, and c_0 are all given. Let $\{\lambda_i, i = 1, 2, ..., n\}$ be the roots of $|P_0 - \lambda P_1| = 0$ and $\{\sigma_i\}$ be the roots of $|P - \sigma P_1| = 0$. Then P must be chosen so that

$$\frac{\lambda_i}{c_0} \geq \frac{\sigma_i}{c} \geq \frac{1}{c_0} \qquad \forall i \in \{1, 2, ..., n\}. \tag{2.29}$$

Let us assume that a P which satisfies the corollary (2.1) or condition (2.9) has been obtained. Then we can readjust the P such that it will also satisfy (2.29).

Let P^* denote the desired gain matrix and let the P^* be decomposed as

$$(P^*)^{-1} = P^{-1} + P_i^{-1} \tag{2.30}$$

then, it follows from (2.22) that the positive (at least semi) definite matrix P_i must satisfy

$$\Gamma'(AP_i^{-1} + P_i^{-1}A')\Gamma = 0. \tag{2.31}$$

The P_i's satisfying (2.31) can be characterized through the following lemma [11].

Lemma: Let Γ_0 be a matrix whose columns form a basis of the Null $(A'\Gamma\Gamma'A)$. Then P_i's defined by (2.32) satisfy (2.31).

$$P_i^{-1} = \Gamma_0 \Lambda \Gamma_0' + B\theta B' \tag{2.32}$$

for any positive (semi) definite matrices, Λ and θ.

3. Example

In this section we study control of a 3-story structure disturbed by an external force $f(t)$ as shown in Figure 2, [11].

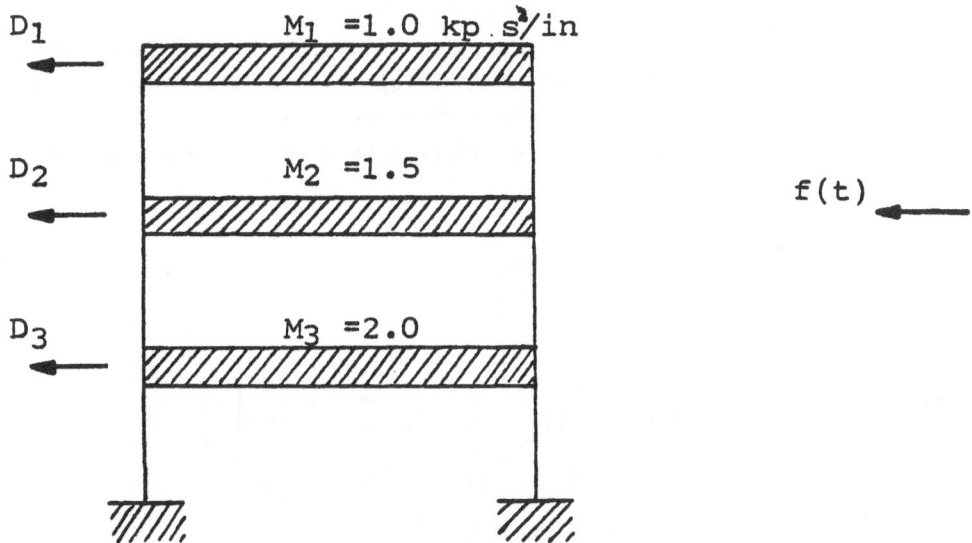

Figure 2

We describe the motion of this linear structure by

$$M\ddot{d} + C\dot{d} + Kd = Ef(t), \tag{3.1}$$

in which

$$M = \begin{bmatrix} 1.0 & 0 & 0 \\ 0 & 1.5 & 0 \\ 0 & 0 & 2.0 \end{bmatrix},$$

$$K = \begin{bmatrix} 600 & -600 & 0 \\ -800 & 1800 & -1200 \\ 0 & -1200 & 3000 \end{bmatrix},$$

$$C = \begin{bmatrix} 2.09 & -0.99 & 0 \\ -0.99 & 4.62 & -1.98 \\ 0 & -1.98 & 7.15 \end{bmatrix} = 1.1M + 0.00165K,$$

and the disturbance input vector is

$$E = \begin{bmatrix} 0.7 \\ 0.2 \\ 0.1 \end{bmatrix}.$$

These numerical values are chosen to characterize a lightly damped structure ($\varsigma = 0.5\%$) and allow decoupling of the equations.

The external disturbance $f(t)$ is assumed to be a linear combination of two slightly damped sinusoidal functions ($\varsigma \doteq 0.05\%$) whose natural frequencies are equal to those of the 1st and 3rd modes of the structure[*].

In what follows we will assume that $f(t)$ can be observed and modeled as:

$$f(t) = Hw = [0, 0.3, 0, 0.7] \begin{bmatrix} w_1 \\ w_2 \\ w_3 \\ w_4 \end{bmatrix}, \tag{3.2a}$$

and

$$\dot{w} = Gw = \begin{bmatrix} 0 & 1 & 0 & 0 \\ -210 & -0.0145 & 0 & 0 \\ 0 & 0 & 0 & 1 \\ 0 & 0 & -2124 & -0.0461 \end{bmatrix} \begin{bmatrix} w_1 \\ w_2 \\ w_3 \\ w_4 \end{bmatrix},$$

with $w_2(0) = w_4(0) = 0$, $|w_1(0)| \leq 1$ and $|w_3(0)| \leq 1$.

(Note that if f can be observed, then w can be reconstructed.)

Introducing the $2n$-dimensional vector $y = [d_1, \dot{d}_1, d_2, \dot{d}_2, d_3, \dot{d}_3]$, one can obtain a linear state form of (3.1) as

---

[*] The undamped natural frequencies of the structure are $\omega_1 = 14.48$, $\omega_2 = 31.05$, and $\omega_3 = 46.1$. These are easily obtained by solving the equation $|K - \omega^2 M| = 0$.

$$\dot{y} = A_0 y + E_0 f, \tag{3.3}$$

where,

$$A_0 = \begin{bmatrix} 0 & 1 & 0 & 0 & 0 & 0 \\ -600 & -2.09 & 600 & 0.99 & 0 & 0 \\ 0 & 0 & 0 & 1 & 0 & 0 \\ 400.02 & 0.66 & -1200.06 & -3.08 & 800.04 & 1.3201 \\ 0 & 0 & 0 & 0 & 0 & 1 \end{bmatrix}, \quad E_0 = \begin{bmatrix} 0 \\ 0.7 \\ 0 \\ 0.133 \\ 0 \end{bmatrix}.$$

By solving $K\phi = \omega^2 M\phi$, one easily obtains the mode shape matrix

$$\phi = \begin{bmatrix} 1 & 1 & 1 \\ 0.644 & -0.601 & -2.57 \\ 0.3 & -0.676 & 2.47 \end{bmatrix}.$$

Since C was assumed so as to satisfy the orthogonality conditions by using the linear modal transformation

$$y = \Phi x,$$

in which

$$\Phi = \begin{bmatrix} 1 & 0 & 1 & 0 & 1 & 0 \\ 0 & 1 & 0 & 1 & 0 & 1 \\ 0.644 & 0 & -0.601 & 0 & -2.57 & 0 \\ 0 & 6.644 & 0 & -0.601 & 0 & -2.57 \\ 0.3 & 0 & -0.676 & 0 & 2.47 & 0 \\ 0 & 0.3 & 0 & -0.676 & 0 & 2.47 \end{bmatrix},$$

we can obtain a modal state form of (3.3).

The augmented modal state equation of the controlled system can be expressed as

$$\dot{\hat{z}} = \hat{A}\hat{z} + \hat{B}u, \tag{3.4}$$

where $\hat{A}$ is obtained by

$$\hat{A} = \begin{bmatrix} \Phi' A_0 \Phi & \Phi' E_0 H \\ 0 & G \end{bmatrix} =$$

$$
= \begin{bmatrix}
0 & 1 & 0 & 0 & 0 & 0 & 0 & 0 & 0 & 0 \\
-210 & -1.45 & 0 & 0 & 0 & 0 & 0 & 0.142 & 0 & 0.33 \\
0 & 0 & 0 & 1 & 0 & 0 & 0 & 0 & 0 & 0 \\
0 & 0 & -966 & -2.693 & 0 & 0 & 0 & 0.063 & 0 & 0.146 \\
0 & 0 & 0 & 0 & 0 & 1 & 0 & 0 & 0 & 0 \\
0 & 0 & 0 & 0 & -2124 & -4.61 & 0 & 0.006 & 0 & 0.013 \\
0 & 0 & 0 & 0 & 0 & 0 & 0 & 1 & 0 & 0 \\
0 & 0 & 0 & 0 & 0 & 0 & -210 & -0.0145 & 0 & 0 \\
0 & 0 & 0 & 0 & 0 & 0 & 0 & 0 & 0 & 1 \\
0 & 0 & 0 & 0 & 0 & 0 & 0 & 0 & -2124 & -0.0461
\end{bmatrix} ,
$$

and for simplicity B is assumed to be:

$$
B = \begin{bmatrix}
0 & 0 & 0 \\
1 & 0 & 0 \\
0 & 0 & 0 \\
0 & 1 & 0 \\
0 & 0 & 0 \\
0 & 0 & 1 \\
0 & 0 & 0 \\
0 & 0 & 0 \\
0 & 0 & 0 \\
0 & 0 & 0
\end{bmatrix} .
$$

Since this example is provided for an illustration, the set of initial conditions and the allowable region are assumed to be described by simple hyperspheres as follows:

$$
\begin{cases}
X_0 = \{x; x' P_0 x \le 1\} \\
X_1 = \{x; x' P_1 x \le 1\},
\end{cases}
\tag{3.5}^{\perp}
$$

in which $P_1 = I_6$ and $P_0 = 2P_1$. That is, it is required to control the structure so that any solution of (3.4) starting at $x_0 \in X_0$ can remain with X_1.

The equation (3.4) is a completely uncoupled form, the design of a state feedback control satisfying the given specifications can be accomplished by considering each mode separately as

$$
\ddot{x}^i = \hat{A}_i \, \dot{x}^i + \hat{B}_i U_i, \quad i = 1, 2, 3.
\tag{3.6}
$$

Considering the 1st mode, we have

$\perp$ The projections of (3.5) onto the i-th modal domain are denoted by $X_0^i = \{x^i; x^{i\prime} V_0^i x^i \le 1\}$ and $X_1^i = \{x^i; x^{i\prime} V_1 x^i \le 1\}$.

$$
\begin{bmatrix} \dot{x}_1 \\ \dot{x}_2 \\ \dot{w}_1 \\ \dot{w}_2 \\ \dot{w}_3 \\ \dot{w}_4 \end{bmatrix} = \begin{bmatrix} 0 & 1 & 0 & 0 & 0 & 0 \\ -210 & -1.45 & 0 & 0.142 & 0 & 0.13 \\ 0 & 0 & 0 & 1 & 0 & 0 \\ 0 & 0 & -210 & -0.0145 & 0 & 0 \\ 0 & 0 & 0 & 0 & 0 & 1 \\ 0 & 0 & 0 & 0 & -2124 & -0.0461 \end{bmatrix} \begin{bmatrix} x_1 \\ x_2 \\ w_1 \\ w_2 \\ w_3 \\ w_4 \end{bmatrix} + \begin{bmatrix} 0 \\ 1 \\ 0 \\ 0 \\ 0 \\ 0 \end{bmatrix} u_1.
\tag{3.7}
$$

The control $u_1 = -\beta_1 \hat{B}_1' P \hat{x}'$ can be directly obtained through the procedures developed in Section 2.

In order to obtain an initial P that satisfies (2.9), we solve an algebraic Riccati equation,

$$
PA_1 + A_1'P - 2PB_1B_1'P + Q = 0.
\tag{3.8}
$$

We choose Q, for example, as the diagonal matrix,

$$
Q = \begin{bmatrix}
.10000e+01 & .00000e+00 & .00000e+00 & .00000e+00 & .00000e+00 & .00000e+00 \\
.00000e+00 & .10000e+01 & .00000e+00 & .00000e+00 & .00000e+00 & .00000e+00 \\
.00000e+00 & .00000e+00 & .10000e-03 & .00000e+00 & .00000e+00 & .00000e+00 \\
.00000e+00 & .00000e+00 & .00000e+00 & .10000e-04 & .00000e+00 & .00000e+00 \\
.00000e+00 & .00000e+00 & .00000e+00 & .00000e+00 & .10000e-02 & .00000e+00 \\
.00000e+00 & .00000e+00 & .00000e+00 & .00000e+00 & .00000e+00 & .10000e-02
\end{bmatrix}.
$$

We obtain from (3.8),

$$
P = \begin{bmatrix}
.60675e+02 & .23809e-02 & .42181e+01 & .11872e-03 & .23144e-01 & .10435e-01 \\
.23809e-02 & .28891e+00 & -.11872e-03 & .20087e-01 & -.10554e+00 & .11166e-03 \\
.42181e+01 & -.11872e-03 & .35543e+02 & .23803e-06 & -.10867e-02 & .72857e-03 \\
.11872e-03 & .20087e-01 & .23803e-06 & .16925e+00 & -.73691e-02 & -.55211e-05 \\
.23144e-01 & -.10554e+00 & -.10867e-02 & -.73691e-02 & .24538e+02 & -.50091e-05 \\
.10435e-01 & .11166e-03 & .72857e-03 & -.55211e-05 & -.50091e-05 & .11536e-01
\end{bmatrix}.
$$

One can then compute the projection of $\hat{X}^1$ onto the controllable subspace as:

$$
X_c^1 = \{x^1; x^{1\prime} V x^1 \le c\},
$$

where

$$
V = P_{11} - P_{12}P_{22}^{-1}P_{21}
$$

$$
= \begin{bmatrix} 60.675 & 0.238 \times 10^{-2} \\ 0.238 \times 10^{-2} & 0.2889 \end{bmatrix}, \quad [11, \text{Sec.3.4}].
$$

Hence, by the condition (2.29), the solutions of the characteristic equation

$$
|V - \sigma V_1| = 0, \text{ in which } V_1 = I_2
$$

must satisfy

$$2 \geq \frac{\sigma_{1,2}}{c} \geq 1 . \tag{3.9}$$

However, there exists no such constant c.

Hence, through the algorithm (2.30) and (2.32) we modify the initial P as

$$P^{*-1} = P^{-1} + \Gamma_0' \Lambda \, \Gamma_0 + \hat{B} \, \theta \, \hat{B} .$$

But

$$\Gamma_0 = \begin{bmatrix} 1 \\ 0 \\ 0 \\ 0 \\ 0 \\ 0 \end{bmatrix} .$$

Hence, with $\Lambda = 0$ and $\theta = -0.2603$, we obtain P^* as:

$$\begin{bmatrix}
.60675e+02 & .23609e-02 & .42181e+01 & .11872e-03 & .23144e-01 & .10435e-01 \\
.23809e-02 & .35288e+02 & -.11872e-03 & .20087e-01 & -.10554e+00 & .11166e-03 \\
.42181e+01 & -.11872e-03 & .35543e+02 & .23803e-08 & -.10867e-02 & .72857e-03 \\
.11872e-03 & .20087e-01 & .22803e-06 & .16925e+00 & -.73681e-02 & -.55211e-03 \\
.23144e-01 & -.10554e+00 & -.10867e-02 & -.73691e-02 & .24538e+02 & -.50091e-05 \\
.10435e-01 & .11166e-03 & .72857e-03 & -.55211e-05 & -.50091e-05 & .11536e-01
\end{bmatrix} .$$

Thus,

$$V^* = P_{11}^* - P_{12}^* P_{22}^{*-1} P_{21}$$

$$= \begin{bmatrix} 60.675 & 0.238 \times 10^{-2} \\ 0.238 \times 10^{-2} & 35.2861 \end{bmatrix} ,$$

satisfying the condition (3.9) with $c = 35$.

To check for $\Gamma'(P^{*-1}A_1' + p^{*-1}A_1)\Gamma < 0$, solving $(BB)'\Gamma = 0$ gives:

$$\Gamma = \begin{bmatrix} 1 & 0 & 0 & 0 \\ 0 & 0 & 0 & 0 \\ 0 & 1 & 0 & 0 \\ 0 & 0 & 1 & 0 \\ 0 & 0 & 0 & 0 \end{bmatrix} .$$

The direct computation yields:

$$\Gamma'(P^{*-1}A_1' + P^{*-1}A_1)\Gamma < 0 .$$

Thus, finally with $\beta_1 = 5$ obtained by (2.21), one arrives at the control gain K_1 for the first mode as:

K1
.119505e-01   .17500e+03   -.59360e-03   .10044e+00   -.5277e+00   .55830e-03.

Through exactly the same procedure one gets the gain matrices K_2 and K_3 for the 2nd and 3rd modes as:

K2
.25880e-02   .42500e+03   .15219e-02   .62800e-04   -.23003e+00   .60875e-03,

K3
.11770e-02   .57500e+03   .34865e-03   .90880e-06   -.30011e-05   .14163e-02.

Displacement and velocity of each floor are plotted in the phase plane in Figures 3, 4, 5.

2. Conclusion

In this manuscript we have presented an approach that allows the design of a linear controller that will maintain the structural response within a suitable prechosen region. Lyapunov function concepts are used. The approach requires a known or assumed linear dynamical model for the external excitations. However, all that is needed is that the external excitations remain within a closed bounded region. In a sense we have a pole placement approach to maintain desired response characteristics in a region. The same procedures can be applied when the structural parameters are known only to within given bounds. If the input dynamics are completely unknown, the concept of inverse dynamical systems can be applied [11]. The technique is easy to compute and appears to have merit. Further study should be made to relax the required bounds on the β coefficient so that less stringent controller actions is required.

Acknowledgement

The research for this paper was partially supported under NSF Grants PFR 78-08811, CEE-8311190.

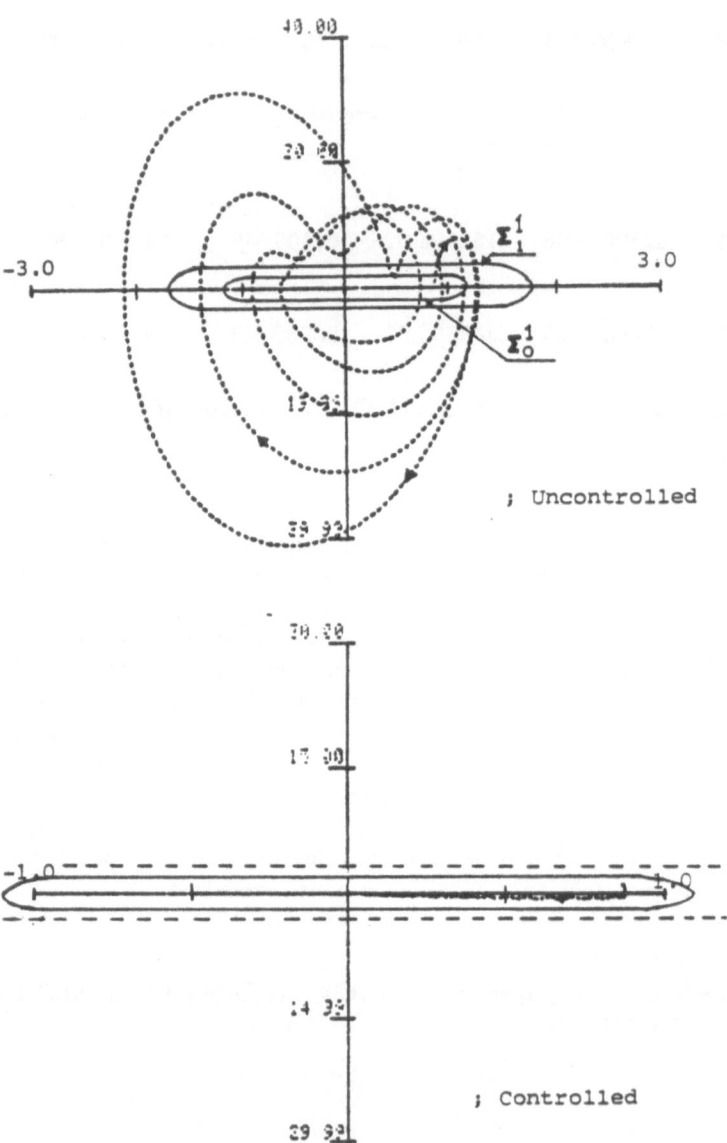

Figure 3: D_1 vs $\dot{D}_1$

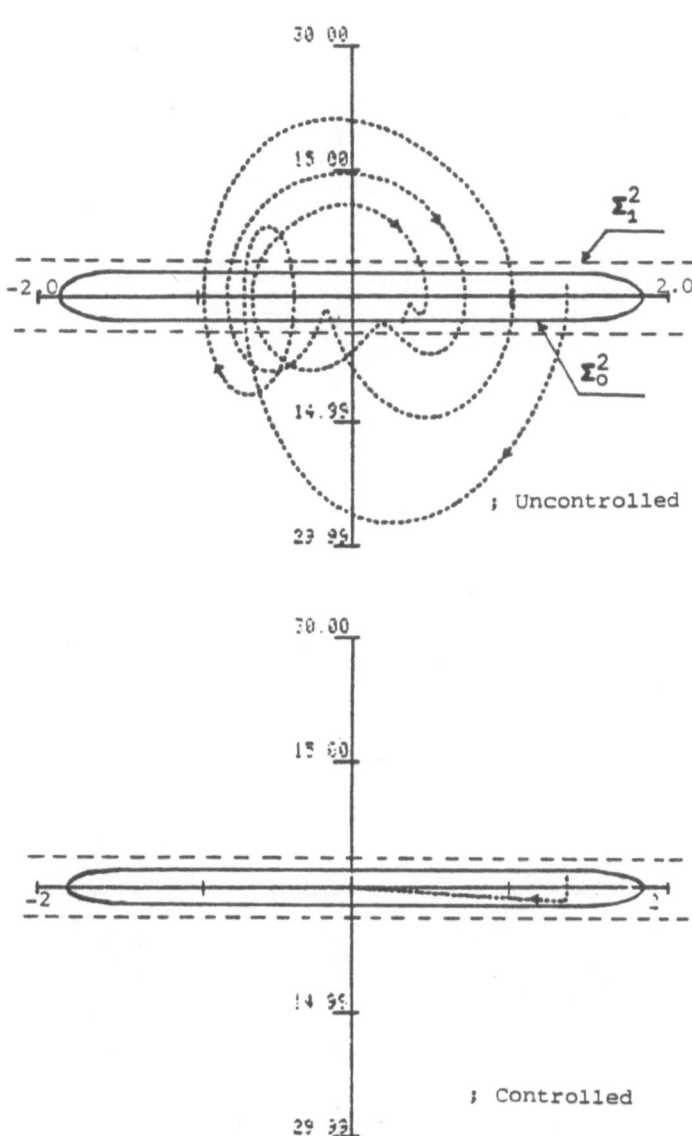

Figure 4: D_2 vs $\dot{D}_2$

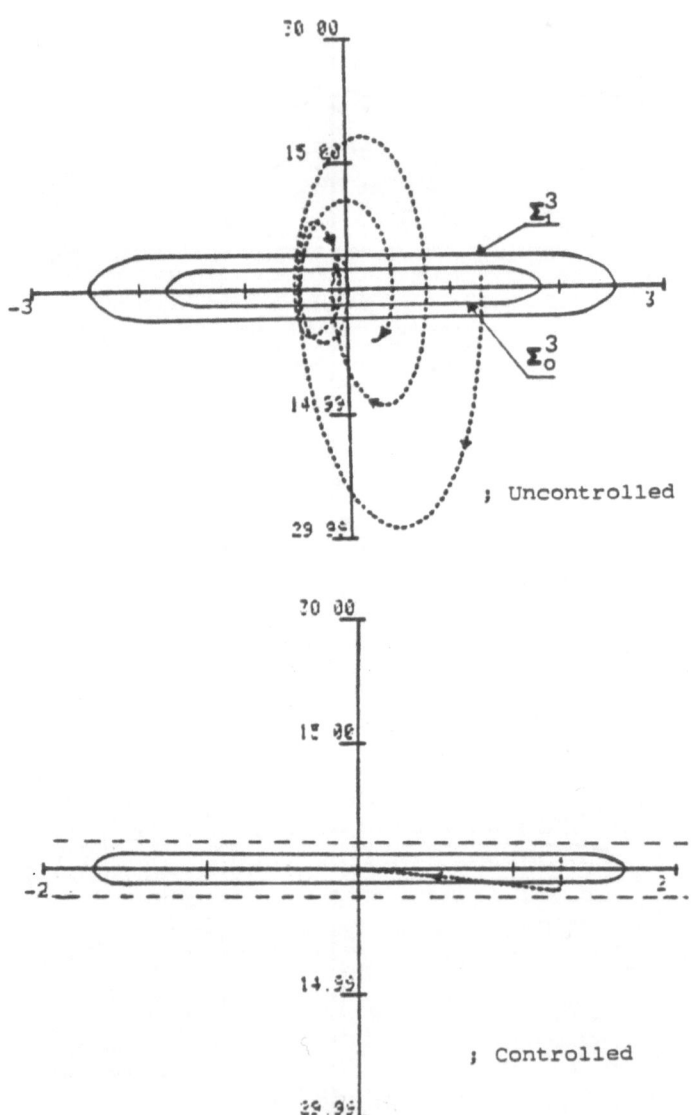

Figure 5: D_3 vs $\dot{D}_3$

References

[1] LaSalle, J. and Lefschetz, S., "Stability by Lyapunov's Direct Method with Applications", Academic Press, New York, N.Y., 1961.

[2] Sarma, G. N. and Kozin, F., "An Active Suspension System Design for Lateral Dynamics of a High-Speed Wheel Rail System", *ASME Jour. Dynamic Systems, Measurement and Control*, Vol. 93, No. 4, Dec. 1971, p. 233.

[3] Yao, J. T. P., "Concept of Structural Control", *ASCE, J. of Structural Div.*, Vol. 98, No. ST 7, Proc. Paper 9048, July 1972, pp. 1567-1574.

[4] Yang, J. N. and Ginnopoulos, F., "Active Tendon Control of Structures", *ASCE, J. of Eng. Mech. Division*, Vol. 104, No. EM3, Proc. Paper 13836, June 1978, pp. 551-568.

[5] Soong, T. T., and Martin, C. R., "Modal Control of Multistory Structures", *ASCE, J. of the Eng. Mech. Division*, Vol. 104, No. EM4, Proc. Paper 12321, August 1976.

[6] Abdel-Rohman, M. and Leipholz, H. H. E., "A General Approach to Active Structural Control", Proc. of the Intl. IUTAM Symposium on Structural Control held at the University of Waterloo, Ontario, Canada, June 1979.

[7] Delfour, M. and Mitter, S., "Reachability of Perturbed Systems and Mini-Sup Problems", *SIAM J. Control*, Vol. 7, Nov. 1969.

[8] Glover, J. D., and Schweppe, F. C., "Control of Linear Dynamic System with Set Constrained Disturbances", *IEEE Trans. on Automatic Control*, Vol. AC-16, October 1975.

[9] Chernousko, F. L., "Elipsoidal Bounds for Sets of Attainability and Uncertainty in Control Problems", *J. Optimal Control Applications and Methods*, Vol. 3, 1983.

[10] Corless, M. and Leitmann, G., "Continuous State Feedback Guaranteeing Uniform Ultimate Boundedness for Uncertain Dynamic Systems", *IEEE Trans. on Automatic Control*, AC-26, No. 5, 1981, pp. 1139-1144.

[11] Lee, S.K., "Bounded State Control of Linear Systems with Application to Active Structural Control", Ph.D. Thesis in System Engineering, Dept. of Electrical Engineering, Polytechnic Institute of New York, Jan. 1985.

ROTOR TO BASE CONTROL
OF ROTATING MACHINERY
TO MINIMIZE TRANSMITTED FORCE

D. W. Lewis
Department of Mechanical and Aerospace Engineering
University of Virginia
Charlottesville, Virginia
P. E. Allaire
Department of Mechanical and Aerospace Engineering
University of Virginia
Charlottesville, Virginia

1. Introduction

One significant source of noise that radiates from submarines originates from vibrations in rotating machinery. The rotating shaft is subject to a number of forces, F_{ext}, such as unbalance,[*] fluid forces on impellers, seals, bearings, et cetera. Shaft vibrations result from these forces which can be transmitted through the machine base to the submersible support, structure, or hull. In a similar fashion, noises in buildings, airplanes, surface ships and large production facilities frequently emanate from rotors and are transmitted to the supports and ultimately cause discomfort to human beings.

Active feedback control for rotating machinery has been studied only recently. Early works concentrated on electromagnetic controls [1,2] for rotors. This has resulted in the development of magnetic bearings by some commercial firms. A multi-mass rotor rig was designed and constructed [3,4] on a large based

---

[*] "Imbalance" and "unbalance" are used interchangeably as Webster's Third International Dictionary, copyright 1971, defines them as "lack of balance" or "absence of balance".

which could essentially be considered as rigid.

Theoretical works related to the control of rotors have been developed from an automatic controls point of view. One example is given in Stanway and Burrows [5]. In these works, the problem is viewed as part identification of system parameters such as bearing stiffness and part as actual control.

An analysis of a single mass rotor with active control on rigid supports was presented in [6] to give the analysis from a machinery dynamics point of view rather than a controls approach. This work was extended to a single mass rotor on flexible supports [7]. A comparison between theory and experiment for a three mass rotor is shown in [8].

Recently several articles have been published which suggest that one method for obtaining the desired form of control is via magnetic bearings. Bearings of this type have been used in high speed rigid rotors as well as flexible shaft compressors [9,10]. Also, they have been employed to reduce vibrations in flexible transmission shafts [11]. As magnetic bearings come to be better understood [12], they may permit the designer to use their controls to implement sophisticated algorithms to reduce transmitted forces.

None of the work currently in the literature relates directly to minimization of forces transmitted (due to rotating machinery) to supports. Other works have discussed applying controls to reduce forces transmitted due to structural vibrations but these are not really relevant here because the control is applied between the structure and the support.

This present work addresses active feedback control for vibration limiting. A rotor model is considered in which a control force is imposed between the rotor and the base supporting the rotor bearings. The force which the base transmits to the surrounding supports is determined according to the form of the active feedback control force acting between the rotor and the base. This is the rotor to base (RTB) force configuration.

2. Rotor Model

Figure 1 shows the geometry of a simple mathematical model with a single rotating disk on a flexible shaft, a machine base, and spring coupling of the base to a support. The position of the shaft relative to the support or hull (in the case of a ship) is defined by the variable x_2. The position of the machinery base relative to the support (or hull) is noted as x_1. The motion of the shaft relative to the base, $x_2 - x_1$, is assumed to be sensed near the disk by non-contacting displacement probes and this signal sent to a controller. Only linear translation in the vertical direction is considered in this analysis. However, the same approach may be made for horizontal motion. Rocking motion is not considered in this work although it could be approached in a similar manner.

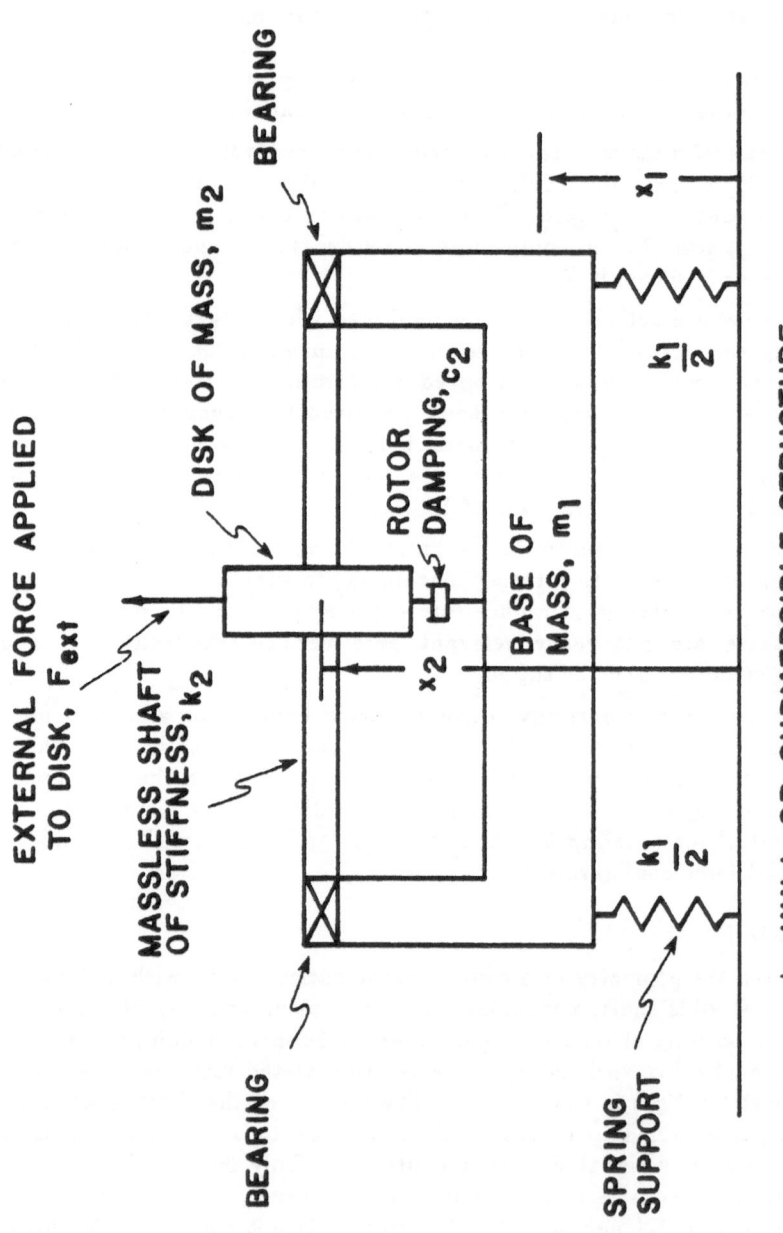

Figure 1: Rotating Machinery Geometry

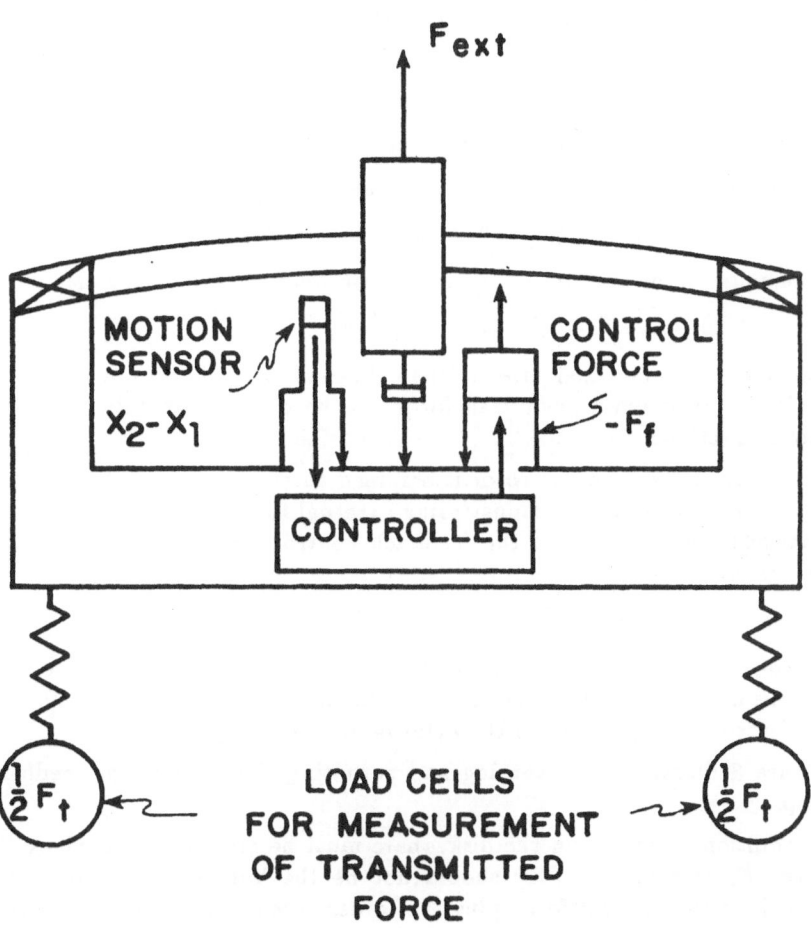

Figure 2: Feedback Control of Rotating Machine to Minimize Transmitted Force - Schematic Diagram

Figure 2 illustrates the automatic feedback control schematic diagram. The controller produces the negative feedback control force through an actuator near the disk. This force, with appropriate amplitude and phase, in one mode of operation may oppose the motion of the rotor shaft and thereby reduce the level of relative vibration, $x_2 - x_1$.

The object of this work is to show that the force transmitted to the support or hull, F_t, may be reduced by the use of RTB control. The force transmitted is:

$$F_t = k_1 x_1 + c_1 \dot{x}_1. \tag{1}$$

In general, F_t is a function of the system parameters as well as the excitation frequency (typically, the angular speed of the rotor).

The equation of motion for the base may be written as:

$$m_1 \ddot{x}_1 + c_1 \dot{x}_1 + k_1 x_1 + c_2(\dot{x}_1 - \dot{x}_2) + k_2(x_1 - x_2) = F_1. \tag{2}$$

Also, the equation of motion for the rotor disk may be formulated as:

$$m_2 \ddot{x}_2 + c_2(cd2 - \dot{x}_1) + k_2(x_2 - x_1) = F_2. \tag{3}$$

Real machines include other effects (than those modeled here) but the basic ideas concerning a minimum transmitted force can be discussed within the context of the above equations.

The control force on the rotor is assumed to act on (or near) the disk in this simple mathematical model. Considering external forces, F_{ext}, such as that resulting from disk unbalance, and the feedback control force, F_f, a force balance on the disk gives:

$$F_2 = F_{ext} - FF. \tag{4}$$

In this equation, the negative sign of the control force is employed in automatic control theory and is called "negative feedback". In this form, the feedback or control force is in opposition to the external forces(s), F_{ext}.

Figure 3 illustrates the sensing and actuating elements of the feedback control loop.

In addition to acting on the disk, there must be another place where the control force F_f is exerted. This should not be the hull or the transmitted force would include the control force plus the forces coming from the base springs and dampers. Thus, the control force will be exerted on the base, thereby defining the rotor to base (RTB) control. The force acting on the base is:

$$F_1 = F_f. \tag{5}$$

This completes the rotor and base equations.

Focus now on the RTB control that operates between the rotor and base. The non-contacting displacements probe, indicated in Figure 2, senses the relative displacement between the shaft at the disk and the base, $x_2 - x_1$. The carrier

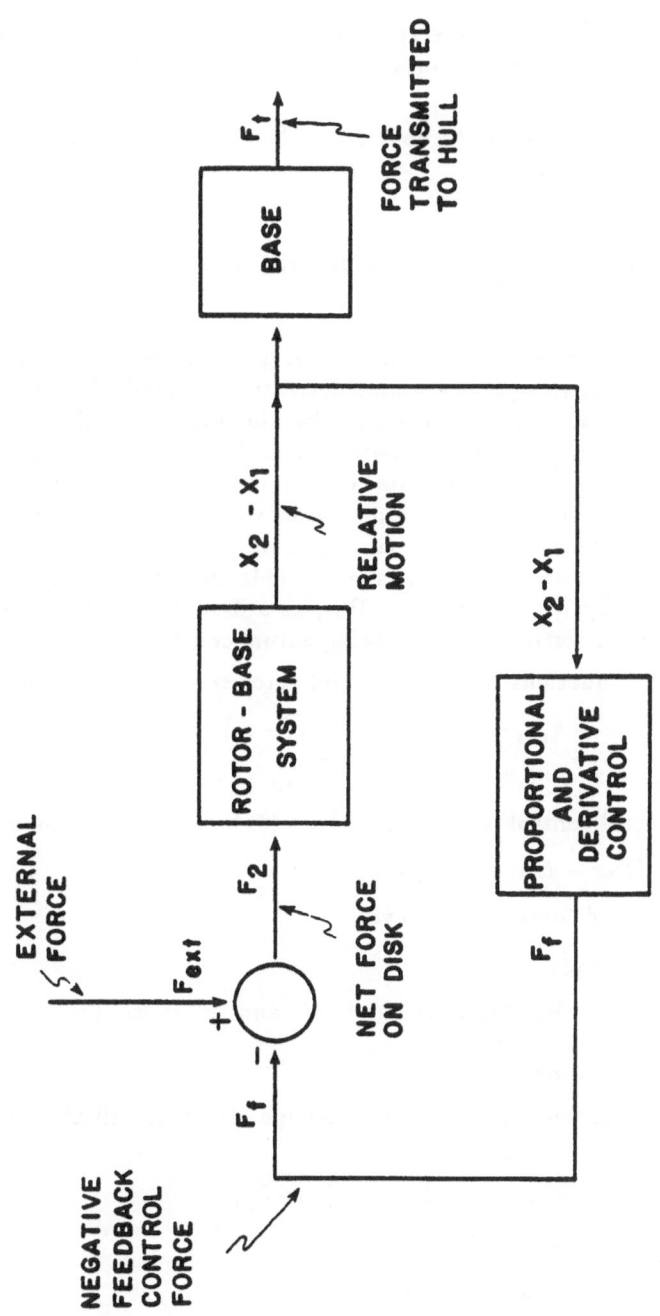

Figure 3: Schematic of Feedback Control Loop

frequency of the typical probe is several megahertz so that there is no significant phase lag in the displacement measurement up to rotor speeds of approximately 100,000 rpm.

If H represents the gain of the controller, the proportional feedback control force will be:

$$F_f = H(x_2 - x_1). \tag{6a}$$

The derivative feedback control force may be written as:

$$F_f = H\alpha(\dot{x}_2 - \dot{x}_1), \tag{6b}$$

in which $H\alpha$ is the derivative feedback control gain. Summing these gives the total RTB control force (proportional plus derivative control). The proportional and derivative feedback control gains may be thought of as simple constants related directly to the setting of a potentiometer. However, with but one small step of the imagination, these "constants" can be replaced by variables that may be functions of rotor speed, as an example. And with one additional step of the imagination, these "constants" can be visualized as variables that change as the physical embodiment of the system changes. This then leads into the general area identified as "adaptive" controllers. But, as a first step, look at the benefits of the proportional and derivative gains being actual constants.

The final set of equations for the base and rotor can then be written as:

$$m_1\ddot{x}_1 + c_1\dot{x}_1 + k_1x_1 + c_2(\dot{x}_1 - \dot{x}_2) + k_2(x_1 - x_2) = F_f, \tag{7}$$

$$-c_2\dot{x}_1 - k_2x_1 + m_2\ddot{x}_2 + c_2\dot{x}_2 + k_2\dot{x}_2 = F_{ext} - F_f. \tag{8}$$

Also, the most general control relation from Eqs. (6a) and (6b) has the form:

$$F_f = -H\alpha\dot{x}_1 - Hx_1 + H\alpha\dot{x}_2 + Hx_2. \tag{9}$$

Finally, the transmitted force is given by:

$$F_t = k_1x_1 + c_1\dot{x}_1 \tag{10}$$

which is to be minimized by the selection of "H" and "α" in Eq. (9).

3. Solution of Equations

Taking the Laplace transforms of Eqs. (7) through (10) with initial conditions set equal to zero yields

$$(m_1s^2 + c_1s + k_1 + k_2)\hat{x}_1 - (c_2s + k_2)\hat{x}_2 = \hat{F}_f, \tag{11}$$

$$-(c_2s + k_2)\hat{x}_1 + (m_2s^2 + c_2s + k_2)\hat{x}_2 = \hat{F}_{ext} - \hat{F}_f, \tag{12}$$

$$\hat{F}_f = -H(\alpha s + 1)\hat{x}_1 + H(\alpha s + 1)\hat{x}_2, \tag{13}$$

$$\hat{F}_f = k_1\hat{x}_1 + c_1s\hat{x}_1. \tag{14}$$

The first three equations are divided by k_2. The result is:

$$z_{11}\hat{x}_1 + z_{12}\hat{x}_2 = \frac{1}{k_2}\hat{F}_f, \tag{15}$$

$$z_{21}\hat{x}_1 + z_{22}\hat{x}_2 = \frac{1}{k_2}(\hat{F}_{ext} - \hat{F}_f), \tag{16}$$

$$\frac{1}{k_2}\hat{F}_f = -z_f\hat{x}_1 + z_f\hat{x}_2, \tag{17}$$

in which

$$z_{11} = M\bar{s}^2 + 2\xi_1\bar{s} + 2\xi_2\bar{s} + 1,$$

$$z_{12} = -(2\xi_2\bar{s} + 1),$$

$$z_{21} = -(2\xi_2\bar{s} + 1),$$

$$z_{22} = \bar{s}^2 + 2\xi_2\bar{s} + 1,$$

$$z_f = 2\xi_f\bar{s} + h_f.$$

Here, all of the z parameters are dimensionless.

Equations (15) and (17) can be combined to yield

$$(z_{11} + z_f)\hat{x}_1 + (z_{12} - z_f)\hat{x}_2 = 0. \tag{18}$$

Solving for $\hat{x}_2$ gives

$$\hat{x}_2 = -\left[\frac{z_{11} + z_f}{z_{12} - z_f}\right]\hat{x}_1. \tag{19}$$

Equation (16) then can be solved for $\hat{x}_1$ as

$$\hat{x}_1 = \frac{1}{k_2}\hat{F}_{ext}\left[\frac{z_{21} - z_f}{(z_{12} - z_f)(z_{21} - z_f) - (z_{11} + z_f)(z_{22} + z_f)}\right] \tag{20}$$

Finally, the transmitted force ratio is obtained from Eq. (14) as:

$$\frac{\hat{F}_t}{\hat{F}_{ext}} = \left[\frac{K(z_{21} - z_f)(K + 2\xi_1\bar{s})}{(z_{12} - z_f)(z_{21} - z_f) - (z_{11} + z_f)(z_{22} + z_f)}\right]. \tag{21}$$

It is this ratio which should be as small as possible.

4. Unbalance Response

As an example, consider an unbalanced disk in which the mass center, G, is a small distance, e_u, away from the geometric center, C, of the disk. If the rotor angular velocity is ω, the unbalanced force acting on the rotor may be written:

$$F_{ext} = m_2 e_u \omega^2 e^{i\omega t},$$

the transmitted force has the form:

$$F_t = A_t e^{i\omega t},$$

in which the magnitude A_t is complex. Define the dimensionless transmitted force as:

$$\bar{F}_t = \frac{F_t}{F_{ext}} = \frac{A_t e^{i\omega t}}{m e_u \omega^2 e^{i\omega t}}. \tag{22}$$

Also, define the dimensionless transmitted force amplitude, A_t, as:

$$\bar{A}_t = \frac{A_t}{m_2 e_u \omega^2}. \tag{23}$$

Note that A_t is a function of the rotor speed, ω, as well as the other system parameters.

The transmitted force is obtained from the general expression Eq. (21) by replacing $\bar{s}$ with fi, in which, the frequency ratio is:

$$f = \frac{\omega}{\omega_{cr}}.$$

The result is:

$$\bar{A}_t = \frac{K(z_{21} - z_f)(K + 2\xi_1 fi)}{[(z_{12} - z_f)(z_{21} - z_f) - (z_{11} + z_f)(z_{22} + z_f)]}$$

in which

$$z_{11} = -Mf^2 + 1 + K + 2\xi_1 fi + 2\xi_2 fi,$$

$$z_{12} = -1 - 2\xi_2 fi,$$

$$z_{21} = -2\xi_2 fi - 1,$$

$$z_{22} = -f^2 + 2\xi_2 fi + 1,$$

$$z_f = 2\xi_f fi + h_f.$$

This shows that the dimensionless transmitted force amplitude is a function of the mass ratio $M = m_1/m_2$, dimensionless base damping $\xi_1 = c_1/(2m_2\omega_{cr})$, dimensionless disk damping $\xi_2 = c_2/(2m_2\omega_{cr})$, stiffness ratio $K = k_1/k_2$, dimensionless feedback damping ratio $\xi_f = H\alpha/(2m_2\omega_{cr})$, dimensionless feedback stiffness ratio $h_f = H/k_2$, as well as the frequency ratio f.

5. Unbalanced Response Example

To lend some feeling to the significance of the previously derived equations, consider an example. Assume:

$$\text{Mass ratio} = M = m_1/m_2 = 5.0,$$

$$\text{Stiffness ratio} = K = k_1/k_2 = 2.0.$$

The significance of these numbers may be realized by noting Figure 1 and the appropriate definitions given in the nomenclature.

The transmitted force amplitude ratio, A_t, is shown as Figure 4 with zero feedback showing the effect of three different values of disk and base damping. The dashed line curve illustrates small damping ($\xi_1 = \xi_2 = 0.01$) showing the two critical speeds. The case of no damping ($\xi_1 = \xi_2 = 0$), dotted line, is not too different but does show that the transmitted force amplitude becomes unbounded as expected. With so-called *critical damping* ($\xi_1 = \xi_2 = 0.5$), the transmitted force amplitude ratio changes form drastically (the solid line of Fig. 4) showing how the second critical has been essentially eliminated.

Figure 4 may be considered a base line case against which other examples should be compared. Figure 4 has no feedback but allows comparisons as it shows clearly the influence of disk and base damping.

Figure 5 depicts the transmitted force amplitude, A_t, versus frequency ratio f with three values of RTB proportional feedback ($h_f = 0.1, 0.4, 0.9$) but with zero derivative feedback. As this feedback increases, the peak amplitudes of the criticals are reduced and the separation frequency likewise increases. These results suggest that one algorithm is to use high gain for running at the lower speeds and then drop the gain as the second critical speed is approached. In short, a constant value of gain is not an optimum approach in using proportional feedback, but the algorithm is not very complex in the case of this system.

Instead of using proportional control, another alternative is to use derivative control in which the force is proportional to the instantaneous velocity difference between the linear vertical motion of the disk and the base. Figure 6 shows the same system ($m = 5$, $K = 2$, $\xi_1 = \xi_2 = 0.01$) with two values of velocity feedback ($\xi_f = 0.1, 0.4$). This type of control is very effective in reducing the transmitted force amplitude normally associated with the second critical speed. As shown by Figure 6, even the response at the fundamental frequency is markedly reduced by this feedback means. Figure 6 should be compared with Figure 4 to appreciate the significance of this type of control.

The next obvious consideration is to employ both proportional and derivative feedback. As can be imagined, the results depend upon the weighting of the mix of these two. The velocity feedback is most significant as this is equivalent to providing a means of extracting energy from the disk. Even a small amount of damping in the system produces very important changes in the dynamic behaviour of the system and in particular, the amount of force transmitted from

D. W. Lewis, P. E. Allaire

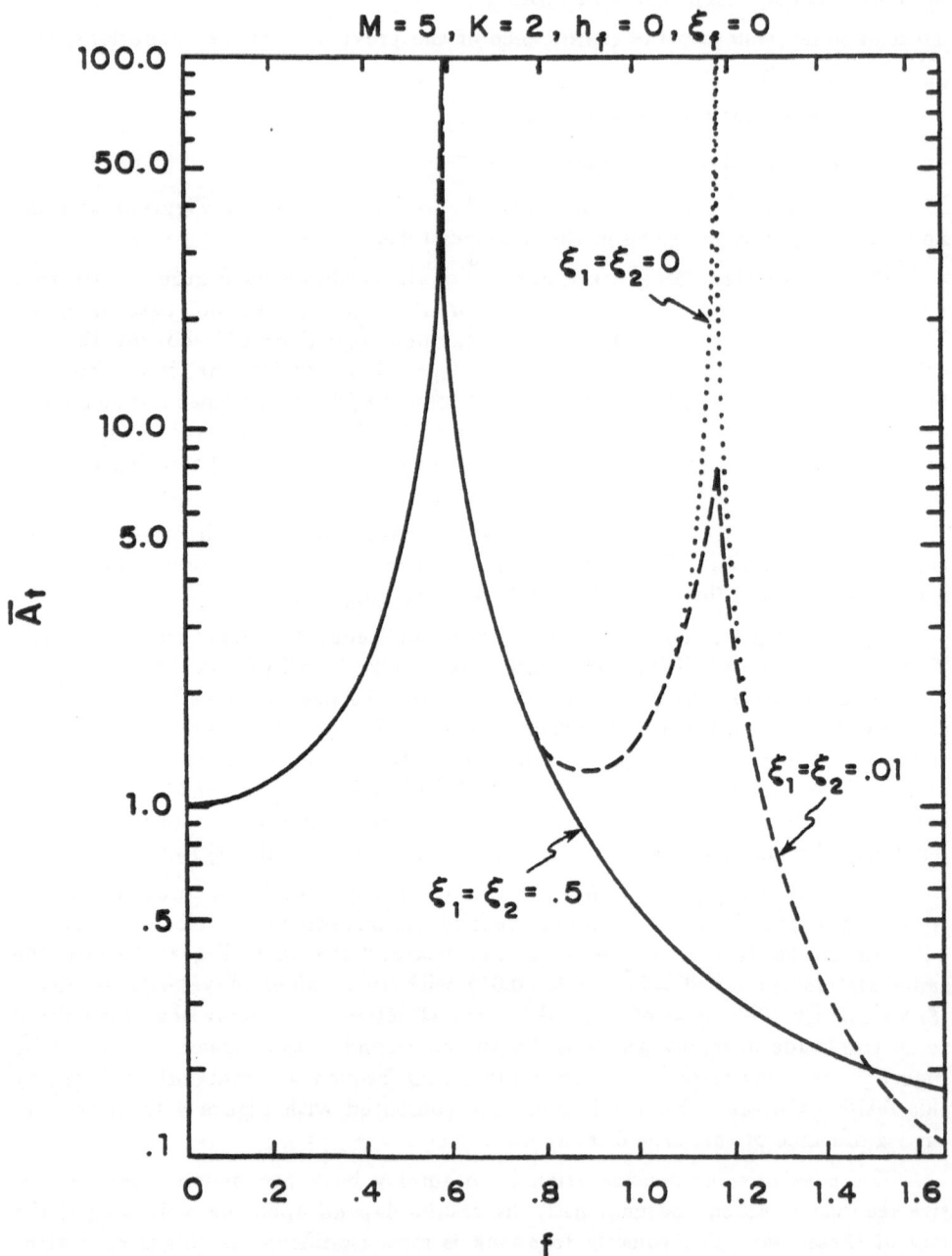

Figure 4: *Transmitted Force Versus Frequency With No Feedback*

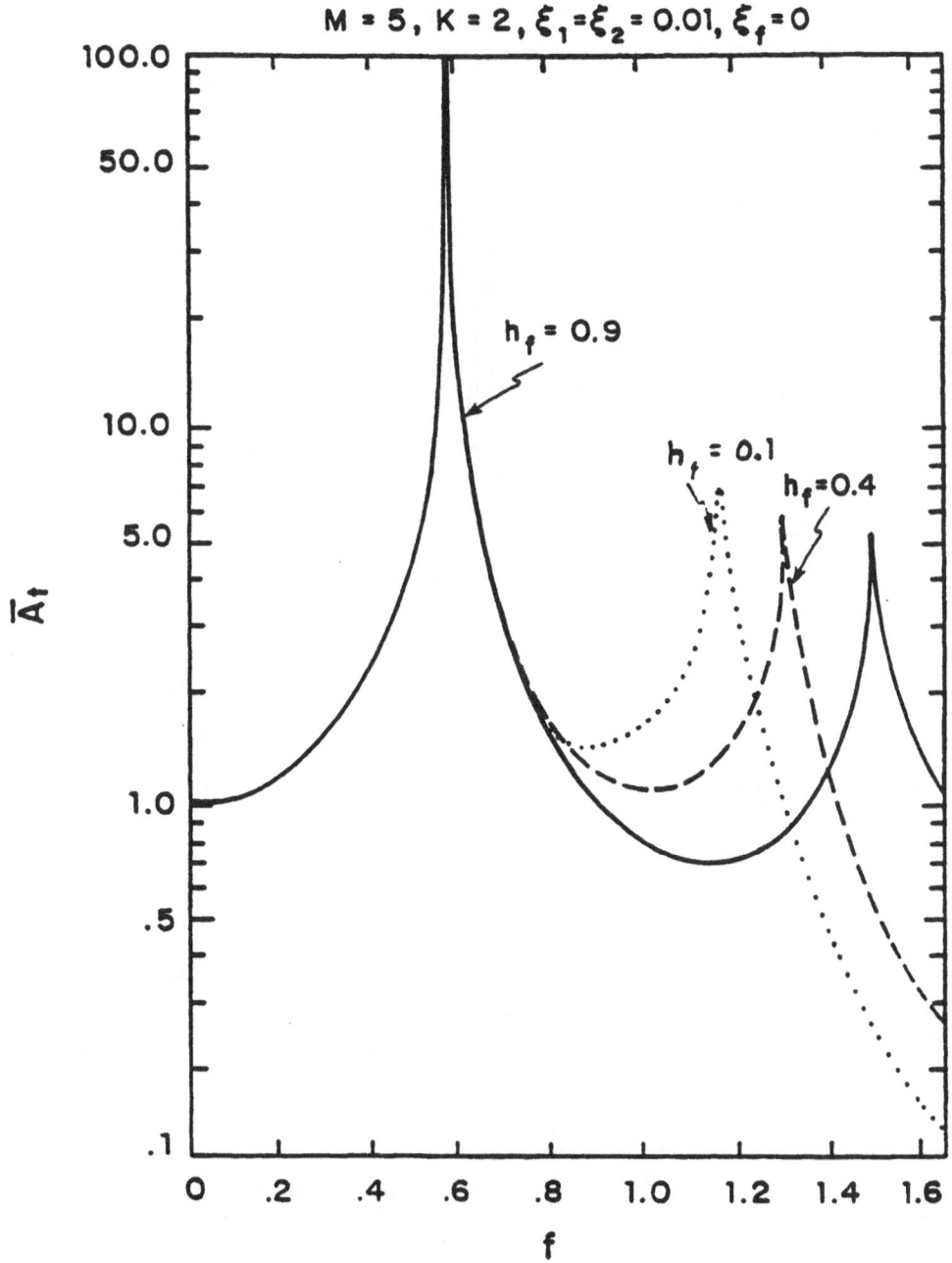

Figure 5: Transmitted Force as a Function of Frequency for Three Values of Proportional Feedback Control

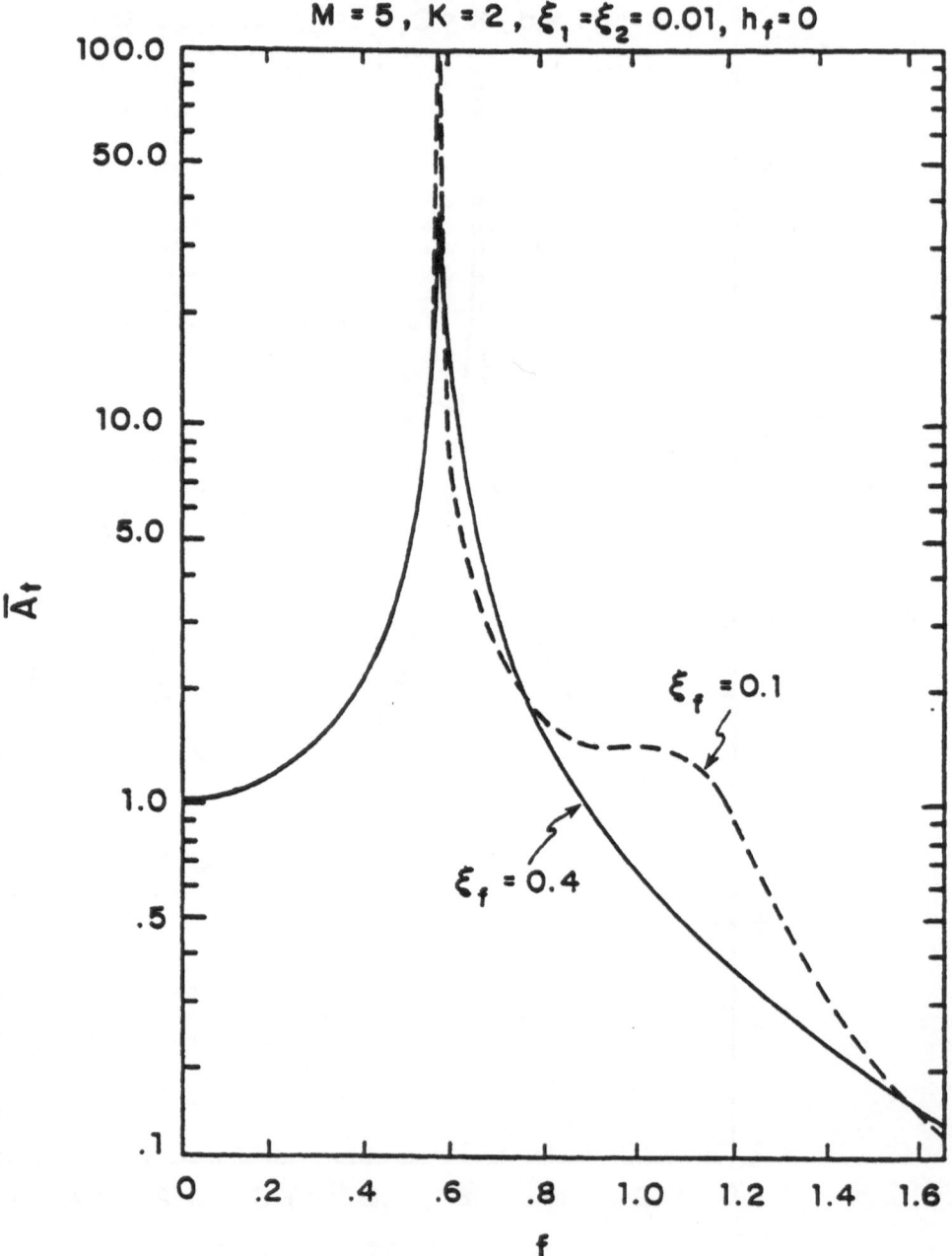

Figure 6: *Transmitted Force Versus Frequency for Three Values of Derivative
Feedback Control*

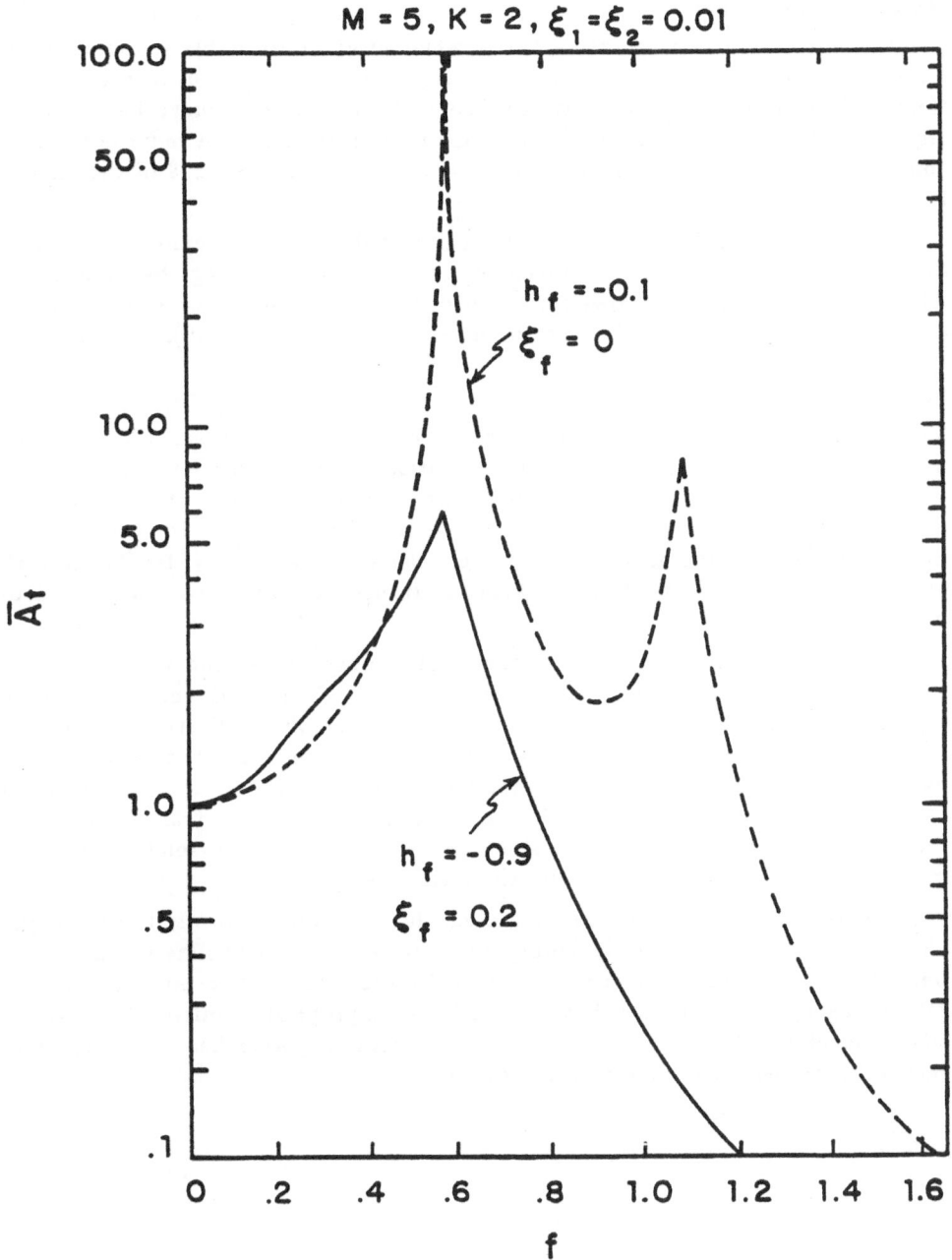

Figure 7: Use of Feedback Forces with Sign Reversal on Proportional Gain

the base to the support.

A not so obvious alternative exists when using controllable forces as illustrated by Figure 7. In this example, the sign has been reversed on the proportional gain yielding some very effective control. The dashed curve has no velocity term and yet the transmitted force amplitude ratio has been reduced so as to compare favourably with Figure 5 (solid curve) even with only 1/9 of the magnitude of the feedback.

The sign of the velocity control term is not very significant. Equivalent damping is most important regardless of whether it leads or lags by 90 degrees. The solid line of Figure 7 shows the effect of the velocity feedback term ($\xi_f = 0.2$) coupled with a large proportional term of reversed sign ($h_f = -0.9$).

6. Conclusions

The system described in this work is a single mass rotor on a base. Changing the mass ratio, $M = m_1/m_2$, and the stiffness ratio, $K = k_1/k_2$, produces wide variations in the force that is transmitted to the support. The use of RTB proportional feedback does not reduce the number of critical speeds. It does afford a means of reducing the transmitted force. Using RTB velocity feedback, with either lag or lead, can effectively reduce the transmitted force even without proportional feedback.

With a sign reversal on the proportional feedback term, the apparent *critical* can be moved somewhat. Even small values of proportional and velocity feedback, if the sign is reversed from the usual controls application, lowers the transmitted force. The *anticipatory* action arising from using the reversed sign on the proportional feedback force has been investigated in this work. Certain combinations of parameters optimize the system, lowering the transmitted force (between base and support) while simultaneously reducing the relative displacement between the disk and the base, the term $x_1 - x_2$.

However, one conclusion can be stated. It is absolutely essential to know the character of a rotor system, including not only the amount of inherent damping, but where in the system it occurs, defined by the mass and spring ratios before attempting to employ feedback measures. The appropriate control of the vibration level of a rotor for a single degree of freedom system has been shown to reduce the transmitted force to the support.

Nomenclature

Symbol	Defintion
A_t	Dimensionless amplitude of force transmitted
c_1	Rotor damping
c_2	Base damping
C	Disk geometric center
e_u	Unbalance eccentricity (distance between G and C)
f	Frequency ratio, $f = \omega/\omega_{cr}$
F_1	Force on base, other than springs and dampers
F_2	Forces on disk, $F_{ext} - F_f$
F_{ext}	External force on disk
F_f	Feedback control force on disk and base
F_t	Force transmitted to support (hull)
G	Disk center of gravity
h_f	Dimensionless feedback stiffness ratio, $h_f = H/k_2$
H	Feedback control gain (stiffness)
k_1	Support stiffness
k_2	Shaft stiffness
K	Stiffness ratio, $K = k_1/k_2$
m_1	Mass of base
m_2	Mass of disk

M	Mass ratio, $M = m_1/m_2$
s	Frequency
$\bar{s}$	Dimensionless Frequency, $\bar{s} = s/\omega_{cr}$
t	Time
x_1	Displacement of base (relative to support or hull)
x_2	Displacement of shaft (relative to support or hull)
α	Velocity feedback constant
ξ_1	Dimensionless base damping $\xi_1 = c_1(2m_2\omega_{cr})$
ξ_2	Dimensionless disk damping $\xi_2 = c_2/(2m_2\omega_{cr})$
ξ_f	Dimensionless feedback damping ratio, $\xi_f = H\alpha/(2m_2\omega_{cr})$
ω	Shaft angular velocity
ω_{cr}	Shaft critical speed, $\omega_{cr} = \sqrt{k_2/m_2}$
$\hat{}$	The Laplace transform of particular variable

Acknowledgement

We wish to thank Josiah Knight for the modeling/calculations and Sandy Smith for typing the manuscript.

References

[1] Schweitzer, G. and Lange, R., "Characteristics of a Magnetic Rotor Bearing for Active Vibration Control," Institution of Mechanical Engineers Conference on Vibrations in Rotating Machinery, Cambridge, 1976, paper C239/76.

[2] Gondhaleker, V. M., Nikolajsen, J. L. and Holmes, R., "Electro-magnetic Control of Flexible Transmission Shaft Vibrations," Proc. IEEE, Vol. 126, 1979, pp. 1008-1010.

[3] Heinzmann, J. D., "The Implementation of Automatic Vibration Control in a High Speed Rotating Test Facility," M.S. Thesis, University of Virginia, January 1980.

[4] Moore, J. W., Lewis, D. W. and Heinzmann, J. D., "Feasibility of Active Feedback Control of Rotor Dynamic Instability," Presented at *Workshop on Rotor Dynamic Instability Problems in High Performance Turbomachinery*, NASA/ARO Conference, Texas A&M University, May 1980, pp. 1-9.

[5] Stanway, R. and Burrows, C. R., "Active Vibration Control of a Flexible Rotor on Flexibly-Mounted Journal Bearings," Joint Automatic Controls Conference, Charlottesville, Virginia, June 1981.

[6] Allaire, P. E., Lewis, D. W. and Jain, V. K., "Feedback Control of a Single Mass Rotor on Rigid Supports," Vol. 132, No. 1, July 1981, pp. 1-11, *Journal of the Franklin Institute*.

[7] Allaire, P. E., Lewis, D. W. and Knight, J. D., "Active Control of a Single Mass Rotor on Flexible Supports," *Journal of the Franklin Institute*, Vol. 315, No. 3, March 1983, pp. 211-222.

[8] Lewis, D. W., Moore, J. W., Bradley, P. L. and Allaire, P. E., "Vibration Limiting of Rotors by Feedback Control," Presented at *Workshop on Rotordynamic Instability Problems in High Performance Turbomachinery*, ARO/NASA Conference, Texas A&M University, May 1982.

[9] Haberman, H., "Le Palier Magnetique Actif ACTIDYNE," AGARD Conference Proceedings, No. 323, 1982.

[10] Haberman, H. and Brunet, M., "The Active Magnetic Bearing Enables Optimum Damping of Flexible Rotor," ASME Paper 84-GT-117, Gas Turbine Conference, Amsterdam, 1984.

[11] Gondhalekar, V. and Holmes, R., "Design of an Electromagnetic Bearing for the Vibration Control of a Flexible Transmission Shaft," *Workshop on Rotordynamic Instability Problems in High Performance Turbomachinery*, Texas A&M University, May 1984.

HOMOGENIZATION AND REINFORCED STRUCTURES

Jacques-Louis Lions
College de France
Paris, France
and
C.N.E.S. (Centre National d'Etudes Spatiales)
Paris

Introduction

Let us consider material with a periodic structure and where the period - say ϵ - is *small* - compared to the size of the material under study.

Examples of such situations are given by *composite materials* or by *perforated materials*.

In this short survey, we want to present some recent results obtained in the study of perforated materials.

Let Ω be an open set $\mathbb{R}^n$ ($n = 2$ *or* 3 in the applications) and let Ω_ϵ be an open set, contained in Ω, obtained from Ω after taking out, in a periodic manner, with period ϵ in all directions (more precise definitions will be given below), holes (or obstacles) O_ϵ[†]:

$$\Omega_\epsilon = \Omega / O_\epsilon$$

[†] Holes correspond to perforated materials, obstacles correspond to flows in porous media or through screens. We refer for this last situation to E. Sanchez-Palencia [1][2], J. L. Lions [1][2], C. Conca·[1], L. Tartar [1] and to the Bibliography therein.

We consider a *boundary value* problem

$$Au_\epsilon = f \text{ in } \Omega_\epsilon \quad \text{†} \qquad (1)$$

where u_ϵ is subject to appropriate boundary conditions on $\partial \Omega_\epsilon$; notice that $\partial \Omega_\epsilon$ consists in two parts: the boundary of Ω minus the intersection with the holes *and* the boundary of the holes: ∂O_ϵ. The situation is *very sensitive* to the nature of the boundary conditions on ∂O_ϵ.

We shall recall first some known results for the case when the boundary conditions on ∂O_ϵ are of the *Neumann type*. Then (cf. D. Cioranescu and J. Saint Jean Paulin [1]) one can show that, for some appropriate topology (to be recalled more precisely in Section 1 below) u_ϵ converges to u as $\epsilon \to 0$ where u is the solution of

$$\mathcal{A} u = f \text{ in } \Omega, \qquad (2)$$

u being subject to "the same" boundary conditions as u_ϵ on $\partial \Omega$; in (2) $\mathcal{A}$ is the so-called *homogenized operator*; formally speaking $\mathcal{A}$ is an operator defined in Ω and which is such that the solution of (2) *approximates* the solution of (1). The interest of such a result is obvious: it allows, at least as a first approximation, to replace the very complicated domain Ω_ϵ by the "usual" domain Ω, at the expense of *computing first the coefficients of $\mathcal{A}$*. We give in Section 1 the (well known) formula for these coefficients.

In Section 2 we recall briefly the *very different* situation where one deals with *Dirichlet conditions* on ∂O_ϵ, according to the Author (J. L. Lions [3]; cf. also J. L. Lions [1] and the bibliography therein). We also mention in this section a very interesting result from M. Vanninathan [1] concerning the *spectrum* of the corresponding operator.

In Section 3 we present some recent results of D. Cioranescu and J. Saint Jean Paulin [1], which extend some of the results given in N. S. Bakhvalov and G. P. Panasenko [1].

These results are concerned with the situation where the holes depend on a second parameter μ: formally speaking (precise hypothesis are given in Section 3 below) the size of the pieces which remain after taking out the holes is $\mu\epsilon$; if one lets first $\epsilon \to 0$, one obtains an homogenized operator *which depends on μ*, say $\mathcal{A}^\mu$; one lets then $\mu \to 0$. At least in some particular geometric situations, one obtains explicit and simple formulae for the limit of $\mathcal{A}^\mu$ as $\mu \to 0$. We refer to the corresponding structures as reinforced, or alveolar, or skeletal structures. Only some hints of the proofs are given: we refer to D. Cioranescu and J. Saint Jean Paulin, loc. cit. for complete proofs.

---

† Here f is given in Ω and in the right hand side of (1), f actually denotes the restriction of f to Ω_ϵ.

The plan is as follows:

1. Homogenization for Perforated Materials. Neumann boundary Conditions on the Boundary of the Holes.

1.1. The set Ω_ϵ.

Let us define

$$ Y =]0,1[^n . $$

A Y-periodic function is a function with period 1 in all directions. In Y we consider an open set O with boundary S (cf. Figure 1) and we set

$$ y = Y \backslash \overline{O} . $$

Let $m = m(y)$ be the Y-periodic function which is defined in Y by

$$ m(y) = \begin{cases} 1 \ in \ y \\ 0 \ in \ \overline{O} \end{cases} . \tag{1.1} $$

We then define

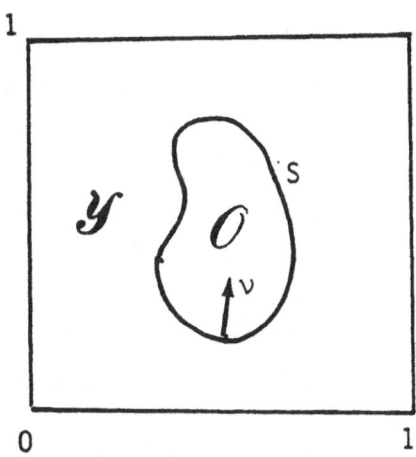

1

$\mathcal{Y}$ S

$\mathcal{O}$

ν

Figure 1   0   1

$$\Omega_\epsilon = \{x \mid x\,\epsilon\,\Omega, \ m(x/\epsilon) = 1\}. \tag{1.2}$$

From a geometrical viewpoint, Ω_ϵ is what remains of Ω after taking out all the holes, obtained by translation (of size ϵ in all directions) of $\epsilon\mathcal{O}$, and which intersect Ω.

We have

$$\begin{cases} \partial\,\Omega_\epsilon = \Gamma_\epsilon \ \bigcup \ S_\epsilon, \\ \Gamma_\epsilon = \Gamma\backslash \text{intersection with the holes}, \\ S_\epsilon = \text{boundary of the holes}. \end{cases} \tag{1.3}$$

1.2. A first example

Let us consider first the simplest boundary value problem one can think of:

$$-\Delta u_\epsilon = f \quad \text{in} \quad \Omega_\epsilon, \tag{1.4}$$

$$u_\epsilon = 0 \quad \text{on} \quad \Gamma_\epsilon, \tag{1.5}$$

$$\frac{\partial u_\epsilon}{\partial \nu} = 0 \quad \text{on} \quad S_\epsilon. \tag{1.6}$$

This is equivalent to finding

$$\inf_v \ J_\epsilon(v), \tag{1.7}$$

$$J_\epsilon(v) = \frac{1}{2} \int\limits_{\Omega_\epsilon} |\nabla v|^2 dx \ - \ \int\limits_{\Omega_\epsilon} fv dx \tag{1.8}$$

when

$$v \in H^1(\Omega_\epsilon) \overset{\pm}{} , \quad v = 0 \text{ on } \Gamma_\epsilon. \tag{1.9}$$

Let us now *homogenize* the problem by using *asymptotic calculus of variations* (cf. J. L. Lions [2]).

We observe first that one can write

$$J_\epsilon(v) = \frac{1}{2} \int_\Omega m(x/\epsilon) |\nabla v|^2 dx - \int_\Omega m(x/\epsilon) f v \, dx \tag{1.10}$$

where one takes now

$$v \in H^1(\Omega), \quad v = 0 \text{ on } \Gamma \tag{1.11}$$

(one can always extend $v \in H^1(\Omega_\epsilon)$ into $v \in H^1(\Omega)$, at least assuming S "not too bad").

We consider next test functions v with the following particular structure (ansatz):

$$v(x) = v_o(x,y) + \epsilon v_1(x,y) + ..., y = x/\epsilon \tag{1.12}$$

where

$$\begin{cases} v_j(x,y) \text{ is defined in } \Omega \times Y \\ v_j(x,y) \text{ is } Y\text{-periodic in } y, \quad \forall \times \in \Omega \\ v_j(x,y) = 0 \text{ for } x \in \Gamma. \end{cases} \tag{1.13}$$

We observe that (with obvious notations)

$$\begin{cases} \nabla v \text{ becomes } \epsilon^{-1} \nabla_y v + \nabla_x v, \text{ where we replace } y \\ \text{by } x/\epsilon \text{ at the end of the computation.} \end{cases} \tag{1.14}$$

We observe next that if ϕ is a *reasonable* function

$$\phi(x,y), \quad x \in \Omega, y \in Y, \quad Y\text{-periodic in } y,$$

then

$$\int_\Omega \phi(x,x/\epsilon) dx \longrightarrow \int\int_{\Omega \times Y} \phi(x,y) \, dx dy.$$

The *general principle* in the asymptotic calculus of variations says that one can *replace* (1.7) by

---

$${}^{\pm} H^1(\Omega_\epsilon) = \{\phi \,|\, \phi, \; \frac{\partial \phi}{\partial x_1}, ..., \; \frac{\partial \phi}{\partial x_n} \in L^2(\Omega_\epsilon)\}.$$

$$\begin{cases} \inf \mathcal{Y}\,(v_o + \epsilon v_1 + \ldots), \\[1mm] \mathcal{Y}(v) = \dfrac{1}{2}\int \int_{\Omega \times Y} m(y)\,|\epsilon^{-1}\nabla_y v + \nabla_z v\,|^2 dx - \int \int_{\Omega \times Y} m(y)fv\ dxdy \ ^{\perp} \\[1mm] v \text{ given as in } (1.12)(1.13), \end{cases} \tag{1.15}$$

and by taking the inf in $v_o, v_1, \ldots$, term after term.

The term in ϵ^{-2} should be zero, i.e.

$$\nabla_y v_o = 0 \tag{1.16}$$

i.e.

$$v_o = v_o(x) \ \textit{does not depend on } y\,. \tag{1.17}$$

Then the term in ϵ^o in $\mathcal{Y}(v_o + \epsilon v_1 + \ldots)$ equals

$$\mathcal{Y}_o(v_o, v_1) = \frac{1}{2}\int\int_{\Omega \times Y} |\nabla_y v_1 + v_x v_o\,|^2\ dxdy - \int\int_{\Omega \times Y} fv_o\ dxdy\,. \tag{1.18}$$

The homogenized problem consists in minimizing (1.18) *where*

$$\begin{cases} v_o \in H^1(\Omega) \ , \ v_o = 0 \text{ on } \Gamma, \\ v_1 = v_1(x,y) \ \text{ is } Y-\text{periodic} \end{cases} \tag{1.19}$$

(and assuming, of course, that the integrals in (1.18) make sense!).

We notice that

$$\int\int_{\Omega \times Y} fv_o\ dxdy = |\mathcal{Y}\,|\int_\Omega fv_o dx\,, \quad |\mathcal{Y}\,| = \text{volume of } \mathcal{Y}\,. \tag{1.20}$$

The homogenized problem admits a unique solution for the minimum in v_o - say u_o - and admits an infinite number of solutions in v_1; if u_1 is *one* of these solutions, then u_1 is defined up to the addition of a function $\hat{u}_1(x)$ of x alone.

Remark 1.1
The function $\hat{u}_1$ ca be defined along with higher order terms of the expansion, but due to *boundary layers* this becomes complicated. $\square$

Euler equations
The Euler equations for inf $\mathcal{Y}_o(v_o, v_1)$ are given by

---

$^{\perp}$ Of course one has also
$$\mathcal{Y}(v) = \frac{1}{2}\int \int_{\Omega \times Y} |\epsilon^{-1}\nabla_y v + \nabla_z v\,|^2\ dxdy - \int \int_{\Omega \times Y} fv\ dxdy\,.$$

$$\iint\limits_{\Omega\times y} (\nabla_y u_1 + \nabla_x u_o)\nabla_y v_1\, dx dy = 0, \tag{1.21}$$

$$\iint\limits_{\Omega\times y} (\nabla_y u_1 + \nabla_x u_o)\nabla_x v_o\, dx dy = |y| \int\limits_\Omega fv_o dx. \tag{1.22}$$

One can easily derive *explicit formulae*. Equation (1.21) is equivalent to

$$\begin{cases} -\Delta_y u_1 = 0 \text{ in } \Omega\times y,\ u_1\ Y-\text{periodic in } y, \\ \dfrac{\partial u_1}{\partial \nu_{(y)}} + \nu_j \dfrac{\partial u_o}{\partial x_j} = 0 \text{ on } S,\ \ \forall\, x \in \Omega; \end{cases} \tag{1.23}$$

in (1.23) $\nu = \nu_{(y)} = \{\nu_j\}$ denotes the normal to S directed towards the exterior of y (cf. Fig. 1).

Let us introduce χ^j by

$$\begin{cases} -\Delta_y \chi^j = 0 \text{ in } y \\ \dfrac{\partial \chi^j}{\partial \nu_{(y)}} = \nu_j(y) \text{ on } S,\ \chi^j\, Y-\text{periodic}. \end{cases} \tag{1.24}$$

Equations (1.24) define χ^j up to an additive constant. We can normalize χ^j by

$$\int\limits_y \chi^j(y) dy = 0$$

but this is by no means essential.

Then

$$u_1(x,y) = -\chi^j(y) \frac{\partial u_o}{\partial x_j}(x) + \hat{u}_1(x). \tag{1.25}$$

We now replace u_1 by (1.25) in (1.22) and we obtain

$$\frac{1}{|y|} \iint\limits_{\Omega\times y} \left[\delta_i^j - \frac{\partial \chi^j}{\partial y_i} \right] \frac{\partial u_o}{\partial x_j} \frac{\partial u_o}{\partial x_i}\, dx dy = \int\limits_{Om} fv_o dx. \tag{1.26}$$

If we define

$$q_{ij} = \frac{1}{|y|} \int\limits_y \left[\delta_i^j - \frac{\partial \chi^j}{\partial y_i} \right] dy \tag{1.27}$$

we have

$$\begin{cases} \int_\Omega q_{ij} \dfrac{\partial u_o}{\partial x_j} \dfrac{\partial v_o}{\partial x_i} dx = \int_{Om} f v_o \ dx \\ \forall \ v_o \in H^1(\Omega) \ , \ v_o = 0 \text{ on } \Gamma, \\ u_o \in H^1(\Omega) \ , \ u_o = 0 \text{ on } \Gamma. \end{cases} \qquad (1.28)$$

This is the *homogenized problem.*

The *homogenized operator* is given by

$$\mathcal{A} = -q_{ij} \frac{\partial^2}{\partial x_i \partial x_j}, \qquad (1.29)$$

and the *homogenized problem* is

$$\mathcal{A}u = f \text{ in } \Omega, \ u = 0 \text{ on } \Gamma. \qquad (1.30)$$

A few remarks are in order.

Remark 1.2

The operator $\mathcal{A}$ is *elliptic.* This is not entirely obvious if one uses (1.27) but it follows from the fact that (1.28) is "equivalent" to inf $\mathcal{Y}_o(v_o, v_1)$. □

Remark 1.3

The homogenized (or effective) coefficients q_{ij} *do not depend on* Ω but only on $\mathcal{Y}$. □

Remark 1.4

The above computations are *formal.* They can be justified by "energy methods". (Cf. D. Cioranescu and J. Saint Jean Paulin [1]).

Remark 1.5

Another way to obtain (1.28) is to use direct asymptotic expansions in (1.4) (1.5) (1.6). One takes the ansatz

$$\begin{cases} u = u_0 + u_1 + \ldots, \\ u_j = u_j(x, y) \ , \ u_j \ Y\text{--periodic in } y: \end{cases} \qquad (1.31)$$

one obtains that $-\Delta$ becomes

$$-\epsilon^{-2}\Delta_y - 2\epsilon^{-1}\Delta_{xy} - \Delta_x \qquad (1.32)$$

where $\Delta_{xy} = \dfrac{\partial^2}{\partial x_i \partial y_i}$, and that $\dfrac{\partial}{\partial \nu}$ becomes

$$\epsilon^{-1}\frac{\partial}{\partial\,\nu_{(y)}} + \nu_j\frac{\partial}{\partial\,x_j}\,.$$ (1.33)

Then by identifying the powers of ϵ^{-2}, ϵ^{-1}, ϵ^o in (1.4) (1.5) (1.8) one obtains:

$$-\Delta_y u_o = 0, \quad \frac{\partial\,u_o}{\partial\,\nu_{(y)}} = 0$$

so that $u_o = u_o(x)$; then

$$-\Delta_y u_1 - 2\Delta_{xy} u_o = -\Delta_y u_1 = 0,$$

$$\frac{\partial\,u_1}{\partial\,\nu_{(y)}} + \nu_j\frac{\partial\,u_o}{\partial\,x_j} = 0 \text{ on } S$$

which is (1.23); then

$$\begin{cases} -\Delta_y u_2 - 2\Delta_{xy} u_1 - \Delta_x u_o = f \\ \dfrac{\partial\,u_2}{\partial\,\nu_{(y)}} + \nu_j\dfrac{\partial\,u_1}{\partial\,x_j} = 0 \text{ on } S, \ \ \forall\, x \in \Omega. \end{cases}$$ (1.34)

Problem (1.34) admits a solution u_2 iff *compatibility conditions* are satisfied: they are obtained by writing that

$$+\int_y (\Delta_y u_2)dy = +\int_S \frac{\partial\,u_2}{\partial\,\nu_{(y)}}dS$$

i.e.

$$\int_y (\Delta_x u_o + 2\Delta_{xy} u_1 + f)dy = -\int_S \nu_j\frac{\partial\,u_1}{\partial\,x_j}dS\,.$$

After some computations, one obtains (1.28) or more precisely, the equivalent form (1.30). □

Remark 1.6

The method given in Remark 1.5 is a little bit more intricated and does not show in an "obvious" manner the elliptic character of the homogenized operator. But it has the advantage to extend to *non symmetric operators* as indicated in Section 1.3 below.

1.3. More general operators

Let us consider, in a much more general manner, an operator

$$A\epsilon = -\frac{\partial}{\partial\,x_i}\left[a_{ij}(x/\epsilon)\frac{\partial}{\partial\,x_j}\right]$$ (1.35)

defined in Ω (and, by restriction, in Ω_ϵ); we assume that

$$\begin{cases} a_{ij}(y) \text{ is } Y-\text{periodic}, \quad a_{ij} \in L^{\infty}(Y) \\ a_{ij}(y)\xi_i\xi_j \geq \alpha\xi_i\xi_i, \quad \alpha > 0. \end{cases} \tag{1.36}$$

(The a_{ij}'s are *not necessarily symmetric*).

Remark 1.7

The case considered so far corresponds to

$$a_{ij}(y) = a_{ij} = \delta^j_i. \tag{1.37}$$

We then consider the boundary value problem

$$\begin{cases} A_\epsilon u_\epsilon = f \text{ in } \Omega_\epsilon, \\ u_\epsilon = 0 \text{ on } \Gamma_\epsilon, \\ \dfrac{\partial u_\epsilon}{\partial \nu_{A_\epsilon}} = 0 \text{ on } S_\epsilon \end{cases} \tag{1.38}$$

where

$$\frac{\partial}{\partial \nu_{A_\epsilon}} = \nu_i a_{ij}(x/\epsilon)\frac{\partial}{\partial x_j}, \quad \nu_i = \nu_i(x/\epsilon).$$

We can again define a *homogenized operator*

$$A = -q_{ij} \frac{\partial^2}{\partial x_i \partial x_j} \tag{1.39}$$

where the q_{ij}'s are given as follows. We introduce

$$A_0 = -\frac{\partial}{\partial y_i}\left[a_{ij}(y)\frac{\partial}{\partial y_j}\right] \tag{1.40}$$

($A_0 = -\Delta y$ in case of Section 1.2) and we define χ^j by

$$\begin{cases} A_0(\chi^j - y_j) = 0 \text{ in } Y^*, \\ \dfrac{\partial}{\partial \nu_{(y)}}(\chi^j - y_j) = 0 \text{ on } S, \\ \chi^j \, Y-\text{periodic}. \end{cases} \tag{1.41}$$

Then

$$q_{ij} = \frac{1}{|Y|}\int_{Y^*}\left[a_{ji} - a_{jk}\frac{\partial \chi^i}{\partial y_k}\right]dy. \tag{1.42}$$

The *homogenized problem* is then

$$A u = f \text{ in } \Omega, \quad u = 0 \text{ on } \Gamma. \tag{1.43}$$

The *convergence result* says that, in some sense, $u_\epsilon \to u$ as $\epsilon \to 0$. In order to make this precise one has *to extend u_ϵ in $P_\epsilon u_\epsilon$ defined* in Ω (such that $P_\epsilon u_\epsilon = u_\epsilon$ in Ω_ϵ, $P_\epsilon u_\epsilon \in H^1(\Omega)$).

Then

$$P_\epsilon u_\epsilon \to u \text{ in } H^1(\Omega) \text{ weakly} \tag{1.44}$$

i.e.

$$P_\epsilon u_\epsilon \to u, \quad \frac{\partial P_\epsilon u_\epsilon}{\partial x_i} \to \frac{\partial u}{\partial x_i} \text{ in } L^2(\Omega) \text{ weakly}, \quad i = 1,\dots,n.$$

(it follows that actually $P_\epsilon u_\epsilon \to u$ in $L^2(\Omega)$ *strongly*).

1.4. Spectral problem

Let us assume that

$$a_{ij} = a_{ji} \quad \forall \ i, j \tag{1.45}$$

and let us consider the *spectral problem*

$$\begin{cases} A_\epsilon u_\epsilon = \lambda(\epsilon) u_\epsilon, \text{ in } \Omega_\epsilon, \\ u_\epsilon = 0 \text{ on } \Gamma_\epsilon, \quad \dfrac{\partial u_\epsilon}{\partial \nu_{A_\epsilon}} = 0 \text{ on } S_\epsilon. \end{cases} \tag{1.46}$$

We have the *discrete spectrum*

$$0 < \lambda_1(\epsilon) \le \lambda_2(\epsilon) \le \lambda_3(\epsilon) \le \dots \quad . \tag{1.47}$$

Then (cf. S. Kesavan [1], D. Cioranescu and J. Saint Jean Paulin [1])

$$\lambda_m(\epsilon) \to \lambda_m \text{ as } \epsilon \to 0 \tag{1.48}$$

where $\{\lambda_m\}$ = spectrum of A in Ω, with Dirichlet condition on Γ.

1.5. Other boundary conditions on Γ

All what has been said readily extends to the case of other boundary conditions of Γ (and on Γ_ϵ), such as

$$\frac{\partial u_\epsilon}{\partial \nu_{A_\epsilon}} + \beta u_\epsilon = 0, \quad \beta = \beta(x) \ge 0. \tag{1.49}$$

If $\beta = 0$, some care should be taken with respect to the compatibility conditions on the right hand side (they disappear if A_ϵ is replaced by $A_\epsilon + \alpha I$, $\alpha > 0$).

2. Homogenization for Perforated Materials.
Dirichlet Boundary Conditions on the Boundary of the Holes.

2.1. Setting of the problem

With the same notations as in Section 1, we consider the problem

$$-\epsilon^2 \Delta u_\epsilon = f \text{ in } \Omega_\epsilon \qquad (2.1)$$

subject to

$$u_\epsilon = 0 \text{ on } \Gamma_\epsilon \qquad (2.2)$$

$$u_\epsilon = 0 \text{ on } S_\epsilon. \qquad (2.3)$$

Remark 2.1

If we compare with (1.4) (1.5) (1.6) we have changed u_ϵ into $\epsilon^2 u_\epsilon$ (in order to simplify notations in the asymptotic expansion; this is a detail!) and we have changed the *Neuman conditions* (1.6) *into the Dirichlet condition* (2.3). As we are going to see, this *changes completely the nature of the expansion*.. □

2.2. Asymptotic expansion

We take the ansatz

$$u_\epsilon = u_o + \epsilon u_1 + \dots , \qquad (2.4)$$

where

$$\begin{cases} u_j = u_j(x,y) \text{ is defined in } \Omega \times Y , \\ u_j(x,y) \text{ is } Y-\text{periodic}, \\ u_j(x,y) = 0 \text{ if } y \in S, \ x \in \Omega, \end{cases} \qquad (2.5)$$

and, if possible (and we shall see below, there are difficulties in general)

$$u_j(x,y) = 0 \text{ if } x \in \Gamma , \ y \in Y . \qquad (2.6)$$

By identification we obtain:

$$\begin{cases} -\Delta_y u_o = f , \\ -\Delta_y u_1 - 2\Delta_{xy} u_o = 0, \\ -\Delta_y u_2 - 2\Delta_{xy} u_1 - \Delta_x u_o = 0 \\ \cdots\cdots\cdots\cdots \end{cases} \qquad (2.7)$$

In these equations, x *plays the role of a parameter,* a situation which is radically different from the previous one in Section 1.

If one introduces $w_o(y)$ by

$$\begin{cases} -\Delta_y w_o = 1 \text{ in } \mathcal{Y}, \\ w_o = 0 \text{ on } S, \ w_o \ Y\text{-periodic} \end{cases} \tag{2.8}$$

then

$$u_o(x,y) = w_o(y)f(x). \tag{2.9}$$

One can proceed; with (2.9), $(2.7)_2$ becomes

$$-\Delta_y u_1 = 2\frac{\partial w_o}{\partial y_i}\frac{\partial f}{\partial x_i}.$$

Therefore if we introduce w_i by

$$\begin{cases} -\Delta_y w_i(y) = 2\dfrac{\partial w_o}{\partial y_i} \text{ in } \mathcal{Y}, \\ w_i = 0 \text{ on } S, \ w_i \ Y\text{-periodic}, \end{cases} \tag{2.10}$$

then

$$u_1 = w_i(y)\frac{\partial f}{\partial x_i}(x), \tag{2.11}$$

and we can proceed in this manner (depending on the smoothness of f).

Since x is a parameter here, *no flexibility is left for the boundary condition on* Γ.

If f (resp. ∇f,...) equals 0 on Γ, then u_o (resp. u_1,...) satisfies (2.6) and *the asymptotic expansion can be justified* (up to an order which depends on the smoothness of f). (cf. J.L. Lions [3]). If not (i.e. if $f \neq 0$ on Γ), then *boundary layers are needed.*

We refer to J. L. Lions [2],[4] for particular geometries; the general case has been studied in O. A. Oleinik, A. S. Shamaev and G. A. Yosifian [1].

2.3. Spectral problem

Let us consider the spectral problem (compare to (1.46)):

$$\begin{cases} -\Delta u_\epsilon = \lambda(\epsilon)u_\epsilon \text{ in } \Omega_\epsilon, \\ u_\epsilon = 0 \text{ on } \Gamma_\epsilon, \ u_\epsilon = 0 \text{ on } S_\epsilon. \end{cases} \tag{2.12}$$

We give here, without proof, the result of M. Vanninathan [1]. The proof uses, among others, an idea of L. Tartar. Formal comments which somewhat "explain" the result are given in J. L. Lions [4].

We introduce the first eigenvalue α of the operator $-\Delta$ in $\mathcal{Y}$:

$$\begin{cases} -\Delta_y \phi = \alpha \phi \ \text{in} \ \mathcal{Y}, \\ \phi = 0 \quad \text{on} \ S, \\ \phi \ Y-\text{periodic}, \end{cases} \tag{2.13}$$

$$\phi(y) > 0 \ \text{in} \ \mathcal{Y} \tag{2.14}$$

(it is (2.14) which defines α as the 1st eigenvalue).

One introduces then

$$B_o = \frac{\partial}{\partial y_i} \left[\phi(y)^2 \frac{\partial}{\partial y_i} \right] \tag{2.15}$$

an elliptic operator which is *degenerated* on S. We define ψ_j by

$$\begin{cases} B_o(\psi_j - y_j) = 0 \ \text{in} \ \mathcal{Y} \\ \psi_j \ Y-\text{period} \end{cases} \tag{2.16}$$

(*no* boundary conditions are needed on S). We then set

$$b_{ij} = \int_{\mathcal{Y}} \phi(y)^2 \left[\delta_i^j + \frac{\partial \psi_j}{\partial y_i} \right] dy \tag{2.17}$$

and

$$B = -b_{ij} \frac{\partial^2}{\partial x_i \partial x_j}. \tag{2.18}$$

One can verify (Vanninathan, loc. cit.) that B is elliptic; B is symmetric. Let $\{\mu_m\}$ be the spectrum of B in Ω:

$$\begin{cases} B\phi_m = \mu_m \, \phi_m \ \text{in} \ \Omega, \\ \phi_m = 0 \ \text{on} \ \Gamma, \quad 0 \le \mu_1 < \mu_2 \le \dots \le \mu_m \le \dots \; . \end{cases} \tag{2.19}$$

Then

$$\lambda_m = \epsilon^{-2}\alpha + \mu_m + 0(\epsilon). \tag{2.20}$$

2.4. Extensions and open questions

All that has been said extends to the higher order elliptic operators, to systems, and to evolution equations - *with the possible exception of Section 2.3.*

If we consider

$$\Delta^2 u_\epsilon \equiv \lambda(\epsilon) u_\epsilon \ \text{in} \ \Omega_\epsilon \tag{2.21}$$

subject to the Dirichlet conditions on $\Gamma_\epsilon \bigcup S_\epsilon$; i.e.

$$u_\epsilon = \frac{\partial u_\epsilon}{\partial \nu} = 0 \quad \text{on} \quad \Gamma_\epsilon \bigcup S_\epsilon, \tag{2.22}$$

then the asymptotic expansion of $\lambda_m(\epsilon)$ (the mth order eigenvalue) *is not known*.

The difficulty seems to arise from the fact that - at least in general - there is no eigenfunction of Δ^2 in $\mathcal{Y}$ (with Y-periodicity and zero Dirichlet boundary conditions on S) which is > 0 in $\mathcal{Y}$.

3. Reinforced or Skeletal Structures

3.1. Setting of the problem

We use the notations of Figure 1 and we assume (cf. Figure 2) that $n = 2^\dagger$ and that $\mathcal{Y} = $ square with same center than Y (i.e. ½, ½) and with size $1 - \mu$, where μ is a new "small parameter".

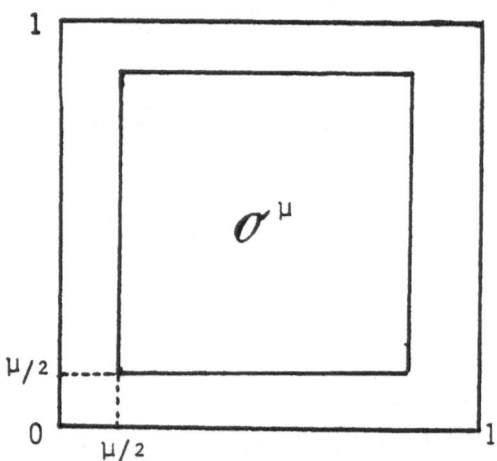

Figure 2

In order to recall the dependence in μ, we shall denote

$$\mathcal{Y} = \mathcal{Y}^\mu, \quad \mathcal{O}^\mu = \text{hole (or obstacle)}. \tag{3.1}$$

Remark 3.1

All that has been said in Section 1 is valid when $\mathcal{Y}$ has corners, hence it applies to the present situation.

---

† To fix ideas. Cf. D. Cioranescu and J. Saint Jean Paulin [2] for higher dimension cases.

Remark 3.2

The domain $\Omega_{\epsilon\mu}$ obtained as in Section 1 (μ fixed for the time being) is what is called a reinforced or a skeleton structure; cf. the shaded part on Figure 3.

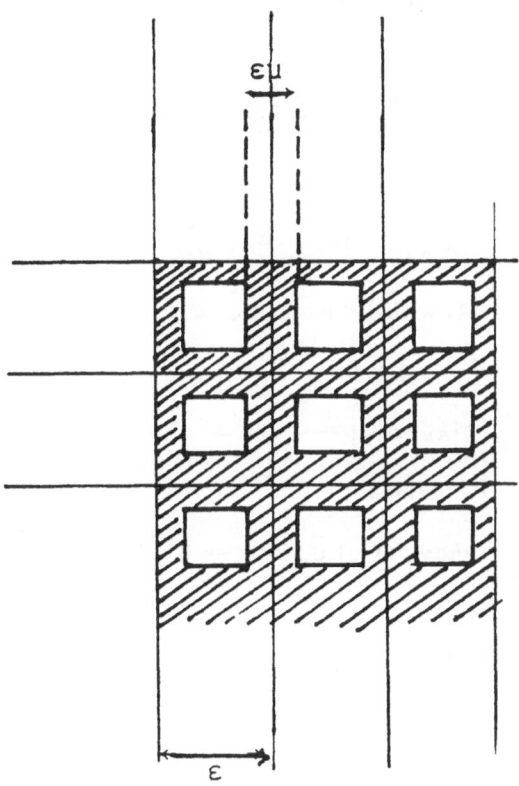

Figure 3

Let A_ϵ now be given as in Section 1.3.

We consider the problem

$$A_\epsilon u_{\epsilon\mu} = f \quad \text{in} \quad \Omega_{\epsilon\mu}, \tag{3.2}$$

$$u_{\epsilon\mu} = 0 \quad \text{on} \quad \Gamma_{\epsilon\mu}, \tag{3.3}$$

$$\frac{\partial u_{\epsilon\mu}}{\partial \nu_{A_\epsilon}} = 0 \quad \text{on} \quad S_{\epsilon\mu}. \tag{3.4}$$

When $\epsilon \to 0$, μ being fixed, one has (with the notations of (1.44))

$$P_\epsilon u_{\epsilon\mu} \rightarrow u_\mu \text{ in } H^1(\Omega) \text{ weakly} \tag{3.5}$$

where u_μ is the solution of

$$\begin{cases} A^\mu u_m = f \text{ in } \Omega, \\ u_\mu = 0 \text{ on } \Gamma. \end{cases} \tag{3.6}$$

In (3.6) A^μ is given as in (1.42):

$$A^\mu = q^\mu_{ij} \frac{\partial^2}{\partial x_i \partial x_j} \tag{3.7}$$

where

$$q^\mu_{ij} = \frac{1}{|y^\mu|} \int_{y^\mu} \left[a_{ji} - a_{jk} \frac{\partial \chi^i_\mu}{\partial y_k} \right] dy ; \tag{3.8}$$

in (3.8) χ^j_μ is given as in (1.14) with y^μ instead of y:

$$\begin{cases} A_o(\chi^j_\mu - y_j) = 0 & \text{in } y^\mu, \\ \dfrac{\partial}{\partial \nu_{(y)}} \left(\chi^j_\mu - y_j \right) = 0 & \text{on } S^\mu \subset \partial \, O^\mu, \\ \chi^j_\mu \ Y\text{--periodic} . \end{cases} \tag{3.9}$$

We want to study the behaviour of u_μ as $\mu \rightarrow 0$.

3.2. Asymptotic formula

Let us assume for simplicity that[+]

$$a_{ij} = \text{constant} . \tag{3.10}$$

We have

$$q^\mu_{ij} \rightarrow q^o_{ij} \text{ as } \mu \rightarrow 0. \tag{3.11}$$

where

$$q^o_{ij} = a_{ij} - \sum_{k=1}^{2} \frac{a_{ik} a_{kj}}{a_{kk}} . \tag{3.12}$$

One can also show (cf. D. Cioranescu and J. Saint Jean Paulin, loc. cit.) that

$$u_\mu \rightarrow u \text{ in } H^1(\Omega) \text{weakly} \tag{3.13}$$

where

[+] Cf. D. Cioranescu and J. Saint Jean Paulin [1] and N. S. Bakhbalov and G. P. Panasenko [1] for more general cases.

$$\begin{cases} \mathcal{A}^\circ u = f \ \text{ in } \Omega, \ \ \mathcal{A}^\circ = -q_{ij}^\circ \ \dfrac{\partial^2}{\partial x_i \partial x_j} \\ u = 0 \ \text{ on } \ \Gamma. \end{cases} \tag{3.14}$$

The proof is technical. The key point is to obtain estimates on q_{ij}.

Since all functions appearing in (3.8) are Y-periodic, one can replace the integration on $\mathcal{Y}^\mu$ by the integration in Z^μ as indicated in shaded on Figure 4, on the square of size 1, and of center $\{1 , 1\}$.

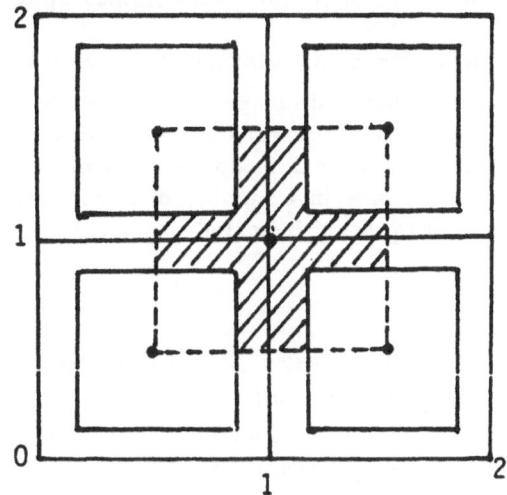

Figure 4

One then uses decompositions of Z^μ in the "horizontal" and in the "vertical" part and after a number of computations, one obtains (3.11), (3.12). □

3.3. Various remarks

Remark 3.3

Similar results hold true for *spectral problems*. If one considers the spectrum $\{\lambda_m^\mu\}$ of $\mathcal{A}^\mu$, it will converge, when $\mu \to 0$, to the spectrum $\{\lambda_m\}$ of $\mathcal{A}$. □

Remark 3.4

For extensions to 3-dimensional cases and with several other geometrical configurations, we refer to the authors already quoted. □

4. Some Variants

4.1. Composite materials

For composite materials with a periodic structure, we refer to A. Bensoussan, J. L. Lions and G. Papanicolaou [1], E. Sanchez-Palencia [1] and to the bibliography therein.

For cases *without* periodic structures and a systematic study of a priori information on the homogenized (or effective) coefficients, we refer to D. Bergman [1], F. Murat and L. Tartar [1] and to the Bibliography therein.

4.2. Control of composite or of reinforced structures

The problem of optimal control of composite or of reinforced structures leads to the following general question: is it valid to "*homogenize first*" and "*to control next*" (so that the control theory can be applied in a much simpler context). This has been proven to be the case in a number of situations; cf. J. L. Lions [5],[6]; some open questions have been indicated in [7], a general account will be presented in [8].

Bibliography

N. S. Bakhbalov and G. P. Panasenko [1] *Averaged Processes in Periodic Media.* Moscow, Nauka, 1984, (in Russian).

A. Bensoussan, J. L. Lions and G. Papanicolaou [1], *Asymptotic Analysis for Periodic Structures.* North-Holland, 1978.

D. Bergman [1] Bulk Physical Properties of Composite Media, in "*Les Méthodes de l'homogénéisation: théorie et applications en Physique*", D. Bergman and al., Eyrolles, 1985, pp. 1-128.

D. Cioranescu and J. Saint Jean Paulin [1] Homogenization in open set with holes. J.M.A.A. 71 (2) (1979), pp. 590-607.

[2] Reinforced and alveolar structures, J.M.P.A., to appear in 1986.

C. Conca [1] Etude d'un fluide traversant une paroi perforée. I. Comportement limite près de la paroi. II. Comportement limite loin de la paroi. To appear.

S. Kesavan [1] Homogenization of elliptic eigenvalue problems. Applied Mathamatics and Optimization, 5, No. 2 and 3, 1979.

J. L. Lions [1] Some Problems connected with Navier Stokes equations. Lectures at ELAM-IV, 1978. Actes 1979, pp. 222-286.

[2] *Some Methods in the Mathematical Analysis of Systems and their Control.* Science Press, Beijing, China and Gordon Breach, New York, 1981.

J. L. Lions [4] Remarques sur les problèmes d'homogénéisation dans les milieux à structure périodique et sur quelques problèmes raides, in *"Les Méthodes de l'homogénéisation: théorie et applications en Physique,"* Eyrolles, 1985, pp. 129-228.

[5] Some asymptotic problems in the optimal control of distributed systems. NASA-J.P.L. Workshop, San Diego, CA., June 1984.

[6] Boundary control of hyperbolic systems and homogenization theory. Vth IFAC Workshop on Control Applications of non-linar programming and optimization. Capri (Italy), June 1985.

[7] Asymptotic problems in distributed systems. Riviere Memorial Lecture, Minneapolis, May 1985.

[8] Book under preparation.

F. Murat and L. Tartar [1] Calcul des variations et homogénéisation, in *"Les Méthodes de l'homogénéisation: théorie et applications en Physique"*, D. Bergman and al. Eyrolles, 1985, pp. 323-369.

O. A. Oleinik, A. S. Shamaev and G. A Yosifian [1] On asymptotic expansions of solutions of the Dirichlet problem for elliptic equations in perforated domains. To appear.

E. Sanchez-Palencia [1] *Non homogeneous media and vibration theory*. Springer, Lecture Notes in Physics, 1980.

[2] Un problème d'écoulement lent d'un fluide incompressible au travers d'une paroi finement perforée. In *"Les Méthodes de l'homogénéisation: théorie et applications en Physique"* by D. Bergman, J. L. Lions, G. Papanicolaou, F. Murat, L. Tartar, E. Sanchez Palencia, Eyrolles, 1985.

L. Tartar [1] Appendix of E. Sanchez Palencia [1].

M. Vanninathan [1] Thesis, Paris, 1979.

CONTROL ISSUES
FOR
PSEUDO DYNAMIC TESTING

N. Harris McClamroch
The University of Michigan
Ann Arbor, Michigan
Robert D. Hanson
The University of Michigan
Ann Arbor, Michigan

1. Introduction

The most realistic method to asses the inelastic response characteristics of a test structure, e.g. a building, subjected to seismic ground motions would be to test the full scale structure on a shaking table. Clearly this is not possible with current experimental facilities, nor will it ever be practical for most large structures. Several alternative approaches have been developed. For example, shaking table tests of reduced scale structural models can be used. The shortcomings of such reduced scale tests are obvious; it is sufficient to point out that many structures cannot be adequately represented by reduced scale structural models [1].

The pseudo dynamic test method has been developed to carry out realistic experimental tests on full scale structures subjected to seismic ground motions.The pseudo dynamic test method (PDTM) uses an on line computer and associated test instrumentation to monitor and control the structure so that the structural displacements closely resemble those that would occur if the structure were subjected to a seismic excitation. The procedure is a self determined experimental technique. Experimental measurements are made of the structural restoring forces during the test; these measured forces are used by the computer,

446

together with a set of mathematical equations for the inertial response characteristics, to determine changes in the structural displacement that should occur
as a consequence of the given ground acceleration. The PDTM differs from classical computer based structural dynamic simulation in that it depends on experimentally measured structural restoring forces rather than on structural restoring
forces computed from a mathematical model. The PDTM makes use of multiple
electrohydraulic actuators to cause the displacement of the structure to track the
computer displacements. Although, in principle, the test could be carried out in
real time, physical limitations on the instrumentation make it necessary to carry
out the test on a step-by-step basis, i.e. pseudo dynamically.

Pseudo dynamic test facilities are in current operation at the Building
Research Institute, Ministry of Construction [15-18] and the University of Tokyo
[14, 23-27] in Japan, and at the University of California, Berkeley [6-8, 19-22] in
the U.S.A. A test facility is currently being constructed at the University of
Michigan.

This paper will review the basic approach for pseudo dynamic testing and its
implementation, giving particular attention to tight control of experimental
errors. Careful implementation of the control logic and the multiple electrohydraulic servo actuators is essential to achieve accurate and robust tracking for
the test structure. Specific forms for the feedback controller logic are suggested
based on certain theoretical results for stabilization of multi degree of freedom
structures using multiple electrohydraulic servo actuators; a control procedure is
also suggested to improve the displacement accuracy of the controlled structure.

The overall effect of experimental errors in the test are strongly influenced
by the accuracy with which the structural displacements are controlled by the
multiple electrohydraulic actuators and controller logic. The final conclusion is
that the PDTM is a viable experimental test method for inelastic response studies
if the experimental procedure is carefully implemented.

2. The Method

The basic assumption in the pseudo dynamic test method is that the dynamic
responses of the structure can be accurately represented by a lumped structural
model which has only a finite number of degrees of freedom. The equations of
motion for an idealized multiple degree of freedom structure can be expressed in
terms of a second-order vector ordinary differential equation, which, in matrix
form, is

$$M\ddot{x} + R(x) = F \tag{1}$$

where $x, \dot{x}$ and $\ddot{x}$ are the structural displacement, velocity and acceleration vectors, respectively; M is the inertial matrix; $R(x)$ is the structural restoring force
vector function, e.g. $R(x) = Kx$ for a linear elastic structure with stiffness matrix
K; and F is the external force vector applied to the structure. Viscous damping
terms have been omitted in (1) only for simplicity. The mass matrix in (1) could

be formulated using standard finite element procedures.

The principal difference between pseudo dynamic testing and computer simulation using numerical integration [2-4,13] is that in the former the computed structural displacements are actually imposed on the test specimen and the resulting restoring forces are measured experimentally from the deformed specimen; in the latter an (inexact) mathematical model is used to specify the restoring forces. Thus, uncertainties associated with the analytical formulation of restoring force characteristics, which are usually the major sources of structural nonlinearities, are not present in a pseudo dynamic test.

A pseudo dynamic test proceeds in a stepwise manner. In particular, (computed) displacements of the structure are determined according to a recursive algorithm; the most common form of the pseudo dynamic test algorithm (PDTA) is based on the central difference formula [4,13] where

$$M\left[\frac{x_{j+1} - 2x_j + x_{j-1}}{T^2}\right] = -R^i + M1a^i \tag{2}$$

Here x_j represents the computed structural displacement at the ith seismic time increment, R^i represents the measured restoring force vector at the ith time increment, a^i represents the value of the seismic ground acceleration at the ith time increment, and the vector $1 = (1, \ldots, 1)$. Finally, T represents the seismic time increment used in (2). Note that the recursive algorithm (2) can be viewed as an approximation to (1) in the case that the external force $F = M1a^i$ is the reaction force due to the seismic ground acceleration. At each step, the computed displacements are quasi-statically imposed on the test specimen by means of computer controlled electrohydraulic actuators. The static restoring forces measured at each step are used to compute the displacement at the next step, based on the specified seismic excitation and equation (2). This step-by-step process is repeated until the desired response history is obtained. The quasi-static approach not only allows a detailed inspection of structural damage during a test, but also permits the use of test equipment available in many structural laboratories.

In theory, the method should be applicable to any structure whose equations of motion can be adequately represented by(1). However, in order to minimize the complexity of the loading apparatus and the number of actuators required, it is desirable that only a limited number of dynamic degrees of freedom be required to describe the overall dynamic response. The most extensive test experience with full scale structures has been at Building Research Institute in Japan, where pseudo dynamic tests of multi-story buildings, with assumed mass concentrations at floor levels, have been completed.

3. Implementation of the Method

A typical implementation scheme for the pseudo dynamic method is shown in Figure 1. A critical part of the implementation is a minicomputer, which performs the following tasks at each step of the test: (i) reads and stores data from a data acquisition unit; (ii) calculates the computed structural response based on the PDTA; and (iii) transmits the computed structural displacement signals to the actuator controllers. These computed displacement signals are commands to the actuators and feedback control system to control the displacement of each structural degree of freedom. The restoring forces developed by the test structure, as well as other measurements of structural response, are measured and fed back to the mini-computer by the data acquisition unit.

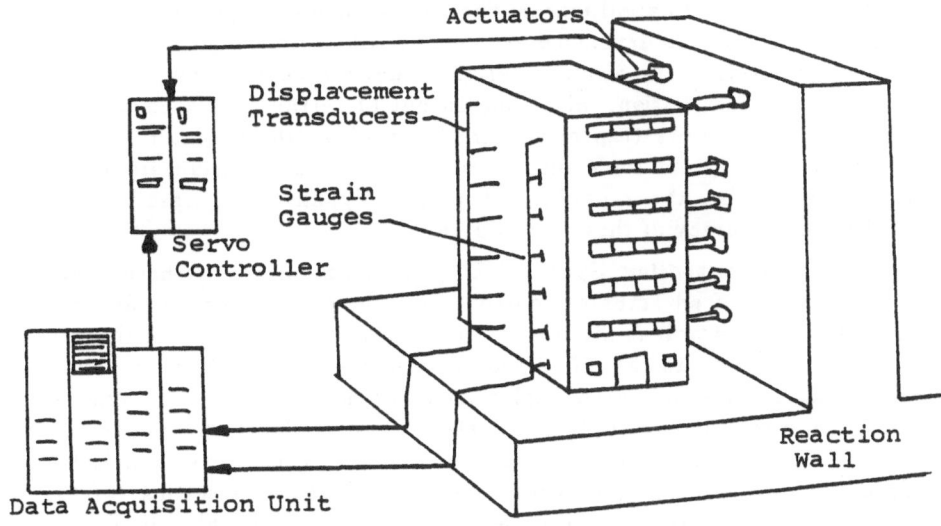

Figure 1

The total time interval required for each step consists of two operational phases: the "hold" and "load" phases. During the hold period, data acquisition and numerical computations are carried out. The load phase is the time period during which the actuators and control systems regulate the structural displacements to track the computed (command) displacements. Data acquisition starts immediately after the load phase is completed. The hold period can be a small fraction of a second if a high speed data acquisition system is used. During the load phase, several auxiliary tasks can be performed, such as recording the acquired data in a disc file, reading the external seismic excitation record, and plotting selected data channels for graphical display. The minimum time for the load phase is limited by the sensitivity of the actuator controller system in responding to the computed command displacement signals. It has been demonstrated that each step of a pseudo dynamic test can be performed in less than one second for simple one or two degree of freedom structures. Tests for multi degree of freedom structures are more involved and require careful coordination between the load and hold phases for each degree of freedom.

The pseudo dynamic test method makes use of standard laboratory equipment. However, since experimental data is a critical part of the self-determined process, pseudo dynamic testing requires high precision control and measurement instruments. The performance characteristics of the major instruments are briefly discussed in the following.

Measurement Transducers: The displacements imposed on a structure and the restoring forces developed are measured by displacement and load transducers, respectively. The accuracy of these devices depends on careful calibration and proper installation. Since the output of certain displacement transducers and load or pressure transducers are fed back to the actuator controller for closed loop control, and force measurements from load transducers are used as part of the PDTA, the reliability of pseudo dynamic test results is directly related to the accuracy of these transducers.

Actuator-Controller System: Displacements of the structure are controlled by means of an actuator-controller system during a pseudo dynamic test. Electronic servo controllers are used to control the displacements of the hydraulic actuators in response to the difference between the structural displacement command signals and the measured displacements of the test structure. The response accuracy of the actuator-controller system to a displacement command signal from the computer depends on the capacity of the servo valves which drive the hydraulic actuators, the servo control loops, and the gain settings of the controller. If the gains are low, the system may respond sluggishly. If the gains are too high, the system may become unstable, and the actuator will overshoot and oscillate about the commanded displacement. Therefore, feedback gains should be carefully selected if an actuator is to respond sensitively and stably to a command signal. This is discussed in more detail in a later section. The maximum response speed of an actuator is limited by the capability of the servo valve and the hydraulic power supply.

4. Importance of the Test Errors

Numerical Integration: The accuracy of the numerical solution obtained from the PDTA (2) depends on the value of integration step size selected. If the method is convergent, the numerical solution should approach the exact solution as the step size goes to zero. On the other hand, to reduce the computational effort, the largest possible value of step size should be selected consistent with the required accuracy. In any event, there are numerical errors inherent in the use of the approximation (2).

Experimental Errors: In addition to the numerical errors mentioned above, experimental errors are introduced during the load phase due to limitations of the actuators and controllers. A computed displacement usually cannot be exactly imposed on the test structure due to errors caused by lack of sensitivity of the actuator-controller system, miscalibration of displacement transducers, or other implementation errors. The structural displacements and restoring forces measured from the test structure may also be different from the actual quantities due to measurement errors.

The effect of these errors in a pseudo dynamic test can be significant; these effects are cumulative since incorrect displacements imposed on the test structure will results in erroneous force feedback; and the errors in force feedback will in turn lead to erroneous displacements being computed at the next step. Due to this error accumulation, a pseudo dynamic test can ultimately be rendered unreliable even though the experimental errors introduced in each step are relatively small [9].

5. Feedback Control of Structure and Electrohydraulic Actuators

Proper implementation of the load phase requires suitable operation of the electrohydraulic actuators used in the tests. Displacement and pressure (or load) feedback is desirable and development of an actuator controller system should take into account the effects of multi degree of freedom and inelastic structural coupling. This section gives particular attention to the selection of an actuator controller which automatically regulates the operation of the electrohydraulic actuators and structure combination as desired. Particular design conditions are suggested; the importance of pressure (or load) feedback in the actuator controller is emphasized so that stability, and hence disturbance and parameter insensitivity, of the controlled system is guaranteed. Such theoretical stabilization issues have been addressed previously in [11]. The emphasis here is on development of a suitable understanding of the load phase control issues.

In order that the test structure actually deform according to the computed structural displacements determined from the PDTA the electrohydraulic actuators must be properly controlled during the load phase at each step; this is the function of the actuator controller (AC). Physically the AC is implemented as an electronic controller. One form for the AC is based on feedback of the displacements and forces of the structure at the actuator locations according to controller relation

$$u = -G^d(x - \underline{x}_i) - G^f F \tag{3}$$

where u is the vector of currents applied to the actuator servo valves, x is the vector of measured structural displacements, F is the vector of forces between the actuators and the structure, and $\underline{x}_i$ is the vector of computed structural displacements (assumed constant during the load phase at each step) obtained from the PDTA. The matrices G^d and G^f represent feedback control gains. Thus the AC electronics generate the input currents to all actuator servo values, based on measured values of the displacements and forces at the actuator locations. Note that either load cells or hydraulic pressure transducers can be used to obtain the force measurements. Leakage of fluid around the actuator pistons has the same effect as force feedback; hence the gain matrix of G^f should take into account the leakage effects.

Knowledge of the structural restoring force function $R(x)$ is not required to implement either the PDTA or the AC. It is assumed that the lumped parameter model for the structure is consistent with the instrumentation in the sense that measurements of the actual structural displacement vector x and of the electrohydraulic force vector F can be directly made. These assumptions are rather restrictive but they could be relaxed at the expense of somewhat greater mathematical complexity.

Conditions for choosing the AC parameters are now presented. In order to complete the description of the testing system it is necessary to include a suitable mathematical model for the hydraulic actuators. It is assumed that the force vector F in (1) is generated by a set of suitably located hydraulic actuators. Assuming linear operation, the dynamics of the hydraulic actuators can be described by [12]

$$\dot{F} = -K_h \dot{x} + K_h K_q u \tag{4}$$

where K_h is a diagonal, positive definite stiffness matrix which depends upon the geometry of the actuator design, the bulk modulus of the hydraulic fluid and the stiffness of the elastic connections between the actuators and the structure [10,11]; and K_q is a nonsingular diagonal current influence matrix which depends upon the geometry of the actuator design and the hydraulic servo valve characteristics.

Thus the actuator and structural dynamics are described by the coupled equations (1) and (4). Substituting the controller equation (3) into equation (4) the closed loop equations for the testing system are obtained

$$M\ddot{x} + R(x) = F$$

$$\dot{F} = -K_h \dot{x} + K_h K_q \left[-G^d (x - \underline{x}_i) G^f F \right].\tag{5}$$

These equations form the basis for selection of the feedback gain matrices G^d and G^f so as to guarantee that the closed loop has desirable response characteristics.

The existing theory suggests that the feedback gain matrices G^d and G^f in the AC should be chosen as diagonal matrices to satisfy the conditions.

$$K_h G^f > G^d > 0 \tag{6}$$

where matrix condition $A > B$ means that $A - B$ is positive definite. As shown in [11], satisfaction of such conditions guarantees that the closed loop is stable for a wide class of realistic inelastic restoring force functions.

In addition, accurate tracking is desired in the sense that at a given step the actual structural displacement vector should tend to the computed displacement vector, as closely as possible. Even for a stable closed loop, perfect tracking is not achievable since the steady state tracking error depends on the *unknown* nonlinear restoring force function. To achieve a small steady state error the feedback gains G^d should be chosen as large as possible in the sense that

$$G^f K \ll G^d \tag{7}$$

where K is the "average elastic stiffness". It is clear from conditions (6) and (7) that there is a tradeoff in achieving satisfaction of the two conditions of stable operation and accurate tracking. This tradeoff is a key feature in achieving a proper implementation of the AC in the PDTM. These observations also suggest that it is important that the electrohydraulic servo valves be selected so that their hydraulic stiffness significantly exceed maximum stiffness of the building to be tested.

Even though the feedback gains may be carefully selected to achieve closed loop stability and to minimize the steady state errors (between the computed displacement vector and the actual structural displacement vector) at each step, the steady state errors at each step may still be excessively large. To further reduce the steady state errors at each step during the load phase the following modified AC, based on integral control with an adaptive parameter β, has been suggested:

$$u = -G^d (x - z) - G^f F$$
$$\dot{z} = \beta (x - \underline{x}_i). \tag{8}$$

In this case, if the control gain matrices G^d and G^f and the adaptive parameter β are chosen appropriately so that the closed loop is stable then necessarily there is zero steady state error between the computed displacement and the actual structural displacement; one such choice is that G^d and G^f satisfy conditions (6) and that β be chosen sufficiently small. This AC has the advantage that, theoretically, the test errors at each step of the pseudo dynamic method can be made arbitrarily small. The disadvantage is that implementation of the AC in (8) is more complicated than that given in (3) and that, in general, closed loop stabilization is more difficult to achieve, possibly resulting in slow convergence toward steady state. Again, it is clear that there is a trade off between the speed and accuracy of the pseudo dynamic test method.

6. Verification Studies

A series of experimental tests have been performed at The University of Tokyo [14,23-27], at The Building Research Institute in Tsukuba, Japan [15-18], and at The University of California in Berkeley [6-8,19-22]. Tests are currently in progress at The University of Michigan at Ann Arbor. Space does not permit a detailed discussion of each study.

In general, studies have compared the results of pseudo dynamic tests and shaking table tests. The general conclusions are that good comparisons were achieved for single and for two degree of freedom tests.

The control issues have not been completely resolved for multiple degree of freedom tests such as used for seven story reinforced concrete building tests and six story steel building tests in Japan. In the first case an equivalent single mode pseudo dynamic test provided excellent results but initial seven degree of freedom tests were unsuccessful. A six story steel structure was successfully tested as a six degree of freedom pseudo dynamic system but required a modification of the PDTA to incorporate 90% of critical damping in the highest three modes. Each of these tests were carried out giving relatively little attention to the AC issues. Our viewpoint is that these tests might have been successful if each step could have been more tightly controlled.

7. Conclusions

1. The pseudo dynamic test method combines well established analytical techniques in structural dynamics with well established experimental methodology.

2. Experimental errors appear to be the major source of inaccuracies in pseudo dynamic testing, due to the cumulative effect of step by step implementation errors. Nevertheless, experimental errors at each step can be reduced to insignificant levels by using high performance test equipment and appropriate control instrumentation.

3. The reliability of the pseudo dynamic method has been verified for low order structures by correlating the results of pseudo dynamic tests with those of shaking table tests.

4. According to current analytical and experimental studies, difficulties will be experienced in testing stiff systems which have large number of degrees of freedom. Such systems are extremely sensitive to experimental errors, but it is expected that improved actuator controller designs may aid to resolve these difficulties.

5. Because of well controlled experimental conditions, pseudo dynamic testing can provide valuable data for increasing our understanding of nonlinear behavior of structures subjected to seismic excitations, and to assess and improve current analytical modeling techniques for analyzing such systems.

Acknowledgements

Financial support of this research by the National Science Foundation through Grants CEE8008583 and CEE8207375 to the University of Michigan is gratefully acknowledged. The results, opinions and conclusions presented in this paper are those of the authors and do not necessarily reflect the views of the National Science Foundation.

References

[1] Bertero, V.V., et al., "Use of Earthquake Simulators and Large-Scale Loading Facilities," Proceedings, Workshop on ERCBC, University of California, Berkeley, 1978.

[2] Clough, R. W. and Penzien, J., *Dynamics of Structures*, McGraw-Hill, New York, 1975.

[3] Clough, R. W. and Wilson, E. L., "Dynamic Analysis of Large Structural Systems with Local Nonlinearities," *Computer Methods in Applied Mechanics and Engineering,* Vol. 17/18, January, 1979, pp. 107-129.

[4] Dahlquist, G., Bjorck, A. and Anderson, N., *Numerical Methods,* Prentice Hall, Englewood Cliffs, NJ, 1974.

[5] Hanson, R. D. and McClamroch, N. H., "Pseudo Dynamic Test Method for Inelastic Building Response," Proceedings, 8th World Conference on Earthquake Engineering, San Francisco, CA, July 1984.

[6] Mahin, S. A., Popov, E. P. and Zayas, V. A., "Seismic Behavior of Tubular Steel Offshore Platforms," Proceedings, Offshore Technology Conference, Houston, TX, May 1980.

[7] Mahine, S.A. and Williams, M.E., "Computer Controlled Seismic Performance Testing," Second ASCE-EMD Specialty Conference on Dynamic Responses of Structures, Atlanta, GA, January, 1981.

[8] Mahin, S.A., Shing, P.B., Dermitzakis, S., Thewalt, C. and Javadian-Gilani, A., "Verification Studies on the Pseudo Dynamic Method," Fouth JTCC Meeting, Tsukuba, Japan, 1983.

[9] McClamroch, N.H., Seraskos, J. and Hanson, R.D., "Design and Analysis of the Pseudo-Dynamic Test Method," Report UMEE 81R3, Department of Civil Engineering, The University of Michigan, September 1981.

[10] McClamroch, N.H. and Haonson, R.D., "Remarks on Some Important Control Issues in Implementation of the PDTM," Fourth JTCC Meeting Tsukuba, Japan, 1983.

[11] McClamroch, N.H., "Displacement Control of Flexible Structures Using Elextrohydraulic Servo Actuators," *ASME Journal of Dynamic Systems, Measurement and Control,* Vol. 107, No. 1, March, 1985, 34-39.

[12] Merritt, H.E., *Hydraulic Control Systems,* John Wiley, 1967.

[13] Newmark, N.M., "A Method of Computation for Structural Dynamics," *Journal of the Engineering Mechanics Division,* ASCE, Vol. 85, No. EM3, July, 1959, pp. 67-94.

[14] Okada, T., Seki, M. and Park, Y.J., "A Simulation of Earthquake Response of Reinforced Concrete Building Frames to Bi-Directional Ground Mtoion by IIS Computer-Actuator On-Line System, " *Proceedings,* 7th World Conference on Earthwuake Engineering, Istanbul, Turkey, September 1980.

[15] Okamoto, S. et al., "A Progress Report on the Full-Scale Seismic Experiment of a Seven Story Reinforced Concrete Building - Part of the U.S. - Japan Cooperative Program," *Research Paper No. 94,* Building Research Institute, Ministry of Construction, Japan, March 1982.

[16] Okamoto, S., Kaminosono, T., Nakashima, M. and Kato, H., "Techniques for Large Scale Testing at BRI Large Scale Structure Test Laboratory," Building Research Institute, Ministry of Construction, Tsukuba, Japan, Research Paper 101, May 1983.

[17] Okamoto, S., Kaminosono, T., Nakashima, M. and Kato, H., "Actuator Control Procedure of BRI Pseudo Dynamic Tesgting System," Fourth JTCC Meeting, Tsukuba, Japan, 1983.

[18] Okamoto, S., et al., "Correlation Between Shaking Table Test and Pseudodynamic Test on Steel Structures", presented at the May, 1983, 15th Joint Meeting, U.S.-Japan Panel on Wind and Seismic Effects, Tsukuba, Japan.

[19] Shing, Pui-Shum and Mahin, Stephen A., "Experimental Error Propagation in Pseudodynamic Testing," Earthquake Engineering Research Report UCB/EERC-83/12, University of California, Berkeley, June, 1983, pp. 168.

[20] Shing, P.B. and Mahin, S.A., "Pseudodynamic Test Method for Seismic Performance Evaluation: Theory and Implementation," UCB/EERC-84/01, Earthquake Engineering Research Center, University of California, Berkeley, January 1984.

[21] Shing, P.B., Mahin, S.A. and Kermitzakis, S.N., "evaulation of On-Line Computer Methods for Seismic Testing," *Proceedings,* 8th World Conference on Earthquake Engineering, San Francisco, CA July, 1984.

[22] Shing, P.B., Javadian-Gilani, A.S. and Mahin, S.A., "Evaluation of Seismic Behavior of a Braced Tubular Steel Structure by Pseudodynamic Testing," *Journal of Energy Resources Technology*, ASME Transactions, September 1984.

[23] Takanashi, K., et al., "Nonlinear Earthquake Response Analysis of Structures by a Computer-Actuator On-Line System," *Bulletin of Earthquake Resistant Structure Research Center*, No. 8, Institute of Industrial Science, University of Tokyo, 1975.

[24] Takanashi, K., Udagawa, K., and Tanaka, H., "Behavior of Bolted Joints in Earthquake Excitation," *Bulletin ERS*, No. 10, 1976, pp. 37-42.

[25] Takanashi, K., Udagawa, K., and Tanaka, H., "Earthquake Response Analysis of a 1-Bay, 2-Story Frame by Computer-Actuator On-Line System," *Bulletin ERS*, No. 11, 1977, pp. 55-60.

[26] Takanashi, K., et al., "Inelastic Response of E-Shaped Columns to Two Dimensional Earthquake Motions," *Bulletin of Earthquake Resistant Structure Research Center*, No. 13, Institute of Industrial Science, University of Tokyo, 1980.

[27] Takanahi, K. and Ohi, K., "Earthquake Response Analysis of Steel Structures by Rapid Computer-Actuator On-Line System," *Bulletin ERS*, No. 16, 1983, pp. 103-109.

CONTROL OF FLUTTER IN BRIDGES

L. Meirovitch and D. Ghosh
Department of Engineering Science and Mechanics
Virginia Polytechnic Institute and State University
Blacksburg, Virginia 24061, USA

1. Introduction

In recent years there has been increasing interest in the application of modern control theory to the problem of vibration suppression in civil structures. The loads on such structures can depend on the environment. For example, the motion of a flexible suspension bridge, such as the original Tacoma Narrows Bridge [Ref. 1], can include unsteady aerodynamic forces. The motion can be described by two partial differential equations for bending and torsional vibration [Ref. 2], with the airstream velocity V as a parameter. The stability of motion is governed by the real part of the system eigenvalues, where the eigenvalues depend on V. For $V = 0$, the system is self-adjoint [Ref. 3] and the eigenvalues consist of pairs of mere imaginary complex conjugates. For small V, the eigenvalues have negative real parts, so that the motion is asymptotically stable. As V increases, some real parts can turn positive, rendering the system unstable. The air speed corresponding to zero real part is known as the critical speed V_{cr}. If the critical speed corresponds to a value for which the imaginary part of an eigenvalue is different from zero, the structure is said to be in *flutter condition* [Ref. 3].

This paper is concerned with the suppression of flutter instability in a suspension bridge deck by means of feedback control forces. To this end, the mathematical model considered consists of the mid-section of the bridge, assumed to be fixed at the towers. The aerodynamic coefficients, which for small oscillations are linear in the bending displacement and torsional rotation and their first derivatives [Ref. 4], are obtained from published wind tunnel tests [Ref. 5]. The partial differential equations of motion in the presence of aeroelastic forces are non-self-adjoint [Ref. 3]. The equations can be converted into a set of coupled second-order ordinary differential equations by means of Galerkin's method [Ref. 3]. For control purposes, it is convenient to transform these equations into a set of first-order state equations.

The control gains for proportional feedback control can be selected in several ways. One well known method is optimal control using a quadratic performance index, which requires the solution of a matrix Riccati equation [Ref. 6]. In general, this entails the solution of a set of nonlinear coupled equations, which for high-order systems (>40) can present computational difficulties. If the independent modal-space control method is used [Ref. 7], then it is only necessary to solve sets of independent second-order Riccati equations, instead of solving a single matrix Riccati equation of high order.

This paper uses concepts presented in Refs. 7 and 8 for optimal modal control of flutter in bridges. As pointed out in Ref. 7, independent control of the modes can be achieved only if control is designed in the modal space. The control implementation is carried out in the actual space, which implies a linear transformation from the modal controls to the actual controls.

2. Equations of Motion

The equations of motion of a bridge deck (Fig. 1) undergoing bending and torsional vibrations in steady airflow can be written as [Ref. 2]:

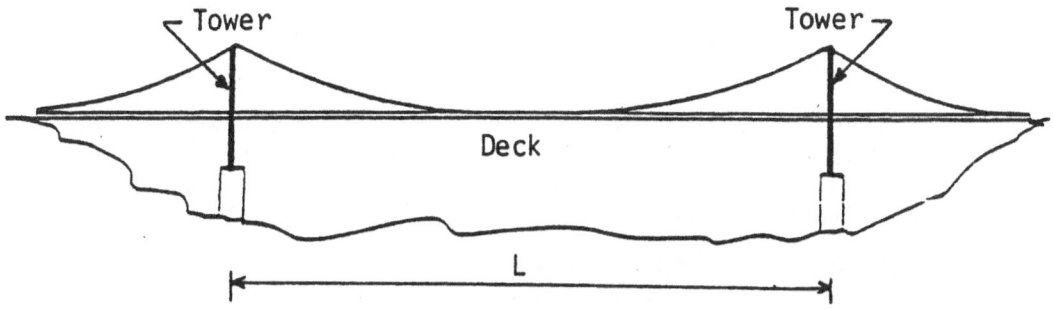

Figure 1: Schematic diagram of a suspension bridge

$$\frac{\partial^2}{\partial x^2}\left(EI\frac{\partial^2 w}{\partial x^2}\right) + m\frac{\partial^2 w}{\partial t^2} + my\frac{\partial^2 \theta}{\partial t^2} + C_w\frac{\partial w}{\partial t} + L_w = F, \qquad (1a)$$

$$-\frac{\partial}{\partial x}\left(GJ\frac{\partial \theta}{\partial x}\right) + my\frac{\partial^2 w}{\partial t^2} + I_2\frac{\partial^2 \theta}{\partial t^2} + C_\theta\frac{\partial \theta}{\partial t} + M_\theta = T, \qquad (1b)$$

where EI is the flexural stiffness, GJ the torsional stiffness, m the mass per unit length, I_2 the mass moment of inertia about the elastic axis, y the distance between elastic axis and inertia axis, C_w the damping coefficient for bending motion, C_θ the damping coefficient for torsional motion, F the distributed control force and T the distributed control torque. The aerodynamic lift force L_w and torque M_θ can be written as [Ref. 4]:

$$L_w = -\frac{1}{2}\rho V^2 2B\left[KH_1^{\cdot}(K)\frac{\dot{w}}{V} + KH_2^{\cdot}(K)B\frac{\dot{\theta}}{V} + K^2H_3^{\cdot}(K)\theta\right], \qquad (2a)$$

$$M_\theta = -\frac{1}{2}\rho V^2(2B^2)\left[KA_1^{\cdot}(K)\frac{\dot{w}}{V} + KA_2^{\cdot}(K)B\frac{\dot{\theta}}{V} + K^2A_3^{\cdot}(K)\theta\right], \qquad (2b)$$

where B is the width of the deck, $K = B\omega/V$ the reduced frequency and ω the frequency of oscillation. Moreover, $H_i^{\cdot}$ and $A_i^{\cdot}$ $(i = 1,2,3)$ are aerodynamic coefficients, which are functions of K. Equations (1) must be satisfied over the domain $0 < x < L$. In addition, if the mid-section of a bridge is fixed at the towers, then w and θ must satify the boundary conditions as:

$$w(0,t) = \frac{\partial w(0,t)}{\partial x} = \theta(0,t) = 0, \qquad (3a)$$

$$w(L,t) = \frac{\partial w(L,t)}{\partial x} = \theta(L,t) = 0. \qquad (3b)$$

Equations (1) can be converted into a set of ordinary second-order differential equations by means of Galerkin's method [Ref. 3]. To this end, $w(x,t)$ and $\theta(x,t)$ can be expressed as:

$$w(x,t) = \sum_{j=1}^{n} a_j(t)\phi_j(x), \quad \theta(x,t) = \sum_{j=n+1}^{n+m} a_j(t)\phi_j(x), \qquad (4a,b)$$

where $\phi_j(x)$ are comparison functions [Ref. 3] and $a_j(t)$ are generalized coordinates. In theory n and m can be infinite, but in practice only a finite number of terms is used. The first n comparison functions must be four times differentiable and satisfy the boundary conditions

$$\phi_j(0) = \phi_j'(0) = \phi_j(L) = \phi_j'(L) = 0, \qquad (5a)$$

where primes denote differentiations with respect to x. The remaining comparison functions must be twice differentiable and satisfy the boundary conditions

$$\phi_j(0) = \phi_j(L) = 0. \qquad (5b)$$

Inserting Eq. (2a) into Eq. (1a) and using Eq. (4a), we obtain

$$\sum_{j=1}^{n} a_j(EI\phi_j'')'' + \sum_{j=1}^{n} \ddot{a}_j m\phi_j + \sum_{j=n+1}^{n+m} \ddot{a}_j my\phi_j + C_w\sum_{j=1}^{n} \dot{a}_j\phi_j$$

$$- \rho VBKH_1^{\cdot}(K)\sum_{j=1}^{n} \dot{a}_j\phi_j - \rho VB^2KH_2^{\cdot}(K)\sum_{j=n+1}^{n+m} \dot{a}_j\phi_j$$

$$- \rho V^2BK^2H_3^{\cdot}(K)\sum_{j=n+1}^{n+m} a_j\phi_j = F. \tag{6}$$

Multiplying Eq. (6) by ϕ_i ($i = 1,...n$), integrating over the domain and using boundary conditions (5a), we can write

$$\sum_{j=1}^{n}\ddot{a}_j\int_0^L m\phi_j\phi_i \ dx + \sum_{j=n+1}^{n+m} a_j\int_0^L my\phi_j\phi_i \ dx + \sum_{j=1}^{n}a_j\int_0^L EI\phi_j{}''\phi_i{}'' \ dx$$

$$+ \sum_{j=n+1}^{n+m} \dot{a}_j\left[\int_0^L \{C_w - \rho VBKH_1^{\cdot}(K)\}\phi_j\phi_i \ dx\right]$$

$$- \sum_{j=n+1}^{n+m} a_j\int_0^L \rho VB^2KH_2^{\cdot}(K)\phi_j\phi_i \ dx$$

$$- \sum_{j=n+1}^{n+m} a_j\int_0^L \rho V^2BK^2H_3\phi_j\phi_i \ dx = \int_0^L F\phi_i \ dx,$$

$$i = 1,2,...,n. \tag{7a}$$

Following a similar procedure, Eq. (1b) yields

$$\sum_{j=1}^{n}\ddot{a}_j\int_0^L my\phi_j\phi_i \ dx + \sum_{j=n+1}^{n+m} \ddot{a}_j\int_0^L I_2\phi_j\phi_i \ dx$$

$$- \sum_{j=1}^{n}\dot{a}_j\int_0^L \rho VB^2KA_1^{\cdot}(K)\phi_j\phi_i \ dx$$

$$+ \sum_{j=n+1}^{n+m} \dot{a}_j\int_0^L \left(C_\theta - \rho VB^3KA_2\right)\phi_j\phi_i \ dx$$

$$+ \sum_{j=n+1}^{n+m} a_j\int_0^L \left(GJ\phi_i{}'\phi_j{}' - \rho V^2B^2KA_3\phi_j\phi_i\right) dx = \int_0^L T\phi_i \ dx,$$

$$i = n+1, n+2,... n+m. \tag{7b}$$

Equations (7) can be written in the matrix form

$$M\ddot{\mathbf{a}} + C\dot{\mathbf{a}} + K\mathbf{a} = \mathbf{F}, \tag{8}$$

where

$$M = \begin{bmatrix} A & B \\ C & D \end{bmatrix}, \tag{9}$$

in which

$$A = [\int_0^L m\phi_i\phi_j \, dx], \qquad i,j = 1,2,...,n; \tag{10a}$$

$$B = [\int_0^L my\phi_i\phi_j \, dx], \quad i = 1,2,...,n; \quad j = n+1, n+2,...,n+m; \tag{10b}$$

$$C = B^T = [\int_0^L my\phi_i\phi_j \, dx], \quad i = 'n+1,n+2,...n+m; \quad j = 1,2,...,n; \tag{10c}$$

$$D = [\int_0^L I_2\phi_i\phi_j \, dx], \quad i,j = n+1,n+2,...,n+m. \tag{10d}$$

Similarly,

$$C = \begin{bmatrix} E & F \\ G & H \end{bmatrix}, \tag{11}$$

where

$$E = \left[\int_0^L \{C_w - \rho VBKH_1^{\cdot}(K)\}\phi_i\phi_j \, dx\right], \quad i,j = 1,2,...,n; \tag{12a}$$

$$F = [\int_0^L \rho VB^2KH_2\phi_i\phi_j \, dx], \quad i = 1,2,...,n; \quad j = n+1,...,n+m; \tag{12b}$$

$$G = F^T = -[\int_0^L \rho VB^2KA_1\phi_i\phi_j \, dx], \quad i = n+1,n+2,...n+m;$$
$$j = 1,...,n; \tag{12c}$$

$$H = [\int_0^L (C_\theta - \rho VB^3KA_2)\phi_i\phi_j \, dx], \quad i,j = n+1,n+2,...,n+m; \tag{12d}$$

and

$$K = \begin{bmatrix} Q & R \\ 0 & T \end{bmatrix}, \tag{13}$$

where

$$Q = [\int_0^L EI\phi_j''\phi_i'' \, dx], \quad i,j = 1,2,...,n; \tag{14a}$$

$$R = -[\int_0^L \rho V^2BK^2H_3\phi_i\phi_j \, dx], \quad i = 1,2,...n; \quad j = n+1,...,n+m; \tag{14b}$$

$$T = [\int_0^L (GJ\phi_i'\phi_j' - \rho V^2B^2KA_3\phi_j\phi_i) \, dx], \quad i,j = n+1,n+2,...,n+m. \tag{14c}$$

Finally,

$$\mathbf{F} = \begin{bmatrix} \mathbf{F}_w \\ \mathbf{F}_\theta \end{bmatrix} = [F_1 \, F_2 \cdots F_{n+m}]^T, \tag{15}$$

where

$$F_i = [\int_0^L F\phi_i \, dx], \quad i = 1,2,...,n; \tag{16a}$$

$$F_i = [\int_0^L T\phi_i \, dx], \quad i = n+1,n+2,...,n+m. \tag{16b}$$

The solution of Eq. (8) can be carried out most conveniently in state form. To transform Eq. (8) to a set of first-order state equations, we introduce the state vector

$$u = \begin{bmatrix} \dot{a} \\ a \end{bmatrix}, \tag{17}$$

consider the identity $K\dot{a} - K\dot{a} \equiv 0$ and obtain

$$M'\dot{u} + K'u = \begin{bmatrix} F \\ 0 \end{bmatrix} = P', \tag{18}$$

where

$$M' = \begin{bmatrix} M & 0 \\ 0 & K \end{bmatrix}, \quad K' = \begin{bmatrix} C & K \\ -K & 0 \end{bmatrix}. \tag{19}$$

The $2(n + m) \times 2(n + m)$ matrix M' is real symmetric and positive definite and the $2(n + m) \times 2(n + m)$ matrix K' is real but otherwise arbitrary. Because M' is positive definite and symmetric it can be decomposed as follows [Ref. 3]:

$$M' = LL^T, \tag{20}$$

where L is a nonsingular lower triangular matrix. Introducing Eq. (18) into Eq. (16) and using the linear transformation [Ref. 7]

$$x(t) = L^T u(t), \quad u(t) = (L^{-1})^T x(t), \tag{21}$$

we obtain

$$\dot{x}(t) = Ax(t) + X(t), \tag{22}$$

where

$$A = -L^{-1}K'(L^{-1})^T, \quad X = L^{-1} \begin{bmatrix} F(t) \\ 0 \end{bmatrix}. \tag{23}$$

3. The Eigenvalue Problem and the Flutter Condition

The eigenvalue problem can be derived by letting $X = 0$ in Eq. (22) and assuming a solution of the homogeneous problem in the form

$$x(t) = e^{\lambda t}x, \tag{24}$$

where λ and x are constants, both quantities in general being complex. Hence, the eigenvalue problem is given by

$$Ax = \lambda x \tag{25}$$

and has $2(n + m)$ solutions λ_s, x_s ($s = 1,2,...,2(n+m)$), where λ_s are the eigenvalues and x_s the eigenvectors. At flutter, damping is zero, so that the eigenvalues are pure imaginary. This implies a pair of complex conjugates

$$\frac{\lambda_r}{\lambda_r} = \pm i\omega_r.$$ (26)

The lowest value of ω_r corresponding to zero real part represents the flutter frequency. The air speed corresponding to this condition is known as the critical air speed V_{cr}.

4. Decoupling of the Equations of Motion

To decouple the equations of motion, we must solve the adjoint eigenvalue problem

$$A^T y = \lambda y.$$ (27)

It possesses the same eigenvalues λ_r but the set of eigenvectors y_r is different from the set of eigenvectors x_s. The two sets of eigenvectors are biorthogonal and can be normalized [Ref. 7] so as to satisfy

$$y_r^T x_s = 2\delta_{rs}, \quad y_r^T A x_s = 2\lambda_s \delta_{rs}, \quad r,s = 1,2,...,2(n+m).$$ (28)

We assume that the eigenvalues and eigenvectors occur in complex conjugate pairs, as follows:

$$x_r = a_r \pm ib_r, \quad y_r = c_r \pm id_r, \quad r = 1,2,...,n+m;$$
$$\lambda_r = \alpha_r + i\omega_r, \quad \bar{\lambda}_r = \alpha_r - i\omega_r, \quad r = 1,2,...,n+m.$$ (29)

Our objective is to formulate the problem in terms of real quantities alone. To this end, we can form the modal matrices [Ref. 7]

$$X = [a_1 \ b_1 \ a_2 \ b_2 \cdots a_{n+m} \ b_{n+m}];$$
$$Y = [c_1 \ -d_1 \ c_2 \ -d_2 \ -c_{n+m} \ d_{n+m}].$$ (30)

In terms of X and Y, the biorthonormality relations can be written as

$$Y^T X = I, \quad Y^T A X = \Lambda,$$ (31)

where

$$\Lambda = block{-}diagonal \ \Lambda_r,$$ (32)

in which

$$\Lambda_r = \begin{bmatrix} \alpha_r & \omega_r \\ -\omega_r & \alpha_r \end{bmatrix}, \quad r = 1,2,...,n+m.$$ (33)

Next, we introduce the linear transformation

$$x = X s$$ (34)

in Eq. (22), premultiply the resulting equation by Y^T and obtain

$$\dot{\mathbf{z}} = \Lambda \mathbf{z} + \mathbf{Z}, \tag{35a}$$

where

$$\mathbf{Z} = Y^T \mathbf{X}. \tag{35b}$$

Equation (35a) represents a set of equations of the form

$$\dot{\mathbf{z}}_r = \Lambda_r \mathbf{z}_r + \mathbf{Z}_r, \qquad r = 1,2,...,n+m; \tag{36}$$

where

$$\mathbf{z}_r(t) = \left[z_{2r-1}(t)\ \ z_{2r}(t) \right]^T, \ \ \mathbf{Z}_r(t) = \left\{ Z_{2r-1}(t)\ \ Z_{2r}(t) \right\}^T, \ \ r = 1,2,...,n+m, \tag{37}$$

are two-dimensional vectors of modal displacements and forces, respectively. If the modal forces $\mathbf{Z}_r(t)$ are functions of $\mathbf{z}_r$ alone, then Eq. (36) represents sets of uncoupled pairs of first-order ordinary differential equations. In the following section, we assume that this is the case and use Eq. (36) to design control forces in the modal space.

5. Optimal Control in Modal Space

Let us consider independent linear control of the type

$$\mathbf{Z}_r(t) = E_r(t)\mathbf{z}_r(t), \qquad r = 1,2,...,n+m. \tag{38}$$

Introducing Eq. (38) into Eq. (36), we obtain the closed-loop equations

$$\dot{\mathbf{z}}_r = [\Lambda_r + E_r]\mathbf{z}_r, \qquad r = 1,2,...,n+m. \tag{39}$$

For optimal control, E_r can be determined by minimizing a cost function of the form

$$J = \sum_{n=1}^{n+m} J_r, \tag{40}$$

where J_r are independent modal cost functions having the expressions

$$J_r = [\mathbf{z}_r(t_f) - \hat{\mathbf{z}}_r]^T H_r[\mathbf{z}_r(t_f) - \hat{\mathbf{z}}_r] + \int_0^{t_f} (\mathbf{z}_r^T Q_r \mathbf{z}_r + \mathbf{Z}_r^T R_r \mathbf{Z}_r) \, dt,$$
$$r = 1,2,...,n+m; \tag{41}$$

where Q_r is a positive semidefinite and symmetric matrix, H_r and R_r are positive definite symmetric matrices and $\hat{\mathbf{z}}_r$ is a reference value to which $\mathbf{z}_r$ is to be driven at final time t_f. For simplicity, we assume that $H_r = 0$ and $Q_r = I$. This leaves us with the problem of selecting R_r.

It is shown in Ref. 6 that the optimal modal control gain matrix has the form

$$E_r(t) = -R_r^{-1}K_r(t), \qquad r = 1,2,...,n+m; \tag{42}$$

where $K_r(t)$ is a real symmetric matrix satisfying the Riccati equation:

$$\dot{K}_r(t) = -K_r\Lambda_r - \Lambda_r^T K_r - I + K_r R_r^{-1}K_r, \quad E_r(t_f) = H_r,$$

$$r = 1,2,...,n+m. \tag{43}$$

Moreover, it is shown in Ref. 7 that R_r can be chosen in the form

$$R_r = \begin{bmatrix} \infty & 0 \\ 0 & R_{r\,22} \end{bmatrix}, \quad r = 1,2,...,n+m; \tag{44}$$

from which it follows that

$$R_r^{-1} = \begin{bmatrix} 0 & 0 \\ 0 & R_{r\,22}^{-1} \end{bmatrix}, \tag{45}$$

so that Z_r can be written as:

$$Z_r = \begin{bmatrix} 0 \\ -R_{r\,22}^{-1}(K_{r\,21}z_{2r-1} + K_{r\,22}z_{2r}) \end{bmatrix}, \tag{46}$$

which leaves us with $R_{r\,22}^{-1}$ for selection. When this is done, components of K_r can be found by solving the Riccati equations:

$$\dot{K}_{r\,11} = -2\alpha_r K_{r\,11} + 2\omega_r K_{r\,12} - 1 + R_{r\,22}^{-1}K_{r\,12}^2,$$

$$\dot{K}_{r\,12} = -\omega_r K_{r\,11} - 2\alpha_r K_{r\,12} + \omega_r K_{r\,22} + R_{r\,22}^{-1}K_{r\,12}K_{r\,22},$$

$$\dot{K}_{r\,22} = -2\omega_r K_{r\,11} - 2\alpha_r K_{r\,12} - 1 + R_{r\,22}^{-1}K_{r\,22}^2,$$

$$r = 1,2,...,n+m. \tag{47}$$

It is shown in Ref. 6 that, if $H_r = 0$ and Q_{r}, R_r are constant matrices, then $K_r(t_f) = K_r = constant$ which implies that $\dot{K}_r = 0$ $(r = 1,2,...,n+m)$. In this paper, we assume that t_f is sufficiently large that steady-state values of K_r can be used in Eq. (42) instead of the time-dependent values. The steady-state values of K_r can be found by Potter's method [Ref. 9].

To select the value of $R_{r\,22}^{-1}$, we return to the closed-loop equations, Eq. (39). Inserting Eqs. (45) and (42) into Eq. (39), we obtain

$$\dot{\mathbf{z}}_r = A_r^*\mathbf{z}_r, \qquad r = 1,2,...,n+m; \tag{48}$$

where A_r^* are nonsymmetric matrices given by:

$$A_r^* = \begin{bmatrix} \alpha_r & \omega_r \\ -(\omega_r + R_{r\,22}^{-1}K_{r\,21}) & (\alpha_r - R_{r\,22}^{-1}K_{r\,22}) \end{bmatrix}, \quad r = 1,2,...,n+m. \tag{49}$$

Letting

$$\mathbf{z}_r = e^{\lambda_r t} \mathbf{z}_r, \qquad r = 1,2,...,n+m; \tag{50}$$

we obtain the eigenvalue problem

$$A_r' \mathbf{z}_r' = \lambda_r \mathbf{z}_r', \qquad r = 1,2,...,n+m; \tag{51}$$

which can be solved for the closed-loop eigenvalues. The values R_{r22}^{-1} are selected such that the closed-loop frequencies are equal to the corresponding open-loop frequencies ω_r. This yields:

$$R_{r22}^{-1} = \frac{4\omega_r K_{r21}}{(K_{r22}^2)}, \qquad r = 1,2,...,n+m. \tag{52}$$

As soon as K_r and R_r^{-1} are determined, the control forces can be found by using $\mathbf{P}^{\cdot} = L\mathbf{X}$. Also the response can be found by using the transformation:

$$u = (L^{-1})^T X \mathbf{z}, \tag{53}$$

where $\mathbf{z}$ is obtained from Eq. (50).

6. Numerical Example

The technique developed in the preceding sections has been applied to a suspension bridge deck having aerodynamic characteristics similar to those of the original Tacoma Narrows Bridge [Ref. 5]. The chosen properties of the bridge deck were $L = 853.0\ m$, $B = 11.89\ m$, $y = 0.01\ m$, $I = 80,000.0\ kg\cdot m$, $m = 8,482.67\ kg/m$, $EI = 2.33 \times 10^5\ kg\cdot m^3/s^2$, $GJ = 9.32 \times 10^9\ kg\cdot m^3/s^2$, $C_w = 50.0\ kg/(m\cdot s)$, $C_\theta = 3.27 \times 10^4\ kg\cdot m/s$, $\rho = 1.12\ kg/m^3$. From wind tunnel tests [Ref. 5], it is observed that the aerodynamic coefficients $A_1^{\cdot}$ and $A_3^{\cdot}$ are negligible. In addition, if we assume that $H_2^{\cdot}$ and $H_3^{\cdot}$ are negligible the aerodynamic terms become decoupled. Then, the coefficients $H_1^{\cdot}(K)$ and $A_2^{\cdot}(K)$ are approximated by:

$$H_1^{\cdot}(K) = -0.1\left[\frac{V}{fB}\right], \quad A_2^{\cdot}(K) = -0.025\left[\frac{V}{fB}\right]\left[2.0 - \frac{V}{fB}\right], \tag{54}$$

where V/fB is the reduced velocity, in which f is the frequency of oscillation.

In this example, only two modes of vibration are considered, one for the bending displacement and one for the torsional displacement. The comparison functions chosen are:

$$\phi_1 = \frac{x}{L}\left[1 - \frac{x}{L}\right]\sin\frac{\pi x}{L}, \quad \phi_2 = \sin\frac{\pi x}{L}. \tag{55}$$

The conditions at flutter were computed and are displayed in Table 1. Results with $V = 20.00\ m/s$ are shown in Tables 2, 3 and 4. From Table 3, it is observed that the real part of the open-loop eigenvalue corresponding to the lowest frequency is positive implying a divergent solution. This mode was targeted for control and the other mode was left uncontrolled. From Table 4. it can be seen that the closed-loop eigenvalues have negative real parts, so that the system has been rendered stable. Figures 2 - 5 show the controlled response of a point at the center of the bridge. The initial conditions were $\dot{w}(L/2,0) = 0.1 \times 10^{-2}\ m/s$, $\theta(L/2,0) = 0.014\ deg/s$ and $w(l/2,0) = \theta(L/2,0) = 0$. Figures 6 and 7 show the required control force and control torque respectively.

Table 1: *Flutter conditions*

Frequency = 1.2563 rad/s
Reduced frequency = 0.7958
Reduced velocity = 7.8950
Velocity = 18.77 m/s

Table 2: *Conditions for V = 20.00 m/s*

Frequency = 1.2563 rad/s
Reduced frequency = 0.7469
Reduced velocity = 8.413
Velocity = 20.00 m/s

Table 3: *Open loop eigenvalues for V = 20.00 m/s*

	1	2	3	4
Real	0.0350	0.0350	-0.0128	-0.0128
Imag.	1.2563	-1.2563	2.5133	-2.5133

Table 4: *Closed loop eigenvalues for V = 20.00 m/s*

	1	2	3	4
Real	-0.0839	-0.0839	-0.0128	-0.0128
Imag.	1.2563	-1.2563	2.5133	-2.5133

Table 5: *Control gain matrices for V = 20.00 m/s*

1		2	
0	0	0	0
-0.0108	-0.2328	0	0

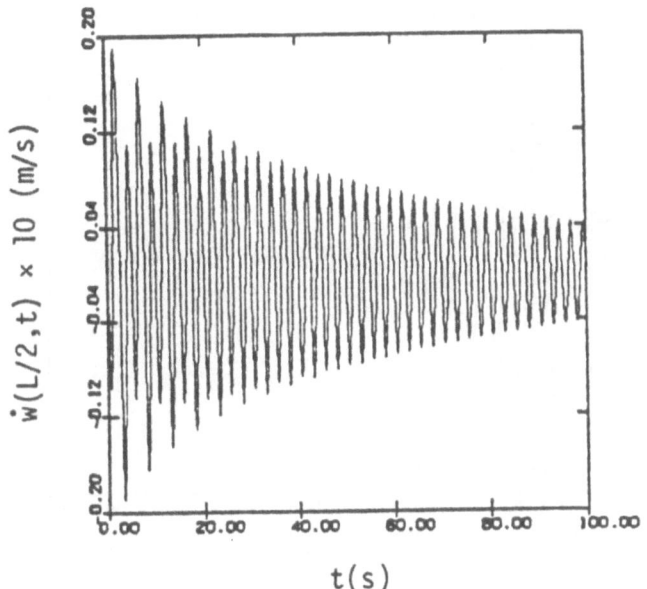

Figure 2: Time history of the velocity $\dot{w}(L/2,t)$

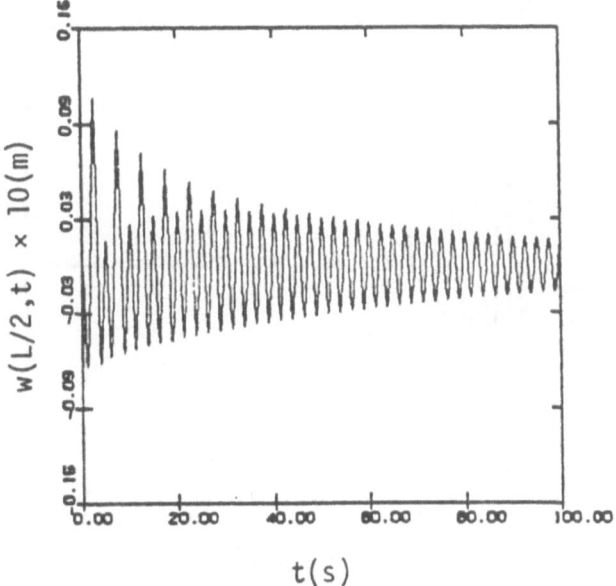

Figure 3: Time history of the displacement $w(L/2,t)$

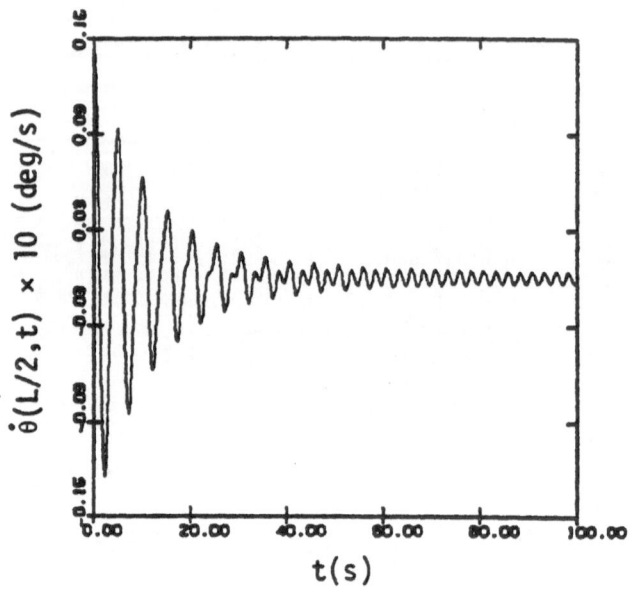

Figure 4: Time history of the angular velocity $\dot{\theta}(L/2,t)$

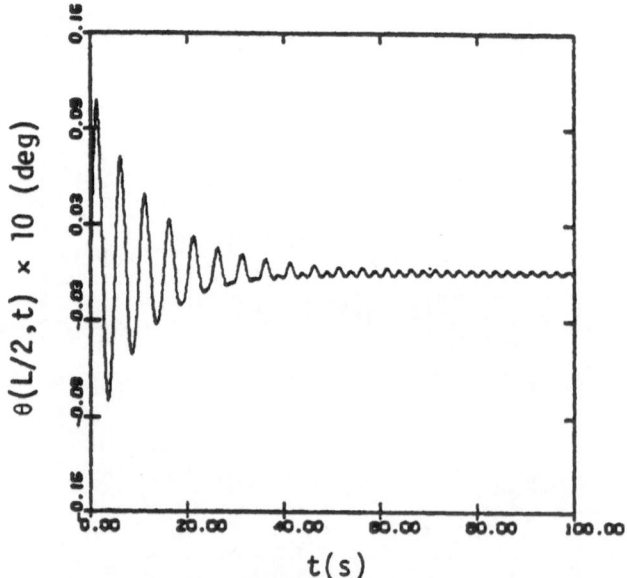

Figure 5: Time history of the angular displacement $\theta(L/2,t)$

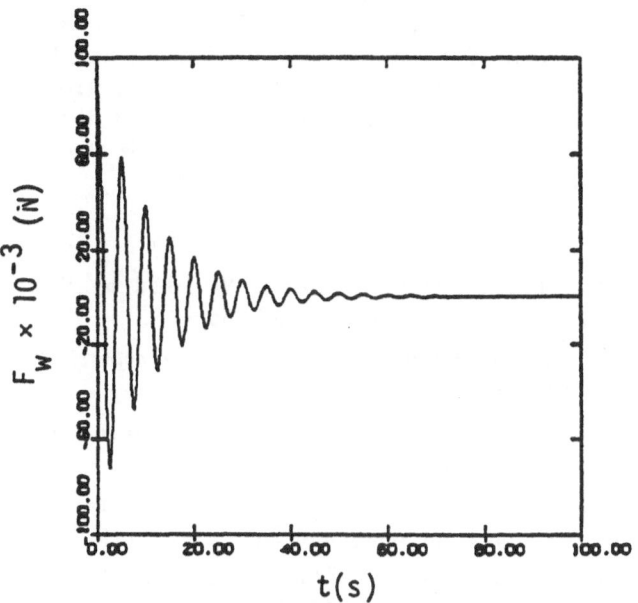

Figure 6: Time history of the control force F_w

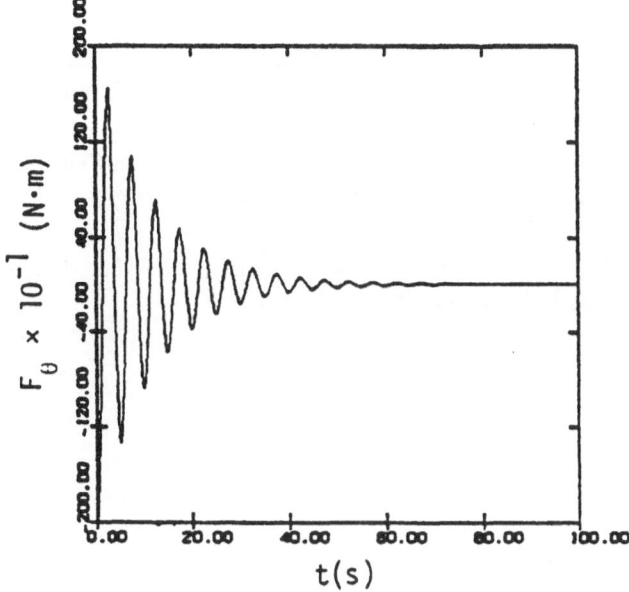

Figure 7: Time history of the control torque F_θ

7. Results and Conclusions

An optimal control scheme for controlling flutter instability in a suspension bridge deck is presented. The optimal control scheme is based on the concept of independent modal-space control, leading to a 2×2 matrix Riccati equation. The control gain matrix (Table 5) is chosen such that the real parts of the closed-loop eigenvalues acquire negative real parts while the open-loop frequencies remain unchanged.

References

[1] Bowers, N.A., "Tacoma Narrows Bridge Wrecked by Wind," *Engineering News Record*, Nov. 14, 1940, pp. 647 and 656.

[2] Fung, Y.C., *An Introduction to the Theory of Aeroelasticity*, Dover Publications, Inc., New Tork, 1969.

[3] Meirovitch, L., *Computational Methods in Structural Dynamics*, Sijthoff-Noordhoff, The Netherlands, 1980.

[4] Simiu, E. and Scanlan, R.H., *Wind Effects on Structures: An Introduction to Wind Engineering*, John Wiley and Sons, New York, 1978.

[5] Scanlan, R.H. and Tomko, J.J., "Airfoil and Bridge Deck Flutter Derivatives," *Journal of Engineering Mechanics (ASCE)*, December 1972, pp. 1717-1737.

[6] Kirk, D.E., *Optimal Control Theory*, Prentice-Hall Inc., Englewood Cliffs, N.J. 1970.

[7] Meirovitch, L. and Baruh, H., "Optimal Control of Damped Flexible Gyroscopic Systems," *Journal of Guidance and Control*, Vol. 4, No. 2, March-April 1981, pp. 157-163.

[8] Meirovitch, L. and Silverberg, L.M., "Control of Non-Self-adjoint Distributed Parameter Systems," *Journal of Optimization Theory and Applications*, (to appear).

[9] Potter, J.E., "Matrix Quadratic Solution," *SIAM Journal of Applied Mathematics*, Vol. 14, pp. 496-501, 1966.

ON THE CONTROL OF INSTABILITIES IN FLUID-STRUCTURE INTERACTION PROBLEMS

V.J. Modi and F. Welt
Department of Mechanical Engineering
The University of British Columbia
Vancouver, B.C. Canada V6T 1W5

1. Introduction

Reponse of aerodynamically bluff bodies when exposed to a fluid stream has been a subject of considerable study for quite some time. The prevention of aeroelastic vibrations of smokestacks, transmission lines, suspension bridges, tall buildings, etc., is of particular interest to engineers. Ever since the pioneering contribution by Strouhal, who correlated periodicity of the vortex shedding with the diameter of a circular cylinder and velocity of the fluid stream, there has been a continuous flow of important contributions resulting in a vast body of literature. This has been reviewed rather adequately by Cermak [1]. In general, the oscillations may be induced by vortex resonance or geometric-fluid dynamic instability called galloping.

Several passive devices such as helical strakes, shrouds, slats, tuned mass dampers, etc., have been proposed over years and have exhibited a varying degree of success in minimizing the effects of vortex induced and galloping types of instabilities [2]. In general, the vibration suppressing devices tend to change aerodynamic characteristics of the structure in such a way as to interfere with and weaken the existing force while the dampers provide a mechanism for dissipating energy. Motivation for the current investigation came from the spacecraft technology where torus shaped partially filled ring-type nutation dampers are

473

frequently used to control very long period (around 1.5 - 24 hours) librational motion. As the frequency of oscillations encountered in wind induced instability of bluff bodies, earthquake response of buildings, and wave excited vibrations of offshore structures is relatively small, it was thought appropriate to explore applicability of nutation dampers to this class of problems.

2. An Approximate Analytical Approach to Evaluate Energy Dissipation

The velocity field within a rigid torus damper oscillating harmonically in translation can be approximated by a potential flow solution. An additional term accounting for the wall boundary layer is introduced to assess energy dissipation through the action of viscous forces. The procedure is similar to the one adopted by Case and Parkinson [3].

2.1. Potential flow solution

The potential function $\phi(r,\theta,z,t)$ represents solution of the differential equation

$$\nabla^2 \phi = 0, \tag{1}$$

with the boundary conditions:

$$\frac{\partial \phi}{\partial n} = V_n \quad at \ the \ wall; \tag{2a}$$

and

$$\frac{\partial^2 \tilde{\phi}}{\partial t^2} + g\frac{\partial \tilde{\phi}}{\partial z} + 2\frac{\partial \tilde{\phi}}{\partial r}\cdot\frac{\partial^2 \tilde{\phi}}{\partial r \partial t} + \dots = 0 \tag{2b}$$

representing kinetic and kinematic conditions at the free surface. Here:

r, θ, z = cylindrical coordinates with origin at the undisturbed free surface (Fig. 1);

t = time;

V_n = damper wall velocity component normal to the boundary;

ϕ = $\tilde{\phi} + \phi_f$;

$\tilde{\phi}$ = potantial function relative to the moving coodinates r,θ,z;

ϕ_f = potential function for the damper motion;

g = acceleration due to gravity.

As the free surface occupies different orientation during the damper motion, a standard Taylor series expansion of Eq. (2b) around $z = 0$ is used and a linear solution obtained by neglecting second higher order terms. Applying the procedure of separation of variables, the linearized system yields a solution in terms of the Fourier-Bessel expansion as:

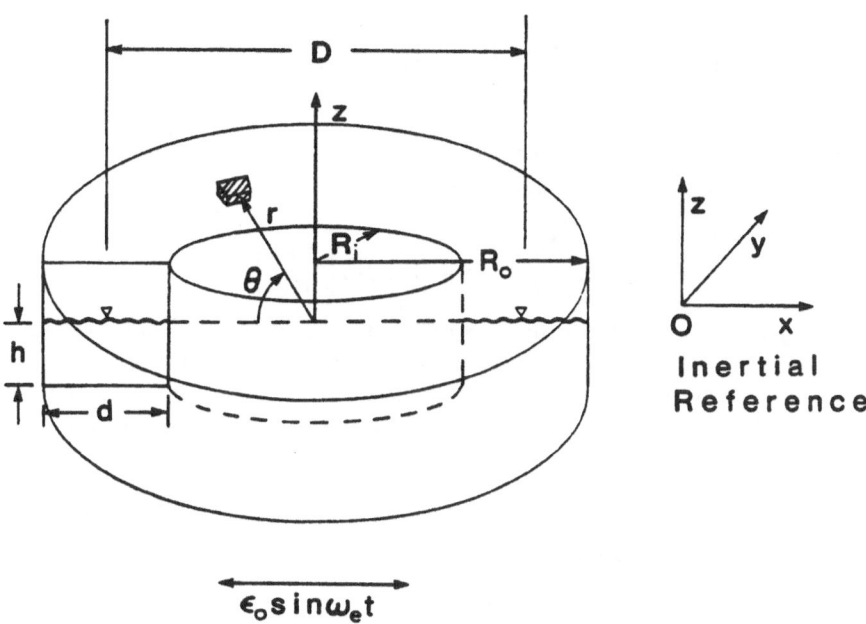

Figure 1: Geometry of the square section damper selected for analytical study

$$\phi(r,\theta,z,t) = \epsilon_o(\cos\omega_e t)(\cos\theta)R_o\left[\frac{r}{R_o} + \sum_i B_i C_1\left(\lambda_{1i}\frac{r}{R_o}\right)\frac{Cosh\left\{\frac{\lambda_{1i}(h+z)}{R_o}\right\}}{Cosh\left(\frac{\lambda_{1i}h}{R_o}\right)}\right]; \quad (3)$$

here:

ϵ_o = amplitude of damper excitation;

ϕ_1 = $\epsilon_o r\cos\omega_e t\cos\theta$;

$$B_i = \frac{2}{\left\{1-\left(\frac{\omega_{1i}}{\omega_e}\right)^2\right\}}\left[\frac{C_1(\lambda_{1i}) - \left(\frac{R_i}{R_o}\right)C_1\left(\frac{\lambda_{1i}R_i}{R_o}\right)}{C_1^2(\lambda_{1i})(\lambda_{1i}^2-1) - C_1^2\left(\frac{\lambda_{1i}R_i}{R_o}\right)\left\{\left(\frac{\lambda_{1i}R_i}{R_o}\right)^2-1\right\}}\right];$$

$f_e, \omega_e =$ damper excitation frequency;

ω_{1i} = liquid natural frequency in the i-th mode,

$$\left\{\left(\lambda_{1i} g/R_o\right)\tanh\left(\lambda_{1i} h/R_o\right)\right\}^{1/2};$$

h = liquid height;

λ_{1i} = eigenvalue for the i-th transverse mode with the first circumferential mode, solution of the equation $C_1'\left(\lambda_{1i} R_i/R_o\right) = 0$;

$$C_1\left(\frac{\lambda_{1i} r}{R}\right) = Y_1'(\lambda_{1i})J_1\left(\frac{\lambda_{1i} r}{R_o}\right) - J_1'(\lambda_{1i})Y_1\left(\frac{\lambda_{1i} r}{R_o}\right).$$

Note, J_1 and Y_1 are Bessel functions of the first and second kind, order one, and prime denotes differentiation with respect to quantities in the brackets. Of course, the linear solution cannot be expected to be valid near liquid resonance, hence a nonlinear solution is obtained by assuming the potential function of the form:

$$\tilde{\phi}(r,\theta,z,t) = \epsilon^{1/3}(\Psi_1\cos\omega_e t + X_1\sin\omega_e t) + \epsilon^{2/3}(\Psi_0 + \Psi_2\cos2\omega_e t + X_2\sin2\omega_e t)$$

$$+ \epsilon(\Psi_3\cos3\omega_e t + X_3\sin3\omega_e t), \tag{4}$$

where Ψ_1, Ψ_2, Ψ_3 and X_1, X_2, X_3 are functions of r,ϕ,z, and $\epsilon = \epsilon_0\omega$. With this, the solution near resonance is given by,

$$\tilde{\phi}(r,\theta,z,t) = \epsilon^{1/3}\left[\gamma\cos\theta\, C_1\left(\frac{\lambda_{11} r}{R_o}\right)\frac{\cosh\left\{\dfrac{\lambda_{11}(z + h)}{R_o}\right\}}{\cosh\left(\dfrac{\lambda_{11} h}{R_o}\right)}\cos\omega_e t\right.$$

$$\left. + \varsigma\sin\theta\, C_1\left(\frac{\lambda_{11} r}{R_o}\right)\frac{\cosh\left\{\dfrac{\lambda_{11}(z + h)}{R_o}\right\}}{\cosh\left(\dfrac{\lambda_{11} h}{R_o}\right)}\sin\omega_e t\right]$$

$$+ \epsilon^{2/3}\left[\sum_{n=1}^{\infty}\frac{\Omega_{on}}{2}(\varsigma^2 - \gamma^2)C_o\left(\frac{\lambda_{on} r}{R_o}\right)\frac{\cosh\left\{\dfrac{\lambda_{on}(z + h)}{R_o}\right\}}{\cosh\left(\dfrac{\lambda_{on} h}{R_o}\right)}\right.$$

$$- \sum_{n=1}^{\infty} \frac{\Omega_{2n}}{2} (\varsigma^2 + \gamma^2) C_2 \left(\frac{\lambda_{2n} r}{R_o} \right) \frac{\cosh\left\{ \dfrac{\lambda_{2n}(z+h)}{R_o} \right\}}{\cosh\left\{ \dfrac{\lambda_{2n} h}{R_o} \right\}} \cos 2\theta \left| \sin 2\omega_e t. \right. \tag{5}$$

Here γ and ς are representative of the liquid amplitude of motion; Ω_{on}, Ω'_{2n} are terms associated with the mode shapes and damper dimensionless number D/d,

$$C_m \left(\frac{\lambda_{mn} r}{R_o} \right) = Y'_m(\lambda_{mn}) J_m \left(\frac{\lambda_{mn} r}{R_o} \right) - J'_m(\lambda_{mn}) Y_m \left(\frac{\lambda_{mn} r}{R_o} \right);$$

and J_m, Y_m are the Bessel functions of the m-th order. It may be pointed out that two solutions can be found: the one where $\varsigma = 0$ corresponds to the 'planar motion' of the liquid and the 'nonplanar solution' for $\varsigma \neq 0$. To simplify the analysis, the third order term in ϵ is omitted in the subsequent development.

2.2. Effect of viscosity

For incompressible flow, the Navier-Stokes equation reduces to:

$$(\bar{u} \cdot \nabla)\bar{u} + \frac{\partial \bar{u}}{\partial t} = -\nabla \left(gz + \frac{p}{\rho} + \frac{\partial V}{\partial t} \right) + \nu \nabla^2 \bar{u}, \tag{6}$$

where $\bar{u}$ is the fluid velocity vector in the moving frame of reference and $\dfrac{\partial V}{\partial t}$ is the acceleration of the frame of reference. Substituting $\bar{u} = \nabla \phi + \bar{u}_2$ and recognizing $\bar{V} = \nabla \phi_f$,

$$\left[(\overline{\nabla \phi} + \bar{u}_2) \cdot \nabla \right] (\nabla \tilde{\phi} + \bar{u}_2) + \frac{\partial}{\partial t} (\nabla \tilde{\phi} + \bar{u}_2)$$

$$= -\nabla \left(gz + \frac{p}{\rho} + \frac{\partial \phi_f}{\partial t} \right) + \nu \nabla^2 (\nabla \tilde{\phi} + \bar{u}_2),$$

i.e.,

$$(\nabla \tilde{\phi} \cdot \nabla)\nabla \tilde{\phi} + \bar{u}_2 \nabla^2 \tilde{\phi} + (\nabla \tilde{\phi} \cdot \nabla)\bar{u}_2 + (\bar{u}_2 \cdot \nabla)\bar{u}_2 + \frac{\partial \bar{u}_2}{\partial t}$$

$$= -\nabla \left(gz + \frac{p}{\rho} + \frac{\partial \phi}{\partial t} \right) + \nu \nabla (\nabla^2 \tilde{\phi}) + \nu \nabla^2 \bar{u}_2. \tag{7}$$

Using Bernoulli's equation, $(\nabla \tilde{\phi})^2 + gz + \dfrac{p}{\rho} + \dfrac{\partial \phi}{\partial t} = $ const., as well as $\nabla^2 \tilde{\phi} = 0$, and neglecting second order terms in $\bar{u}_2$, the above equation reduces to:

$$\frac{\partial \bar{u}_2}{\partial t} = \nu \nabla^2 \bar{u}_2. \tag{8}$$

The corresponding continuity equation is $\nabla \cdot (\nabla \tilde{\phi} + \bar{u}_2) = 0$, i.e.,

$$\nabla \cdot \bar{u}_2 = 0. \tag{9}$$

As $\nabla \tilde{\phi} + \bar{u}_2 = 0$ at the wall,

$$\bar{u}_2 = -\nabla \tilde{\phi}. \tag{10}$$

Taking $\bar{u}_2$ to be harmonic, i.e., $\bar{u}_2 = \bar{U}_2 e^{i\omega_e t}$, Eq. (8) becomes:

$$i\omega_e \bar{U}_2 = \nu \left\{ \frac{1}{r} \frac{\partial}{\partial r} \left[r \frac{\partial \bar{U}_2}{\partial r} \right] + \frac{1}{r^2} \frac{\partial^2 \bar{U}_2}{\partial \theta^2} + \frac{\partial^2 \bar{U}_2}{\partial z^2} \right\},$$

or neglecting the curvature effects,

$$i\omega_e \bar{U}_2 = \nu \left\{ \frac{\partial^2 \bar{U}_2}{\partial r^2} + \frac{1}{r^2} \frac{\partial^2 \bar{U}_2}{\partial \theta^2} + \frac{\partial^2 \bar{u}_2}{\partial z^2} \right\}. \tag{11}$$

Now $\bar{U}_2$ is considered to be significantly different from zero close to the boundary only. Near the damper vertical walls, it is assumed that:

$$\frac{\partial^2 \bar{U}_2}{\partial r^2} \gg \frac{\partial^2 \bar{U}_2}{\partial z^2}; \quad \frac{\partial^2 \bar{U}_2}{\partial r^2} \gg \frac{1}{r^2} \frac{\partial^2 \bar{U}_2}{\partial \theta^2}.$$

Therefore, Eq. (13) becomes:

$$\bar{U}_2 = \frac{\nu}{i\omega_e} \frac{\partial^2 \bar{U}_2}{\partial r^2} \quad \text{near } r = R_o \text{ and } r = R_i. \tag{12}$$

Similarly, near the bottom wall, it is assumed that:

$$\frac{\partial^2 \bar{U}_2}{\partial z^2} \gg \frac{\partial^2 \bar{U}_2}{\partial r^2}, \quad \frac{1}{r^2} \frac{\partial^2 \bar{U}_2}{\partial \theta^2};$$

therefore,

$$\bar{U}_2 = \frac{\nu}{i\omega_e} \frac{\partial^2 \bar{U}_2}{\partial z^2} \quad \text{near } z = -h. \tag{13}$$

Resolution of Eqs. (9), (10), (12) and (13) provides an analytical solution for $\bar{u}_2$ in terms of the known potential function $\tilde{\phi}$.

2.3. Energy dissipation

The expression for energy dissipation in a viscous liquid can be shown to be as [4]:

$$\frac{dE}{dt} = \mu\left[\int_v |\nabla\times\overline{u}|^2\, dv + \int_s (\overline{n}\cdot\nabla |\overline{u}|^2\, ds - 2\int_s \overline{n}\cdot\overline{u}\times(\nabla\times\overline{u})\, ds\right]$$

where: μ = absolute viscosity;

v = liquid volume;

s = area of liquid boundary;

$\overline{n}$ = unit vector normal to the boundary.

Setting $\overline{u} = \nabla\tilde{\phi} + \overline{u}_2$, cosidering $\nabla\times\nabla\tilde{\phi} = 0$ (irrotational flow) and $\overline{u} = 0$ at the boundaries,

$$\frac{dE}{dt} = \mu\left[\int_v |\nabla\times\overline{u}_2|^2\, dv + \int_{s=F_s} \overline{n}\cdot\nabla |\nabla\tilde{\phi}|^2\, ds\right].$$

Here F_s represents instantaneous free surface elevation. The integral over a cycle permits calculation of an equivalent damping ratio η.

Figure 2 shows typical variation of the damping ratio with excitation amplitude at two different frequencies (f_ℓ = natural frequency of liquid, M_ℓ = mass of liquid, M = total mass of the oscillating system). For $f_e/f_\ell = 1.23$, the damping ratio remains essentially constant until a step jump at $\epsilon_o/d \approx 0.145$. Note, with a further increase in ϵ_o/d the damping ratio diminishes, however, with a reduction in the excitation amplitude enormous dissipation levels can be attained. The change in excitation frequency ($f_e/f_\ell = 1.48$) only affected location of the step without substantially changing the overall behaviour. A few experimental data points indicated on the chart also show the same trend.

3. Experimental Determination of Damping Ratio and Energy Dissipation

To evolve a rational test program, a dimensional analysis of the variables was carried out which led to twenty five π numbers: dimensionless amplitude, dimensionless liquid height, Reynolds number (Re), Weber number, Froude number, etc. Damping characteristics were assessed through the use of two criteria: the damping ratio η and the dimensionless energy dissipation parameter.

To assess effectiveness of the damper during nutational motion, a static stand was designed. Essentially it consisted of a 1.2 m swing arm pivoted by a ball bearing at one end and supported by two springs whose stiffness and locations can be changed so as to provide a desired range of natural frequency. A family of thirty dampers was designed to carry out the parametric study. The details of some of the models tested are given in Table 1.

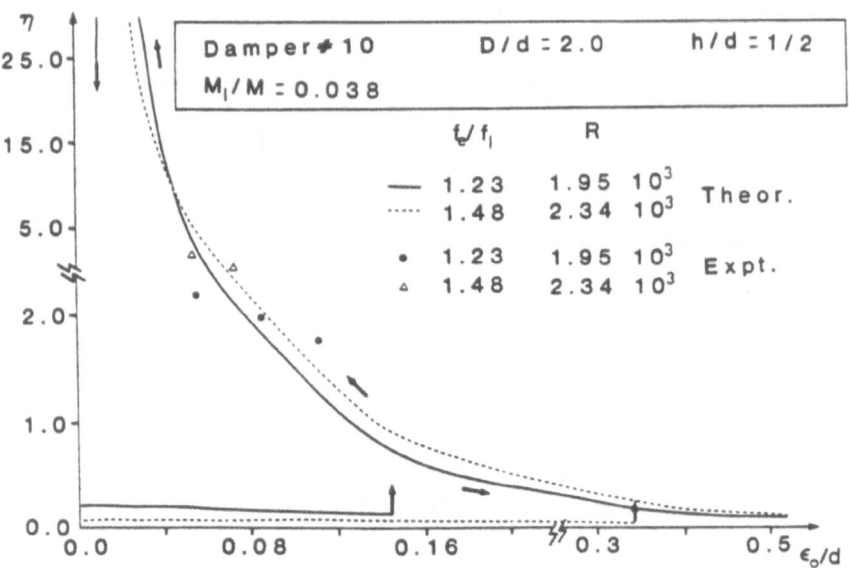

Figure 2: Theoretically predicted variation of the damping ratio with ampli-
tude of two excitation frequencies for model 10. A sample of
corresponding experimental results is also indicated. $M_1/M = 0.038$

3.1. Damper characteristics

The amount of information collected through a planned variation of the system
parameters is rather extensive; however, for conciseness, only a sample of data
suggesting trends is presented here.

3.1.1. Effect of height/section diameter (h/d) ratio

Damping ratio η and energy dissipation rate $\overline{E}_{a,f}$ are highly dependent on the
liquid height. For the circular cross-section damper (damper 1 in Fig. 5),
optimum value of the liquid height parameter leading to a maximum η was found
to be in the range 3/4 - 1. The parameters $\rho d^3 f^2/\sigma$ (Weber number) and $f^2 d/g$
(Froude number) have only a slight effect on this optimum value. However, $\overline{E}_{a,f}$
has a tendancy to be maximum for low values of h/d, typically equal to 1/8. In
other words, the increase in damping η with h/d, as observed in Figs. 3a and 3b,
is at a cost of an increase in the amount of liquid.

The square cross-section damper (Damper 2) showed similar trends at rela-
tively high values of $\rho d^2 f/\mu$, $\rho d^3 f^2/\sigma$ and $f^2 d/g$ (i.e., higher frequencies), where
η reached a maximum for h/d close to 1 (Fig. 3c). However, at lower values of

Table 1:   *Details of some of the damper models used in the test program*

Damper	Dimensions		Description				
	d (cm)	D (cm)	Internal Configuration	Cross-Section	Material	Weight (gms)	Capacity (m)
1	2.4	33.1	plain	circular	rubber	914	422
2	3.5	15.9	plain	square	plexiglas	1710	640
7	2.6	6.65	plain	circular	plastic	26	167
7A	3.0	8.3	plain	circular	plastic	27	184.1
7B	2.8	7.6	plain	circular	plastic	18	157.6
7C	2.4	5.5	plain	circular	plastic	12	78
8	1.9	28.9	plain	circular	copper	842	235
9	1.55	23.6	plain	circular	polyethylene	215	136
10	2.8	5.6	plain	square	plexiglas	50	140
11	2.8	5.6	baffles	square	plexiglas	72	140
12	2.8	5.6	perforated tube	square	plexiglas	80	126
16	3.5	15.9	perforated tube	square	plexiglas	1740	635
17	3.5	15.9	baffles	square	plexiglas	1710	640
18	3.5	15.9	middle layer	square	plexiglas	1771	589
19	3.5	15.9	pieces of wood in flow	square	plexiglas	1710	640

$\rho d^3 f^2/\sigma$ and $f^2 d/g$ (lower frequencies), η and $\overline{E}_{a,f}$ are both optimal for h/d around 1/8 (Fig. 3d). It should be mentioned that experiments with viscous engine oils (i.e., very small $\rho d^2 f/\mu$) showed similar trends, suggesting that the optimal damping as a function of h/d is essentially unaffected by the $\rho d^2 f/\mu$ parameter.

In summary, there is a value of h/d for which the damping is optimal. It depends on $\rho d^3 f^2/\sigma$, $f^2 d/g$ (frequency), D/d and cross-sectional shape (geometry). This value is typically in the range 3/4 - 1, although it is shifted towards the lower end in some cases due to reduced contribution of the top sur-face at lower frequency and liquid height. Energy dissipation rate was found to be generally maximum at low h/d, typically $< 1/8$, when there is high contact surface area per unit volume of water.

The experimental program systematically assessed effects of the other π-numbers. The results are recorded and discussed at length in reference [5]. Some of the more important conclusions may be summarized as follows:

(i) In general, at higher Reynolds numbers, an increase in amplitude of motion results in higher energy dissipation rate due to the corresponding increase in the turbulence level. Although higher turbulence level is beneficial, viscous forces are more efficient in dissipating energy. Thus lower values of $\rho d^2 f/\mu$ is desirable. Recognizing that f cannot be controlled and variation in d would

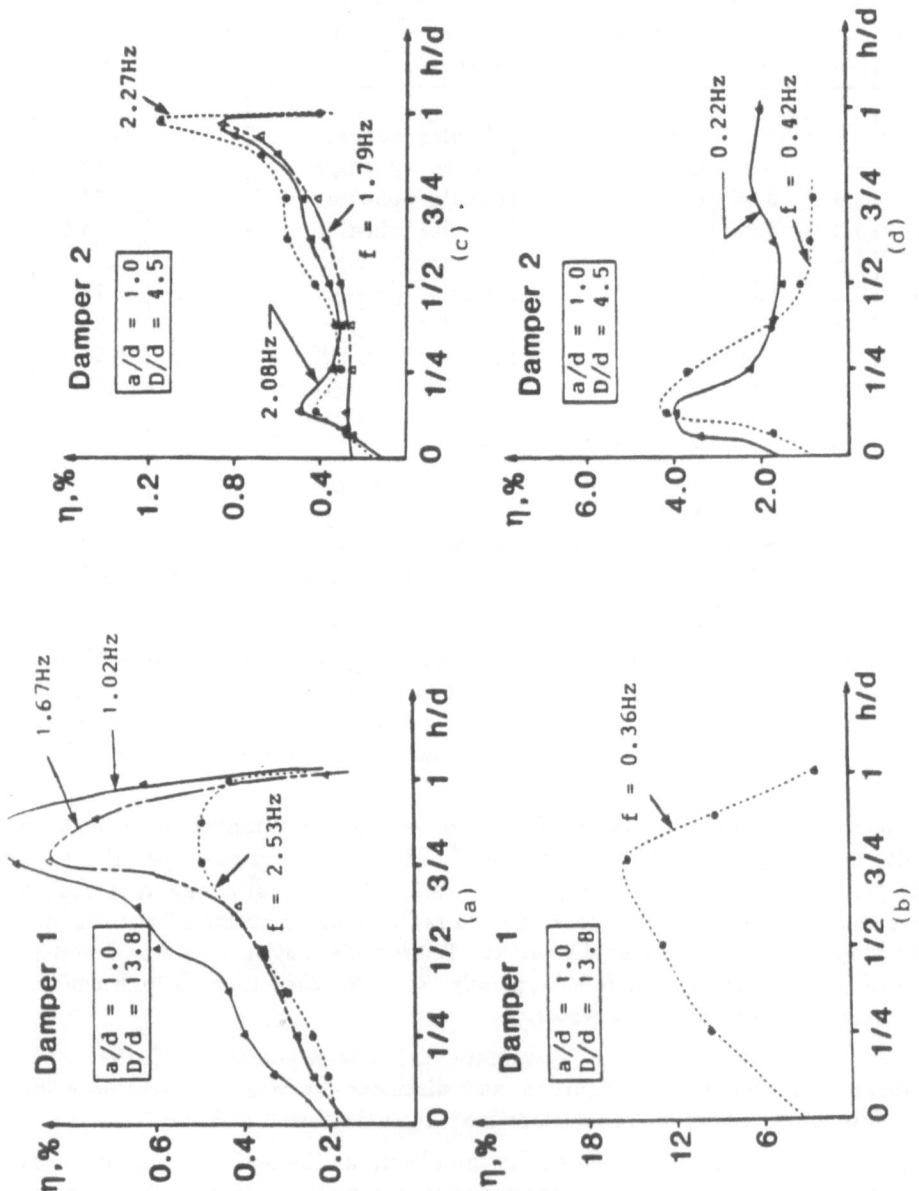

Figure 3: Variation of the damping ratio with the liquid height

affect several other dimensionless parameters, the easiest way to realize this condition is through the use of high kinematic viscosity fluid.

(ii) Damping is primarily dependent on h/d (liquid height), D/d (damper geometry), and f^2d/g (frequency of oscillations) parameters. There is a relationship between optimal damping and these three parameters that corresponds to the situation when the liquid frequency reaches resonance. The information should prove useful in the damper design when the structural frequency of oscillation is known. For instance, low structural frequencies imply use of dampers with high D/d ratio for a given h/d.

3.1.2. Modified internal configurations

A large number of tests aimed at assessing the influence of modifications in the internal geometry on the energy dissipation process were undertaken. They included introduction of: (i) surface roughness; (ii) spheres in flow; (iii) screens; (iv) perforated tube; (v) baffles; (vi) horizontal partition; (vii) floating pieces of wood. Some of the more promising results are summarized here (Fig. 4). In the high frequency range, perforated inside tube device and floating pieces of wood exhibit highest peak values of η, while baffles show more consistent improvement over the entire range of h/d. The two layer damper is not competitive at high frequency, but is most successful in the low frequency range ($f = 0.22\ Hz$). As can be expected, the energy dissipation rate $\overline{E}_{a,f}$ showed peak values in the low range of h/d where this class of nutation dampers are most efficient, whereas the figure emphasizes high values of h/d where damping ratios are maximum. Thus, design of a damper will be a compromise dictated by the structural frequency together with the physical constraint represented by D/d. The ultimate objective would be to minimize the damper mass and yet attain the liquid resonance for optimal damping.

4. Wind Induced Instability Studies

Effectiveness of the dampers in controlling vortex resonance and galloping instabilities for a family of two and three dimensional models was studied in both laminar and turbulent flows. A closed circuit laminar flow wind tunnel with a test-section of $0.69 \times 0.91 \times 2.44\ m$ is able to produce a stable flow with velocity ranging from 0.3 - 30 m/s and turbulence level of less than 0.1%. Turbulent flow tests were conducted in a boundary layer tunnel ($1.58 \times 2.44 \times 24.4\ m$) with 20.74 m of roughness board installed in the test-section upstream of the model to produce desired boundary layer thickness and turbulence intensity.

The dampers were mounted on a velocity of bluff body aerodynamic models during the tests. The models had either circular or square cross-section. The two-dimensional models essentially spanned the height of the tunnel (0.69 m) while the models of finite span simulated three dimensional conditions (Table 2).

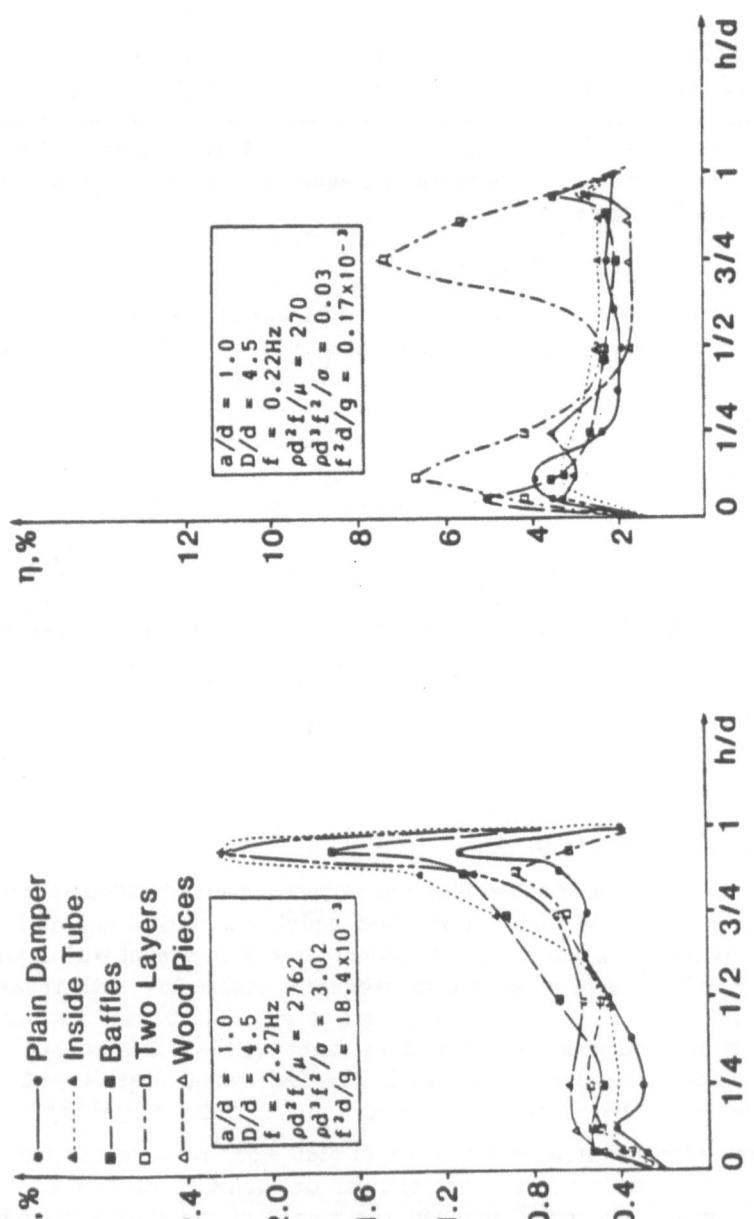

Figure 4: Comparative performance of several more promising damper configurations

Table 2: Physical description of aerodynamic models

MODEL	I, 2-D	II, 2-D	III, 3-D	IV, 3-D	V, 3-D	VI, 2-D
CROSS-SECTIONS	○ 76 mm	□ 32 mm	□ 51 mm	□ 51 mm	□ 102 mm	○ 102 mm
LENGTH (mm)	686	686	483	705	483	686
MATERIAL	ALUMINUM	BALSA	BALSA	BALSA	BALSA	BALSA
MASS (g)	383	234	253	*VARIABLE	376	124

MODEL	VII, 2-D	VIII, 2-D	IX, 2-D	X, 3-D	XI, 3-D	XII, 3-D
CROSS-SECTIONS	○ 102 mm	□ 102 mm	□ 51 mm	○ 102 mm	□ 102 mm	□ 51 mm
LENGTH (mm)	673	673	673	508	508	508
MATERIAL	PCV	BALSA & ALUMINUM	BALSA & ALUMINUM	PCV	BALSA & ALUMINUM	BALSA & ALUMINUM
MASS (g)	786	745	349	644	515	244

4.1. Response during vortex resonance

The aerodynamic instability for a circular cylinder exposed to a fluid stream is due to vortex shedding. In this set of tests, Models VII and X were used in conjunction with several dampers at different natural frequencies of the model and system mass parameter $M/L_m \rho_a d_m^2$ (M = total system mass, L_m = model length, ρ_a = air density, d_m = model diameter or width). Typical results are shown in Figs. 5, 6, 7. Here U, the dimensionless free stream velocity is defined as $V_\infty/\omega d_m$ where V_∞ represents the free stream velocity and ω is the system's natural frequency.

Figure 5a shows the effect of system frequency in relation to the liquid natural frequency with and without the damping liquid (water in all cases). In the absence of nutation damping, the dimensionless amplitude Y (y/model width) at resonance is around 0.6 when the frequency ratio is 0.52 and reduces to around 0.32 as the frequency ratio increases to 1.94. Note, even the plain nutation damper (#10, $D = 56$ mm, $d = 26$ mm) appears to be quite effective. As suggested by the discussion in Sec. 3, energy dissipation can be expected to be maximum when the excitation frequency coincides with the liquid natural frequency ($f_e/f_l = 1$). Although for the results presented here this condition is not satisfied, the nutation damper continues to be quite successful reducing the resonance response by around 40% at the frequency ratio of 0.52 and a dramatic reduction by 96% at $f_e/f_l = 1.94$. As can be expected, at $f_e/f_l \approx 1$, the damper is effective in virtually suppressing the oscillations irrespective of its geometry (damper models 10, 11, 12, Fig. 5b).

As can be expected, the damper diameter ratio D/d has a significant effect on the energy dissipation. Obviously this is reflected in the dynamical response of the models during vortex resonance as shown in Fig. 6. Two cases of relatively low (1 Hz, Fig. 6a) and high (2.5 Hz, Fig. 6b) system frequency are considered. Response of the system in absence of the damper is also included to facilitate comparison.

At the low system frequency, the damper with $D/d = 15.2$ is relatively more effective compared to that with $D/d = 2$. However, the behaviour is reversed at the higher system frequency (Fig. 6b). The governing parameter again appears to be f_e/f_l. Ideally, one would like it to be unity. However, when that condition is not satisfied, it was observed that:

(i) dampers with $f_e/f_l > 1$, in general, perform better than that with $f_e/f_l < 1$, perhaps because of the liquid lagging behind the excitation and thus resisting motion (Fig. 6a);

(ii) as f_e/f_l approaches 1, the damper performance improves (Fig. 6b).

Figure 7 presents results of a two-dimensional test to assess the effect of liquid height in the damper. Note, by an appropriate selection of h/d, the dynamical response during vortex resonance can be limited to any desired value. Thus in practical application to industrial aerodynamics problems, there is

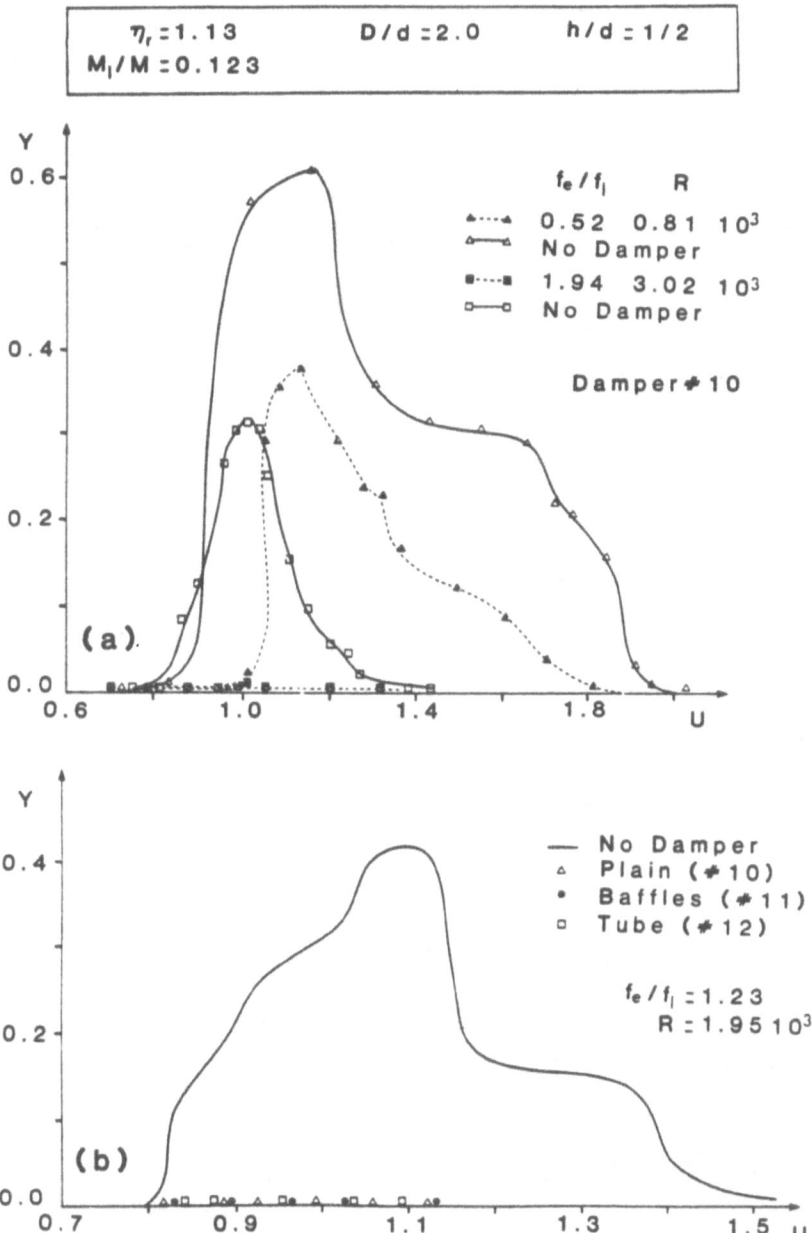

Figure 5: *Vortex resonance response of a three dimensional model showing the effect of: (a) system frequency, (b) damper configurations*

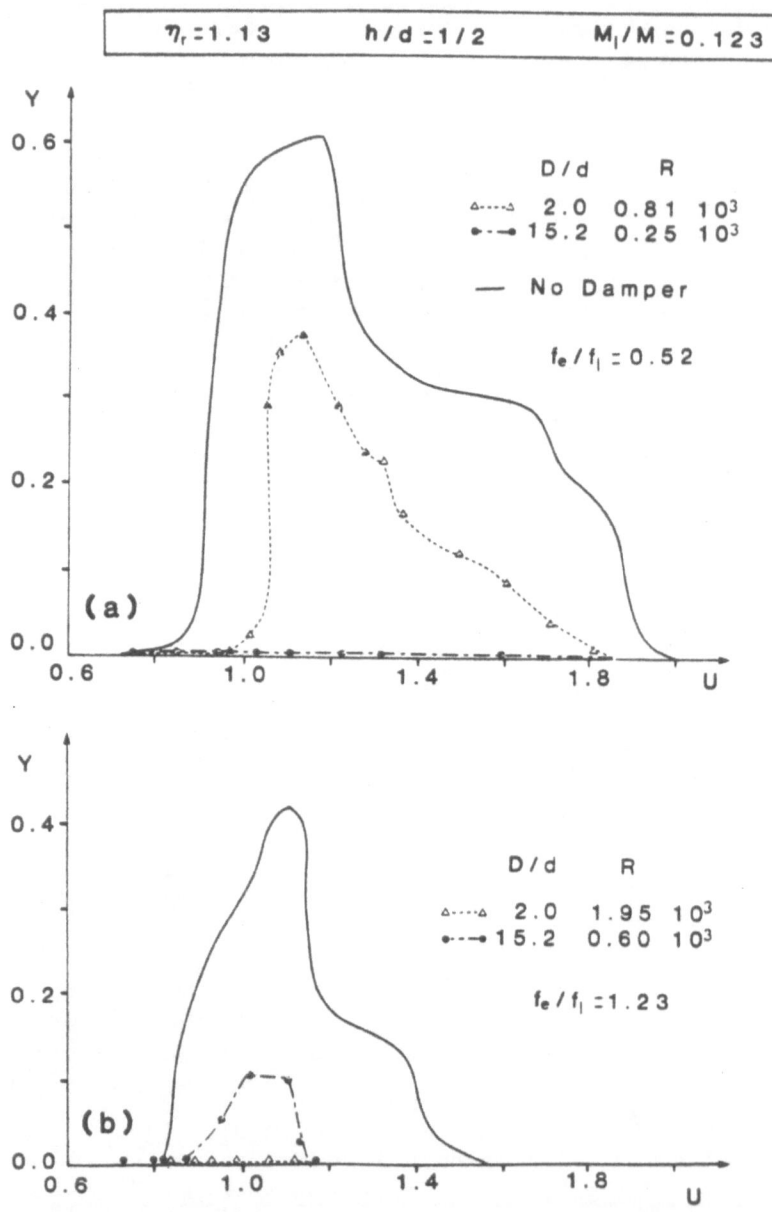

Figure 6: *Effect of the damper diameter ratio D/d during vortex resonance of a three dimensional model: (a) low system frequency of 1 Hz; (b) high system frequency of 2.5 Hz*

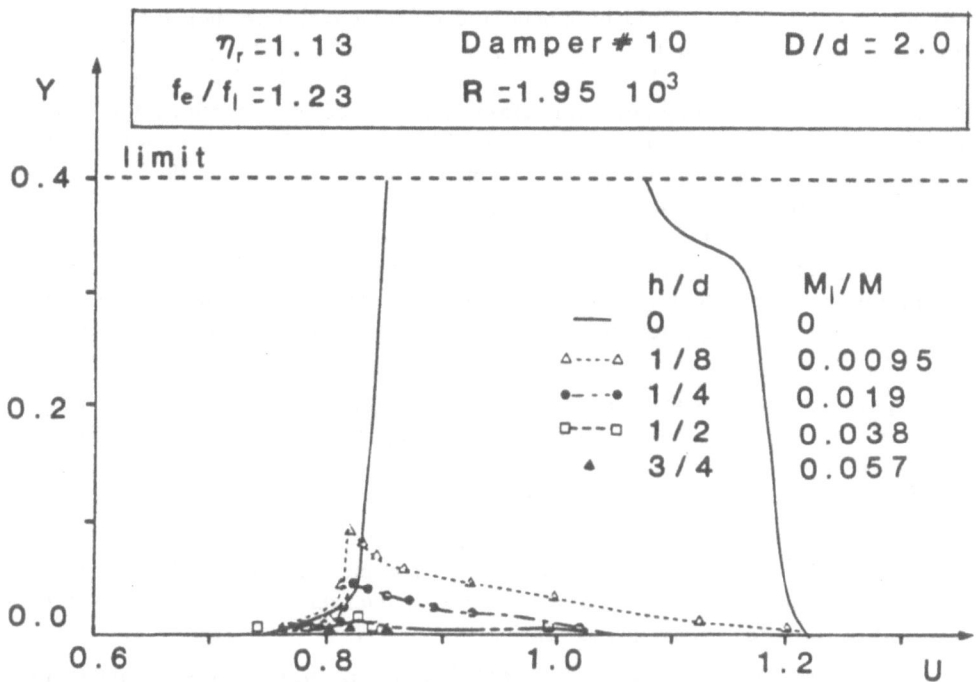

Figure 7: Influence of liquid height on vortex resonance response of a two-dimensional model

considerable freedom in designing an effective damper that would perform close to $f_e/f_c = 1$ through a suitable combination of D/d and h/d.

4.2. Galloping response

Galloping represents a type of self-excited vibration. Geometrically bluff sections may exhibit galloping type of oscillations because of the nature of the fluid forces and moments. If the aerodynamic loading shows an increase with model attitude and the amount of resulting energy input by the fluid forces exceeds that dissipated through various forms of damping, then oscillations would continue to grow until a net energy balance is established.

For square prisms, galloping occupies a more important position than vortex resonance so far as the aerodynamic instabilities are concerned. Effectiveness of nutation dampers in controlling galloping motion is illustrated in Figs. 8 - 11. In general, results show the dampers to be remarkably effective even with their modest size and h/d as small as 1/8.

Typical data showing the effect of the frequency ratio, f_e/f_c, and damper configurations are presented in Fig. 8. Note, in absence of the damping fluid, the classical vortex resonance peak appears to merge with the onset of galloping at around $U = 1.2$. This should be considered coincidental here, being governed by the mass-damping parameter. However, with the nutation damper the resonance peak can be identified followed by a hysteresis loop indicating two equilibrium positions. In this region, equilibrium state of the model jumps from a low level to a high value when subjected to an external disturbance. In absence of a nutation damper, large amplitude oscillations set in which quickly exceed the limit of the test arrangement at $U \approx 2$. However, the damper (model 10) is able to delay the onset of instability to U as high as 4 (Fig. 8a, $f_e/f_l \approx 1$). As in the vortex resonance case, ratio of the excitation frequency to the liquid frequency continues to be the dominant parameter. Note by an appropriate combination of the system variables, critical speed for the onset of galloping can be further delayed to $U > 12$ (Fig. 9)!

Figure 8b attempts to assess influence of modifications in the damper's internal geometry. It is apparent that for $h/d = 1/2$, each damper configuration contributes positively, however, the plain damper (i.e., the one without any internal modifications) proves to be most promising in raising the critical wind speed.

Figure 9 indicates effects of the liquid height parameter and the damper diameter ratio. The trends in both cases are as expected. For the system parameters chosen, an increase in h/d to 1/2 and a decrease of D/d to 2 leads to a favourable value for the frequency ratio.

The mass damping parameter (reduced damping, η_r) is proportional to product of the damping ratio and inertia of the system, both of which tend to resist motion. Figure 10a shows galloping response for models VIII and IX, representing two different values of the reduced damping, with the plain damper model 10. As anticipated, for larger value of $\eta_r = 11.7$, galloping amplitude remained negligibly small throughout the tunnel speed range.

It has been reported in the literature that presence of end gaps between the model and the tunnel walls affects two dimensional character of the flow along the model span. To assess gap effects on galloping, the wind tunnel model VIII was fitted with two end plates. As expected it improved spanwise coherence of the fluctuating loading leading to a lower value for the critical U (Fig. 10b).

FIgure 11 attempts to simulate three dimensional character of engineering structures commonly encountered in practice. Using the aerodynamic model XI in conjunction with dampers 9 and 10, it studies the effect of liquid height and damper diameter ratio on the galloping response during the condition of three dimensional flow. The general trend continues to be the same as that observed earlier in Fig. 9. As explained before relative effectiveness of h/d and D/d continues to be governed by the favourable value of f_e/f_l.

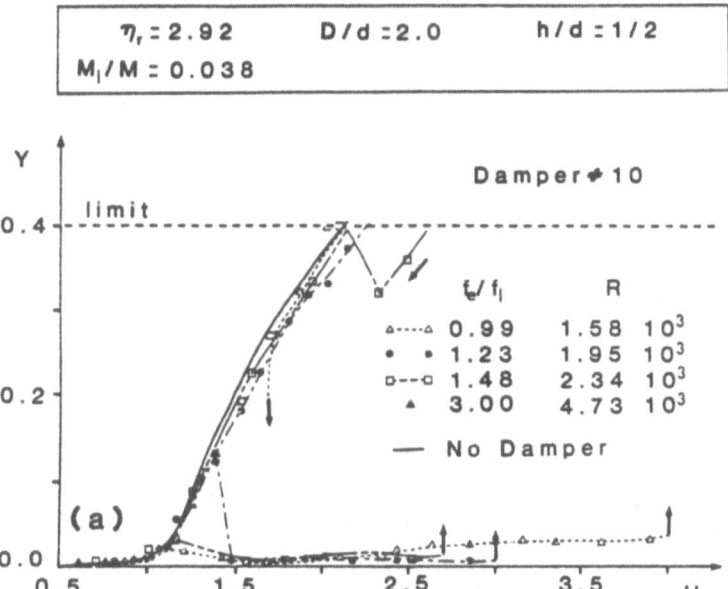

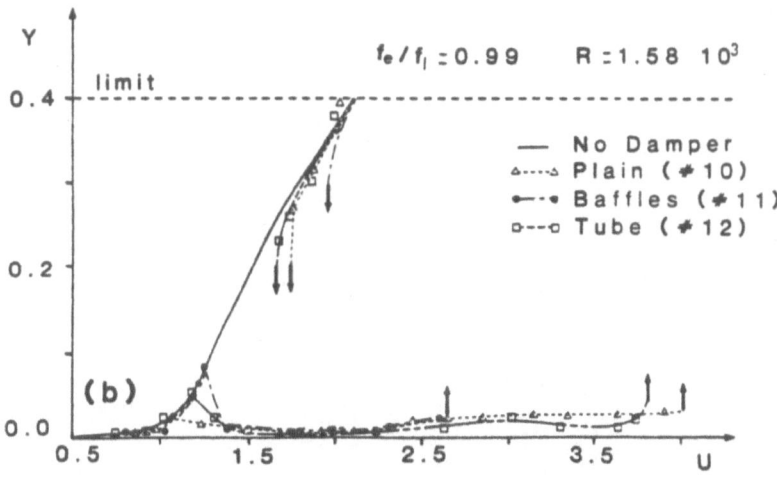

Figure 8: Galloping response of a two dimensional square cyclinder (model VIII) showing the effect of: (a) frequency ratio f_e/f_i; (b) damper configurations

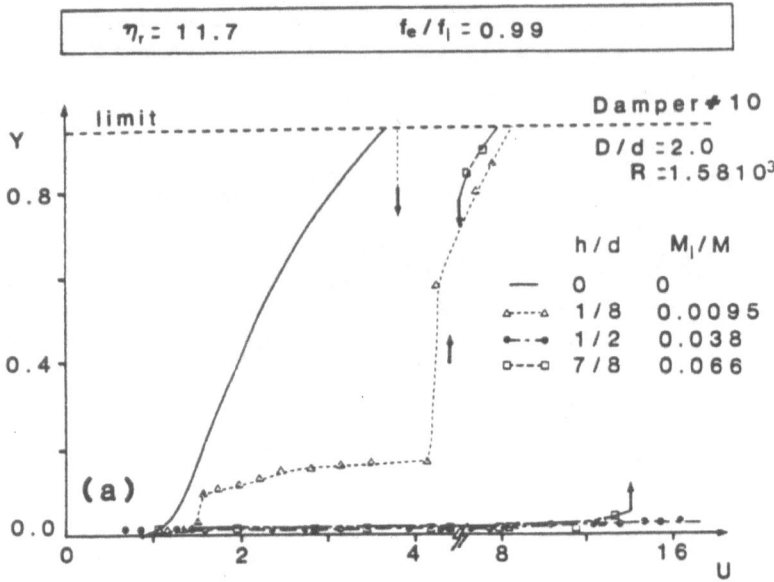

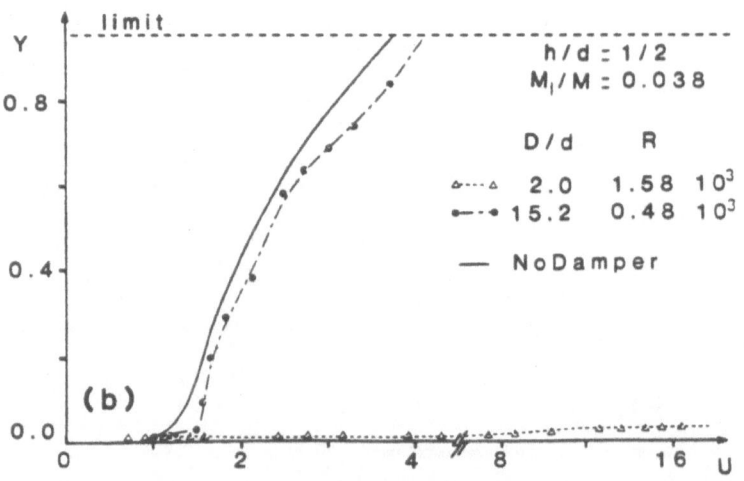

Figure 9: Effect of liquid height and damper diameter ratio on the galloping response of a two-dimensional model (model IX)

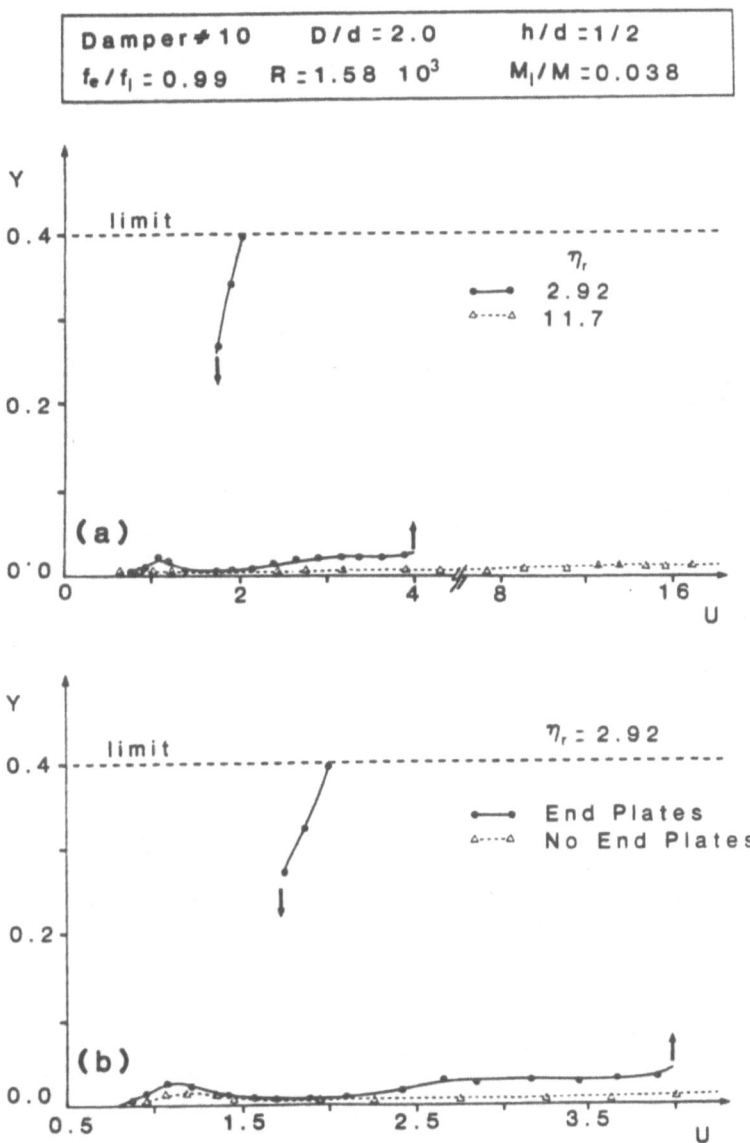

Figure 10: *Influence the mass-damping parameter and end-plates on the galloping of two dimensional models*

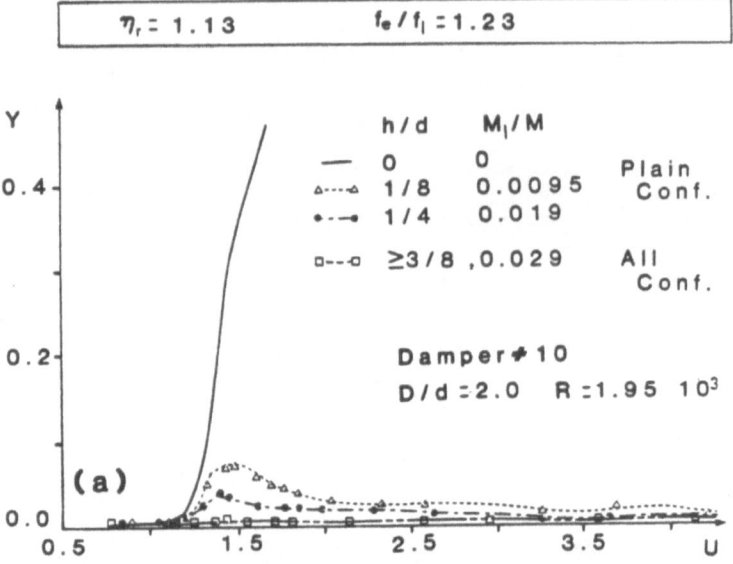

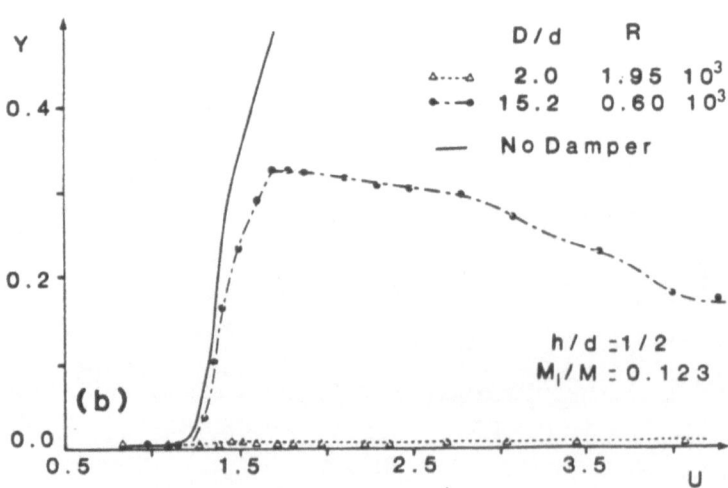

Figure 11: Galloping response of a three-dimensional model. Effect of: (a) h/d; (b) D/d

5. Concluding Remarks

The experimental program has attempted to assess the influence of more significant parameters on the energy dissipation process. The amount of information
obtained is rather extensive and would be useful in the design of efficient
dampers for a wide range of applications. The proposed nutation dampers
represent a useful concept in minimizing the effects of fluid induced instabilities.
They showed encouraging performance during both vortex induced and self-
excited galloping type of oscillations. The experiments also demonstrated effectiveness of the dampers for either two dimensional plunging motion or pivoted
rigid body rotation for three dimensional models. The simple analytical model
appears promising in that it suggests the right trends. However, experimental
measurement of damping during forced excitation is necessary to assess its validity. Such an experiment is in the planning stage.

Acknowledgement

This investigation was supported by the Natural Sciences and Engineering
Research Council of Canada, grant No. A-2181.

References

[1] Cermak, J.E., "Application of Fluid Mechanics to Wind Engineering," Free-
     man Scholar Lecture, *Transactions of ASME, Journal of Fluid Engineer-
     ing*, Vol. 97, No. 1, March 1975, pp. 9-38.

[2] Zdravkovich, M.M., "Review and Assessment of Effectiveness of Various
     Aero and Hydromechanic Means for Suppressing Vortex Shedding," *Proceed-
     ings of the 4th Colloquium on Industrial Aerodynamics*, Aachen, West
     Germany, Part 2, Editors: C. Kramer et al., 1980, pp. 29-46.

[3] Case, K.M. and Parkinson, W.C., "Damping of Surface Waves in an
     Incompressible Liquid," *Journal of Fluid Mechanics*, Vol. 2, Part 2, March
     1957, pp. 172-184.

[4] Lamb, H., *Hydrodynamics*, Dover Publication Inc., 1945, pp. 581.

[5] Welt, F., "A Parametric Study of Nutation Dampers," M.A.Sc. Thesis,
     University of British Columbia, July 1983.

A MESS-LIKE COMPENSATOR
FOR ACTIVE CONTROL
OF SELECTED MODES
OF A LARGE SPACE STRUCTURE

A. Nayak

HR Textron Inc.

2485 McCabe Way

Irvine, California 92714

1. Introduction

While controlling a large structure, such as a future space structure, it is prudent to control only certain modes for various reasons. Most of the time, only a few modes contribute the most to performance degradation. Thus, only these modes, henceforth called controlled modes, need to be controlled. One needs to control these modes carefully. If it is not done properly, control spillover and observation spillover can cause the closed-loop system to be unstable.

The Model Error Sensitivity Suppression (MESS) technique deals effectively with the control spillover and the observation spillover problem as described by Sesak and Likins in Reference [1]. They formulate an LQG problem in which the minimization criterion contains the control and the observation spillover terms. Thus, the resultant compensator tries to suppress the control and observation spillover. This method is conceptually easy but needs the solution of Riccati equations for the control gains and the filter gains. This can be formidable if the dimensions are high.

To circumvent the problem of solving Riccati equations in the LQG approach, researchers have tried to devise simple output feedback types of control. In this control, a feedback gain is devised which tries to give desirable performance characteristics to the closed loop system (see Canavin, Reference [2],

Lin et al, Reference [3], Elliott, et al, Reference [4]).

Our research tries to combine these two techniques and is motivated by the problem of controlling the VPI pendulous plan grid test article with five velocity sensors and five actuators. The test article is a combination of highly flexible aluminum grid beams, which hang from a horizontal, steel, top-beam assembly supported in nearly frictionless bearing. Details of its design and theoretical modeling are given in References [5] and [6].

The section on Preliminaries sets up the general framework for the problem, and the sections on Control and Estimation Laws develop MESS-like control and estimation laws. The Closed Loop Analysis section derives the sufficiency condition for the stability of the closed loop system. The sections on Collocated Actuators and Sensors and Simulation Results contain the numerical results for the pendulous plane grid.

2. Preliminaries

The dynamics of the pendulous plane grid are assumed to be given by

$$[M]\ddot{\mathbf{x}} + [D]\dot{\mathbf{x}} + [K]\mathbf{x} = [B]\mathbf{u}, \tag{1}$$

$$\dot{\mathbf{y}} = [C]\dot{\mathbf{x}}. \tag{2}$$

$[M]$, $[D]$, and $[K]$ are mass, damping and stiffness matrices, respectively. $\mathbf{x}$ represents the vector of physical displacements of the nodes. $[B]$ and $[C]$ are the control and the observation matrices, respectively. $\dot{\mathbf{y}}$ is the observation vector, and $\mathbf{u}$ is the control vector.

Equation (2) represents the fact that only velocity sensors are used.

The transformation between modal coordinates, $\mathbf{q}$, and physical coordinates is

$$\mathbf{x} = [\phi]\mathbf{q}, \tag{3}$$

which has the following properties:

$$[\phi^T][M][\phi] = [I], \tag{4}$$

$$[\phi^T][K][\phi] = [\omega^2], \tag{5}$$

$[I]$ is an identity matrix. The columns of $[\phi]$ represent the mode shapes, and $[\omega^2]$ is a diagonal matrix whose diagonal elements are squares of the natural angular frequencies of the modes.

Using the above transformation, Eqs. (1) and (2) can be written as follows:

$$\ddot{\mathbf{q}} + [2\varsigma\omega]\dot{\mathbf{q}} + [\omega^2]\mathbf{q} = [\phi^T][B]\mathbf{u}, \tag{6}$$

$$\dot{\mathbf{y}} = [C][\phi]\dot{\mathbf{q}}, \tag{7}$$

where $[2\varsigma\omega] = [\phi^T][D][\phi]$ and is assumed to be diagonal. ς is the damping ratio.

Define

$$[\phi_A] \triangleq [\phi^T][B], \tag{8}$$

and

$$[\phi_o] \triangleq [C][\phi]. \tag{9}$$

Equations (6) and (7) can be written in terms of the modes to be controlled and the residual modes (or left over modes)

$$\ddot{q}_c + [2\varsigma\omega]_c\dot{q}_c + [\omega^2]_c q_c = [\phi_A]_c u, \tag{10}$$

$$\ddot{q}_\gamma + [2\varsigma\omega]_\gamma\dot{q}_\gamma + [\omega^2]_\gamma q_\gamma = [\phi_A]_\gamma u, \tag{11}$$

$$\dot{y} = [\phi_o]_c\dot{q}_c + [\phi_o]_\gamma\dot{q}_\gamma. \tag{12}$$

The subscripts c and γ refer to the modes that are to be Controlled and to the Residual modes. Thus, q_c refers to the modal coordinates of the modes to be controlled and q_γ refers to the modal coordinates of the residual modes. The matrices $[2\varsigma\omega]$, $[\omega^2]$, $[\phi_A]$, and $[\phi_o]$ are partitioned accordingly.

The purpose of the control u is to induce active damping in controlled modes, i.e.,

$$[\phi_A]u = -[2\varsigma_A\omega]_c\dot{q}_c, \tag{13}$$

where $[2\varsigma_A\omega]_c$ is a diagonal matrix, and ς_A is the desired active damping ratio.

Thus, the purpose here is to generate control u which is as close to Eq. (13) as possible and based on $\dot{y}$, the measurements.

3. Control Law

The following form for the control will be assumed:

$$u = -[K_c]\dot{q}_c. \tag{14}$$

The control gain $[K_c]$ should have the following property:

$$[\phi_A]_c[K_c] = [2\varsigma_A\omega]_c. \tag{15}$$

This will satisfy the Eq. (13), the control requirement.

The control forces, will "spillover" into the residual modes. Thus, to suppress that, (this is the MESS principle, Reference [1]), $[K_c]$ should be

$$[\phi_A]_\gamma[K_c] = [0]. \tag{16}$$

Combining Eqs. (15) and (16) produces

$$[\phi_A][K_c] = \left[\frac{[2\varsigma_A\omega]_c}{[0]} \right]. \tag{17}$$

$[K_c]$ can be obtained as a least square solution of Eq. (17) as:

$$[K_c] = \left\{ [\phi_A^T][\phi_A] \right\}^{-1} [\phi_A^T] \left[\frac{[2\varsigma_A \omega]_c}{[0]} \right]. \tag{18}$$

It is assumed that the rank of ϕ_A is equal to the dimension of the control u.

The above control law may be obtained by the Linear Quadratic Regulator method by proper choices of weighting matrices, as suggested in Reference [7].

4. Estimation Law

The following form for the estimate will be assumed:

$$\hat{\dot{q}}_c = [K_E]\dot{y}, \tag{19}$$

$\hat{\dot{q}}_c$ is the estimate of $\dot{q}_c$.

Substituting Eq. (12) for $\dot{y}$, one obtains

$$\hat{\dot{q}}_c = [K_E][\phi_o]_c \dot{q}_c + [K_E][\phi_o]_\gamma \dot{q}_\gamma. \tag{20}$$

For a good estimate $[K_E]$ should be such that

$$[K_E][\phi_o]_c = [I], \tag{21}$$

where $[I]$ is an identify matrix.

The term $[K_E][\phi_o]_\gamma$ represents the observation spillover, and to suppress that a MESS-like approach will be used, i.e., one needs,

$$[K_E][\phi_o]_\gamma = [0] \tag{22}$$

Combining Eqs. (21) and (22), one gets

$$[K_E][\phi_o] = \left[[I] \vdots [0] \right]. \tag{23}$$

$[K_E]$ can be obtained as a least square solution of Eq. (23) as:

$$[K_E] = \left[[I] \vdots [0] \right] [\phi_o^T] \left\{ [\phi_o][\phi_o^T] \right\}^{-1}. \tag{24}$$

It is assumed that the rank of $[\phi_o]$ is equal to the dimension of the observations $\dot{y}$.

5. Closed Loop Analysis

From the implementation point of view, the control u will be generated by

$$u = -[K_c]\hat{\dot{q}}_c \tag{25}$$

where $[K_c]$ is obtained from Eq. (18) and $\hat{\dot{q}}_c$ is obtained from Eqs. (19) and (24).

Substituting Eqs. (25), (19), (7) and (8) in Eq. (6), produces

$$\ddot{q} + [2\varsigma\omega]\dot{q} + [\omega^2]q = -[\phi_A][K_c][K_E][\phi_o]\dot{q}. \tag{26}$$

Define

$$[2\varsigma\omega]_I \triangleq [\phi_A][K_c][K_E][\phi_o]. \tag{27}$$

The matrix $[2\varsigma\omega]_I$ will dictate the stability of the closed loop system. If the matrix,

$$[2\varsigma\omega]_M = [2\varsigma\omega] + [2\varsigma\omega]_I \tag{28}$$

is positive definite, the closed loop system will be stable. For details see Appendix A.

In the case of collocated actuators and sensors, closed loop stability is assured for the above control and estimation laws. This is shown in the next section.

When actuators and sensors are not collocated, i.e., $[\phi_A^T] \neq [\phi_o]$, stability condition may not be satisfied. The system may be stable in two ways; (i) change the locations of actuators and sensors, and/or (ii) change $[K_c]$ and $[K_E]$ in the following way:

$$[K_c] = \left\{ [\phi_A^T][Q_c][\phi_A] \right\}^{-1} [\phi_A^T][Q_c]\left[\frac{[2\varsigma_A\omega]}{[0]} \right], \tag{29}$$

and

$$[K_E] = \left\{ [I] \vdots [0] \right\} [Q_E][\phi_o]^T \left\{ [\phi_o][Q_E][\phi_o^T] \right\}^{-1}. \tag{30}$$

Equations (29) and (30) are generalized pseudo inverse solutions of Eqs. (17) and (23), respectively. The arbitrary matrices $[Q_c]$ and $[Q_E]$ must be positive-- definite and symmetric. They can be varied such that $[2\varsigma\omega]_M$ is a positive definite matrix, thus stabilizing the closed loop system.

6. Collocated Actuators and Sensors

Collocation of actuators and sensors implies

$$[\phi_o] = [\phi_A^T]. \tag{31}$$

Simplifications occur in this special case. Equation (24) now can be written as

$$[K_E] = \left[[I] \,\vdots\, [0] \right] [\phi_A] \left\{ [\phi_A^T][\phi_A] \right\}^{-1}. \tag{32}$$

Equation (27) can now be written using Eqs. (31), (32) and (18) as:

$$[2\varsigma\omega]_I = [\phi_A] \left\{ [\phi_A^T][\phi_A] \right\}^{-1} [\phi_A^T] \left[\begin{array}{c} [2\varsigma_A\omega]_C \\ [0] \end{array} \right] \left[[I] \,\vdots\, [0] \right] [\phi_A] x \left\{ [\phi_A^T][\phi_A] \right\}^{-1} [\phi_A^T]. \tag{33}$$

$[2\varsigma\omega]_I$ can be seen to be a positive semidefinite matrix. Since $[2\varsigma\omega]$ is a positive definite matrix, $[2\varsigma\omega]_M$ is a positive definite matrix. Thus, the closed loop system is stable.

The implication of $[2\varsigma\omega]_I$ being positive semidefinite is that almost all the modes of the system can get active damping. Thus, the closed loop system is more robust than the open loop system.

7. Numerical Analysis

The above problem of collocated actuators and sensors was numerically simulated using the data provided by VPI in Reference [8]. Five actuator-sensor pairs are placed at node locations 1, 4, 10, 13 and 22 of the dynamic model of the test article. The first 20 modes are used to model the dynamics of the test article. It is desired to provide active damping to the first six modes.

Figure 1 shows the schematic of the closed loop system.

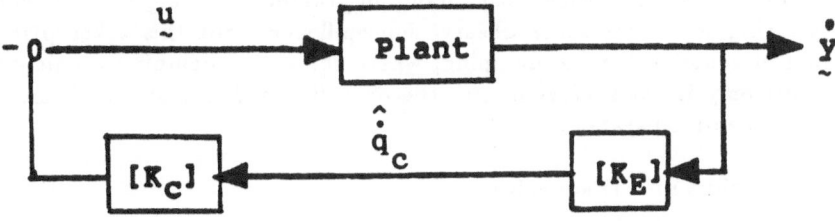

Figure 1

The control gain matrix $[K_c]$ and the estimation gain matrix $[K_E]$ were obtained using Eqs. (18) and (32), respectively.

Table 1 contains the open-loop eigenvalues and closed-loop eigenvalues for two cases. The results confirm the analytical observation that collocated actuators and sensors induce active damping in almost all system modes.

Table 1: System performance

OPEN LOOP		CASE A*				CASE B*			
		CLOSED LOOP NO ESTIMATOR		CLOSED LOOP WITH ESTIMATOR		CLOSED LOOP NO ESTIMATOR		CLOSED LOOP WITH ESTIMATOR	
Frequency (Hz)	ς	Frequency (Hz)	ς	Frequency (Hz)	ς	Frequency (Hz)	ς	Frequency (Hz)	ς
.58385	.043	.5848	.077	.5856	.109	.5849	.077	.5887	.153
.8898	.05	.8898	.081	.8897	.077	.8899	.081	.8881	.084
1.3731	.035	1.3724	.104	1.3712	.1115	1.3734	.104	1.370	.137
3.2792	.01	3.2877	.0563	3.2859	.0354	3.3197	.102	3.3118	.0595
3.593	.009	3.582	.0556	3.5840	.0363	3.5469	.103	3.555	.0618
4.9572	.013	4.955	.0599	4.9557	.03667	4.950	.107	4.9526	.0594
5.4848	.0035	5.4848	.0035	5.4846	.004	5.485	.0035	5.484	.0044
5.6535	.003	5.6535	.003	5.6535	.0032	5.654	.003	5.6533	.0033
6.1941	.006	6.1941	.006	6.1938	.0085	6.1942	.006	6.193	.0105
7.9967	.005	7.9966	.005	7.9966	.0077	7.9966	.005	7.997	.0102
8.3988	.005	8.3988	.005	8.3970	.0098	8.3988	.005	8.3912	.0139
9.393	.005	9.393	.005	9.390	.0113	9.3930	.005	9.386	.0175
9.8065	.005	9.8065	.005	9.806	.0058	9.8065	.005	9.8052	.0065
11.634	.005	11.634	.005	11.634	.0059	11.634	.005	11.633	.0065
13.273	.005	13.273	.005	13.272	.0073	13.273	.005	13.27	.0095
20.556	.005	20.556	.005	20.556	.0050	20.556	.005	20.556	.0050
24.730	.005	24.729	.005	24.729	.0053	24.729	.005	24.73	.0054
26.959	.005	26.959	.005	26.958	.0054	26.959	.005	26.958	.0058
28.856	.005	28.854	.005	28.865	.0054	28.855	.005	28.85	.0057
30.748	.005	30.747	.005	30.747	.0052	30.747	.005	30.747	.0054

*
Case A: Desired active damping ratio in first six modes (.1, .1, .1, .1, .1, .1)
Case B: Desired active damping ratio in first six modes (.1, .1, .1, .2, .2, .2)

Active damping is provided only to the modes to be controlled when the controller is used without the estimator. When the estimator is also used, almost all the residual modes get some active damping, thereby showing that the compensator is trying to suppress the observation spillover. But this takes place at the expense of controlled mode damping, where the active damping predicted by the controller only is not achieved. But the overall closed-loop system is more robust than the open-loop system.

8. Conclusion and Discussion

MESS-like control and estimation laws are developed to control the first six modes of the VPI pendulous plane grid. A general case has been analyzed to study the stability of the closed-loop system. If the closed-loop system is found to be unstable, that can be remedied in two ways: (i) change the actuator and sensor locations, and/or (ii) change the control gain and/or estimates gain using generalized pseudoinverses.

It has been shown that for collocated actuators and sensors, almost all modes - controlled and residual - can get some active damping; thus, the closed-system is always stable. The numerical simulation confirms this result.

The estimator developed here does not have any dynamics. This reduces complexity. There is no guarantee of closed-loop stability in the general situation. However, the system will be always stable when actuators and sensors are collocated. A Kalman filter may be advisable, but it will introduce estimator dynamics and complicate implementation somewhat.

Acknowledgements

The author wants to thank Prof. Hallauer of VPI and R. Quartararo of HR Textron for their suggestions and help.

References

[1] Sesak, J. R. and Likins, P. W., "Model Error Sensitivity Suppression: Quasi-Static Optimal Control for Flexible Structures," Proc. of 18th IEEE Conference on Decision and Control, Fort Lauderdale, FL., Dec. 1979.

[2] Canavin, J. R., "The Control of Spacecraft Vibrations Using Multivariable Output Feedback," Proc. of AIAA/AAS Astrodynamics Conference, Palo Alto, CA., Aug. 1978.

[3] Elliott, L. E., Mingori, D. L., and Iwens, R. P., "Performance of Robust Output Feedback Controller for Flexible Spacecraft," Proc. of the 2nd VPI & SU/AIAA Symposium on Dynamics and Control of Large Flexible Spacecraft, Editor Meirovitch, June 1979.

[4] Lin, J. G., Hegg, D. R., Lin, Y. H., and Keat, J. E., "Output Feedback Control of Large Space Structures: An Investigation of Four Design Methods," Proc. of 2nd VPI & SU/AIAA Symposium on Dynamics and Control of Large Flexible Spacecraft, Editor Meirovitch, June 1979.

[5] Masse, M. A., "A Plane Grillage Model for Structural Dynamics Experiments: Design, Theoretical Analysis, and Experimental Testing," M.S. Thesis, Virginia Polytechnic Institute and State University, 1983.

[6] Gehling, R. N., "Experimental and Theoretical Analysis of a Plane Grillage Structure with High Modal Density," M.D. Thesis, Virginia Polytechnic Institute and State University, 1984.

[7] Nayak, A., "Analysis of the VPI Active Structural Damping Experiment," HR Textron Report, July 18, 1984.

[8] Numerical modal data provided to HR Textron by VPI in April 1984, 56 degree of freedom modal.

[9]   Hefner, R. D. and Hallman, W. P., "Space Structure Control via a Frequency-Shaped KTC Approach," paper presented at AAS/AIAA Astrodynamics Specialist Conference, Lake Tahoe, Nevada, 1981.

APPENDIX A

Consider the linear time-invariant system

$$[A]\ddot{\mathbf{x}} + [D]\dot{\mathbf{x}} + [G]\dot{\mathbf{x}} + [K]\mathbf{x} = 0, \tag{A.1}$$

where $\mathbf{x}$ is a vector, $[A]$ and $[D]$ are symmetric, positive definite matrices, $[K]$ is a symmetric matrix and $[G]$ is a skew symmetric matrix.

The null solution of (A.1) will be asymptotically stable if all of the eigenvalues of $[K]$ are positive and unstable if $[K]$ has at least one negative eigenvalue.

This theorem is known as the Kelvin-Tait-Chataev theorem. For more details, see Reference [9].

Note that any matrix can be represented as the sum of a symmetric matrix and a asymmetric matrix, e.g.,

$$[E] = \frac{\left[E + E^T\right]}{2} + \frac{\left[E - E^T\right]}{2}. \tag{A.2}$$

Thus, positive definiteness of the matrix implies positive definiteness of the symmetric part.

FUNDAMENTAL ASPECTS
OF
STRUCTURE CONTROL

H. Öz
The Ohio State University
Department of Aeronautical and Astronautical Engineering
Columbus, Ohio 43210

1. Introduction

Within the last decade control problems of flexible systems have attracted increasingly more interest and importance. By now, the well-known problems associated with computational difficulties, spillover, parameter uncertainties, robustness, high-dimensionality, actuator/sensor numbers and locations have been addressed by numerous investigators [1-4]. According to the way controls are designed, control techniques have been classified such as coupled-control (CC) and Independent Modal-Space Control (IMSC) and comparisons have been made regarding their advantages and disadvantages [5]. Mostly linear state feedback approaches dominated the field by utilizing pole-allocation techniques and linear optimal control theory, although output feedback schemes also apppeared rather frequently. The highlight of the difficulties has been the control and observation spillover phenomena which invited a host of approaches to eliminate or minimize their effects. Almost across the board, whatever the control technique, the emphasis in controlling structural systems has been on control theories and approaches and on how the principles of these theories could be applied properly to the structural system. However, an important point has been and is being overlooked: it is that the controlled system, being a structurally dynamic system, does constitute a separate branch of science which has a particular discipline in its own right with its deep-rooted foundation in the ages-old "Principles of Mechanics." In controlling such a physical system, which requires interdisciplinary attention, it is imperative to make sure not only how the principles of control theory are applied properly but also how the principles of mechanics are to be paid due respect and not violated by the control approach. If a control approach violates the principles of structural mechanics, regardless of how well the principles of its own discipline are served, the desired response of the structure will never be obtained. The ultimate objective of a control problem is to elicit a desired response. The desired response finds its raison d'etre in the satisfaction of laws of nature which are expressed as principles of mechanics which in turn are reflected by the structural equations of motion. No response of a structural system (whether it is uncontrolled or controlled) which does not find its origin in these principles can be realized. Any control approach which violates the principles of mechanics will be an ill-posed problem. The major theme of this paper is to review typical control approaches and establish whether the structural control problem is well-posed from the structural mechanics viewpoint.

2. Review of Basic Principles of Mechanics

In 1874 Gustav Robert Kirchoff gave the definition of mechanics. He then said [6] "Mechanics is the science of motion; its task is to describe completely (and accurately), in the simplest manner the motions occurring in nature " under whatever loads may be acting within and/or on the boundaries of a system.

To mathematically define a system, loads must be defined in terms of the displacements, derivatives of the displacements (both spatial and time), and independent variables (space and time). These loads may be natural such as aerodynamic, gravitational and inertia forces or they may be applied by a control system. A motion can be analyzed (response problem) or synthesized (control problem). Definition of loads in terms of displacements and their derivatives can be known directly through observation of and experiments with natural phenomena (such as viscous friction forces) or can be known indirectly by deriving the equations of motion by techniques such as Hamilton's Law (for example, Newton's second law, which can be obtained from Hamilton's Law, is a description of loads in terms of displacements). Some loads can be deliberately designed as known functions of displacements and their derivatives. Such is the case in a feedback control problem. The *displacements* of a system are *essential quantities* and the *loads* on that system are *derived quantities of mechanics*. The essential and derived quantities have a bonding feature in that for every load at a point, there exists a compatible displacement. The compatibility is determined by the fact that the product of a load and its associated displacement must yield an energy term [7-8]. How the essential and derived quantities interact with each other is always governed by the *laws of nature*. The natural laws are completely expressed by *the fundamental postulate of mechanics* [9]: "A particle, a system of particles and/or a continuum will follow a path and/or assume a configuration for which the time integral of the total work of the system is a minimum for whatever forces (whatever these forces may be functions of) that may act on or within the system [10]." The forces include applied and inertia forces. The applied forces may certainly include control forces. Nature will supply those forces in the form of inertia forces and/or internal forces, that will cause a system to simultaneously follow a path or assume a configuration of least work and be in equilibrium, regardless of the prescribed forces or of the other natural forces that may act on the system. Mathematical expressions of this postulate are represented in the forms such as "Hamilton's Law" [9-11] and the "principle of virtual work" [12-13]. Starting from these mathematical forms in which embedded is the fundamental postulate one can derive the "rules of the game" of the motion of a structural system. To recall how the laws of nature demand and are paid due respect in the formulation of structural problems, we may consider the bending motion of a beam and derive descriptions of loads in terms of displacements. The result is [14]:

$$\frac{\partial}{\partial x}\left[k'GA\left(\frac{\partial u}{\partial x} - \psi\right)\right] - m\frac{\partial^2 u}{\partial t^2} + f(x,t) = 0, \tag{2.1a}$$

$$0 < x < L$$

$$\frac{\partial}{\partial x}\left(EI\frac{\partial \psi}{\partial x}\right) + k'GA\left(\frac{\partial u}{\partial x} - \psi\right) - k^2 m\frac{\partial^2 \psi}{\partial t^2} = 0, \tag{2.1b}$$

with the boundary conditions (B.C.):

$$\left(EI\frac{\partial \psi}{\partial x}\right)\delta\psi\Big|_0^L = 0, \tag{2.2a}$$

$$k'GA\left(\frac{\partial u}{\partial x} - \psi\right)\delta u\Big|_0^L = 0. \tag{2.2b}$$

$u(x,t)$, $\psi(x,t)$ are the translational and angular displacements due to bending. EI, $k'GA$ are the flexural and shear rigidities, respectively, k is a radius of gyration and m and f are the mass and externally applied load densities over the domain $0 < x < L$. To accentuate the implications of natural laws both shear and rotatory effects have been included. Equations (2.1-2) are manifestations of the working of natural laws and can be obtained via Hamilton's Law or the principle of virtual work. Based on this, one of the first principles of mechanics as reflected by Eq. (2.1-2) is that: at the points of a *given structure* where applied loads (derived quantities) are prescribed one cannot prescribe apriori what the corresponding compatible displacements (essential quantities) will be at those points; conversely, at the points where displacements are specified one cannot prescribe apriori what the corresponding compatible loads will be at those points. If the loads (displacements) are specified at some points then the corresponding displacements (loads) have to be what they have to be to satisfy the natural laws regardless of what one may wish them to be. Formally, they can be found by

solving Eqs. (2.1-2) viz, they will be dictated on the system by nature which
requires equilibrium of all loads (inertial, elastic, external or control) on or within
the system. Alternately, they can be found directly by using Hamilton's Law
[11]. This stated principle of mechanics is implicit in the B.C.s, Eqs. (2.2). Equation (2.2a) means that either the derived quantity $EI\,\partial\psi/\partial x$ = bending
moment (the load) or the variation of the rotation $\delta\psi$ (associated compatible
essential quantity) must vanish at both ends. If the rotation is specified or
known (through observation of geometric conditions at the ends) at either or
both ends, $\delta\psi = 0$ and then the associated moment load $EI\,\partial\psi/\partial x$ at that point
is unknown and cannot be specified in advance arbitrarily. It can only be found
after one solves for ψ explicity from Eqs. (2.1) which are the statements of equilibrium of forces and moments in nature. Conversely, if the bending moment
$EI\,\partial\psi/\partial x$ vanishes (known through the observation of the nature of supports)
at either or both ends, then the corresponding $\delta\psi$ is unknown and does not have
to be zero. $\delta\psi$ has to be what it has to be as the solution of Eqs. (2.1) will
require. In any case the B.C. Eq. (2.2a) will be satisfied. Similarly, Eq. (2.2b)
tells us that either the derived quantity $k'GA(\partial u/\partial x - \psi)$ = shear force or the
variation δu of the compatible displacement must vanish at each end. In other
words, only the shear force or the displacement is known (prescribed) at either
end, the remaining quantity must be found via Eqs. (2.1). Note that the B.C.s
Eqs. (2.2) have the units of energy. Hence the natural laws are at work at the
boundaries by allowing apriori knowledge or specification of either a derived
quantity or a compatible essential quantity, but not of both. In structural
mechanics, along with the information obtained by observing the nature of supports, in order to state the correct B.C.s, one has to make sure via Eqs. (2.2) that
the aforementioned fundamentals are not violated so that the problem is mechanically well-posed and a meaningful solution can be obtained. The fundamentals
are also at work within the interior domain $0 < x < L$ through Eqs. (2.1).
Indeed, for a given structure one can either specify the derived quantity $f(x,t) =$
load distribution in Eqs. (2.1) and has to accept whatever the essential quantity
displacement u has to be so that the equilibrium of loads (differential equations)
are satisfied at every point (response problem) or one can specify the displacement u and has to discover what the load distribution $f(x,t)$ has to be so that
the equilibrium of loads (differential equations) will be satisfied while realizing
exactly the specified displacement (control problem). But one must realize that
both $[f(x,t), u(x,t)]$ can never be specified simultaneously at every point or over
subdomains of the structure and expect the natural balances of loads to be
insured at every point or over those subdomains. Similar statements can also be
made for the rotation $\psi(x,t)$ and the external moment distribution. Hence:
*simultaneous arbitrary specification of both the essential quantities (displace-
ments) and the compatible derived quantities (loads) at a point, at some points
or at every point of a structural system violates the natural laws of mechanics,
constitutes an overspecification of the structure dynamically and therefore is
not allowable.* Otherwise, if overspecifications occur, the system dynamics will be

altered by nature inspite of the specified quantities so as to create an event that is fully in accordance with the natural laws of mechanics. However, a subtle point must not be misunderstood: controlling the response of a structural system will not constitute an ability to change the working of natural laws of mechanics. On the contrary, the control system itself together with the original system cannot result in an aggregate system which violates the principles of mechanics; the closed-loop system will be under the jurisdiction of nature also. No control objective which violates the laws of mechanics can take place in reality. If such controls are contemplated and applied to the system, the laws of mechanics will dominate to yield a process different from the objectives of the control design. Nature will always manage to produce a process that insures compatibility with its laws. This is where the phenomenon known as control spillover will find its origin.

Traditionally, to control a given structural system, the problem is posed by specifying certain control objectives most of which can be translated into a desired displacement behavior (essential quantity). To accomplish this, one also specifies the load (control forces and/or moments) distribution, (derived quantities) well in advance by assuming a number of spatially discretized (such as pointwise) inputs. However, from the point of view of principles of mechanics, some relevant questions arise: are the specified control objectives (which can be translated into specified (desired) behavior of essential quantities) and the prespecified derived quantities (by the virtue of the assumed spatially discrete input distribution) consistent with the laws of nature? May the problem have been overspecified, underspecified, or specified properly in regards to the principles of mechanics? From this perspective, if the control problem is not well formulated, the attempted solution will not be correct or it can at best be regarded as an approximation to the desired behavior of the essential quantities. Starting in Sec. 4 we shall address these questions. However, before then, in Sec. 3, we shall review an alternative approach to structural dynamics. To this end, for the remainder of the paper, for simplicity, we shall disregard the shear and rotatory effects reducing Eqs. (2.1-2) to:

$$\frac{\partial}{\partial x^2}\left[EI\frac{\partial^2 u}{\partial x^2}\right] + m\frac{\partial^2 u}{\partial t^2} = f(x,t), \tag{2.3a}$$

$$EI\frac{\partial^2 u}{\partial x^2}\delta\left(\frac{\partial u}{\partial x}\right)\Big|_0^L = 0, \tag{2.3b}$$

and in particular, we shall consider a pinned-pinned beam with uniform properties. The observed (geometric) B.C.s are $u(0,t) = u(L,t) = 0$. The remaining B.C.s are obtained from Eq. (2.3b) in the form $EI\,\partial^2 u/\partial x^2 = 0$, $x = 0, L$ since for a pinned-pinned beam the variation of the rotation $\delta\psi/\delta(\partial u/\partial x)$ at the end points cannot be known apriori.

3. Integral Formulation of Structural Dynamic Systems

In this section, we shall present an alternate formulation of the problems of structural mechanics which brings the importance of and the relation between the essential and derived quantities to a sharper focus. The formulation is the integral formulation of system dynamics via the influence function concept of structural mechanics. The interested reader is referred to Ref. [15] on this subject. Let us consider the simply supported beam of Fig. 1.

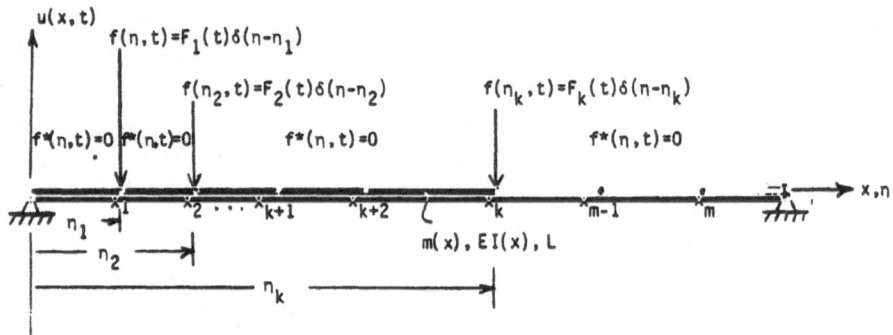

Figure 1: A flexible structure with a typical control force distribution

The displacement $u(x,t)$ at any point x due to a distributed load $f'(\eta,t)$ at η can be written in the form:

$$u(x,t) = \int_0^L C(x,\eta)f'(\eta,t)\,d\eta, \tag{3.1}$$

in which the function $C(x,\eta)$ is known as the influence function which gives the displacement at x due to a unit load at η. The function $C(x,\eta)$ is a characteristic of the structural system and it depends on the geometry, material properties and the boundary conditions of the structural system. The derivation of the influence functions $C(x,\eta)$ already takes into account the equilibrium of loads acting on and within the system along with the B.C.s and therefore Eq. (3.1) constitutes a complete and accurate formulation from the viewpoint of structural principles [15]. We shall find it more convenient to use Eq. (3.1) instead of Eqs. (2.3) to illustrate our points. Eq. (3.1) is valid for static and dynamic problems of mechanics. In a dynamic setup the load distribution $f'(\eta,t)$ on the structure consists of:

$$f'(\eta,t) = -m(\eta)\ddot{u}(\eta,t) + f(\eta,t), \qquad 0 < \eta < L, \tag{3.2}$$

where the first and the second terms on the right are the inertia and external loads, respectively, acting at η. Using Eq. (3.2) in Eq. (3.1) and rearranging we have an alternate formulation of the structural dynamic problem:

$$u(x,t) + \int_0^L C(x,\eta)m(\eta)\ddot{u}(\eta,t)\,d\eta = \int_0^L C(x,\eta)f(\eta,t)\,d\eta, \qquad (3.3)$$

in which $C(x,\eta)$ is assumed known. We shall consider the external load distribution in the form $f(\eta,t) = \sum_{j=1} F_j \delta(\eta - \eta_j)$ in which δ is the Dirac delta function. If the external load is continuously distributed then $j \to \infty$ otherwise for finite j one has a spatially discrete load distribution. In addition, we shall concentrate on a number of points, x_1, x_2, ..., x_k, x_{k+1}, ..., x_m (m may be infinite). First let us consider a static case: according to Eq. (3.3) the displacements at $x_i(i = 1,...,m)$ and the loads at $\eta_j(j = 1,....)$ are related by:

$$
\begin{bmatrix} u_1 \\ u_2 \\ u_k \\ u_{k+1} \\ \cdot \\ \cdot \\ \cdot \\ u_m \end{bmatrix}
=
\begin{bmatrix}
c_{11} & c_{12} & c_{1k} & c_{1,k+1} & \ldots & c_{1m} & c_{1,m+1} & c_{1\infty} \\
c_{21} & c_{22} & c_{2k} & c_{2,k+1} & \ldots & c_{2m} & c_{2,m+1} & c_{2\infty} \\
c_{k1} & c_{k2} & c_{kk} & c_{k,k+1} & \cdots & c_{km} & c_{k,m+1} & c_{k\infty} \\
c_{k+1,1} & c_{k+1,2} & c_{k+1,k} & c_{k+1,k+1} & \cdots & c_{k+1,m} & c_{k+1,m+1} & c_{k+1\infty} \\
\cdot & \cdot & \cdot & \cdot & & \cdot & \cdot & \cdot \\
\cdot & \cdot & \cdot & \cdot & & \cdot & \cdot & \cdot \\
\cdot & \cdot & \cdot & \cdot & & \cdot & \cdot & \cdot \\
c_{m,1} & c_{m,2} & c_{mk} & c_{m,k+1} & \cdots & c_{mm} & c_{m,m+1} & c_{m\infty}
\end{bmatrix}
\begin{bmatrix} F_1 \\ F_2 \\ F_k \\ F_{k+1} \\ \cdot \\ \cdot \\ F_m \\ F_{m+1} \\ F_\infty \end{bmatrix}, \qquad (3.4)
$$

in which $u_i = u(x_i)$ and $c_{ij} = C(x_i,\eta_j)$ represents an influence coefficient. We arranged the elements of the force vector such that F_1, F_2, ..., F_m correspond to the displacements at x_1, x_2, ..., x_m. Let us assume now that $f(\eta,t)$ is discrete in-space such that F_{m+1}, F_{m+2}, ..., F_∞ are all zero. Equation (3.4) reduces to:

$$
\begin{bmatrix} \mathbf{u}_{1,k} \\ \mathbf{u}_{k+1,m} \end{bmatrix}
=
\begin{bmatrix} C_{11} & C_{12} \\ C_{21} & C_{22} \end{bmatrix}
\begin{bmatrix} \mathbf{F}_{1,k} \\ \mathbf{F}_{k+1,m} \end{bmatrix}, \quad or \quad \mathbf{u} = C\mathbf{F}, \qquad (3.5)
$$

in which C_{11} etc. are the corresponding submatrix partitions in Eqs. (3.4) and $\mathbf{u}_{1,k}^T = [u_1, u_2, ..., u_k]$, $\mathbf{F}_{1,k}^T = [F_1, F_2, ..., F_k]$, $\mathbf{u}_{k+1,m}$ and $\mathbf{F}_{k+1,m}$ are defined similarly. C is the complete matrix of influence coefficients. We note that u_i and $F_i(i = 1, ..., m)$ are associated essential and derived quantities of mechanics, respectively. Furthermore, Eq. (3.5) reflects the relationship between these quantities compatibly from the point of view of natural laws. In accordance with its principles, in Eqs. (3.5) one can specify apriori all of the components of $\mathbf{F}^{*T} = [F_1^*, F_2^*, ..., F_m^*]$ and find the displacements $u_1, u_2, ..., u_m$ at the corresponding points (static response problem) i.e., $\mathbf{u} = C\mathbf{F}^*$. An (*) denotes prescribed quantities. Alternately, one can specify apriori a set of desired displacements at all points $u_i = u_i^*(i = 1, ..., m)$ and find the required set of forces $\mathbf{F}$ by an inversion process:

$$\mathbf{F} = C^{-1}\mathbf{u} = K\mathbf{u}^*, \tag{3.6}$$

in which K is the matrix of stiffness coefficients. For a specified u, Eq. (3.6) is the solution of a *static shape control problem*. Note that the components of **F** have to be as Eq. (3.5) requires them to be consistent with the natural equilibrium laws. One can also specify parts of the vectors **F** and u such that $\mathbf{F}_{k+1,m} = \mathbf{F}_{k+1,m}^*$ and $\mathbf{u}_{1,k} = \mathbf{u}_{1,k}^*$ while the associated compatible quantities $\mathbf{u}_{k+1,m}$ and $\mathbf{F}_{1,k}$ are left as unknowns. In this case Eqs. (3.5) appear in the form:

$$\begin{bmatrix} \mathbf{u}_{1,k}^* \\ \mathbf{u}_{k+1,m} \end{bmatrix} = \begin{bmatrix} C_{11} & C_{12} \\ C_{21} & C_{22} \end{bmatrix} \begin{bmatrix} \mathbf{F}_{1,k} \\ \mathbf{F}_{k+1,m}^* \end{bmatrix}. \tag{3.7}$$

From the upper part of (3.7) one can solve for $\mathbf{F}_{1,k}$ in terms of prescribed quantities $\mathbf{u}_{1,k}^*$ and $\mathbf{F}_{k+1,m}^*$ and then substitute the result into the lower part to solve for $\mathbf{u}_{k+1,m}$ in terms of $\mathbf{u}_{1,k}^*$ and $\mathbf{F}_{k+1,m}^*$. This way, all prescribed and computed quantities will be compatible with governing natural laws reflected in Eq. (3.5). As an example, one may consider a highly accurate finite element formulation of the mechanics of the beam of Fig. 1 and regard points $i = 1, ..., k, k+1, ..., m$ as the nodal points of the finite elements. A basic feature of obtaining solutions to structural problems by this popular technique is that at the nodal points where displacements are prescribed the corresponding compatible nodal displacements are left as unprescribed in keeping with the principles of mechanics.

The situation is no different in the case of dynamics of structures. The same principles must be adhered to. This time the total displacement at a point is the sum of a *quasi-static displacement* component due to the external applied forces and dynamic component due to inertia loads acting on the system. Quasi-static displacement is defined to be the displacement that would be obtained at a point had all the external load distribution at instant t been frozen as static loading on the structure. The quasi-static loading is given by the value of the integral on the right hand side of Eq. (3.3). Clearly, then from the left of Eq. (3.3) the quasi-static displacement is also the difference of the total displacement and the dynamic displacement. We shall denote the quasi-static displacement by $\bar{u}(x,t)$ and express it as:

$$\bar{u}(x,t) = u(x,t) + \int C(x,\eta)m(\eta)\ddot{u}(\eta)\,d\eta. \tag{3.8}$$

With this notation one may write an equation similar to Eq. (3.1):

$$\bar{u}(x,t) = \int C(x,\eta)f(\eta,t)\,d\eta, \tag{3.9}$$

in which the dynamic nature of the problem is embedded in the characterization of a quasi-static displacement on the left. Considering again a set of points $i = 1, ..., k, k+1, ..., m$, the counterpart of Eqs. (3.5) in the case of dynamics is that:

$$\begin{bmatrix} \bar{u}_{1,k} \\ \bar{u}_{k+1,m} \end{bmatrix} = \begin{bmatrix} C_{11} & C_{12} \\ C_{21} & C_{22} \end{bmatrix} \begin{bmatrix} F_{1,k}(t) \\ F_{k+1,m}(t) \end{bmatrix}, \quad or \quad \bar{u}(t) = CF(t), \text{ for } all \ t > 0, \quad (3.10)$$

in which $\bar{u}_i(t) = u_i(t) + \int C(x_i,\eta)m(\eta)\ddot{u}(\eta,t)\,d\eta$, $0 \le x_i \le L$. Again the principles of mechanics require that if the displacements $\bar{u}(x,t)$ are specified at some points, the associated compatible loads at those points must remain unprescribed and vice versa, at all t. If one has $\bar{u}_{1,k} = \bar{u}_{1,k}^*$ and $F_{k+1,m} = F_{k+1,m}^*$ then Eqs. (3.10) become:

$$\begin{bmatrix} \bar{u}_{1,k}^*(t) \\ \bar{u}_{k+1,m}(t) \end{bmatrix} = \begin{bmatrix} C_{11} & C_{12} \\ C_{21} & C_{22} \end{bmatrix} \begin{bmatrix} F_{1,k}(t) \\ F_{k+1,m}^*(t) \end{bmatrix}. \quad (3.11)$$

As special cases, if all of the components of $F(t)$ are specified on the right hand side, none of the $\bar{u}(t)$ can be specified arbitrarily, but they must be found for given $F(t)$ via Eq. (3.10) (dynamic response problem). Conversely, if all of the elements of $\bar{u}(t)$ are specified on the left hand side of Eq. (3.10), then none of the components of $F(t)$ can be specified arbitrarily, but they must be found, for given $\bar{u}(t)$, from Eq. (3.10) (dynamic control problem) if a meaningful solution is to be obtained from a structural mechanics point of view. Simultaneous arbitrary specification of essential, $\bar{u}_i(t)$, and derived, $F_i(t)$, quantities at a number of or at all points i will constitute an overspecification of the problem mechanically and Eqs. (3.10) will yield an inconsistent set. Equations (3.10) describe facts of life and they must hold regardless of the source of external load distribution $f(\eta,t)$ and no matter what variables $f(\eta,t)$ may be a function of. If $f(\eta,t)$ is generated by a control system, it too together with the resulting displacements must satisfy the principles of mechanics associated with Eqs. (3.10, 3.11). In the next section, we shall address the control problem of the structural system represented by Eqs. (3.10, 3.11) from this perspective.

4. A Common Control Approach to the Structural Dynamic System

To control the response $u(x,t)$ of the system of Fig. 1 and of Eqs. (2.3a) or (3.9,10), it is most practical to consider a modal transformation based on the associated eigenvalue problem. The details are much too familiar to be repeated here. The displacement is represented in the form $u(x,t) = \sum\limits_{r=1}^{\infty} \phi_r(x)\xi_r(t)$ where $\phi_r(x)$ and $\xi_r(t)$ represent an eigenfunction and a modal coordinate, respectively. Next, utilizing the orthogonality of the eigenfunctions, the system is transformed to a set:

$$\ddot{\xi}_r(t) + \omega_r^2\xi_r(t) = \int \phi_r(\eta)f(\eta,t)\,d\eta = f_r(t), \qquad r = 1,...,\infty, \quad (4.1)$$

in which f_r is a modal input. Equations (2.3a) and (4.1) are completely equivalent based on the rigor of mathematics of transform theory, therefore one can consider either one as the analysis warrants without sacrificing mathematical equivalency. The premise of control is based on the fact that by controlling the response of modal coordinates $\xi_r(t)$, it is possible to elicit a desired response $u(x,t)$. Thus instead of Eq. (2.3a) and relevant B.C.s one concentrates on the set (4.1) leading to the concept of so-called Modal Control Theory. Manipulation of responses $\xi_r(t)$ can be done directly by Eqs. (4.1). However, it is customary in control theory to transform Eqs. (4.1) to modal-state-space by introducing the new variable $\eta_r = \omega_r^{-1}\dot{\xi}_r$. For every r one obtains the pair:

$$\begin{bmatrix} \dot{\xi}_r \\ \dot{\eta}_r \end{bmatrix} = \begin{bmatrix} 0 & \omega_r \\ -\omega_r & 0 \end{bmatrix} \begin{bmatrix} \xi_r \\ \eta_r \end{bmatrix} + \begin{bmatrix} 0 \\ \omega_r^{-1}f_r = \omega_r^{-1}\int \phi_r(\eta)f(\eta,t)\,d\eta \end{bmatrix}, \qquad r = 1,2.... . \quad (4.2)$$

To control the distributed parameter system (DPS) via Eqs. (4.2) or (4.1) *the most common FIRST STEP* is to assume the external control input distribution in the form:

$$f(\eta,t) = \sum_{i=1}^{k} F_i(t)\delta(\eta - \eta_i), \qquad 0 \le \eta \le L, \qquad (4.3)$$

in which $\delta(\eta - \eta_i)$ is the Dirac delta function. Equation (4.3) describes a k-point input distribution (Fig. 1). The justification for assuming a point input distribution of the form of Eq. (4.3) is the economical concern that a continuously distributed input cannot be realized so that it is replaced by a finite number of point inputs. The objective is to keep the number of inputs as small as possible as long as the system remains controllable. However, the assumed form of input distribution, Eq. (4.3), is not based on any principles of mechanics and it has nothing to do with the nature of mechanics exhibited by the specific structural system (other than making sure that $F_i(t)$ are not put at particular controlled mode shape nodes). However, Eq. (4.3) has a subtle but a very significant implication from the structural mechanics point of view. A mere gaze at Fig. 1 and a mathematical view of Eq. (4.3) reveal that:

$$f(\eta,t) = 0 = f^*(\eta,t) = \sum_{i=1}^{\infty} 0\delta(\eta - \eta_i), \quad 0 < \eta < L, \quad \eta \neq \eta_1, \eta_2, ..., \eta_k. \quad (4.4)$$

From the principles of mechanics view, it is not so much important that the input has the form of Eq. (4.3) but it is very important that it has the form of Eq. (4.4), because, the assumption of point inputs prescribes (constraints) apriori as zero the force distribution (a derived quantity) at almost infinitely many points of the structural domain before the control problem is solved. On the other hand, only at a small number of points $\eta_1, \eta_2, ..., \eta_k$ the input distributions $F_1(t), F_2(t), ..., F_k(t)$ have not been specified (constrained) as these will be unknowns to be found by the particular control technique. Now, recall the concepts presented in Sec. 3 and regard points $i = 1, 2, ..., k$ as being where point control inputs $F_i(t)(i = 1, 2, ..., k)$ are located and points $i = k+1, ..., m$ as an arbitrary number of points at which there are no control inputs and hence the constraints $F_{k+1}(t) = F_{k+2}(t) = \cdots F_m(t) = 0$. Corresponding to Eqs. (3.10) and (3.11) one has:

$$\mathbf{F}^T = \left[\mathbf{F}_{1,k}^T \; \mathbf{F}_{k+1,m}^T\right] = \left[\mathbf{F}_{1,k}^T \; \mathbf{0}_{k+1,m}^T\right] = \left[\mathbf{F}_{1,k}^T \; \mathbf{F}_{k+1,m}^{*T}\right],$$

and from Eqs. (3.11), whatever the subsequent control theory is, if the solution is to be compatible with the laws of mechanics that will govern the system, it must not arbitrarily specify the essential displacements $\overline{\mathbf{u}}_{k+1,m}$ versus the apriori specified derived quantities $\mathbf{F}_{k+1,m}^*$. This is where we shall be critical of the subsequent treatment of the control problem. The *SECOND STEP* to the control problem which is intimately related to step one is to consider in Eqs. (4.2) a finite number of modes to determine the inputs based on:

$$\dot{\mathbf{w}} = \Lambda\mathbf{w} + B\mathbf{F}^M, \tag{4.5a}$$

$$\mathbf{w} = \left[\xi_1, \eta_1, \xi_2, \; \cdots \; \xi_n, \eta_n\right]^T, \tag{4.5b}$$

$$\mathbf{F}^M = \left[F_1(t), F_2(t), ...F_k(t)\right]^T, \tag{4.5c}$$

in which the elements of Λ and B should be evident by comparison to the first $2n$ of Eqs. (4.2). We shall view Eqs. (4.5) as the *control design model* and the input vector $\mathbf{F}^M$ as found via Eqs. (4.5) will be regarded as the *control model force input*. To denote this distinction a superscript M is used. Without necessarily transforming to the state-space, the control problem can also be handled in the modal configuration space via:

$$\ddot{\xi}(t) + [\omega^2]\xi(t) = B'\mathbf{F}^M, \quad B_{ij} = \phi_i(\eta_j), \quad i = 1,...,n, \quad j = 1,...,k,$$
$$[\omega^2] = \text{diag}[\omega_1^2 \; \omega_2^2 \; \cdots \; \omega_n^2]. \tag{4.6}$$

The object of modern control theory is to design $\mathbf{F}^M$ as $\mathbf{F}^M = \mathbf{F}^M(\mathbf{w}) = \mathbf{F}^M(\xi, \dot{\xi})$ such that the coordinates $\mathbf{w}$ (or ξ) will behave as desired. The control design models, Eqs. (4.5) and (4.6) imply that the distributed displacement is expressed in the form:

$$u(x,t) = \sum_{r=1}^{n}\phi_r(x)\xi(t) = \sum_{r=1}^{n}[\phi_r \; 0]\mathbf{w}_r(t), \quad \mathbf{w}_r = \left[\xi, \eta_r\right]^T. \tag{4.7}$$

The two most common methods of dealing with the control of Eqs. (4.5) are the pole-allocation technique and the linear-optimal control theory. In the first technique, one tries to find a gain matrix G for an assumed linear state feedback $\mathbf{F}^M = G\mathbf{w}$ such that the closed-loop control model $\dot{\mathbf{w}} = (\Lambda + BG)\mathbf{w}$ has desired eigenvalues $\rho[\Lambda_{CL} = \Lambda + BG]$ where $\rho[.]$ denotes the set of eigenvalues of $[.]$. In the second technique, the idea is to find a gain matrix G such that the performance index $J = 1/2 \int_0^{\infty} (\mathbf{w}^T Q\mathbf{w} + \mathbf{F}^{MT}R\mathbf{F}^M) \, dt$ is minimized. The solution to this problem is obtained by solving a steady-state matrix Riccati equation[16]. For given weighting matrices (Q,R) with proper signs and the satisfaction of observability and controllability, the solution G is stabilizing. Corresponding to the optimal gain G the closed-loop control model Λ_{CL} has a set of eigenvalues which can be found readily after the matrix G is computed. The implication is that the linear *optimal control theory itself merely serves as an indirect pole-allocation technique.* In view of this, it is appropriate to describe the modern control objectives unifiedly as "eigenvalue allocation." Satisfaction of the eigenvalues' allocation by a particular technique, such as technique "1" results in a closed-loop matrix $\Lambda_{CL}^{(1)} = \Lambda + BG^{(1)}$ from which for an initial condition $w(0) = w_0$, the modal trajectories, corresponding to technique "1" and assumed input distribution, are indirectly arbitrarily specified in the form: $w^{(1)} = \Phi_{CL}^{(1)} w_0 = w^{*(1)}$. Φ_{CL} is the closed-loop transition matrix. That choosing "eigenvalue allocation" as a control objective, for a given number and locations of inputs, constitutes a purely arbitrary specification of modal trajectories can be seen if it is recalled that the gains that produce a desired set of eigenvalues are not unique [17]. If a different technique such as technique "3" is used for the same eigenvalues of technique "1" a different gain $G^{(3)}$ will be obtained to yield a different $\Phi_{CL}^{(3)}$. Thus, for the same w_0 the new controlled trajectories will be $w^{(3)} = \Phi_{CL}^{(3)} w_0 = w^{*(3)} \neq w^{*(1)}$. This shows the arbitrariness of the specification of modal responses by eigenvalue allocation schemes of control theory. Next, because expansion (4.7) is assumed, the solution w^* of the control problem (4.5) also prescribes the distributed displacement arbitrarily:

$$u(x,t) = \sum_{r=1}^{n} \phi_r(x)\xi_r^*(t) = \sum_{r=1}^{n} \left[\phi_r(x)\ 0\right] w_r^*(t) = u^*(x,t). \qquad (4.8a)$$

This in turn implies that the quasi-static displacement field of the structural system is also prescribed arbitrarily by means of the assumed control technique as:

$$\bar{u}(x,t) = u^*(x,t) + \int C(x,\eta)\eta\ddot{u}^*(\eta,t)\ d\eta = \bar{u}^*(x,t). \qquad (4.8b)$$

In Eq. (4.8b) $\ddot{u}^*(\eta,t)$ has been prescribed according to the response solution $w^*(t)$ of Eqs. (4.5 or 4.6). In view of Eq. (4.8b), the control design model specifies arbitrarily the complete set of quasi-static displacements at every point of the structural domain including points $x_1, x_2, ..., x_m$ on the left of structural dynamic system (3.10 or 3.11). Indeed, one has $\bar{u}_{k+1,m} = [\bar{u}_{k+1}^*, \bar{u}_{k+2}^*, ..., \bar{u}_m^*]^T = \bar{u}_{k+1,m}^*(t)$ and similarly $\bar{u}_{1,k}(t) = \bar{u}_{1,k}^*(t)$, ...

5. Violation of Principles of Mechanics by the Control Approach

· · · But the structural principles laid out in Secs. 2 and 3 require that the essential quantities, $\bar{u}_{k+1,m}$ must not be specified arbitrarily, because the corresponding compatible loads, derived quantities $F_{k+1,m}$, have been specified apriori as $F_{k+1,m}(t) = F_{k+1,m}^* = 0$ by the FIRST STEP of the control approach which considered a point-input distribution. It is also clear from the first and second steps of the control approach that no attention has been paid to such structural mechanics principles discussed in Secs. 2 and 3. Unless one is clairvoyant, $\bar{u}_{k+1,m}^*$, specified indirectly by the control problem and the specified input distribution $F_{k+1,m}^* = 0$, will not be compatible mechanically and mathematically. The structural system, Eq. (3.11) takes the form:

$$\begin{bmatrix} \bar{u}_{1,k}^* \\ \bar{u}_{k+1,m}^* \end{bmatrix} = \begin{bmatrix} C_{11} & C_{12} \\ C_{21} & C_{22} \end{bmatrix} \begin{bmatrix} F_{1,k} \\ F_{k+1,m}^* \end{bmatrix}, \tag{5.1}$$

and at least the lower set of these equations is overspecified by the control strategy. Presumably, since the vector $F_{1,k}$ is to be the designed control vector F^M, if we put $F_{1,k} = F^M$ and $F_{k+1,m}^* = 0$, we must check whether the equalities:

$$\bar{u}_{1,k}^* \overset{?}{=} C_{11} F^M, \qquad \bar{u}_{k+1,m}^* \overset{?}{=} C_{21} F^M,$$

$$\bar{u}^* \overset{?}{=} \overline{C} F^M, \qquad \overline{C} = \begin{bmatrix} C_{11}^T & C_{21}^T \end{bmatrix}^T, \tag{5.2a,b,c}$$

are satisfied to insure compatibility with the structural principles. To this end, it will be more instructive to evaluate $\bar{u}_i(t)$ at a point $x = x_i$ and determine the degree of inconsistency that may exist in Eqs. (5.2). Combining Eqs. (4.8) and noting that $\xi_r = \omega_r \eta_r$ and considering the second of every pair of Eqs. (4.2), $r = n$, we obtain:

$$\bar{u}^*(x_i,t) = \bar{u}_i^*(t) = \sum_{r=1}^{n} \left[\phi_r(x_i) - \omega_r^2 \int C(x_i,\eta)\phi_r(\eta)m(\eta)\, d\eta \right] \xi_r^*(t)$$

$$+ \sum_{r=1}^{n} \sum_{j=1}^{k} \int C(x_i,\eta)\phi_r(\eta)m(\eta)\, d\eta\ \phi_r(x_j) F_j^M(t). \tag{5.3a}$$

It can be shown from the integral formulation of the eigenvalue problem of the structural system and the orthogonality of the eigenfunctions [15] that:

$$\int C(x,\eta)\phi_r(\eta)m(\eta)\, d\eta = \omega_r^{-2}\phi_r(x). \tag{5.3b}$$

Utilizing Eq. (5.3b) in Eq. (5.3a) we obtain:

$$\bar{u}_i^*(t) = \sum_{r=1}^{n} \sum_{j=1}^{k} \omega_r^{-2}\phi_r(x_i)\phi_r(x_j) F_j^M(t) = \bar{u}^*(x,t), \quad x = x_i, \quad i = 1,2,\dots . \tag{5.3c}$$

This equation gives the quasi-static displacement at point $i = x_i$ as implied by the control design model which provides $F_j^M(t)$. The prescribed displacements at a number of points $i = 1, \dots, m$ can be written in compact matrix form:

$$\bar{u}^*(t) = Z F^M(t) = \begin{bmatrix} \bar{u}_{,k}^* \\ \bar{u}_{k+1,m}^* \end{bmatrix} = \begin{bmatrix} Z_1 \\ Z_2 \end{bmatrix} F^M(t), \tag{5.4}$$

$$Z_{ij} = \sum_{r=1}^{n} \omega_r^{-2}\phi_r(x_i)\phi_r(x_j), \quad i = 1, 2, \dots m, \quad j = 1, \dots, k,$$

where Z_{ij} play the role of influence coefficients. Now refer to Eqs. (5.2) and use Eq. (5.4) to obtain:

$$\epsilon = \bar{u}^M - \bar{u}^* = E F^M, \quad E = \overline{C} - Z, \quad \bar{u}^M = \overline{C} F^M, \tag{5.5}$$

in which ϵ is the difference (measure of incompatibility) between the quasi-static displacement $\bar{u}^*(t)$ prescribed by the control technique and the actual quasi-static displacement $\bar{u}^M(t)$ that will be obtained if the control model input F^M is applied to the structural system. We shall refer to E as the incompatibility (error) matrix. Note that E is an $m \times k$ matrix and m is allowed to go to infinity depending on how fastidious one may wish to be to discover the extent of incompatibility. The control vector is given by $F^M = G\Phi_{CL}w_0$ and depends on the control technique and the initial state arbitrarily. If Eqs. (5.2) are to be consistent at every point x_i in $0 < x < L$ regardless of the number of control inputs and their magnitudes at all times, one must have $E = \bar{C} - Z = 0$, $E_{ij} = C_{ij} - Z_{ij} = 0$, $i = 1, 2, \cdots$; $j = 1, 2, ..., k$; so that $\epsilon_i = 0$ identically in Eqs. (5.5). However, it is never possible to have $E = 0$ as we show next: it can be shown that the influence function has the form [15]:

$$C(x,\eta) = \sum_{r=1}^{\infty} \omega_r^{-2}\phi_r(x)\phi_r(\eta), \tag{5.6}$$

Using the form of $\bar{C}$ as given by (5.2c) along with (5.4) and (5.6) one gets:

$$E_{ij} = C_{ij} - Z_{ij} = \sum_{r=n+1}^{\infty} \omega_r^{-2}\phi_r(x_i)\phi_r(\eta_j), \quad i = 1, 2, .. \text{ arbitrary in } 0 < x < L,$$

$$j = 1, 2, .., k \quad \text{input locations.} \tag{5.7}$$

It is clear that $E_{ij} = 0$ can never be realized for all and arbitrary i, j, and k. Therefore $\epsilon \neq 0$ and Eqs. (5.2) will not be consistent. Consequently, the particular control approach will always violate the fundamental laws of mechanics embodied in Eqs. (3.10). In different terminology: *the displacement field $u^*(x,t)$ and therefore the acceleration (specific inertia load) field $\ddot{u}^*(x,t)$ implied by the pole-allocation solution to control model are not compatible with the apriori assumed point input distribution. The natural laws of mechanics, requiring equilibrium of all loads at every point in space and in time, cannot be maintained by the proposed common control strategy.*

In view of these results it is important to discover the consequences of the incompatibility. The error ϵ can be written in the form $\epsilon = EG\Phi_{CL}w_0$ and it *reflects a randomness in the degree of inconsistency of the control strategy with the laws of mechanics* depending on the initial conditions and the control technique used to compute G and Φ_{CL}. Our thrust is not to quantify a measure of this inconsistency. However, it is important to ascertain the qualitative nature of the effect of ϵ if $\mathbf{F}^M$ is used on the structural system. The structural system will behave in another way as the natural laws will not permit inconsistent impositions of the common control strategy on its behavior, and choose to eliminate it regardless of whatever it takes to do so. The structural system may get rid of the overspecification imposed by the control method by relaxing a set of the specified quantities through the action of natural laws so that the equilibrium of loads will always be maintained by the closed-loop system. There are two ways of viewing what may happen in reality. The first possibility is that the system, by producing additional loads by itself over the domain by an amount $f_\epsilon(\eta,t)$, will render the entire right hand side of Eqs. (5.1) unprescribed so that the specified left hand side can be maintained as $\bar{u}^*(t)$ without violating the structural principles. The second possibility is that the specified displacements $\bar{u}^*(t)$ may be relaxed by the system by producing an additional apriori unknown displacement $\bar{u}_e(t)$ such that the entire left hand side of (5.1) will be rendered unprescribed in the form $\bar{u}^*(t) + \bar{u}_e(t)$. Then the right hand side of Eqs. (5.1) with prescribed forces as the control strategy dictates will not be in violation of structural principles. It turns out that both of these possibilities will have to sustain each other, i.e. *be self-equilibriating* for their existence in order to insure compatibility with the laws of mechanics. According to the first possibility, $f_e(\eta,t)$ must be such that the equality:

$$\bar{u}^* = \sum_{j=1}^{k} C(x,\eta_j)F_j^M + \int C(x,\eta)f_e(\eta,t)\,d\eta \qquad (5.8a)$$

is maintained. Upon rearranging one obtains:

$$\int C(x,\eta)f_e(\eta,t) = \bar{u}^* - \bar{u}^M = -\epsilon(x,t), \qquad (5.8b)$$

which implies that $f_e(\eta,t)$ must cause a displacement $-\epsilon(x,t)$ thereby annihilating the inconsistency ϵ caused by $\mathbf{F}^M$. The load $f_e(\eta,t)$ can be found by inverting the integral equation (5.8b):

$$f_e(\eta,t) = -\int K(\eta,x)\epsilon(x,t)\,dx \qquad (5.9)$$

where $K(\eta,x)$ is the inverse operator of $C(x,\eta)$. The function $K(\eta,x)$ is known as the stiffness function of the given structural system and laws of mechanics are represented by it completely [15].

Next using Eqs. (5.5,7) in Eq. (5.8b) multiplying it by $\Phi_s(x)m(x)$ and integrating over x and η we obtain $\int \Phi_s(\eta)f_e(\eta,t) = 0$, $s = 1, 2, ..., n$. Therefore $f_e(\eta,t)$ will be orthogonal to the subspace spanned by the control model and will have no effect on the modes of the control model. This indicates that the displacement $\bar{u}^*$ specified by the control model will be preserved as designed and the modal coordinates $\xi_r(t)$, $r = 1, 2,...,n$, will not be altered by $f_e(\eta,t)$. According to the second possibility, the additional required $\bar{u}_e(x,t)$ must be such that:

$$\bar{u}^*(x,t) + \bar{u}_e(x,t) = \sum_{j=1}^{k} C(x,\eta_j)F_j^M = \bar{u}^M(x,t), \quad or,$$

$$\bar{u}_e(x,t) = \bar{u}^M - u^* = \epsilon(x,t), \qquad 0 < x < L, \qquad (5.10a)$$

is satisfied to insure compatibility. Note that the error quasi-static displacement corresponds to an error $u_e(x,t)$ in the actual displacement and will be given by an equation similar to (4.8b):

$$\bar{u}_e(x,t) = u_e(x,t) + \int C(x,\eta)m(\eta)\ddot{u}_e(\eta,t)\,d\eta = \epsilon(x,t). \qquad (5.10b)$$

Comparing Eqs. (5.10) and (5.8b) we conclude that both possibilities are causal, and one must have:

$$\bar{u}_e = u_e(x,t) + \int C(x,\eta)m(\eta)\ddot{u}_e(\eta,t)\,d\eta = -\int C(x,\eta)f_e(\eta,t)\,d\eta = \epsilon(x,t). \quad (5.11)$$

Considering that $\int\int C(x,\eta)K(\eta,s)u(s,t)\,ds\,d\eta = u(x,t)$, one can write from Eqs.
(5.11): $\quad \int C(x,\eta)\Big[\int K(\eta,s)u_e(s,t)\,ds + m(\eta)\ddot{u}_e(\eta,t)\Big]d\eta = -\int C(x,\eta)f_e(\eta,t)\,d\eta$
which yields:

$$f_e(\eta,t) = -\int K(\eta,s)u_e(s,t)\,ds - m(\eta)\ddot{u}_e(\eta,t), \quad\quad\quad\quad (5.12)$$

that is, $f_e(\eta,t)$ needed to preserve compatibility is the net elastic and inertia load
at η solely sustained by the error displacement $u_e(\eta,t)$. It remains to find out
now how f_e and u_e are developed.

From Eqs. (5.5,7 and 8b) we recognize:

$$\int C(x,\eta)f_e(\eta,t)\,d\eta = -\sum_{r=n+1}^{\infty}\sum_{j=1}^{k}\omega_r^{-2}\phi_r(x)\phi_r(\eta_j)F_j^M, \quad\quad (5.13)$$

and rewrite Eq. (5.13) in view of Eq. (5.3b):

$$\int C(x,\eta)f_e(\eta,t)\,d\eta = -\sum_{r=n+1}^{\infty}\int \phi_r(\eta)C(x,\eta)m(\eta)\sum_{j=1}^{k}\phi_r(\eta_j)F_j^M(t)\,d\eta,$$

from which we obtain:

$$f_e(\eta,t) = -\sum_{r=n+1}^{\infty} m(\eta)\phi_r(\eta)\sum_{j=1}^{k}\phi_r(\eta_j)F_j^M(t)$$

$$= -\sum_{r=n+1}^{\infty} m(\eta)\phi_r(\eta)f_{er}(t), \quad\quad\quad\quad (5.14)$$

in which we defined the coordinates $f_{er} = \sum_{j=1}^{k}\Phi_r(\eta_j)F_j^M(t)$. In order to insure
compatibility with the natural laws one must have $f_e(\eta,t) \neq 0$ and since the basis
functions $m(\eta)\Phi_r(\eta)$ in the expansion of Eq. (5.14) are independent, it follows
that some (at least one) of the coefficients $f_{er}(t)$, $r = n + 1,, ..., \infty$, must be
non-zero. But the reader will recognize the load coordinates f_{er} precisely as the
modal inputs on the modes $r = n + 1, ...$ beyond the control model. Hence, we
showed that in order to insure satisfaction of natural laws some f_{er} must be nontrivial regardless of the number, location, and magnitude of control inputs $F_j^M(t)$
obtained from the control model (4.5). We conclude that for compatibility, the
closed-loop structural dynamic system will derive the necessary loads in the form
of additional distributed elastic and inertia loads (5.12) from the discrete design
control input $\mathbf{F}^M$ according to Eq. (5.14). Equation (5.14) indicates that in order
to compute the load $f_e(\eta,t)$ one should have knowledge of the many modes
$r > n$ beyond the control model. At first this may look disturbing. However, in
the following we shall show that the *mean convergence* of the expansion of (5.14)
can be found in terms of the parameters of the control model without any

knowledge of the uncontrolled system. Consider the general differential equation of motion of the structure, with point inputs:

$$m(\eta)\ddot{u} + L(u) = \sum_{j=1}^{k} F_j \delta(\eta - \eta_j),$$ (5.15a)

in which $L(\ )$ is the differential stiffness operator. The displacement $u(\eta,t)$ which is compatible with Eq. (5.15a) can be written as $u = u^* + u_\epsilon$ where u^* is the displacement desired by way of construction of the control design model and u_ϵ is the error displacement from it. Substituting the expression for u into Eq. (5.15a) and rearranging we obtain:

$$m(\eta)\ddot{u}_\epsilon + L(u_\epsilon) = \sum_{j=1}^{k} F_j \delta(\eta - \eta_j) - \left[m(\eta)\ddot{u}^* + L(u^*) \right],$$ (5.15b)

where one may designate the terms on the right hand side as the *differential equation error*. From a comparison of differential and integral formulation of system dynamics one can also see that for a particular structure $\int K(\eta,s) u(s)\, ds = L(u(\eta))$ with the implied boundary conditions on $u(\eta)$ and its spatial derivatives. With this last property in mind, we conclude from Eqs. (5.12) and (5.15b) that:

$$f_\epsilon(\eta,t) = -\left[m(\eta)\ddot{u}_\epsilon + L(u_\epsilon) \right]$$

$$= -\left[\sum_{j=1}^{k} F_j \delta(\eta - \eta_j) - \left[m(\eta)\ddot{u}^* + L(u^*) \right] \right].$$ (5.16)

The additional load required to insure compatibility is the negative of the differential equation error. From the control model we can show that:

$$m(\eta)\ddot{u}^* + L(u^*) = \sum_{r=1}^{n} \left[m(\eta)\phi_r(\eta)\ddot{\xi}_r^* + L\left[\phi_r(\eta) \right]\xi_r^* \right].$$

Recalling the eigenvalue problem $\omega_r^2 m(\eta)\Phi_r(\eta) = L(\Phi_r)$, we obtain:

$$m(\eta)\ddot{u}^* + L(u^*) = \sum_{r=1}^{n} m(\eta)\phi_r(\eta)\left[\ddot{\xi}_r^* + \omega_r^2 \xi_r^* \right]$$

$$= \sum_{r=1}^{n} m(\eta)\phi_r(\eta)\sum_{j=1}^{k} \phi_r(\eta_j)F_j.$$ (5.17)

Substituting Eq. (5.17) into Eq. (5.16) we obtain an alternate expression for the load $f_\epsilon(\eta,t)$:

$$f_\epsilon(\eta,t) = -\left[\sum_{j=1}^{k} F_j \delta(\eta - \eta_j) - \sum_{r=1}^{n} m(\eta)\phi_r(\eta)\sum_{j=1}^{k} \phi_r(\eta_j)F_j \right].$$ (5.18)

The significance of Eq. (5.18) over Eq. (5.14) is that it refutes the observation that one needs to know the modes beyond the control model in order to obtain some measure of the load f. Equation (5.18) implies the contrary, that the load f_ϵ can be determined completely from the control model itself, one need not go any further. In view of Eqs. (5.14) and (5.18) one is inclined to write:

$$f_\epsilon(\eta,t) = - \sum_{r=n+1}^{\infty} m(\eta)\phi_r(\eta)f_{\epsilon r}$$

$$= -\sum_{j=1}^{k} F_j\delta(\eta - \eta_j) + \sum_{r=1}^{n} m(\eta)\phi_r(\eta)f_r, \qquad (5.19a)$$

where $f_r = \sum_{j=1}^{k}\Phi_r(\eta_j)F_j$ is the modal input as usual, and conclude that Eq. (5.14) converges to Eq. (5.18) pointwise. Rearranging Eq. (5.19a) one has:

$$\sum_{j=1}^{k} F_j\delta(\eta - \eta_j) = \sum_{r=1}^{n} m(\eta)\phi_r(\eta)f_r + \sum_{r=n+1}^{\infty} m(\eta)\phi_r(\eta)f_{\epsilon r}, \qquad (5.19b)$$

which represents an expansion of the discontinuous input function $\sum F_j\delta(\eta - \eta_j)$. Transposing the controlled terms to the left of Eq. (5.19b) it follows that the remaining terms $r > n + 1$, ..., ∞ represent a Fourier expansion to the discontinuous closed-form function:

$$\sum_{j}^{k} F_j\delta(\eta - \eta_j) - \sum_{r=1}^{n} m(\eta)\phi_r(\eta)f_r = -f_\epsilon(\eta,t). \qquad (5.19c)$$

Hence we conjecture that Eq. (5.14) converges to Eq. (5.18) in the mean. Equation (5.19b) is not surprising, however, the relationship of its terms to the mechanical incompatibility and the resulting load $f_\epsilon(\eta,t)$ we discussed heretofore is worth noting.

A third expression for f_ϵ can be given by spatial discretization of the integral Eq. (5.9), leading to :

$$\mathbf{F}_\epsilon = -C^{-1}\epsilon = -C^{-1}E\mathbf{F}^M, \qquad f_{i\epsilon} = \frac{F_i}{\Delta\eta_i}, \qquad (5.20)$$

where $f_{i\epsilon}$ is an approximation to $f_\epsilon(\eta_i)$ and $\Delta\eta_i$ is the stepsize between points i and $i + 1$ over which the integrals are assumed to remain constant.

The error displacement u_ϵ associated with the load $f_\epsilon(\eta,t)$ can now be obtained by solving Eq. (5.11) for u_ϵ in conjunction with Eq. (5.14) or (5.18). To this end one may utilize the fact that $f_\epsilon(\eta,t)$ is orthogonal to the subspace spanned by Φ_s, $s = 1, 2, ..., n$ and expand u_ϵ in the form $u_\epsilon(s,t) = \sum_{r=n+1}^{\infty} \Phi_r\xi_\epsilon$ in which ξ_ϵ are the solutions of:

$$\ddot{\xi}_\epsilon + \omega_r^2\xi_\epsilon = -\int \phi_r(\eta)f_\epsilon(\eta,t)\,d\eta = f_{\epsilon r}(t), \qquad r = n+1,... . \qquad (5.21)$$

The question is whether one should use Eq. (5.14) or Eq. (5.18) for f_ϵ since we stated above that the convergence of Eqs. (5.14) and (5.18) was not necessarily pointwise. As it turns out from a response point of view, it is immaterial which expression is used for f_ϵ, since both of them yield the same modal inputs on the modes $r > n + 1$, ... Indeed, by using either Eq. (5.14) or (5.18) in Eq. (5.21) we obtain $f_{\epsilon r} = \sum_j \Phi_r(\eta_j) F_j^M$, $r = n + 1,...$, from which the displacement u_ϵ can be computed by modal synthesis via Eq. (5.21). In this sense, the loads Eqs. (5.14) and (5.18) are "equipollent." The total displacement exhibited by the structure will be $u = u^* + u_\epsilon \neq u^*$, therefore the control objective will never be achieved.

The error phenomenon we discussed above from the point of view of structural mechanics is known in the field of control theory as the *control spillover phenomenon*. The literature of control is abound with examples of this phenomenon showing its significance [1-4]. Specifically, the modal inputs

$$f_{\epsilon r}(t) = \sum_{j=1}^{k} \Phi_r(\eta_j) F_j^M = f_r(t), \; r = n + 1,..., \text{ are known as control spillover in}$$

control theory. With this, referring back to Eqs. (5.14) and (5.21), the connection between the error load distribution $f_\epsilon(\eta,t)$ and the control spillover is abundantly clear. *It is important to realize that in the above derivations, based solely on the fundamental laws of mechanics, we showed why control spillover must exist. No apriori assumptions of its existence was made. In particular, we did not assume that not all $f_r = f_{\epsilon r} = \sum_j \Phi_r(\eta_j) F_j^M$, $r = n + 1,...$, were non-zero,*

on the contrary we proved why some of them had to be non-zero. In addition, we gave a closed-form expression for the (spillover) load $f_\epsilon(\eta,t)$ in terms of the control model only, Eq. (5.18). The closed-loop structural system will develop these (spillover) loads at every point enough to insure conformity with the fundamental postulate of mechanics.

Returning to Eq. (5.19c), we shall refer to the function:

$$\sum_{r=1}^{n} m(\eta)\phi_r(\eta)f_r(t) = \sum_{r=1}^{n} m(\eta)\phi_r(\eta)\sum_{j=1}^{k} \phi_r(\eta_j)F_j(t) \qquad (5.22)$$

as the *effective control* or *control effectiveness* since it represents only that portion of the input $\sum_{j=1}^{k} F_j\delta(\eta - \eta_j)$ which can be made reach the subspace of the distributed parameter system spanned by the control design model and therefore it is the only useful part of the input for control. The remaining part of the input is wasted in the form of the terms $r = n + 1$, ... in Eq. (5.19b). Hence, we recognize that *control effectiveness, Eq. (5.19c), is complementary to control spillover."* As mathematical as Eq. (5.19b) seems, it is more important to focus on its physical significance. Essentially, Eq. (5.19b) is an expression of a *principle of input conservation* which can be stated as: *the sum of control effectiveness and control spillover at any instant must be equal to the magnitude of the total physical input on the system at that instant.* Because the input distribution and the control effectiveness are completely determined by the solution of the control design model, the amount of spillover can be ascertained by using the input conservation principle without any explicit reference to what is beyond the control design model.

6. The Unique Compatible Method of Approach to Structural Control

Because it will always lead to error ϵ, F^M is not the correct solution for the structural system subject to the constraints (control objectives) $u^*(x,t) = \sum_{r=1}^{n} \left[\Phi_r(x) \, 0 \right] w_r^*$, $f^*(\eta,t) = 0$, $\eta \neq \eta_1,...,\eta_k,...,$. When F^M is applied to the structure, laws of mechanics make it a solution to a different dynamic system with $u = u^* + u_\epsilon$ and $f^*(\eta,t) = 0$, $\eta \neq \eta_1, ..., \eta_k$ where u_ϵ is the displacement which is the degradation of the displacement objective. In fact, *because it is overspecified, mechanically there is no (compatible) solution to the structural control problem under the constraints u^* and $f^*(\eta,t)$.* The question we pose now is what it takes to obtain a solution which is also compatible with the natural equilibrium laws. One has to relax the constraints to make the problem properly stated. The objective $u^*(x,t)$ is essential and it must be retained, after all it is the theme of the control problem. It follows then that *the other constraint $f^*(\eta,t) = 0$ ($\eta \neq \eta_1, ..., \eta_k$) must be done away with. This is precisely the overspecification introduced into the control formulation via the FIRST STEP in Sec. 4.* This is not surprising because apriori imposition of this constraint on the input distribution has no structural dynamic basis, and as we proved, is ill-founded. On the other hand $u^*(x,t)$ does have a mechanical base and is the main control objective. When the control problem imposes a constraint on this essential quantity over the whole domain, no more room is left for specifying the input distribution $f(\eta,t)$ at any point of the structural domain. $f(\eta,t)$ is the derived quantity of the mechanical system, it has to be derived literally from the specified time history of the essential quantity and cannot be specified in advance. Hence, in order *to obtain solutions to the structural control problem that are compatible with the fundamental principles of mechanics one must start solving the control problem without any apriori assumed input distribution.* The control model input distribution $f^M(\eta,t)$ must be treated as unknown both in spatial and time dependence and it must be found both in space and time dependence in order to achieve $u = u^*(x,t)$. Again, it should be remembered that the controlled behavior $u^*(x,t)$ is specified via the modal pole-allocation for the modal equations. Eqs. (4.2) must now be achieved without any knowledge of the η dependence of the input $f(\eta,t)$ and consequently, spatial integrations over $f(\eta,t)$ in Eq. (4.2) cannot be performed. An inspection of Eqs. (4.2) reveals that under this condition the only way to realize the control objective (pole-allocation) is by *first designing the modal control inputs $f_r(t)$, $r = 1,2, ..., n$, directly.* Let us assume that this is done and $f_r = f_r^M(t) = f_r^M(w)$, $r = 1, 2, ..., n$, are the *modal inputs* that yield desired eigenvalue behavior for the control design model. The desired displacement behavior specified (indirectly) by the control objective is then:

$$\bar{u} = u + \int C(x,\eta)m(\eta)\ddot{u}(\eta,t)\, d\eta$$

$$= \sum_{r=1}^{n} \left[\phi_r(x)\xi_r + \int C(x,\eta)m(\eta)\phi_r(\eta)\ddot{\xi}_r\, d\eta \right],$$

but since $\ddot{\xi}_r(t) = -\omega_r^2 \xi_r + f_r^M(t)$ we obtain by substituting into the above:

$$\bar{u}(x,t) = \int C(x,\eta)\sum_{r=1}^{n} m(\eta)\phi_r(\eta)f_r^M(t)\, d\eta = \bar{u}^*(x,t), \qquad (6.1)$$

in which $f_r^M(t)$ is known for any given initial conditions from the solution of the pole-allocation problem. $f_r^M(t)$ do not represent physical control inputs however, and one must search for a physical input $f^M(\eta,t)$ which will be compatible with the equilibrium laws of mechanics for the desired $\bar{u}^*(x,t)$ in Eq. (6.1). Hence, for compatibility we write the integral equation of the first kind:

$$\epsilon(x,t) = \bar{u}^M - \bar{u}^* = \int C(x,\eta)f^M(\eta,t)\, d\eta - \bar{u}^* = 0, \quad 0 \leq x \leq L, \qquad (6.2)$$

for the unknown $f^M(\eta,t)$, $\bar{u}^*$ is known according to Eq. (6.1). Note that this last equation is the same as Eq. (3.9) with $\bar{u} = \bar{u}^*$ and $f(\eta,t)$ unknown. Substituting Eq. (6.1) into Eq. (6.2):

$$\int C(x,\eta)f^M(\eta,t)\, d\eta = \int C(x,\eta)\sum_{r=1}^{n} m(\eta)\phi_r(\eta)f_r^M(t)\, d\eta, \qquad (6.3)$$

from which we immediately obtain a solution for $f^M(\eta,t)$:

$$f^M(\eta,t) = \sum_{r=1}^{n} m(\eta)\phi_r(\eta)f_r^M(t) = \int K(\eta,x)\bar{u}^*(x,t)\, dx. \qquad (6.4a,b)$$

It is seen that *the physical input is determined uniquely and harmoniously by the control theory which first provides the time dependence of the input via modal controls $f_r^M(t)$ and by the structural mechanics theory which yields the spatial dependence of the input via (6.1-4) so that the control objective $u^*(x,t)$ can be realized without violating the natural laws of mechanics.* It is also clear that *the consistent solution to the structural control problem requires a separation of tasks* in the sense that control design must be carried out directly for the modal control inputs independently of the determination of (or any apriori impositions on) the spatial input distribution and the spatial input distribution must be determined independently of the control design process. In the solution for $f^M(\eta,t)$, Eqs. (6.4) and the spatial and time dependencies are clearly separated. *Because the modal control inputs f_r^M must be determined first, finding a compatible solution is truly a genuine modal-space design and modal synthesis process.* Most importantly, the form of the solution, Eq. (6.4) is a *must* for conformity with natural laws, it is *not a choice*. Recall now the approach to the control problem given in Sec. 4. There, the *separation of tasks* is not observed, the apriori assumed input distribution and control design are combined by attempting to design the components of a physical input $\mathbf{F}^M$ directly which leads to the violation of natural equilibrium laws of mechanics.

Finally, in view of Eqs. (6.4) we note that the modal model truncation becomes a non-issue. *The naturally compatible solution does not require n → ∞ nor does it require any knowledge of modes beyond n.* Consequently, *model truncation is not the source of incompatibility* encountered in Sec. 5. On the other hand, when the controller is truncated via spatially discrete inputs, the model truncation becomes a concern also, indeed, the unmodelled modes provide the only means of relief for the satisfaction of natural laws at the expense of dissatisfaction of control objective $u = u^*$. Natural laws of mechanics do have the priority above all.

7. A Revisit: Independent Modal-Space Control (IMSC) and Coupled Control (CC) Techniques

The common method of control given in Sec. 4 which is not based on the separation of tasks and which, therefore, is not compatible with structural principles, is referred to in the literature as Coupled Control (CC) technique [5]. On the other hand, another control technique based on the separation of tasks as defined in the previous section has also been proposed for some years. The technique is known as Independent Modal-Space Control (IMSC) [17-23]. The IMSC designs first the modal controls f_r, $r = 1, 2, ...,n$, independently of each other and obtains the solution to the structural control problem theoretically in the form of Eq. (6.4) by modal synthesis regardless of the number n [21]. In other words, the foundation of IMSC is composed of what it takes to produce a solution consistent with the most fundamental laws of mechanics. The compatibility of the IMSC lies not in the fact that the modal controls are designed independently but in the fact that modal control laws are designed first without any apriori assumed physical input distribution as we concluded in Sec. 6.

The structural mechanics perspective that is presented heretofore is quite significant indeed as the concept of a "globally optimal control for flexible systems" has been pursued recently in [17, 22-25]. Proofs of "globally optimal properties" of IMSC have been given in [17, 22, 23]. In particular, a new concept of optimality for control of flexible systems which is more consistent with the physical nature of flexible structures has been proposed in [22, 26]. References [22, 26] show that the modal performance measure of the IMSC discussed therein is a weighted quadratic performance measure of the control input precisely given by Eq. (6.4). Consequently, the modal performance measure of the IMSC (referred to as design performance quality (DPQ) measure in [22, 26]) represents a fundamental quantity that is uniquely compatible with the laws of mechanics. Coupled Control techniques violate the laws of mechanics because the formulation of the control problem is mechanically constrained (overspecified) from the beginning. It follows that their associated coupled performance measures are also overconstrained and therefore they are suboptimal relative to comparable IMSC designs. In light of concepts forwarded in [17, 22, 23, 26], the IMSC and CC designs are also designated as "natural" and "unnatural" controls, respectively. In light of the structural principles that we presented in this paper, such designation of the two approaches to control becomes all the more appropriate. Reference [22] defines an index of unnaturality for all CC designs, the index essentially being a degree of suboptimality relative to the modal performance measure of IMSC. The global optimal properties of IMSC are intrinsic in its method of approach which is compatible with the fundamental principles of mechanics (therefore naturality). The control approaches failing to satisfy conformity with these principles (therefore unnaturality) are necessarily suboptimal.

Having pointed out the fundamental traits of IMSC and CC techniques, it is now possible to address some practical issues of structural control with better physical insight. Note from the form of solution, Eq. (6.4), that how the inputs $f_r^M(t)$ are designed has no consequences in regards to structural principles, but it has significant computational implications. Designing them in the form $f_r = f_r^M(w_r)$ is computationally most attractive. This reduces the system to a collection of n second-order subsystems resulting in IMSC. There is no curse of high dimensionality in IMSC. Alternately, Eq. (6.4) does not prohibit modal inputs of the form $f_r^M = f_r^M(w)$, but this is unnecessary computational burden leading to a coupled modal-space control (a phrase which should not be confused with the CC technique discussed above and in the references). As a result, independent control of the modes in modal-space is preferred. Theoretically compatible solution, Eq. (6.4), to the structural control problem is a spatially continuous input distribution and its form is unique. The present state of the art does not permit continuously distributed inputs so that the solution (6.4) has to be implemented approximately by spatially discrete--such as point or piecewise continuous--inputs. An incompatible solution is inevitable in practice when discrete inputs are used and one obtains a structural control system exhibiting degradation of (specified) control objectives. Implementation of IMSC by spatially discrete point and piecewise continuous inputs is addressed in [23, 26]. One of the fundamental aspects of the implemented IMSC is that the "separation of tasks" is still maintained and the design of modal control laws f_r^M are kept independent of the spatially discrete actuator locations [5]. When point inputs are used by IMSC to approximate Eq. (6.4) via modal synthesis, n inputs are required so that the design modal controls f_r^M, $r = 1, 2, ..., n$, can be realized exactly despite the spatial approximation of the controller. The important practical consequence of this is that *the modal performance of the control design model of IMSC implemented by a finite number of point inputs is identical to the modal performance measure of the continuously distributed input, Eq. (6.4). Consequently, the implemented IMSC preserves quantitatively in the control design model the global properties of the compatible solution (6.4)* [29]. In the CC technique, the modal control laws do depend on the apriori selected actuator number and locations, the technique lacks the "separation of tasks." If the actuator number and locations are changed the control design must be repeated. An important objective of CC technique is that the designs strive to keep the actuator number $k < n$ (possibly $k \ll n$). If n is too large, computability of control inputs becomes a major obstacle. Because the number of inputs used in CC can be less than n, (the required number of point inputs to implement IMSC), a CC design is more constrained relative to the implemented IMSC from the viewpoint of structural principles. The specified input distributions of CC and implemented IMSC have the forms:

$$f^M(\eta,t) = f^*(\eta,t) = 0, \qquad \eta \neq \eta_1, \eta_2, \ldots, \eta_k;$$

and,

$$f^M(\eta,t) = f^*(\eta,t) = 0, \qquad \eta \neq \eta_1, \eta_2, \ldots \eta_k, \eta_{k+1}, \ldots \eta_n, \quad 0 \leq \eta \leq L,$$

respectively. This shows that the CC inputs are specified (constrained) as zero at a larger number of points than the implemented IMSC. The excess number of inputs of implemented IMSC relaxes that many more constraints on the input distribution, and the controls approximate more closely the correct solution Eq. (6.4), thus making it mechanically less constrained. The result is that corresponding to similar control objectives such as discussed in Ref. [17] within the context of dynamically similar closed-loop systems, an IMSC implemented via point inputs is more compatible with the laws of mechanics and (therefore) its performance will always be superior to that of a CC design. Another way of understanding why a CC design is more constrained relative to an IMSC design is to consider the modal-space spanned by the modal control inputs of the two techniques. The ranks of modal gain matrices of the two techniques are the number of inputs used by the respective techniques. Therefore, the dimension of the subspaces spanned by the modal controls of the CC and IMSC designs are k and n, respectively. Since $n > k$, the CC designs are carried out in a smaller subspace than the IMSC designs. Hence, CC is a more constrained design also from the perspective of linear analysis. The IMSC design is done in the largest possible modal space of dimension n for the control model providing the controller with more degrees of freedom. This is mechanically meaningful. It follows from the above facts that the number of constraints that matters structurally is inversely proportional to the number of inputs used by the IMSC and CC designs. For the chosen control design model the IMSC is referred to in Ref. [17] as a full-rank control as opposed to a CC design which is rank-deficient. Further implications of these are studied in Ref. [17] in detail and it is proven rigorously that the IMSC design is a globally minimum control gain design among all dynamically similar closed-loop control designs.

In practice, it will be desirable to keep some measure of incompatibility $\epsilon(x,t)$ at an acceptable level in controlling a structural system. Different candidate designs can be evaluated relative to each other in their measures of $\epsilon(x,t)$ and a best design can be selected. Most conveniently, rather than using a direct measure of ϵ, one may work with suitably defined quadratic performance measures for the designs. Comparison of such performance measures will be a comparison of the incompatibilities indirectly. Comparisons of different designs with each other is most meaningful if they can all be compared to a known control design which has no incompatibility ($\epsilon(x,t) = 0$) associated with it to accomplish the same eigenvalue objectives. In this case, comparisons to such an ideal design will be absolute and global in nature with undisputable fundamental concepts behind them. For specified control objectives a theoretical closed-form solution does not exist for CC designs and no base design for an absolute measure of incompatibilities $\epsilon(x,t)$ can be provided from amongst the CC designs. Even in the case when the inputs are increased indefinitely in CC ($k \rightarrow \infty$) so as to have a continuous input there is no guarantee that the incompatibility $\epsilon(x,t)$ will be eliminated. In this sense a CC represents an open-loop character with respect to manipulation of $\epsilon(x,t)$. The inherent incapability of CC techniques to eliminate this incompatibility (hence spillover) even as $k \rightarrow \infty$ is discussed in [22, 26] within the context of performance quality measures. On the other hand, for the same objectives set for the CC designs a dynamically similar IMSC design can be obtained readily [17] from which the closed-form compatible solution (6.4) can be determined by modal synthesis. Note that as pointed out earlier, the implemented IMSC also has the same modal performance in its control design model as the distributed compatible input (6.4). The performance measure associated with such an IMSC solution will provide an absolute reference by which other control designs can be evaluated to assess their properties relative to each other. Most importantly, this way one will be comparing the designs to a globally optimum form of solution for the control objectives that is fully compatible with the fundamental laws of mechanics. In this sense, comparison of CC designs to a dynamically similar IMSC design will not be a matter of cosmetics but will be a matter of evaluating in a most fundamental way how well the natural laws are respected by the CC designs. References [5, 17, 22, 26, and 29] provide examples of such comparisons.

8. Numerical Results

A detailed numerical example for a pinned-pinned beam of Fig. 1 is presented in Ref. 30 illustrating the perspective of this paper. Because of space limitations, the reader is referred to Ref. [30].

9. Conclusions

Approaches to control of structural systems have been discussed from the point of view of principles of mechanics. To this end, basic incompatibilities are shown and the unique compatible method of approach to structural control is presented. While the presentation may have been tutorial in certain aspects, between the lines lies an interdisciplinary assessment of current methodology for structural control.

References

[1-4] Proceeding of the First, Second, Third and Fourth VPI & SU/AIAA Symposia on Dynamics and Control of Large Flexible Spacecraft, Blacksburg, VA, 1977, 1979, 1981, 1983.

[5] Meirovitch, L., Baruh, H., and Öz, H., "A Comparison of Control Techniques for Large Flexible Systems," *Journal of Guidance, Control and Dynamics*, Vol. 6, No. 4, 1983.

[6] Von Laue, M., *History of Physics*, Academic Press, inc., New York, 1950.

[7] Bruhn, E. F., *Analysis and Design of Flight Vehicle Structures*, S. R. Jacobs, Indianapolis, 1973.

[8] Öz, H., "Structural Analysis and Design of Aerospace Vehicles," Lecture Notes, Dept. of Aeronautical and Astronautical Engineering, The Ohio State University (Manuscript in Preparation).

[9] Bailey, C. D., "A New Look at Hamilton's Principle," *Foundation of Physics*, Vol. 5, No. 3, 1975.

[10] Bailey, C. D., "An Introduction to the Displacement of Deformable Bodies." Notes of Aircraft Structures II, Dept. of Aeronautical and Astronautical Engineering, The Ohio State University, Jan. 1981.

[11] Bailey, C. D., "Hamilton, Ritz and Elastodynamics," *ASME Journal of Applied Mechanics*, Vol. 98, No. 4, 1976.

[12] Fung, Y. C., *Foundations of Solid Mechanics*, Prentice-Hall, New Jersey, 1965.

[13] Meirovitch, L., *Methods of Analytical Dynamics*, McGraw-Hill, N.Y., 1970.

[14] Meirovitch, L., *Analytical Methods in Vibrations*, McMillan Co., N.Y., 1969.

[15] Bisplinghoff, R. L., Ashley, H., and Halfman, R. L., *Aeroelasticity*, Addison-Wesley Pub. Co., Reading, Mass., 1955.

[16] Kirk, D. E., *Optimal Control Theory*, Prentice-Hall Inc., New Jersey, 1970.

[17] Öz, H., "Dynamically Similar Control Systems and a Minimum Gain Control Technique: IMSC," Ref. 4. *Journal of Optimization Theory and Applications* (to appear)

[18] Meirovitch, L. and Öz, H., "Computational Aspects of the Control of Large Flexible Structures," Proc. of the 18th IEEE Conference on Decision and Control, Dec. 1979, Ft. Lauderdale, FL, pp. 220-229.

[19] Meirovitch, L. and Öz, H., "Modal-Space Control of Distributed Gyroscopic Systems," *Journal of Guidance and Control*, Vol. 3, No. 2, 1980, pp. 140-150.

[20] Öz, H. and Meirovitch, L., "Optimal Modal-Space Control of Flexible Gyroscopic Systems," *Journal of Guidance and Control*, Vol. 3, No. 3, 1980, pp. 218-226.

[21] Öz, H. and Meirovitch,L., "Stochastic Independent Modal-Space Control of Distributed Parameter Systems," *Journal of Optimization Theory and Applications*, Vol. 40, No. 1, May 1983.

[22] Öz, H., "Another View of Optimality for Control of Flexible Systems," Ref. 4.

[23] Meirovitch, L. and Silverberg, L., "Globally Optimal Control of Self-Adjoint Systems," Ref. 4.

[24] Floyd, M. A., Lindberg, R. E., and Lyons, M. G., "Comments on A Comparison of Control Techniques for Large Flexible Systems," *Journal of Guidance Control and Dynamics*, Vol. 7, No. 5, 1984.

[25] Meirovitch, L., Baruh, H., and Öz, H., "Reply to Authors to M. A. Floyd, R. E. Lindberg and M. G. Lyons," *Journal of Guidance Control and Dynamics*, Vol. 7, No. 5, 1984.

[26] Öz, H., "A New Concept of Optimality for Control of Flexible Structures," A revised and extended version of Ref. 22, (to appear).

[27] Meirovitch, L., Baruh, H., Montgomery, R. D. and Williams, J. P., "Nonlinear Natural Control of An Experimental Beam," *Journal of Guidance, Control and Dynamics*, Vol. 7, No. 4, 1984.

[28] Skidmore, G. R., Hallauer, W. L., and Gehling, R. N., "Experimental-Theoretical Study of Modal-Space Control," Second International Modal Analysis Conference, Orlando, FL, Feb. 1984.

[29] Öz, H. and Adiguzel, E., "Generalized Natural Performance Charts for Control of Flexible Systems," Paper No. 84-1951 CP AIAA Guidance and Control Conference, Seattle, WA, Aug. 1984.

[30] Öz, H., "Control of Flexible Systems and Principles of Structural Mechanics," AIAA-84-2001, AIAA/AAS Astrodynamics Conference, Seattle, WA, Aug. 1984.

GRAPHICAL FEEDBACK STABILITY CRITERION FOR UNDAMPED CONTINUOUS ELASTIC SYSTEMS

R. Piche
Department of Civil Engineering
University of Waterloo

1. Introduction

A "frequency domain" approach to the analysis of feedback control systems has a number of advantages over a state space approach; among several others the following may be mentioned:

- the graphical techniques associated with frequency domain methods (e.g. Nyquist plots) are powerful design aids, allowing spatial intuition to guide the designer in modifying a system to meet various performance criteria.

- mathematical models for many distributed parameter systems can be derived in closed form and written in the frequency domain as transcendental transfer functions; in contrast, a state space model requires the use of infinite-order matrices [1] or abstract operators [9].

The Nyquist stability criterion for feedback systems is applicable to a wide variety of linear time-invariant systems, including distributed parameter systems [3]. However, undamped continuous elastic systems cannot be handled by the Nyquist criterion, because they have an infinite number of 'unstable' poles (i.e. poles with zero real part). Even when linear damping is assumed to be present, the sharpness of the resonance peaks in the frequency response leads to Nyquist diagrams that are very sensitive within narrow frequency bands [7, chpt.5].

In [10, 11], a feedback stability criterion is derived that is applicable to genaral convolution systems. This criterion is based on the factorization of the (possibly unstable) transfer function into a ratio of two stable transfer functions. This factorization makes it possible to use a Nyquist-like plot to study the input-output stability of feedback systems, provided a factorization can be found. The theory for the factorization method is outlined in Sec. 2 of this paper.

In Sec. 3, the factorization method is applied to a specific problem in structural control: the stabilization of an undamped one-dimensional wave equation system using a single actuator/sensor pair at the boundary. This is perhaps the simplest conceivable example of a continuous elastic system, and provides insight into this class of problems. It is shown that a graphical, frequency domain approach to the problem can be used to study the effects of controller dynamics and delays in a straightforward manner.

A brief description of work in progress concludes the paper.

2. Theory

2.1. The algebra $\mathcal{A}$ of stable causal convolution operators

Consider a system for which the zero-intial-state output is the convolution of the input with some 'impulse-response' generalized function. Such a 'convolution system' with impulse-response g is said to be *causal* [14, §10.3] if $g(t) = 0$ for all $t < 0$, and *bounded-input-bounded-output (BIBO) stable* if $g * y \in L_\infty$ whenever $y \in L_\infty$ (L_∞ denotes the Lebesgue space of bounded functions.)

A large class of causal BIBO-stable impulse response functions consists of generalized functions of the form [4, 12]

$$g(t) = g_0(t) + \sum_{k=1}^{\infty} g_k \delta(t - \tau_k), \tag{1}$$

where $\delta(\,\cdot\,)$ denotes the delta distribution, and the following conditions are imposed:

$$g_0(t) = 0, \quad \text{for } t < 0, \tag{2}$$

$$\tau_k \geq 0, \quad \text{for } k = 1,2,..., \tag{3}$$

$$\int_0^\infty |g_0(t)|\, dt < \infty, \tag{4}$$

$$\sum_{k=1}^{\infty} |g_k| < \infty. \tag{5}$$

The class of impulse-response functions described above is denoted $\mathcal{A}$. A norm for elements g of $\mathcal{A}$ is defined as :

534   R. Piche

$$\|g\|_A := \int_0^\infty |g_0(t)| \, dt + \sum_{k=1}^\infty |g_k|. \tag{6}$$

The causality of elements of A follows directly from conditions (2) and (3). The BIBO-stability follows from conditions (4) and (5), which can be used to prove the inequality

$$\|g * y\|_A \le \|y\|_{L_\infty} \cdot \|g\|_A \qquad \text{for all } y \in L_\infty \quad g \in A \tag{7}$$

The Laplace transform of an impulse-response function is termed the 'transfer function' of the system. Let $\hat{A}$ denote the set of Laplace transforms of elements of A. The norm of an element of $\hat{A}$ is taken to be equal to the norm of the corresponding element in A. If $\hat{g} \in \hat{A}$, then according to [4] the function $\Gamma{:}\omega \to \hat{g}(i\omega)$ is uniformly continuous and bounded for all $\omega \in \mathbb{R}$. Also, $\hat{g}(s)$ is analytic and bounded on $\text{Re}s > 0$.

Clearly, the sum of two elements of A is in A. It can also be shown that the convolution of two elements of A is in A. Thus, A forms an 'algebra'. Another important property of A is the following [5]: if $g \in A$, then g has a convolution inverse in A (i.e. $1/\hat{g} \in \hat{A}$) if and only if

$$\inf_{\text{Re}s \ge 0} |\hat{g}(s)| > 0 \,. \tag{8}$$

A graphical Nyquist-type test to verify condition (8) is the following. First, note that $\hat{g}$ maps the domain $\{\text{Re}s > 0\}$ onto a bounded domain in the complex plane; boundary of the image domain is the curve $\Gamma := \{\hat{g}(i\omega), \omega \in \mathbb{R}\}$. If Γ encircles the origin, it is because $\hat{g}$ has at least one zero in $\text{Re}s > 0$; if Γ crosses the origin, it is because $\hat{g}$ has at least one zero on $\text{Re}s = 0$; if Γ approaches the origin, it is because there is a sequence of points $s_k \in \{\text{Re}s \ge 0\}$ such that $\hat{g}(s_k) \to 0$. Thus, a graphical criterion for invertibility of g in A is that the curve Γ should not encircle, cross, or approach the origin in the complex plane. (Note that $g(\cdot)$ is real, so it is only necessary to plot the curve for $\omega \ge 0$, since $\hat{g}(-i\omega) = \hat{g}(i\omega)$, i.e. the curve for negative ω is simply the reflection about the real axis of the curve for positive ω.)

2.2. BIBO stability of feedback systems

Consider the feedback configuration shown in Fig. 1. The convolution operator (impulse-response function) for the system to be controlled (the "plant") is denoted p; c is the compensator impulse-response function, u is the input, v is the output, e is the compensator input ($=$ difference $u-v$), and f is the cmpensator output. The hat '$\hat{}$' denotes the Laplace transform. The equations for the closed-loop system of Fig. 1 can be solved to yield:

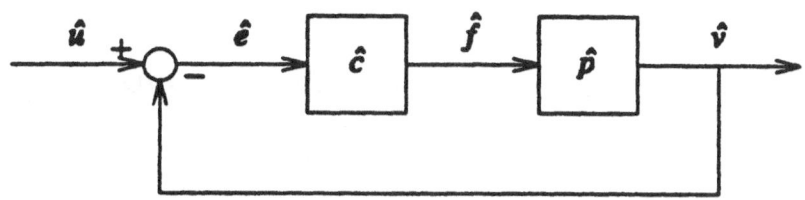

Figure 1: Feedback system configuration

$$\hat{v} = \frac{\hat{p}\hat{c}}{1 + \hat{p}\hat{c}}\hat{u}, \tag{9a}$$

$$\hat{e} = \frac{1}{1 + \hat{p}\hat{c}}\hat{u}, \tag{9b}$$

and

$$\hat{f} = \frac{\hat{c}}{1 + \hat{p}\hat{c}}\hat{u}. \tag{9c}$$

It will be assumed that the compensator impulse-response c is BIBO stable, i.e. $c \in \mathcal{A}$. With this assumption, $\hat{p}/(1 + \hat{p}\hat{c}) \in \hat{\mathcal{A}}$ is a sufficient condition for *all* of the response in the closed-loop system (e, f, v) to be bounded for any bounded input u.

Proof:

By assumption, $\hat{p}/(1 + \hat{p}\hat{c}) \in \hat{\mathcal{A}}$ and $\hat{c} \in \hat{\mathcal{A}}$, thus $\hat{p}\hat{c}/(1 + \hat{p}\hat{c})$ is the product of two elements of $\mathcal{A}$. Since $\hat{\mathcal{A}}$ is closed under multiplication, then $\hat{p}\hat{c}/(1 + \hat{p}\hat{c}) \in \hat{\mathcal{A}}$. Similarly, since $\hat{\mathcal{A}}$ is closed under addition and $1 \in \hat{\mathcal{A}}$, $1/(1 + \hat{p}\hat{c}) = 1 - \hat{p}\hat{c}/(1 + \hat{p}\hat{c}) \in \hat{\mathcal{A}}$. Multiplying by $\hat{c}$ gives $\hat{c}/(1 + \hat{p}\hat{c}) \in \hat{\mathcal{A}}$. ∎

The feedback system will be termed BIBO stable whenever $\hat{p}/(1 + \hat{p}\hat{c}) \in \hat{\mathcal{A}}$.

Assume that $\hat{p}$ can be expressed as a ratio of two causal stable transfer functions (elements of $\mathcal{A}$) as follows:

$$\hat{p}(s) = \frac{\hat{n}(s)}{\hat{d}(s)}, \quad \hat{n}, \hat{d} \in \hat{\mathcal{A}} \tag{10}$$

A sufficient condition for BIBO stability of the feedback system is

$$\frac{1}{\hat{n}\hat{c} + \hat{d}} \in \hat{\mathcal{A}}. \tag{11}$$

This is due to the fact that $\hat{p}/(1 + \hat{p}\hat{c}) = \hat{n}/(\hat{n}\hat{c} + \hat{d})$ would then be a product of elements of $\hat{\mathcal{A}}$, and would thus itself be in $\hat{\mathcal{A}}$. Since $\hat{n}\hat{c} + \hat{d}$ is itself an element of $\hat{\mathcal{A}}$, condition (11) is equivalent to requiring that $\hat{n}\hat{c} + \hat{d}$ be invertible in $\hat{\mathcal{A}}$; one may therefore use the graphical test described earlier to verify condition (11). The graph of $\Gamma{:}\omega \rightarrow (\hat{n}\hat{c} + \hat{d})(i\omega)$ will be termed the *characteristic curve* for the feedback system.

Two elements $\hat{n},\hat{d} \in \hat{A}$ are said to be *coprime in* $\hat{A}$ if there exist elements $\hat{n}_1,\hat{d}_1 \in \hat{A}$ such that

$$\hat{n}_1(s)\hat{n}(s) + \hat{d}_1(s)\hat{d}(s) = 1 \qquad \text{for all Res} \geq 0 \tag{12}$$

It has been shown [11] that (11) is also a necessary condition for BIBO stability of the feedback system, provided the factorization in (10) is coprime.

proof:

If $\hat{p}/(1 + \hat{p}\hat{c}) = \hat{n}/(\hat{n}\hat{c} + \hat{d}) \in \hat{A}$, then also $1/(1 + \hat{p}\hat{c}) = \hat{d}/(\hat{n}\hat{c} + \hat{d}) \in \hat{A}$ (see previous proof). If $\hat{n},\hat{d}$ are coprime in $\hat{A}$, there exist elements $\hat{n}_1,\hat{d}_1 \in \hat{A}$ such that $\hat{n}_1\hat{n} + \hat{d}_1\hat{d} = 1$ in Res ≥ 0. Then write $1/(\hat{n}\hat{c} + \hat{d}) = \hat{n}_1\hat{n}/(\hat{n}\hat{c} + \hat{d}) + \hat{d}_1\hat{d}/(\hat{n}\hat{c} + \hat{d})$; since the right-hand side of this last equation is made up of elements of $\hat{A}$, it is an element of $\hat{A}$. ∎

Note that condition (11) ensures BIBO stability of the feedback system, whether or not $\hat{n}$ and $\hat{d}$ are coprime. However, if it is not known whether $\hat{n}$ and $\hat{d}$ are coprime, then a violation of (11) will not necessarily imply that the system is unstable.

3. Boundary Feedback Control on One-Dimensional Wave Equation System

3.1. Derivation of transfer functions

Consider a uniform undamped elastic shaft in torsion. (Since they are governed by the same equation, one could as well have chosen a rod with axial vibration or a taut string with lateral vibration or a shear beam or a finite transmission line etc.) The equation of motion is given by the familiar one-dimensional wave equation:

$$\mu r^2 \frac{\partial^2 w}{\partial t^2}(x,t) - GJ\frac{\partial^2 w}{\partial x^2}(x,t) = 0, \tag{13}$$

where μ is the mass per unit length, r^2 is the radius of gyration, GJ is the torsional stiffness, ℓ is the length, and $w(x,t)$ is the angular rotation at position x and time t. Taking Laplace transforms with zero initial conditions gives

$$\mu r^2 s^2 \hat{w}(x,s) - GJ\frac{d^2\hat{w}}{dx^2}(x,s) = 0. \tag{14}$$

Let the end of the shaft $x = \ell$ be free, and let the other end $x = 0$ be subjected to a concentrated applied torque $f(t)$, so that the boundary conditions are

$$\frac{d\hat{w}}{dx}\bigg|_{x=0} = -\frac{\hat{f}(s)}{GJ}, \qquad \frac{d\hat{w}}{dx}\bigg|_{x=\ell} = 0. \tag{15}$$

With these boundary conditions, the solution of (14) is found to be

$$\hat{w}(x,s) = (\coth\beta s \cosh\frac{\beta s x}{\ell} - \sinh\frac{\beta s x}{\ell})\frac{\hat{f}}{\gamma s}, \tag{16}$$

where $\beta^2 := \mu r^2 \ell^2/GJ$ and $\gamma^2 := \mu r^2 GJ$ are positive constants.

Let $\hat{v}(s)$ denote the angular velocity at the point $x = 0$. The response $\hat{v}$ to applied torque $\hat{f}$ is found by substituting $\hat{v} := s\hat{w}(0,s)$ into (16), which yields

$$\hat{v} = \frac{\coth\beta s}{\gamma}\hat{f}. \tag{17}$$

The plant transfer function for this system is thus given by

$$\hat{p} := \frac{\hat{v}}{\hat{f}} = \frac{\coth\beta s}{\gamma}. \tag{18}$$

Since there is no damping, the system is not BIBO stable. For example, a unit step applied torque ($\hat{f} = 1/s$) produces a zero-state velocity response given by

$$v(t) = \mathcal{L}^{-1}\left[\frac{1}{s}\cdot\frac{\coth\beta s}{\gamma}\right] = \begin{cases} 1/\gamma & \text{for } 0 < t < 2\beta \\ 3/\gamma & \text{for } 2\beta < t < 4\beta \\ 5/\gamma & \text{for } 4\beta < t < 6\beta \\ \cdots & \cdots \end{cases} \tag{19}$$

i.e. the velocity grows without bound. Note that, although unstable in the BIBO sense, the system is stable in the usual sense of structural mechanics (i.e. there is no buckling or flutter).

Assume now that a mechanism (active or passive) is added to the end of the shaft which acts in such a way that the applied torque is proportional to the angular velocity at that point, according to a linear control law of the form

$$\hat{f} = \hat{c}(\hat{u} - \hat{v}) \tag{20}$$

where $\hat{c}$ is a transfer function to be determined, and $u(t)$ is a noise or tracking signal. The resulting feedback system is represented by the block diagram shown in Fig. 1; the physical situation is shown in Fig. 2. The stabilization problem consists of finding a compensator transfer function $\hat{c}$ such that the feedback system is BIBO stable. Due to the infinite number of unstable poles in the plant transfer function of Eq. (18) (namely, $\pm n\pi i$, $n=0,1,2,...$), this problem cannot be dealt with using the Nyquist criterion; however, the factorization method will be seen to be effective in analyzing this feedback system.

Figure 2: Physical realization of shaft boundary feedback

3.2. BIBO-stabilization condition

A coprime factorization of the plant transfer function (18) is $\hat{p}(s) = \hat{n}(s)/\hat{d}(s)$ with

$$\hat{n}(s) = \frac{1}{\gamma}(1 + e^{-2\beta s}), \quad \hat{d}(s) = 1 - e^{-2\beta s}. \tag{21}$$

To verify that $\hat{n}$ and $\hat{d}$ are elements of $\widehat{A}$, one need only consider the corresponding impulse-response functions (inverse Laplace transforms):

$$n(t) = \frac{1}{\gamma}[\delta(t) + \delta(t - 2\beta)],$$

$$d(t) = \delta(t) - \delta(t - 2\beta), \tag{22}$$

which are clearly of the form (1). To verify the coprimeness of $\hat{n}$ and $\hat{d}$, note that

$$\frac{\gamma}{2}\hat{n}(s) + \frac{1}{2}\hat{d}(s) = 1. \tag{23}$$

Since the constants $\gamma/2$ and $1/2$ are both elements of $\widehat{A}$, the coprimeness condition (12) is satisfied.

The condition for feedback stability is found by substituting (21) into (8) to yield

$$\inf_{\mathrm{Re}s \geq 0} \left| \frac{\hat{c}(s)}{\gamma}(1 + e^{-2\beta s}) + 1 - e^{-2\beta s} \right| > 0. \tag{24}$$

If $\hat{c}$ is a *strictly proper* transfer function (i.e. $\lim_{\substack{\mathrm{Re}s \geq 0 \\ |s| \to \infty}} |\hat{c}(s)| = 0$), it will not stabilize the shaft system.

proof:

Consider the points $s_r = r\pi i/\beta$, $r=1,2,...$ The image of these points under the characteristic mapping is given by

$$(\hat{n}\hat{c} + \hat{d})(s_r) = \frac{\hat{c}(s_r)}{\gamma}(1 + e^{-2\pi ri}) + 1 - e^{-2\pi ri} = \frac{2}{\gamma}\hat{c}(s_r). \tag{25}$$

If $\hat{c}(s)$ is strictly proper, the image of the points $\{s_r\}$ approaches the origin in the complex plane as $r \to \infty$, and the BIBO feedback stability condition (24) is violated. Since (21) is a *coprime* factorization, this means the feedback system will be unstable in the BIBO sense. ∎

Roughly speaking, the above result is due to the fact that the effectiveness of a strictly proper compensator diminishes at higher frequencies, so that the unstable high frequency modes are not affected by the controller. Thus, it is necessary for a stabilizing compensator to have an "infinite bandwidth". This is a consequence of the plant being modelled as having a non-strictly-proper transfer function.

3.3. Global stability

If cndition (24) is satified, this guarantees that for any bounded input u, the angular velocity v at the end $x = 0$ will remain bounded. But what about the velocity at any other point along the shaft? It turns out that if the feedback system is BIBO stable, then the velocity response at *any* point along the shaft will remain bounded for bounded input u, and a bound on this response can be calculated.

proof:

Substitute (17) into (16) to get

$$\hat{w}(x) = \frac{1}{s}\frac{\cosh(1 - x/\ell)\beta s}{\cosh\beta s}\hat{v}. \tag{26}$$

Let $\hat{v}_\xi$ denote the angular velocity response at a point $x = \xi\ell$ ($0 \leq \xi \leq \ell$), i.e. $\hat{v}_\xi := s\hat{w}(\xi\ell)$. Substituting this into (26) gives

$$\hat{v}_\xi = \hat{h}_\xi\hat{v} \tag{27}$$

with

$$\hat{h}_\xi := \frac{\cosh(1 - \xi)\beta s}{\cosh\beta s}. \tag{28}$$

Substituting (9a) into (27) gives the velocity $\hat{v}_\xi$ in terms of the input $\hat{u}$ as:

$$\hat{v}_\xi = \hat{h}_\xi\left(\frac{\hat{p}\hat{c}}{1 + \hat{p}\hat{c}}\right)\hat{u} = \frac{\hat{h}_\xi\hat{n}\hat{c}}{\hat{n}\hat{c} + \hat{d}}\hat{u}. \tag{29}$$

If $\hat{c} \in \hat{\mathcal{A}}$ and $(\hat{n}\hat{c} + \hat{d})^{-1} \in \hat{\mathcal{A}}$, then a sufficient condition for $\hat{h}_\xi\hat{n}\hat{c}(\hat{n}\hat{c} + \hat{d})^{-1}$ to be in $\hat{\mathcal{A}}$ is that $\hat{h}_\xi\hat{n}$ be in $\hat{\mathcal{A}}$. Expanding $\hat{h}_\xi\hat{n}$ using (28) and (21) gives:

$$\hat{h}_\xi\hat{n} = \frac{\cosh(1 - \xi)\beta s}{\cosh\beta s}\left[\frac{1 + e^{-2\beta s}}{\gamma}\right] = \frac{1}{\gamma}(e^{-\xi\beta s} + e^{-(2-\xi)\beta s}), \tag{30}$$

which can readily be verified to be in $\hat{\mathcal{A}}$. A bound on the velocity response at $x = \xi\ell$ is found by applying inequality (7) to (29) to get

$$\|\hat{v}_{\xi}\|_{L_{\infty}} \le \|\hat{h}_{\xi}\hat{n}\|\|\hat{\imath}\cdot\|\frac{\hat{c}}{\hat{n}\hat{c} + \hat{d}}\|\hat{\imath}\cdot\|\|\hat{u}\|_{L_{\infty}} = \frac{2}{\gamma}\|\frac{\hat{c}}{\hat{n}\hat{c} + \hat{d}}\|\hat{\imath}\cdot\|\|\hat{u}\|_{L_{\infty}}. \qquad (31)$$

3.4. Proportional compensation (ideal viscous damper)

Consider the compensator defined by

$$\hat{c}_0(s) := \gamma\kappa_0, \qquad (32)$$

where κ_0 is a dimensionless constant, and γ is the positive structural constant defined in (§3.1). Since $\hat{c}_0$ is a constant, it is an element of $\hat{\mathcal{A}}$. In physical terms, since the compensator torque is here assumed to be proportional to the velocity, the transfer function corresponds to an ideal viscous damper.

Using the definitions (21) and (32), the characteristic mapping is given by

$$(\hat{n}\hat{c}_0 + \hat{d})(i\omega) = \kappa_0(1 + e^{-2\beta i\omega}) + 1 - e^{-2\beta i\omega}. \qquad (33)$$

The characteristic curve is a circle in the complex plane (see Fig. 3). The centre of the circle is at $\kappa_0 + 1$, and the radius is $|\kappa_0 - 1|$. The condition for BIBO feedback stability is that no part of the circle or its interior cross the origin; by geometry the condition for this is simply

$$\kappa_0 > 0, \qquad (34)$$

that is, *any* passive ideal viscous damper on the boundary will stabilize *all* the modes of the system.

The stability condition (34) can be verified using inverse Laplace transforms. The impulse response function corresponding to the closed-loop operator $\hat{p}/(1 + \hat{p}\hat{c}_0)$ is given by

$$\mathcal{L}^{-1}\left[\frac{\hat{p}}{1 + \hat{p}\hat{c}_0}\right] = \mathcal{L}^{-1}\left[\frac{1/\gamma}{\kappa_0 + \tanh\beta s}\right]$$

$$= \frac{1/\gamma}{1 + \kappa_0}\left[\delta(t) + \frac{2}{1 + \kappa_0}\sum_{r=1}^{\infty}\left(\frac{1 - \kappa_0}{1 + \kappa_0}\right)^{r-1}\delta(t - 2r\beta)\right]. \qquad (35)$$

The norm is

$$\left\|\frac{\hat{p}}{1 + \hat{p}\hat{c}_0}\right\|_{\hat{\mathcal{A}}} = \frac{1/\gamma}{1 + \kappa_0}\left[1 + \frac{2}{1 + \kappa_0}\sum_{r=1}^{\infty}\left|\frac{1 - \kappa_0}{1 + \kappa_0}\right|^{r-1}\right]. \qquad (36)$$

The series in (36) converges to $[1 - |(1 - \kappa_0)/(1 + \kappa_0)|]^{-1}$ if and only if condition (34) is satified, in which case the norm of $\hat{p}/(1 + \hat{p}\hat{c}_0)$ is

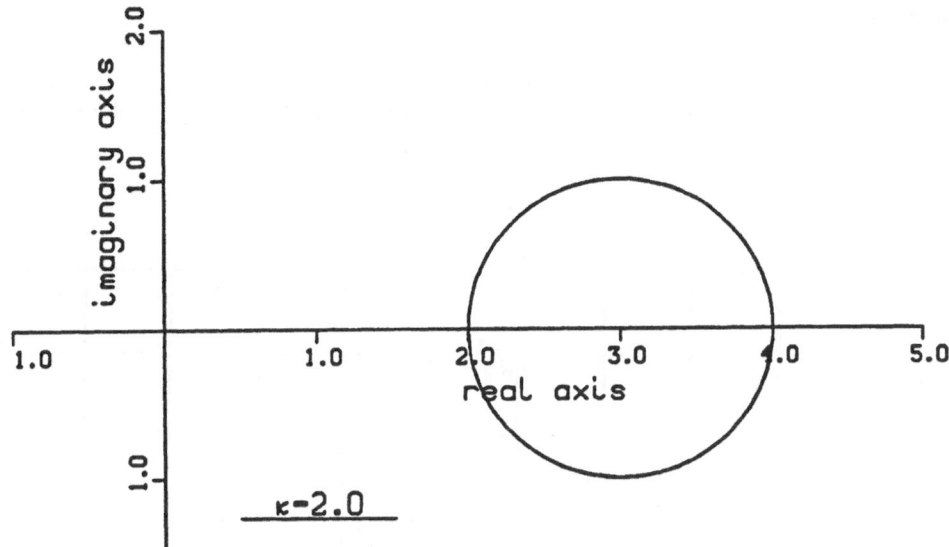

Figure 3: Characteristic curve, proportional compensator

$$\left\| \frac{\hat{p}}{1 + \hat{p}\hat{c}_0} \right\|_{\hat{A}} = \frac{1/\gamma}{1 + \kappa_0} \left[1 + \frac{2}{(1 + \kappa_0)\left[1 - \left| \dfrac{1 - \kappa_0}{1 + \kappa_0} \right| \right]} \right]. \tag{37}$$

Thus (34) is verified as a stability condition.

Recall that the $\hat{A}$ norm ($= A$ norm) provides a bound on the magnitude of the response to any bounded input. For a *tracking* problem (where $u(t)$ is a tracking signal), one is interested in minimizing the error signal e. The closed-loop transfer function for $\hat{e}$ as a response to $\hat{u}$ is $1/(1 + \hat{p}\hat{c}_0)$. Assuming $\kappa_0 > 0$, the $\hat{A}$ norm is found to be

$$\left\| \frac{1}{1 + \kappa_0 \coth\beta s} \right\|_{\hat{A}} = \frac{1}{1 + \kappa_0} \left[1 + \frac{2\kappa_0}{(1 + \kappa_0)\left[1 - \left| \dfrac{1 - \kappa_0}{1 + \kappa_0} \right| \right]} \right]. \tag{38}$$

For a *disturbance rejection* problem (where the input $u(t)$ is a noise signal), one is interested in minimizing the boundary velocity v. The closed-loop transfer function for $\hat{v}$ as a response to $\hat{u}$ is $\hat{p}\hat{c}_0/(1 + \hat{p}\hat{c}_0)$; since this is only $\hat{p}/(1 + \hat{p}\hat{c}_0)$ times the constant $\hat{c}_0 = \kappa_0\gamma$, the $\hat{A}$ norm is simply the right-hand side of Eq. (37) times the constant $\kappa_0\gamma$. The $\hat{A}$ norms defined above are upper bounds on the outputs e and v for any unit-bounded input u. These bounds are plotted as functions of the compensator gain constant κ_0 in Fig. 4. Note that the values of the norms lie in $[1,2)$, with the common minimum occuring at $\kappa_0 = 1$. The compensator gain $\kappa_0 = 1$ corresponds to perfect impedance matching at the boundary, that is, the boundary condition is such that no reflection of incident waves occurs.

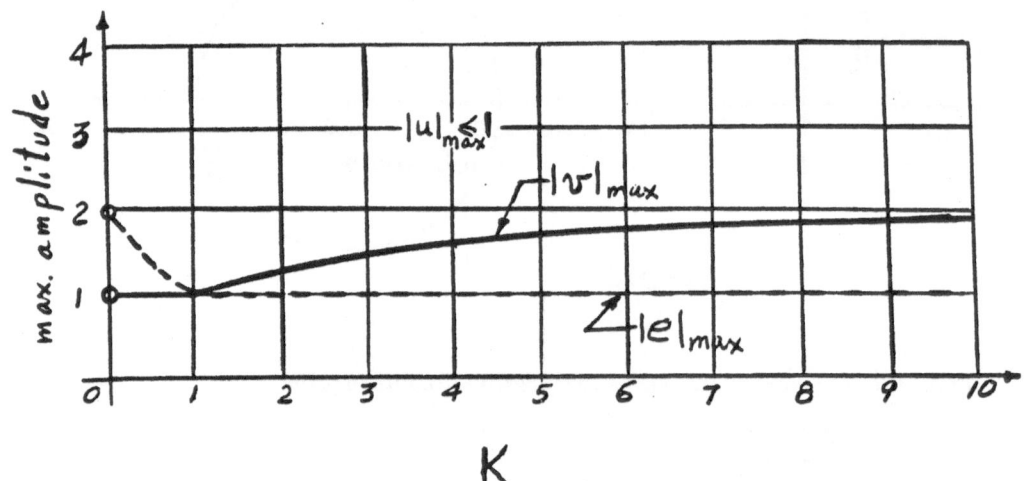

Figure 4: Input-output norms, shaft boundary velocity-proportional feedback

3.5. Lag compensator (Maxwell damper)

Consider the compensator defined by

$$\hat{c}_1(s) := \frac{\gamma \kappa_1}{\tau s + 1}, \tag{39}$$

where κ_1 is the compensator gain, τ is the lag time constant, and γ is the positive structural constant defined in §3.1. This compensator corresponds to an ideal Maxwell damper on the boundary (see Fig. 5), with spring stiffness $\gamma \kappa_1 / \tau$ and damper constant $\gamma \kappa_1$. In circuit theory it is termed a 'lag' compensator.

The compensator impulse response function $c_1(t)$ is given by

$$c_1(t) := L^{-1}[\hat{c}_1(s)] = \frac{\gamma \kappa_1}{\tau} e^{-t/\tau}, \tag{40}$$

thus the compensator is stable (i.e. $c_1 \in A$) provided $\tau > 0$.

Figure 5: *Mechanical realization of lag compensator*

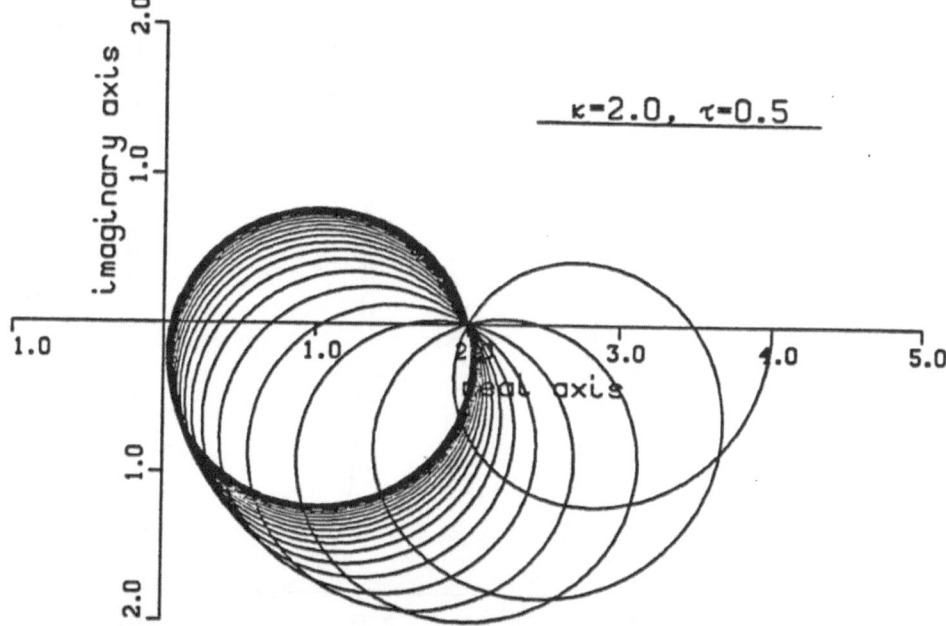

Figure 6: *Characteristic curve, lag compensator*

Plotting the characteristic curve for this compensator and the factorized shaft transfer function (21) gives a curve such as the one shown in Fig. 6. The curve does not encircle the origin, but does approach it as $\omega \to \infty$. This is due to $\hat{c}_1$ being a strictly proper transfer function; as shown in §3.2, such a compensator cannot stabilize the feedback system, since the resonant response of higher modes becomes arbitrarily large. For example, consider the steady-state amplitude response v when a unit-amplitude sinusoidal input of frequency ω is applied:

$$|\hat{v}(\omega)| = \left| \left(\frac{\hat{p}\hat{c}_1}{1 + \hat{p}\hat{c}_1} \right)(i\omega) \right|. \tag{41}$$

A typical graph of this amplitude response function, Fig. 7, shows how the size of the resonance peaks grows with frequency. The response to high-frequency resonant inputs becomes unbounded as the input frequencies increase. Theoretically, then, a bounded input with an 'infinite frequency' component (e.g. white noise) would lead to a response with 'infinite' amplitude. This form of instability could be termed "resonance at infinity".

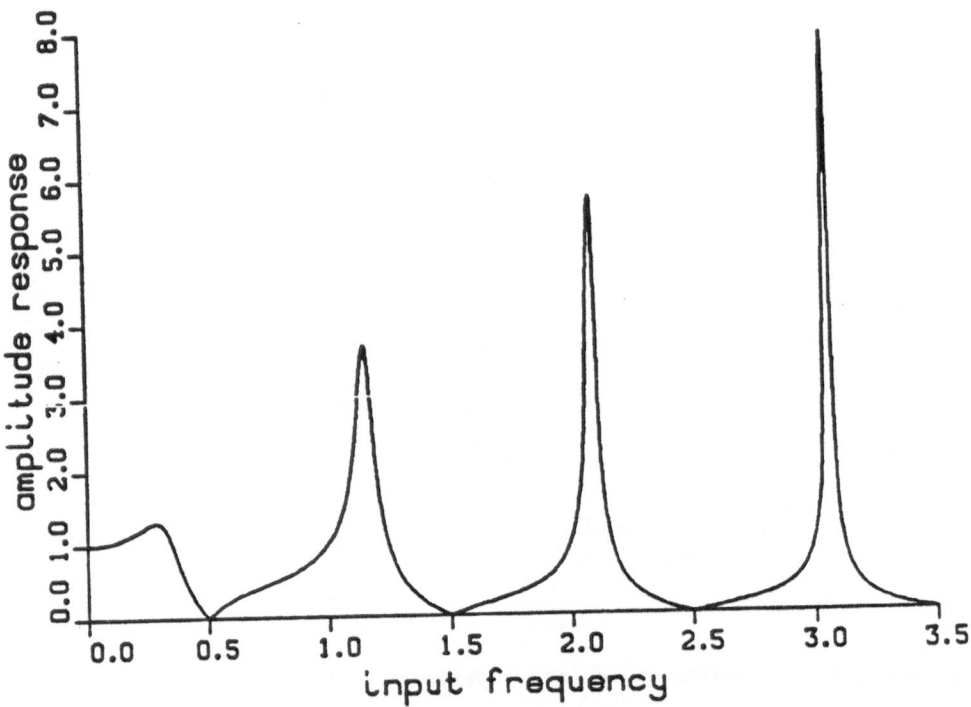

Figure 7: Tip velocity amplitude response to unit sinusoidal input

It is interesting to note that an analysis of this system using a finite-dimensional lumped parameter model of the shaft would not predict this type of instability, because the transfer function would be strictly proper. It can be shown [7] that, in contrast to the above results, lag feedback compensation of a lumped-parameter version of the shaft transfer function can lead to a BIBO stable system. The difference in results is due to the loss of higher-order unstable modes in the lumped-parameter model.

3.6. Lag-lead compensator (Three-element viscous damper)

Consider a compensator defined by

$$\hat{c}_2 := \gamma\kappa_2\frac{1 + \tau_1 s}{1 + \tau_2 s}, \tag{42}$$

where κ_2 is the gain constant, τ_1 and τ_2 are lead and lag time constants, and γ is the positive structural constant defined in §3.1. In circuit theory this is termed a 'lag-lead' compensator. In mechanics, this transfer function corresponds to a 'three element viscous damper' [2]; Fig. 8 shows one possible mechanical realization using ideal dampers and an ideal spring.

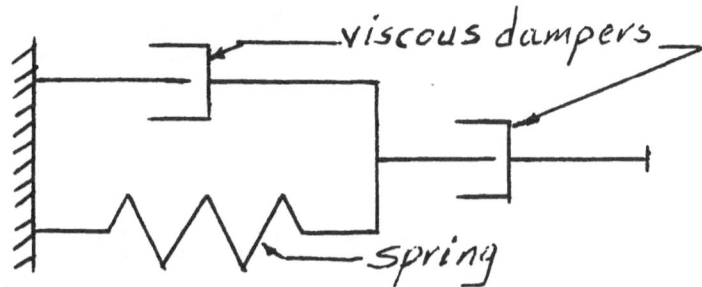

Figure 8: Mechanical realization of lag-lead compensator

The compensator impulse response function $c_2(t)$ is given by

$$c_2(t) := L^{-1}[\hat{c}_2(s)] = \gamma\kappa_2\frac{\tau_1}{\tau_2}\left[(\frac{1}{\tau_1} - \frac{1}{\tau_2})e^{-t/\tau_2} + \delta(t)\right], \tag{43}$$

thus $c_2 \in A$ provided $\tau_2 > 0$.

Figure 9 shows a characteristic curve for the feedback system with lag-lead compensator. Starting ($\omega = 0$) at a point along the positive real axis, the curve spirals clockwise, crossing another point on the positive real axis repeatedly, and gradually approaches a circle.

Some general characteristics of the characteristic curve $(\hat{n}\hat{c}_2 + \hat{d})(i\omega)$ curve can be derived analytically. The starting point ($\omega = 0$) is given by

$$(\hat{n}\hat{c}_2 + \hat{d})(0) = 2\kappa_2. \tag{44}$$

A common point for each spiral loop occurs at $\beta\omega = \pi/2, 3\pi/2,...$, for which the characteristic curve value is

$$(\hat{n}\hat{c}_2 + \hat{d})[\frac{(2r - 1)\pi i}{2\beta}] = 2, \qquad r = 1,2,... \ . \tag{45}$$

To assess the asymptotic behaviour of the curve for high frequencies, let $\omega = (\pi r + \phi)/\beta$, where $0 \leq \phi \leq \pi$ and r is an integer. Then

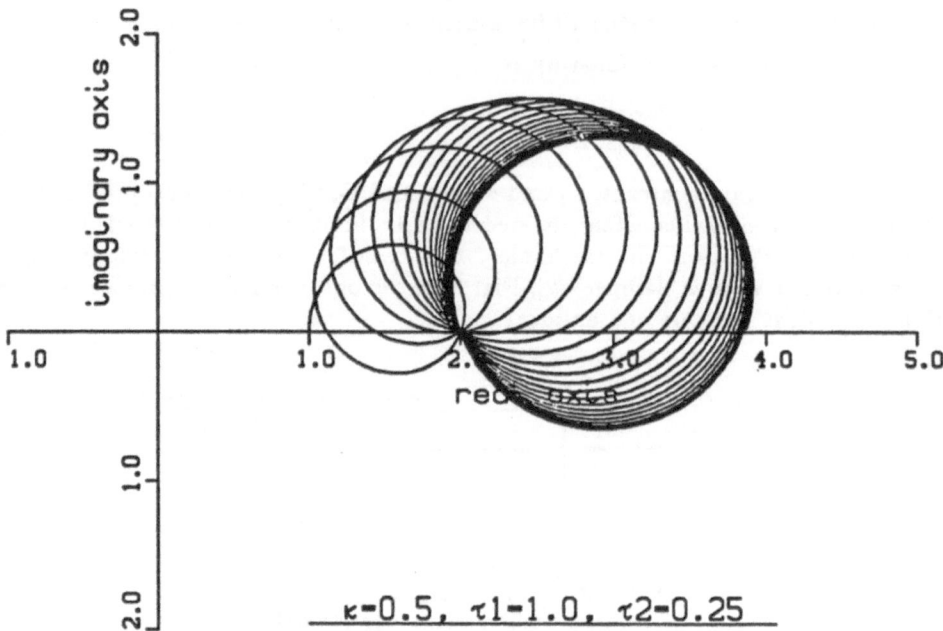

Figure 9: Characteristic curve, lag-lead compensator

$$\lim_{r \to \infty}(\hat{n}\hat{c}_2 + \hat{d})[\frac{i(\pi r + \phi)}{\beta}] = \kappa_2\frac{\tau_1}{\tau_2}(1 + e^{-2i\phi}) + 1 - e^{-2i\phi}. \tag{46}$$

In the limit, the characteristic curve for the feedback system with the lag-lead compensator tends to the same as for the system with the proportional compensator in §3.4. As $s \to \infty$, the compensator transfer function (42) approaches the constant value $\hat{c}_2 \to \kappa_2\tau_1/\tau_2$, so that at high frequencies the lag-lead compensator behaves like the proportional compensator (32) with proportionality constant $\kappa_2\tau_1/\tau_2$ instead of κ_0. One condition for stability is therefore $\kappa_2\tau_1/\tau_2 > 0$, analogous to (34). Also, the starting point defined in (44) should lie to the right of the origin; this gives the condition $\kappa_2 > 0$. Since $\tau_2 > 0$ (by assumption, for stability of c_2) and $\kappa_2 > 0$, the proportionality constant condition reduces to $\tau_1 > 0$. Three necessary conditions for BIBO stability of the feedback system with stable lag-lead compensator $\hat{c}_2$ are therefore simply: $\kappa_2 > 0$, $\tau_1 > 0$, and $\tau_2 > 0$.

In general, one can expect similar results for other non-strictly-proper rational compensator transfer functions: the asymptotic high-frequency behaviour of the feedback system will resemble that of the system with proportional compensator, due to the fact that the compensator transfer function approaches a constant value.

3.7. Delayed feedback

If the controller is an active, digitally controlled device, there will be a certain finite delay between the sensor measurement and the compensator action. Consider a velocity-proportional feedback compensator, as in §3.4, delayed by a given time $\theta > 0$. The transfer function for this delayed compensator is

$$\hat{c}_3(s) := \gamma\kappa_3 e^{-\theta s}. \tag{47}$$

The impulse-response function corresponding to (47) is

$$c_3(t) = \gamma\kappa_3\delta(t - \theta), \tag{48}$$

so $c_3 \in \mathcal{A}$.

A typical characteristic curve for delayed proportional feedback compensation of the shaft system is shown in Fig. 10. The curve encircles the origin repeatedly, thus the system is unstable.

Many graphs such as Fig. 10 have been drawn, but no combination of positive parameters for Eq. (47) has been found to yield a stable system. This leads to the conjecture that the undamped shaft system cannot be stabilized using a delayed compensator, and that *any* finite delay will lead to instability. This conjecture should of course be verified analytically.

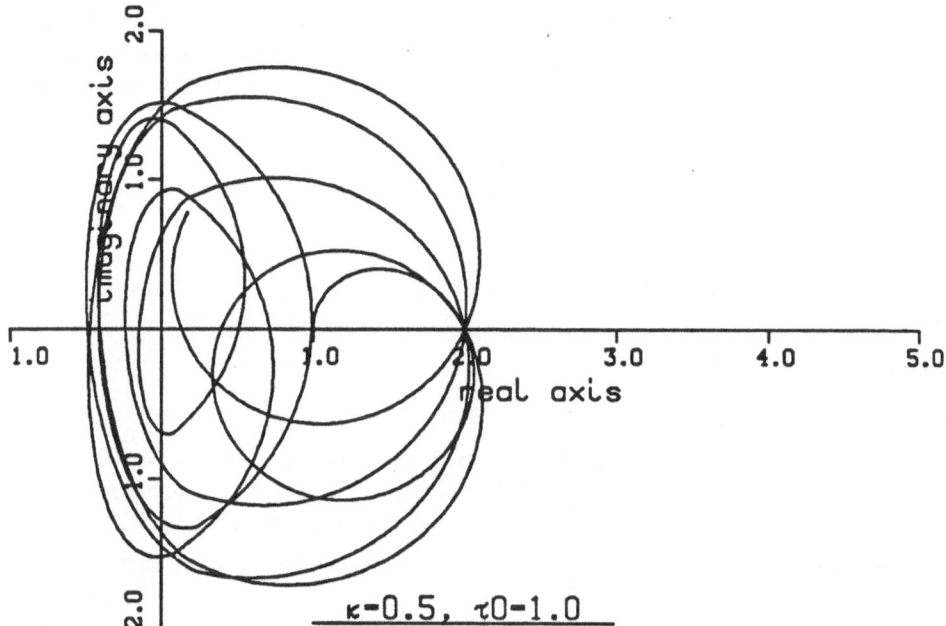

Figure 10: Characteristic curve, delayed compensation

4. Work in Progress

The purpose of this paper has been to introduce the coprime factorization
method as a tool in the analysis of feedback systems involving elastic structures,
especially for continuum or 'distributed parameter' models of such structures.
Several extensions of the work presented here are in progress, and will be
reported on in later communications. Some of these are described below.

- This paper has only shown the factorization of the closed-form transfer function for a torsional shaft. Closed-form transfer functions can also be written
  for systems involving membrance, beams, plates, and shells; analysis of feedback control can be done using the factorization method.

- The dynamic stiffness matrix method can be used to derive the closed-form
  transfer functions for complex structures made up of continuous elastic
  members. It is gaining popularity as an efficient method for computing the
  natural frequencies and frequency responses of structures [15, 6, 8, 13]. The
  dynamic stiffness matrix method can be modified to yield stable factorizations of transfer functions. This opens up the possibility of using the factorization method for a wide class of structures of practical interest.

- The transfer functions for structures with linear structural damping can be
  written by replacing the elastic constant E by a linear operator $\hat{E}(s)$. The
  characteristic curves drawn using stable factorizations are found to be much
  less sensitive to computational errors than Nyquist diagrams. Also, preliminary results indicate that the "resonance at infinity" for strictly proper compensators and the instability for "any" finite delay in the feedback do not
  occur in systems with damping.

Acknowledgements

The support and encouragement of Prof. H.H.E. Leipholz are gratefully ack-
nowledged. Part of this work was carried out under a research contact for the
Communications Research Centre (Space Technology and Applications Branch),
Dept. of Communications, Ottawa.

References

[1] Berkman, F. and Karnopp, D., "Complete Response of Distributed Systems
    Controlled by a Finite Number of Linear Feedback Loops," *ASME Journal
    of Engineering for Industry*, November 1969, pp. 1063-1068.

[2] Bland, D.R., *The Theory of Linear Viscoelasticity*, Pergamon, 1960.

[3] Desoer, C.A., "A General Formulation of the Nyquist Criterion," *IEEE
    Trans. on Circuit Theory*, June 1965, pp. 230-234.

[4] Desoer, C.A. and Vidyasagar, M., *Feedback Systems: Input-Output Proper-
    ties*, Academic Press, 1975.

[5] Hille, E. and Phillips, R.S., *Functional Analysis and Semi-Groups*, American Mathematical Society, 1957.

[6] Kärnä, T., "A Computer Model for Wind-Induced Vibrations of Guyed Masts," pp. E79-80 in *Proceeding of Tenth Canadian Congress of Applied Mechanics (CANCAM)*, University of Western Ontario, London, Ontario, June 2-7 1985.

[7] Piche, R., "Frequency-Domain Continuum Modelling and Control of Third-Generation Spacecraft," Contractor Report DOC-CR-85-032, Dept. of Communications, Ottawa, May 1985.

[8] Poelaert, D., "DISTEL, A Distributed Element Program for Dynamic Modelling and Response Analysis of Flexible Structures," pp. 319-338 in *Proc. 4th VPI&SU/AIAA Symp. on Dynamics and Control of Large Structures*, Blacksburg, Virginia, June 6-8 1983.

[9] Russell, D.L., "Control Canonical Structure for a Class of Distributed Parameter Systems," *Third IMA Conference on Control Theory*, Academic Press, 1981, pp. 794-825.

[10] Vidyasagar, M., "Input-Output Stability of a Broad Class of Linear Time-Invariant Multivariable Systems," *SIAM J. Control*, 10(1), February 1972, pp. 203-209.

[11] Vidyasagar, M., "Coprime Factorizations and Stability of Multivariable Distributed Feedback Systems," *SIAM J. Control*, 13(6), November 1975, pp. 1144-1155.

[12] Willems, J.C., *The Analysis of Feedback Systems*, MIT Press, 1971.

[13] Williams, F.W. and Wittrick, W.H., "Exact Buckling and Frequency Calculations Surveyed," *J. Structural Engineering*, 109(1), January 1983, pp. 169-187.

[14] Zemanian, A.H., *Distribution Theory and Transform Analysis*, McGraw-Hill, 1965.

[15] Akesson, B. and Friberg, P., "Exact Buckling and Frequency Calculations Surveyed - Discussion," *ASCE Journal of Structural Engineering*, Vol. 110, 1984, pp. 186-188.

EIGENVALUE SPECIFICATION USING FEEDBACK FOR A CLASS OF INFINITE DIMENSIONAL CONTROL SYSTEMS

R.L. Rebarber
Department of Mathematics and Statistics
University of Nebraska-Lincoln

1. Introduction

In this paper we discuss the use of control canonical forms to help construct feedback controls solving the eigenvalue specification problem for a class of infinite dimensional linear control systems.

$$\dot{x}(t) = Ax(t) + bu(t)$$

$$x(0) = x_0 \tag{1}$$

where $x(t) \epsilon H$, a Hilbert space, and $u(t)$ is a scalar control. We assume that A has a Riesz basis of eigenvectors $\{\phi_k\}_{k\epsilon I}$ with associated eigenvalues $\{\lambda_k\}_{k\epsilon I}$. b is an "admissible input element", as defined in Ho and Russell [4], which can be written in the form

$$b = \sum_{k\epsilon I} b_k \phi_k.$$

Let $\{\psi_k\}_{k\epsilon I}$ be the Riesz basis of H which is biorthogonal to $\{\phi_k\}_{k\epsilon I}$.

We will be primarily interested in the case where the eigenvalues lie between two curves of the form

$$\Gamma_a := \{a \pm x e^{\pm i\sigma} | x \in (0,\infty)\}$$

where $\sigma \in [0,\pi/2]$. In this case A generates a holomorphic semigroup in

$$\tilde{\Omega} := \{z \,||\arg(z)| < \sigma\}.$$

Systems of this sort come up in models for structural damping of a vibrating system. A class of examples can be described as follows:

Let B be a positive, self adjoint operator on a Hilbert space X with a Riesz basis of eigenvectors $\{\chi_k\}$ and associated eigenvalues $\{\mu_k\}$. The associated undamped, uncontrolled vibrating system is $\ddot{x} + Bx = 0$. To model structural damping, we could use

$$\ddot{x} + 2\rho B^{\cdot} \dot{x} + Bx = 0, \quad 0 < \rho < 1.$$

Let $x_1 = B^{\cdot}x$ and $x_2 = \dot{x}$. Then, the equivalent first order system is

$$\begin{bmatrix} x_1 \\ x_2 \end{bmatrix} = \begin{bmatrix} 0 & B^{\cdot} \\ -B^{\cdot} & -2\rho B^{\cdot} \end{bmatrix} \begin{bmatrix} x_1 \\ x_2 \end{bmatrix} := A \begin{bmatrix} x_1 \\ x_2 \end{bmatrix}.$$

A has eigenvalues $\{\pm i\sqrt{\mu_k}(-\rho \pm i(1-\rho^2)^{\cdot})\} =: \{\pm i\sqrt{\mu_k}(e^{\pm i\theta})\}$, and eigenvectors

$$\begin{bmatrix} \chi_k \\ ie^{i\theta}\chi_k \end{bmatrix},$$

which form a Riesz basis for $H = X \otimes X$.

Chen and Russell show that if the damping term is $2\rho C$, where C is "close" to B, then A still generates a holomorphic semigroup. For example, this happens if $C = B^{\cdot} + DB^{\cdot}$, where $||D|| < 1$.

In the case of the vibrating beam, B is the beam operator $\partial^4/\partial s^4$ on a domain $\mathcal{D}(B)$ determined by the boundary conditions. For some domains, $B^{\cdot} = -\partial^2/\partial s^2$ on $\mathcal{D}(B)$. However, there are choices for $\mathcal{D}(B)$ for which $-\partial^2/\partial s^2$ is not "close" to $B^{\cdot}$ in any useful sense. In some of these cases it can still be shown that the associated first order operator still has a Riesz basis of eigenvectors and genarates a holomorphic semigroup.

2. Canonical Forms

We consider a control canonical form to be a scalar functional equation and an associated canonical state space equation. These two equations are equivalent in the sense that we can recover the solution of one from the solution of the other. To use the canonical form, we map the original system to the canonical state space equation, and then use the scalar equation to analyze the properties of the solution. Canonical forms relevant to this discussion have been constructed for infinite dimensional systems by Russell [7,8], Teglas [12] and Rebarber [5].

This study of canonical forms for these systems is dependent upon the placement of the eigenvalues $\{\lambda_k\}_{k\in I}$. Different eigenvalue configurations require somewhat different methods, although there is a unifying structure which we will try to indicate. All of these methods hinge upon the generalization of the characteristic polynomial, called a cardinal function, which is an entire or meromorphic function. The cardinal function will represent the functional equation in the frequency domain, and the properties of the canonical state space are analyzed using the properties of the cardinal function.

The first work on canonical forms along these lines was done by Russell. In [7] a class of augmented and deficient hyperbolic distributed parameter systems is shown to be equivalent to nth order neutral integral equations. In this case the eigenvalues lie in a strip the complex plane of the form $\{\,|Re(z)| < a\}$, and are in clusters which are asymptotically linearly spaced. The cardinal functions here will be entire and have certain growth properties.

In [3], Ho studies a class of degenerate hyperbolic systems. Frequency domain techniques are used to construct the canonical state space. In this case the eigenvalues lie on a curve in the left half plane. Ho also studies a class of retarded functional equations, and argues that these can serve as a canonical form for these systems. The methods used in Rebarber [5] can be easily adapted to rigorously show this equivalence. These methods also complete an argument in [8], where Russell considers systems which have negative real eigenvalues spaced like integers squared.

In all of the above cases, the eigenvalue placement allows an entire function with zeros at the eigenvalues and useful growth conditions. This is not the case for the systems described above governed by holomorphic semigroups. In this case the cardinal function is meromorphic. In this paper, we will be primarily interested in describing the canonical form and control results for these systems.

We begin a more detailed examination of infinite dimensional canonical forms in the next section, where we briefly discuss the canonical form for hyperbolic distributed parameter systems done by Russell in [7,9], and then give some indication of the general structure for canonical forms of this sort.

3. Example of an Infinite Dimensional Canonical Form

We assume that the eigenvalues are of the form $\{2\pi ji + \epsilon_j + \alpha\}_{j\in Z}$ where $\{e^{\lambda_j \cdot}\}_{j\in Z}$ is a Riesz basis for $L^2[-1,1]$. This space will serve as the canonical state space. The canonical state space equation in $L^2[-1,1]$ will be

$$\dot{x}(t) = Dx(t) + \hat{b}u(t). \tag{2}$$

Here D is a "differentiation" operator in the sense that $D(\sum_{k\in Z} x_k e^{\lambda_k s}) = \sum_{k\in Z} x_k \lambda_k e^{\lambda_k s}$. The domain of D, $\mathcal{D}(D)$, is the subspace of $H^2[-1,1]$ which contains all elements of the form $\sum_{k\in Z} x_k e^{\lambda_k s}$, where $\{x_k \lambda_k\}_{k\in Z}$ is square summable. $\hat{b}$ is the canonical input element, which can be represented by $\sum (1/p'(\lambda_k))\phi_k$, where p is the associated cardinal function. $\hat{b}$ doesn't belong to $L^2[-1,1]$, but it is an admissible input element [3].

The canonical functional equation is of the form

$$(\eta * y)(t) := <\eta, y(t+\cdot)> = u(t), \tag{3}$$

where the "generating functional" η is an element of the dual space of $H^2[-1,1]$. $\mathcal{D}(D)$ has the alternate characterization as all x in $H^2[-1,1]$ such that $<\eta,x> = 0$, For these systems Eq. (3) is a zero order neutral equation. The convolution operator can be defined as in Eq. (3), but it can also be defined in the frequency domain by using $p(\lambda) = F\eta(\lambda) := <\eta, e^{\lambda \cdot}>$. To solve this functional equation, we can use the two-sided Laplace transform theory developed in [8]. It can be shown that if $w(t,s) := y(t+s)$ for real t, then $w(t,s) = x(t,s)$, the solution of the canonical state space equation, for $-1 < s < 1$.

In general, it is not possible to find a useful state space which has $\{e^{\lambda_k \cdot}\}_{k \in I}$ as a Riesz basis. In the general approach, the canonical state space E will be a space of equivalence classes of a space X, where the equivalence class containing x is denoted by $\{x\}$. E will have $\{\{e^{\lambda_k \cdot}\}\}_{k \in I}$ as a Riesz basis. We can put the above canonical form in this framework as follows: let

$$X^{-a} := \left\{ x \, |e^{-\rho|\cdot|}x \in L^2(-\infty,\infty), \, \forall \, \rho > a \right\}.$$

We then write $x \sim y$ if $x = y$ in $L^2[-1,1]$. In this way we can think of the canonical state space as being the space of all equivalence classes of X^a. The canonical state space equation will be an appropriate generalization of Eq. (2).

More generally, we start with space X and Y, and a bilinear form $<\cdot,\cdot>$ defined on $Y \otimes Y$. Then we say that $x_1 \sim x_2$ if $<y,x_1> = <y,x_2>$ for all $y \in Y$. In the above hyperbolic case, X is X^a, Y is $L^2[-1,1]$, and $<y,x>$ is $\int_{-1}^{1} x(t)y(t) \, dt$. Furthermore, since we would like to map the original state space equation to the canonicl state space equation, it will be useful to have a Riesz basis of Y which is dual to $\{\{e^{\lambda_k \cdot}\}\}_{k \in I}$. In the hyperbolic case, the dual basis is $\{g_k\}_{k \in I}$, where $g_k = L^{-1}p_k$, and

$$p_k(\lambda) := \frac{p(\lambda)}{\left[p'(\lambda_k)(\lambda - \lambda_k)\right]}. \tag{4}$$

We will illustrate this approach by describing the canonical form in the case where A generates a holomorphic semigroup. All of the following results can be found in Rebarber [5], except as noted. We will only present those results necessary to discuss the construction of feedback controls in detail.

4. Holomorphic Semigroup Case

4.1. A Laplace transform

The canonical state space and functional equation are constructed and analyzed in the frequency domain, so a Laplace transform relevent to holomorphic semigroups is needed. This is done in Rebarber [5].

Let $\ell_\theta := \{xe^{i\theta} | x \epsilon (0,\infty)\}$. Let X^a be the space of all functions $x(s)$ which are holomorphic in $\tilde{\Omega}$, with L^2 boundary values, such that for all $\rho > a$ and all $|\theta| < \sigma$, $e^{-\rho \vdash} | x \epsilon L^2[\ell_\theta]$.

Let $z > \Gamma_a$ mean that z is to the right of Γ_a. Let $\Omega_a := \{z > \Gamma_a\}$. Then Ψ^a will be the set of all Ψ holomorphic in Ω_a which satisfy certain growth properties in Ω_a.

For $f \epsilon X^a$, let $||f||_\rho$ be defined as the L^2 norm of $e^{-\rho} f$ on $\ell_\sigma \bigcup \ell_{-\sigma}$. For $\psi \epsilon \Psi^a$, let $||\psi||_\rho$ be defined as the L^2 norm of ψ on Γ_ρ.

We can then define a Laplace transform $L:X^a \rightarrow \Psi^a$. The relationship between x and Lx is similar to the relationship between the semigroup generated by A and the resolvent of A in the classical theory of holomorphic semigroups [2].

Theorem 1: L is one to one and onto. When X^a and Ψ^a are given the above norms, with $\rho > a$, L is an isomorphism.

4.2. The canonical state space

Let $X^{a,b}$ be the set of all f in X^a such that Lf is meromorphic and has growth properties in $\{z < \Gamma_b\}$ similar to the growth properties in $\{z > \Gamma_a\}$. We choose $a > b$ so that $\{\lambda_k\}_{k\epsilon I}$ lies between Γ_a and Γ_b. The canonical state space will be a set of equivalence classes of $X^{a,b}$.

Let p be a cardinal function with zeros at $\{\lambda_k\}_{k\epsilon I}$ and poles at $\{\mu_k\}_{k\epsilon I}$. The precise growth conditions on p and spacing conditions on $\{\lambda_k\}_{k\epsilon I}$ and $\{\mu_k\}_{k\epsilon I}$ are given in [5]. A class of cardinal functions are constructed in [5].

Let $\{\alpha_k\}_{k\epsilon I}$ be a set of complex numbers between Γ_a and Γ_b.

$$X\{\alpha_k\} := \left\{ \sum_{k\epsilon I} x_k e^{\alpha_k \cdot} \Big| \sum_{k\epsilon I} |x_k|^2 < \infty \right\}$$

and

$$\Psi\{\alpha_k\} := \left\{ \sum_{k\epsilon I} x_k (\lambda - \alpha_k)^{-1} \Big| \sum_{k\epsilon I} |x_k|^2 < \infty \right\}.$$

The following theorem can be found in [6].

Theorem 2: Let $\{\alpha_k\}_{k \epsilon I}$ be the zero set or the pole set of a cardinal function. For all $x = \sum_{k \epsilon I} x_k e^{\lambda_k \bullet}$ and $r > a$ there exists positive M_r and m_r such that

$$m_r \, ||x\,||_r \leq \left[\sum_{k \epsilon I} |x_k|^2 \right]^{\frac{1}{2}} \leq M_r \, ||x\,||_r.$$

This Theorem shows that $X\{\alpha_k\}$ is a Hilbert space with norm $||\cdot||_r$. This also shows that $\{e^{\alpha_k \bullet}\}_{k \epsilon I}$ is a Riesz basis for $X\{\alpha_k\}$, so $\{(\lambda - \alpha_k)^{-1}\}_{k \epsilon I}$ is a Riesz basis for $\Psi\{\alpha_k\}$.

We will define a duality relationship between $X := X^{a,b}$ and $Y := X\{\mu_k\}$. We then take equivalence classes of X with respect to Y. Let $M := LX$ and $\Phi := LY$. For $\psi \, \epsilon \, M$ and $\phi \, \epsilon \, \Phi$ we define the bilinear form

$$<\phi,\psi> \; = \; \frac{1}{2\pi i} \int_{-\Gamma_r} \phi(z)\psi(z) \, dz,$$

which is easily seen to be independent of $r \, \epsilon \, (c,b)$, where c is chosen so that $\{\mu_k\}_{k \epsilon I}$ are to the left of Γ_c. For $x \, \epsilon \, X$ and $y \, \epsilon \, Y$, define $<y,x> \; = \; <Ly,Lx>$. We write that $x_1 \sim x_2$ if $<y,x> \; = \; <y,x>$ for every $y \, \epsilon \, Y$. Let E be the space of equivalence classes with respect to this relation.

It can be shown that, if p_k is given by Eq. (4), then $\{p_k\}_{k \epsilon I}$ is a Riesz basis for Φ. Let $g_k := L^{-1}p_k$, so $\{g_k\}_{k \epsilon I}$ is a Riesz basis for Y. E is clearly a Hilbert space with norm

$$||\{x\}\,||_r \; := \; \sup_{y \epsilon Y} \frac{<y,x>}{||y\,||_r}. \qquad \qquad \circ$$

In fact, we have the following result, which shows that E is isomorphic to $X\{\lambda_k\}$:

Theorem 3: $\{\{e^{\lambda_k \bullet}\}\}_{k \epsilon I}$ is a Riesz basis for E, with dual basis $\{g_k\}_{k \epsilon I}$.

4.3. The canonical state space equation

The semigroup generator for the canonical form will be a "differentiation" operator D on E, which is defined on an appropriate subspace of "differentiable" elements of D. We will be necessarily vague about details here, to conserve space, referring the reader to [5].

Let $\tilde{X}$ be the space of functions in X which also have derivatives in X, and let $\tilde{M} = L(\tilde{X})$. Let $\tilde{E}$ be the set of equivalence classes of $\tilde{X}$. It can be shown that if $f_1 \sim f_{2,}$ and $f_1, f_2 \, \epsilon \, X$, then $f_1{}' \sim f_2{}'$. Therefore, we can define differentiation on $\tilde{E}$ as $D\{f\} = \{f'\}$. If we restrict D to

$$\mathcal{D}(D) = \left\{ \sum_{k \epsilon I} x_k \{e^{\lambda_k s}\} \Big| \sum_{k \epsilon I} |x_k \lambda_k|^2 < \infty \right\},$$

then D will be the infinitessimal generator of the strongly continuous holomorphic semigroup on E given by

$$S(t)\left[\sum_{k \epsilon I} x_k \{e^{\lambda_k s}\} \right] = \sum_{k \epsilon I} x_k e^{\lambda_k t} \{e^{\lambda_k s}\}$$

for $t \epsilon \tilde{\Omega}$.

Let $\hat{b} = \sum_{k \epsilon I} \left[1/p'(\lambda_k) \right] \{e^{\lambda_k s}\}$. Then $\hat{b}$ can be shown to be an "admissible input element" in the sense given in [3]. The canonical state space equation will be

$$\{\dot{x}(t)\} = D\{x(t)\} + \hat{b}u(t),$$
$$\{x(0)\} = \{x_0\}. \tag{5}$$

4.4. The canonical functional equation

We now define a functional equation on X which is equivalent to the above state space equation. We first define a convolution type operator on X related to p. For $z > \Gamma_a$ or $z < \Gamma_b$, and $\psi \epsilon M$ we define

$$(p * \psi)(\lambda) := \frac{1}{2\pi i} \int_{\Gamma_r \cup -\Gamma_\rho} \frac{p(\varsigma)\psi(\varsigma)}{\lambda - \varsigma} \, d\varsigma,$$

which is independent of $r > a$ and $\rho < b$. It can be shown that $p * \psi \epsilon M$ for $x \epsilon X$, $(\eta * x)$ is defined by $L^{-1}(p * Lx)$. There is an associated bounded linear functional η on X for which $(\eta * x)(t) = (\eta, x(t + \cdot))$, and $<\eta, e^{\lambda \cdot}> = p(\lambda)$.

Consider the following eqation, for $u \epsilon X$:

$$\eta * y = u$$
$$y \sim x_0$$
$$\{x(t,s)\} = \{y(t + s)\} \tag{6}$$

We need some convenient machinery to solve Eq. (6). This is given by the next two results.

Theorem 6: Let $\psi = Lx$. Then $L(\eta * x) = p\psi + \phi$, where $\phi \epsilon \Phi$.

Theorem 7: If $Lu \epsilon M$, then $Lu/p \sim 0$.

Thus, if $u \epsilon X$, the solution of Eq. (6) is $y = L^{-1}\psi$, where $\psi = \psi_0 + Lu/p$, with $\psi_0 \epsilon \Psi\{\lambda_k\}$ and $\psi_0 \sim Lx_0$.

Now we have the canonical state space equation, which is of the same form as the original state space equation, and a functional equation, which we can solve easily.

Theorem 8: The solutions of Eq. (5) and Eq. (6) are the same.

Thus, we can use the functional equation to help solve the original state equation. The evolution of the system is much clearer from the Eq. (6) than it is from Eq. (5).

5. Control Results

We now have the machinery necessary to use the canonical form to help construct feedback controls for Eq. (1). We are interested in the eigenvalue placement problem: Given a set of complex numbers $\{\alpha_k\}_{k \epsilon I}$, can a feedback control $u(t)$ be found for which the closed loop system has eigenvalues at $\{\alpha_k\}_{k \epsilon I}$? In [11], Sun considers systems which include those we are considering here, except that he insists that b be an element of H. He proves that a necessary and sufficient condition that there exists a bounded functional h^* such that $A + bh^*$ has eigenvalues at $\{\alpha_k\}_{k \epsilon I}$ is

$$\sum_{k \epsilon I} \left| \frac{\alpha_k - \lambda_k}{b_k} \right|^2 < \infty, \tag{7}$$

where $\{\lambda_k\}_{k \epsilon I}$ are the eigenvalues of A. One drawback of this result is that the eigenvectors of the closed loop operator are not explicitly demonstrated, although they can be found by solving an infinite dimensional linear system of equations. Another restriction is that b must be a bounded input, which rules out many interesting cases, including boundary control. Furthermore, this result is restricted to bounded feedback elements h^*, which is why we cannot move the eigenvalues uniformly away from the original eigenvalues. Sun's sufficient condition has been duplicated using control canonical forms of this sort in [5, 7, 12].

To get control results, we first discuss control in the canonical state space system, and then translate to the original system. We therefore need a map from the original system to the canonical system. First, let

$$T_0 : E \longrightarrow H : \{e^{\lambda_k s}\} \longrightarrow \phi_k.$$

Then T_0 is an isomorphism which takes $\mathcal{D}(A)$ onto $\mathcal{D}(D)$. Let

$$T_1 : E \longrightarrow E : \{e^{\lambda_k s}\} \longrightarrow \{e^{\lambda_k s}\} p'(\lambda_k) b_k.$$

If there exists m, $M > 0$ such that

$$m < b_k < M, \quad k \epsilon I, \tag{8}$$

then T_1 is an isomorphism. If there exists $M > 0$ such that

$$b_k \neq 0 \quad \text{and} \quad |b_k| < M, \quad k \,\epsilon\, I, \tag{9}$$

then T_1 is 1–1, but not necessarily onto. If $x(t)$ solves Eq. (1) and $x(t) = T_1 T_0\{z(T)\}$, then $\{z(t)\}$ solves

$$\{\dot{z}(t)\} = D\{z(t)\} + \hat{b}u(t),$$

$$\{z(0)\} = (T_0 T_1)^{-1} x_0 =: z_0, \tag{10}$$

Let $T_2 := (T_0 T_1)^{-1}$.

5.1. Admissible feedback elements

We would like to consider feedback output elements which are not necessarily bounded. We first need to define an extension of A. We can represent the dual space of $D(A^*)$ as

$$W := D(A^*)' = \left\{ \sum_{k \,\epsilon\, I} x_k \phi_k \,\Big|\, \sum_{k \,\epsilon\, I} |x_k/\lambda_k|^2 < \infty \right\}.$$

In our cases the admissible input elemnet $\hat{b}$ is an element of W, which can be thought of as a Hilbert space with norm $||\sum x_k \phi_k|| = (\sum_{k \,\epsilon\, l} |x_k/\lambda_k|^2)^{\cdot\cdot}$. Let

$$\hat{A}{:}H \longrightarrow W{:}\sum x_k \phi_k \longrightarrow \sum x_k \lambda_k \phi_k,$$

so $\hat{A}$ is bounded. If h^* is a functional defined on $D(h^*) \subset H$, we can compute $Ax + bh^*x$ as an element of W for $x \,\epsilon\, D(h^*)$.

Definition 9. h^* is an admissible feedback element if:
(1) $D(A + bh^*)$ is dense in H.
(2) There exists an extension of $A + bh^*$, called A_h, such that A_h is the infinitessimal generator of a strongly continuous semigroup $S_h(t)$, which is holomorphic in Ω_θ for some $\theta \,\epsilon\, (0,\pi/2)$.
(3) For all $x \,\epsilon\, H$ and $T \,\epsilon\, \tilde{\Omega}_\theta$,

$$||h^*S_h(t)x||_{L^2[0,T]} < \infty.$$

This definition is a modification of a more general definition given in Salamon [10].

5.2. Control of the canonical state space equation

Theorem 10: If $\{\alpha_k\}_{k\epsilon l}$ is the zero set of a cardinal function q with poles at $\{\mu_k\}_{k\epsilon l}$, the pole set of p, then there exists an admissible feedback element d^* such that $D + \hat{b}d^*$ has eigenvalues at $\{\alpha_k\}_{k\epsilon l}$ and a Riesz basis of associated eigenvectors $\{\{e^{\alpha_k \cdot}\}\}_{k\epsilon l}$.

Proof: Let μ be the functional for which $F\mu = q$. Let z solve

$$\mu * z = 0,$$

$$z \approx z_0,$$

and let $\{y(t)\} := \{z(t + \cdot)\}$. Then, by the results in Sec. 3, $\{y(t)\}$ satisfies

$$\{y(t)\} = D_\alpha\{y(t)\},$$

$$\{y(0)\} = \{z_0\},$$

where D_α is a different "differentiation" operator given by $D\left[\sum_{k\epsilon I} x_k\{e^{\alpha_k \cdot}\}\right] = \sum_{k\epsilon I} x_k \alpha_k\{e^{\alpha_k \cdot}\}$ on $\mathcal{D}(D) = \left\{\sum_{k\epsilon I} x_k\{e^{\alpha_k \cdot}\} \,|\, \sum_{k\epsilon\infty} |x_k \; \alpha_k|^2 < \infty\right\}$. Let $u(t) = (\eta * z)(t)$, so $\{y(t)\}$ also solves Eq. (10).

Let $q_k(\lambda) = q(\lambda)/q'(\alpha_k)(\lambda - \alpha_k)$, and $g_k^\alpha = L_{q_k}^{-1}$. $\{g_k^\alpha\}_{k\epsilon I}$ is a Riesz basis of Y which is dual to $\{\{e^{\alpha_k \cdot}\}\}$, which is a Riesz basis of E. Then

$$z = \sum_{k\epsilon I} <g_k^\alpha, z_0>e^{\alpha_k \cdot} \epsilon \; X\{\alpha_k\},$$

$$u = \sum_{k\epsilon I} <g_k^\alpha, z_0>p(\alpha_k)e^{\alpha_k \cdot} \epsilon \; X\{\alpha_k\},$$

and

$$\{y(t)\} = \sum_{k\epsilon I} <g_k^\alpha, z_0>e^{\alpha_k t}\{e^{\alpha_k \cdot}\}.$$

We can write $u(t)$ in feedback form, i.e. as a linear function of $\{y(t)\}$: let $d^* = \sum_{k\epsilon I} p(\alpha_k)g_k^\alpha$, with $\mathcal{D}(d^*) = \left\{\sum_{k\epsilon I} x_k\{e^{\alpha_k \cdot}\} \,|\, \sum_{k\epsilon I} |x_k| < \infty\right\}$. Then $u(t) = d^*\{y(t)\}$. To show that d^* is an admissible feedback element, it will suffice to show that $D + \hat{b}d^*$ has a Riesz basis of eigenvectors. We now need the following result.

Lemme 11: Let $\{\alpha_k\}_{k\epsilon I}$ be the zero set of some cardinal function (not necessarily with poles at $\{u_k\}_{k\epsilon I}$) such that $e^{\alpha_k \cdot} \epsilon \; X$ for all $k \epsilon I$, and let $d^* = \sum_{k\epsilon I} p(\alpha_k)g_k^\alpha$. Then $\{e^{\alpha_k \cdot}\}$ is an eigenvector of $D + bd^*$ with associated eigenvalue α_k.

Proof of Lemma:

$$\{e^{\alpha_k \cdot}\} = \sum_{j\epsilon I} <g_j, e^{\alpha_k \cdot}>\{e^{\lambda_j \cdot}\}.$$

For $\rho \; \epsilon \; (c,b)$,

$$<g_j, e^{\alpha_k{}'}> = \frac{1}{2\pi i} \int_{-\Gamma_{\rho}} \frac{p(\lambda)d\lambda}{p'(\lambda_j)(\lambda - \lambda_j)(\lambda - \alpha_k)},$$

so

$$\{e^{\alpha_k{}'}\} = \sum_{j\epsilon I} \frac{p(\alpha_k)}{p'(\lambda_j)} \frac{1}{\alpha_k - \lambda_j} \{e^{\lambda_j{}'}\},$$

$$D\{e^{\alpha_k{}'}\} = \sum_{j\epsilon I} \frac{p(\lambda_k)}{p'(\lambda_j)} \frac{\lambda_j}{\alpha_k - \lambda_j} \{e^{\lambda_j{}'}\},$$

$$\hat{b}d^*\{e^{\alpha_k{}'}\} = \hat{b}p(\alpha_k) = \sum_{j\epsilon I} \frac{p(\alpha_k)}{p'(\lambda_j)} \{e^{\alpha_j{}'}\}. \tag{11}$$

Hence, $(D + bd^*)\{e^{\alpha_k{}'}\} = \alpha_k \sum_{j\epsilon I} \frac{p(\alpha_k)}{p'(\lambda_k)} \frac{1}{\alpha_k - \lambda_j} \{e^{\lambda_j{}'}\} = \alpha_k\{e^{\alpha_k{}'}\}$, proving the Lemma.

Since $\{\{e^{\alpha_k{}'}\}\}_{k\epsilon I}$ is a Riesz basis for Y, by the results in iec. 3, this shows that $D + \hat{b}d^* = D_\alpha$. Furthermore, $d^*S_d \epsilon X$, so d^* satisfies all of the conditions in Definition 9. Hence, d^* is admissible, finishing Theorem 10.

We can also get a formula for the feedback of element in terms of $\{g_k\}_{k\epsilon I}$:

$$g_j^\alpha = \sum_{k\epsilon I} <g_j^\alpha, e^{\lambda_k{}'}>g_k = \sum_{k\epsilon I} \frac{q(\lambda_k)}{q'(\alpha_j)} \frac{1}{\lambda_k - \alpha_j} g_k, \tag{12}$$

so

$$d^* = \sum_{j\epsilon I} p(\alpha_j) \sum_{k\epsilon I} \frac{q(\lambda_k)}{q'(\alpha_j)} \frac{1}{\lambda_k - \alpha_j}. \tag{13}$$

5.3. Control of the original equation

In the case where b satisfies Eq. (8), T_2 is an isomorphism between H and E which takes Eq. (1) to (5). We then get

Corollary 12: If b satisfies Eq. (8) and $\{\alpha_k\}_{k\epsilon I}$ is the zero set of a cardinal function q with poles at $\{\mu_k\}_{k\epsilon I}$, then there exists an admissible feedback element h^* such that $A + bh^*$ has eigenvalues $\{\alpha_k\}_{k\epsilon I}$, and a Riesz basis of associated eigenvectors.

We can get formulas for the new eigenvectors $\{\chi_k\}_{k\epsilon I}$, the new dual basis $\{h_k^\alpha\}$, and the feedback element h^* in terms of the original eigenvectors $\{\phi_k\}_{k\epsilon I}$ and the original dual basis $\{\psi_k\}_{k\epsilon I}$. Since $\chi_k = T_2^{-1}\{e^{\alpha_k{}'}\}$, using Eq. (11) we get that

$$\chi_k = \sum_{j \in I} \frac{p(\alpha_k)}{p'(\lambda_j)} \frac{1}{\alpha_k - \lambda_j} p'(\lambda_j) b_j \phi_j. \tag{14}$$

Since $h_k^\alpha = T_2^* g_k$, it can be shown that

$$h_j^\alpha = \sum_{k \in I} \frac{q(\lambda_k)}{q'(\alpha_k)(\lambda_k - \alpha_j) b_k p'(\lambda_k)} \psi_k. \tag{15}$$

Then

$$h^* = T_2^* d^* = \sum_{k \in I} p'(\alpha_k) h_k^\alpha. \tag{16}$$

In Section 5, we summarize some results about the zero and pole placement of a class of cardinal fnctions which we have constructed. Combining those results with Theorem we get solutions to the eigenvalue placement problem for some specific systems and choices of $\{\alpha_k\}_{k \in I}$.

Suppose now that b only satisfies Eq. (9), so T_2 is not necessarily an isomorphism. This occurs, for instance, when $b \in H$. We can prove

Theorem 13: Suppose b is an admissible input element which satisfies Eq. (9). Let $\{\alpha_k\}_{k \in I}$ be the zero set of a cardinal function q which has poles at $\{\mu_k\}_{k \in I}$, and satisfies Sun's condition Eq. (7). Then, h^*, given by Eqs. (15) and (16), is a bounded feedback element such that $A + bh^*$ has eigenvalues at $\{\alpha_k\}_{k \in I}$ and a Riesz basis of eigenvectors $\{\chi_k/p'(\lambda_k) b_k\}_{k \in I}$, with χ_k given by Eq. (14).

Theorem 14: If $\{(\alpha_k - \lambda_k)/b_k\}_{k \in I}$ is bounded, then h^* is an admissible feedback element and the eigenvectors form a basis for H in the sense of being a complete, strongly independent set.

We can also construct feedback controls to realize different eigenvalues than those realized above, by using "almost cardinal" functions. Almost cardinal functions have a different distribution of zeros with respect to poles than cardinal functions do, but still have some of the properties of cardinal functions which make them useful for control. These properties can be stated in terms of conditions on the biorthogonal set $\{q_k\}_{k \in I}$. Some examples of almost cardinal functions are given in [5].

Theorem 15: Let $\{\alpha_k\}_{k \in I}$ be the zero set of an almost cardinal function $q(z)$ with poles at $\{\mu_k\}_{k \in I}$. Let d^* be constructed as in Eq. (13). Then d^* is an admissible feedback element, and $D + bd^*$ has spectrum at $\{\alpha_k\}_{k \in I}$. The associated eigenvectors $\{\{e^{\alpha_k t}\}\}_{k \in I}$ form a basis in the sense of being a complete, strongly independent set in E. If h^* is constructed as in Eqs. (15) and (16), and b satisfies Eq. (8), then h^* is an admissible feedback element, $A + bh^*$ has spectrum at $\{\alpha_k\}_{k \in I}$ and $\{\chi_k\}_{k \in I}$, given by Eq. (14), is a basis for H.

6. Cardinal functions

In [5] we construct two classes of cardinal functions. In one class the zeros and pole are space asymptotically linearly. These functions have poles at

$$\left\{e^{i\sigma}(-\frac{\rho_k}{a_1} + \gamma_1 + \delta_1 i)\right\}_{k \in I_1} \cup \left\{e^{-i\sigma}(-\frac{\xi_k}{a_2} + \gamma_2 + \delta_2 i)\right\}_{k \in I_2} \qquad (17)$$

and zeros at

$$\left\{e^{i\sigma}(-\frac{\eta_k}{a_1} + \gamma_1 + \alpha_1 i)\right\}_{k \in I_1} \cup \left\{e^{-i\sigma}(-\frac{\beta_k}{a_2} + \gamma_2 + \alpha_2 i)\right\}_{k \in I_2}, \qquad (18)$$

where $a_1, a_2, \gamma_1, \gamma_2, \alpha_1$, and α_2 are all real, with $\delta_1 > \alpha_1$, $\delta_2 < \alpha_2$, and a_1 and a_2 are positive. I_1 and I_2 are subsets of Z^+ which contain all but finitely many elements of Z^+. ρ_k, ξ_k, η_k and β_k are all of the form

$$k + \epsilon_k, \qquad \sum |\epsilon_k|^2 < \infty. \qquad (19)$$

We also obtain a class of cardinal functions with zeros and poles spaced asymptotically quadratically. These functions have zeros and poles as in Eqs. (17) and (18), except that Eq. (19) is replaced by

$$k^2 + k\epsilon_k, \qquad \sum |\epsilon_k|^2 < \infty.$$

"Almost cardinal" functions have some, but not all, of the growth properties of cardinal functions. A class of these functions can be constructed using perturbed Gamma functions. These have zeros and poles of the form Eqs. (17), (18), with ρ_k and ξ_k of the form Eq. (19), η_k of the form

$$k - d_1 + \epsilon_k, \qquad \sum |\epsilon_k|^2 < \infty$$

and β_k of the form

$$k - d_2 + \epsilon_k, \qquad \sum |\epsilon_k|^2 < \infty,$$

with $0 \leq d_1 + d_2 < \frac{1}{2}$.

Acknowledgements

This research was supported by the National Science Foundation under Grant No. MCS-82-15064, the Army Research Office under Contract No. DAAG29-80-C-0041, and the Air Force Office of Scientific Research under Grant No. AFOSR-84-0088.

References

[1] Chen, G. and Russell, D.L., "A mathematical model for linear elastic systems with structural damping," Technical Summary Report No. 2089. Mathematics Research Center, University of Wisconsin-Madison, June 1980.

[2] Hille, E. and Phillips, R.S., *Functional Analysis and Semigroups*, Amer. Math. Soc. Colloq. Publ., Vol. 31, Amer. Math. Soc., Providence, 1957.

[3] Ho, L.F., "Controllability and soectral assignability of a class of hyperbolic control systems with retarded control canonical forms," Thesis, University of Wisconsin-Madison, July 1981.

[4] Ho, L.F. and Russell, D.L., "Admissible input elements for systems in Hilbert space and a Carleson measure criterion," *SIAM J. Control*, Vol. 21, 1983, pp. 614-639.

[5] Rebarber, R., "Control of holomorphic semigroups generated by a class of spectral operators," Thesis, University of Wisconcin-Madison, August 1984.

[6] Rebarber, R., "A Laplace transform relevant to holomorphic semigroups," (to appear).

[7] Russell, D.L., "Closed loop eigenvalue specification for infinite dimensional systems: Augmented and deficient hyperbolic cases," Technical Summary Report No. 2021, Mathematics Research Center, University of Wisconsin-Madison, August 1979.

[8] Russell, D.L., "Functional equations as control canonical forms for distributed parameter control systems and a state space theory for certain differential equations of infinite order," in *Volterra and Functional Differential Equations*, Hannsgen, Herdman, Stech, Wheeler, Ed., Marcel Dekker, Inc., New York, 1974.

[9] Russell, D.L., "Uniform bases of exponentials, neutral groups and a transform theory for $H[a,b]$," Technical Summary Report No. 2149. Mathematics Research Center, University of Wisconsin-Madison, December 1980.

[10] Salamon, D., *Control and Observation of Neutral Systems*, Pitman, Boston, 1984.

[11] Sun, S.H., "On spectrum distribution of completely controllable linear systems," translated by Ho, L.F., *SIAM J. Control*, Vol. 19, 1981, pp. 730-743.

[12] Teglas, R., "On the control canonical structure of a class of scalar input systems," *SIAM J. Control*, Vol. 22, 1984, pp. 552-569.

AN ON-LINE CONTROL ALGORITHM
FOR INELASTIC STRUCTURES

A.M. Reinhorn, G.D. Manolis and C.Y. Wen
Department of Civil Engineering
State University of New York
Buffalo, N.Y. USA

1. Introduction

Active control has traditionally been applied to linear systems [1], and there is very little information on its extension to non-linear systems [2]. In this work, an algorithm for the control of large displacements in structures undergoing inelastic deformations has been developed using an active pulse system. The rationale for such a development comes from the field of earthquake engineering. During intense shaking, a structure develops inelastic deformations in an effort to dissipate the energy input from the earthquake. If these inelastic deformations are excessive, extensive repairs may be required, or it may even become necessary to demolish the structure. If, on the other hand, the structure is allowed to enter the non-linear range but we subsequently intervene with an active control mechanism to ascertain that the kind of deformations that would render the structure unserviceable do not materialize, then a rational and economical method of earthquake resistant design will result.

There are three basic types for the control of flexible structures such as extremely tall buildings or long span bridges. They are:

(a) Passive control, where energy dissipating devices are introduced into the structure to absorb energy and can be replaced if extensively damaged. Such devices are metal U-shaped devices [3,4], rubber bearing [5], tuned mass dampers [6], and the expandable top story approach [7].

(b) Semi-active control, where passive devices such as the tuned mass dampers are augmented by simple control mechanisms [8].

(c) Active control, where the system's response (displacements, velocities, and accelerations) and the input (loads) are continuously monitored and corrected through the input of external forces. Among the active control devices used today we have electromagnetic or servocontrolled hydraulic devices for aircraft [9], tuned mass dampers coupled with external power supplies [10], appendages for tall buildings [11], tendons or cables [12,13], and pulses applied through either gas jets or prestressed cables [14]. Based on the way data is acquired, active control systems can be classified as closed loop, open loop, or open closed loop [15]. Finally, based on the type of control rule, we have optimal control, sub-optimal control and non-optimal control [15].

The suggested control algorithm for discrete systems is a closed loop non-optimal algorithm based on the standard Newmark beta numerical integration method in conjunction with the Newton-Raphson method. The state vector U_{i+1} at the end of a time interval Δt is computed. If U_{i+1} does not exceed a certain tolerance dictated by serviceability conditions, the next state vector U_{i+2} is computed. If U_{i+1} exceeds the tolerance, then a pulse is computed and applied to the structure at the beginning of the time interval. The intensity of the pulse is such that the new state vector U_{i+1} at the end of the time interval will not exceed the tolerance. The duration of the pulse Δt_p ranges from 0.1 Δt to 0.2 Δt.

This control algorithm is applicable to any inelastic structure whose stiffness is represented by a piecewise curve. Furthermore, it can easily be extended to multi-degree of freedom systems. The advantage of this control algorithm is that it results in a minimum number of on-line computations. Since the algorithm is based on a predicted response at the end of the computational interval [14], a larger time step Δt than the ones commonly used in other control algorithms can be used. As a result, the algorithm is more suitable for on-line computations arising from practical applications.

The applicability of this algorithm is exemplified for a structure subjected to an harmonic (sine) force and to an earthquake accelerogram with random characteristics. The response of the structure in both cases is maintained within either elastic or inelastic bounds based on initially prescribed limits.

2. Development of Control Algorithm

2.1. Review of the numerical algorithm

The control algorithm that will be used essentially derives from a reformulation of the Newmark β numerical integration algorithm. Although there are quite a few methods for the numerical integration of the equations of motion resulting from a multi-degree-of-freedom (MDOF) representation of a structural system [16], the Newmark β algorithm has been shown to be one of the best [17].

Focusing, for the time being, on a single-degree-of-freedom (SDOF) system, we have the equation of motion in the form:

$$My(t) + Cy(t) + Ky(t) = R(t). \tag{1}$$

As usual, M, C, and K are the mass, damping, and stiffness, respectively, y the displacement, dots indicate time derivatives, and R the external force. If the stiffness is a constant, then, (1) can be rewritten in the following incremental form

$$\hat{K}y_{i+1} = \hat{R}_{i+1}, \tag{2}$$

where $\hat{K}(K,M,C)$ is the effective stiffness and $\hat{R}_{i+1}(R_{i+1},y_i,\dot{y}_i,\ddot{y}_i,M,C)$ is the residual force during time step t_{i+1}. Upon solution of (2), the accelerations and velocities at $i+1$ are found as functions of y_{i+1}, y_i, $\dot{y}_i$, and $\ddot{y}_i$. If the stiffness K is variable, then iterations are necessary during the duration of the time increment Δt, where $t_{i+1} = t_i + \Delta t$. Combining, therefore, the Newton-Raphson method with Newmark's β algorithm gives

$$\hat{K}_i\Delta y^{(k)} = \hat{R}_{i+1}^{(k)} - \hat{R}_{i+1}^{(k-1)} = DF^{(k)}, \tag{3}$$

where k is the number of iterations and

$$y_{i+1}^{(k)} = y_{i+1}^{(k-1)} + \Delta y^{(k)}. \tag{4}$$

Furthermore, $\hat{K}_i$ is a function of the current stiffness K_i, M, and C. The need to iterate arises when the new displacement computed using the current stiffness K_i exceeds the yield deformation y_{yield} in a bilinear inelastic model. The iterative process stops when the difference between $DF^{(k)}$ and $DF^{(k-1)}$ is less than some prescribed tolerance. In order to expedite the iterative process, the current stiffness K_i is updated as

$$K_i^{(k)} = \frac{DF^{(k)}}{(y_{i+1}^{(k)} - y_i)}, \tag{5}$$

Two different expressions for $DF^{(k)}$ exist, depending on whether we have loading $(R_i \geq 0)$ or unloading $(R_i < 0)$.

2.2. Reformulation of the numerical integration algorithm

In order to be able to estimate a SDOF system's state variables at the end of a time interval in the manner required by the control algorithm, the Newmark β algorithm described previously, needs to be reformulated. First, the displacement, velocity and acceleration at $i+1$ are written in their usual form [16]:

$$y_{i+1} = y_i + \dot{y}_i\Delta t + \left[(0.5 - \alpha)\ddot{y}_i + \alpha\ddot{y}_{i+1}\right]\Delta t^2,$$

$$\dot{y}_{i+1} = \dot{y}_i + \left[(1 - \delta)\ddot{y}_i + \delta\ddot{y}_{i+1}\right]\Delta t, \tag{6}$$

$$\ddot{y}_{i+1} = \left(\frac{1}{\alpha \Delta t}\right)\dot{y}_i - \left(\frac{1}{2\alpha} - 1\right)\ddot{y}_i.$$

Note that for $\alpha = 1/6$ and $\beta = 1/2$, the linear acceleration algorithm is recovered. Denoting $a_o = 1/(\alpha \Delta t^2)$, $a_1 = \delta/(\alpha \Delta t)$, $a_2 = 1/(\alpha \Delta t)$, $a_3 = 1/(2\alpha) - 1$, $a_4 = \delta/\alpha - 1$, $a_5 = (\delta/\alpha - 2)\Delta t/2$, $a_6 = \Delta t(1 - \delta)$, $a_7 = \delta \Delta t$, and using Eq. (6) in conjunction with the equation of motion (1) yields

$$\hat{K} = K + a_o M + a_1 C, \tag{7}$$

and

$$\hat{R}_{i+1} = R_{i+1} + M\left(a_o y_i + a_2 \dot{y}_i + a_3 \ddot{y}_i\right) + C\left(a_1 y_i + a_4 \dot{y}_i + a_5 \ddot{y}_i\right), \tag{8a}$$

or

$$\hat{R}_{i+1} = R_{i+1} + \underline{D}^T \underline{U}_i, \tag{8b}$$

where the vector $\underline{D}_i^T = [Ma_o + Ca_1, Ma_2 + Ca_4, Ma_3 + Ca_5]$ is time independent and the state vector $\underline{U}_i^T = \left[y_i, \dot{y}_i, \ddot{y}_i\right]$ contains the state variables at time step i. Also, the operation T denotes transposition. Using the above notation, the equilibrium Eq. (2) reads

$$\hat{K} y_{i+1} = R_{i+1} + \underline{D}^T \underline{U}_i \tag{9}$$

for which

$$y_{i+1} = \hat{K}^{-1} r_{i+1} + \hat{K}^{-1} \underline{D}^T \underline{U}_i. \tag{10}$$

The remaining state variables at $i+1$ are obtained as

$$\dot{y}_{i+1} = a_7 \ddot{y}_{i+1} + \underline{G}^T \underline{U}_i,$$
$$\ddot{y}_{i+1} = a_o y_{i+1} - \underline{E}^T \underline{U}_i, \tag{11}$$

where $\underline{G}^T = [0,1,a_6]$ and $\underline{E}^T = [a_o, a_1, a_2]$. Equations (10) and (11) may be combined to yield the state vector at $i+1$:

$$\underline{U}_{i+1} = \left\{\begin{matrix}1\\a_1\\a_o\end{matrix}\right\} K^{-1} R_{i+1} + \left[\begin{matrix}K^{-1}\underline{D}^T\\a_o K^{-1}\underline{d}^T - a_7\underline{E}^T + \underline{G}^T\\a_o K^{-1}\underline{D}^T - \underline{E}^T\end{matrix}\right]\underline{U}_i, \tag{12a}$$

or

$$\underline{U}_{i+1} = \underline{S} R_{i+1} + \underline{Q} \underline{U}_i. \tag{12b}$$

The Newton-Raphson method can now easily be introduced in Eq. (12) to account for a variable stiffness.

568 A.M. Reinhorn, G.D. Manolis, C.Y. Wen

2.3. Step-by-step feedback control

The time stepping algorithm developed above (12) can now be used for the computation of a corrective force to be applied to the structure during the computation interval Δt, if the computed state vector $\underline{U}_{j+1}$ at the end of the interval is found to exceed a prescribed limit $\underline{U}_{\lim}$. At first, Eq. (12) is used to forecast the state vector at $t_i + \Delta t$ from the measured values of $\underline{U}_i$. The only problem is that R_{i+1} is not yet known. If, however, the forcing function is assumed to remain constant during the computation interval, then the measured value of R at i can be used. The validity of this assumption has been substantiated by Prucz et al [14]. For earthquake-type motions,

$$\underline{U}_{j+1} = \underline{S}Mu_{gi} + \underline{Q}\underline{U}_i, \tag{13}$$

where $\ddot{u}_g$ is the continuously monitored ground acceleration.

Two types of corrective measures are suggested for the case $\underline{U}_{j+1} > \underline{U}_{\lim}$:

(a) *Corrective-force*: This force is continuously applied to the structure through-out the interval $t_i \leq t \leq t_i + \Delta t$ and its magnitude $\underline{F}_j$ at time i is computed from the inequality

$$\left|\underline{U}_{\lim}\right|_j \geq \left|\underline{U}_{j+1} + \underline{S}\underline{F}_i\right|_j, \tag{14}$$

where only one component, j, (either, displacement, velocity or acceleration) is used for control design. Therefore, using Eq. (12):

$$\underline{F}_i = \begin{Bmatrix} \hat{K}\left(U_{\lim} - U_{i+1}\right) \\ \hat{K}\left(\dot{U}_{\lim} - \dot{U}_{i+1}\right)/a_1 \\ \hat{K}\left(\ddot{U}_{\lim} - \ddot{U}_{i+1}\right)/a_o \end{Bmatrix}. \tag{15}$$

Only one of the three components of $\underline{F}_i$ will be used for correction, depending on the nature of the prescribed component of $\underline{U}_{\lim}$.

(b) *Corrective Pulse*: This pulse is a force applied over a short interval of time, which ranges from 10% to 20% of Δt [14]. The effect of the pulse is to introduce an initial velocity which is directly added to the actual velocity of the system. Thus,

$$\dot{y}_i(total) = \dot{y}_i + \frac{P_i \Delta t_p}{M}, \tag{16}$$

where P_i is the magnitude of the pulse at time i, and Δt_p is the duration of its application. Inserting Eq. (16) into Eq. (13) results in

$$\underline{U}_{j+1} = \underline{S}M\ddot{u}_{gi} + \underline{QU}_j + \{Q_{2j}\underline{P}_j\}\frac{\Delta t_p}{M}, \tag{17}$$

where $Q_{2j} = [Q_{21}, Q_{22}, Q_{23}]$ is the second row of $\underline{Q}$. Solving for the intensity of the pulse gives

$$\underline{P}_j = \begin{cases} (M/\Delta t_p)|U_{\text{lim}} - U_{i+1}|/Q_{21} \\ (M/\Delta t_p)|\dot{U}_{\text{lim}} - \dot{U}_{i+1}|/Q_{22} \\ (M/\Delta t_p)|\ddot{U}_{\text{lim}} - \ddot{U}_{i+1}|/Q_{23} \end{cases}. \tag{18}$$

Again, depending on which component of $\underline{P}_j$ is selected, displacement, velocity or acceleration control results. For an inelastic system, adjustment of $\underline{Q}$ is required since its coefficients are stiffness dependent [2].

2.4. Control algorithm implementation

The control procedure is best described by reference to the block diagram shown in Fig. 1. The following steps are distinguished: (a) Monitoring of the state vector and recording of its values by the digital computer at time t_i. (b) Estimation of the values of the state vector at the end of the time interval Δt, i.e., at $t_{i+1} = t_i + \Delta t$. (c) Checking to see if the threshold levels for the response are exceeded. (d) If threshold levels are exceeded, the corrective force or pulse is computed using Eq. (15) or (18). (e) Generation of the corrective signal in the tendons or pulse generators.

The above procedure is repeated continuously. Since the state vector is continuously monitored, the corrective signal is computed based on 'fresh' (actual) information. Thus, the accumulation of errors that usually appears in similar on-line systems such as the pseudo-dynamic testing technique [18] is altogether eliminated.

3. Numerical Examples

Based on the numerical algorithm presented in Sec. 1, two numerical simulations of control for a simple structure have been done.

3.1. Control during harmonic excitation

Figure 2 shows the uncontrolled and controlled response of both an elastic and an elastoplastic system with fundamental frequency $f_o = 2.5Hz$ and 5% of critical damping. The system is under sinusoidal excitations of frequency $f = 0.5Hz$. Control was achieved by applying pulses (Eq. (18)) to the system once the response exceeded 0.8 of the static displacement for both the elastic and the inelastic system. It is observed that the controlled response remains well within the prescribed bounds. Figure 3 plots the hysteresis loop traced by the same inelastic system as above, both prior to and after pulse control is applied. The effect of the control is to limit the amount of energy dissipated by the system, which is

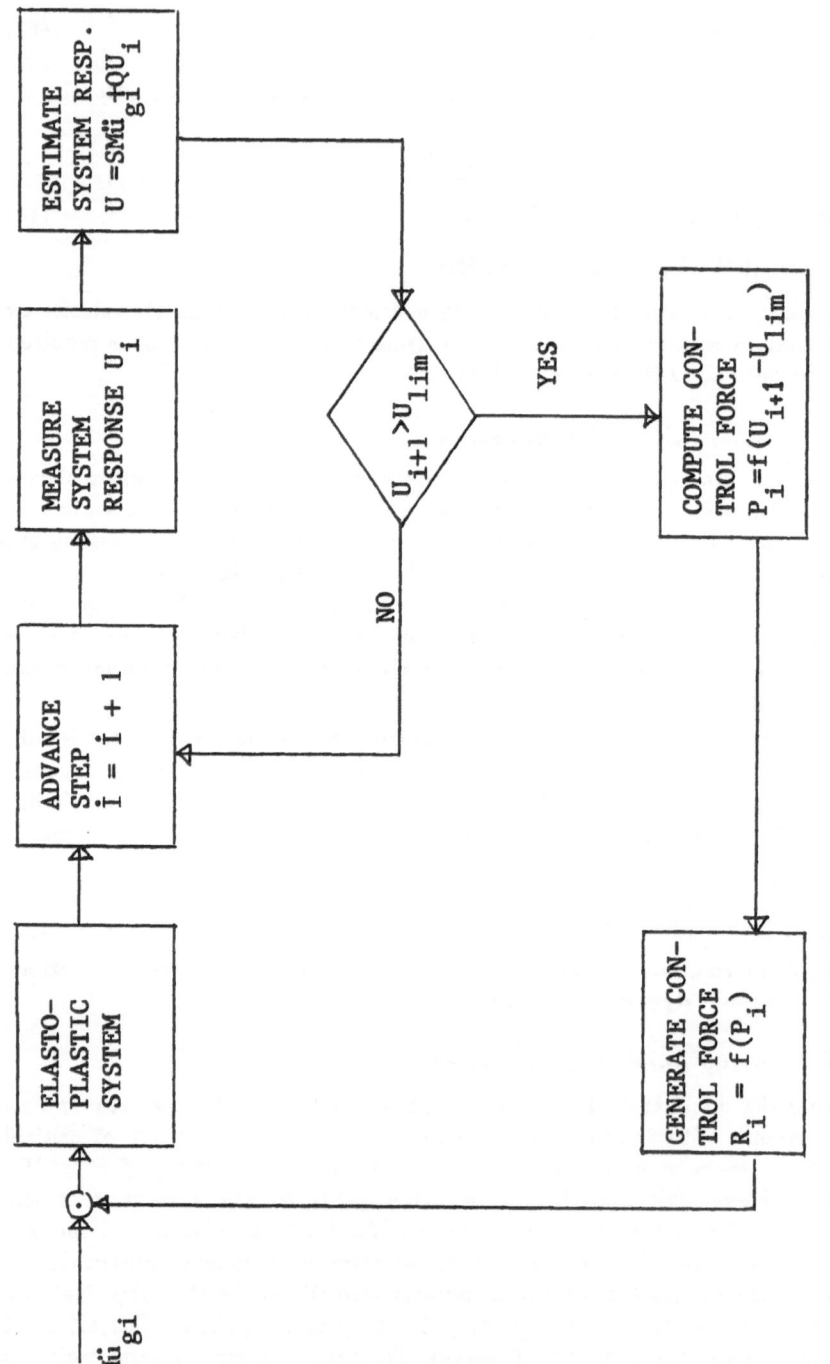

Figure 1: Block diagram of the control algorithm

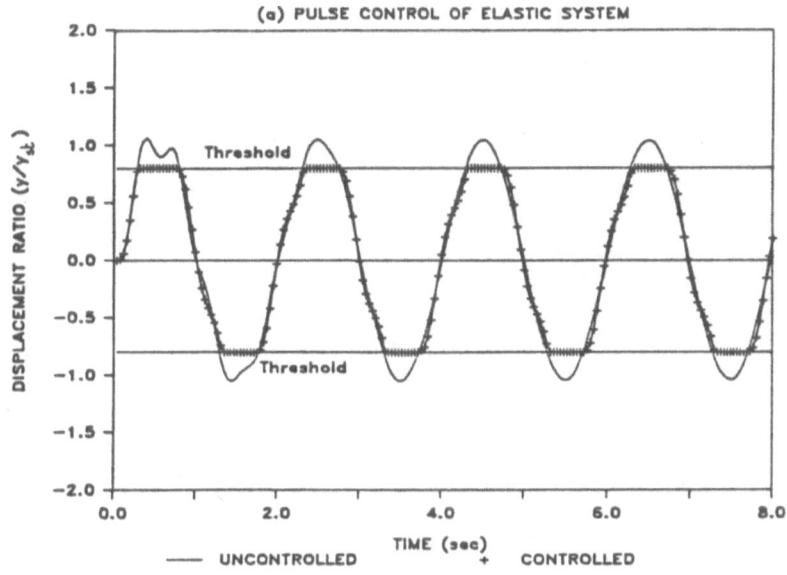

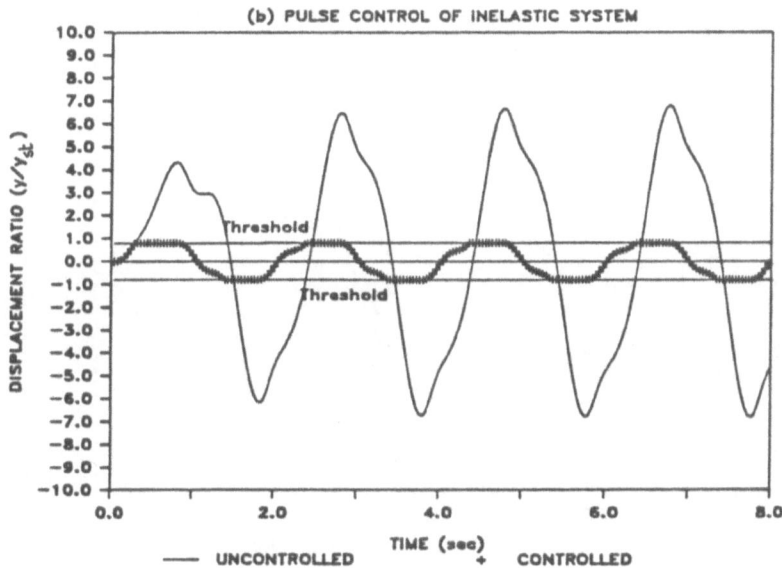

Figure 2: Pulse control of systems under harmonic loads

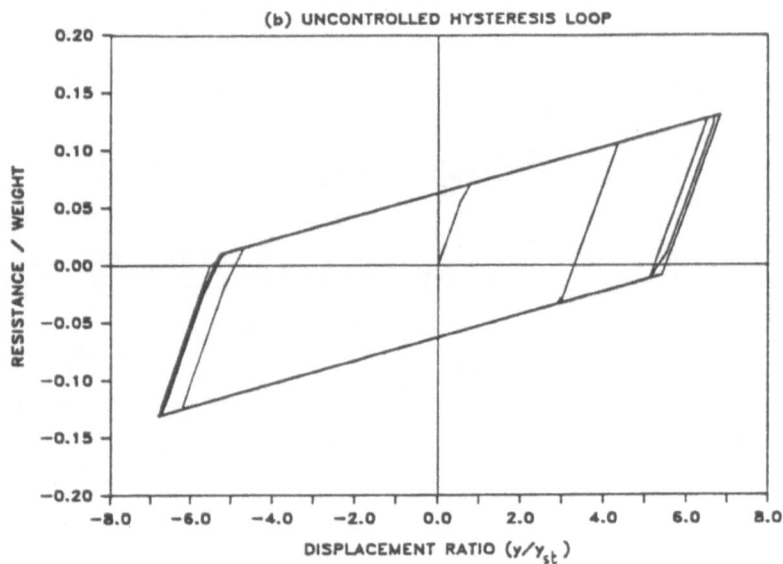

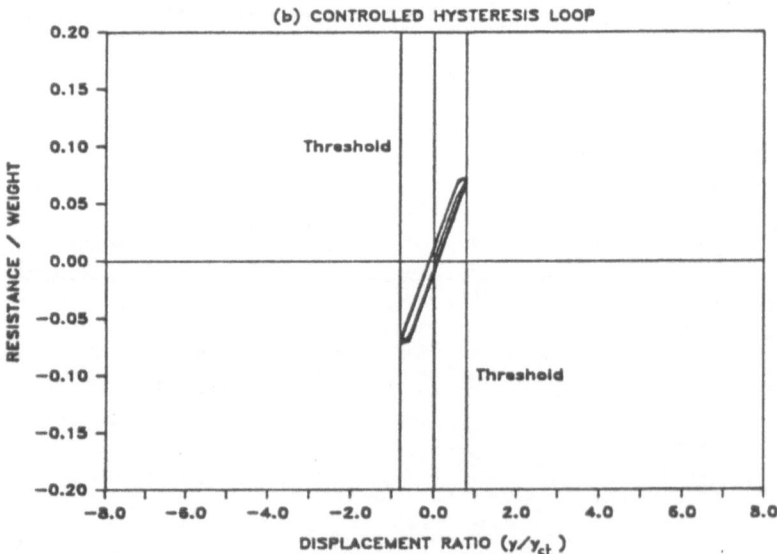

Figure 3: Hysteresis loop of inelastic system under harmonic loads

manifested by the narrowing of the hysteresis loop.

3.2. Control during earthquake excitation

The same system that was described above is now subjected to base motion. For this purpose, a recorded accelerogram with random characteristics is used. The accelerogram, which is the horizontal North-South component of the 1940 El-Centro earthquake, normalized to a maximum of 10 percent of the gravity, is shown in Fig. 4. The systems' elastic response, both uncontrolled and controlled, is shown in Fig. 5(a). Similarly, its inelastic uncontrolled and controlled response is shown in Fig. 5(b). Again it is seen that the controlled response remains within the 0.6 and 0.8 of the 'static' response bounds prescribed for the elastic and ine-lastic systems, respectively. Since there is no energy absorbed by yielding in the elastic system, more and stronger pulses are necessary to limit the response, as shown in Fig. 6(a). In contrast, the inelastic system yields at 0.6 of the 'static displacement' and the threshold maintained at 0.8 of the 'static' displacement, a substantial amount of the energy input from the earthquake is dissipated. There-fore, fewer and less intense pulses are needed for the control, as shown in Fig. 6(b). Finally, Fig. 7 depicts the uncontrolled and controlled inelastic system's hysteresis loops. It is again evident that control forces are viewed by the system as input energy resulting in narrowing of the hysteresis loop.

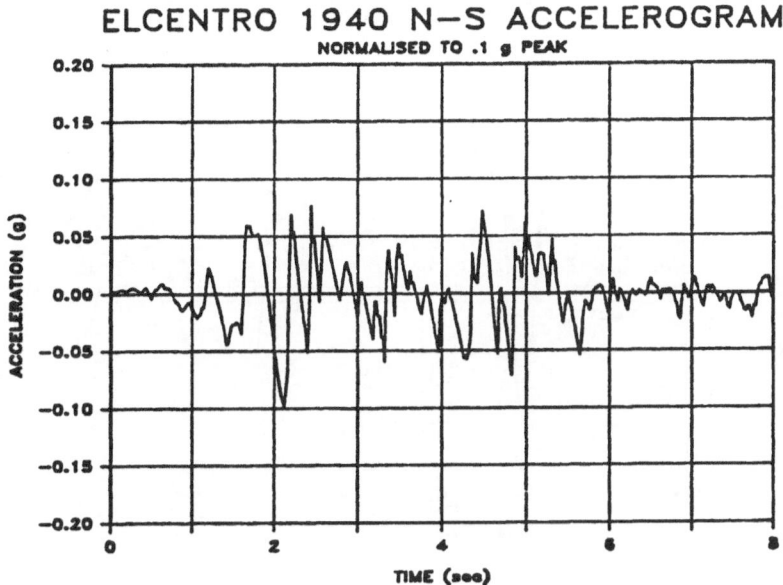

Figure 4: Earthquake accelerogram normalized

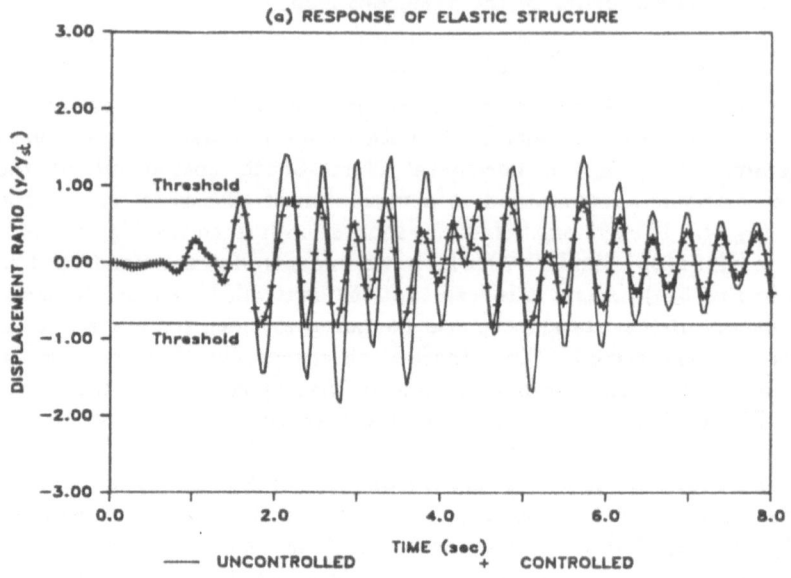

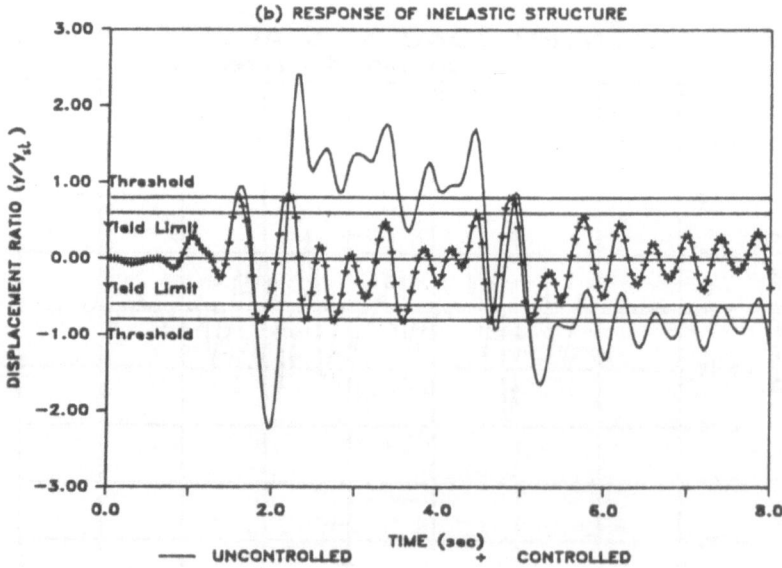

Figure 5: Pulse control of systems under earthquake excitation

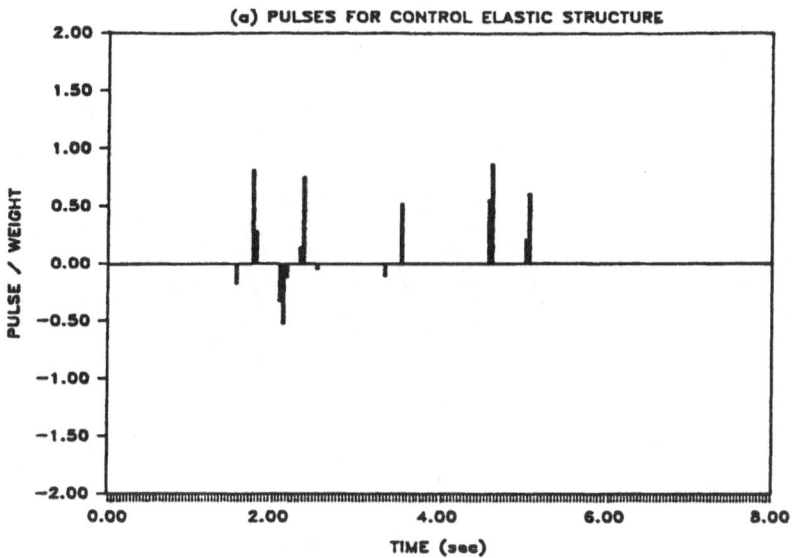

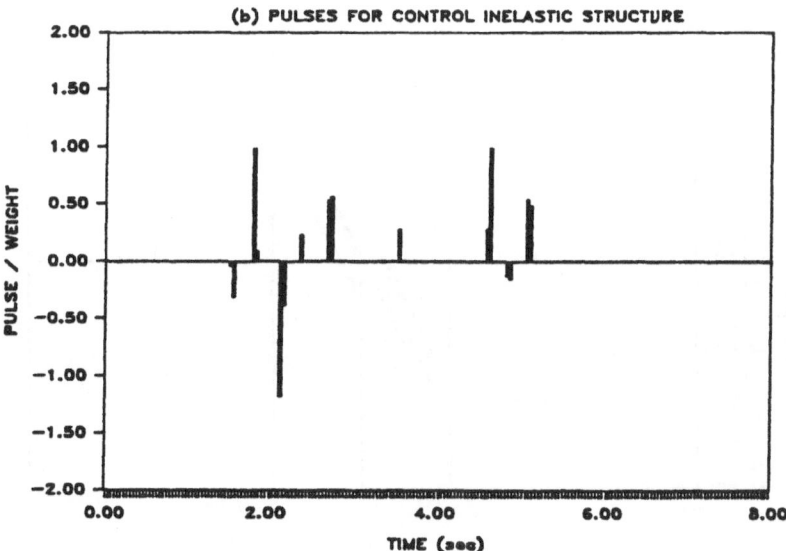

Figure 6: Pulses required for control under earthquake excitation

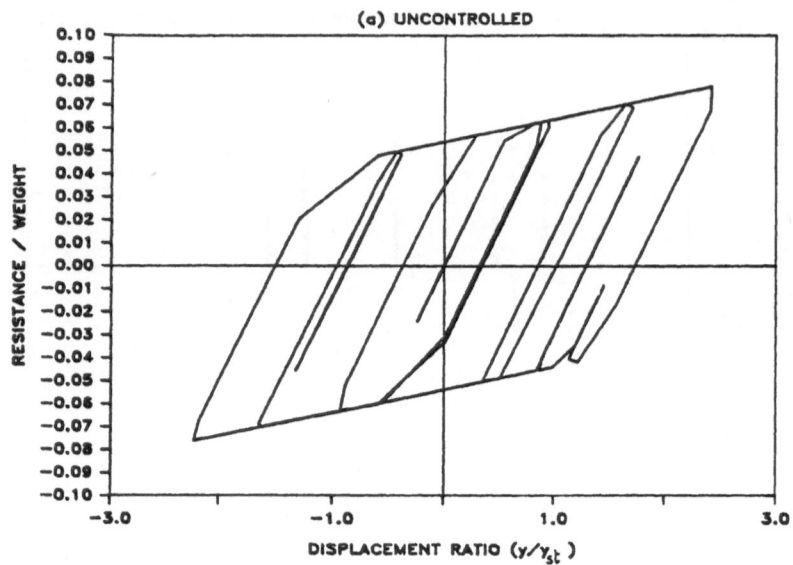

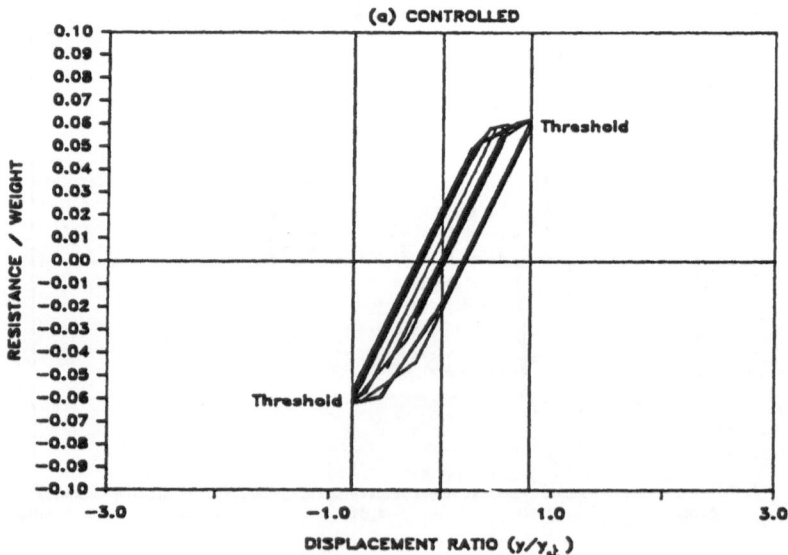

Figure 7: Hysteresis loops of inelastic system under earthquake excitation

4. Conclusion Remarks

The active, non-optimal control algorithm described so far was developed with elastic and inelastic (bilinear) SDOF systems in mind. The present formulation, therefore, needs to be augmented in order to include general nonlinear system characteristics, multi-degree-of-freedom system (MDOF), and on-line identification of system characteristics. More specifically, the following steps are distinguished:

(a) Development of a control algorithm for MDOF systems exhibiting known inelastic behavior, i.e., $\underline{K}_i = \underline{K}(y_i)$.

(b) Refinement of the above control algorithm so that a parallel identification of the system's current stiffness characteristics by extrapolating from the stiffness characteristics measured during the previous time steps is possible.

(c) Error analysis of both control algorithm developed in (a) and (b).

Upon completion of the above three steps, the final form of the augmented control algorithm will be adopted for digital control so as to additionally include the influence of the time delay in the controller's response and the influence of friction in the system. The control algorithm can then be tested on a simulated real time system to observe the error propagation. The tests can be done using numerical simulation in the digital controller and real time harmonic signals.

Acknowledgements

The authors would like to thank Professor T.T. Soong for his support. Also, the University Computing Center of SUNY/Buffalo is acknowledged for making its facilities available to the authors.

References

[1] Leipholz, H.H.E., Editor, *Structural Control*, Proceeding of the International IUTAM Symposium on Structural Control held at University of Waterloo, Ontario, Canada, 1979.

[2] Masri, S.F., Bekey, G.A., and Caughey, T.K., "On-Line Control of Non-Linear Flexible Structures," *Journal of Applied Mechanics*, Vol. 49, pp. 871-884, 1981.

[3] Kelly, J.M., "Control Devices for Earthquake Resistant Structural Design," in Structural Control, Leipholz, H.H.E., Editor, North Holland, Amsterdam, 1980, pp. 391-414.

[4] Skinner, R.L., "Base Isolated Structures in New Zealand," in Proceedings, 8th World Conference on Earthquake Engineering, Vol. V, San Francisco, California, 1984, pp. 927-934.

[5] Robinson, W.H. and Tucker, A.G., "A Lead Rubber Shear Damper," *Bulletin of New Zealand National Society for Earthquake Engineering*, Vol. 10, No. 3, 1977, pp.151-153.

[6] Warburton, G.B., "Optimum Absorber Parameters for Various Combinations of Response and Excitation Parameters," *International Journal of Earthquake Engineering and Structural Dynamics*, Vol. 10, 1982, pp. 381-401.

[7] Jagadish, K.S., Raghu Prasad, B.K., and Vasudeva Rao, P., "The Inelastic Vibration Absorber Subjected to Earthquake Ground Motions," *International Journal of Earthquake Engineering and Structural Dynamics*, Vol. 7, 1977, pp. 317-326.

[8] Hrovat, D., Barak, P., and Robins, M., "Semi-Active vs. Passive or Active Tuned Mass Dampers for Structural Control," *Journal of Engineering Mechanics Division, ASCE*, Vol. 109, No. EM3. 1983, pp. 691-701.

[9] Gravelle, A., "Active Flutter Control in Transonic Conditions," in Structural Control, Leipholz, H.H.E., Editor, North Holland, Amsterdam, 1980, pp. 297-312.

[10] Chang, J.C.H. and Soong, T.T., "Structural Control Using Active Tuned Mass Dampers," *Journal of Engineering Mechanics Division, ASCE*, Vol. 106, No. EM6, 1980, pp. 1091-1098.

[11] Soong, T.T. and Skinner, G., "Experimental Study of Active Structural Control," *Journal of the Engineering Mechanics Division, ASCE*, Vol. 107, No. EM6, 1981, pp. 1057-1067.

[12] Roorda, J., "Tendon Control in Tall Structures," *Journal of the Structural Division, ASCE*, Vol. 101, No. ST3, 1975, pp. 505-521.

[13] Yang, J.N. and Giannopoulos, F., "Active Control and Stability of Cable-Stayed Bridge," *Journal of the Engineering Mechanics Division, ASCE*, Vol. 105, No. EM4, 1979, pp. 677-694.

[14] Prucz, Z., Soong, T.T., and Reinhorn, A.M., "An Analysis of Pulse Control for Simple Mechanical Systems," *Journal of Dynamic Systems, Measurement and Control, ASME*, Paper #B-122, June, 1985 (in print).

[15] Reinhorn, A.M. and Manolis, G.D., "Current State of Knowledge on Structural Control," *Journal of the Shock and Vibration Digest, Vibration Institute*, 1985 (in print).

[16] Bathe, K.J., *Finite Element Procedures in Engineering Analysis*, Prentice-Hall, England Cliffs, 1982.

[17] Goudreau, G.L. and Taylor, R.L., "Evaluation of Numerical Integration Methods in Elastodynamics," *Journal of Computer Methods in Applied Mechanics and Engineering*, Vol. 2, No. 1, 1973, pp. 303-317.

[18] Mahin, S.M. and Williams, E.M., "Computer Controlled Seismic Performance Testing," Second ASCE-EMD Specialty Conference on Dynamic Response of Structures, Atlanta, Georgia, 1981.

PREDICTIVE STRUCTURAL CONTROL

J. Rodellar
School of Civil Engineering
Universidad Politécnica de Cataluña
Barcelona, Spain

J. Martín-Sánchez
Department of Chemical Engineering
University of Alberta
Edmonton, Alberta, Canada

1. Introduction

In recent years, different methods have been taken from Control Theory in order
to develop active control systems for civil engineering structures. Classical feedback control techniques have been used in [1-3]. The interest in improving the
performance given by classical feedback control has led to the application of
optimal control theory for structural control as considered in [4-7]. Since a
matrix Ricatti equation has to be solved, the optimal control law applied in references [4-7] may not be useful for on-line control when the structure has many
degrees of freedom. Active control by modal synthesis has been proposed in [8].
This technique consists of controlling independent specific dynamic modes after
decoupling the structure's equations of motion. An example, in which a linear
optimal control is applied to each mode, may be found in [9]. However decoupling is possible only when the damping matrix verifies the orthogonality conditions.

Structural control by pole placement has been considered in [10-11]. This
method is based on a feedback control law in which the gain is chosen in such a
way that the set of closed-loop poles have a specified value in the stable half of
the s-plane. Although at least one set of stable poles exists if the open-loop

system is completely controllable, it may not be possible to find suitable locations that guarantee a good performance for civil engineering structures. Within this context, a least-square algorithm is used in [12] to estimate the control law gain that reduces the structural response in a factor specified a priori.

In this paper a predictive control method is proposed as a new approach to structural control. This approach is a subset of a more general adaptive-predictive control system (APCS) introduced in [13-15] and used in a number of control applications [16-18]. It is important to indicate that APCS methodology is supported by results of global stability when the plant is under external disturbances [19], which is generally the case of interest in structural control. The purpose of this paper is to describe the concepts in which the method is based and show its potential and advantage for structural control. The problem of reducing the vibration level induced in a building structure by a seismic excitation is considered as an illustrative numerical example.

Since the method is conceived to be implemented on a digital computer, it is mathematically formulated in discrete time.

2. Conceptual Description of the Predictive Control Method

The main concept of the method is that the control or input applied to the plant, at consecutive control instants, is the one that makes the predicted plant output equal to a dynamic desired plant output. To develop this concept into a practical scheme, the input and output variables of the plant, which will in this paper be a structure, should be appropriately selected in a previous stage of the implementation. Then, as shown in Fig. 1, a predictive model and a driver block are used, the specific functions of which at each control instant are explained as follows:

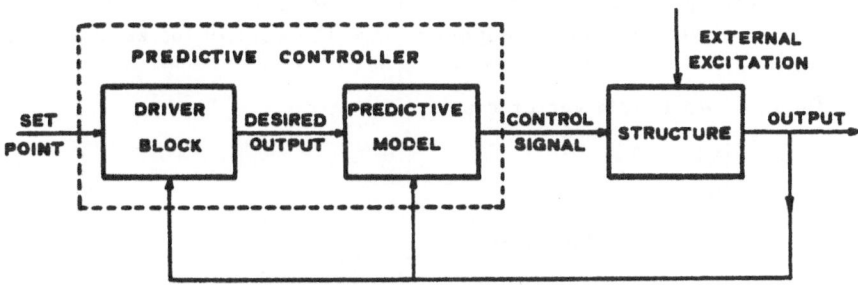

Figure 1: Basic scheme of the predictive controller

- The driver block takes into account the desired steady state (set point), which in structural control is generally the equilibrium state of the structure, to generate a future desired output value of the structure that belongs to a desired output trajectory. By the way of example, the desired output trajectory may start from the current value of the measured output and reach the set point according to some chosen dynamics.

- The predictive model is used to generate a control signal that makes the predicted output of the structure equal to the desired output given by the driver block.

According to the above explanation, the key factors in the implementation of predictive structural control are the choice of the predictive model and the driver block design. Since the method is designed to be implemented on a digital computer, the differential equations of the structure's motion will be at first discretized in the next section, where the considered discrete-time structural control problem will be defined. Then, using the discrete-time equations of the structure, the specific choices of the predictive model and the driver block design, used in this implementation of predictive method, will be discussed in the subsequent section.

3. A Discrete-Time Formulation for a Structural Control Problem

Consider a civil engineering structure modelled as a lumped-mass n-degree-of-freedom mechanical system subjected to a horizontal ground acceleration $\ddot{d}_o(t)$. Its horizontal motion can be described by the equation:

$$M\ddot{y}+C\dot{y}+Ky=u-m\ddot{d}_o \tag{1}$$

where M, C and K are respectively the $n \times n$ mass, damping and stiffness matrices; y, $\dot{y}$ and $\ddot{y}$ are respectively the $n \times 1$ displacement, velocity and acceleration vectors relative to the ground and m is the $n \times 1$ mass vector; u is the $n \times 1$ control vector whose components are the control forces applied on the structure.

Defining a $2n \times 1$ state vector x as

$$x = \begin{pmatrix} y \\ \dot{y} \end{pmatrix} \tag{2}$$

Eq. 1 can be expressed in the form:

$$\dot{x} = Fx + Gu + v \tag{3}$$

where F and G are matrices defined by

$$F = \begin{pmatrix} 0 & I \\ -M^{-1}K & -M^{-1}C \end{pmatrix}, \quad G = \begin{pmatrix} 0 \\ M^{-1} \end{pmatrix} \tag{4}$$

O and I being respectively the $n \times n$ null and identity matrices; v is a vector representing the external excitation which is defined as

$$v = \begin{pmatrix} 0 \\ -M^{-1}m\ddot{d}_o \end{pmatrix} \tag{5}$$

Eq. 3 represents the continuous time history of the controlled structure in a simple, compact and general state variable form.

Since the structure will be controlled by using a digital computer, the digital control scheme of Fig. 2 must be considered.

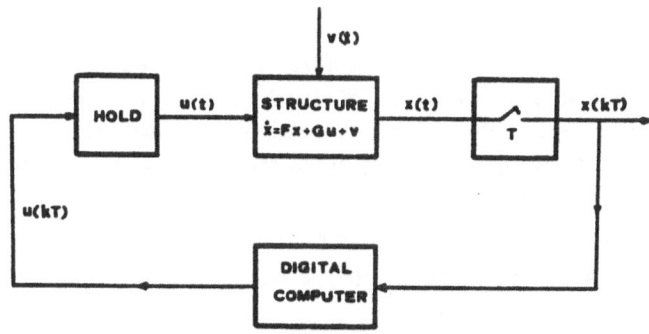

Figure 2: Basic scheme for digital structural control

At each sampling instant kT, where T is the sampling period and k is an ordinal integer, a set of sensors will measure the displacement and velocity components of the state vector $x(kT)$. Then the digital computer, implementing the predictive control method, will calculate the value of the control vector $u(kT)$. A hold device will convert the sequence of consecutive $u(kT)$ into a continuous control signal $u(t)$ which will be applied to the structure by means of a set of appropriate actuators. Generally a zero-order hold is used in control applications which converts the sequence of $u(kT)$ into a piecewise function $u(t)$ by holding constant the value of $u(kT)$ between instant kT and the consecutive instant $kT+T$. In this case, a difference equation, relating the sequence of $x(kT)$ as a function of the control sequence $u(kT)$ and the excitation signal $v(t)$, may be defined as follows:

$$x(k+1) = Ax(k) + Bu(k) + v(k) \tag{6}$$

where k indicates the sampling instant kT and A, B and $v(k)$ are given by [20]:

$$A = e^{FT} \tag{7a}$$

$$B = \int\limits_{0}^{T} e^{Ft} G \, dt \tag{7b}$$

$$v(k) = \int\limits_{kT}^{kT+T} e^{F(kT+T-t)} v(t) \, dt \tag{7c}$$

Using Eq. 2, the displacement vector $y(k)$ may be expressed as a function of the sampled state vector in the form:

$$y(k) = Hx(k) \tag{8}$$

where $H = [I \,|\, 0]$ \hfill (9)

The control objective will be to reduce the displacement vector $y(k)$ with a non excessive control action $u(k)$. Obviously $y(k)$ is considered as the output vector of the structure. The following section illustrates the predictive control method as applied to the previously defined structural control problem under the following assumptions:

1. Matrices A and B defined by Eq. 7 are considered known.

2. The external excitation vector $v(k)$ is an unknown and unmeasured perturbation acting on the structure.

4. Implementation of Predictive Structural Control

Since the value of the external excitation is assumed unknown, the following predictive model is chosen:

$$\hat{x}(k+1\,|\,k) = Ax(k) + Bu(k) \tag{10}$$

where $\hat{x}(k+1\,|\,k)$ represents the predicted value, at time k, of the state vector at time $k+1$. It can be observed that Eq. 10 uses all the available information known at time k. This model can also be used to predict, at sampling instant k, a future sequence of state vectors as a function of a future sequence of control vectors, as it is indicated by the equation:

$$\hat{x}(k+j\,|\,k) = A\hat{x}(k+j-1\,|\,k) + B\hat{u}(k+j-1\,|\,k) \tag{11}$$

$$(j = 1, \ldots, \lambda)$$

where $\hat{x}(k+j\,|\,k) \; (j=1,\ldots,\lambda)$ is the predicted sequence of state vectors, $\hat{u}(k+j-1\,|\,k) \; (j=1,\ldots,\lambda)$ is the future sequence of control vectors and λ is a positive integer that defines the finite prediction horizon $[k, k+\lambda]$. Obviously $\hat{x}(k\,|\,k) = x(k)$. The way in which the above model is used in the present implementation of predictive structural control is explained as follows.

In the case in which no further change in the control action were made after instant k, the future sequence of state vectors could be computed from Eq. 11 under the condition:

$$\hat{u}(k\,|k)=\hat{u}(k+1\,|k)=...=\hat{u}(k+\lambda-1\,|k) \tag{12}$$

Using Eqs. 11 and 12, the value of the predicted state vector at sampling instant $k+\lambda$ can be expressed in the form:

$$\hat{x}(k+\lambda\,|k)=\alpha(\lambda)x(k)+\beta(\lambda)\hat{u}(k\,|k) \tag{13}$$

where $\alpha(\lambda)$ and $\beta(\lambda)$ are matrices given by

$$\alpha(\lambda)=A^{\lambda} \tag{14a}$$

$$\beta(\lambda)=(A^{\lambda-1}+A^{\lambda-2}+...+A^{2}+A)B+B \tag{14b}$$

Therefore, using Eqs. 8 and 13, the predicted output at instant $k+\lambda$ can be written as

$$\hat{y}(k+\lambda\,|k)=H\alpha(\lambda)x(k)+H\beta(\lambda)\hat{u}(k\,|k) \tag{15}$$

Using Eq. 15, the control vector $u(k)$, that will make the predicted output $\hat{y}(k+\lambda\,|k)$ equal to a desired output $y_d(k+\lambda)$, can be computed by

$$u(k)=\hat{u}(k\,|k)=(H\beta(\lambda))^{-1}y_d(k+\lambda)+(H\beta(\lambda))^{-1}H\alpha(\lambda)x(k) \tag{16}$$

As new sampling instants arrive, new future desired values of the output vector will be defined, which may require a change in the control vector. Therefore, the previously indicated procedure for the computation of the control vector is repeated at every sampling instant.

In the present implementation, the driver block is designed to generate, at each sampling instant k, a desired output $y_d(k+\lambda)$ that belongs to a reference trajectory defined by

$$y_r(k+j)=\theta_1 y_r(k+j-1)+\theta_2 y_r(k+j-2) \tag{17}$$

$$(j=1,\ldots,\lambda)$$

This reference trajectory is redefined, at each sampling instant k, in such a way that it starts from the previous and present output vectors of the structure and approaches the steady-state output (null displacement in this case) with a chosen desired dynamics defined by matrices θ_1 and θ_2. This redefinition of the reference trajectory implies that, at each instant k:

$$y_r(k)=y(k) \tag{18a}$$

$$y_r(k-1)=y(k-1) \tag{18b}$$

The desired output $y_d(k+\lambda)$ is made equal to $y_r(k+\lambda)$, which is calculated using Eqs. 17 and 18.

Therefore, the implementation of the predictive structural control scheme requires the execution, at each sampling instant, of the following two steps:

1. Computation of the desired output $y_d(k+\lambda)$ by Eqs. 17 and 18.
2. Computation of the control vector by Eq. 16.

The above control scheme has illustrated the choice and use of a digital predictive model of the structure and has presented a particular driver block that can be casted into the general driver block design methodology. Under this general methodology, the desired output may be obtained through the minimization of a cost function of the predicted process input/output sequences over a finite time horizon $[k,k+\lambda]$ without requiring the solution of a matrix Ricatti equation [21-22].

5. Simulation Example

Consider a four-degree-of-freedom shear building structure, as shown in Fig. 3, with parameters:

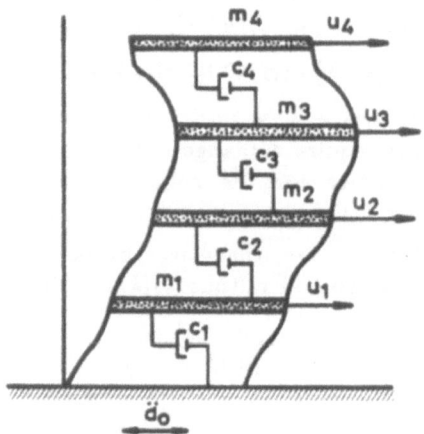

Figure 3: Shear building structure

$$m_i = 400\,N s^2/cm$$

$$c_i = 640\,N s/cm$$

$$k_i = 6400\,N/cm$$

$$(i = 1, \ldots, 4)$$

The structure is excited by an earthquake ground acceleration $\ddot{d}_o$ which is represented in Fig. 4.

The displacement response of each floor of the structure, when no control action is taken, is shown in Fig. 5.

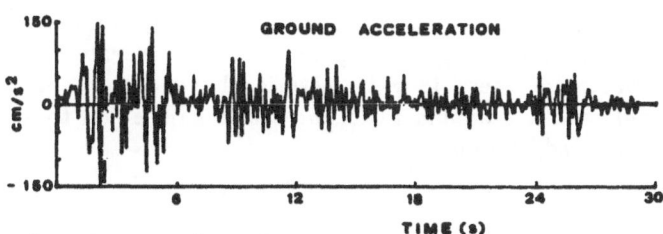

Figure 4: *Earthquake ground acceleration*

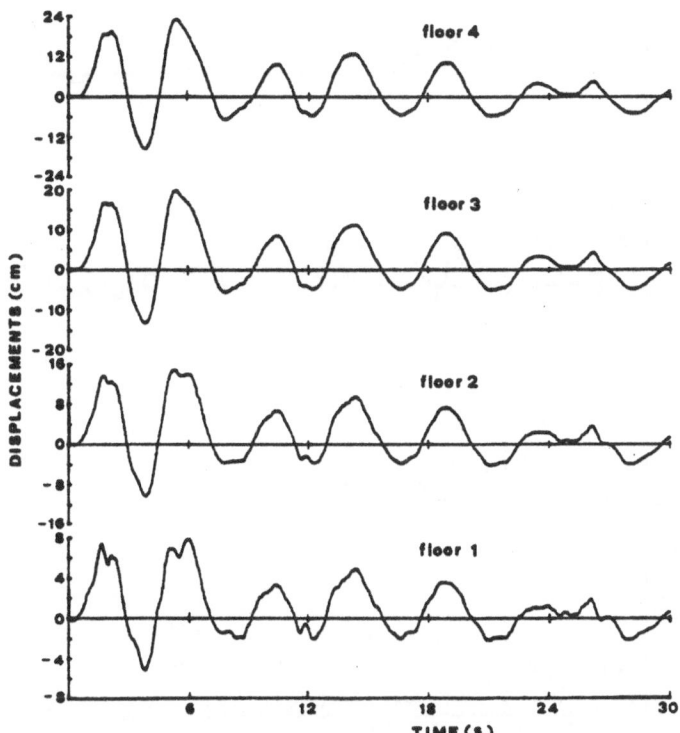

Figure 5: Displacement response of the uncontrolled structure

The predictive method has been applied to reduce the response of the structure. Figs. 6a and 6b represent the displacement and the control force
corresponding to each floor when the following parameters have been used:

-    Sampling period: $T=0.05$ seconds.

- Length of the prediction horizon: $\lambda = 4$.
- Matrices θ_1 and θ_2 of the desired output trajectory defined by Eq. 17 have been chosen diagonal, what corresponds to four separate scalar trajectories. The parameters of these four trajectories have corresponded to a second order model with a damping ratio and static gain equal to 1 and a natural frequency of 4 rad/second.

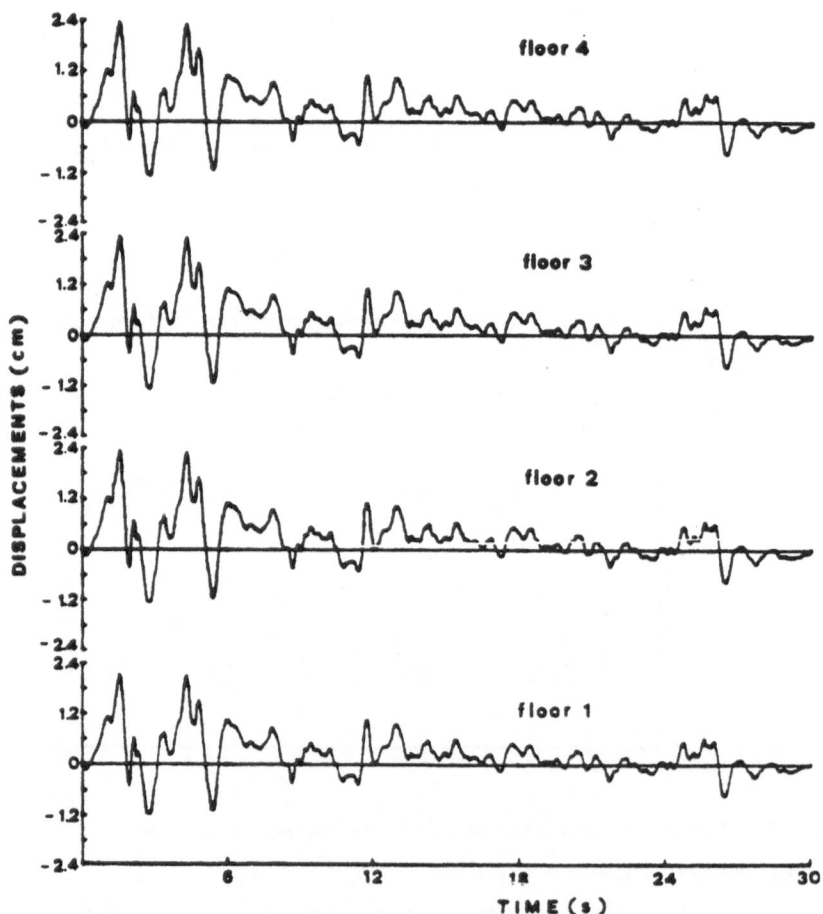

Figure 6a: Controlled response using a prediction horizon $\lambda = 4$

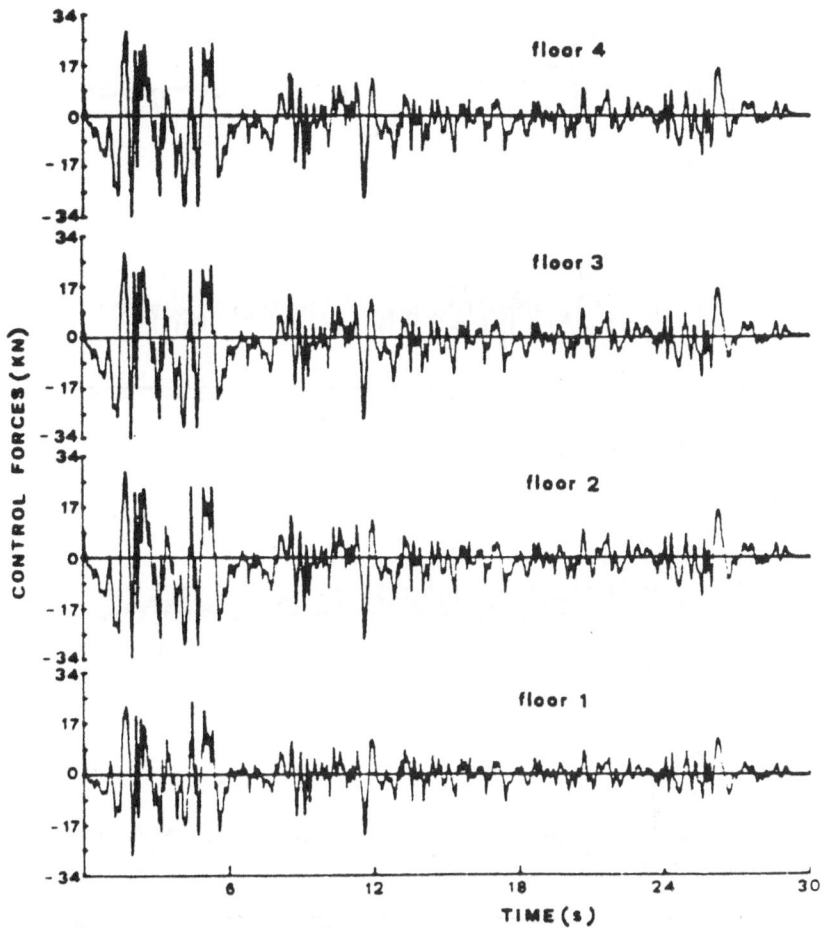

Control forces using a prediction horizon $\lambda = 4$

.er to analyze the effects on the control of different choices in the
k parameters, the results of two additional experiments are presented.
ιows the response and control force of the fourth floor with a reference
identical to the one used in the experiment shown in Fig. 6 but with a
horizon defined by $\lambda = 2$. The experiment shown in Fig. 8 considers a
horizon with $\lambda = 4$, but in this case the reference trajectory discontinu-
s from the measured output value to the desired null displacement.

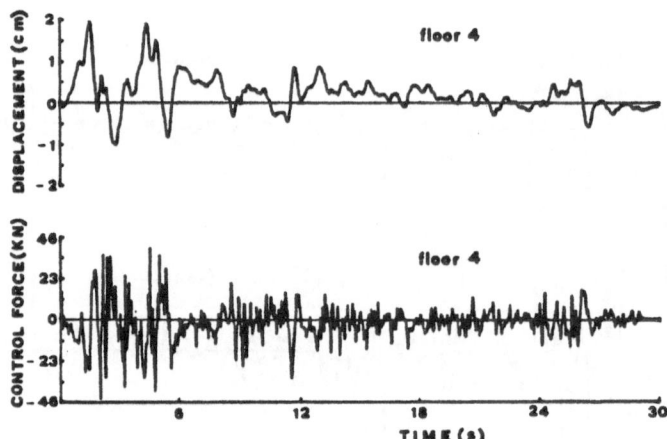

Figure 7: Controlled response using λ=2

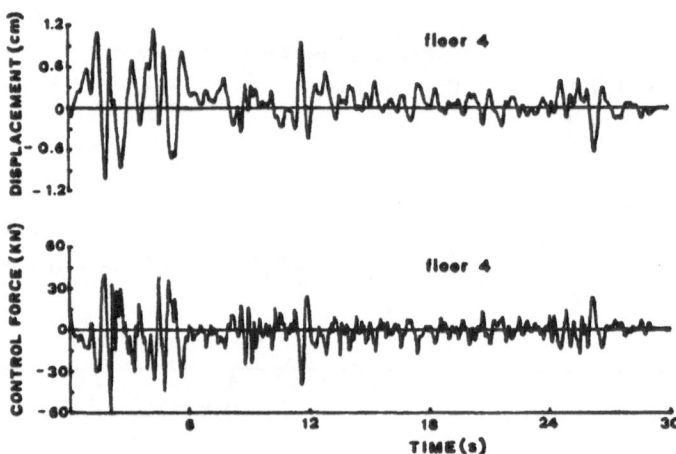

Figure 8: Controlled response with a discontinuous reference trajectory and
        λ=4

6. Discussion of the Results

A comparative analysis of Fig. 5 and Figs. 6, 7 and 8 shows that the predictive
control scheme drastically reduces the displacement response of the structure
induced by the earthquake ground acceleration. For instance, it can be observed
that, when the structure is uncontrolled (Fig. 5), the amplitude of the displacement reaches approximately 24 cm in the fourth floor. However, when the structure is under predictive control, the maximum amplitude of the displacement in
the same fourth floor is reduced at approximately 2.4, 2 and 1.2 cm as respectively shown in Figs. 6, 7 and 8.

Also, it is of interest to note that the maximum amplitude of the control force in Fig. 6b is 34 KN, while in Figs. 7 and 8 this value is 46 and 60 KN respectively. This indicate that, when the factor of reduction of the displacement is already 10, an additional reduction of approximately 0.4 cm requires an added increment of the maximum control force of approximately 12 KN. Therefore, an excessively demanding reduction in the displacement may require too large control forces, which are undesirable.

The results obtained in the experiment of Fig. 6 show an efficient reduction of the displacement response by means of a non-excessive control effort. Fig. 7 shows that, if the value of λ is reduced to 2, a greater reduction in the displacement is obtained at the cost of an undesirable large increment in the control action. Fig. 8 shows that the elimination of a "smooth" dynamics for the reference trajectory of the driver block also results in an excessively large control effort. Clearly in these two last cases the reduction of λ and the elimination of the desired dynamics determine a more demanding desired output vector. The designer will be in general able to choose in a simple manner the driver block parameters that will yield the desired results as it has been the case in the discussed simulation example.

7. Conclusions

This paper has presented the conceptual basis and the implementation issues of a digital predictive control method for structural control. The method has been illustrated as applied to reduce the displacement response of a simulated structure subjected to a seismic excitation. The conclusions are that predictive control may be particularly well suited for on line structural control applications due to the following reasons:

1) It is mathematically simple and easy to use through a digital computer. It does not need the decoupling of the equations of motion, as modal synthesis techniques, and the calculation of the control vector does not require a great computational effort, as optimal control methods.

2) It can reduce the displacement response of the structure in a significant factor with not excessive control forces.

References

[1] Roorda, J., "Tendon Control in Tall Structures", *Journal of the Structural Division, ASCE*, Vol. 101, No. ST3, 1975, pp. 505-521.

[2] Roorda, J., "Experiments in Feedback Control of Structures", in *Structural Control*, edited by H.H. Leipholz, North-Holland Pub. Co., New York, 1980, pp. 629-662.

[3] Yang, J.N. and Giannopoulos, F., "Active Tendon Control of Structures", *Journal of the Engineering Mechanics Division, ASCE*, Vol. 104, No. EM3, 1978, pp. 551-568.

[4] Abdel-Rohman, M. and Leipholz, H.H., "A General Approach to Active Structural Control", *Journal of the Engineering Mechanics Division, ASCE*, Vol. 105, No. EM6, 1979, pp. 1007-1023.

[5] Abdel-Rohman, M. and Leipholz, H.H., "Automatic Active Control of Structures", in *Structural Control*, edited by H.H. Leipholz, North-Holland Pub. Co., New York, 1980, pp. 29-56.

[6] Chang, J. and Soong, T.T., "Structural Control Using Active Tuned Mass Dampers", *Journal of the Engineering Mechanics Division, ASCE*, Vol. 106, No. EM6, 1980, pp. 1091-1098.

[7] Yang, J.N., "Application of Optimal Control Theory to Civil Engineering Structures", *Journal of the Engineering Mechanics Division, ASCE*, Vol. 101, No. EM6, 1975, pp. 819-838.

[8] Meirovitch, L. and Oz, H., "Active Control of Structures by Modal Synthesis", in *Structural Control*, edited by H.H. Leipholz, North-Holland Pub. Co., New York, 1980, pp. 505-521.

[9] Meirovitch, L. and Silverberg, L.M., "Control of Structures Subjected to Seismic Excitation", *Journal of the Engineering Mechanics Division, ASCE*, Vol. 109, No. 2, 1983, pp. 604-618.

[10] Martin, C.R. and Soong, T.T., "Modal Control of Multistory Structures", *Journal of the Engineering Mechanics Division, ASCE*, Vol. 102, No. EM4, 1976, pp. 613-623.

[11] Abdel-Rohman, M. and Leipholz, H.H., "Structural Control by Pole Assignement Method", *Journal of the Engineering Mechanics Division, ASCE*, Vol. 104, No. EM5, 1978, pp. 1159-1175.

[12] Basharkhah, M.A. and Yao, J.T.P., "Some Recent Developments in Structural Control", *Journal of Structural Mechanics*, Vol. 11(2), 1983, pp. 137-152.

[13] Martín-Sánchez, J.M., "Contribution to Model Reference Adaptive Systems from Hyperstability Theory" (in Spanish), Doctoral Dissertation, Universidad Politécnica de Cataluña, Spain, 1974.

[14] Martín-Sánchez, J.M., "A New Solution to Adaptive Control", *Proceedings of the IEEE*, Vol. 64, No. 8, 1976, pp. 1209-1218.

[15] Martín-Sánchez, J.M., "Adaptive-Predictive Control Method", USA Patent No. 4,197,576, priority date August 4, 1976.

[16] Martín-Sánchez, J.M., *Modern Control Theory; Adaptive-Predictive Control Method: Theory and Applications* (in Spanish), book published by March Foundation of Spain, Madrid, 1977.

[17] Martín-Sánchez, J.M., Shah, S.L. and Fisher, D.G., "Application of a Multivariable Adaptive-Predictive Control System", *Proceedings of the Workshop on Applications of Adaptive Systems Theory*, Yale University, 1981, pp. 138-148.

[18] Martín-Sánchez, J.M. and Shah, S.L., "Multivariable Adaptive-Predictive Control of a Binary Distillation Column", *Automatica*, Special issue on adaptive systems, September 1984.

[19] Martín-Sánchez, J.M., "A Globally Stable APCS in the Presence of Bounded Noises and Disturbances", *IEEE Transactions on Automatic Control*, Vol. AC-29, No. 5, 1984, pp. 461-464.

[20] Franklin, G. and Powell, J.D., *Digital Control of Dynamic Systems*, Addison-Wesley Publishing, 1980.

[21] Martín-Sánchez, J.M., Reply to "Comments to a New Solution to Adaptive Control", Proceedings Letters, *Proceedings of the IEEE*, April 1977.

[22] Rodellar, J., "Optimal Design of Driver Block in the Adaptive-Predictive Control System" (in Spanish), Doctoral Dissertation, Universidad de Barcelona, Spain, 1982.

APPROXIMATE METHODS FOR MULTIOBJECTIVE DISTRIBUTED CONTROL OF A VIBRATING BEAM

I. S. Sadek
Department of Mathematics and Statistics
Queen's University
Kingston, Ontario, Canada K7L 3N6

1. Introduction

The progress on the application of modern Control methods to vibrating structures is relatively slow due to the inherent complexities of the subject. Indeed, complications arise when the behavior of vibrating elastic structures are described by partial differential equations. In contrast, standard optimal control techniques are applicable only to systems described by ordinary differential equations [4]. However, the subject has gained a timely importance and even some urgency in recent years due to the possibility of constructing large stations in space, the oscillation of which should be actively controlled in a weightless vacuum.

The research activities in the area of structural control through the employment of distributed forces as the control variable have been increased over the last few years. A review of the work on the control of vibrating elastic systems was given by [5] in 1978. Several related articles appeared in the proceedings of 1979 Symposium on structural control [9]. The control of torsional vibrations of a shaft through an applied moment force and of a beam to bring it to a specified state in a given time are considered in [3] and [8] respectively. In [6,16], the authors studied the optimal control of the transverse vibrations of a beam where the total energy are chosen as the cost functional. Moreover, optimal control of plates by means of control forces were investigated in [10,11]. Narukava [12]

594

studied the control of string vibrations with upper and lower bounds on the control functions. Also, a number of control problems related to one dimensional structures were investigated in [7].

In this present paper, we consider the problem of minimizing the dynamic response of a damped simply supported beam in a specified time with a minimum expenditure of force. The dynamic response of the beam is defined as the dynamic deflection and/or velocity of the beam. Thus, this control problem involves the objectives of minimizing the deflection, velocity, and force spent and is formulated as a multiobjective control problem by using a pareto optimal approach [1,2,13]. The optimal control of the formulated problem is determined by three suggested methods of solutions: analytic and two computational approaches.

The analytic method is based on the maximum principle derived in [14]. The method consists of converting the basic control problem of minimizing a weighted sum objective functional under given constraints to solving an initial-terminal boundary-value problem in partial differential equations (PDE) by using the maximum principle [14]. The solution of the problem is expressed in terms of explicit formulae by using the eigenfunction expansion method. In this case, the method of solution of the posed problem using maximum principle is analytic. We also suggest two computational methods for the solution of the present problem. The first method leads to a mathematical programming problem while the second one leads to an integral equation which can be solved by a number of computational techniques. These computational methods give the exact solutions when closed-form solutions are available. In general, however, they yield only approximate results.

The approximate methods considered in this study are computationally easier to handle than the analytic method for solving an optimal control problem, since we do not need to solve PDE. In particular, the variational method (integral equation approach) yields integral equation which are always easier to handle than PDE. Furthermore, a direct method (mathematical programming approach) gives us great freedom as to choice of our objective functional, so that we can obtain a final solution that is even better than the one found by analytical approach (maximum principle).

2. Formulation of the Control Problem

We consider a simply supported beam of length ℓ and of mass m per unit length. The distributed control force applied on the beam is $\bar{F}(X, t^{*})$ where X is the position along the beam and t^{*} is the time. The equation governing the vibrations of a simply supported beam is given by

$$mW_{t^{*}t^{*}} + C^{*}W_{t^{*}} + EIW_{XXXX} = F(X, t^{*}), \qquad (2.1)$$

$$0 < X < \ell, t^{*} \geq 0,$$

in which C^{*} is the damping coefficient, $W(X, t^{*})$ is the transverse deflection, and EI is the flexural rigidity.

We introduce the following dimensionless quantities:

$$x = X/\ell, w = W/\ell \quad , \quad t = \frac{t^{*}}{\ell^{2}} \left[\frac{EI}{m} \right]^{1/2},$$

$$C = \frac{C^{*}\ell^{2}}{\sqrt{mEI}}, \quad f = \frac{\ell^{3}F}{EI}. \tag{2.2}$$

Using (2.2), the dimensionless form of the governing equation is obtained as:

$$w_{tt} + Cw_{t} + L[w] = f(x, t), \tag{2.3}$$

where $L = [\quad]_{xxxx}$.

For a simply supported beam the boundary conditions are:

$$w(0, t) = w_{xx}(0, t) = w(1, t) = w_{xx}(1, t) = 0. \tag{2.4}$$

The displacement and velocity at $t = 0$ is given by:

$$w(x, 0) = \phi(x) \quad , \quad w_{t}(x, 0) = \Psi(x). \tag{2.5}$$

Equations (2.3)-(2.5) give the nondimensional formulation of the vibrating beam problem subject to a distributed control force.

The objective of the study is to determine the control force $f(x, t)$ such that the deflection $w(x, t)$ and the velocity $w_{t}(x, t)$ of the beam will be minimized in a specified time $0 \leq t \leq T$ with a minimum expenditure of control energy. Thus the present control problem has the multiple objectives of (a) minimizing w, (b) minimizing w_{t}, and (c) minimizing the amount of force spent for this purpose. Consequently, the control problem can be stated in the following form:

Determine the optimal control function $f(x, t)$, $0 \leq x \leq 1$ and $0 \leq t \leq T$ which minimizes the cost functionals:

$$J_{1} = \int_{0}^{1} w^{2}(x, T)dx, \quad J_{2} = \int_{0}^{1} w_{t}^{2}(x, T)dx, \quad J_{3} = \int_{0}^{T}\int_{0}^{1} f^{2}(x, t)dxdt, \tag{2.6}$$

with $w(x, t)$ subject to (2.3)-(2.5).

The quadratic functions of w, w_{t} and f in (2.6) were chosen in such a way to obtain an explicit control law for the problem.

The objective of the posed problem is to determine an optimal control function which will minimize the deflection, the velocity and the force function simultaneously. Unfortunately, this is impossible, since the objectives are normally in conflict with each other. It is therefore necessary to seek some other concept of optimality which is both physically meaningful and easy to apply. We feel that pareto optimality is well suited to this problem. For further details on pareto optimality, we refer the reader to [1,2,13] where this concept was used for optimality design of multiobjective structures. We compute the solutions by minimizing a single performance index obtained by combining the multiple objectives in a weighted sum

$$J(f) = \sum_{i=1}^{3} \mu_i J_i(f), \quad \mu_i \geq 0, \quad \mu_1 + \mu_2 > 0, \quad \mu_3 > 0, \tag{2.7}$$

where μ_i are the weighting constants and reflect the relative weight attached to minimizing J_i. Consequently the optimal control to be solved is the following:

$$\min_{f \in U} \ J(f) \tag{2.8}$$

subject to (2.3)–(2.5) ,

where U is the set of all functions measurable and bounded on $(0,1) \times (0,T)$.

3. Analytic Solution via the Maximum Principle

In order to investigate this problem we will use the following maximum principle [14].

Theorem 1: If the control function $f^o(x,t), 0 \leq x \leq 1, 0 \leq t \leq T$

(i) is a solution of the optimal control problem (2.8), and

(ii) the differential operator L together with appropriate boundary conditions is self-adjoint and has a complete orthogonal set of eigenfunctions, and

(iii) Range of L together with the appropriate boundary conditions and initial conditions contains U, then the following maximum principle [14] holds

$$\underset{f \in U}{\text{Max}} \ H(x,t,v^o,f) = H(x,t,v^o,f^o) \text{a.e. in } (0,1) \times [0,T], \tag{3.1}$$

where

$$H(x,t,v^o,f) = v^o(x,t)f(x,t) - \mu_3 f^2(x,t),$$

and v° satisfies

$$v_{tt}^o - C v_t^o + L[v^o] = 0 \quad \text{on} \ (0,1) \times (0,T) \tag{3.2}$$

with boundary conditions (2.4) and the terminal conditions:

$$v^o(x,T) = -2\mu_2 w_t^o(x,T) \ ,$$

$$v_t^o(x,T) = 2\mu_1 w^o(x,T) - 2C\mu_2 w_t^o(x,T), \tag{3.3}$$

where w° is the state function corresponding to the optimal control.

Remark: In [14] sufficiency and uniqueness follow from a convexity assumption which is automatically satisfied in the case here.

3.1. Verification of maximum principle hypotheses

In order to apply the maximum principle it is necessary to verify that hypotheses of theorem 1 are satisfied, where we need only to verify (ii).

ii(a). L together with boundary conditions is self-adjoint. To see this note that

$$vL[w] - wL[v] = \frac{d}{dx}\beta(v,w), \tag{3.4}$$

where

$$\beta(v,w) = vw'' - v'w'' + v''w' - v''w .$$

By using the boundary conditions, it follows that $\beta(v,w) = 0$ and hence L is self-adjoint.

ii(b). L has a complete set of orthogonal eigenfunctions. The eigenfunctions of the problem is given by

$$W_n(x) = \sin(n\pi x), \quad n = 1,2,\ldots,\infty. \tag{3.5}$$

3.2. Method of solution

We obtain the solution of the control problem formulated in the previous section by employing the maximum principle. Let the adjoint variable $v(x,t)$ satisfying the differential equation (3.2) be

$$v(x,t) = \sum_{n=1}^{\infty} \phi_n(t)\sin(n\pi x), \tag{3.6}$$

where $\phi_n(t)$ satisfies

$$\ddot{\phi}_n(t) - C\dot{\phi}_n(t) + \lambda_n\phi_n(t) = 0, \quad \left[\dot{\phi}_n = \frac{d\phi_n}{dt}\right], \tag{3.7}$$

and

$$\lambda_n = (n\pi)^4, \quad n = 1,,2,3,\ldots,\infty.$$

The general solution of (3.7) is

$$\phi_n(t) = a_n M_n(t) + b_n N_n(-t), \tag{3.8}$$

where

$$M_n(t) = e^{\frac{1}{2}(P_n + C)t}, \quad N_n(t) = e^{\frac{1}{2}(P_n - C)t}, \quad P_n = \left(C^2 - 4\lambda_n\right)^{1/2},$$

and a_n and b_n are constants to be determined.

Now we invoke the maximum principle, theorem 1. It follows that a necessary and sufficient condition for the control f to be optimal f° is that

$$f^\circ = \frac{1}{2\mu_3} v^\circ, \tag{3.9}$$

where

$$v^\circ(x,t) = \sum_{n=1}^{\infty} \phi_n^\circ(t)\sin(n\pi x)x, \tag{3.10}$$

$$\phi_n^\circ(t) = a_n^\circ M_n(t) + b_n^\circ N_n(-t). \tag{3.11}$$

Next solve

$$w_{tt}^\circ + Cw_t^\circ + L[w^\circ] = f^\circ \tag{3.12}$$

with boundary conditions (2.4) and initial conditions (2.5). The solution is given by:

$$w^\circ(x,t) = \sum_{n=1}^{\infty} Z_n^\circ(t)\sin(n\pi x), \tag{3.13}$$

where

$$\ddot{Z}_n^\circ(t) + C\dot{Z}_n^\circ(t) + \lambda_n Z_n^\circ(t) = \frac{1}{2\mu_3} \phi_n^\circ(t). \tag{3.14}$$

The general solution of (3.14) with initial conditions (2.5) is given by:

$$Z_n^\circ(t) = \alpha_n^\circ N_n(t) + \beta_n^\circ M_n(-t) + \frac{1}{2\mu_3 P_n} \int_0^t \left[N_n(t-\tau) - M_n(\tau-t) \right] \phi_n^\circ(\tau) d\tau, \tag{3.15}$$

$$\dot{Z}_n^\circ(t) = \alpha_n^\circ \dot{N}_n(t) + \beta_n^\circ \dot{M}_n(-t) + \frac{1}{2\mu_3 P_n} \int_0^t \left[\dot{N}_n(t-\tau) - \dot{M}_n(\tau-t) \right] \phi_n^\circ(\tau) d\tau, \tag{3.16}$$

where α_n° and β_n° are constants to be determined. Equations (3.15) and (3.16) can be written explicitly at $t = T$, as:

$$Z_n^\circ(T) = D_{1n} + a_n^\circ D_{2n} + b_n^\circ D_{3n}, \tag{3.17}$$

$$\dot{Z}_n^\circ(T) = D_{1n}' + a_n^\circ D_{2n}' + b_n^\circ D_{3n}', \tag{3.18}$$

where

$$D_{1n} = \alpha_n^\circ N_n(T) + \beta_n^\circ M_n(-T), \tag{3.19}$$

$$D_{2n} = \frac{1}{2\mu_3 P_n C} \left(M_n(T) - N_n(T) \right) - \frac{1}{2\mu_3 P_n (P_n + C)} \left(M_n(T) - M_n(-T) \right), \tag{3.20}$$

$$D_{3n} = \frac{-1}{2\mu_3 P_n(P_n - C)} \left[N_n(-T) - N_n(T) \right] - \frac{1}{2\mu_3 P_n C} \left[N_n(-T) - M_n(-T) \right], \quad (3.21)$$

$$D'_{1n} = \frac{1}{2}\alpha_n^o(P_n - C)N_n(T) - \frac{1}{2}\beta_n^o(P_n + C)M_n(-T), \quad (3.22)$$

$$D'_{2n} = \frac{1}{4\mu_3 P_n C} \left[(P_n + C)M_n(T) - (P_n - C)N_n(T) \right]$$

$$- \frac{1}{4\mu_3 P_n} \left[M_n(T) + M_n(-T) \right], \quad (3.23)$$

$$D'_{3n} = \frac{1}{4\mu_3 P_n} \left[N_n(-T) + N_n(T) \right] - \frac{1}{4\mu_3 P_n C} \left[-(P_n - C)N_n(-T) + (P_n + C)M_n(-T) \right], \quad (3.25)$$

Expanding the $\psi(x)$ and $\Psi(x)$ of (2.5) in their Fourier series

$$\phi(x) = \sum_{n=1}^{\infty} \phi_n \sin(n\pi x), \quad (3.26)$$

$$\Psi(x) = \sum_{n=1}^{\infty} \Psi_n \sin(n\pi x), \quad (3.27)$$

with

$$\phi_n = 2\int_0^1 \phi(x)\sin(n\pi x)dx, \quad (3.28)$$

$$\Psi_n = 2\int_0^1 \Psi(x)\sin(n\pi x)dx. \quad (3.29)$$

Hence the initial conditions become for (3.14):

$$Z_n^o(0) = \phi_n, \quad \dot{Z}_n^o(0) = \Psi_n. \quad (3.30)$$

Solving for α_n^o and β_n^o give

$$\alpha_n^o = \frac{1}{2P_n} \left[\phi_n(P_n + C) + 2\Psi_n \right], \quad (3.31)$$

$$\beta_n^o = \frac{1}{2P_n} \left[\phi_n(P_n - C) - 2\Psi_n \right]. \quad (3.32)$$

The only unknown constants that remain to be evaluated are a_n^o and b_n^o appearing in the solution of (3.11). Using (3.3), (3.10) and (3.13), we obtain

$$\phi_n^o(T) = -2\mu_2 \dot{Z}_n^o(T), \quad (3.33)$$

$$\dot{\phi}_n^o(T) = 2\mu_1 Z_n^o(T) - 2C\mu_2 \dot{Z}_n^o(T). \quad (3.34)$$

Inserting (3.8), (3.14) into (3.33) and (3.34), we have:

$$\delta_{1n}a_n^o + \delta_{2n}b_n^o = \gamma_{1n},$$

$$\delta_{3n}a_n^o + \delta_{4n}b_n^o = \gamma_{2n},$$

or,

$$a_n^o = \Delta^{-1}(\gamma_{1n}\delta_{4n} - \gamma_{2n}\delta_{2n}), \qquad (3.35)$$

$$b_n^o = \Delta^{-1}(\gamma_{2n}\delta_{1n} - \gamma_{1n}\delta_{3n}), \qquad (3.36)$$

where

$$\Delta = \delta_{1n}\delta_{4n} - \delta_{2n}\delta_{3n},$$

$$\delta_{1n} = M_n(T) + 2\mu_2 D'_{2n},$$

$$\delta_{2n} = N_n(-T) + 2\mu_2 D'_{3n},$$

$$\gamma_{1n} = -2\mu_2 D'_{1n},$$

$$\delta_{3n} = \frac{1}{2}(P_n + C)M_n(T) - 2\mu_1 D_{2n} + 2C\mu_2 D'_{2n}, \qquad (3.37)$$

$$\delta_{4n} = -\frac{1}{2}(P_n - C)N_n(-T) - 2\mu_1 D_{3n} + 2C\mu_2 D'_{3n},$$

$$\gamma_{2n} = +2\mu_1 D_{1n} - 2C\mu_2 D'_{1n},$$

where $D_{1n}, D_{2n}, D_{3n}, D'_{1n}, D'_{2n}, D'_{3n}$ are given by (3.19) - (3.25), respectively.

The control function $F^o(x,t)$ is computed from (3.9), (3.10) and (3.11) to be

$$f^o(x,t) = \frac{1}{2\mu_3}\sum_{n=1}^{\infty}\left[a_n^o M_n(t) + b_n^o N_n(-t)\right]\sin(n\pi x), \qquad (3.38)$$

where a_n^o and b_n^o are given by (3.35) and (3.36).

4. Solution of the problem by the direct method

In this section we shall apply a direct method (mathematical programming) to the problem (2.8) and reduce it to the solution of a mathematical programming problem.

We, in particular, assume that the control function can be expressed in the form:

$$f(x,t) = \sum_{n=1}^{\infty}\phi_n(t)\sin(n\pi x) \qquad (4.1)$$

where $\phi_n(t)$ is the only unknown. We can approximate $\phi_n(t)$ by using splines or global functions and subsequently minimize J over the unknowns. To illustrate the equivalence of the analytic approach with the present one, we choose the approximating functions in the following form:

$$\phi_n(t) = H_n M_n(t) + K_n N_n(-t) \tag{4.2}$$

where H_n and K_n are constants to be determined.

We know that the objective functional evaluated at $t = T$ is given by

$$J(f) = \frac{1}{2} \sum_{n=1}^{\infty} \left\{ \mu_1 Z_n^2 + \mu_2 \dot{Z}_n^2 + \mu_3 \overline{@}_0^T \phi_n^2(t)dt \right\}, \tag{4.3}$$

where $Z(T)$ and $Z'(T)$ are given by:

$$Z(T) = D_{1n} + H_n \overline{D}_{2n} + K_n \overline{D}_{3n},$$
$$\dot{Z}(T) = D'_{1n} + H_n \overline{D}'_{2n} + K_n \overline{D}'_{3n},$$
$$\overline{D}_{in} = 2\mu_3 D_{in}, \quad \overline{D}'_{in} = 2\mu_3 D'_{in}, \quad i = 2,3.$$

Thus,

$$J(f) = \frac{1}{2} \sum_{n=1}^{\infty} \left\{ \mu_1 (D_{1n} + H_n \overline{D}_{2n} + K_n \overline{D}_{3n})^2 + \mu_2 (D'_{1n} + H_n \overline{D}'_{2n} + K_n \overline{D}'_{3n})^2 \right.$$

$$+ \mu_3 \frac{H_n^2}{(P_n + C)} (M_n^2(T) - 1) - \frac{\mu_3 K_n^2}{P_n - C} (N_n^2(-T) - 1)$$

$$\left. + \frac{2\mu_3 H_n K_n}{C} (Q^2(T) - 1) \right\}, \tag{4.4}$$

where $Q(T) = e^{\frac{1}{2}CT}$.

Now we have to solve the following mathematical programming problem:

$$\min_{H_n, K_n} \quad J, \tag{4.5}$$

where H_n, K_n are the unknown control parameters. Indeed, a vast number of solution techniques is available to (4.5). The method of setting

$$\frac{\partial J}{\partial H_n} = 0, \quad \frac{\partial J}{\partial K_n} = 0, \tag{4.6}$$

for $n = 1, 2, \ldots$, and solving the resulting algebraic equations for H_n and K_n is one of the available methods and we can use it here because of the special form of $\phi_n(t)$ and it follows that

$$\overline{\delta}_{1n} H_n + \overline{\delta}_{2n} K_n = \overline{\gamma}_{1n}, \tag{4.7}$$

$$\overline{\delta}_{3n} H_n + \overline{\delta}_{4n} K_n = \overline{\gamma}_{2n}, \tag{4.8}$$

where

$$\bar{\delta}_{1n} = \mu_1 \bar{D}_{2n}^2 + \mu_2 \bar{D'}_{2n}^2 + \frac{\mu_3}{(P_n + C)}(M_n^2(T) - 1),$$

$$\bar{\delta}_{2n} = \mu_1 \bar{D}_{3n} \bar{D}_{2n} + \mu_2 \bar{D'}_{3n} \bar{D'}_{2n} + \frac{\mu_3}{C}(Q^2(T) - 1),$$

$$\bar{\gamma}_{1n} = -(\mu_1 D_{1n} \bar{D}_{2n} + \mu_2 D'_{1n} \bar{D'}_{2n}),$$

(4.9)

$$\bar{\delta}_{3n} = \mu_1 \bar{D}_{2n} \bar{D}_{3n} + \mu_2 \bar{D'}_{2n} \bar{D'}_{3n} + \frac{\mu_3}{C}(Q^2(T) - 1),$$

$$\bar{\delta}_{4n} = \mu_1 \bar{D}_{3n}^2 + \mu_2 \bar{D'}_{3n}^2 - \frac{\mu_3}{(P_n - C)}(N^2(-T) - 1),$$

$$\bar{\gamma}_{2n} = -(\mu_1 D_{1n} \bar{D}_{3n} + \mu_2 D'_{1n} \bar{D'}_{3n}).$$

Thus, the forcing function is

$$f(x,t) = \sum_{n=1}^{\infty} \left[\frac{1}{E_n} \left(\bar{\gamma}_{1n} \bar{\delta}_{4n} - \bar{\gamma}_{2n} \bar{\delta}_{2n} \right) M_n(t) \right.$$

$$\left. + \frac{1}{E_n} \left(\bar{\gamma}_{2n} \bar{\delta}_{1n} - \bar{\gamma}_{1n} \bar{\delta}_{3n} \right) N_n(-t) \right] \sin(n\pi x),$$

(4.10)

$$E_n = \bar{\delta}_{1n} \bar{\delta}_{4n} - \bar{\delta}_{2n} \bar{\delta}_{3n},$$

where, $\bar{\gamma}_{in}(i = 1,2)$, $\bar{\delta}_{jn}(j = 1,2,,3,4)$ are given by (4.9). Thus (4.10) is the exact solution of the problem and identical to (3.38). This is not surprising since we have assumed the form $\phi_n(t)$ to be the same as the one obtained by the analytic method. Any other approximating function of splines would yield an approximate result.

It is possible to obtain a solution by the direct method which is even better than one found by the analytical approach (maximum principle) where the choice of the objective functional is restricted. Here we use the expression *better* in the sense that at the time T the deflection, the velocity and the total force spent become smaller than what we have found from the analytical approach. Obviously, this is quite an important advantage in favor of the direct method.

4.1. Numerical example

Consider the undamped vibrating beam problem of length π as an example to justify the use of the direct method. The control problem under consideration is of the following form:

Determine the optimal control function $f^o(x,t)$ in $0 \le x \le \pi$ and $0 \le t \le \pi/2$ which minimizes the cost functional

$$J(f)=\int\limits_0^\pi \left| w\left(x,\frac{\pi}{2}\right)\right| dx + \int\limits_0^\pi |w_t(x,\frac{\pi}{2})| dx + \lambda \int\limits_0^{\frac{\pi}{2}} \int\limits_0^\pi f^2(x,t)dxdt, \quad (4.11)$$

where λ is weight attached to force term and with $w(x,t)$ satisfying the differential equation

$$w_{tt} + w_{xxxx} = f(x,t), \quad 0<x<\pi, \quad 0\leq t \leq \pi/2, \qquad (4.12)$$

and the boundary and the initial conditions

$$w(0,t)=w_{xx}(0,t)=w_{xx}(\pi,t)=w(\pi,t)=0, \qquad (4.13)$$

$$w(x,0)=\sin x, \quad w_t(x,0)=-\sin x .$$

Now let us choose the control function to be the simplest one, for example we take

$$\phi(t)=\sigma . \quad \text{(i.e. constant force)} \qquad (4.15)$$

Then the deflection of the beam is given by:

$$w(x,t)=\sin x(\cos t - \sin t + \sigma \int\limits_0^t \sin(t-\tau)d\tau)$$

$$= \sin x(\cos t - \sin t + \sigma(1-\cos t)), \qquad (4.16)$$

and its velocity is

$$w_t(x,t)=\sin x \left[(\sigma-1)\sin t - \cos t\right]. \qquad (4.17)$$

Next compare w and w_t at $t = \pi/2$, thus

$$w(x,\frac{\pi}{2})=\sin x(-1+\sigma), \qquad (4.18)$$

$$w_t(x,\frac{\pi}{2})=\sin x(-1+\sigma). \qquad (4.19)$$

Inserting (4.18), (4.19) and (4.15) into (4.11) we obtain

$$J(f)=4|\sigma-1|+\frac{\lambda\sigma^2\pi^2}{4}. \qquad (4.20)$$

Now we determine σ by using the condition $\frac{\partial J}{\partial \sigma}=0$. If $\sigma \geq 1$, $\frac{\partial J}{\partial \sigma}=4+\frac{\lambda\sigma\pi^2}{2}=0$ then $\sigma = \frac{-8}{\lambda\pi^2}\leq 1$ which is unacceptable. But if $\sigma \leq 1$, $\frac{\partial J}{\partial \sigma}=-4+\frac{\lambda\sigma\pi^2}{2}=0$ then $\sigma = \frac{8}{\lambda\pi^2}$ where $0<\lambda<\infty$. Thus $\sigma=\frac{8}{\lambda\pi^2}\leq 1$ implies that $\lambda \geq \frac{8}{\pi^2}$.

The following results are easily obtained:

$$\max_z w\left(x,\frac{\pi}{2}\right)=w\left(\frac{\pi}{2},\frac{\pi}{2}\right)=-1+\sigma=-1+\frac{8}{\lambda\pi^2} \; , \tag{4.21}$$

$$\max_z w_t\left(x,\frac{\pi}{2}\right)=w_t\left(\frac{\pi}{2},\frac{\pi}{2}\right)=-1+\sigma=-1+\frac{8}{\lambda\pi^2} \; . \tag{4.22}$$

Note that in equations (4.21) and (4.22) we took the value of σ to be $8/\lambda\pi^2$. With $f(x,t)=\sigma\sin x$, we obtain

$$F=\int_0^{\frac{\pi}{2}}\int_0^{\pi}|f|\,dx\,dt=\sigma\pi=\frac{8}{\lambda\pi} \; , \tag{4.23}$$

where F gives an indication of the amount of force spent in controlling the beam up to the terminal time $T=\pi/2$.

We now focus our attention at the solution of the problem (4.12)-(4.14) in case of $\phi(t)=\sigma$ with the following cost functional:

$$J(f)=\int_0^{\pi} w^2\left(x,\frac{\pi}{2}\right)dx+\int_0^{\pi} w_t^2\left(x,\frac{\pi}{2}\right)dx+\int_0^{\frac{\pi}{2}}\int_0^{\pi} f^2(x,t)\,dx\,dt. \tag{4.24}$$

Note that we have chosen a convex functional (4.24) in order to obtain sufficient conditions. In the direct method there is no such restriction, since we are not going to deal with optimality conditions.

The solution of the problem (4.12)-(4.14) with (4.24) by analytical method is given by:

$$w(x,t)=(\cos t-\sin t)\sin x+\frac{4\sin x}{6+\pi}\int_0^{t}\sin(t-\tau)(\sin\tau+\cos\tau)\,d\tau, \tag{4.25}$$

$$f^{\circ}(x,t)=\frac{8}{6+\pi}(\sin t+\cos t)\sin x. \tag{4.26}$$

Thus,

$$w\left(x,\frac{\pi}{2}\right)=\left(\frac{-4}{6+\pi}\right)\sin x, \tag{4.27}$$

$$w_t\left(x,\frac{\pi}{2}\right)=\left(\frac{-4}{6+\pi}\right)\sin x. \tag{4.28}$$

It follows from (4.26) and (4.28) that

$$\max_{x} w\left(x,\frac{\pi}{2}\right) = w\left(\frac{\pi}{2},\frac{\pi}{2}\right) = \frac{-4}{6+\pi} \, , \tag{4.29}$$

$$\max_{x} w_t\left(x,\frac{\pi}{2}\right) = w\left(\frac{\pi}{2},\frac{\pi}{2}\right) = \frac{-4}{6+\pi} \, . \tag{4.30}$$

Using equation (4.26) we compute

$$F = \int_0^{\frac{\pi}{2}} \int_0^{\pi} |f| \, dx \, dt = \frac{32}{\pi+6} \, . \tag{4.31}$$

Figures 1 and 2 give the graphs of the equations (4.21), (4.23), (4.29) and (4.31).

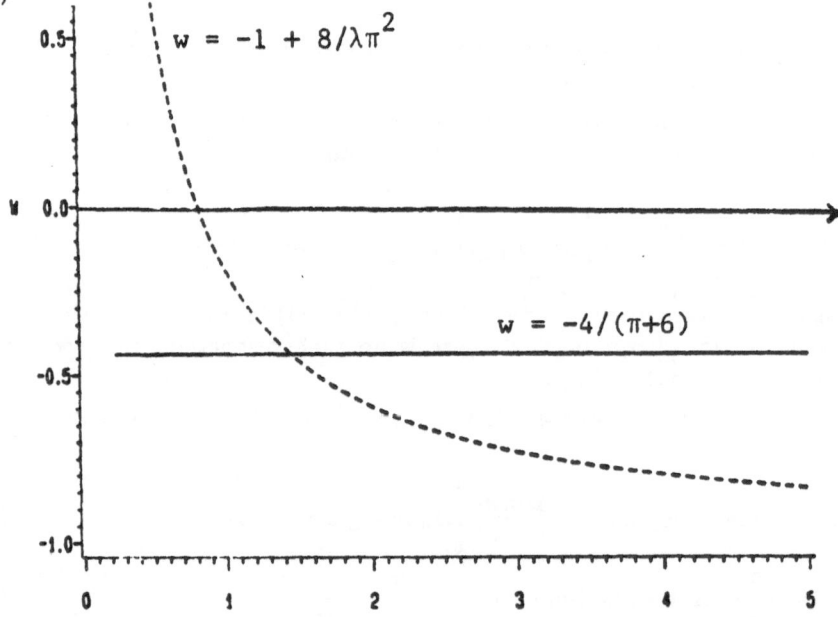

Figure 1: Graph of $\max_{x} w(x,\frac{\pi}{2})$

For example by choosing $\lambda = 1$ we obtain a solution which has the following properties:

$$w\left(\frac{\pi}{2},\frac{\pi}{2}\right) = w_t\left(\frac{\pi}{2},\frac{\pi}{2}\right) = -1 + \frac{8}{\pi^2} \, ,$$

and thus

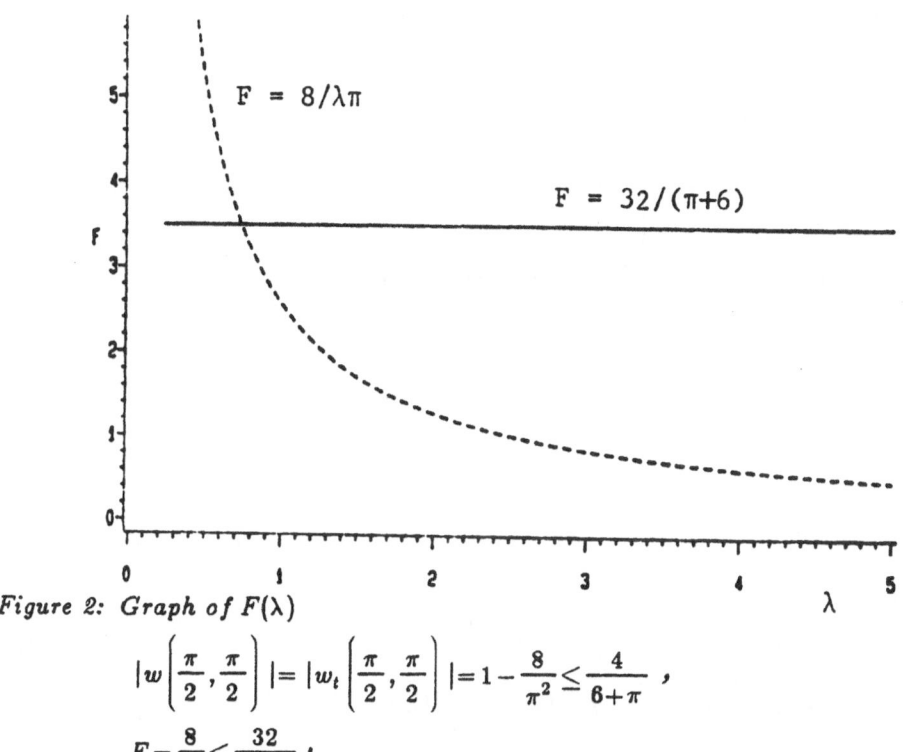

Figure 2: Graph of $F(\lambda)$

$$\left| w\left(\frac{\pi}{2},\frac{\pi}{2}\right) \right| = \left| w_t\left(\frac{\pi}{2},\frac{\pi}{2}\right) \right| = 1 - \frac{8}{\pi^2} \le \frac{4}{6+\pi} \ ,$$

$$F = \frac{8}{\pi} \le \frac{32}{\pi+6} \ .$$

Thus, a control force $f = \frac{8}{\pi^2}\sin x$, obtained by choosing the objective functional (4.11), yields smaller value for $w\left(\frac{\pi}{2},\frac{\pi}{2}\right)$, $w_t\left(\frac{\pi}{2},\frac{\pi}{2}\right)$ and spends less force as compared to the control function (4.15) which was obtained by choosing the objective functional (4.24).

Thus, the freedom provided by the direct approach in specifying the cost functional can in some cases lead to better control functions.

5. Solution of the Problem by Integral Equation Approach

In this section, we shall formulate the optimal control problem considered in section 2 as a variational problem and shall derive an integral equation as necessary condition by using the Euler-Lagrange multiplier theorem [15]. It turns out that the integral equation can be solved by an algebraic manipulation and is easier to handle than differential. Thus, the availability of integral equation as necessary condition provides an effective computational tool for the solution of the present optimal control problem.

Now assume the control function can be expanded in the form:

$$f(x,t) = \sum_{n=1}^{\infty} \phi_n(t)\sin(n\pi x),$$ (5.1)

where $\phi_n(t)$ is the new unknown control function.

From (2.7), (3.13) and (5.1), it follows that

$$J(\phi_n) = \frac{1}{2}\sum_{n=1}^{\infty}\left\{\mu_1 Z_n^2(T) + \mu_2 \dot{Z}_n^2(T) + \mu_3 \int_0^T \phi_n^2(t)dt\right\},$$ (5.2)

where

$$Z_n(t) = r_n(t) + \int_0^t \phi_n(\tau)S_n(\tau-t)d\tau,$$ (5.3)

$$\dot{Z}_n(t) = \dot{r}_n(t) + \int_0^t \phi_n(\tau)\dot{S}_n(\tau-t)d\tau,$$ (5.4)

in which

$$r_n(t) = \alpha_n N_n(t) + \beta_n M_n(-t),$$ (5.5)

$$\dot{r}_n(t) = \frac{1}{2}\alpha_n(P_n - C)N_n(t) - \frac{1}{2}\beta_n(P_n + C)M_n(-t),$$ (5.6)

$$S_n(t-\tau) = (N_n(t-\tau) - M_n(\tau-t))/P_n,$$ (5.7)

$$\dot{S}_n(t-\tau) = ((P_n - C)N_n(t-\tau) + (P_n + C)M_n(\tau-t))/(2P_n).$$ (5.8)

Next we proceed with the derivation of necessary conditions of optimality using the Euler-Lagrange theorem [15]. It follows that

$$\delta_{\phi_n(t)} J = \lim_{\epsilon\to0}\frac{J(\phi_n(t) + \epsilon\Delta\phi_n(t)) - J(\phi_n(t))}{\epsilon} = 0,$$ (5.9)

for sufficiently small $\epsilon > 0$. Where $\Delta\phi_n(t)\epsilon C(0,T)$ is an arbitrary function.

Using (5.2) - (5.8) in (5.9) and then making use of the Fundamental Theorem of Calculus of Variations, we obtain

$$\mu_1 D_{1n}S_n(T-\tau) + \mu_1 S_n(T-\tau)\int_0^T S_n(T-\varsigma)\phi_n(\varsigma)d\varsigma + \mu_2 D_{1n}'\dot{S}_n(T-\tau)$$

$$+ \mu_2\dot{S}_n(T-\tau)\int_0^T \dot{S}_n(T-\varsigma)\phi_n(\varsigma)d\varsigma + \mu_3\phi_n(\tau) = 0, \quad \text{for all } n \geq 1,$$ (5.10)

where $D_{1n} = r_n(T)$, $D_{1n}' = \dot{r}_n(T)$ are given by (3.19) and (3.22) respectively.

We note that (5.10) is an integral equation and valid for any T. This integral equation (5.10) can be solved explicitly as follows: Let

$$g_n(\tau) = S_n(t-\tau), \qquad h_n(\tau) = \dot{S}_n(T-\tau). \tag{5.11}$$

By the use of (5.11), (5.10) becomes

$$\mu_1 D_{1n} g_n(\tau) + \mu_1 g_n(\tau) \int_0^T g_n(\varsigma)\phi_n(\varsigma)d\varsigma + \mu_2 D'_{1n} h_n(\tau)$$

$$+ \mu_2 h_n(\tau) \int_0^T h_n(\varsigma)\phi_n(\varsigma)d\varsigma + \mu_3 \phi_n(\tau) = 0, \tag{5.12}$$

and consequently,

$$\phi_n(\tau) = -\frac{1}{\mu_3}\left[\mu_1 D_{1n} g_n(\tau) + \mu_1 g_n(\tau) x_n + \mu_2 D'_{1n} h_n(\tau) + \mu_2 h_n(\tau) y_n\right], \tag{5.13}$$

in which

$$x_n = \int_0^T g_n(\varsigma)\phi_n(\varsigma)d\varsigma, \qquad y_n = \int_0^T h_n(\varsigma)\phi_n(\varsigma)d\varsigma, \tag{5.14}$$

where x_n and y_n are unknowns to be determined.

Multiplying (5.12) by $g_n(\tau)$ and $h_n(\tau)$ respectively and then integrating over $[0, T]$, we obtain

$$(\mu_1 d_n + \mu_3)x_n + \mu_2 e_n y_n = -(\mu_1 D_{1n} d_n + \mu_2 D'_{1n} e_n), \tag{5.15}$$

$$\mu_1 e_n x_n + (\mu_2 \ell_n + \mu_3)y_n = -(\mu_1 D_{1n} e_n + \mu_2 D'_{1n} \ell_n), \tag{5.16}$$

where

$$d_n = \int_0^T g_n^2(\tau)d\tau = \left[-(P_n - C)^{-1}(1 - N_n^2(T)) + (P_n + C)^{-1}(1 - M_n^2(-T))\right.$$

$$\left. - 2C^{-1}(1 - Q^2(-T))\right]/P_n^2, \tag{5.17}$$

$$e_n = \int_0^T h_n(\tau)g_n(\tau)d\tau = \left[\frac{1}{2}(N_n^2(T)-1) + (2C)^{-1}(P_n - C)(Q^2(-T)-1)\right.$$

$$\left. + (2C)^{-1}(P_n + C)(1 - Q^2(-T)) + \frac{1}{2}(M_n^2(-T)-1)\right]/P_n^2, \tag{5.18}$$

$$\ell_n = \int_0^T h_n^2(\tau)d\tau = \left[\frac{1}{4}(P_n - C)(N_n^2(T) - 1)\right.$$

$$+ \frac{1}{4}(P_n + C)(1 - M_n^2(-T)) + (2C)^{-1}(P_n - c)(P_n + C)(1 - Q^2(-T)) \Bigg] \bigg/ P_n^2, \qquad (5.19)$$

and therefore,

$$x_n = \frac{1}{R_n} \left[-(\mu_2 \ell_n + \mu_3)(\mu_1 D_{1n} d_n + \mu_2 D'_{1n} e_n) + \mu_2 e_n (\mu_1 D_{1n} e_n + \mu_2 D'_{1n} \ell_n) \right], \qquad (5.20)$$

$$y_n = \frac{1}{R_n} \left[-(\mu_1 d_n + \mu_3)(\mu_1 D_{1n} e_n + \mu_2 D'_{1n} \ell_n) + \mu_1 e_n (\mu_1 D_{1n} d_n + \mu_2 D'_{1n} e_n) \right], \qquad (5.21)$$

where

$$R_n = (\mu_1 d_n + \mu_3)(\mu_2 \ell_n + \mu_3) - \mu_1 \mu_2 e_n^2.$$

Thus, the forcing function is

$$f(x,t) = \sum_{n=1}^{\infty} -\frac{1}{\mu_3} \left[\mu_1 (D_{1n} + x_n) g_n(t) + \mu_2 (D'_{1n} + y_n) h_n(t) \right] \sin n\pi x, (5.22)$$

where $g_n(t)$, $h_n(t)$, x_n, y_n, D_{1n}, and D'_{1n} are given by (5.11), (5.20), (5.21), (3.19), (3.22), respectively.

In view of (5.22), if follows that $f(x,t)$ is identical after simplication to the solutions found by other methods, that is identical to (3.38) and (4.10).

6. Conclusion

Analytic solution approach is valuable in the sense that it provides the necessary conditions of optimality and sometimes sufficiency too; but direct approaches are easier to handle in order to obtain explicit results. Obviously, there is a certain price that has been paid for using the direct approaches. The price is simply that in the direct approaches, we cannot claim that what we have found is definitely the best solution. That is why we called them approximate methods. But in the special case, it is shown that explicit results are obtained for certain choices of the approximating control functions.

References

[1] Adali, S., "Multiobjective Design of an Antisymmetric Angle-ply Laminate by Nonlinear Programming," *American Society of Mechanical Engineers Journal of Mechanisms*, 1983, pp. 214-219.

[2] Adali, S., "Pareto Optimal Design of Beams Subjected to Support Motions," *Computers and Structures*, Vol. 16, 1983, pp. 297-303.

[3] Akulenko, L. D., and Bolotnik, N. N., "On the Control of Systems with Elastic Elements," *Journal of Applied Mathematics and Mechanics*, Vol. 44, 1980, pp. 13-18.

[4] Breakwell, J. A., "Optimal Control of Distributed Systems," *The Journal of the Astronautical Sciences,* Vol. XXIX, No. 4, 1981, pp. 343-372.

[5] Köhne, M., "Distributed Parameter Systems (Editors W. H. Ray and D. G. Lainiotis)," The Control of Vibrating Structures, New York: Marcel Decker Inc., pp. 387-456.

[6] Komkov, V., "The optimal Control of a Transverse Vibration of a Beam " *SIAM J. Control,* Vol. 6, 1968, pp. 401-421.

[7] Krabs, W., "Optimal Control of Processes Governed by Partial Differential Equations," *Part II: Vibrations,* Zeitschrift Für Operations Research, Vol. 26, 1982, pp. 63-86.

[8] Ishmukhametov, A. Z., "Optimal Control of Transverse Vibrations of a Rod," Moscow University Computational Mathematics and Cybernetics, Vol. 15(4), 1981, pp. 59-64.

[9] Leipholz, H. H. E. (Editor), "Structural Control," Amsterdam: North-Holland Publishing Company, 1980.

[10] Leipholz, H. H. E., "Distributed Control of Elastic Plates," Mechanics Research Communications, Vol. 9, 1982, pp. 133-136.

[11] Luzzato E., and Jean M., "Mechanical Analysis of Active Vibration Damping in Continuous Structures," *Journal of Sound and Vibrations,* Vol. 86, 1983, pp. 455-473.

[12] Narukawa, K., "Admissible Controllability of Vibrating Systems with Constrained Controls," *SIAM Journal on Control and Optimization,* Vol. 20, 1982, pp. 770-782.

[13] Sadek, I. S., Adali, S., "Control of the Dynamic Response of A Damped Membrane by Distributed Forces" *Journal of Sound and Vibration,* Vol. 96(3), 1984, pp. 391-406.

[14] Sadek, I. S., Bruch, J. C., Jr., and Sloss, J. M., "A Maximum Principle for Problems Governed by Systems of Partial Differential Equations," submitted.

[15] Smith, D. R., "Variational Methods in Optimization," Prentice Hall Inc., Englewood Cliffs, New Jersey, 1974.

[16] Yavin, Y., "Optimal Control of the Transverse Vibrations of a Beam with a Bound on the Potential Energy," *Journal of Optimization Theory and Applications,* Vol. 5(5), 1970, pp. 376-381.

ACTIVE TENDON CONTROL SYSTEM FOR WIND-EXCITED TALL BUILDING

B. Samali, J.N. Yang
Dept. of Civil, Mechanical and Environmental Eng.
The George Washington University
Washington, D.C. 20052.
C.T. Yeh
Graduate Institute of Civil Engineering
Tamkang University
Tamshui, Taipei
Taiwan, R.O.C.

1. Introduction

Under strong wind environments, most buildings do not have safety problems. However, the acceleration response of buildings causing discomfort of tenants is of considerable concern. To alleviate such undesirable building motion under strong wind gusts different passive or active control systems have been proposed and investigated [e.g., Refs. 1-3, 8, 9, 11-16, 19, 24, 25, 27]. Recently, an exploratory study for the application of an active mass damper control system to tall buildings had been considered in Ref. [16], in which the coupled lateral-torsional motions of the wind-excited tall building was considered. The importance of the torsional motion has been substantiated experimentally in Ref. [7] and analytically, in Refs. [10, 15, 16, 21]. It is the purpose of this paper to investigate the effectiveness of an active tendon control system for tall buildings subjected to strong wind turbulences. A transfer matrix formulation recently proposed to analyze the stochastic response of tall buildings under earthquake and wind loads [15-17, 19-23] is used.

The wind turbulence is modeled as a stochastic process with a mean wind velocity that varies with the building height (nonhomogeneous) plus a random fluctuation with a cross-power spectral density suggested in the literature. A random vibration analysis along with a closed-loop control algorithm is used to determine the stochastic response of tall buildings with or without an active tendon control system. The statistics of the required active control forces have been determined. Further, a Monte Carlo simulation technique has been employed to illustrate the building response behavior under an active tendon control system. A numerical example of a forty-story building under strong wind excitations is worked out to demonstrate the effectiveness of the proposed active tendon control system.

2. Formulation

Consider a structural model shown in Fig. 1a for an idealized N-story building, with the following assumptions: (1) the inertia of a typical floor is lumped at the floor level characterized by a mass, m, and a mass moment of inertia, I, about the mass center, C, and (2) linear elasticity is provided by the massless columns or shear walls in each story and it is characterized by three stiffness constants, K_{ex}, K_{ey}, and K_{et}. For simplicity it is further assumed that all story units are identically constructed, i.e., the mass, the mass moment of inertia, as well as the stiffness constants are identical for every story.

The floor response is described by three displacement variables, i.e., translations along the x- and y-axes and rotation about the z-axis, and three force variables, i.e., shear forces along the x- and y-axes and torsional moment about the z-axis. The state vector $\{Z\}_j$ for the jth floor is defined as:

$$\{Z\}_j = \{\overline{u}_j, \overline{v}_j, \overline{\theta}_j, \overline{Q}_j, \overline{V}_j, \overline{U}_j\}', \tag{1}$$

where u_j, v_j and θ_j are, respectively, the displacements of the elastic center, E, along the x- and y-axes and the rotation about the vertical z-axis, of the jth floor, and U_j, V_j, and Q_j are the corresponding shear forces and torsional moment immediately above the jth floor. An upper-bar denotes a Fourier transformation and a prime indicates the transpose of a vector or matrix. In Eq. (1) each variable is frequency dependent, which is the Fourier transform of the corresponding time-dependent variable.

An active tendon control system can be installed in any story unit as shown in Fig. 1. A total of four pairs of active tendons are installed in each story unit, and each pair of tendons is connected to a controller as shown in Figs. 1b and 2. Each controller is regulated by sensors placed on the floors immediately above and below the story unit. Displacement, velocity or acceleration sensors may be used. By controlling each pair of active tendons in the opposite frames of the building independently, the torsional motion of the building can be controlled. For simplicity, it is assumed that each controller is identical, thus simplifying the transfer matrix properties as will be seen later. The controllers considered herein

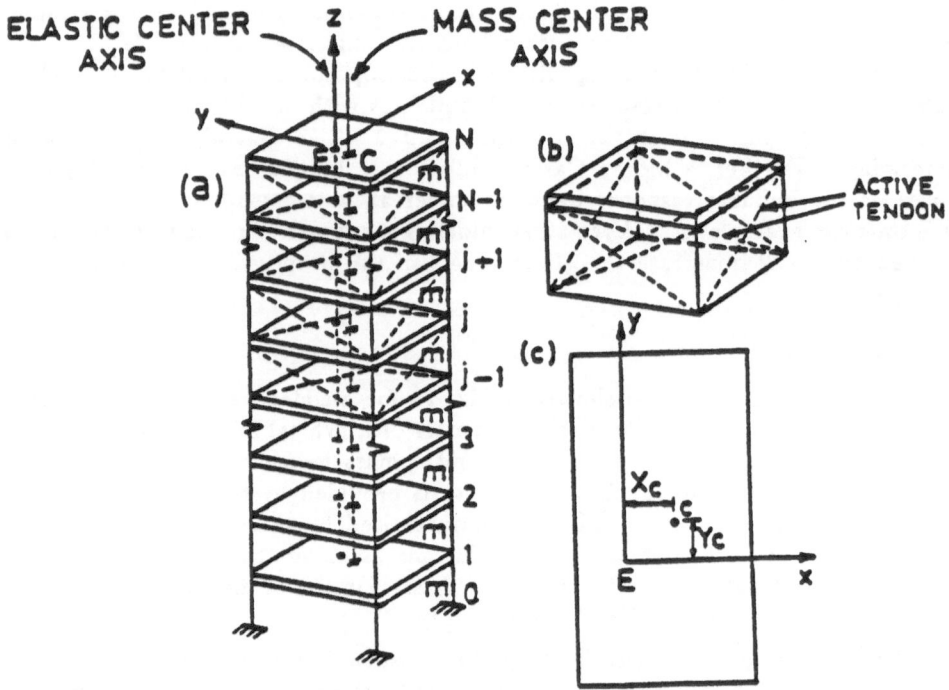

Figure 1: Structural model of the multi-story building; (a) structural model; (b) typical story unit; and (c) top view of the jth floor

are electrohydraulic servomechanisms.

The response of the jth floor of a wind-excited building equipped with the active tendon control system shown in Fig. 1 is related to the response of the $(j-1)$th floor through [15]

$$\{Z\}_j = [A]_j\{Z\}_{j-1} - \{0,0,0,\overline{F}_{\theta j},\overline{F}_{yj},\overline{F}_{xj}\}',\tag{2}$$

where $[A]_j$ is the 6×6 transfer matrix from $(j-1)$th floor to jth floor for the story unit with active tendon control, i.e.,

$$[A]_j = [T]_j + [\tilde{C}]_j,\tag{3}$$

in which $[T]_j$ is the transfer matrix for the story unit without control and $[\tilde{C}]_j$ is the transfer matrix contributed by the active tendon control system. The 6×6 matrix $[T]_j$ only involves the structural properties, e.g., floor mass, centroidal mass moment of inertia, eccentricities along the x- and y-directions, elastic translational stiffnesses along x- and y-directions, elastic torsional stiffness about the z-axis and the external as well as the internal damping coefficients. The 6×6 matrix $[\tilde{C}]_j$ involves control parameters τ_x, τ_y, ϵ_x and ϵ_y, denoting the normalized

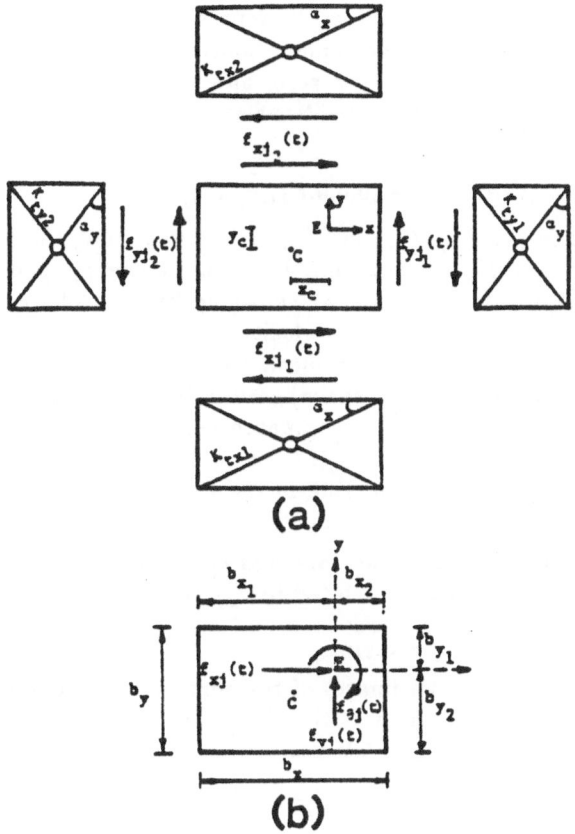

Figure 2: Active tendon control system of the jth story unit. (a) jth floor top view and its four exterior frames equipped with active tendons and their corresponding control force; (b) jth floor top view and the resultant control forces acting at the elastic center

loop gains and feedback gains of control systems installed along the x- and y-directions, respectively. It also involves tendon properties and geometry, e.g., stiffnesses of tendons in x- and y-directions, and the angles of inclination of tendons; type of sensor used; and building dimensions. The 6×6 transfer matrix $[A]_j$ is given in Refs. [15, 17].

Since the structural properties and controllers employed for each story unit are identical, the subscript j will be dropped from $[A]_j$, $[T]_j$ and $[\tilde{C}]_j$, i.e., $[A]_j$, $[T]_j$ and $[\tilde{C}]_j$ will be replaced by $[A]$, $[T]$ and $[\tilde{C}]$, respectively. Note that $[A]$ represents the transfer mechanism from one floor to another for all story units equipped with active tendon control system whereas $[T]$ represents the transfer mechanism from one floor to another for all story units equipped with active tendon control system whereas $[T]$ represents the transfer mechanism from one floor to another without the tendon control system. In the absence of the active control system, matrix $[\tilde{C}]$ is zero, and matrix $[A]$ reduces to matrix $[T]$. (Also see Ref. [15]). Therefore, the transfer relation without an active control system is given by

$$\{Z\}_j = [T]_j\{Z\}_{j-1} - \{0,0,0,\overline{F}_{\theta j},\overline{F}_{yj},\overline{F}_{zj}\}'. \tag{4}$$

In Eqs. (2) and (4), $\overline{F}_{\theta j}$, $\overline{F}_{yj}$, and $\overline{F}_{zj}$ are, respectively, the Fourier transforms of $F_{\theta j}(t)$, $F_{yj}(t)$ and $F_{zj}(t)$, denoting the wind loads applied to the jth floor through an aerodynamic center, A, as shown in Fig. 3. The random wind forces are produced by the random wind velocity, that consists of a mean value, and two random fluctuating parts. The mean wind velocity that varies with respect to the height (inhomegeneous in space) results in the static wind loads, and the induced static building response is computed separately (see Refs. [15, 21]). The wind forces $F_{zj}(t)$, $F_{yj}(t)$ and $F_{\theta j}(t)$, as well as the corresponding Fourier transforms $\overline{F}_{zj}$, $\overline{F}_{yj}$ and $\overline{F}_{\theta j}$ appearing in the following formulation are those due to the fluctuating parts of the random wind velocity.

The Fourier transform of the torsional component of wind loads applied to the jth floor, denoted by $\overline{F}_{\theta j}$, is related to $\overline{F}_{zj}$ and $\overline{F}_{yj}$ thorugh

$$\overline{F}_{\theta j} = -y_A\overline{F}_{zj} + x_A\overline{F}_{yj}, \tag{5}$$

where x_A and y_A are the coordinates of the aerodynamic center referenced to the coordinate system with its origin at the elastic center E, as shown in Fig. 3. The above wind forces are directly related to the drag and lift forces defined as:

$$\text{Drag} = \frac{1}{2}\rho bhV_{\text{rel}}^2 C_D, \qquad \text{Lift} = \frac{1}{2}\rho bhV_{\text{rel}}^2 C_L, \tag{6}$$

in which ρ = air density; b = characteristic width of the building; h = story height; C_D and C_L = the drag and lift coefficients, respectively, which are functions of the angle of attack α and V_{rel} = wind velocity relative to the structure. The wind forces $F_{zj}(t)$, $F_{yj}(t)$ and $F_{\theta j}(t)$, can be expressed in terms of the fluctuating wind velocities following the approach given in Refs. [10, 15, 16, 21].

Let the controllers be installed in all story units above the ℓth floor. Then Eq. (2) can be applied repeatedly for $j = \ell + 1$ to N to obtain a relation between $\{Z\}_\ell$ and $\{Z\}_N$ as follows:

$$\{Z\}_N = [A]^{N-\ell}\{Z\}_\ell - \sum_{j=\ell+1}^{N} [A]^{N-j}\{0,0,0,\overline{F}_{\theta j},\overline{F}_{yj},\overline{F}_{zj}\}'. \tag{7}$$

Similarly, one can obtain the relation between $\{Z\}_\ell$ and $\{Z\}_0$ by repeated applications of Eq. (4),

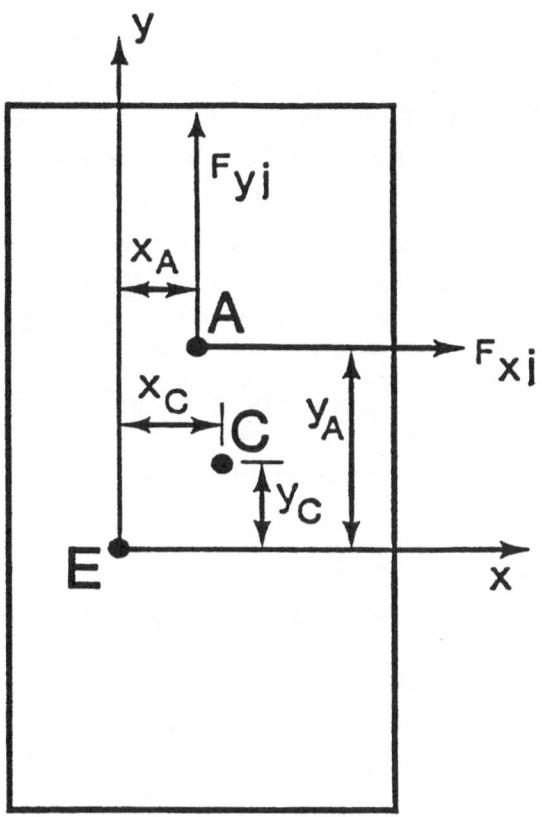

*Figure 9: jth floor of the multi-story building under wind excitations; elastic
center E, mass center C and aerodynamic center A*

$$\{Z\}_\ell = [T]' \{Z\}_0 - \sum_{j=1}^{\ell} [T]^{\ell-j} \{0,0,0,\bar{F}_{\theta j}, \bar{F}_{yj}, \bar{F}_{zj}\}'. \tag{8}$$

Substitution of Eq. (8) into Eq. (7) yields the following relation between $\{Z\}_N$ and $\{Z\}_0$:

$$\{Z\}_N = [B(N,\ell)]\{Z\}_0 - [A]^{N-\ell} \sum_{j=1}^{\ell} [T]^{\ell-j} \{0,0,0,\bar{F}_{\theta j}, \bar{F}_{yj}, \bar{F}_{zj}\}'$$

$$- \sum_{j=\ell+1}^{N} [A]^{N-j} \{0,0,0,\bar{F}_{\theta j}, \bar{F}_{yj}, \bar{F}_{zj}\}', \tag{9}$$

where $[B(N,\ell)]$ is a 6×6 matrix defined as:

$$[B(N,\ell)] = [A]^{N-\ell}[T]^{\ell}. \tag{10}$$

The matrix $[B(N,\ell)]$ is the overall transfer matrix. To compute the matrix $[B(N,\ell)]$, elements of matrices $[A]^{N-\ell}$ and $[T]^{\ell}$ are needed. An analytical method for computing $[A]^n$ and $[T]^n$ for different values of n was proposed in Ref. [23].

The boundary conditions at the base and the top floor of the building are given by

$$\{Z\}_N = \{\bar{u}_N, \bar{v}_N, \bar{\theta}_N, 0, 0, 0\}', \qquad \{Z\}_0 = \{0, 0, 0, \bar{Q}_0, \bar{V}_0, \bar{U}_0\}'. \tag{11}$$

By substituting the boundary conditions into Eq. (9) and considering the last three rows of the resulting matrix equation, the unknown quantities $\bar{Q}_0$, $\bar{V}_0$ and $\bar{U}_0$ can be solved. The results are given as follows:

$$\{\bar{Q}_0, \bar{V}_0, \bar{U}_0\}' = [B_{22}(N,\ell)]^{-1} \sum_{j=1}^{N} [D_1(j,\ell)]\{\bar{F}_{\theta j}, \bar{F}_{yj}, \bar{F}_{zj}\}', \tag{12}$$

where $[D_1(j,\ell)]$ is a 3×3 complex matrix defined as:

$$[D_1(j,\ell)] = [A_{21}(N-\ell)][T_{12}(\ell-j)] + [A_{22}(N-\ell)][T_{22}(\ell-j)], \quad \text{for } j \leq \ell,$$

$$= [A_{22}(N-j)], \quad \text{for } j > \ell, \tag{13}$$

in which matrices $[A_{21}(N-\ell)]$, $[T_{12}(\ell-j)]$, $[A_{22}(N-\ell)]$, $[T_{22}(\ell-j)]$ and $[A_{22}(N-j)]$ are given by Eq. (17).

With the quantities $\bar{Q}_0$, $\bar{V}_0$, $\bar{U}_0$ obtained in Eq. (12), the unknown quantities $\bar{u}_N$, $\bar{v}_N$ and $\bar{\theta}_N$ can be determined from the first three rows of the aforementioned matrix equation. The results are given in the following:

$$\{\bar{u}_N, \bar{v}_N, \bar{\theta}_N\}' = \sum_{j=1}^{N} [D_2(j,\ell)]\{\bar{F}_{\theta j}, \bar{F}_{yj}, \bar{F}_{zj}\}', \tag{14}$$

in which $[D_2(j,\ell)]$ and $[D_3(j,\ell)]$ are 3×3 complex matrices defined by

$$[D_2(j,\ell)] = [B_{12}(N,\ell)][B_{22}(N,\ell)]^{-1}[D_1(j,\ell)] - [D_3(j,\ell)], \tag{15}$$

$$[D_3(j,\ell)] = [A_{11}(N-\ell)][T_{12}(\ell-j)] + [A_{12}(N-\ell)][T_{22}(\ell-j)], \quad \text{for } j \leq \ell,$$

$$= [A_{12}(N-j)], \quad \text{for } j > \ell. \tag{16}$$

In Eqs. (13), (15) and (16), the 3×3 matrices $[B_{12}(N,\ell)]$, $[B_{22}(N,\ell)]$, $[A_{11}(n)]$, $[A_{12}(n)]$, $[A_{21}(n)]$, $[A_{22}(n)]$, $[T_{12}(n)]$, $[T_{22}(n)]$ are given in the following

$$[B_{12}(N,\ell)] = \begin{bmatrix} b_{14}(N,\ell) & b_{15}(N,\ell) & b_{16}(N,\ell) \\ b_{24}(N,\ell) & b_{25}(N,\ell) & b_{26}(N,\ell) \\ b_{34}(N,\ell) & b_{35}(N,\ell) & b_{36}(N,\ell) \end{bmatrix},$$

$$[B_{22}(N,\ell)] = \begin{bmatrix} b_{44}(N,\ell) & b_{45}(N,\ell) & b_{46}(N,\ell) \\ b_{54}(N,\ell) & b_{55}(N,\ell) & b_{56}(N,\ell) \\ b_{64}(N,\ell) & b_{65}(N,\ell) & b_{66}(N,\ell) \end{bmatrix},$$

$$[A_{11}(n)] = \begin{bmatrix} a_{11}(n) & a_{12}(n) & a_{13}(n) \\ a_{21}(n) & a_{22}(n) & a_{23}(n) \\ a_{31}(n) & a_{32}(n) & a_{33}(n) \end{bmatrix}, \quad [A_{12}(n)] = \begin{bmatrix} a_{14}(n) & a_{15}(n) & a_{16}(n) \\ a_{24}(n) & a_{25}(n) & a_{26}(n) \\ a_{34}(n) & a_{35}(n) & a_{36}(n) \end{bmatrix},$$

$$[A_{21}(n)] = \begin{bmatrix} a_{41}(n) & a_{42}(n) & a_{43}(n) \\ a_{51}(n) & a_{52}(n) & a_{53}(n) \\ a_{61}(n) & a_{62}(n) & a_{63}(n) \end{bmatrix}, \quad [A_{22}(n)] = \begin{bmatrix} a_{44}(n) & a_{45}(n) & a_{46}(n) \\ a_{54}(n) & a_{55}(n) & a_{56}(n) \\ a_{64}(n) & a_{65}(n) & a_{66}(n) \end{bmatrix},$$

$$[T_{12}(n)] = \begin{bmatrix} \tau_{14}(n) & \tau_{15}(n) & \tau_{16}(n) \\ \tau_{24}(n) & \tau_{25}(n) & \tau_{26}(n) \\ \tau_{34}(n) & \tau_{35}(n) & \tau_{36}(n) \end{bmatrix}, \quad [T_{22}(n)] = \begin{bmatrix} \tau_{44}(n) & \tau_{45}(n) & \tau_{46}(n) \\ \tau_{54}(n) & \tau_{55}(n) & \tau_{56}(n) \\ \tau_{64}(n) & \tau_{65}(n) & \tau_{66}(n) \end{bmatrix}, \qquad (17)$$

where $b_{ij}(N,\ell)$, $a_{ij}(n)$ and $\tau_{ij}(n)$ are, respectively, the (i,j) element of matrices $[B(N,\ell)]$, $[A]^n$ and $[T]^n$.

The state vector for the mth floor when $m > \ell$, i.e., $\{Z\}_m = \{\bar{u}_m, \bar{v}_m, \bar{\theta}_m, \bar{Q}_m, \bar{V}_m, \bar{U}_m\}'$, is obtained from Eq. (9) by relacing N by m in Eqs. (9) and (10). When $m \leq \ell$, the corresponding $\{Z\}_m$ is obtained from Eq. (8) by relacing ℓ by m.

If the tendon controllers are installed below the ℓth floor, the transfer relations between $\{Z\}_N$ and $\{Z\}_0$ can be obtained conveniently from Eq. (9) by interchanging matrices $[A]$ and $[T]$, In such a case, the equations corresponding to Eq. (10) and Eqs. (12)-(16) are obtained by interchanging matrices $[A]$ and $[T]$.

The vector of the Fourier transforms of the forces at the jth floor is a function of control parameters ($\tau_x, \tau_y, \epsilon_x$ and ϵ_y), tendon properties and geometry, type of sensor used, and building dimensions. This vector is given in Refs. [15] and [17].

2. Root Mean Squares of Response and Control Forces

With the exclusion of the static mean response, the 3×3 matrices of the cross-spectral densities for the force type responses $\bar{U}_0$, $\bar{V}_0$ and $\bar{Q}_0$ at the base, and the displacement type responses, $\bar{u}_N$, $\bar{v}_N$ and $\bar{\theta}_N$, at the top floor can be obtained using Eqs. (12) and (14) as follows (see Ref. [15]):

$$[\Phi_0] = \begin{bmatrix} \Phi_{Q_0 Q_0} & \Phi_{Q_0 V_0} & \Phi_{Q_0 U_0} \\ \Phi_{V_0 Q_0} & \Phi_{V_0 V_0} & \Phi_{V_0 U_0} \\ \Phi_{U_0 Q_0} & \Phi_{U_0 V_0} & \Phi_{U_0 U_0} \end{bmatrix}$$

$$= [B_{22}(N,\ell)]^{-1} \left\{ \sum_{j=1}^{N} \sum_{k=1}^{N} [D_1(j,\ell)][\Phi_F]_{jk}[D_1(k,\ell)]^{\prime *} \right\} \left([B_{22}(N,\ell)]^{-1} \right)^{\prime *}, \tag{18}$$

$$[\Phi_N] = \begin{bmatrix} \Phi_{u_N u_N} & \Phi_{u_N v_N} & \Phi_{u_N \theta_N} \\ \Phi_{v_N u_N} & \Phi_{v_N v_N} & \Phi_{v_N \theta_N} \\ \Phi_{\theta_N u_N} & \Phi_{\theta_N v_N} & \Phi_{\theta_N \theta_N} \end{bmatrix}$$

$$= \sum_{j=1}^{N} \sum_{k=1}^{N} [D_2(j,\ell)][\Phi_F]_{jk}[D_2(k,\ell)]^{\prime *}, \tag{19}$$

in which the asterick represents the complex conjugate.

A 3×3 cross-spectral matrix for the top floor acceleration response, denoted by $[\Phi_{\ddot{N}}]$, is directly related to $[\Phi_N]$ and the frequency ω as follows:

$$[\Phi_{\ddot{N}}] = \omega^4 [\Phi_N]. \tag{20}$$

In Eqs. (18) and (19), $[\Phi_F]_{jk}$ is the cross-spectral matrix of the wind loads applied to the jth and kth floors, which has been derived in Refs. [15, 16, 21]. In deriving the matrix $[\Phi_F]_{jk}$, the wind velocity is assumed to consist of a mean wind velocity, that varies with respect to the height from the ground according to a power law [4, 5], and fluctuating wind velocities, that are stationary random processes defined by the cross-spectral densities [5, 6]. Fuethermore, the mean wind velocity that results in static mean responses of the building is separated from the random fluctuating wind velocities for analysis purposes. Similar expressions to those given by Eqs. (18) and (19) can be obtained for spectral matrices of the control forces, $[\Phi_{cf}]$.

Let $\{\Phi_0\}_D$, $\{\Phi_N\}_D$, $\{\Phi_{\ddot{N}}\}_D$ and $\{\Phi_{cf}\}_D$ be, respectively, vectors containing all diagonal elements of matrices $[\Phi_0]$, $[\Phi_N]$, $[\Phi_{\ddot{N}}]$ and $[\Phi_{cf}]$. Furthermore, let $\{\sigma_0\}$, $\{\sigma_N\}$, $\{\sigma_{\ddot{N}}\}$ and $\{\sigma_{cf}\}$ be, respectively, standard deviation vectors of force-type responses U_0, V_0 and Q_0 at the base, top floor displacements (including rotation), top floor accelerations and active tendon control forces. These quantities can be evaluated through numerical integrations as follows:

$$\{\sigma_0\} = \left[\int_{-\infty}^{\infty} \{\Phi_0\}_D \, d\omega \right]^{1/2}, \quad \{\sigma_N\} = \left[\int_{-\infty}^{\infty} \{\Phi_N\}_D \, d\omega \right]^{1/2},$$

$$\{\sigma_N\} = \left[\int_{-\infty}^{\infty} \{\Phi_{\ddot{N}}\}_D \, d\omega \right]^{1/2} = \left[\int_{-\infty}^{\infty} \omega^4 \{\Phi_N\}_D \, d\omega \right]^{1/2}, \tag{21}$$

$$\{\sigma_{cf}\} = \left[\int_{-\infty}^{\infty} \{\Phi_{cf}\}_D \, d\omega \right]^{1/2}.$$

Then, the corresponding root mean squares can be obtained from the standard deviation and the static mean value determined from the mean wind velocity.

4. Simulation of Structural Response and Control Forces

Monte Carlo simulations of structural response quantities and the required active tendon control forces under wind excitations are also carried out to illustrate the structural behavior under wind excitations. The method of Monte Carlo simulation is performed using the Fast Fourier transform (FFT) technique described in Refs. [15, 16, 18, 26].

5. Numerical Example

A forty-story building, in which all stories are identically constructed, is considered for illustrative purposes. The properties of each floor are: (1) floor mass = 1000.00 $tons$; (2) centroidal mass moment of inertia = 2.4×10^5 $tons - m^2$; (3) eccentricity along the x-direction = 2.4 m (about 10% of structural dimension along the x-axis); (4) eccentricity along the y-direction = 4.8 m (about 10% of structural dimension along the y-axis); (5) mass moment of inertia about the elastic center = 2.688×10^5 $tons - m^2$; (6) elastic translational stiffness along the x-direction = 7.0×10^6 kN/m; (7) elastic translational stiffness along the y-direction = 1.05×10^7 kN/m; (8) elastic torsional stiffness about the z-axis = 3.5×10^9 $kN/rad.$; (9) external damping = 20 $kN/m/sec$; and (10) the internal damping coefficients are all zero.

The building is assumed to have dimensions of 24 and 48 meters along the x- and y-directions, respectively, with a story height of 4 meters. Foundamental natural frequencies of the uncontrolled building aling the x- and y-directions are, respectively, 3.12 and 3.92 rad/sec.

The properties of the active tendon control systems are as follows: total stiffness of tendons in the x-direction = 1.4×10^5 kN/m; total stiffness of tendons in the y-direction = 2.1×10^5 kN/m; angles of inclination of tendons in the x- and y-directions = 0.1651 and 0.0831 radians, respectively. The same active tendon controllers using electrohydraulic servomechanisms described in Refs. [15-17, 19, 20, 24 , 25] are used.

A maximum reduction of response quantities is achieved if the tendon controllers are installed between every adjacent floors for the entire building; however, such an arrangement may neither be feasible nor economical. For practical applications, a few active tendon controllers may be installed above or below the ℓ th floor. For illustrative purpose, tendon controllers are installed below the 20th floor.

A search has been conducted for the best combination of control parameters, including the normalized loop gains and feedback gains, which provide a better response reduction while requiring smaller control forces. They are found to be $\epsilon_x = 2$, $\epsilon_y = 2$, $r_x = 10$ and $r_y = 10$ and the most effective type of sensor is found to be the velocity sensor. Hence, velocity sensors are used in this example.

622   B. Samali, J.N. Yang, C.T. Yeh

The aerodynamic data used are as follows: gradient height $= 300\ m$; gradient velocity (velocity at gradient height) $= 90.0\ m/sec$; air density $= 1.23\ kg/m^3$; drag coefficient $= 1.6$; lift coefficient $= 0.6$; reference mean wind velocity at 10 meter height $= 23.1\ m/sec$; exponent for the mean-wind-profile power law $= 0.4$; constants in the wind spectrum $C_1 = 7.7$ and $K_0 = 0.03$ [15, 16, 21]; angle of attack of the mean wind $= 45\ °$; and coordinates of the aerodynamic center, $x_A = -0.5\ m$ and $y_A = 4.8\ m$.

In order to demonstrate the effectiveness of the active tendon control system in reducing the structural response under strong wind gusts, the power spectral densities of the top floor displacement and acceleration along the x-direction, and the base torsional moment are presented in Figs. 4, 5 and 6, respectively. Since the mean values of these response quantities, with the exception of acceleration response, are non-zero, a δ-function type singularity should appear at $\omega = 0$ in each of these spectra (see Ref. [21]). However, these δ-singularities are not included in Figs. 4-6; therefore, the spectra shown are actually the frequency distributions of the variances.

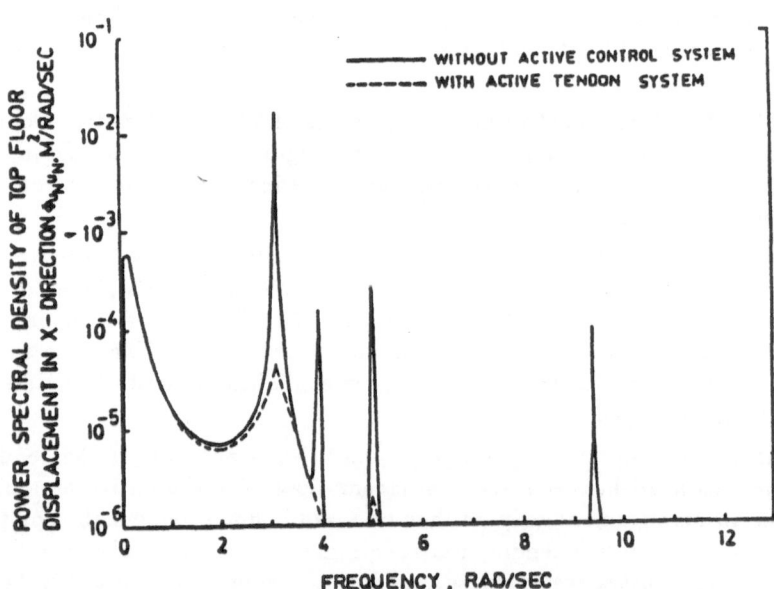

Figure 4: Power spectral density of top floor displacement in x-direction

It is noted that each of these spectra has a few dominant peaks. For the force and displacement type responses, the first peak located at $\omega = 0.08\ rad/sec$ is attributed to the peak of the wind load spectra [15, 16, 21]. Other peaks are associated with the natural frequencies of the building. It is observed from these figures that the implementation of the active tendon control system cannot control the peak of the input wind spectra; however, other peaks

Active Tendon Control System

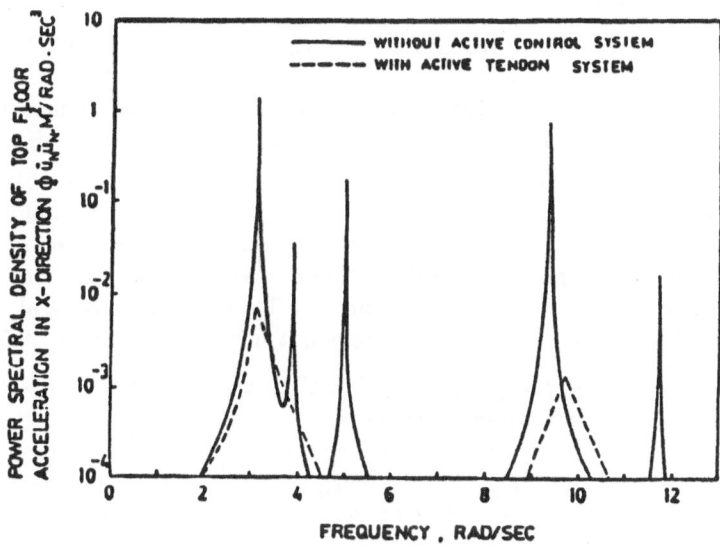

Figure 5: Power spectral density of top floor acceleration in x-direction

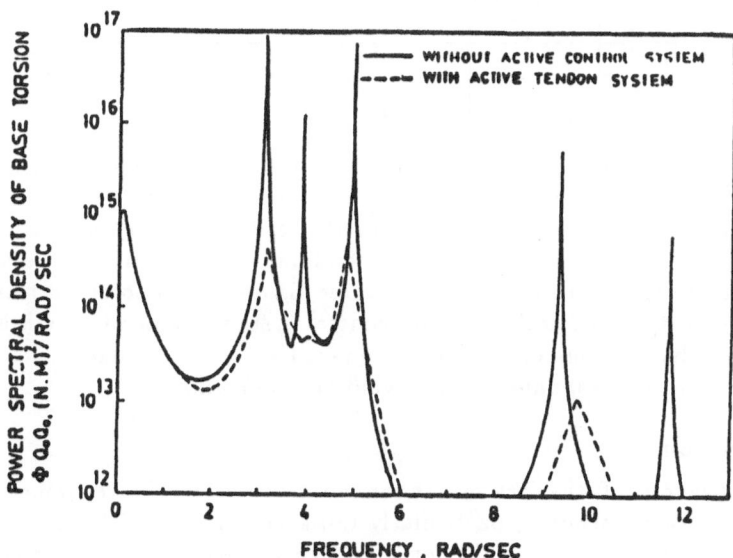

Figure 6: Power spectral density of base torsion

are suppressed considerably.

Unlike the force and displacement spectra, the spectra of the top floor accelerations do not have a peak at $\omega = 0.08$ rad/sec resulting from the wind spectra. This is because the low frequency displacement contributes little to the acceleration response. However, the higher modes contribute significantly to acceleration type responses.

The statistics of the response quantities are presented in Table 1 for the forty-story building with and without the active tendon control system. The mean values of the response quantities shown in the first row of Table 1 are the static responses and they are not influenced by the active control system. In other words, the active control system does not influence the static deflection.

Table 1:   *Response of the forty-story building with and without active control systems to strong wind gusts*

		u_N cm	v_N cm	θ_N rad	$\ddot{u}_N$ cm/sec²	$\ddot{v}_N$ cm/sec²	$\ddot{\theta}_N$ rad/sec²	U_0 kN	V_0 kN	Q_0 kN-m	cf_{x1} kN	cf_{y1} kN
	Mean Value	7.6	2.3	$.69\times10^{-3}$	0	0	0	20,441	9,291	93,470	-	-
Without Active Control System	Standard Deviation	3.8	1.7	$.10\times10^{-2}$	43.5	28.6	$.22\times10^{-1}$	10,689	6,965	139,280	-	-
Active Tendon System	Standard Deviation	1.9	0.8	$.27\times10^{-3}$	9.7	4.9	$.39\times10^{-2}$	5,471	3,444	37,054	528	335

Finally, simulated sample functions of (i) the top floor displacement and acceleration in the x-direction, (ii) the base torsional moment, and (iii) the active control force of the lowest controller in the x-direction are presented in Figs. 7-10. The response behaviors of the tall building with or without the active control system are clearly demonstrated in these figures. It is observed from Table 1 and Figs. 4-10 that the active tendon control system proposed herein is effective in reducing the tall building response quantities, in particular the acceleration and torsional responses, under strong wind environments.

6. Conclusion

The effectiveness of the active tendon control system in reducing the wind-induced structural response, particularly the acceleration and torsional responses, has been demonstrated. For the force and acceleration type responses, the contribution of higher modes is not insignificant and considering only one mode in the analysis may lead to misleading and nonconservative results. The active control system is capable of reducing the acceleration and torsional responses substantially.

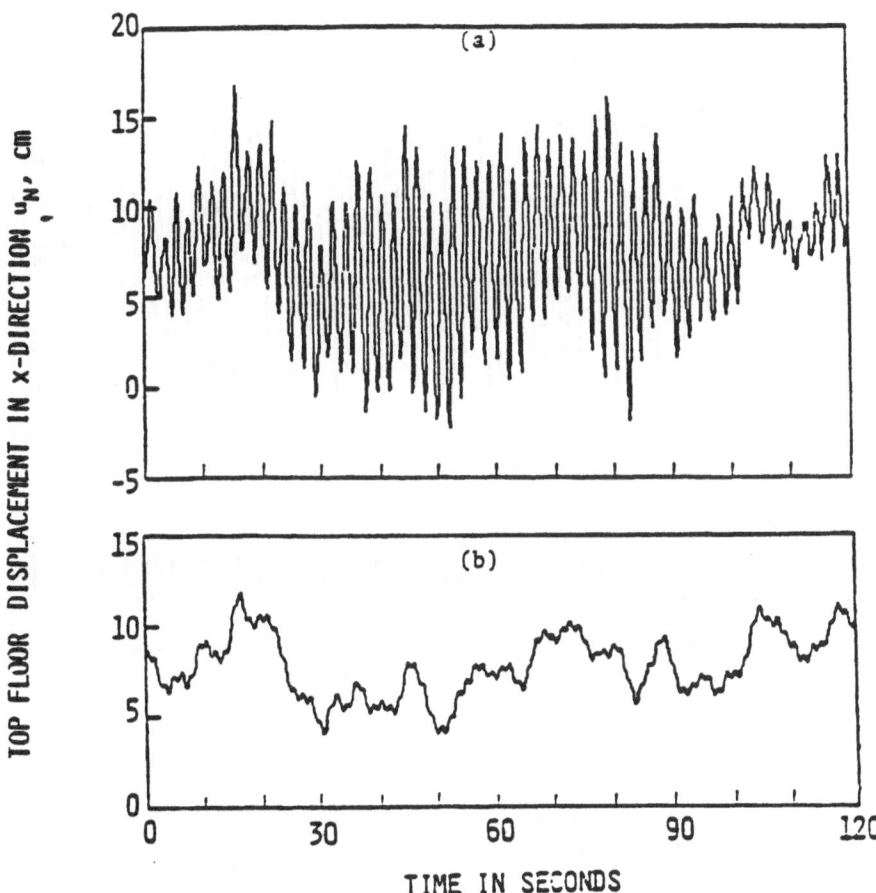

Figure 7: Top floor displacement in x-direction. (a) without active control system; (b) with active tendon control system

The analytical expression of the wind load suggested in the literature and used in the present investigation may not necessarily represent the natural wind in all cases. However, improved or different analytical expression can be incorporated easily in the analysis procedure presented herein. The acceleration response contributed by higher vibrational modes requires further experimental investigations, since the mechanism of wind-structure interaction in higher mode responses may be different from that in the first mode response. Finally, the effectiveness of the active tendon control system presented herein requires an experimental demonstration of practical applications.

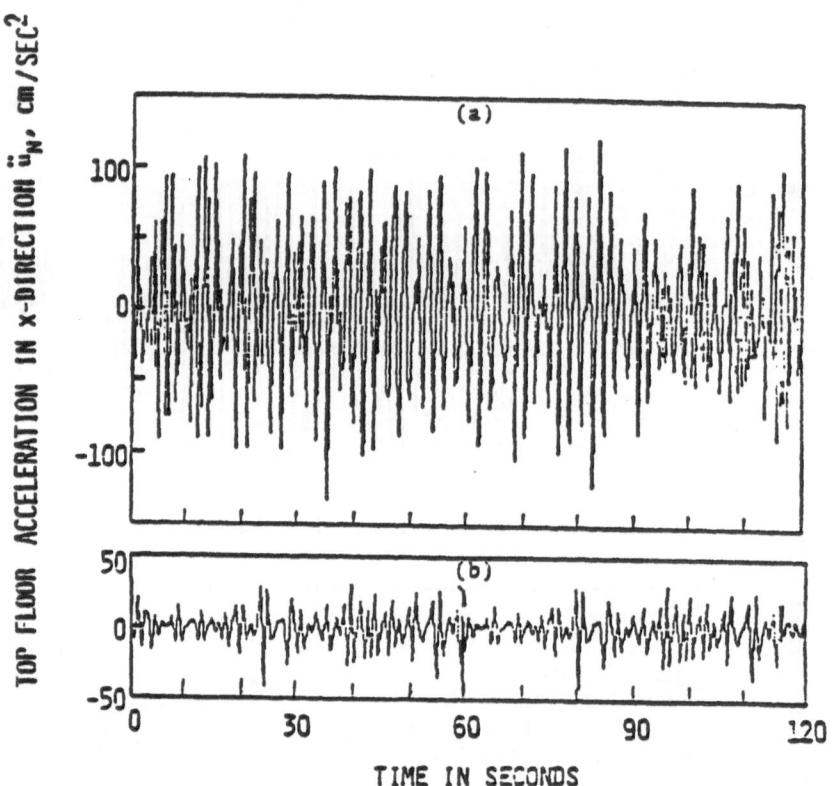

Figure 8: Top floor acceleration in x-direction. (a) without active control system; (b) with active tendon control system

Acknowledgment

This research was supported by the National Science Foundation Grant No. NSF-CEE-81-05307.

References

[1] Abdel-Rohman, M. and Leipholz, H.H., "Active Control of Tall Buildings," *Journal of the Structural Division*, ASCE, Vol. 109, No. 3, March 1983, pp. 628-645.

[2] Abdel-Rohman, M. and Leipholz, H.H., "Active Control of Flexible Structures," *Journal of the Structural Division*, ASCE, Vol. 104, No. ST8, August 1978, pp. 1251-1266.

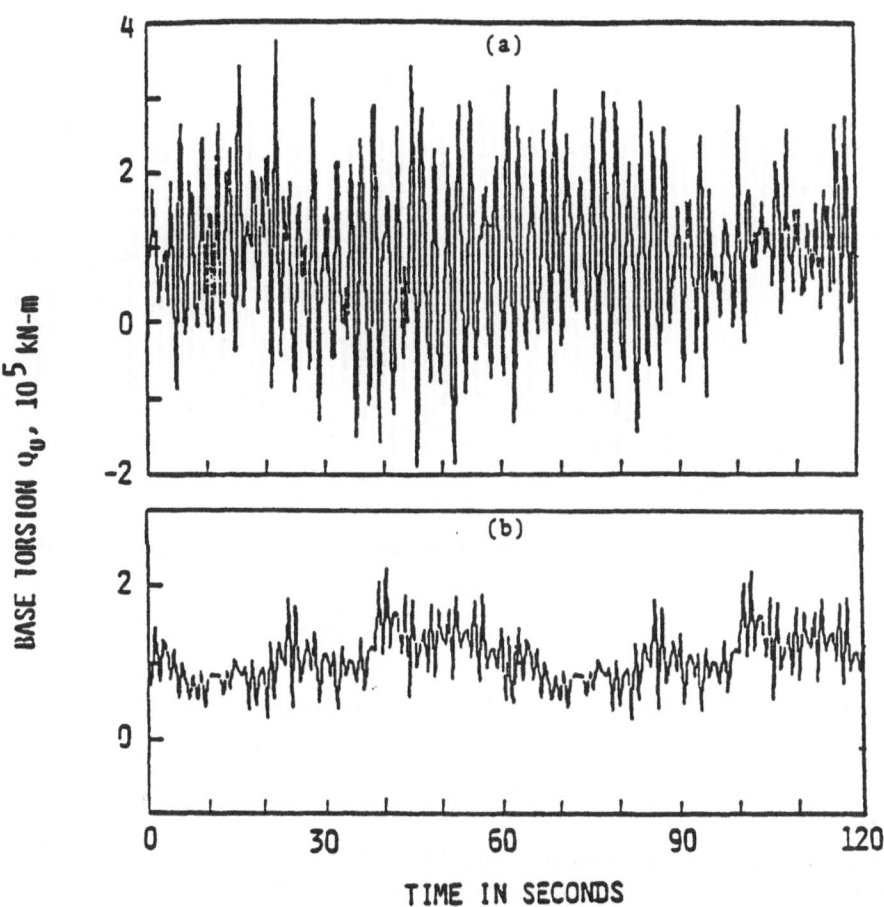

Figure 9: Base torsion. (a) without active control system; (b) with active tendon control system

[3] Chang, J.C.H. and Soong, T.T., "Structural Control Using Active Tuned Mass Damper," *Journal of the Engineering Mechanics Division*, ASCE, Vol. 106, No. EM6, December 1980, pp. 1091-1098.

[4] Davenport, A.G., "The Treatment of Wind Loading on Tall Buildings," *Proceeding of Symposium on Tall Buildings*, University of Southampton, Pergamon Press, Inc., New York, N.Y., 1966.

[5] Davenport, A.G., "The Application of Statistical Concepts to the Wind Loading of Structures," *Proceedings, Institution of Civil Engineers*, London, England, Vol. 19, August 1961, pp. 449-472.

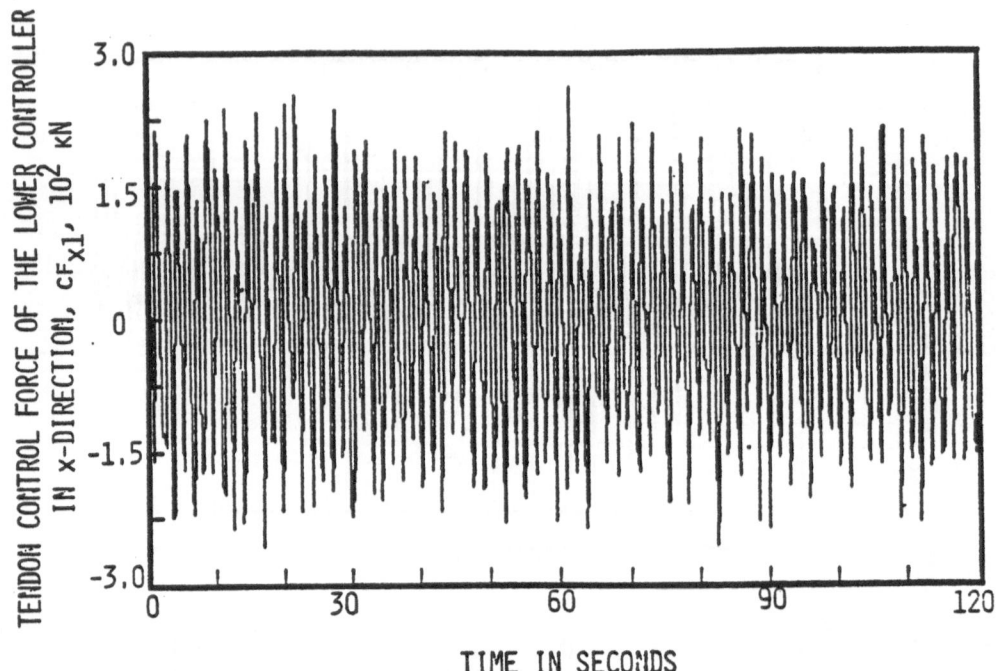

Figure 10: Tendon control force of the lower controller in x-direction

[6] Davenport, A.G., "The Spectrum of Horizontal Gustiness Near the Ground in High Winds," *Quarterly Journal, Royal Meteorological Society*, Vol. 87, April 1961, pp. 194-211.

[7] Hart, G.C., DiJulio, R.M., Jr., and Lew, M., "Torsional Response of High-Rise Buildings," *Journal of Structural Division*, ASCE, Vol. 101, No. ST2, Proc. Paper 11128, Feb. 1975, pp. 397-416.

[8] Lund, R., "Active Damping of Large Structures in Winds," *Structural Control*, ed. H.H.E. Leipholz, IUTAM 1979, North-Holland, pp. 459-470.

[9] Martin, C.R. and Soong, T.T., "Modal Control of Multi-Story Structures," *Journal of Engineering Mechanics Division*, ASCE, Vol. 104, No. EM4, August 1978, pp. 613-632.

[10] Patrickson, C.P. and Friedman, P., "A Study of the Coupled Lateral and Torsional Response of Tall Buildings of Wind Loadings," Technical Report, UCLA-ENG-7616, University of California at Los Angeles, Dec. 1976.

[11] Petersen, N.R., "Design of Large Scale Tuned Mass Damper," *Structural Control*, ed. H.H.E. Leipholz, IUTAM 1979, North-Holland, pp. 581-596.

[12] Roorda, J., "Tendon Control in Tall Structures," *Journal of Structural Division*, ASCE, Vol. 101, No. ST3, pp. 505-521.

[13] Roorda, J., "Active Damping in Structures," Granfield Report Aero., No. 8, Granfield Institute of Technology, Granfield, U.K., 1971.

[14] Sae-Ung, and Yao, J.T.P., "Active Control of Building Structures," *Journal of Engineering Mechanics Division*, ASCE, Vol. 104, No. EM2, April 1978, pp. 335-350.

[15] Samali, B., "Control of Coupled Lateral-Torsional Motion of Buildings Under Environmental Loads," Doctoral Dissertation, CMEE Department, The George Washington UNiversity, February 1984.

[16] Samali, B., Yang, J.N., and Yeh, C.T., "Control of Lateral-Torsional Motion of Wind-Excited Buildings," *Journal of the Engineering Mechanics*, ASCE, June 1985.

[17] Samali, B., Yang, J.N., and Liu, S.C., "Control of Lateral-Torsional Motion of Buildings under Seismic Loads," to appear in the *Journal of the Structural Engineering*, ASCE, 1985.

[18] Yang, J.N. and Liu. M.J., "Building Critical-Mode Control: Nonstationary Earthquake," *Journal of Engineering Mechanics Division*, ASCE, Vol. 109, No. 6, December 1983, pp. 1375-1389.

[19] Yang, J.N. and Samali, B., "Control of Tall Building in Along-Wind Motion," *Journal of Structural Division*, ASCE, Vol. 109, No. 1, January 1983, pp. 50-68.

[20] Yang, J.N.,"Control of Tall Buildings Under Earthquake Excitation," *Journal of Engineering Mechanics Division*, ASCE, Vol. 108, No. EM5, Oct. 1982, pp. 833-849.

[21] Yang, J.N., Lin, Y.K., and Samali, B., "Coupled Motion of Wind-Loaded Multi-Story Building," *Journal of Engineering Mechanics Division*, ASCE, Vol. 107, No. EM6, Dec. 1981, pp. 1209-1226.

[22] Yang, J.N. and Lin, Y.K., "Along-Wind Motion of a Multi-Story Building," *Journal of Engineering Mechanics Division*, ASCE, Vol. 107, No. EM2, April 1981, pp. 295-307.

[23] Yang, J.N., Lin, Y.K., and Sae-Ung, S., "Tall Building Response to Earthquake Excitations," *Journal of Engineering Mechanics Division*, ASCE, Vol. 106, No. EM4, August 1980, pp. 801-817.

[24] Yang, J.N. and Giannopoulos, F., "Active Control of Two Cable-Stayed Bridge," *Journal of Engineering Mechanics Division*, ASCE, Vol. 105, No. EM5, Oct. 1979, pp. 795-810.

[25] Yang, J.N. and Giannopoulos, F., "Active Tendon Control of Structures," *Journal of Engineering Mechanics Division*, ASCE, Vol. 102, No. EM4, August 1976, pp. 613-632.

[26] Yang, J.N., "Simulation of Random Envelope Processes," *Journal of Sound and Vibration*, Vol. 21, No. 1, 1972, pp. 73-85.

[27] Yao, J.T.P. and Tang, T.P., "Active Control of Civil Engineering Structures," Technical Report No. CE-STR-73-1, Purdue University, 1973.

IDENTIFICATION AND MODEL ADJUSTMENT OF A HANGING PLATE DESIGNED FOR STRUCTURAL CONTROL EXPERIMENTS

Bernd Schäfer
Hans Holzach
German Aerospace Research Establishment (DFVLR)
D-8031 Oberpfaffenhofen, Federal Republic of Germany

1. Introduction

Future spacecraft systems are planned having large flexible elements and requiring advanced attitude control systems. In recent years much emphasis has been placed upon theoretical investigations of controller designs rather than their experimental verification. The present study therefore contributes to the experimental research of the attitude control of large flexible structures via ground testing methods for representative structural models.

The laboratory test element has to be properly designed in order to meet actual dynamical characteristics associated with planned large flexible spacecraft. The most challenging problems for advanced controller to be designed arise from the flexibility of the large sized structures which are distributed parameter systems having, consequently, an infinite number of degrees of freedom. Very often, critical structural frequencies are very low, mostly below 1 Hz, which in general appear in closely spaced clusters. They are expected to fall within the bandwidth of the attitude control system and hence can no longer be ignored. Their natural damping typically is very poorly known and extremely small, often below 1% of

631

critical damping, which mostly is very poorly known. Moreover, the prediction of the dynamical behaviour in space via ground testing remains quite limited [1].

The very low damping of the structural modes is a critical problem which has involved new controller design philosophy. Here, damping augmentation of the structural modes (termed as "Low-Authority Control" [2,3]) by either direct velocity feedback (based on modal controller design) or, more recently, by wave-absorbing compensators proved to be very effective in order to ease the task of primary attitude controller. Basic experimental investigations on one-dimensional structures like various beam configurations [4-6] showed satisfactory performance, although not being very characteristic for large complex structures. More typical test elements have been proposed and investigated, using two-dimensional planar structures like a circular homogeneous plate [7], or quadrilateral grids [8,9].

The objective of the present study is the design of a suitable ground test element to permit the verification of candidate control systems. An accurate dynamical model is desirable in order to study the basic problems inherent in the interactive system structure/controller thus avoiding the strong impact possibly given by poorly predicted dynamics. But otherwise, since dynamical modeling for structures set into space is obviously less reliable, robust or adaptive controllers may be envisaged to largely overcome model uncertainties. Demonstration of the performance of such controllers in laboratory, on the basis of poorly known dynamics, is an alternate but very important study objective which easily may be treated by successive deterioration of the originally very accurate plate model, either analytically or experimentally or both.

The test element will be a rectangular homogeneous plate, suspended vertically by two parallel light wires. The suspension mechanism allows for low-frequent quasi-rigid-body oscillations of the whole configuration, which serve for necessary attitude degrees of freedom.

Proper dimensioning of the geometrical and material parameters is required to achieve the characteristic dynamical behaviour. The modeling will be performed analytically including the expected strong impact of gravity as could be observed recently in an equivalent beam experiment [5,10]. This model will be backed up by a finite element approach based on the analytically derived plate parameters. A modal survey testing of the configuration will be performed for identifying the modal parameters experimentally and, importantly, for checking the reliability of the model. However, model corrections are expected to be necessary, and the model adjustment will be a complementary process, theoretically as well as experimentally in order to meet the desired structural behaviour.

2. Hanging Plate Dynamical Model

The wire suspended plate configuration in its final design is already anticipated at this stage of description and is shown in Figure 1. Additionally, it carries two concentrated masses, the purpose of which will be explained below in section 4. For the present, the dynamical influence of these masses will be ignored, but added later on. For the analytical approach the configuration is subdivided in two separate models: (a) a pendulous, rigidly assumed plate involving three oscillation modes with very low frequencies if designed adequately, and (b) a purely elastic plate without wires but subjected to gravity force. This important separation greatly reduces the analytical effort, but it is expected to yield satisfactory results if the quasi-rigid-body modes are well separated from the structural ones. As a consequence, this depends strongly on the proper design of the geometrical parameters like the length of wires and plate sides and the plate thickness.

More recently, a plate dynamical model for an isotropic assumed material behaviour has been presented [11] without experimental verification. The rigid body modeling has been described there in detail and will be skipped here in favour of the more dominant problem of a careful model adjustment. Moreover, the experimental tests of section 4 exhibited clearly, that an isotropic material behaviour can no longer be assumed. Furthermore, during the test set-up phase, imperfect manufacturing processing of the plate material had been observed which caused remarkable deformations of the hanging plate from the vertical plane. To overcome these effects, plates of greater thickness have been envisaged, which evidently alter the dynamics. For the purposes of the analytical modeling this dynamical aspect has to be dealt with only to the extent of keeping the lowest structural frequencies below a preset value. Different constraints mainly due to digital implementation of the control algorithms, required the lowest 10 to 15 modal frequencies to be below approximately 10 Hz, but some of them must be closely spaced.

The structural plate model is treated on the basis of the above required items. No work in literature has been found for the case of rectangular orthotropy including variable in-plane loads. Therefore, an approximate model is developed regarding both effects. The analytical approach is based on the Rayleigh-Ritz procedure assuming classical thin plate theory with small deflections only, which is expected to hold true even in the case of in-plane gravity loads. To be more general, orthotropic material behaviour is admitted. The four sides of the rectangular plate are completely free. The origin of the rectangular coordinate system (x,y) is regarded to be centered at the middle of the plate such that x- and y-axes are falling together with the two symmetry lines of the plate. Both axes are thought to lie in the undeformed middle surface of the plate. The x-axis is oriented upwards and is parallel to the downwards oriented direction of the gravity field.

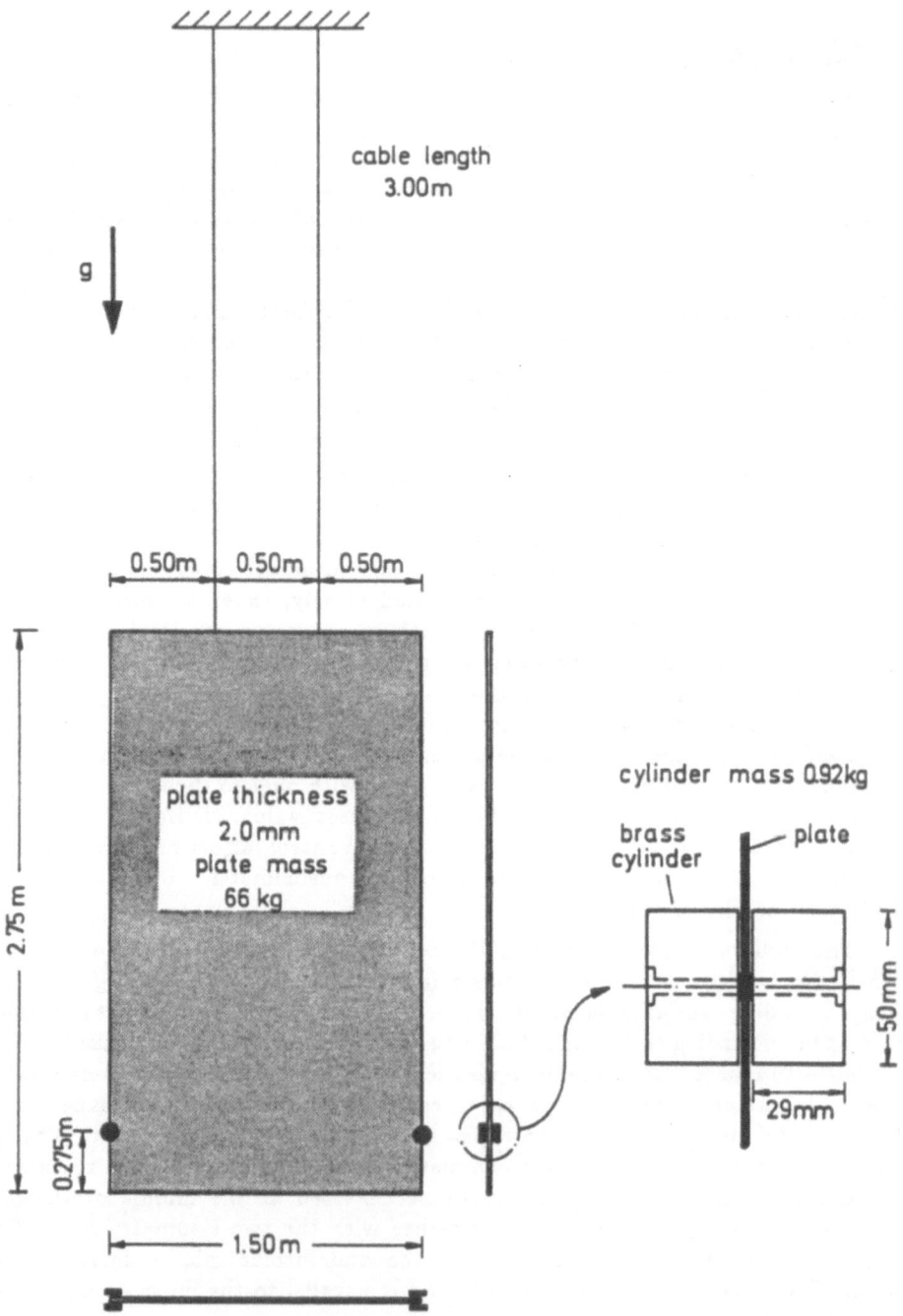

Figure 1: Hanging plate configuration

The resultant in-plane forces per unit length, N_x, N_y and $N_{xy} = N_{yx}$, are assumed not to vary when the plate is vibrating. Then they are independent from the deflection function, w, and can be calculated from the static state of equilibrium, giving

$$N_x = \rho g h \left(x + \frac{a}{2} \right),$$

(1)

where ρ is density, g is acceleration due to gravity, and the geometrical parameters are denoted by a, b and h (i.e. the vertical and the horizontal plate length, and its thickness). Both other in-plane loads vanish in the special case to be considered here. The governing partial differential equation of motion for the transverse bending is given by (see e.g. Ref. 12,13)

$$D_x \frac{\partial^4 w}{\partial x^4} + 2(D_1 + 2D_{xy}) \frac{\partial^4 w}{\partial x^2 \partial y^2} + D_y \frac{\partial^4 w}{\partial y^4} - \frac{\partial}{\partial x}\left[N_x \frac{\partial w}{\partial x} \right] + \rho \frac{\partial^2 w}{\partial t^2} = 0,$$

(2)

where t is time, and the four elastic constants, being independent, are

$$D_x = \frac{E_x}{1 - \nu_x \nu_y} \frac{h^3}{12}, \quad D_y = \frac{E_y}{1 - \nu_x \nu_y} \frac{h^3}{12},$$

(3a)

$$D_1 = \nu_y D_x = \nu_x D_y, \quad D_{xy} = G \frac{h^3}{12}.$$

(3b)

Here, E_x, E_y, ν_x, ν_y are Young's moduli and Poisson's ratios in the respective directions, G is the modulus in shear. Since it follows from Eqs. (3a), (3b) that $\nu_x / E_x = \nu_y / E_y$, because of the required symmetry of the stress-strain relations, only four of these elastic constants are independent from each other. The well-known classical boundary conditions for unloaded, isotropic, completely free plates are modified in presence of in-plane loads and orthotropy, but they are skipped here, since no exact solution exists even in this much simpler case.

The Rayleigh-Ritz method takes advantage of the maximum potential, U, and kinetic energy, T, of the system. Assuming a solution of the form

$$w(x,y,t) = W(x,y)\sin \omega t$$

(4)

for the free vibration analysis, where ω is the frequency of vibration and $W(x,y)$ is the corresponding mode shape, the energy expressions are then given by

$$U = \frac{1}{2} \int_{-\frac{a}{2}}^{\frac{a}{2}} \int_{-\frac{b}{2}}^{\frac{b}{2}} \left[D_x \left(\frac{\partial^2 W}{\partial x^2} \right)^2 + 2D_1 \frac{\partial^2 W}{\partial x^2} \frac{\partial^2 W}{\partial y^2} + D_y \left(\frac{\partial^2 W}{\partial y^2} \right)^2 + 4D_{xy} \left(\frac{\partial^2 W}{\partial x \partial y} \right)^2 \right.$$

$$+ N_x \left(\frac{\partial W}{\partial x}\right)^2 \Bigg] dx \, dy \,, \tag{5}$$

accounting for both the strain energy of bending and twisting and the energy of the gravitational force, and

$$T = \frac{1}{2} \rho h \omega^2 \int\limits_{-\frac{a}{2}}^{\frac{a}{2}} \int\limits_{-\frac{b}{2}}^{\frac{b}{2}} W^2 dx \, dy \,. \tag{6}$$

The principle of energy conservation then leads to the Rayleigh quotient which may be expressed in the form $\omega^2 = V/T^*$, where $T^* = T/\omega^2$, and V and T^* are both functionals depending upon the spatial function W only. According to the Ritz method [14] the unknown function W is expanded in a series of admissible functions (note that the boundary conditions are solely of dynamical and not of geometrical character). Moreover, it is by far more convenient to separate by a product of two functions, say $X(x)$ and $Y(y)$, the one depending on the spatial coordinate x only, the other on y. A further advantage is yielded by choosing the orthogonal eigenfunctions of the free-free beam [13], the one-dimensional analog of the completely free plate, which finally gives

$$W(x,y) = \sum_{k=0}^{n_x} \sum_{l=0}^{n_y} c_{kl} \, X_k(x) Y_l(y). \tag{7}$$

The constants c_{kl} have to be chosen such as to minimize Rayleigh's quotient. This gives $n_{xy} = (n_x + 1)(n_y + 1)$ upper bounds of the lowest eigenfrequencies and corresponding mode shapes. The finding of minimum values then leads to the special eigenvalue problem:

$$(A - \Lambda I)c = 0, \tag{8}$$

where A is a square matrix, I is the identity matrix, both of order n_{xy}, c is the eigenvector, and $\Lambda = \omega^2/\omega_B^2$ with $\omega_B^2 = D_x/(\rho h a^4)$ is the non-dimensional eigenvalue. Matrix A contains the plate material and geometrical properties. Moreover, A is symmetric and positive semidefinite because of the self-adjoint and positive semidefinite boundary value problem, which gives real, positive eigenvalues ($\Lambda \geq 0$).

Some remarks shall be left to the isotropic case. The corresponding dynamical equations can be easily deduced from the above equations by simply setting $\nu_x = \nu_y \equiv \nu$, $E_x = E_y \equiv E$ giving $G = E/(2(1 + \nu))$. This means, only two elastic constants are necessary in isotropic plate bending theory.

3. Numerical Results and Model Adjustment

Equation (8) has been solved by standard numerical methods [15]. Both the non-dimensional eigenfrequency quantity $\sqrt{\Lambda} = \omega / \omega_B$ and the corresponding mode shape depend altogether upon six parameters, which are the four elastic constants, the aspect ratio a/b, and a parameter τ, which defines the gravity effect, $\tau = \rho h g a^3 / D_z$. The plate thickness enters just in both parameters ω_B and τ. Extensive numerical studies have been performed in order to design a suitable plate configuration. Two different plate materials, i.e. steel and aluminium, have been considered, but both exhibit no remarkable difference in the dynamics since the material parameters appear in the combination E/ρ and ν itself (in isotropic case) which are nearly the same for both materials. Hence, parameter studies on the aspect ratio a/b and on the plate thickness h proved to be sufficient for the final design. For the sake of reducing computer times for the numerical calculations, the truncation numbers n_z and n_y have been limited to $n_z = n_y = 5$, which results in 36 terms in the expansion of Equation (7) and hence in 36 upper bounds of the exact eigenfrequencies and the corresponding mode shapes. Since accuracy of the approximated modal data decreases for higher mode numbers, the first 20 modes are seen to be approximated with sufficient accuracy, as could be shown by an increase of truncation number.

However, once the final plate parameters had been determined, the truncation number was increased to 11, giving 144 terms in the expansion. This was done in order to be much more accurate in the final model.

3.1. Isotropic case

Since no severe anisotropy could be expected before the final test configuration was set up, the parameter studies have been performed on the basis of an isotropic material behaviour. During the parameter studies the material constants have been adopted to $\sqrt{E/\rho} = 5000$ m/s and $\nu = 0.3$. For the studies, the plate thickness has been limited to 1.0, 1.5 and 2.0 mm, which are values for commercially available plates. Because of expected imperfect manufacturing processing, mainly caused by the rolling mills, the value of 2.0 mm thickness has been preferred. Suspending the plate such that the direction of rolling is parallel to the vertical, then would largely overcome remarkable plate deformations by means of its own weight, which otherwise are possibly raised by the rolling process.

Finally, values for the horizontal and the vertical plate side are obtained, which provide the required dynamical characteristics: $a = 2.75\ m$, $b = 1.50\ m$. Furthermore, the decision of the plate material has been brought on in favour of steel (of non-magnetic type (austenitic steel) because of the dedicated electrodynamic actuator system) rather than aluminum. The respective eigenfrequencies and the corresponding mode shapes will be given later in conjunction with the experimental results.

3.2. Orthotropic case and parameter adjustment

The experimentally obtained structural frequencies of the isotropically designed plate showed great deviations from the predicted ones (cf. Table 1). Nevertheless, the experimentally identified structural frequencies fortunately also exhibited intended dynamical plate behaviour, meaning low-frequent modes with several clusters. The next step of analytical modeling then consisted of adjusting the model to the observed system behaviour.

Table 1:   Experimental and theoretical eigenfrequencies (in Hz) and measured modal damping coefficients

mode number	plate without masses 2.0 mm × 2.75 m × 1.50 m cable length 3.00 m				plate with masses 2.0 mm × 2.75 m × 1.50 m cable length 3.00 m		measured modal damping coeff. in % of crit. damping
	exp.	anal. orth.	anal. isot.	FEM isot.	exp.	FEM orth.	
1	0.16	0.17	0.17	0.16	0.16	0.16	4.51
2	0.23	0.23	0.23	0.23	0.23	0.23	3.34
3	0.77	0.90	0.90	0.76	0.77	0.77	1.01
4	2.05	2.08	1.89	1.70	1.94	1.96	0.34
5	2.07	2.09	1.98	2.04	2.07	2.15	0.29
6	4.61	4.56	3.74	3.69	4.49	4.37	0.25
7	4.78	4.74	4.37	4.41	4.77	4.80	0.36
8	7.16	7.18	4.75	4.81	6.38	6.48	0.30
9	7.45	7.78	5.81	5.86	7.09	7.70	0.36
10	8.43	8.39	6.50	6.47	8.38	7.94	0.32
11	8.99	9.01	7.94	8.15	8.87	9.07	0.34
12	11.15	11.37	8.48	8.53	11.10	11.1	0.22
13	12.50	12.07	10.26	10.34	12.32	12.1	0.27
14	12.65	14.44	11.89	12.07	12.73	14.5	-
15	13.25	15.59	13.16	13.44	14.00	15.7	-

Since the plate rolling direction was oriented parallel to the vertical it was hoped that the gravitational force would compensate largely for the deformed shape. This proved to be rather successful, but still non-negligible, small deviations from the vertical plane remained (less than 1 cm of the bottom plate edge). In an additional preceding remedy the plate had been stretched in the direction of rolling by means of a stretcher device which is able to provide for uniformly distributed tensile forces along opposite plate edges. This improvement was carried out at the steel manufacturer's laboratories.

Because of processes, which influence the plate material as well as its geometry, the assumption of an ideally isotropic behaviour could no longer be maintained. Especially, the deviations from the vertical plane which consist

Table 2:   Plate characteristics

material	austenitic stainless steel
density ρ	8000 kg/m^3
cable length ℓ	3.00 m
vertical plate side a	2.75 m
horizontal plate side b	1.50 m
thickness h	2.00 mm
vertical Young's modulus E_z	$2.364 \cdot 10^{11}$ N/m^2
horizontal Young's modulus E_y	$4.376 \cdot 10^{11}$ N/m^2
shear modulus G	$1.252 \cdot 10^{11}$ N/m^2
vertical Poisson ratio ν_z	0.293
horizontal Poisson ratio ν_y	0.542
weight of attached mass, each	0.918 kg
total weight of plate without masses	66.0 kg

almost entirely of curvature about horizontal axes, add much more stiffness to
the plate in the horizontal direction than in the vertical. This effect can be
favourably treated by adopting a virtual Young's modulus in the horizontal direction much greater than in the vertical one. Moreover, stretching the plate in its
vertical direction, may also alter the elastic constants with respect to the vertical,
and a possibly strong increase of E_z has to be considered. The expected orthotropic behaviour is foreseen to arise much more by geometrical deformations than
by different material characteristics in different directions.

Because of the above discussed dynamical influences it seems obvious, that
the material parameters are hardly to be determined experimentally. Moreover,
applying e.g. shell theory to the deformed plate, would pass the effort to the
strong demand of measuring very precisely the geometrical deformations deviating from the vertical plane; and even a subsequent finite element approach would
not overcome the addressed difficulties.

Therefore, the four independent orthotropic elastic constants, say E_z, E_y, ν_z
and G are derived by means of the analytical model while fitting the theoretical
eigenfrequencies f_i^{theor} to the corresponding experimental f_i^{exp}. The equivalent
measured frequencies of the plate without the additionally mounted two masses
are listed in column 3 of Table 1. The lowest 9 structural frequencies have been
considered for parameter optimization. The optimization is based on a least
squared error fit of the type

$$\Phi(p_1,...,p_r) = \sum_{i=4}^{12} a_i \big(f_i^{theor}(p_1,...,p_r) - f_i^{exp}\big)^2, \qquad (9)$$

for which the error function Φ, depending on the parameters (or variables)
$p_1,...,p_r$ (here $r = 4$), is to be minimized. The theoretical eigenfrequencies are
calculated by the Rayleigh-Ritz model. The summation starts with $i = 4$ since
the first three modes correspond to quasi rigid body oscillations, which are not

accounted for in this structural plate approach disregarding the suspending cables. Additionally, different weighting factors a_i are admitted in order to obtain a better fit for the lower frequencies than for the higher ones. These factors are not part of the r variables p. Their values have to be chosen before the minimization process begins. Then a quick numerical standard routine has been employed [15] which determines the minimum of a function Φ of r variables using the quasi-Newton method.

Since the elastic constants E_z, E_y and G usually have large numbers on the order of $10^{11} N/m^2$, derived quantities as defined by the Rayleigh-Ritz model are used for variables p in the optimization routine. Hence, the results for the so optimized parameters are as follows:

$$
\begin{aligned}
p_1 &= \sqrt{E_z/\rho} &&= 5.436.0 \ m/s, \\
p_2 &= \sqrt{E_y/E_z} &&= 1.851, \\
p_3 &= G/G_0 &&= 1.369, \\
p_r &= \nu_z &&= 0.293,
\end{aligned}
\tag{10}
$$

where $G_0 = E_z/(2(1 + \nu_z))$ is the modulus in shear in the isotropic case. Moreover, the weighting factors $\alpha_4 = \alpha_5 = 30$, $\alpha_6 = \alpha_7 = 15$, $\alpha_k = 1(k \geq 8)$ have been chosen. Now, the values of the five elastic constants can be derived from Equation (10). They are comprised in Table 2 together with other material and geometrical parameters already defined before, which completely define the hanging plate configuration (with mounted masses, to be discussed in section 4).

The eigenfrequencies of this orthotropic model have already been listed in Table 1, column 2. There, they are compared to the equivalent frequencies of the isotropic case (column 4). Comparison of the respective elastic constants in both cases show, that Young's modulus in the vertical direction is just by 18.2% larger. The reason for this increase may be details of the rolling process and the subsequent stretching of the plate in this direction.

Moreover, regarding now the horizontal elastic modulus, this value exceeds that of E_z by an amazing 85%. This behaviour can be explained by the small plate deformations which evidently enhance the stiffness in the horizontal direction and which have been discussed above. Hence, the high value for E_y is caused much more by geometrical reasons than by orthotropic material properties. Although this large value is more virtual rather than physically real, its use is justified by the success in developing a dynamical model of high fidelity.

At long last it should be mentioned that the parameter fit is indeed based on a comparison of experimental and modelled eigenfrequencies, but it does not account for the corresponding mode shapes. That is to say, there is no proper reliability in the true ordering of corresponding mode shapes. For example, exchange in ordering may be expected from modes lying close together. But this is just the case for all of the lowest 8 structural modes where four closely spaced pairs are given. An appropriate extension of Equation (9) is possible, in principle, in order to avoid mode exchanges. Nevertheless, this aspect has not been considered here, since comparison with the experiments indicated no difference at all, except for modes No. 12 and 13 (see section 4).

3.3. Finite element approach

The final analytical model has been backed up by a finite element model using the program system ASKA [16]. In this approach the influence of the suspension wires and the optionally added concentrated masses have been accounted for. The model uses triangular bending elements with a complete fifth order displacement field. The membrane forces being necessary to include the in-plane gravity loads by means of a geometrical stiffness matrix, are computed prior to the vibration analysis while using triangular membrane elements. Furthermore, a stability analysis preceded the dynamical analysis in order to ensure that no buckling of the thin plate occurred. The model contains 117 nodal points with 3 degrees of freedom for each point, which then are reduced by static condensation to $3 \times 35 = 105$ degrees of freedom. The resulting eigenfrequencies are listed in Table 1 for the isotropic case (column 5) and for the orthotropic case including both attached concentrated masses (column 7).

4. Identification of Modal Parameters

A modal survey testing has been performed in order to establish the validity of the structural plate model, and if necessary, to correct the model. The modal data of the test element, i.e. the resonant frequency, damping factor and mode shape of each of several modes of vibration have been identified by mans of the transfer function technique.

4.1. Test facilities and test procedure

A structural dynamics analyzer of type Hewlett Packard HP 5423A has been used for force commanding, data sampling, data processing, and modal data identification. The plate has been stimulated using the quasi white noise signal output of the spectral analyzer. The excitation signal has been commanded to the force actuation system which is of non-contacting electrodynamical type, while sensing has been provided by a non-contacting infrared laser diode based system [5,6] of high resolution (0.05%), a range of ±16 mm and a frequency response range up to 2 kHz. The respective performance data for the actuator system refer to a displacement range of ±12 mm, a maximum force amplitude of 1 N and a frequency response range up to 200 Hz.

Before the response signals are sampled by the spectral analyzer, they have been filtered by a relatively broad bandwidth filter of 1 kHz and by the analyzer's anti-aliasing filter which is adapted automatically to the selected sampling rate. Since the analyzer possesses only one single input and output channel, the transfer behaviour between the driving point and only a single other plate location can be processed. Consequently, this procedure for modal identification has proved to be very time-consuming.

As a necessary preliminary for the complete modal survey testing, the resonant frequencies have to be identified carefully, this is also necessary in order to provide a first verification of the theoretical model. Different driving points were investigated to find the best locations for excitation, such that all of the lowest modes below about 12 Hz could be observed. Evidently, these locations are the four plate corner points. Also, different locations for sensing were selected in order to get, already in the preliminary tests, rough estimates for the correspondence of resonance frequency and mode shape by comparison with those of the theoretical model. By this procedure, 13 resonance frequencies below 12.5 Hz could be detected (cf. column 6 of Table 1). In addition, in order to get more reliable information on frequencies, further measurements were performed using sinusoidal excitation technique. Here, the single vibration modes were stimulated by sweeping over the range of the resonant frequency to be identified.

The maximum peak-to-peak deflection of the plate was limited to about 3 cm. These maximum values occurred at the two bottom plate corner points, mainly caused by oscillations of the first two modes (torsional and first pendulum oscillation). Applied maximum forces did not exceed 0.3 N. For measurement of the various transfer functions the spectral analyzer was used to sample the deflection sensor data at three different sample rates which correspond to three different values of maximum frequency to be detected: 3.125, 6.25 and 12.5 Hz bandwidth. This was done because of the existence of four pairs of closely spaced frequencies in the range up to 12.5 Hz: Thee pairs are in the neighbourhood of 2.0, 4.6, 6.7 and 8.6 Hz. Consequently, since the analyzer's frequency resolution depends directly on the selected sampling rate, the lower modes, having the smallest frequency spacing, would be detected with very poor accuracy while admitting high bandwidth values.

For the modal survey testing, thirty almost equally spaced locations were defined on the plate surface which coincided with equivalent grid points of the finite element net. For the present, it was decided to consider only one fixed driving point, which is one of the two upper plate corner points. And since, after all, the mode shape results showed good agreement compared to the theoretical ones, no further attempt was made to obtain frequency responses from other driving points. However, in order to obtain reliable transfer function data for one single input/output path, many successive measurements (up to 100) were carried out, and their average was taken for further data processing.

The modal parameters are identified by mathematically curve fitting an analytical form of a transfer function to measured data. The curve fitting method uses a single mode (or single degree-of-freedom) method, which means that the function fit to the data is for a single mode of vibration.

4.2. Plate modifications

The frequency results showed that just the lowest two structural modes at about 2 Hz were separated by only 0.02 Hz, compared to a predicted value of 0.3 Hz for the isotropic plate model. This extremely small frequency spacing would have provided severe difficulties in separating the two corresponding mode shapes experimentally. Hence, an increase in frequency spacing was desired which favourably had been achieved with low technical expense by adding two small masses made of brass to the structure, symmetrically arranged with respect to the vertical plate symmetry axis. By a series of experiments the best positions for mounting with respect to frequency spacing was found to be near the lower plate corners. The weight of both masses was kept as low as possible (here 0.92 kg for each mass) in order to avoid enhanced structural modifications of the originally proposed configuration. By this a spacing of 0.13 Hz was achieved which proved to be adequate for proper mode shape isolation.

Both masses have two opposing influences on the plate resonant frequencies: first, the kinetic energy will be increased and second, the in-plane loads increase slightly due to the additional weight. While the first effect reduces the frequencies to some extent, the second one acts contrarily by increasing the effective plate stiffness. These effects have been accounted for only in the finite element model rather than in the analytical one. The results show that the frequencies are influenced much more than the mode shapes which behave almost as before except that neighbouring modes possibly exchange in ordering. Furthermore, such modes having vertical nodal lines are affected more strongly than others, since here the added modal mass is much higher.

4.3. Experimental and theoretical results

Resonance frequencies have been measured up to 50 Hz. The lowest 15 of them are listed in Table 1 for the plate without (column 2) and including the two concentrated masses attached to the plate (column 6). No attempt at modal identification for resonant frequencies beyond 12.5 has been made. Typical graphs of measured transfer functions showing the plate frequencies up to 42 Hz are presented in Figure 2 and 3. Due to the considerably large residue values at these high resonances it seems obvious that eigenfrequencies far beyond this limit are detectable by the employed measurement technique.

In addition, Table 1 comprises the measured modal damping coefficients (last column). They include material damping as well as damping effects due to the environmental air and the suspension devices. Compared to the rigid body modes the damping of the structural modes is very low, less than 0.4% of critical damping.

Figures 4 and 5 show the nodal pattern of the mode shapes for the isotropic analytical model and the corresponding finite element model considering the orthotropic effects and the attached concentrated masses. The equivalent

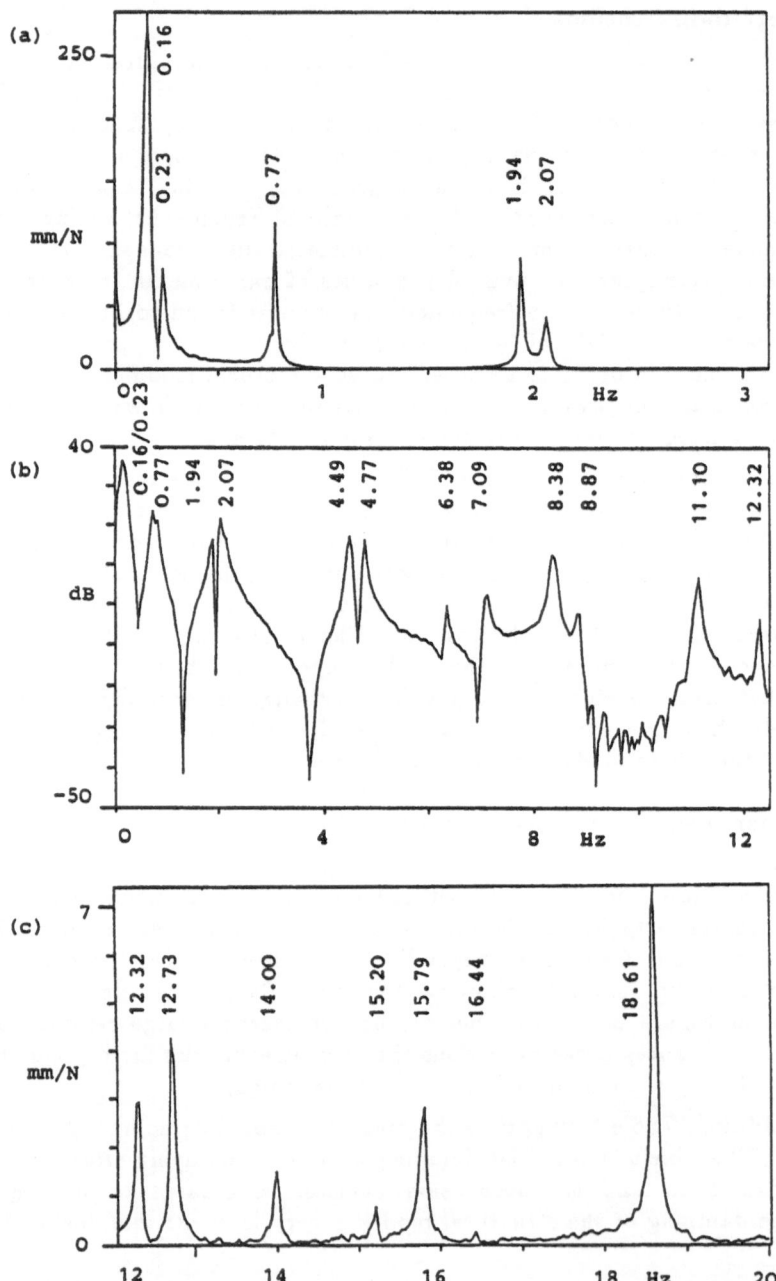

Figure 2: *Typical graphs of transfer function magnitude (driving point input/output channel) for different sampling rates: (a) 3.125 Hz, (b) 12.5 Hz, (c) 8.0 Hz, showing low-resonant frequencies up to 20 Hz*

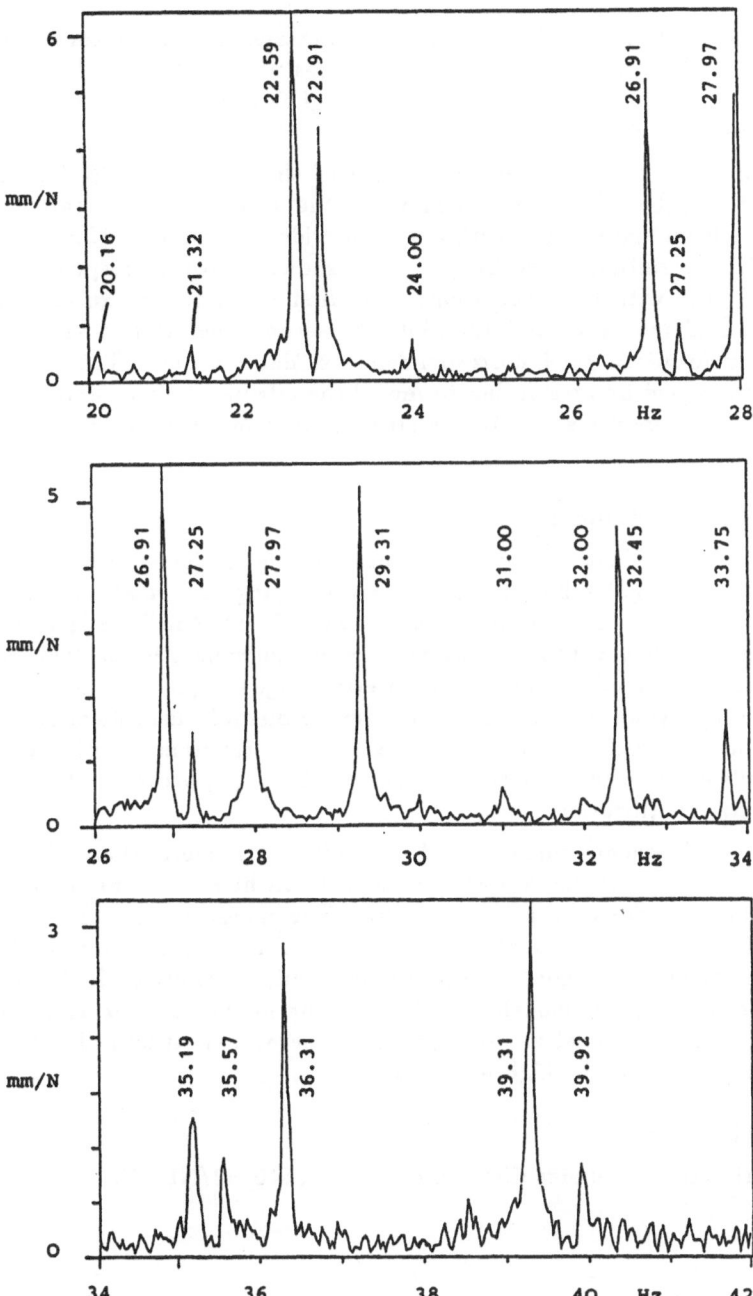

Figure 3: Transfer function magnitude (input/output channel is driving point) for 8 Hz sampling rate (resolution 0.03 Hz) showing the resonant frequencies between 20 and 42 Hz

orthotropic analytical model results without masses are not presented, since there is no remarkable deviation in the mode shapes with respect to those of the finite element approach, except that just two neighbouring modes exchange in ordering, modes 4 and 5.

In order to obtain the nodal pattern of the experimentally identified mode shapes, a bi-cubic spline interpolation of the residues of the thirty plate grid points has been applied [15], with a user provided software routine for finding the zeros. The resulting mode shapes (not shown here) correspond well to the predicted ones with two exceptions. The first one refers to an exchange of the closely spaced modes 12 and 13, while the second one refers to modes 8 and 10 where the formerly good correspondence is deteriorated. This deviation may have been caused by attempting to model the effects of plate curvature with simple orthotropy; modes 8 and 10 are strongly influenced, both having two vertical nodal lines.

5. Concluding Remarks

Both the modal identification technique and the model adjustment for a wire-suspended rectangular homogeneous plate carrying two small masses have been described in detail. Mainly due to non-negligible but small static plate deformations, the modeling approach accounts for orthotropic effects. The good agreement between experimental and theoretical eigenfrequencies and mode shapes justify the dynamic modeling and experimental identification techniques employed. The technique of measuring transfer functions at different sampling rates while exciting the structure by quasi white noise, proved to be successful for modal data identification.

Measured eigenfrequencies did not differ by more than 8% from their predicted values with the adjusted model. Even high resonant frequencies up to 50 Hz could be detected experimentally. The measured modal damping coefficients proved to be very small for the structural modes, less than 0.4% of critical damping, which was a goal for the present study. Definitely, it should be stated that both the dynamic modeling and measuring process complemented each other by correcting the model in order to meet the actual structural behaviour, that itself is characterized by high modal density.

Acknowledgement

This work was done under ESA contract No. 5310/82/NL/BI.

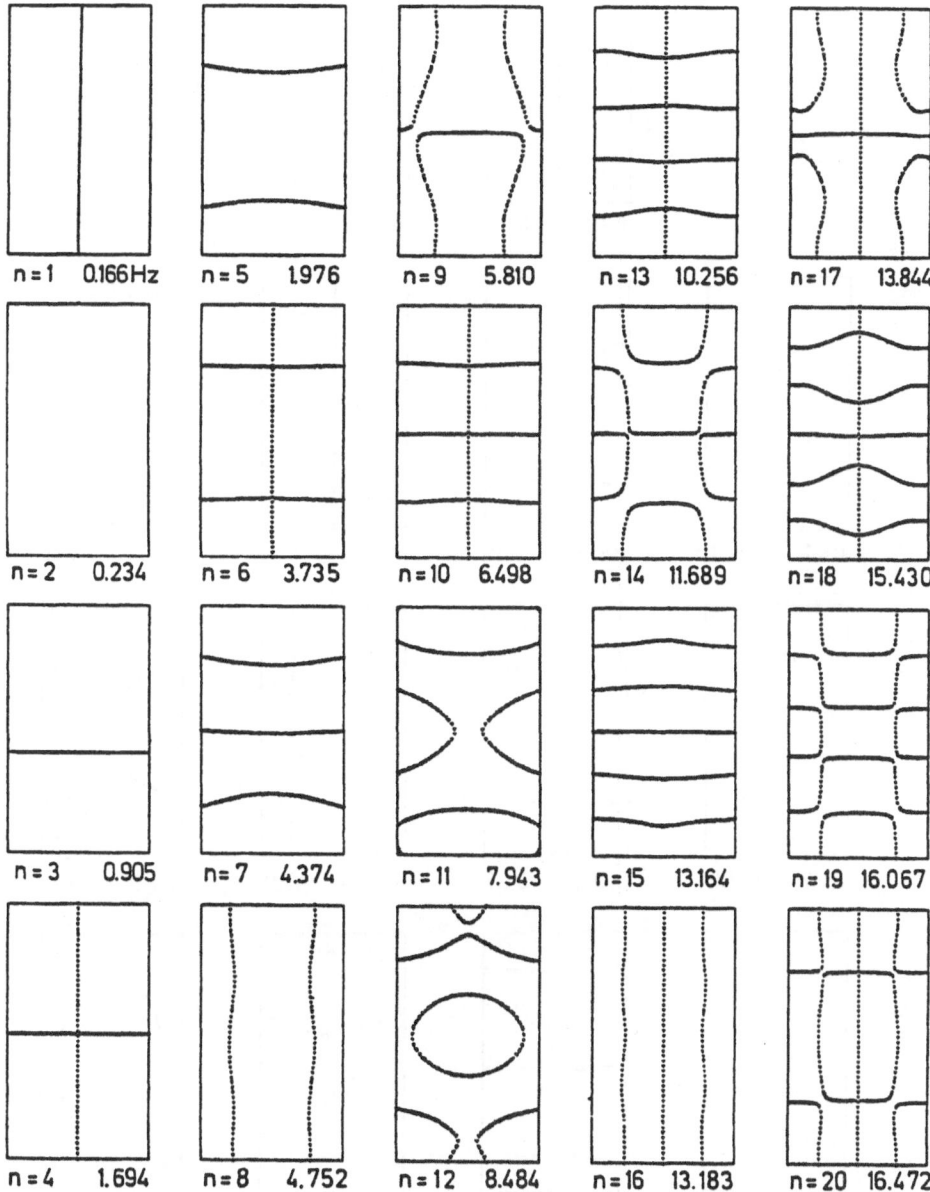

Figure 4: **The lowest 20 mode shapes with corresponding eigenfrequencies for the isotropic Rayleigh-Ritz model without the two concentrated masses**

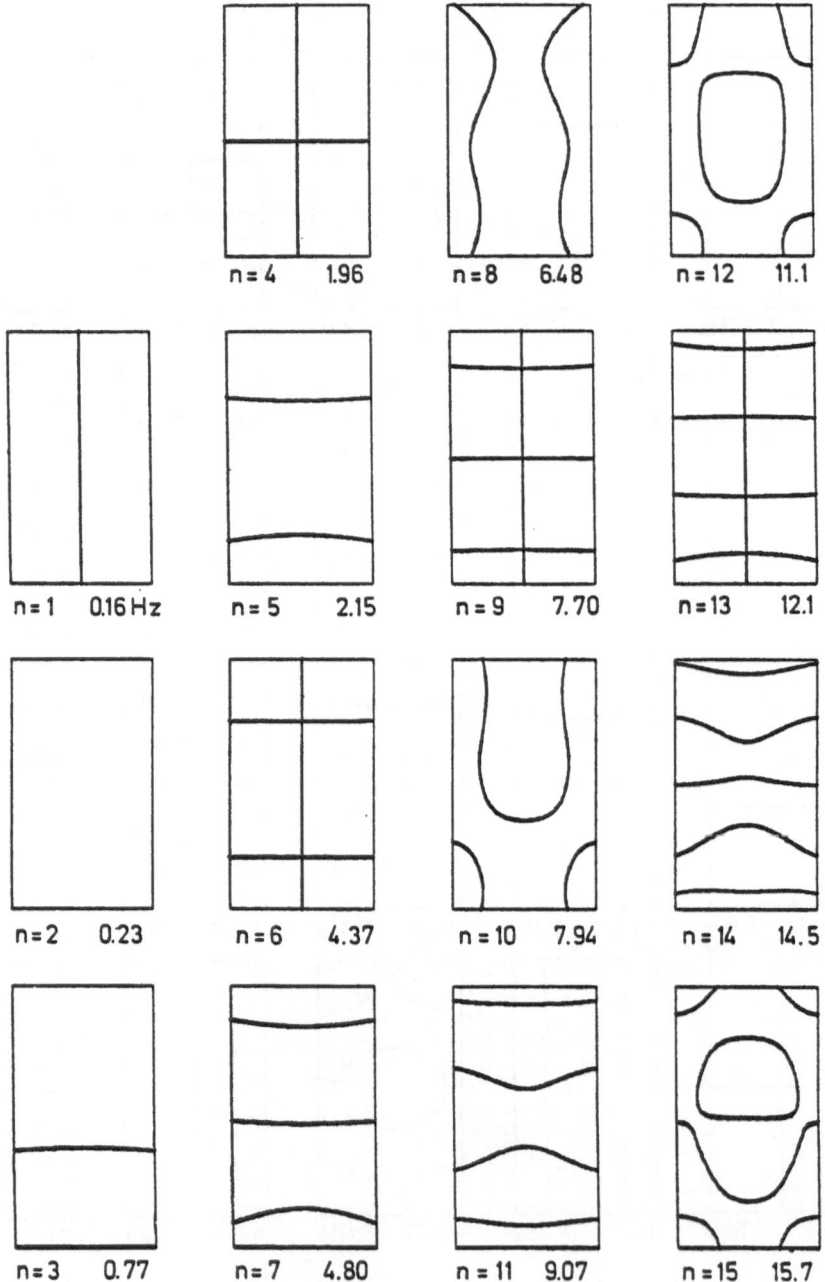

Figure 5: The lowest 15 mode shapes (nodal pattern) of the finite element model with the two concentrated masses

References

[1] Balas, M. J., "Trends in Large Space Structure Control Theory: Fondest Hopes, Wildest Dreams", *IEEE Trans. Automatic Control,* Vol. 27, 1982, pp. 522-535.

[2] Aubrun, J.-N., "Theory of the Control of Structures by Low-Authority Controllers", *AIAA J. Guidance and Control,* Vol. 3, Sept.-Oct. 1980, pp. 444-451.

[3] DFVLR/Dornier System, "Study on Investigation of the Attitude Control of Large Flexible Spacecraft", *ESTEC Contract No. 5310/82/NL/BI, Phase 1 and 2 Final Report, 1984 and 1985.*

[4] Schaechter, D.B., "Hardware Demonstration of Flexible Beam Control", *AIAA J. Guidance and Control,* Vol. 5, 1982, pp. 48-53.

[5] Schäfer, B., Holzach, H., "Experimental Research on Felxible Beam Modal Control", *AIAA Dynamics Specialists Conference,* Paper No. 84-1020 CP, Palm Springs, Ca., May 17-18, 1984. (Appears also in *J. Guidance, Control and Dyanmics,* July-Aug. 1985.)

[6] Flotow, A. H. von, Schäfer, B., "Experimental Comparison of Wave-Absorbing and Modal-Based Low-Authority Controllers for a Flexible Beam", *AIAA Guidance, Navigation and Control Conference,* Snowmass, Colorado, Aug. 19-21, 1985.

[7] Aubrun, J.N., Ratner, M.J., Lyons, M.G., "Structural Control for a Circular Plate", *J. Guidance, Control and Dynamics,* Vol. 7, Sept.-Oct. 1984, pp. 535-545.

[8] Montgomery, R.C., Sundararajan, N., "Experiments in Structural Dynamics and Control Using a Grid", Workshop on Identification and Conrol of Flexible Space Structures, San Diego, Ca., June 4-6, 1984.

[9] Hallauer, W.L., Skidmore, G.R., Gehling, R.N., "Modal-Space Active Damping of a Plane Grid: Experiment and Theory", *AIAA Dynamics Specialists Conference,* Paper No. 84-1018 CP, Palm Springs, Ca., May 17-18, 1984.

[10] Schäfer, B., "Free Vibrations of a Gravity-Loaded Clamped-Free Beam", *Ingenieur-Archiv,* Vol. 55, 1985, pp. 66-80.

[11] Schäfer, B., "Dynamical Modelling of a Gravity-Loaded Rectangular Plate as a Test Configuration for Attitude Control of Large Space Structure", 35th Congress of the Int. Astronautical Federation, Paper No. IAF-84-391, Lausanne, Switzerland, Oct. 7-13, 1984.

[12] Timoshenko, S., Woinowsky-Krieger, S., "Theory of Plates and Shells", McGraw-Hill Int. Book Comp., 2nd Edition, Tokyo, 1981.

[13] Leissa, A.W., "Vibration of Plates", NASA Report SP-160, 1969.

[14] Collatz, L., "Eigenwertaufgaben mit technischen Anwendungen", Akademische Verlagsgesellschaft, Leipzig, 1949.

[15] IMSL, "Library Reference Manual", Customer Relation Houston, Texas, Edition 9.2, Nov. 1984.

[16] ASKA, "Automatic System for Kinematic Analysis, User's Reference Manual, Part I, II and III", Stuttgart, FRG, 1979.

ACTIVE CONTROL OF FLOATING STRUCTURES

M. Shinozuka, E. Samaras and C. Paliou
Civil Engineering and Engineering Mechanics
Columbia University
New York, New York, 10027, USA

1. Introduction

Many floating and semi-floating structures exist at present, from small heliports to large offshore drilling platforms. Furthermore, a wide variety of floating structures have been proposed, including offshore nuclear power plants, high-rise housing, and even airports. These applications have been proposed for floating structures since they would utilize space not otherwise available, and they would be isolated from earthquakes. Such structures are, however, subject to waves, which provide a continuous disturbance.

The purpose of the research reported here is to demonstrate experimentally the effectiveness of active control to reduce the influence of waves on the motion of such floating structures. The type of structure chosen was an open-bottom structure supported by a cushion of trapped air underneath. This configuration is particularly natural for active control using a system that can sense the structure height and add or remove air from the open-bottom chamber to correct for any heave motion. Use of more than one open-bottom chamber allows control of not only heave motion, but pitch and roll motions as well.

The experimental demonstration developed here involves a cylindrical open-bottom object floating in a water tank and being controlled to decrease heave motion in the presence of wave disturbances. A wave generator is used to subject the object to various different wave conditions and evaluate the effectiveness of various potential control system designs. In designing the control system,

system identification is performed on the plant and actuator hardware using impulse response, step response, and frequency response techniques. Because of several undesirable properties of the determined actuator dynamic behavior, considerable effort was expended to develop control compensation techniques to improve the actuator characteristics. Some and perhaps all of this effort could have been avoided by searching for different hardware with different properties, but this of course requires more experiments for identification and was not done. With the compensated actuator, a control law was developed with good phase and gain margins and good disturbance response characteristics. The final controller design is demonstrated in the experiments to be an effective method of controlling the influence of heave disturbances to the structure.

2. Description of Experimental Apparatus

Heave control for structures such as heliports or offshore drilling platforms requires obtaining height measurements for feedback. In some applications it might be possible to make direct height measurements relative to a fixed body or the ocean floor. Otherwise, all measurements can be taken totally from instruments on board. An expensive approach is the use of an inertial platform; a more practical method is to use some of the wave height measurement instrumentation, developed for use in buoy sensors, which are pendulum-like devices that are nearly neutrally buoyant in a liquid-filled chamber. In either case, the device would automatically give feedback data that could be used for pitch and roll control in addition to the data needed for heave control. For purposes of the laboratory experiment, the heave motion was sensed quite simply by an LVDT connected to a fixed base.

The method considered for effecting the height of the floating structure is to adjust the air pressure in the open-bottom chamber supporting the structure. In actual implementation, the structure would have on-board a compressor and perhaps a high pressure air reservoir to add pressure to the chamber when the control system needs to raise the structure, and air would be vented to the atmosphere or to a low pressure reservoir when the structure needs to be lowered. In such a system no air would have to be pumped through the system when there are no waves producing heaving motion.

Because of the small scale of the laboratory experiment, it was more convenient to use the fixed air pressure supply available in the laboratory. This pressure supply was connected directly to the open-bottom chamber beneath the floating object. The air from the chamber was then vented to the atmosphere through a butterfly or venturi type valve whose angle was adjusted to obtain the desired height of the body. This arrangement conforms to the basic concept of heave control of floating structures, but differs slightly from the most likely configuration described above in that air is continually flowing through the chamber when there are no waves producing heaving motion.

Figure 1 shows the experimental system. The floating structure is a 10 lb. cylindrical plastic object, 17 inches high and 10 inches in diameter, with an air chamber 12 inches high and 6 inches in diameter. It floats in a cylindrical iron water tank, 2 foot square by 30 inches tall. The wave generator is a styrofoam ring with outer diameter 22 inches and inner diameter 12 inches. As the ring is moved up and down, heaving motion is imparted to the water in the tank, which in turn affects the floating structure.

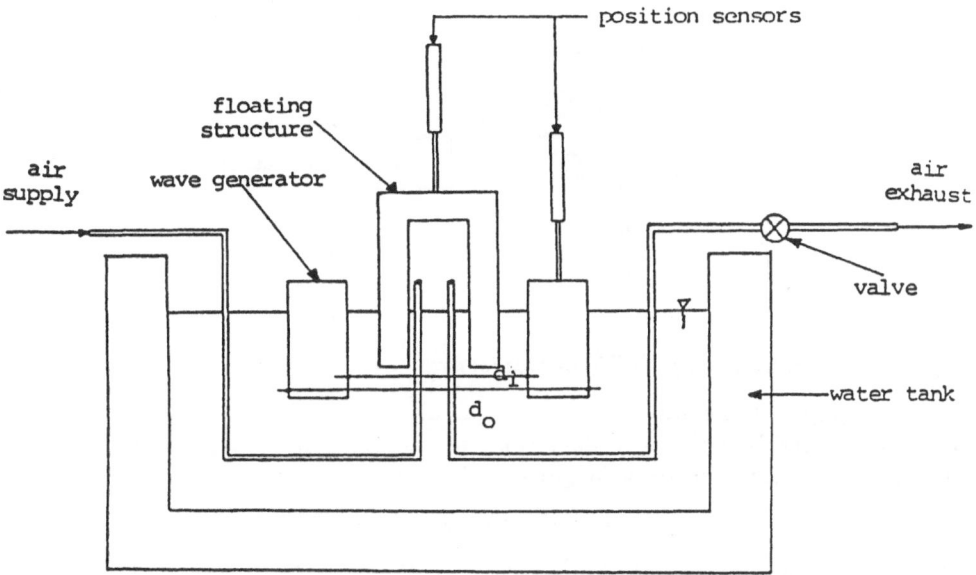

Figure 1: Experimental apparatus

The LVDT that sensed the height of the floating object was mounted as shown in Figure 2. The object was capable of some rolling or pitching motion as well as horizontal translation. Tethers prevented extreme translation, but were maintained slack during all experiments. To prevent binding of the LVDT shaft due to any such motions, even though small in amplitude, the point of contact between the LVDT stem and the floating object was on a glass plate with minimal friction between the two.

The output of the LVDT is a voltage that is proportioned to the input height when operating in the linear range. The size of this range as well as the proportionality constant depends on the input impedance of the device to which the LVDT is connected. In this application, the proportionality constant was

$$k_f = 2.3 \text{volts/inch} \tag{1}$$

and the linear range was sufficiently generous to cause no problems when operating about 0 volts.

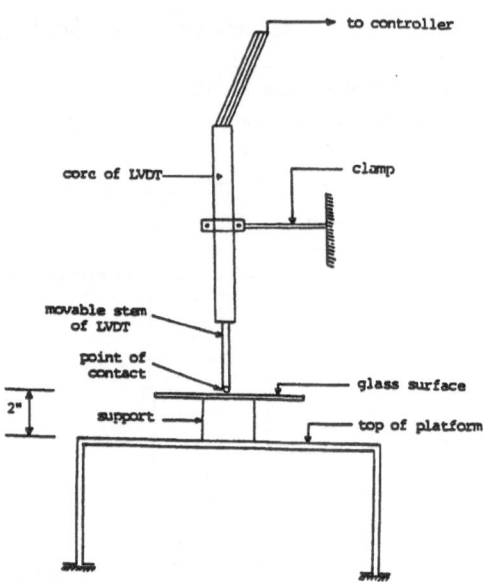

Figure 2: Attachment of LVDT to platform

The height of the wave generator was also sensed by an LVDT with a low pass filter added in series to reduce some of the noise in the signal, producing an overall transfer function from height to voltage:

$$H_1(s) = k_w \left(1 + s/2000\right)/\left(1 + s/54\right) \; ,$$

with gain $k_w = 5$ volts/inch and a linear range of -0.5 volts to +4.0 volts.

The butterfly valve on the exhaust line from the open-bottom chamber was actuated with an AC servo motor described in Figure 3. A DC command signal from the controller is sent to the amplifier which transforms the signal to AC to supply the motor. The shaft of the motor is coupled to the shaft of the butterfly valve, whose angular position is sensed by a rotary potentiometer, which in turn sends a DC feedback signal back to the amplifier to compare with the command signal. This servomechanism was conveniently available to the authors, and would appear on first examination to be ideal for the purpose at hand. During the course of the research it was found to have several undesirable dynamic properties which are described later, that complicated the control system design process.

The butterfly valve itself contains a geometric nonlinearity relating valve angle and area of the valve opening. The nominal valve opening for a given height of the floating object is not zero, since there is a steady flow of air through the system. For reasonable variations in valve angle about this point, the relationship can be linearized and linear control theory used to design the controller. An improved design would use a different type of valve to eliminate

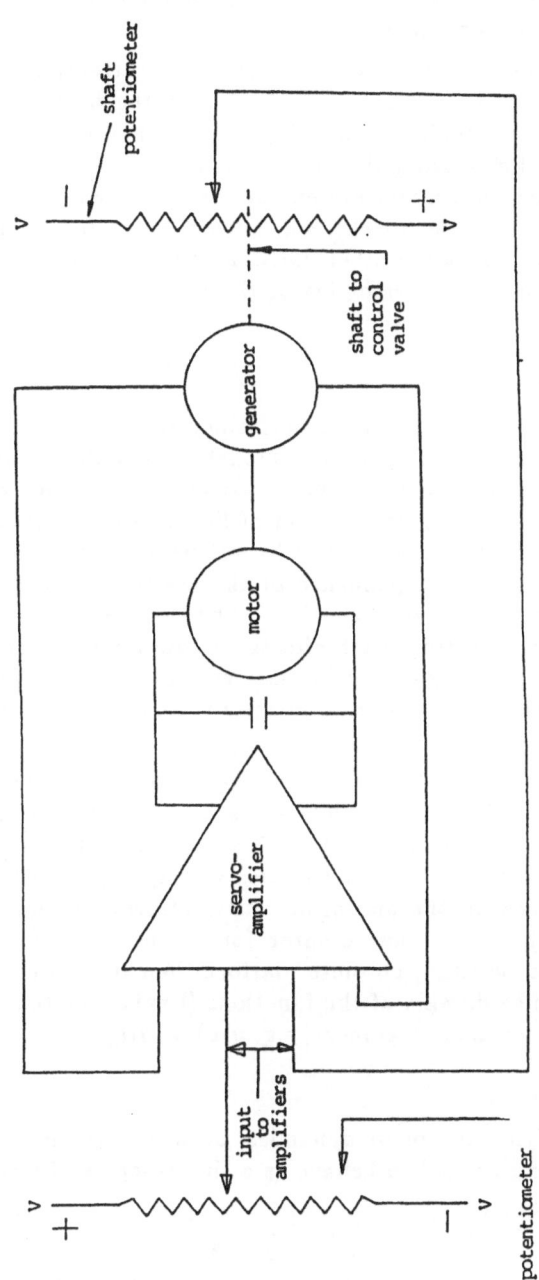

Figure 3: Motor-Amplifier Circuit

nonlinear effects from this source. Some nonlinear behavior in the overall system
was observed, but most of this was during large amplitude motions that one
might expect to go outside the linear range.

Another nonlinear phenomenon that required some attention during experi-
mentation is the fact that there are two angles that the butterfly valve could
assume for a given valve opening, depending on whether the valve is rotated
clockwise or counterclockwise from the closed position to reach the needed area
of valve opening. Hence, a control system designed for control about one of
these positions (with negative feedback), will have positive feedback when inad-
vertently used to control about the other valve position as a result of the method
of start-up of the system. This scrambling of the error signal can make the system unstable.

3. System Identification

This section is devoted to the experiments performed to identify the actuator and
plant dynamics, and obtain a mathematical model of how the inputs to the AC
servo motor, and wave disturbance inputs, affect the floating object. It is
assumed that this model can be given as a set of linear differential equations with
constant coefficients. Experimental studies by others have shown the equations
of floating objects to exhibit a dependence of the coefficients of the differential
equations on a frequency of wave excitation. The assumption here is that the
frequency dependence arising from such effects as added mass is negligibly small
over the range of frequencies involved in the experiments, and this assumption
allows us to apply the methods of classical control theory to the problem.

Figure 4 gives a block diagram of the complete feedback system together
with indications of the meaning of the variables r, u, d and y. The parts of the
system to be identified are the blocks labelled actuator and platform dynamics.
The actuator dynamics, denoted by the transfer function $H_a(s)$, includes the AC
servo motor and the butterfly valve, and relates voltage inputs u to the servo to
the resulting force applied on the platform. The platform dynamics, denoted by
transfer function $H_p(s)$, relates the actuator forces and wave forces d, on the
floating platform to the resulting absolute platform height y, which is the quantity to be controlled. The domain of the functions (Laplace or time) will be indicated, where necessary, by their argument, e.g. $y(s)$ or $y(t)$.

3.1. Plant identification

For purposes of identifying the plant dynamics, both impulse and step responses
were used. The plant is assumed to behave as a single-degree-of-freedom system.

$$\ddot{y} + 2\varsigma_p \omega_p \dot{y} + \omega_p^2 y = \frac{1}{m}F,$$

where F is the total applied force and m is the mas. This expression would be
considerably more complicated if added mass effects were considered and a

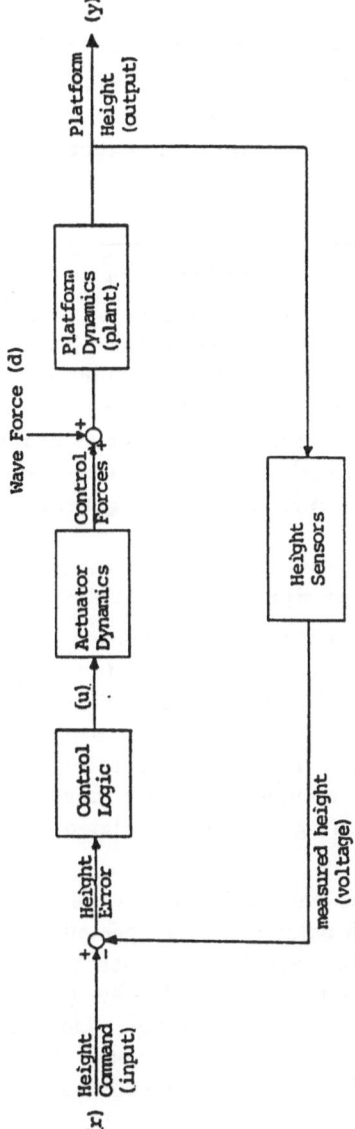

Figure 4: Block Diagram of Experimental Feedback Control System

frequency response method of identification would then be more natural.

The plant was studied from our wave generator as disturbance input to the platform height as output. An impulsive disturbance was applied yielding a decaying sinusoidal output whose ith peak R_i and peak time t_i were recorded for nine successive peaks. From this the undamped natural frequency ω_p (and the damped natural frequency ω) and damping ratio can be determined. The results were $\omega_p = 8.4$ rad/sec and $\varsigma_p = 0.03$.

To complete the determination of the transfer function treating the height of the wave generator ring as the input:

$$H_p(s) = \frac{k_p \omega_p^2}{s^2 + 2\varsigma_p \omega_p s + \omega_p^2} \, , \tag{2}$$

required obtaining the DC gain k_p which was done by applying a step distubance to the system and observing the final value of the platform height $y(t)$. Using the final value theorem this height is k_p, and was found to be 2.8 inches of output height per inch of ring motion. Note that the stability properties of the feedback system are influenced by the product of the gains of the plant and actuator, and this product is measured directly in future sections. Hence, the actual value of k_p is of little importance, and in later computer generation k_p was set to 1 giving unity DC gain.

3.2. Identification of actuator dynamics

Since it is difficult to directly test the transfer function from AC motor input voltage to force on the floating platform, the transfer function from input voltage to platform response (H_s) was tested and identified. Then, knowing H_p from above means that the desired actuator transfer function is obtainable as $H_a = H_s / H_p$. Since H_p was a transfer function from wave generator ring position to platform height, the $H_a(s)$ that results is actually a transfer function from voltage input of the AC motor to an equivalent ring position associated with the applied force to the platform.

Frequency response tests as well as DC tests were made for $H_s(s)$. Figure 5 displays the DC behavior which exhibits an unexpected characteristic - when the input goes from 0 to -0.19 volts, back to zero, then to +0.19 volts and back to zero, the output fails to return to its original value (-0.29 volts vs. -0.19 volts) and therefore depends on the history of the input. This effect was traced to the motor/valve system, but the exact cause was unclear. Since the remainder of the experiments were performed about a fixed set point, and since the main aim of the research was to demonstrate controller attenuation of wave disturbances, the DC dependence on the history of the input was not considered further in the control design.

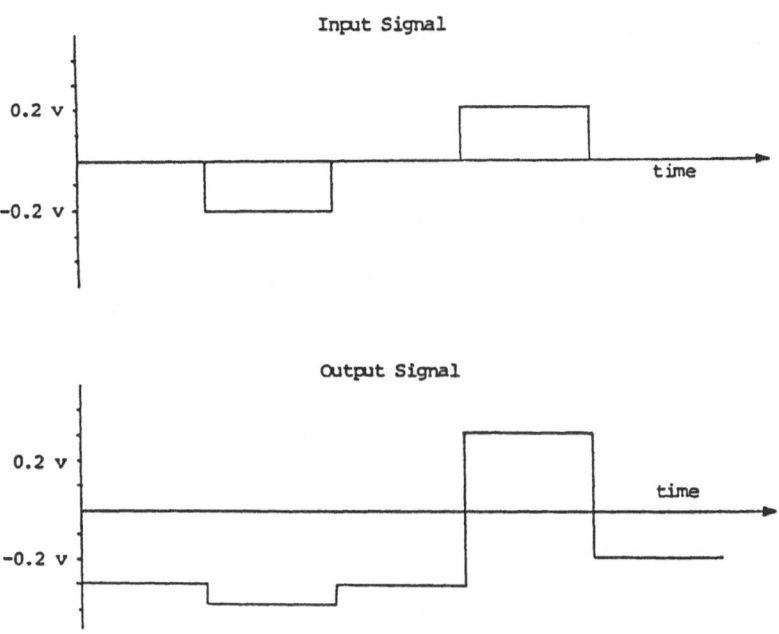

Figure 5: Nonlinear system DC behavior (Transients not shown)

Identification of $H_s(s)$ about the set point was made by experimentally determining the Bode plot of output-over-input amplitude ratio for sinusoidal inputs over the range of interest. The transfer function is then

$$H_s(s) = \frac{k_s}{(s + p_a)(s^2 + 2\varsigma_m \omega_m s + \omega_m^2)} . \qquad (3)$$

The gain $k_s = 1.26$. The two complex poles are due to the motor and can be determined more accurately, including the damping ratio ς_m, by tests on the motor alone. This is done in the next section for purposes of designing a compensator for the motor. It is not clear where the pole p_a is from (there may be some delay from the valve piping), but the presence of additional dynamics is not unusual.

Having determined $H_s(s)$, the actuator dynamics are then given by

$$H_a(s) = \frac{H_s(s)}{H_p(s)} = \frac{k_a(s^2 + 2\varsigma_p \omega_p s^2 + \omega_p^2)}{(s+p_a)(s^2 + 2\varsigma_m \omega_m s^2 + \omega_m^2)} \ , \tag{4}$$

where, $\omega_m = 25.1$ radians/sec (4 Hz), $p_a = 1.9 \ \text{sec}^{-1}$ (0.3 Hz), and

$$k_a = k_s /(k_p \omega_p^2).$$

This contains the surprising result that the resonant peak of the platform frequency response is cancelled by the zeros of the above actuator transfer function. This is an unfortunate circumstance that complicates the controller design. (Note that in many of the computer simulations made, the zeros of $H_a(s)$ had an undamped natural frequency of 8.2 radians/sec, instead of 8.4 for the platform, and a damping ratio of 0.03. These numbers came from a previous test of the actuator system, and do not produce exact cancellation, although the effect on the results is not significant.)

4. Compensation and Controller Design

The actuator, consisting of the AC motor (Figure 4) and the butterfly valve with high pressure supply, has two very undesirable characteristics. One is that the motor resonance is at low frequency and poorly damped, producing oscillatory responses to command signals, and the second is the presence of zeros that cancel the platform dynamics. The motor system was an off-the-shelf item available to us in the laboratory. We entertained the question of whether to search for a commercially available product with better characteristics, but due to the time invested in the present set-up, decided to continue with the equipment at hand.

The coinciding of the zeros of $H_a(s)$ and the poles of $H_p(s)$ means that the actuator attenuates controller actions at the platform natural frequency. Wave disturbances at this frequency become very hard to control since the transfer function relating disturbances to outputs (from Fig. 4) is

$$\frac{H_p(s)}{1+H_i(s)H_s(s)} = \frac{(k_p \omega_p^2)(s+p_a)(s^2 + 2\varsigma_m \omega_m s + \omega_m^2)h_2(s)}{(s^2 + 2\varsigma_p \omega_p s + \omega_p^2)[(s+p_a)(s^2 + 2\varsigma_m \omega_m s + \omega_m^2)h_2(s) + Kk_s h_1(s)]} \ ,$$

where $H_i(s) = kh_1(s)/h_2(s)$ is the control logic block of Figure 4 which is yet to be specified. The transfer function still contains the plant's characteristic equation as one of the factors of its denominator (or closed loop characteristic equation) in spite of the feedback. Due to this effect, the control objective for the experiment was limited to controlling the influence of wave disturbances at frequencies below the plant natural frequency. There are, however, methods of increasing the controller bandwidth to include frequencies above the plant natural frequency. For example, $h_2(s)$ can contain the plant dynamics as one of its factors to build up the control signal at the resonant frequency of the platform, and simultaneously allow one to affect all the (non-cancelled) roots of the characteristic equation. This could be accomplished with either a passive filter or an active filter containing operational amplifiers.

The other undesirable property of the actuator, the poorly damped oscillations of the AC motor, was eliminated by adding a compensating feedback circuit around the motor, as described below.

4.1. Compensation of motor dynamics

Tests of the motor alone (Figure 3) showed that the transfer function from input voltage command u to butterfly valve angle θ_v as output could be modeled as a second-order system

$$H_m(s) = \frac{k_m \omega_m^2}{s^2 + 2\zeta_m \omega_m s + \omega_m^2} \cdot \tag{5}$$

The parameter values depend on the voltage supplied to the rotary potentiometer. The larger voltages gives more corrective action for a given input signal, but makes the system less stable and makes it exhibit a higher natural frequency. A potentiometer voltage of ± 10 volts was used.

A cascade compensator with a feedback loop was used which cancels the undesirable resonant behavior, and substitutes some better system dynamics (since no physical system can supply pure zeros without supplying poles as well). This is shown in Figure 6a where the cascade compensator transfer function is given by

$$H_c(s) = \frac{k_c(s^2 + 2\zeta_m \omega_m s + \omega_m^2)}{(s + p_1)(s + p_2)} \, , \tag{6}$$

which is realized with passive elements in a bridged T-circuit (see Figure 6b) followed by an amplifier. The resistor and capacitor values used were $C = 1.5$ μf, $R_1 = 1.5 \times 10^3 \Omega$, $R_2 = 10^5 \Omega$, and $R_L = \infty$. Then the expected closed-loop transfer function for the compensated AC motor, Figure 6a, is given by

$$H_{cm}^e(s) = \frac{H_c(s)H_m(s)}{1 + H_c(s)H_m(s)}$$

$$= \frac{k_c k_m \omega_m^2}{(s + p_1)(s + p_2) + k_c k_m \omega_m^2} \cdot$$

After assembling the hardware for the compensated motor, the magnitude vs. frequency Bode plot was obtained experimentally to identify the time transfer function for the new system (Figure 7). The extra compensation has accentuated some part of the system behavior that was not observed in previous experiments, so that the transfer function model obtained from the plot is now third-order

$$H_{cm}(s) = \frac{k_{cm}}{(s + p_{cm})(s^2 + 2\zeta_{cm} \omega_{cm} s + \omega_{cm}^2)} \, , \tag{7}$$

where $p_{cm} = 20$ rad/sec, $\omega_{cm} = 11$ rad/sec (1.75 Hz), $\zeta_{cm} = 0.45$, and $k_{cm} = 7.93$. Then the compensated actuator dynamics are given by

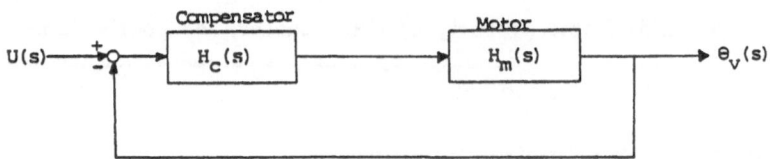

a)   Compensated Valve Control System

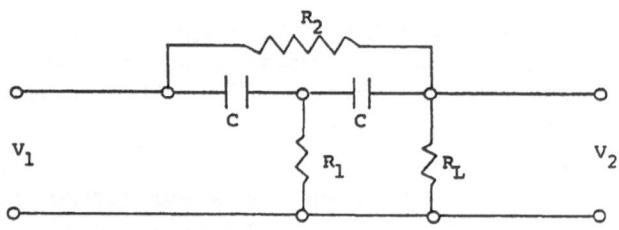

b)   Bridged T-Circuit for $H_c(s)$

Figure 6: AC motor compensation

$$H_{ca}(s) = \frac{k_{ca}(s^2 + 2\varsigma_p \omega_p s + \omega_p^2)}{(s + p_a)(s + p_{cm})(s^2 + 2\varsigma_{cm} \omega_{cm} s + \omega_{cm}^2)} \qquad (8)$$

from Eqs. 7 and 4.

The aim of the compensation was to improve the dynamic behavior of the motor, and in this the design was successful, considerably improving the motor's damping.

4.2. Controller design

The compensated actuator transfer function, Eq. 8, represents the new actuator dynamics block for Figure 4. The platform dynamics are given by Eq. 2, and the LVDT sensor feedback block by Eq. 1. It remains to fill the control logic block with a transfer function $H_l(s)$ which is effective at reducing the influence of wave disturbances on the platform height.

The transfer function of the closed loop system of Figure 4, relating wave disturbance inputs to their influence on platform height, is given by

$$H_d(s) = H_p(s)/[1 + k_l H_l(s) H_{ca}(s) H_p(s)]$$

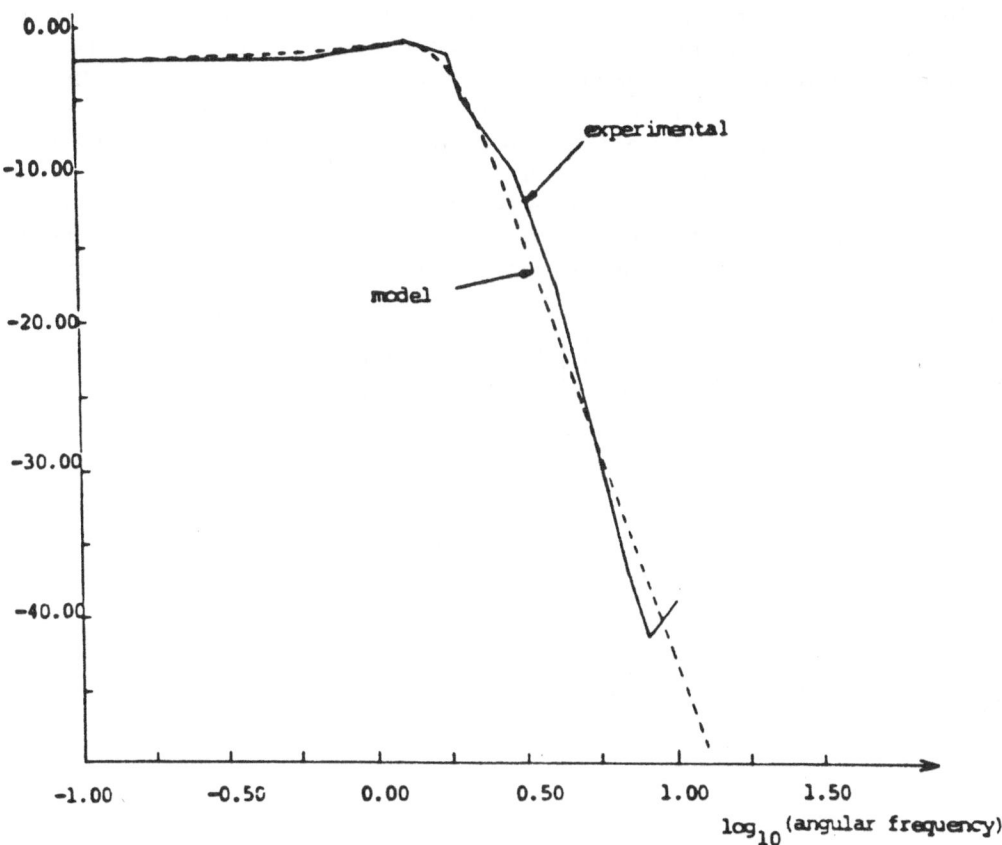

Figure 7: Magnitude vs. frequency bode plot of compensated AC motor

$$= \frac{k_p \omega_p^2 (s+p_a)(s+p_{cm}) g_{cm} h_2(s)}{g_p [(s+p_a)(s+p_{cm}) g_{cm} h_2(s) + K k_f k_{ca} k_p \omega_p^2 h_1(s)]} \tag{9}$$

where g_p and g_{cm} represent the second-order polynomial $s^2 + 2\varsigma \omega s + \omega^2$ with appropriate subscripts supplied. For sinusoidal wave disturbances of radian frequency ω, the ratio of the amplitude of the steady state sinusoidal output to the amplitude of the input is given by $M(\omega) = |H_d(i\omega)|$, and hence the platform height is disturbed less in the steady state as the control gain K is increased. The design process using classical control theory is largely trial and error, relying on the experience of the designer, and on certain general guidelines indicating how proportional, integral, and PID controllers, and lead and lag compensators are likely to affect the transient and the forced response of a system.

Several candidate designs have been proposed for the controller. When a proportional controller is introduced it is found to be inadequate. The use of a lead compensator improves the response by removing the closed loop zero $s + p_a$ in $H_d(s)$. Then an additional lead compensator is introduced to reduce the peak height. The final design adds integral control action which reduces the effect of constant disturbance to zero in the steady state, and has a similar good effect at low frequencies. But it has a detrimental effect on stability, and as a result additional lead compensators were required to maintain stability. The final design frequency response curve in Figure 8 shows the behavior of the controller which is outstanding at low frequencies. In addition, the gain and phase margins, as measures of the degree of stability, were found to be quite good. The transfer function of the controller is:

$$ H_, = K \left[\frac{s+2}{s+20} \right] \left[\frac{s+7}{s+70} \right]^2 \left[\frac{s+7}{s} \right], \qquad K = 150. $$

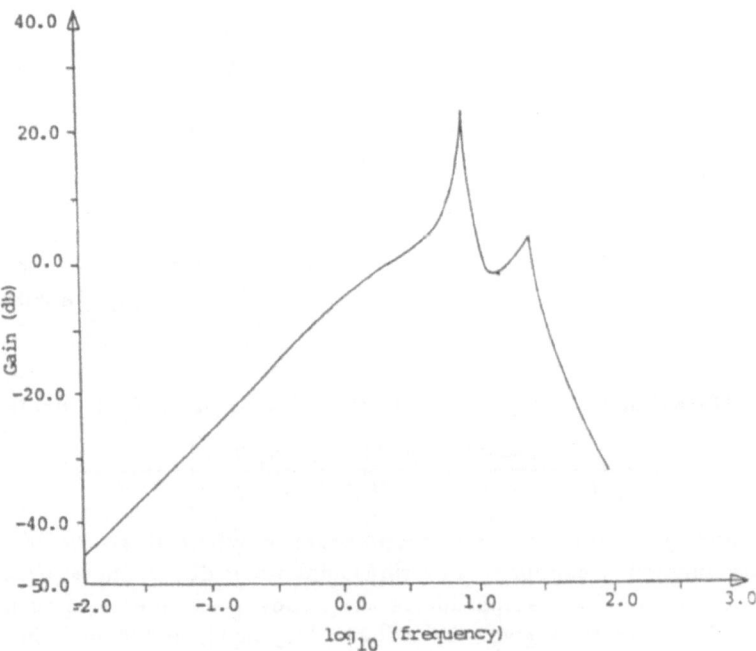

Figure 8: Disturbance response for final control design

5. Experimental Results

The control design of the previous section was implemented using analog electronic devices (operational amplifiers, resistors and capacitors). A diagram of the control circuitry is given in Figure 9. The symbols A1 through A3 represent inverting operational amplifiers with gains of 1, 1/2 or 10, and 2, respectively. AL represents an active lead compensator with transfer function $(s+2)/(s+20)$, while PL represents a passive lead compensator $(s+6.67)/(s+73.3)$. I represents a proportional plus integral term $(s+6.67)/s$. LPF is an active low pass filter $100/(s+200)$ added to reduce the 60 Hz noise in the system.

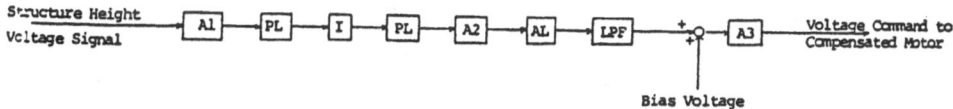

Figure 9: Analog circuitry for control logic block

The experimental results are shown in Figures 11 and 12. The gain indicated is the experimetal output amplitude to the input amplitude. Both input and output signals are roughly sinusoidal with the same frequency. The two control cases A and B, correspond to the A2 amplifier gains of 1/2 and 10, respectively. Both control gains were predicted to have good stability properties based on computer simulations.

The reduction is more pronounced at low frequencies where a reduction of 55-65% of the uncontrolled amplitude is possible at 0.1 Hz. At frequencies close to the natural frequency, the reduction disappears as expected. Control A has less effect on the disturbances, but is more stable than Control B. Control B, with larger control gain, appeared to have poorer stability properties than predicted analytically. With sufficiently large motions it could go into a limit cycle behavior - a nonlinear phenomenon, which cannot be predicted based on a linear model.

The experimental and theoretical gain and phase Bode plots appear in Figure 10. The theoretical gain and phase margins were determined to be 4 and 90° respectively, which are quite good values. Experiments, however, show that the hardware behaves roughly as expected, but differs in the phase angle plot. At the floating structure's natural frequency, the actual phase lead introduced is only 90°, much less than the 125° shift expected. This suggests that a better hardware implementation might produce the expected behavior.

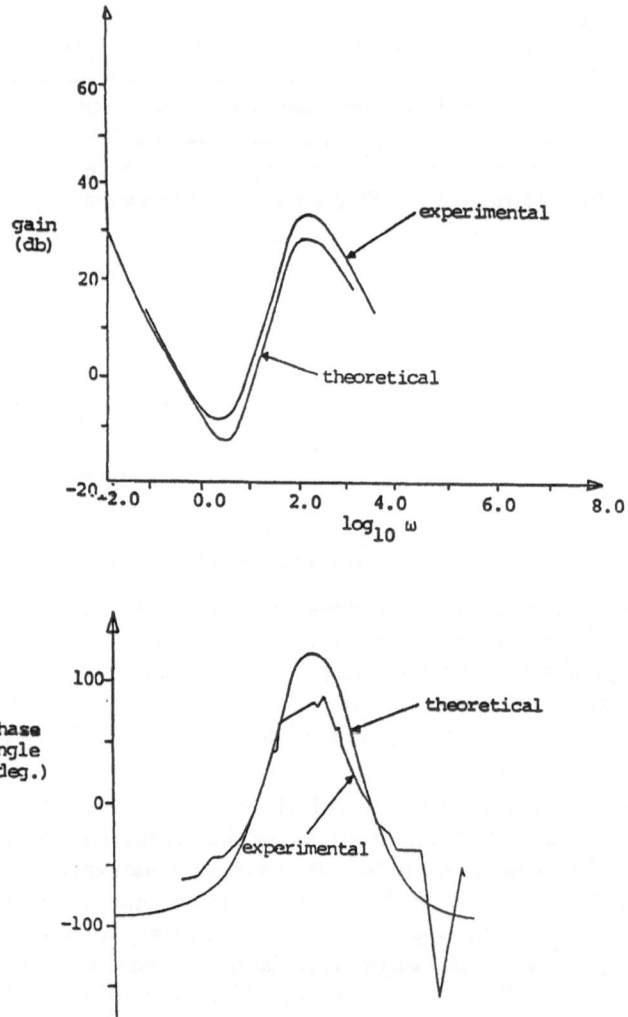

Figure 10: Gain and phase bode plot for controller

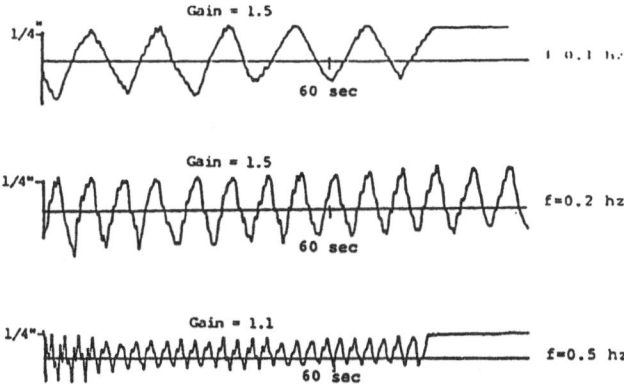

Figure 11: Uncontrolled response to waves

6. Conclusions

It has been experimentally demonstrated that active control of floating structures can substantially reduce wave induced motion at frequencies lower than the structure's natural frequency. Due to the complex nature of the system however, the required control design was not simple. The motor/valve actuator system is clearly a vital part of the control system. In order to further reduce the disturbance effects, it seems likely that still more circuitry is necessary. This could be done using analog devices, but changing to a digital system could be simpler in the long run as it is much more flexible and easier to modify, although it is harder to set up initially. The experiments should also be extended to use a zero steady state flow rate through the open-bottom chamber during undisturbed operation, and extended to more than one chamber in order to simultaneously control heave, pitch and roll. In this process the nonlinear DC behavior should disappear.

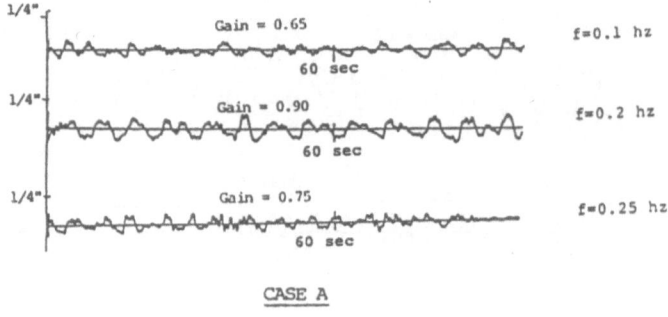

CASE A

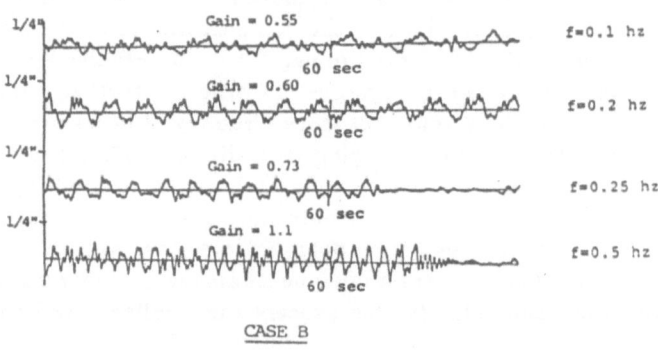

CASE B

Figure 12: Controlled response to waves

A STANDARDIZED MODEL FOR STRUCTURAL CONTROL EXPERIMENTS AND SOME EXPERIMENTAL RESULTS

T. T. Soong, A. M. Reinhorn
Department of Civil Engineering
State University of New York at Buffalo
Buffalo, New York 14260 U.S.A.
J. N. Yang
Department of Civil Engineering
The George Washington University
Washington, D.C. 20234 U.S.A.

1. Introduction

Structural control in civil engineering is an exciting concept. It not only provides an attractive means of enhancing structural safety and serviceability during large loading episodes, but also leads to the notions of 'active' structures whereby structures are designed with active elements in mind - a fundamental departure from past and current passive structural design practices. As Zuk [1] puts it, one sees in active structural control 'a basic change in kind, not just in amounts; as different as an automobile is to a wagon'.

It is gratifying to see that more and more investigators are becoming involved in civil engineering structural control research. One measure of this increase in interest and activity can be seen from annual grant support figures in this area reported by the U.S. National Science Foundation, which shows an increase from a funding level of approximately $40,000 in 1975 to a figure close to $2.5 million in1984. It is also clear from voluminous publications which have appeared over the last few years that much advance has been made. One should note, however, a major portion of these contributions goes to analytical and numerical simulation of control algorithms as applied to control of structures.

As in the normal progression from conceptualization to implementation of a scientific idea, laboratory model analysis is now needed which is based on carefully planned experiments and which can, on the one hand, correlate experimental and anlytical results and, on the other, provide useful extrapolation to a real full-scale structural setting. To be sure, a few experiments have been carried out to date. These include:

(a) Testing of a light beam with control supplies by an on-line microprocessor connected to a mass damper feedback loop [2].

(b) Testing of simple beams using tendons connected to servo-hydraulic actuators [3]. A nonoptimal feedback algorithm was employed. A clamped-clamped beam has also been tested using optimal state feedback implemented by an on-line digitial programmable computer [4].

(c) Wind tunnel testing of an active appendage placed on top of a scaled tall building model and activated by a solenoid coupled to an anlog control circuit operating on an optimal feedback algorithm [5].

(d) Optimal and nonoptimal control experiments of beams using a system of noncontacting optical displacement sensors and electromagnetic actuators [6,7].

(e) Pulse control experiments using servocontrolled air jet generators on a six-story model structure and other flexible mechanical models [8,9].

While results of these experiments shed considerable light on the dynamic behavior associated with an actively controlled structural model, a number of important questions remain unanswered because of one or more of the following:

(a) Models are too simple to infer real structural behavior.

(b) Models are not calibrated or parameters are not properly identified to permit comparison with analytical predictions.

(c) Models are not dynamically similar to 'real' structures, thus making extrapolation to real structural behavior difficult.

(d) Excitations and control inputs are unrealistic. For example, control forces generated in most of these experiments are small, being of the order of one pound or less. On the other hand, one of the distinctive characteristics of civil engineering structural control is that large forces must be generated and they can give rise to a host of problems in a laboratory model experiment.

2. The Standardized Model

This paper describes a research program currently underway. The main objective is to obtain experimental results using carefully selected and calibrated model structures with the aim of evaluating the implementability of active control to structures of this type. More specific goals are:

(a) The development and construction of a structural model. The model is used to simulate single-degree-of-freedom (SDOF) as well as multi-degree-of-freedom (MDOF) dynamic behavior through proper bracing.

(b) A critical examination of control algorithms for experimental purposes using several practical control laws so that hardware requirements and algorithm limitations can be assessed.

(c) Design and experimental verification of an efficient on-line control system for the model structure.

(d) Control optimization from viewpoints of control force requirement, location of sensors and controllers, computing requirements, control and observation spillover compensation, and control effectiveness.

(e) Correlation of experimental and analytical results.

(f) Examination and analysis of experimental results. Assessment of active control applications to real structures.

To accomplish these objectives, considerable effort has been expanded on the development of model fabrication, dynamic test facility and instrumentation.

2.1. Model fabrication and calibration

The model structure was developed with two goals in mind. First, it was to be carefully fabricated and calibrated in order to permit comparisons between analytical and experimental results. Second, it was to be dynamically similar to a prototype whose structural characteristics and response history are accurately known, thus permitting a direct correlation between model and prototype.

A three-story steel frame model structure utilizing artificial mass simulation was selected on the basis of the above-mentioned considerations. It is similar in geometry⁺, material properties and boundary conditions to structural models extensively tested at other institutions [10,11] and it is an approximately 1/4-scale model of a prototype structure (1/2 scale of Berkeley model) which was also extensively tested under seismic conditions [12,13].

A schematic diagram of the model structure is shown in Figure 1. The model is rigidly braced in the direction perpendicular to the motion to prevent lateral side effects. As shown in Fig. 1, the upper floors can also be rigidly

---

⁺One of the dimensions was scaled differently than the overall scale in order to create a model compatible to the operational range of the controlling instrumentation.

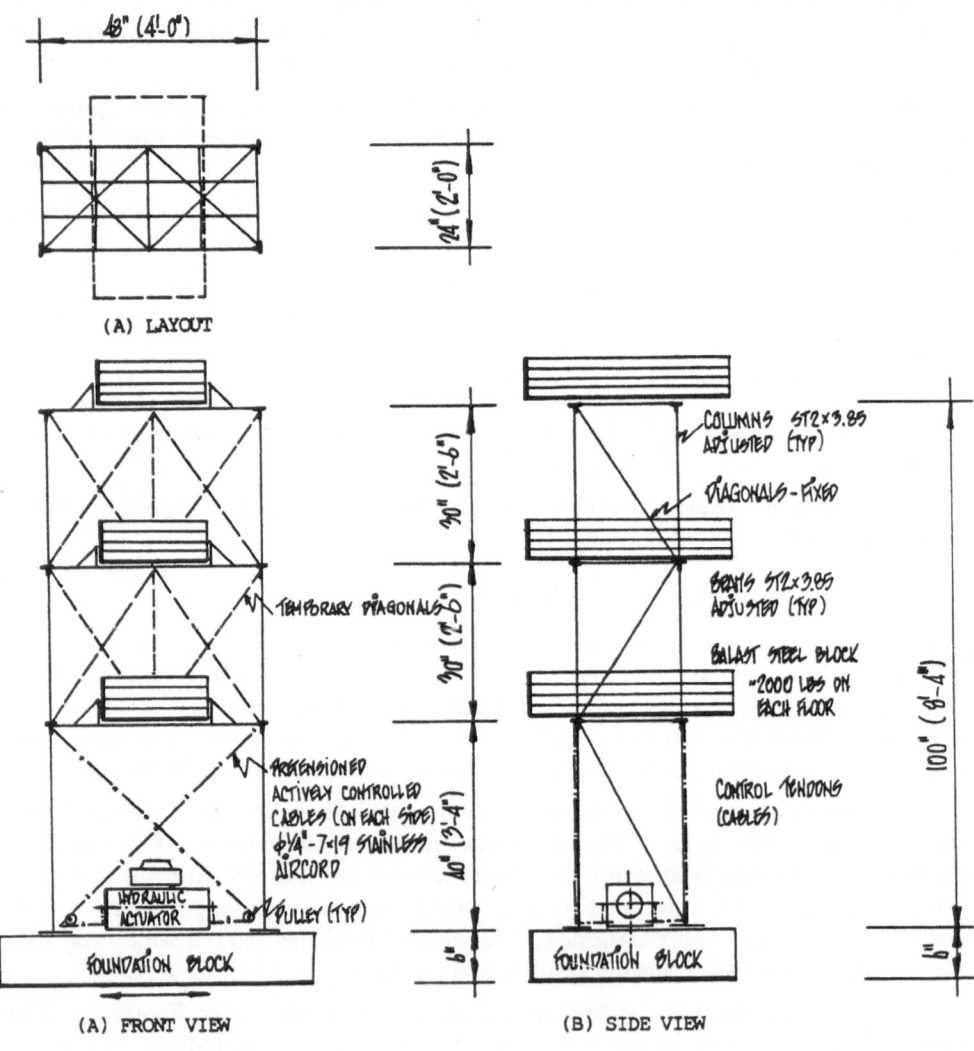

Figure 1: Model Structure - Elevations

braced for modeling a single-degree-of-freedom structure. Figure 2 shows its SDOF configuration in the laboratory.

Some dynamic characteristics of the model structure are shown in Fig. 3b. It can be observed that the SDOF model indeed behaves as a rigid mass supported by first-story columns; higher modes are the result of local beam vibrations.

Figure 2: Braced Model Structure (S.D.O.F.)

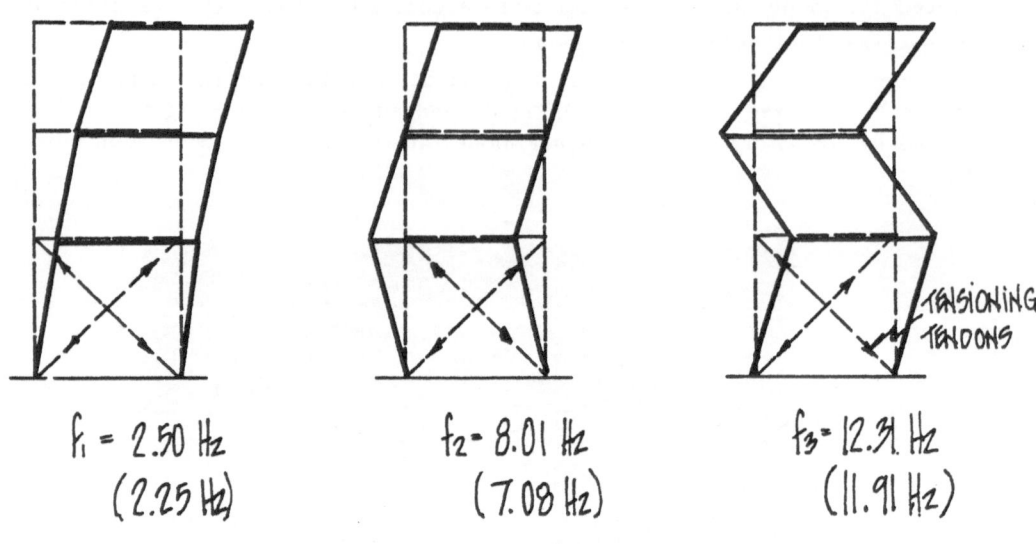

$f_1 = 2.50$ Hz $f_2 = 8.01$ Hz $f_3 = 12.31$ Hz
$(2.25$ Hz$)$ $(7.08$ Hz$)$ $(11.91$ Hz$)$

(A) FREE MODEL – FOR M.D.O.F. TEST

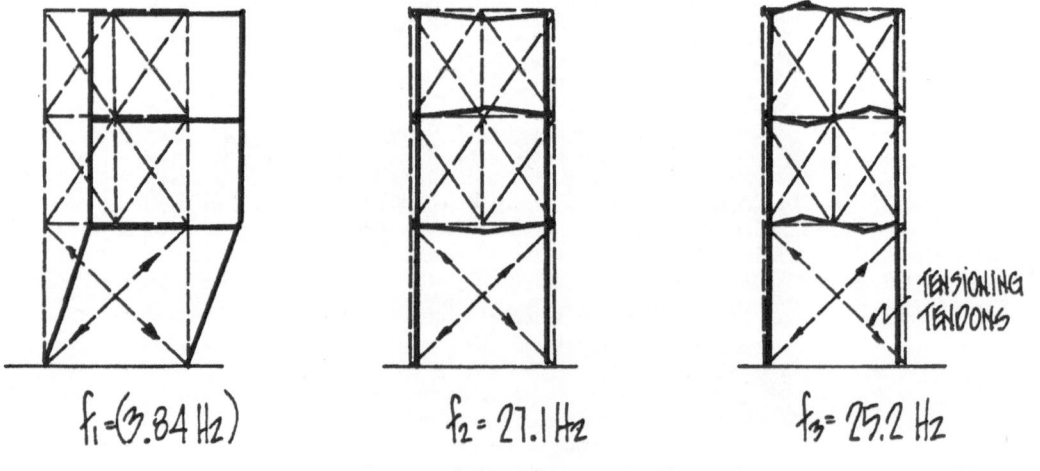

$f_1 = (3.84$ Hz$)$ $f_2 = 27.1$ Hz $f_3 = 25.2$ Hz

(B) BRACED MODEL – FOR S.D.O.F. TEST

Figure 3: *Modal Shapes of Model Structure*
(The values in parenthesis are the testing results.)

Table 1: Parameter values of model structure - MDOF and SDOF

Characteristic	Values	Remarks
(1) (2)	(3)	(4)
1. Height	8'-4"	3'-4"+2'-6"+2'-6"
2. Weight	6449 LBS.	1950 Lbs. per floor+Weight of Structure
3. Natural Frequencies		Braced Model with Cables ('SDOF')
1st Mode	3.84Hz	
2nd Mode	25.20Hz	Local Modes Out of Operation Range
3rd Mode	27.10Hz	
4. Natural Frequencies		Unbraced Model with Cables ('MDOF')
1st Mode	2.25Hz	
2nd Mode	7.08Hz	
3rd Mode	11.91Hz	
5. Structure's Stiffness	9725.84 Lbs./in.	Braced Model ('SDOF')
6. Cables' Stiffness		All Cables
	5958.88 Lbs./in.	Inclined
	3854.50 Lbs./in.	Horizontal
7. Cables' Inclination	36.46° (0.6353 Radians)	
8. Critical Damping	0.5%	Braced Model

Calibration of the model structure was carried out using white noise input and a series of sinusoidal inputs supplied by the base motion. Table 1 shows parameter values of the model structure which are used for analytical calculations. The model has been designed with adjustments which allow the elimination of transverse and torsional vibrations. During the calibration process the natural frequencies of these modes were increased above the operating range.

2.2. Dynamic test facility

An integrated testing system was developed to permit suitable model excitation as well as measurement, data processing, recording and manipulation of control and response parameters. The centerpiece of the facility is a 12' × 12' shaking table. The table has 5 degree of freedom, of which 3 (vertical, lateral and roll) can be individually programmed. The other two are controlled only for correction, and the sixth is constrained by two hydrostatic slide bearings.

Structural models of up to 50 metric tons (110,000 bl) can be shaken with a maximum acceleration of .55g. At less than full load, the maximum acceleration is considerably greater. Figure 4 shows the shaking table system with its four vertical and two lateral actuators and the two slide bearings. The table is driven by two hydraulic power supplies with a combined capacity of 280 gallons per minute. Several on-line accumulators supply high flow demands at peak velocities. The system is capable of developing velocities of 34 in/sec and 71 in/sec in the horizontal and vertical directions, respectively. At low frequencies, the system is limited by strokes of 6 in. laterally and 3 in. vertically. A multigraph of the maximum system performance is given in Figure 5.

The analog control system can simultaneously handle all five independent degrees of freedom. As mentioned previously, only three are programmable and the remaining two are active for correcting any cross coupling between the various degrees of freedom. The analog control system is based on the state-of-the-art, three-variable feedback system illustrated in Fig. 6. This unique feedback arrangement leads to excellent tracking between the table and the command signal.

The frequency limit of the system is defind by the natural frequency of the table and of the supporting oil column, both at approximatley 60 Hz. This allows operations over a wide band of frequencies without substantial error. Figure 7 shows the transfer function for the table, which is essentially flat up to 50 Hz. Besides the analog feedbacks, the system has a dedicated software for prior adjustment of the desired motion by the inverse transfer function of table-model system. The result of the use of the adjusted motion history is a better reproduction of the desired motion with an almost total compensation for the model-table interface. Inputs or command signals to the table can be of the following types: Harmonic motions (sinusoidal, square, triangular), random motions, and any recorded earthquake motion from the PDP 11/34 library of 3000 recorded accelerograms obtained from the Caltech data base. Additional software is available for the collection and processing of test data. Fourier analysis, time analysis, and other on-line procedures can be requested and the results can either be printed or plotted. Data can also be transferred via the TE-16 tape unit to other computers for further processing. Figure 8 illustrates the control and data acquisition system of the earthquake simulation facility.

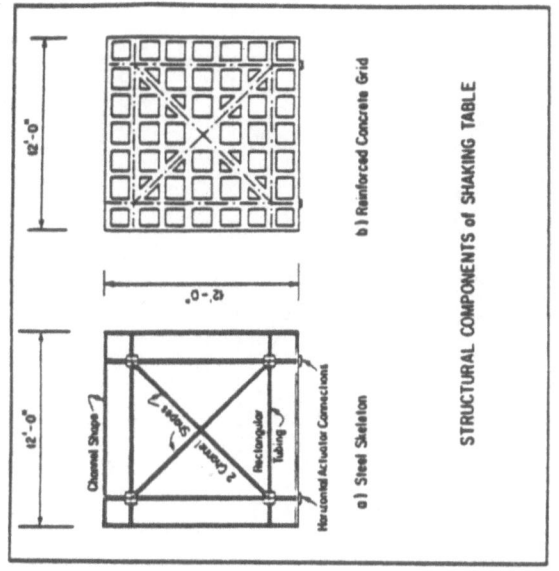

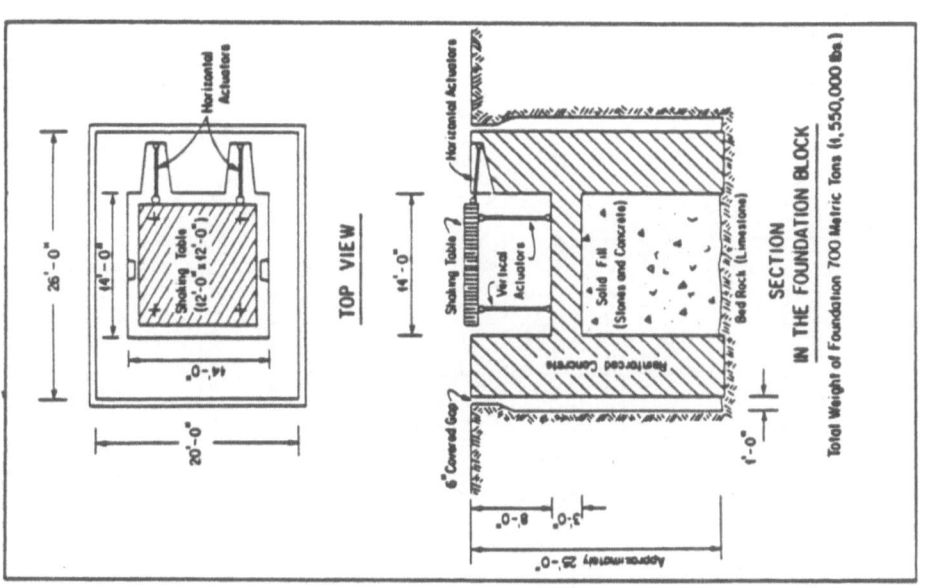

Figure 4: State University of New York at Buffalo earthquake simulation facility

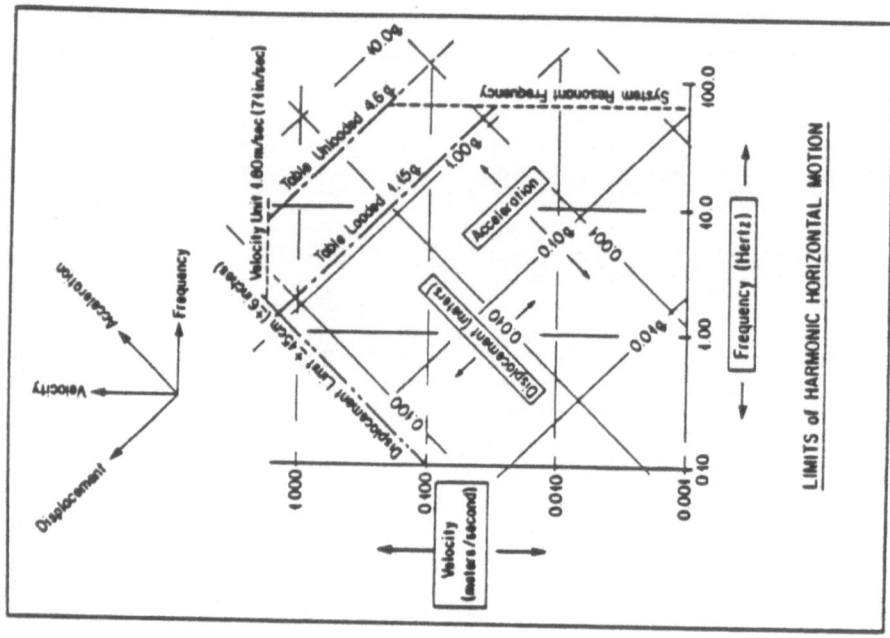

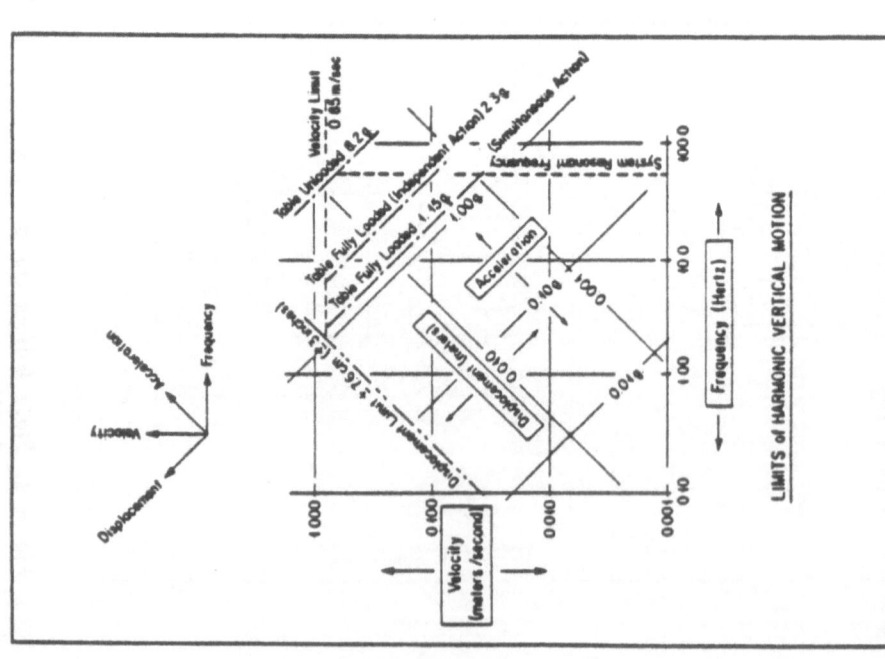

Figure 5: Maximum performance multigraphs for the shaking table in the vertical and horizontal directions

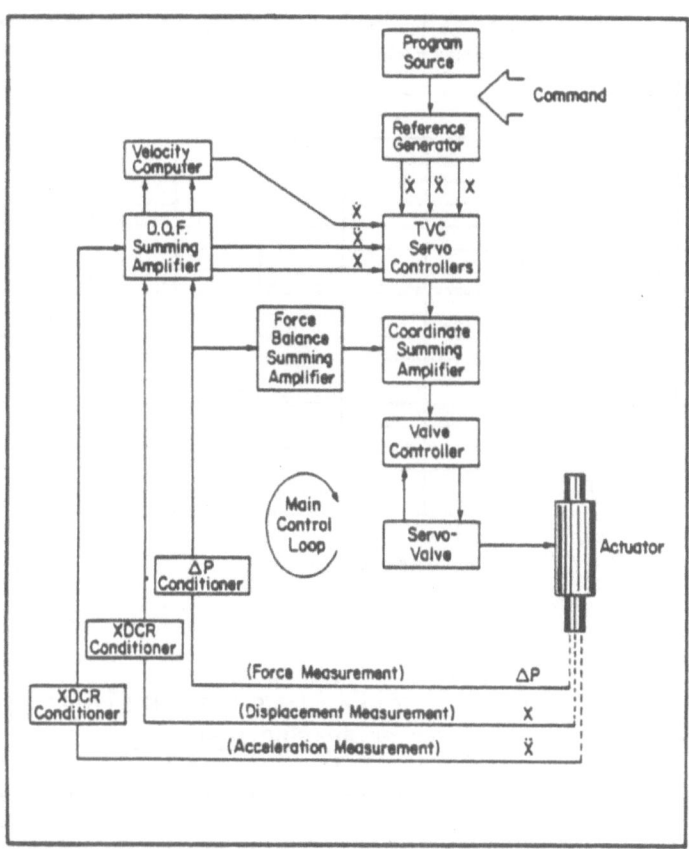

Figure 6: Simplified three-variable control loop

2.3. Actuator and sensing equipment

Control forces are transmitted to the structure through tendon cables operating either in a continuous or discontinuous (pulse) mode. The cables are made of stainless Aircord 7×9-ϕ 1/8 in. with a cyclic stiffness ranging from 73.85 (kips/in.) in to 75.79 (kips/in).in. A 500 lb-capacity hydraulic actuator with an attached servovalve having a maximum flow of 10 gpm was chosen to supply the control forces to active tendons. The maximum speed of the above actuator allows it to develop a force rate of 12.3 kips/sec when the maximum expected force rate is 6 kips/sec. The actuator is equipped with both force and stroke controllers to allow a better stabilization of control forces. The actuator and tendon configuration for the SDOF model test is shown in Figs. 9 and 10.

The sensing equipment for the measurements of displacement, velocity, acceleration and tendon forces have been assembled and calibrated. The instrumentation includes

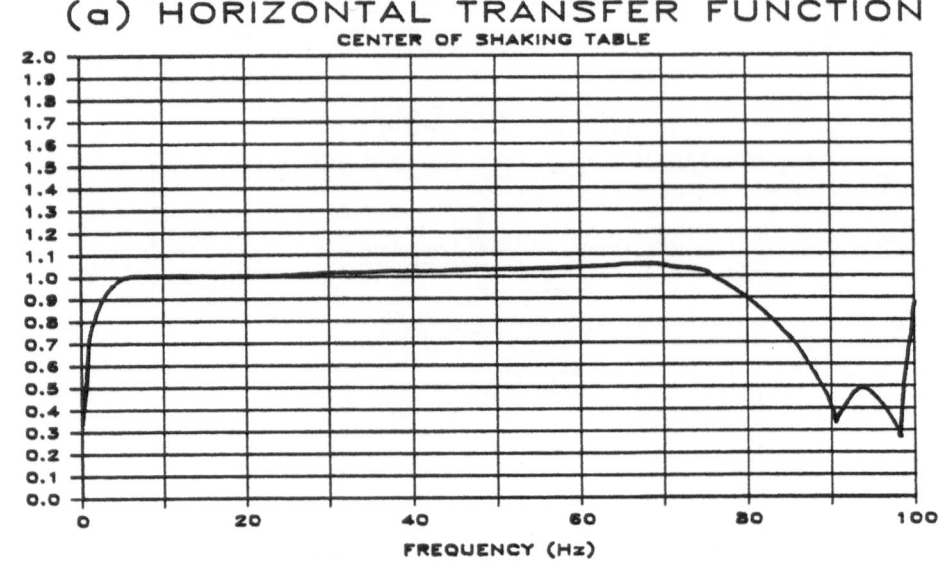

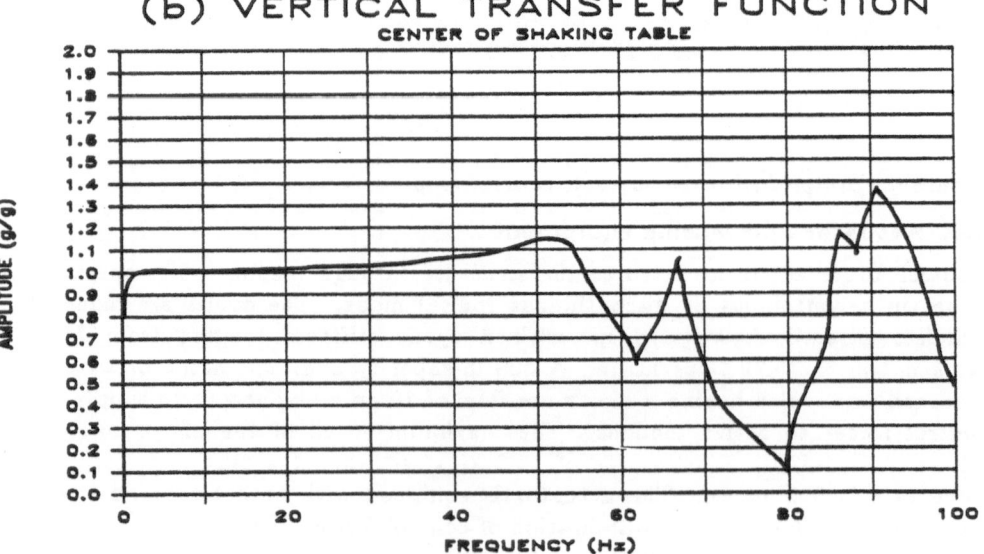

Figure 7: Transfer functions for shaking table.

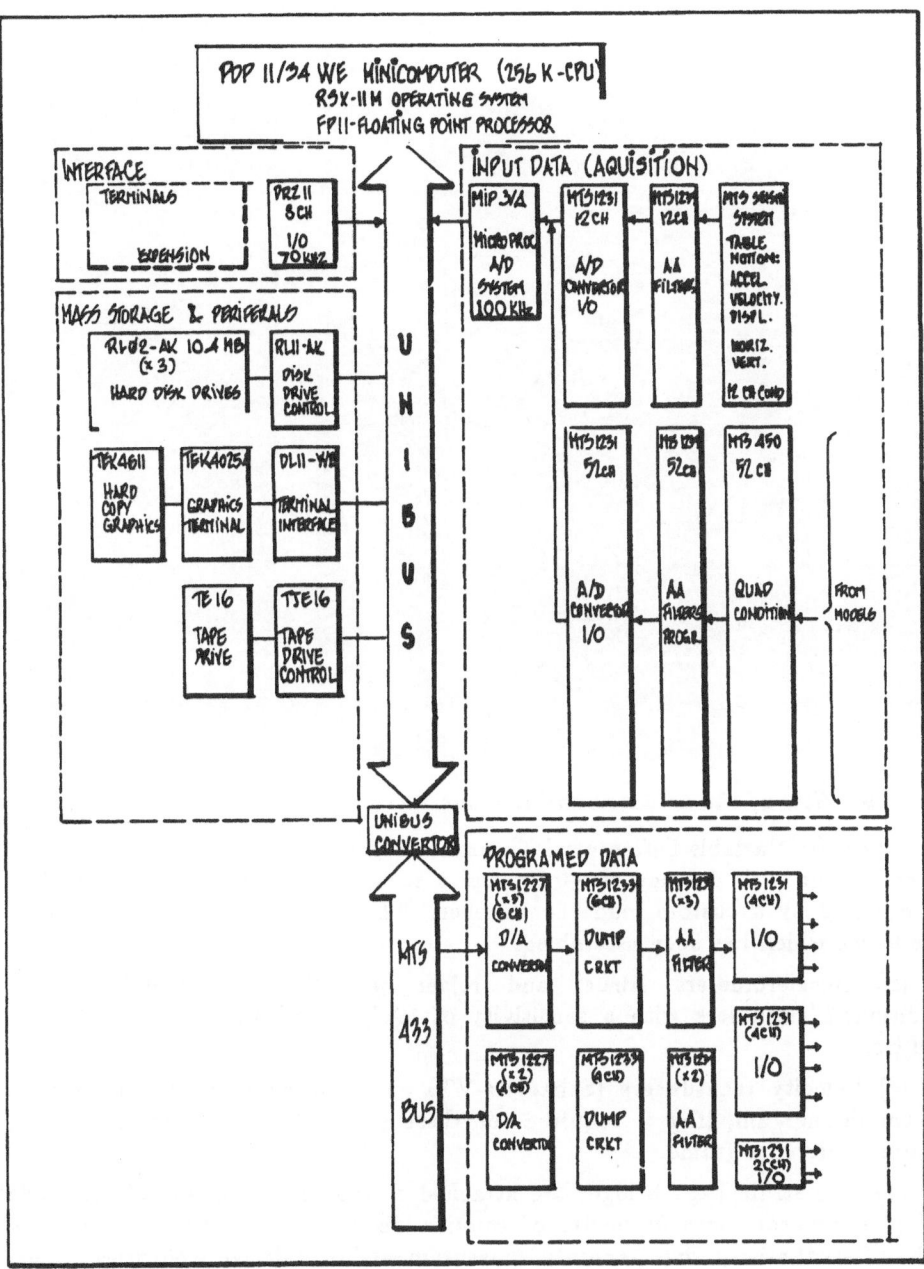

Figure 8: Control and data acquisition block diagram for earthquake simulation system

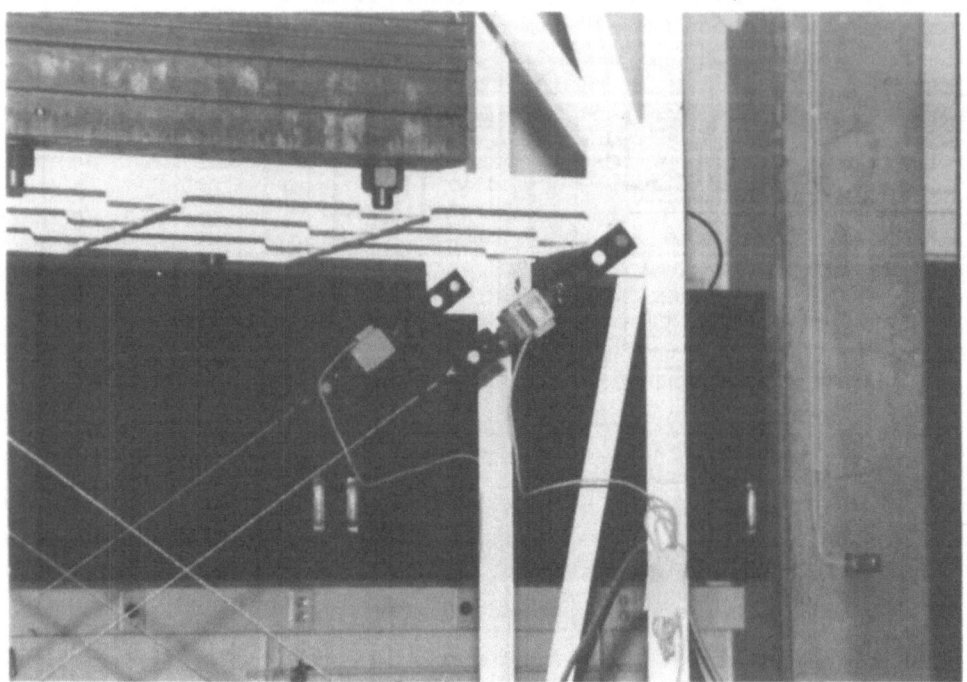

Figure 9: View of the actuator and tendons

(a) Linear Variable Differential Transducers (Schaevitz-LVDT) operated with direct current with a range of ±0.5in and a sensitivity of 10 V/in. The LVDTs are powered by a custom made conditioner. These transducers require a reference frame which has been assembled.

(b) Accelerometers (Bruel and Kjaer #4702) conditioned by PCB-conditioners/amplifiers with a sensitivity of 10 V/g and with a low pass filter 0-50 Hz.

(c) Velocity transducers (Schaevitz-VT) operated by magnetic inductance. No conditioner/amplifier is required for these transducers and, as in (a), they require a reference frame.

(d) Two strain gage bridges are attached to the first floor legs of the model and give accurate measurements of relative displacements and, when coupled with differentiators, give accurate measurements of relative velocities. Using these custom made transducers, the need for the reference frame is eliminated.

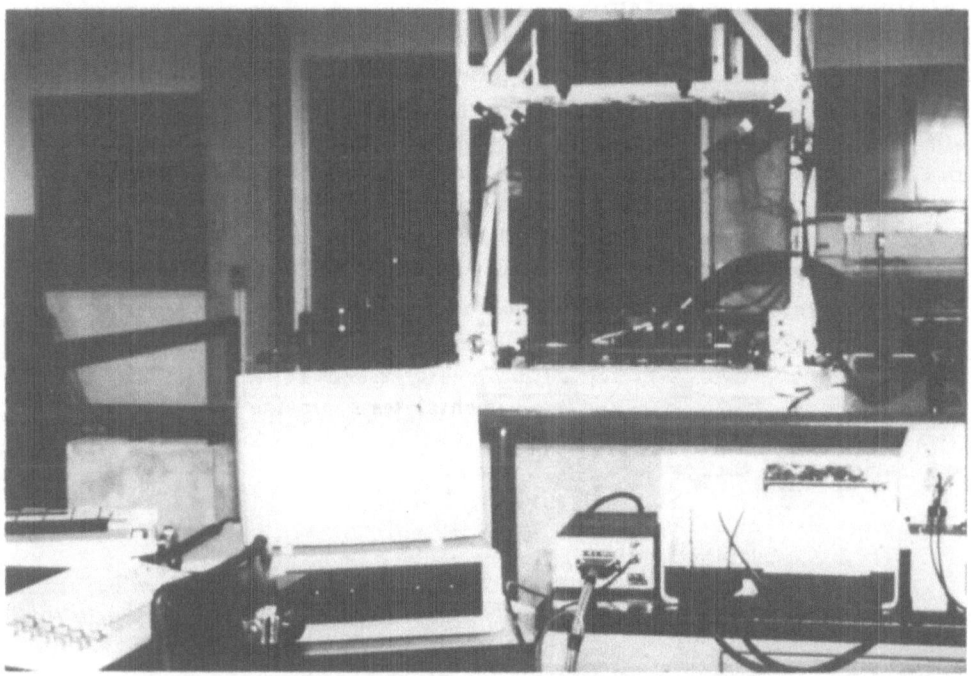

Figure 10: Details of tendons connection and load cells

(e) Load cells (Vishay 1000) with a capacity of 1000-lb conditioned through a Vishay 2200 conditioner/amplifier with a maximum sensitivity of 100 lb/volt.

2.4. Processing unit

The processing unit is based on an IBM-PC computer (256K) operated with PC-DOS 2.1 (real time) operating system. The unit is complemented by 8 floating A/D conversion channels (Data Translation™) with anti-aliasing (PCB-Piezontronics) filters for the incoming signal with 3400 samples per second of acquisition. The unit has a D/A convertor from which an analog control signal is transmitted to the actuator. All the control algorithms are computed on-line by the processing unit using measured data and sending output to the control actuator.

2.5. Data acquisition system

The data acquisition system consists of 52 channels of conditioning, amplification and low-pan-frequency filters. Data acquisition was carefully calibrated and verified for accurate recording. Signals from the conditioners are recorded on magnetic disc after initial processing for frequency response functions, transfer functions and mode identifications. The data acquisition system was updated by the addition of a close circuit video system with slow-motion reviewing for better understanding of vibration effects. A frequency analyzer (HP#3582A) is used for structural identification of the model structure necessary for control calculations.

The integrated testing facility is illustrated in Fig. 11.

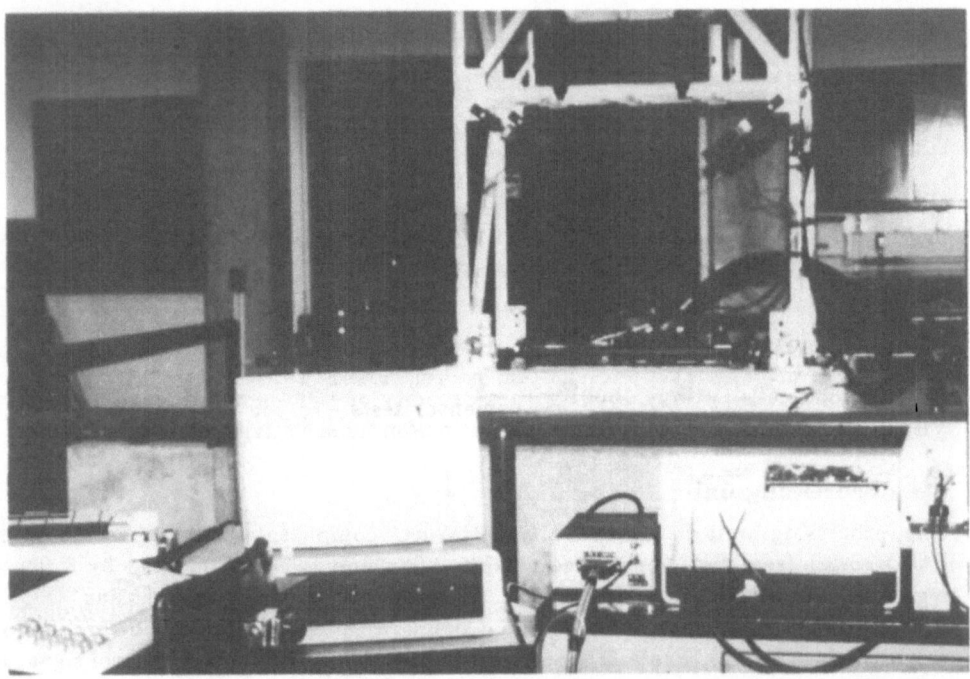

Figure 11: Experiment set-up

3. Control Algorithms

It is seen that the experimental test facility is designed with sufficient versatility to permit the performance of a variety of active control experiments. In terms of control algorithms, the following can be critically examined from the point of view of sensor and controller hardware requirements, speed and quality of data processing, on-line computation and potential application to real structures.

3.1. Optimal closed-loop control

Classical linear optimal control theory is employed using the familiar quadratic performance index. Sensors are required at each floor for measuring displacements and velocities relative to the structural base. Tendons are tensioned to levels proportional to the structural response through a gain matrix. The above algorithm requires the computation of the system's gain matrix based on the solution of Ricatti equation. This approach intends to yield a system which is controlled to operate in the elastic range on the basis of optimal feedback control.

3.2. Optimal open-loop control

Time dependent performance index is used and sensors are required for measurement of base acceleration. The required on-line computation involves only the algebraic manipulation of matrices. However, displacement and velocity sensors at each floor are not required [14].

3.3. Sub-optimal control

This control algorithm requires motion measurements at each floor as well as at the structural base. Since the experimental tests provide both hardware and data processing capabilities required by this control algorithm, it is constructive to consider its implementation. It is expected to be more effective than either of the above-mentioned control algorithms.

3.4. Sub-optimal pulse control

This control scheme, based on some recent development on pulse control theories [16,17], calls for the tendons to be tnesioned for short periods of time (pulses) at various intervals to act against excessive build-up of controlled response. The control scheme is thus a step-by-step procedure in which the structural response is monitored at the beginning of each step of control and the adjustment pulse, if needed, is determined and applied. An advantage of this approach is the possibility of savings in control energy required and of operating a control system in the inelastic, as well as elastic, range of the structure.

While for the optimal closed loop control sensors are required at each floor to measure displacements and velocities, this control involves only a limited number of sensors installed in some preselected floors, including base of the structure. The gain matrix is determined by a search procedure [15]. Velocity sensors are found to be more efficient than the displacement sensors.

4. Experimental Results

The testing program started with the validation of efficiency of linear optimal feedback control for the SDOF system. For this purpose the shaking table has been excited (a) by a sinusoidal motion with different frequencies; (b) by a banded white noise (0 - 10Hz) random motion; and (c) by a scaled ground motion accelerogram (El-Centro 1940 - N-S component).

The well known closed loop optimal algorithm has been reformulated to produce actual control forces including the influence of cable's dynamics. The control force has been programmed using Microsoft BASIC Compiler and the resulting algorithm executed in real time by the IBM-PC microcomputer. For convenience the relative displacements and velocities were measured by the specially designed strain-gage bridges. The microcomputer's digital to analog (D/A) convertor generated a constant voltage proportional to the required control force. The computation cycle of the microcomputer (analog to digital conversion, force computations, and digital to analog conversion) was performed in less than 8 msec. considered adequate for control of up to 15Hz vibrations.

The measured frequency response function for the transient range for a sinusoidal base motion of 5Hz is shown in Fig. 12. The influence of the feedback control was a 20% reduction of the peak response at 5Hz and an immediate damping of the eigen component at 3.84Hz (due to the small damping, 0.5%, no substantial reduction was noticed during the controlled transient response).

When the model was subjected to a banded white noise random base motion (between 0 to 10Hz), the freqency response function showed a large peak of .0087 in at the natural frequency of 3.48Hz as seen in Fig. 13. When the control was applied, the frequency response function showed only .0005 in at a frequency of 4.32Hz. A reduction of 95% of the initial response was obtained. The shift of the peak of the controlled response to a frequency of 4.32Hz shows that the effect of controlling it manifests as a stiffening of the structure of 25% compared with the structural stiffness of the original model (uncontrolled), as expected from theory.

The efficiency of the optimal control can be noticed from the ratio of peak relative displacements of the controled model to the uncontrolled model as shown in Fig. 14. When the feedback gain is set to its optimal values (1.38%), a reduction of approximately 80% is obtained. Any additional increase of gain (= increase of the control force) does not substantially decrease the displacement ratio (increase of 5% of the gan renders a decrease of the response by an additional 15% only).

The response of the model to a typical earthquake accelerogram (EL-Centro 1940 - N-S component) scaled to 25% of its maximum peak is shown in Figs. 15 and 16. The controlled response with optimal control parameters shows a

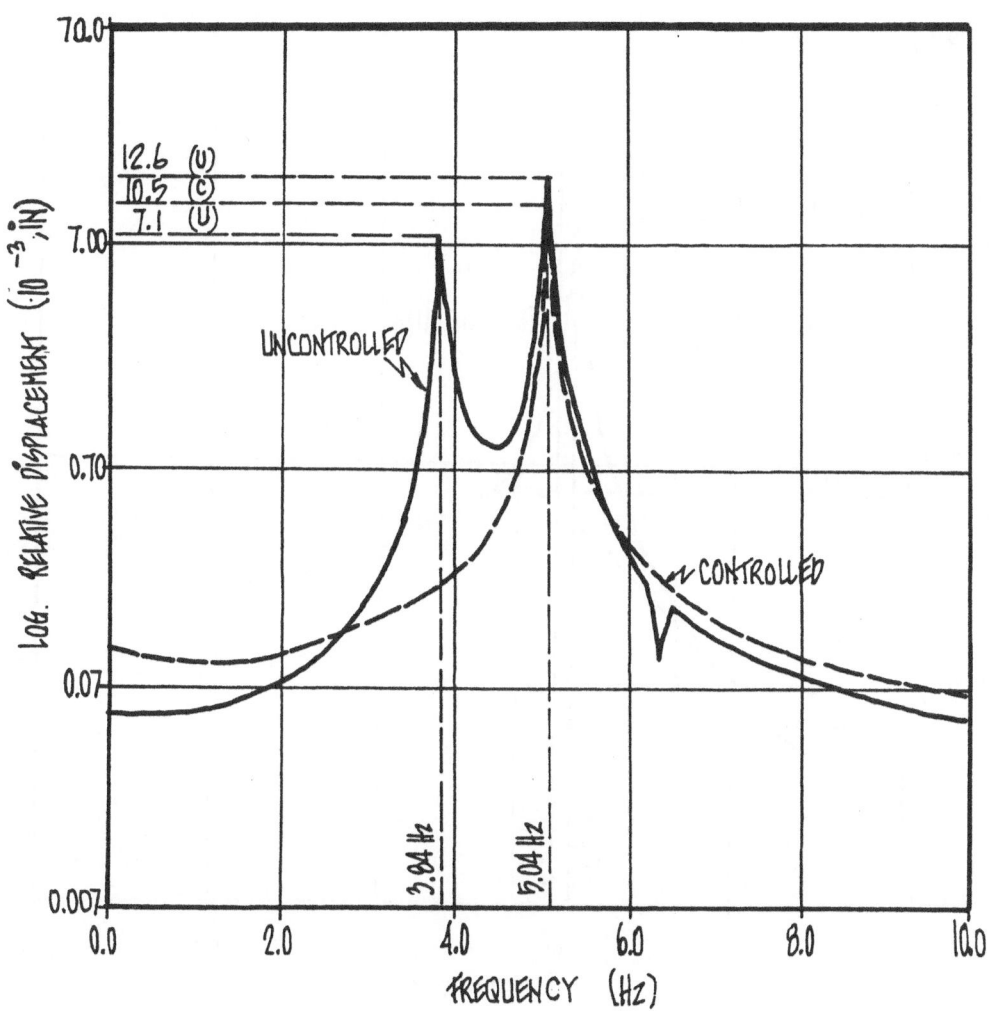

Figure 12:   Frequency response function of relative displacement of first floor under sinusoidal excitation (5Hz)

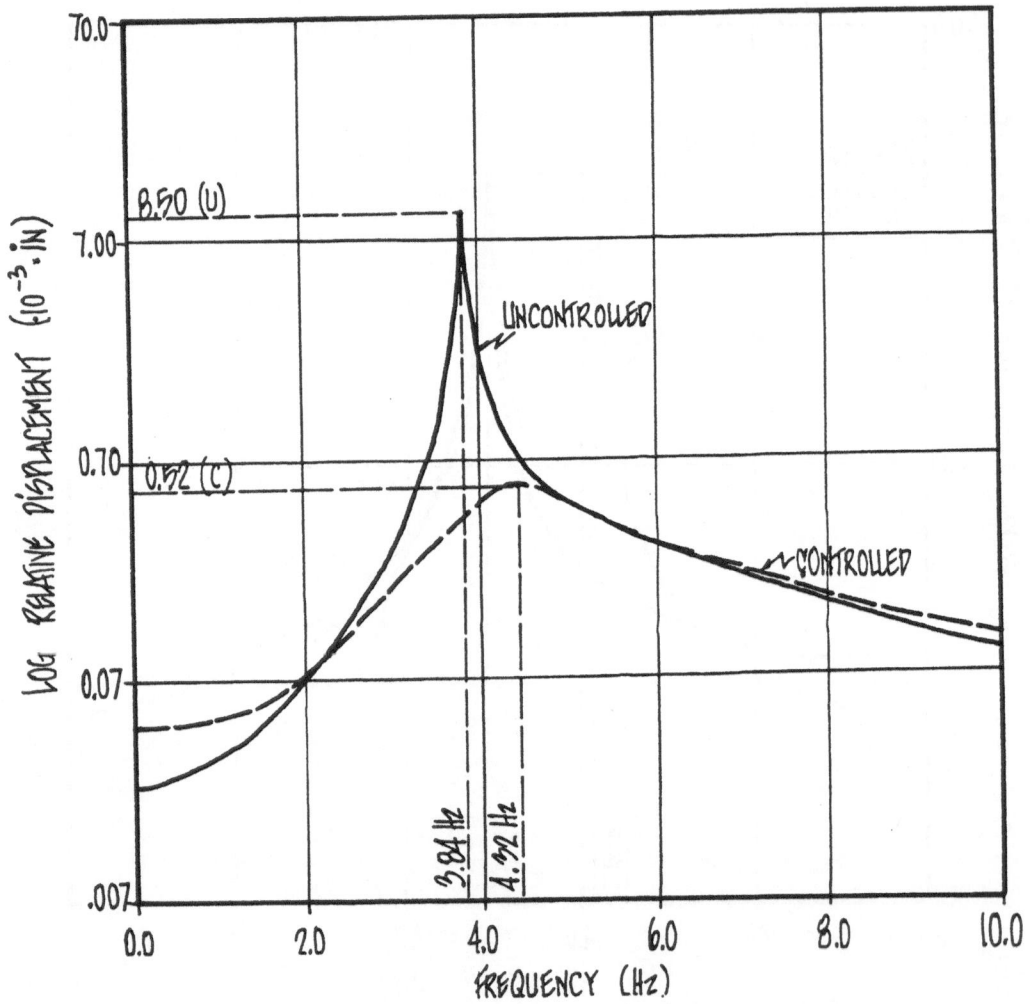

Figure 13: Frequency response function of relative displacement of first floor under random excitation

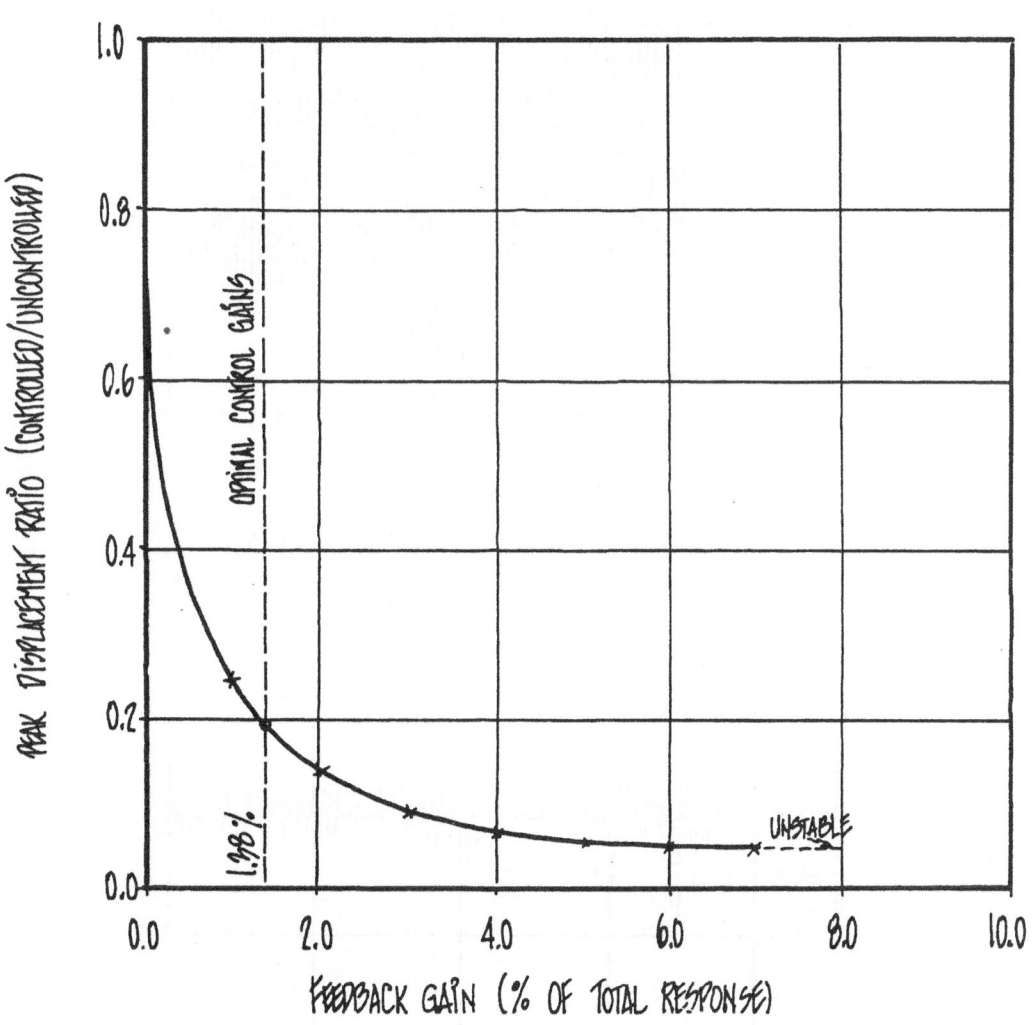

Figure 14: Influence of feedback gain to peak displacement ratio

(a) UNCONTROLLED RESPONSE

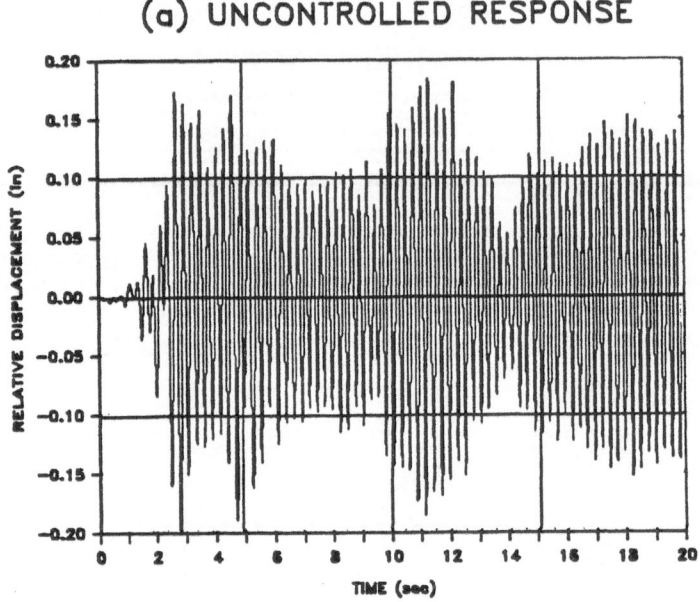

(b) CONTROLLED RESPONSE

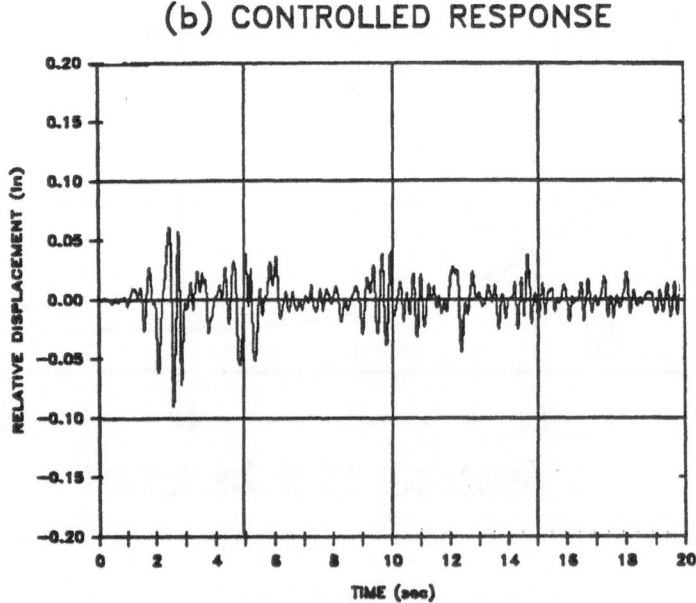

Figure 15: Time response functions under earthquake excitations (25% El Centro accelerogram)

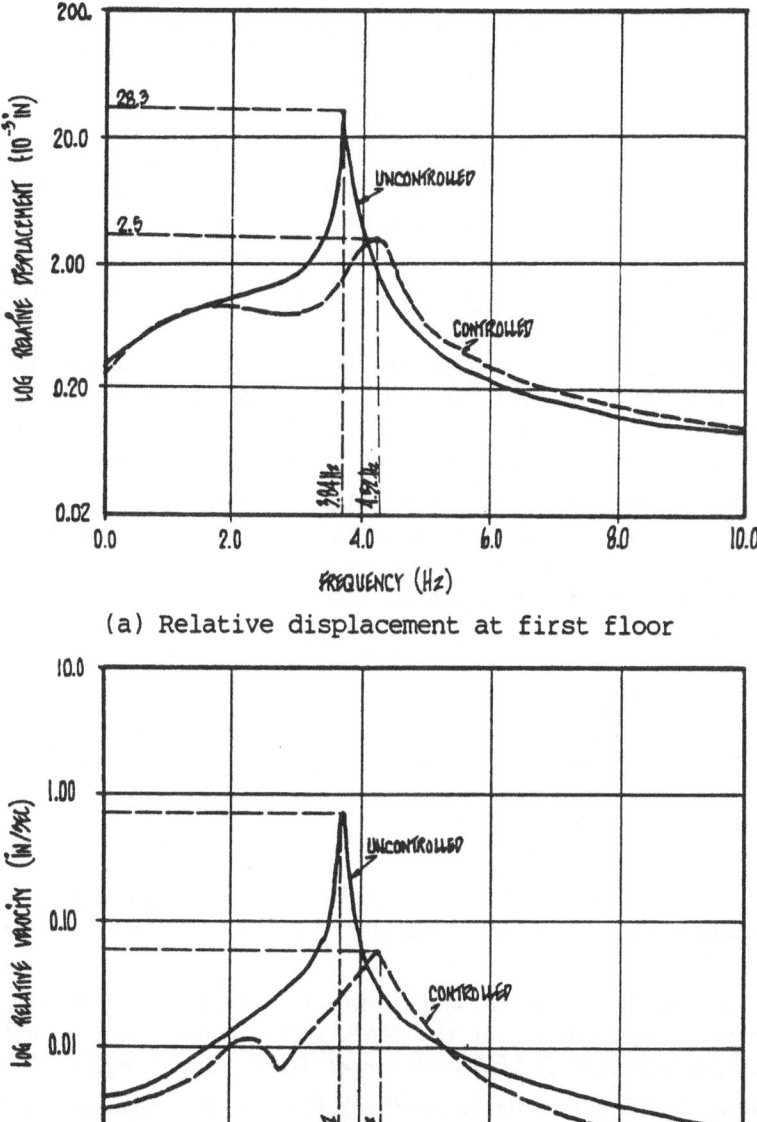

(a) Relative displacement at first floor

(b) Relative velocity at first floor

Figure 16: Frequency response functions under earthquake excitation (25% El Centro accelerogram)

reduction of 90 percent of the maximum peak which occurs at a frequency of 4.32Hz, proving again the stiffening trend during feedback control.

5. Conclusions

The experimental research underway in the SUNY/Buffalo laboratory produced a reliable model with well defined characteristics and a variety of adjustments for sound calibration to be used in active control of structures.

The active control is applied by tendons capable of transferring a few hundred pounds of control forces produced by a servocontrolled hydraulic device which insures their accuracy.

The experiments performed on the braced model simulating one degree of freedom show the applicability of feedback control to structures with emphasis on optimal control. Further testing and analysis will show the efficiency of active control versus theoretical analyses for the SDOF and MDOF structure.

Preliminary studies on other algorithms revealed that some adjustments are necessary before the open loop and open-close loop control can be formulated for real-time operation by the digital controller (microcomputer); a communication on this subject is to be presented by one of the co-authors later during this conference.

Acknowledgements

This work was supported by the National Science Foundation under Grant No. CEE 8311879. The authors are indebted to Mr. Mark Pitman, laboratory technician, for his dedicted work and his valuable suggestions.

References

[1] Zuk, W. "The Past and Future of Active Structural Control Systems", in *Structural Control*, H.H.E. Leipholz (ed.), North Holland, Amsterdam, 779-810, 1980.

[2] Zimmerman, D., Inman, D. and Horner, G., "Dynamic Characterization and Microprocessor Control of the NASA/UVA Proof Mass Actuator", *Proc. AIAA/ASME/ASCE/AHS 25th Struct. Dyna. Mat. Conf.*, 2, 573-577, 1984.

[3] Roorda, J., "Experiments in Feedback Control of Strucutres," in *Structural Control,* H.H.E. Leipholz (ed.), North Holland, Amsterdam, 505-521, 1980.

[4] Herrik, D., Canavin, J. and Strunce, J., "An Experimental Investigation of Modern Modal Control," Paper 79-0199, 17th Aerospace Meeting, New Orleans, 1979.

[5] Soong, T. T. and Skinner, G., "Experimental Study of Active Structural Control," *J. EMD, ASCE,*, 107, 1057-1067, 1981.

[6] Schafer, B. and Hozach, H., "Experimental Research on Flexible Beam Modal Control," *Proc. AIAA/ASME/ASCE/AHS 25th Struct. Dyna. Mat. Conf.,* 2, 317-326, 1984.

[7] Strunce, R. R. and Carman, R. W., "Active Control of Space Structures (ACROSS): A Status Report", *ibid.,* 84-1027, 1984.

[8] Dehghanyar, T. J., Masri, T. J., Miller, S. F., Beckey, R. K. and Caughy, T.K., "An Analytical and Experimental Study into Stability and Control of Nonlinear Felxible Structures", *Proc. 4th VPI and SU/AIAA Symp. Dyna. Contr. Large Struct.,* Blacksburg, VA, 1983.

[9] Dehghanyar, T. J., Masri, T. J., Miller, S. F., Bekey, R. K. and Caughy, T. K., "Sub-optimum Control of Nonlinear Flexible Space Structures", *NASA/ACC Workship on Ident. Contr. of Flexible Space Struct.,* San Diego, 1984.

[10] Miller, R. S., Krawinkler, H. and Gere, J. M., *Model Tests on Earthquake Simulators Development and Implementation of Experimental Procedures,* Report No. 39, Department of Civil Engineeringg, Stanford University, CA, 1979.

[11] Moncarz, P. M. and Krawinkler, H., *Theory and Application of Experimental Model Analysis in Earthquake Engineering,* Report No. 39, Department of Civil Engineering, Stanford University, CA, 1981.

[12] Clough, R. W. and Tang, D. T., *Earthquake Simulator Study of a Steel Frame Structure, Vol. I: Experimental Results,* EERC Report No. 75-6, Berkeley, CA, 1975.

[13] Tang, D. T., *Earthquake Simulator Study of a Steel Frame Structure, Vol. II: Analytical Results,* EERC Report No. 75-36, Berkley, CA, 1985.

[14] Yang, J. N. et al., "On-Line Computation for Active Control of Structures Under Seismic Loads," Paper to be presented at *2nd Int. Conf. on Struct. Contr.,* to appear.

[15] Yang, J. N., "Control of Tall Buildings under Earthquake Excitation", *J. EMD., ASCE,* 108, 50-68, 1982.

[16] Prucz, Z., Soong, T. T. and Reinhorn, A. M., "An Analysis of Pulse Control for Simple Mechanical Systems", *J. Dyn. Syst. Meas. Contr., ASME,* 107, 123-131, 1985.

[17] Reinhorn, A. M. and Manolis, G. D., "An On-Line Control Algorithm for Inelastic Structures", *Proc. 2nd Int. Conf. on Struct. Contr.,* to appear.

A GENERAL THEORY
OF NON-DESTRUCTIVE
DAMAGE DETECTION IN STRUCTURES

N. Stubbs
College of Architecture and Environmental Design, and
Department of Civil Engineering
Texas A & M University
College Station, Texas

1. Introduction

The design characteristics of a passive or an active controller depend upon the mechanical properties of the structure to be controlled. If a change in mechanical properties accurs at any point in the structure, a corresponding adjustment should be made in the control system if the original control objectives are to be continually satisfied. Therefore, as part of the overall control program, it is practicable to nondestructively monitor changes in the mechanical properties of controlled structures. Furthermore, if it is possible to specify the probable location of the change in mechanical properties and estimate the magnitude of the change, such valuable information may provide the basis for decisions regarding recommendations concerning the future use of, or repairs to, the structure.

As damage accumulates in a structure (or structural element), the stiffness of that structure changes [1, 6, 7, 12]. This stiffness change can be used as a quantitative record of damage growth. Changes in stiffness are reflected in changes in eigenfrequencies of the structure (or element) [7]. The measurement of natural frequencies is attractive because such quantities can be measured at a single point on the structure and are independent of the position selected.

694

Offshore structures and structural elements made of composite materials have been a frequent subject of nondestructive damage detection studies [4, 9, 11, 16, 17, 18]. In most of these studies, structural damage, i.e., the reduction in structural stiffness at one or more points of the structure, has been inferred via changes in the dynamic characteristics of the structure. However, a review of existing damage evaluation methods indicates a significant variation in experimental approaches and analytical procedures. For example, in offshore applications some methods excite the platform to generate transfer functions [4, 11] while others rely on the random forces of wind, waves and currents to provide the excitation [17]. In all cases the motion was sensed using accelerometers; however, a variety of placement criteria were used. In fact, no two structures utilized the same placement philosophy. The analytical procedures used to analyze the transduced signal were equally varied. The procedures include the Random Decrement Signature Technique (which essentially extracts the deterministic part of the response from the full response) [17, 18], Fast Fourier Transforms [4], spectral analysis [9] and engineering judgement [16]. The result is a collection of eigenfrequencies, mode shapes, or damping ratios. However, how all (or section of) this information is used depends upon the particular investigator selected. For example, the Random Decrement technique computes a reference signature for the undamaged structure and compares it with signature from later recording [17]; while in the study by Kenly and Dodds [9], damage location and detection was achieved by simulating the effect of stiffness and mass change using a finite element analysis.

Furthermore, each of the methods make a different claim regarding the effectiveness and practicability of structural integrity monitoring using vibration techniques. Nataraja [11] concluded that such techniques comprising accelerometers at the surface and deck can only detect 'global changes'; Yang et al. [17] claimed that the Random Decrement technique was able to identify all the change and non-damage situations with the usage of only four accelerometers mounted on each of the legs of the structure, and Crohas and Lepert [4] concluded that such a method has demonstrated that the techniques (Vibro detection) possesses considerable capability for proposes of structural integrity monitoring and for potential use in design.

It is interesting to observe that although all of the above methods of nondestructive damage evaluation have the same objective (namely, to predict the location and severity of damage in an offshore platform), and that the experimental phase of each method produces the same modified dynamic properties, such a broad variation in damage prediction techniques exists. It is quite reasonable to assume that the existing variations in the conclusions and claims made for each method reflect the nuances of the prevaling models for damage evaluation rather than the capabilities of the method of nondestructive damage evaluation using vibration data. In short, we need a basic theory of structural damage evaluation in which changes in the dynamic properties of a structure (i.e., eigenvalues, mode shapes and modal damping) are theoretically related to changes (at one or more locations) in the mass, damping and stiffness of the structure itself.

This paper presents a theory of nondestructive damage detection that is applicable to any mechanical system that may be damaged in one or more locations. The approach takes into consideration variations in mass, damping, and stiffness. A first difference is applied to the homogeneous equations of motion of a damaged structural system to yield expressions for changes in modal stiffness in terms of modal masses, modal damping, eigenfrequencies and eigenvectors, and their respective changes. From matrix structural analysis, expressions relating variations in the stiffnesses of structural elements to the variations in modal stiffness are generated. Assuming that the changes in eigenfrequencies, masses, and damping can be experimentally determined, the result is a system of algebraic equations with a load vector of changes in modal stiffness and unknowns of changes in member stiffnesses. Methods of solution for the resulting system of equations are discussed along with special cases in which the structure may be modelled to yield unique predictions. Furthermore, how the reliability of the prediction is affected by statistical uncertainty in material properties and dynamic measurements is addressed. The method is demonstrated using a numerical example of a multistory structure damaged at arbitrary locations and with arbitrary severities.

2. Theory of the Method

2.1. Sensitivity equations for damped free motion

Consider at time t_o an undamaged structure described by the system of equations:

$$\mathbf{M}\left[\frac{d^2Y}{dt^2}\right] + \mathbf{C}\left[\frac{dY}{dt}\right] + \mathbf{K}[Y] = 0, \tag{1}$$

where $\mathbf{M}$, $\mathbf{K}$, and $\mathbf{C}$ represent the mass, stiffness, and damping matrices, respectively, and $\mathbf{Y}$ is a matrix of the structural displacements. At some later time t_1 the structure is damaged in one or more locations and the resulting equation of motion becomes (see Fig. 1):

$$\mathbf{M}^*\left[\frac{d^2Y^*}{dt^2}\right] + \mathbf{C}^*\left[\frac{dY^*}{dt}\right] + \mathbf{K}^*[Y^*] = 0, \tag{2}$$

where the asterisk denotes the appropriate quantities of the damaged structure.

During this intervening period, the structural properties are ralated by the equations:

$$\dot{\mathbf{M}}^* = \mathbf{M} + \Delta\mathbf{M}, \quad \mathbf{K}^* = \mathbf{K} + \Delta\mathbf{K}, \quad \text{and} \quad \mathbf{C}^* = \mathbf{C} + \Delta\mathbf{C}. \tag{3}$$

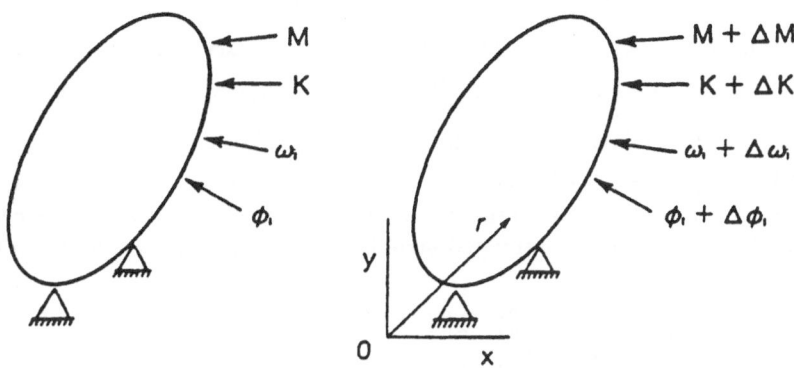

a) Undamaged Structure   b) Damaged Structure

Figure 1: Undamaged and damaged structures

Assuming the usual orthogonality restrictions on the damping matrix, Eq. (1) can be uncoupled to give

$$M_n \left[\frac{d^2 Z_n}{dt^2} \right] + C_n \left[\frac{dZ_n}{dt} \right] + K_n[Z] = 0, \tag{4}$$

in which the modal quantities:

$$M_n = \{\Phi\}_n^T[M]\{\Phi\}_n, \quad K_n = \{\Phi\}_n^T[K]\{\Phi\}_n \quad \text{and} \quad C_n = \{\Phi\}_n^T[C]\{\Phi\}_n, \tag{5}$$

where $(n = 1,2,...,NF)$ and the transformation required to uncoupled generalized coordinates is defined by

$$[Y] = \{\Phi\}[Z]. \tag{6}$$

Note that $\{\Phi\}$ represents the matrix of modal vectors for the system, superscript T represents the transpose of the matrix, and subscript n denotes the mode shape number.

The system of uncoupled equations for the damaged state can be developed using similar reasoning. Using the transformation

$$[Y^*] = \{\Phi^*\}[Z^*] = \{\Phi + \Delta\Phi\}[Z^*] \tag{7}$$

and the appropriate definitions of the modal properties, the uncoupled version of Eq. (2) becomes

$$M_n^* \left[\frac{d^2 Z_n^*}{dt^2} \right] + C_n^* \left[\frac{dZ_n^*}{dt} \right] + K_n^*[Z_n^*] = 0, \tag{8}$$

in which the modal quantities for the damaged structure are given by

$$M_n' = M_n + \Delta M_n, \quad K_n' = K_n + \Delta K_n \quad \text{and} \quad C_n' = C_n + \Delta C_n, \quad (9)$$

where, for example, the change in modal stiffness, ΔK_n, is given by the equation

$$\Delta[K_n] = \{\Phi\}_n^T[\Delta K]\{\Phi\}_n + \{\Delta\Phi\}_n^T[K]\{\Phi\}_n + \{\Phi\}_n^T[K]\{\Delta\Phi\}_n + [O^2]. \quad (12)$$

Analogous expressions exist for the change in modal masses and modal damping.

Defining $\varsigma_n = C_n/2M_n\omega_n$ as the modal damping ratio and assuming only underdamped motion, the frequencies of damped oscillations for the nth mode in the undamaged and the damaged structure (ω_{nd} and ω_{nd}', respectively) are given by

$$\omega_{nd}^2 = \omega_n^2(1 - \varsigma_n^2) \quad \text{and} \quad \omega_{nd}'^2 = \omega_n'^2(1 - \varsigma_n'^2). \quad (13)$$

If we set $\omega_n'^2 = \omega_n^2 + \Delta\omega_n^2$ and $\varsigma_n'^2 = \varsigma_n^2 + \Delta\varsigma_n^2$, then subtracting Eq. (13) results in an expression for the changes in natural damped frequency in terms of the changes in the undamped natural frequencies and the changes in the modal damping ratios. Namely;

$$\frac{\Delta\omega_{nd}^2}{\omega_{nd}^2} = \frac{\Delta\omega_n^2}{\omega_n^2} + \frac{\Delta\varsigma_n^2}{(1 - \varsigma_n^2)}. \quad (14)$$

Repeating the same analysis as above, but this time neglecting any damping in the system, it can be shown that

$$\frac{\Delta\omega_n^2}{\omega_n^2} = \frac{\Delta K_n}{K_n} - \frac{\Delta M_n}{M_n}. \quad (15)$$

Finally, on substituting Eq. (15) into Eq. (14) and solving for only change in the modal stiffness that is directly attributable to the change in the structural stiffness, we get

$$\frac{\{\Phi\}_n^T[\Delta K]\{\Phi\}_n}{K_n} = \frac{\Delta\omega_{nd}^2}{\omega_{nd}^2} + \frac{\Delta M_n}{M_n} + \frac{\Delta\varsigma_n^2}{(1 - \varsigma_n^2)}$$
$$- \frac{[\{\Delta\Phi\}_n^T[K]\{\Phi\}_n + \{\Phi\}_n^T[K]\{\Delta\Phi\}_n + \cdots]}{K_n}. \quad (16)$$

If all the terms on the right hand side of Eq. (16) can be determined directly or via experimental measurements, then the remaining change in the modal stiffness resulting only from the changes in the structural stiffness can be estimated. This last equation is the basis for the method of nondestructive damage evaluation presented here.

2.2. Modal stiffness - structural stiffness relationship

Let the portion of the change in modal stiffness resulting from structural degradation be given by ΔK_{ns}, then,

$$\Delta K_{ns} = \{\Phi\}_n^T [\Delta K]\{\Phi\}_n. \tag{17}$$

Now note that from matrix structural analysis, the change in the global system stiffness matrix $[\Delta K]$ may be given by

$$\Delta K = N^T \Delta K'' N, \tag{18}$$

where N is a joint-member incidence matrix and K'' is the primitive stiffness matrix consisting only of diagonal elements or diagonal elements that are partitioned matrices [14]. Substituting Eq. (18) into Eq. (17) and defining $\{B\}_n = N\{\Phi\}_n$ we obtain

$$\Delta K_{ns} = \{B\}_n^T [\Delta K'']\{B\}_n, \tag{19}$$

where $\{B\}_n$ is a partitioned column matrix with the number of elements equal to the number of structural members.

Assuming that the elements of matrix $\{B\}_n$ are $\{b_{n1}, \ldots, b_{nNE}\}^T$, where NE is the total number of structural members (e.g. in a truss or a beam), then Eq. (19) is a quadratic form [10] in the variables b_{nj} $(j = 1,1,\ldots, NE)$ and Eq. (17) can be rewritten as:

$$\Delta K_{ns} = \sum_{i=1}^{NE} b_{ni}^T \Delta k_i \, b_{ni}, \tag{20}$$

where k_i is the appropriate structural stiffness of member i.

2.3. Basic equation for damage prediction

Let $z_n = \Delta K_{ns}/K_n$, furthermore, let $\Delta k_i = a_i k_i$ (where a_i is the fractional change in stiffness of element i with respect to the original stiffness), then for a one-force member system:

$$z_n = \sum_{i=1}^{NE} \frac{(k_i b_{ni}^2 a_i)}{K_n} = \{F_n\}[A], \tag{21}$$

where $\{F_n\} = \{f_{n1}, \ldots, f_{nj}, \ldots, f_{nNE}\}$, $[A]^T = [a_1, \ldots, a_{NE}]$ and

$$f_{nj} = \frac{k_i b_{nj}^2}{K_n}. \tag{22}$$

Finally, if we define two more matrices, $\{F\}^T = \{F_1^T, \ldots, F_{NE}^T\}$ and $[Z]^T = [z_1, z_2, \ldots, z_{NE}]$, the governing matrix equation for damage prediction in the structural system becomes

$$[Z] = \{F\}[A]. \tag{23}$$

Thus if the elements z_i are experimentally determined (see L.H.S. of Eq. (16)) and the elements of the F matrix are developed using Eq. (22), then Eq. (23) may be solved to estimate the location and severity of the damage. Recall that Eq. (23) will only yield a unique solution if the number of damage locations considered is equal to the number of eigenfrequencies (NF) available. If the number of damage locations or the number of available eigenfrequencies are not equal, then the determination of the location of the damage can proceed in several ways. First, we may resort to generalized inverse techniques (13). For example, as a special case a least squares estimate of A may be obtained using the equation:

$$\mathbf{A} = (\mathbf{F}^T\mathbf{F})^{-1}\mathbf{F}^T\mathbf{Z}. \tag{24}$$

In a second approach external information regarding the state of damage for the elements of interest may be provided. For example, on the basis of engineering judgement or the result of a local nondestructive evaluation scheme, it may be decided that the $(NF-NE)$ conditions (which render the system non-unique) are known. On incorporating these conditions into Eq. (23), F is reduced to a square matrix.

A third approach may be to model the structure such that $NE=NF$. Then solve Eq. (23) to find undamaged locations. The structure is then remodelled using the information from the previous analysis. For example, a multistory structure may be initially modelled as a shear beam in which the level of damage prediction is limited to that of a story. Knowing the approximate location of damage the problem may now be reformulated in terms of damage parameters related to specific members on that floor.

2.4. Uncertainties in structural response and experimental measurements

The preceding analysis is entirely deterministic. In practical situations, however, both the experimentally determined quantities (e.g. $\Delta\omega_{nd}^2$, ΔM, and ΔC) and the structural quantities derived from analysis (e.g. M_n, K_n, and ς_n) are really random variables. Consequently, the location and the magnitude of the damage (a_i) is also a random variable. Thus in the real world one should speak of the probability that the damage at a particular location exceeds a prescribed value, or the probability that the damage at a particular location is between certain limits. Thus we feel that it is fitting to extend the proposed formulation to incorporate these areas of uncertainty.

A *priori*, the exact nature of the probability density functions associated with the problem variables of interest are unknown. However, it is often possible to estimate the values of the first and second moments of such variables. The first and second moments of the experimental values can be estimated directly from a collection of measurements. Analytical techniques also exist and are being developed to estimate the first and second moments of the modal quantities (5, 8, 15). Thus armed with the statistics of the basic quantities, the problem is to estimate the statistics of the damage variable (i.e., the probable locations and severities).

Consider a nonlinear function $f(\overline{X}) = f(X_1, \ldots, X_p)$ of p random variables $(X_1, \ldots, X_p)$ with means $(\mu_1, \ldots, \mu_p)$ and variances $(\sigma_1^2, \ldots, \sigma_p^2)$. Then, approximate values for the function mean (μ_f) and variance $(\mathrm{Var}[f] = \sigma_f^2)$ are determined by

$$\mu_f = f(\mu_1, \ldots, \mu_p), \tag{25}$$

and

$$\sigma_f^2 = \sum_{i=1}^{p} \sum_{j=1}^{p} \left(\frac{\partial f}{\partial X_i} \right) \left(\frac{\partial f}{\partial X_j} \right) Cov[X_i, X_j], \tag{26}$$

where the partial derivatives are evaluated at the mean values. Equating the left hand sides of Eq. (16) and (21) (and neglecting the contribution of first order terms in eigenvectors) leads to the equation:

$$\sum_{i=1}^{NE} \frac{(k_i b_{ni}^2 a_i)}{K_n} = \frac{\Delta \omega_{nd}^2}{\omega_{nd}^2} + \frac{\Delta M_n}{M_n} + \frac{\Delta \varsigma_n^2}{(1 - \varsigma_n^2)} \ . \tag{27}$$

If first we treat the variables in Eq. (27) as random variables, then secondly, approximate the means of the LHS and RHS, and finally, equate the two means, we obtain Eq. (21) evaluated at the mean value of the variables. Similarly, if we approximate the variance of the LHS and the RHS of Eq. (27) using Eq. (26), equate the two expressions, then rearrange the result (while keeping all terms in $\mathrm{Var}[a_i]$ on one side of the equation), we obtain (ignoring any correlation):

$$\sum_{i=1}^{NE} \left\{ \frac{(k_i b_{ni}^2)}{K_n} \right\}^2 \mathrm{Var}[a_i] \simeq \mathrm{Var}[z_n] - \sum_{i=1}^{NE} \left\{ \frac{(b_{ni}^2 a_i)}{K_n} \right\}^2 \mathrm{Var}[k_i]$$

$$- \sum_{i=1}^{NE} \left\{ \frac{(k_i a_i)}{K_n} \right\}^2 \mathrm{Var}[b_{ni}^2] - \sum_{i=1}^{NE} \left\{ \frac{(k_i b_{ni}^2 a_i)}{K_n^2} \right\}^2 \mathrm{Var}[K_n], \tag{28}$$

where

$$\mathrm{Var}[z_n] = \mathrm{Var}\left[\frac{\Delta \omega_{nd}^2}{\omega_{nd}^2} \right] + \mathrm{Var}\left[\frac{\Delta M_n}{M_n} \right] + \mathrm{Var}\left[\frac{\Delta \varsigma_n^2}{(1 - \varsigma_n^2)} \right]. \tag{29}$$

Assuming that all quantities on the RHS of Eq. (28) are known, then NF linear equations in NE unknowns of $\mathrm{Var}[a_i]$ become available. On solving for $\mathrm{Var}[a_i]$ and setting $\sigma_{a_i}^2 = \mathrm{Var}[a_i]$, the probability that the damage in member i exceeds some predetermined magnitude a_{i0} (i.e., $P[a_i \geq a_{i0}]$) may be readily estimated.

3. A Numerical Example

3.1. Problem description

The example structure, a four-story building modelled as a shear beam, is shown in Fig. 2. The appropriate masses and stiffnesses for each level of the structure are indicated on the diagram (see Ref. [3]). Given the eigenfrequencies and mode shapes for the undamaged structure and the eigenfrequencies for the undamaged structure, the problem is to predict the magnitude and location of any damage to the structure.

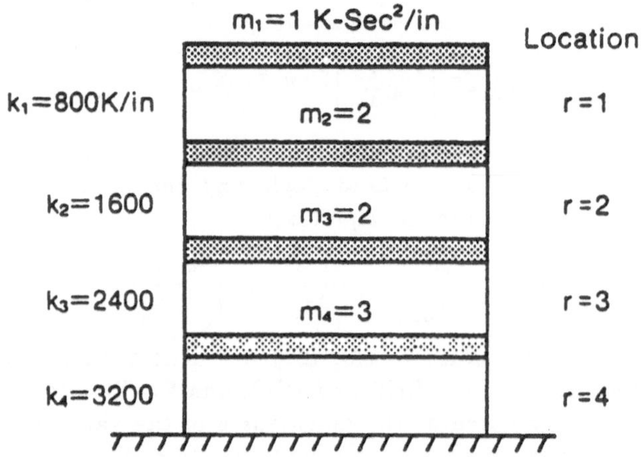

Figure 2: Shear beam model of structure

3.2. Numerical data

The eigenfrequencies for the undamaged structure are as follows: $\omega_1 = 13.294$, $\omega_2 = 29.660$, $\omega_3 = 41.079$, $\omega_4 = 55.882$. Similarly the modal masses are given by $M_1 = 2.87288$, $M_2 = 2.17732$, $M_3 = 4.36658$ and $M_4 = 3.64239$. Finally, the matrix of mode shape vector X is given by:

$$X = \begin{bmatrix} 1.00000 & 1.00000 & -0.90145 & 0.15436 \\ 0.77910 & -0.09963 & 1.00000 & -0.44817 \\ 0.49655 & -0.53989 & -0.15859 & 1.00000 \\ 0.23506 & -0.43761 & -0.70797 & -0.63688 \end{bmatrix}.$$

The example structure was subjected to a total of 38 damage scenarios. The scenarios were selected to a) investigate the accuracy of the proposed formulation and b) demonstrate the capability of damage location and estimation in multiple locations. The first 24 cases were limited to the structure damaged in one location. Cases 25-34 were restricted to the structure damaged at two locations.

Finally, cases 35-38 considered the structure damaged at more than two locations.

The associated eigenfrequencies and the corresponding magnitudes and damage locations for the damaged structure are summarized in Tables 1 and 2. Note that these modified frequencies were obtained by resolving the structure with the appropriate damage inflicted.

Table 1: Eigenfrequencies for structure damaged in one location

DAMAGE CASE	DAMAGE LOCATION	DAMAGE MAGNITUDE	EIGENFREQUENCIES			
			ω_1	ω_2	ω_3	ω_4
1	1	0.010	13.288	29.584	40.998	55.875
2	1	0.030	13.278	29.428	40.840	55.861
3	1	· 0.075	13.252	29.056	40.493	55.831
4	1	0.100	13.236	28.835	40.307	55.815
5	1	0.200	13.162	27.847	39.614	55.756
6	1	0.400	12.927	25.325	38.478	55.657
7	2	0.010	13.277	29.635	41.019	55.800
8	2	0.030	13.242	29.585	40.896	55.637
9	2	0.075	13.159	29.465	40.613	55.280
10	2	0.100	13.110	29.394	40.450	55.087
11	2	0.200	12.886	29.065	39.767	54.355
12	2	0.400	12.254	28.146	38.289	53.087
13	3	0.010	13.272	29.658	41.058	55.724
14	3	0.030	13.227	29.654	41.016	55.408
15	3	0.075	13.122	29.644	40.912	54.698
16	3	0.100	13.060	29.638	40.850	54.305
17	3	0.200	12.781	29.612	40.552	52.744
18	3	0.400	12.029	29.538	39.631	49.788
19	4	0.010	13.270	29.612	41.034	55.850
20	4	0.030	13.222	29.515	40.945	55.787
21	4	0.075	13.108	29.287	40.746	55.650
22	4	0.010	13.041	29.156	40.636	55.575
23	4	0.020	12.737	28.591	40.208	55.290
24	4	0.040	11.907	27.271	39.410	54.784

Table 2: Eigenfrequencies for structure damaged in multiple locations

DAMAGE CASE NO	DAMAGE LOCATION AND MAGNITUDE				EIGENFREQUENCIES			
	(1)	(2)	(3)	(4)	ω_1	ω_2	ω_3	ω_4
25	0.050	0.00	0.00	0.00	13.266	29.267	40.684	55.847
26	0.048	0.00	0.013	0.00	13.240	29.280	40.674	55.642
27	0.043	0.00	0.025	0.00	13.216	29.318	40.689	55.455
28	0.035	0.00	0.035	0.00	13.198	29.381	40.730	55.302
29	0.025	0.00	0.043	0.00	13.185	29.458	40.790	55.183
30	0.013	0.00	0.048	0.00	13.180	29.551	40.873	55.113
31	0.000	0.00	0.050	0.00	13.182	29.650	40.971	55.092
32	0.050	0.00	0.05	0.00	13.155	29.255	40.583	55.053
33	0.050	0.05	0.00	0.00	13.179	29.158	40.356	55.446
34	0.000	0.05	0.00	0.05	13.088	29.282	40.564	55.310
35	0.010	0.02	0.03	0.04	13.096	29.337	40.637	55.105
36	0.020	0.02	0.02	0.02	13.160	29.362	40.666	55.320
37	0.020	0.02	0.02	0.00	13.206	29.456	40.756	55.385
38	0.000	0.02	0.02	0.02	13.170	29.510	40.830	55.335

3.3. Determining elements of Z and F matrices

Since damping and changes in mass have been neglected in this example, $Z_n = \Delta\omega_n^2/\perp_n^2$, where $\Delta\omega_n^2$ is the eigenfrequency change from the undamaged state and ω_n is the original value of the nth eigenfrequency.

The elements of the **F** matrix are given by Eq. (22). In this example if the primitive stiffness matrix for the shear beam model is given by $K'' = Diag[k_1, k_2, k_3, k_4]$, then the bar-node incidence matrix **N**, for the structure becomes:

$$N = \begin{bmatrix} 1 & -1 & 0 & 0 \\ 0 & 1 & -1 & 0 \\ 0 & 0 & 1 & -1 \\ 0 & 0 & 0 & 1 \end{bmatrix}.$$

Thus the matrix $\{B\}_n = \{\Phi_{n1} - \Phi_{n2},\ \Phi_{n2} - \Phi_{n3},\ \Phi_{n3} - \Phi_{n4},\ \Phi_{n4}\}^T$ were Φ_{ni} is the ith component of the nth eigenvector. The elements of the **F** matrix are then obtained according to Eq. (28).

3.4. Results

The error between the actual value of the inflicted damage and the predicted value for the damage in one location (i.e., the accuracy of the technique) as a function of the inflicted damage is shown in Figs. 3 to 6. The cmparison between the inflicted damage values and the predicted values for the structure damaged in two locations is shown in Figs. 7 to 8. Analogous results for the structure damaged in 3 or 4 locations are shown in Figs. 9 - 10.

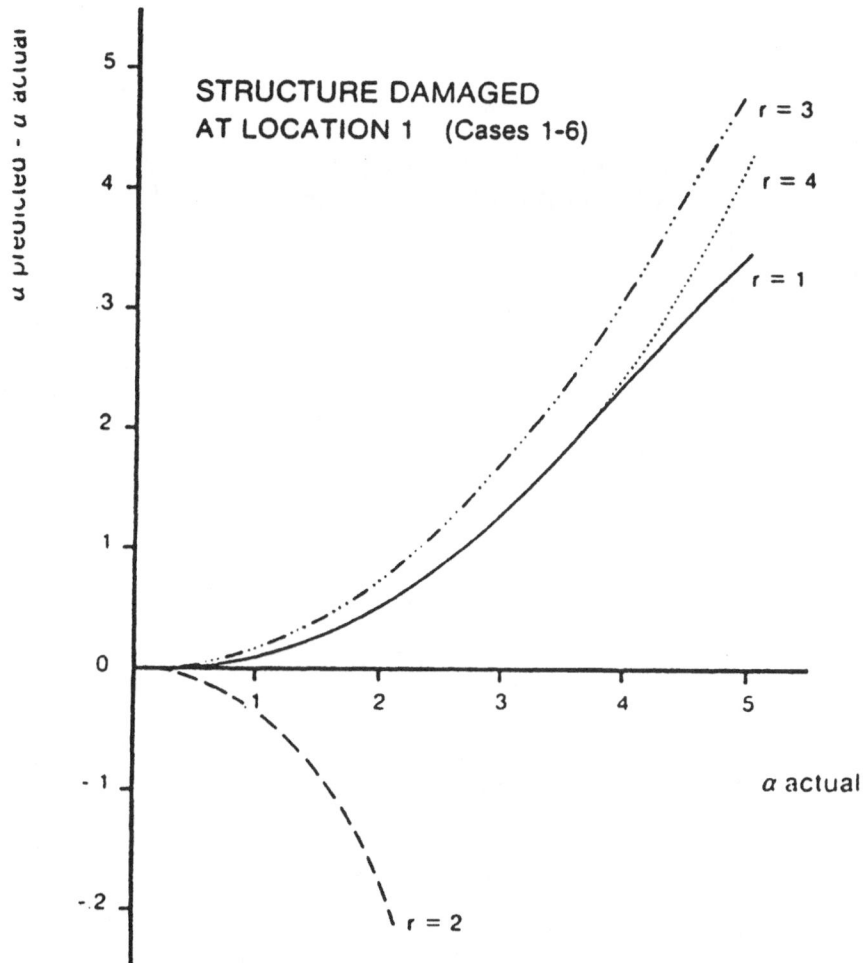

Figure 3: Prediction errors for structure damaged at location 1

3.5. Discussion of results

By studying the cases of the structure damaged in one location (Figs. 3 to 6), two trends can readily be observed. First, if the magnitude of inflicted damage at a single location is less than one tenth the original magnitude of the stiffness at that location, then the error in the prediction is approximately less than 10 percent. Thus for damage predictions of these magnitudes, the present formulation seems to provide favorable results. The second, and obvious, trend is that the error in the value of the predicted magnitude of damage rapidly increases as the

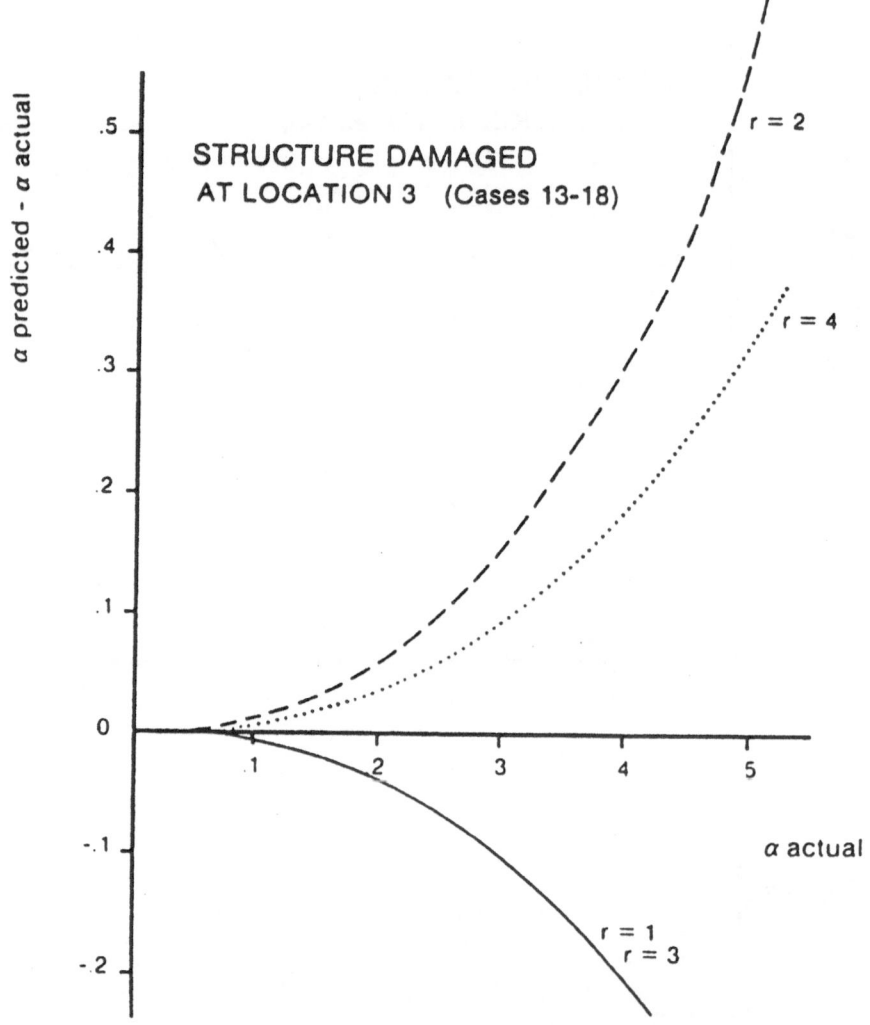

Figure 4: Prediction errors for structure damaged at location 3

magnitude of the inflicted damage increases. As expected, this trend follows from the fact that this formulation considers only first order effects in addition to the limitation that the first order changes resulting from damages in eigenvectors have been ignored.

However, the magnitude of the error can be reduced in several ways. First, the formulation can be extended to include the effects of first order changes in eigenvectors (see for example Eq. (21)). Secondly, the linearized model can be

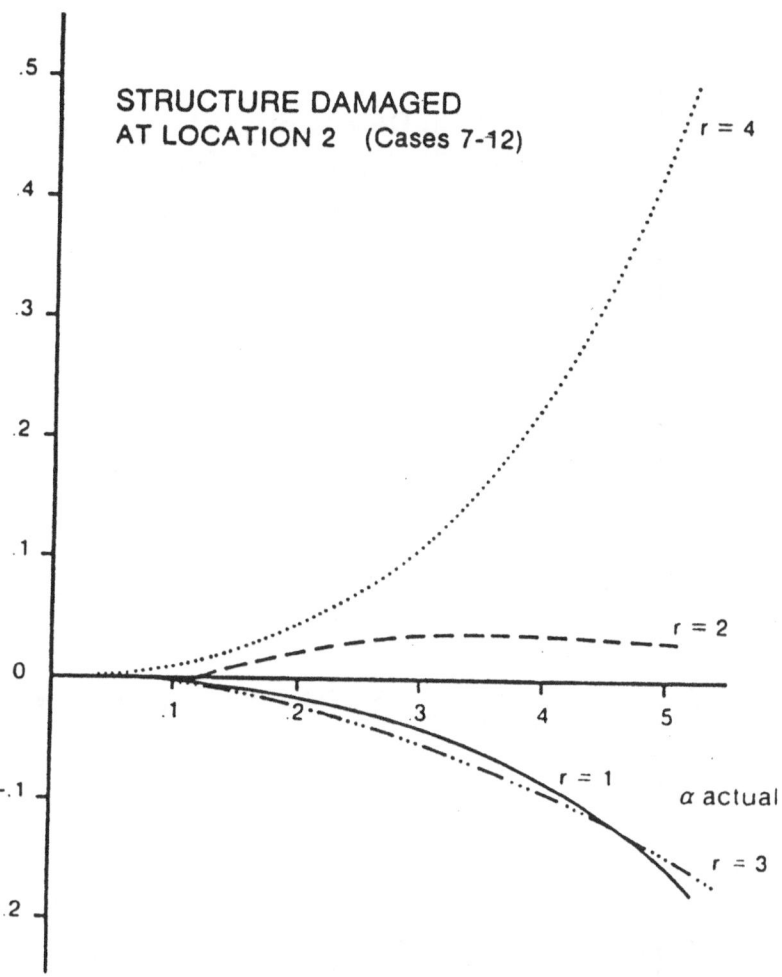

Figure 5: Prediction errors for structure damaged at location 2

applied iteratively. For example, given a fractional change in frequency represented by the matrix $[Z]$, we may divide $[Z]$ into p parts such that $[Z] = \sum_{j=1}^{p} [Z]_j$. For each $[Z]_j$, a corresponding level of damage for the system is computed using the result:

$$[A]_j = \{F_{j-1}\}^{-1}[Z]_j, \tag{30}$$

where

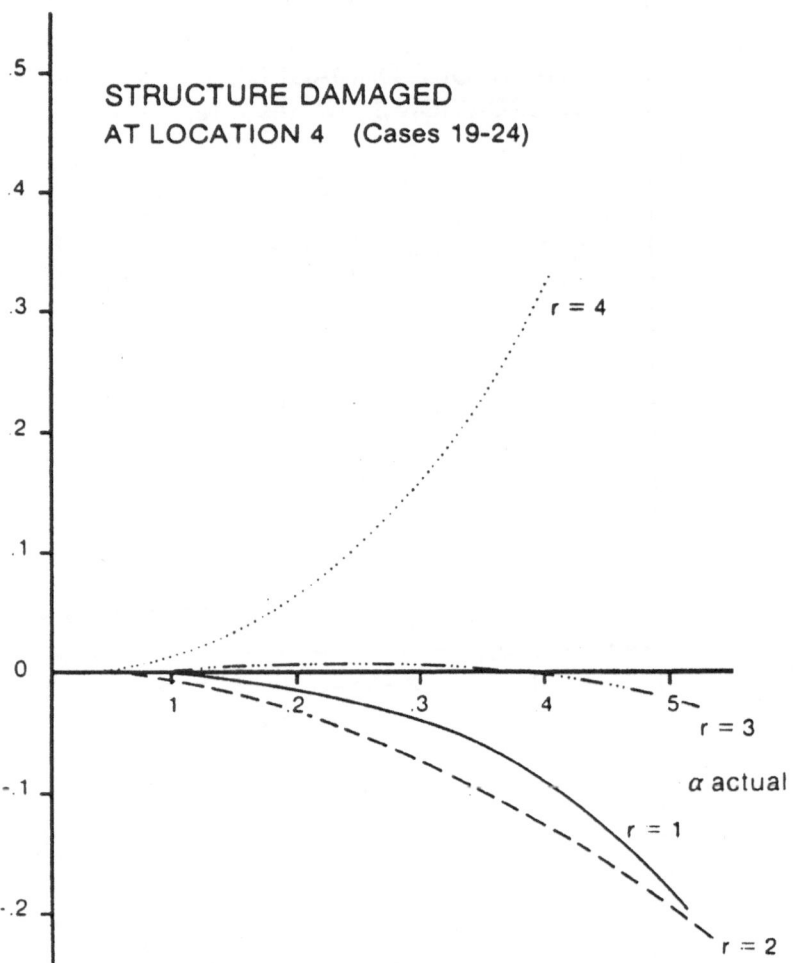

Figure 6: Prediction errors for structure damaged at location 4

$$F_{j-1} = F_{j-1}([F_{j-1}], [A_{j-1}]). \tag{31}$$

The total magnitude of the damage is thus given by:

$$[A] = \sum_{j=1}^{p} [A]_j \tag{32}$$

For the cases in which the structure has been damaged at two locations (Figs. 7 to 9), the predictions using the present formulation closely reproduce the actual damage inflicted on the structure. It should be pointed out that the level of damage in these and future cases was kept below 0.1 of the original stiffness of each location as suggested by the findings for the structure damaged in one location. However, note that in at least four cases (Fig. 7), when the damage was zero at location 2, the corresponding prediction showed an actual increase (albeit small) in stiffness for that location. This condition may be improved using one of the techniques discussed above. However, it should be noted that the condition disappears as the damage magnitude at location 1 decreases. Excellent agreement is observed for the cases in which the structure is damaged in three locations (see Figs. 7 to 10). For example, in damage cases 29-32, the deviation between the inflicted and the predicted damage is insignificant. Similarly, the error in the prediction is insignificant in cases 35-38.

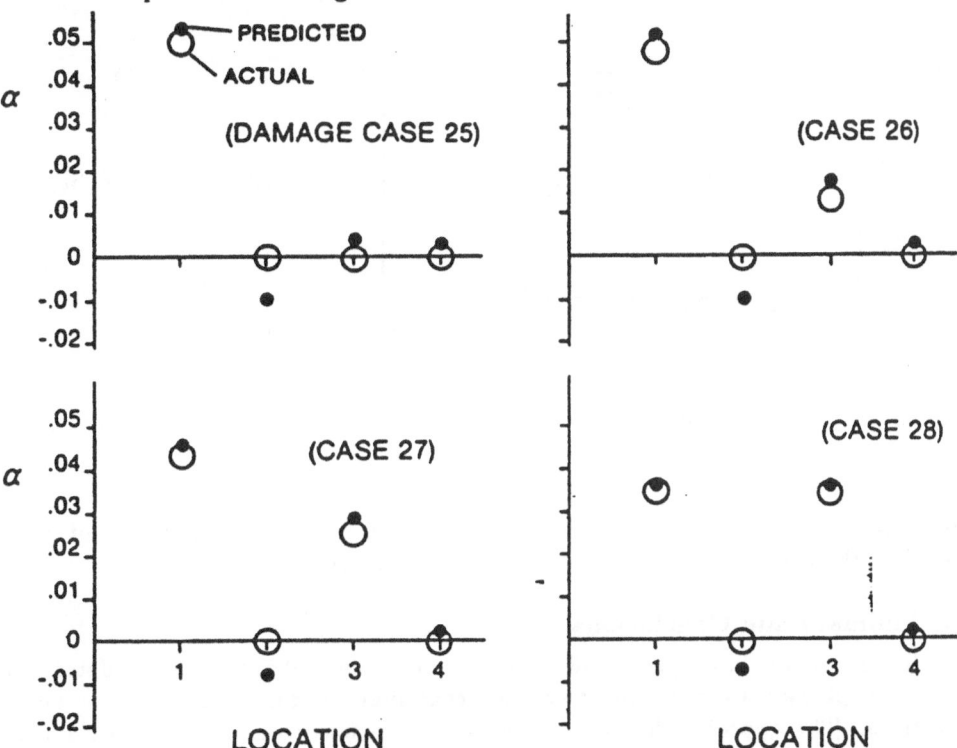

Figure 7: Comparison between actual damage and predicted damage for cases 25-28

Thus we may tentatively conclude from this example that if the structure is damaged in one location, the proposed formulation accurately predicts the location of damage for magnitudes less than approximately one tenth the value of the original stiffness. Secondly, when the magnitude of damage is kept to less than one tenth the value of the original stiffness in a member, the present formulation accurately predicts the location and magnitude of damage in a structure

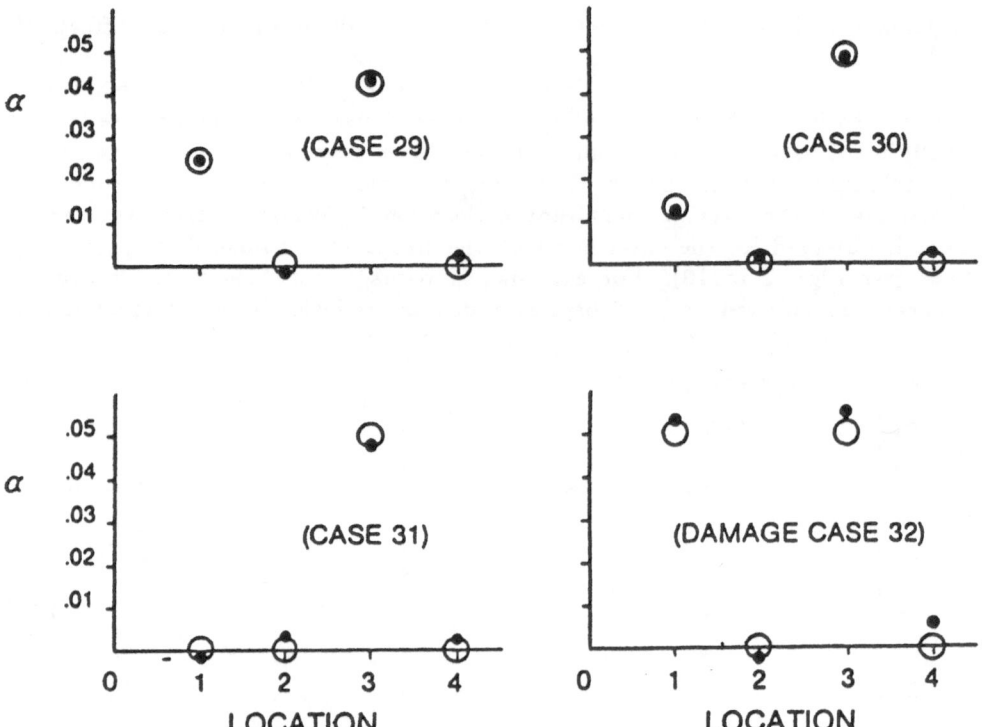

Figure 8: Comparison between actual damage and predicted damage for cases 29-32

damaged at multiple locations. Finally, when the magnitude of damage in one location is greater than approximately one tenth the original stiffness of that location, the present formulation may have to be corrected.

4. Summary and Conclusions

In this paper we have presented a theory of nondestructive damage detection that is applicable to mechanical systems that may be damaged in one or more locations. The approach has taken into consideration variations in mass, damping, and stiffness. A first difference was applied to the homogeneous equations of motion of a damaged structural system to yield expressions for changes in modal stiffnesses in terms of modal masses, modal damping, eigenfrequencies, eigenvectors, and their respective changes. From matrix structural analysis, expressions relating variations in stiffness of structural elements to the variations in modal stiffness were generated. Assuming that the changes in eigenfrequencies, masses, and damping can be experimentally determined, the result was a system of algebraic equations with a load vector of changes in modal stiffness and unknowns of changes in member stiffnesses. Methods of solution for the resulting system of equations were discussed along with special cases in which the structure may be

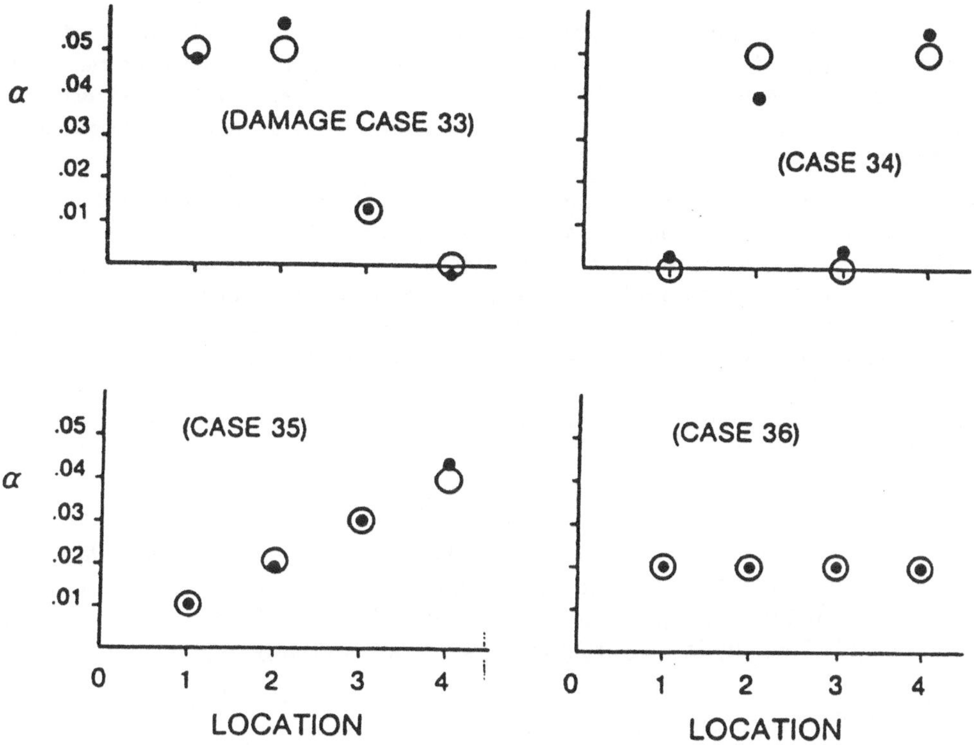

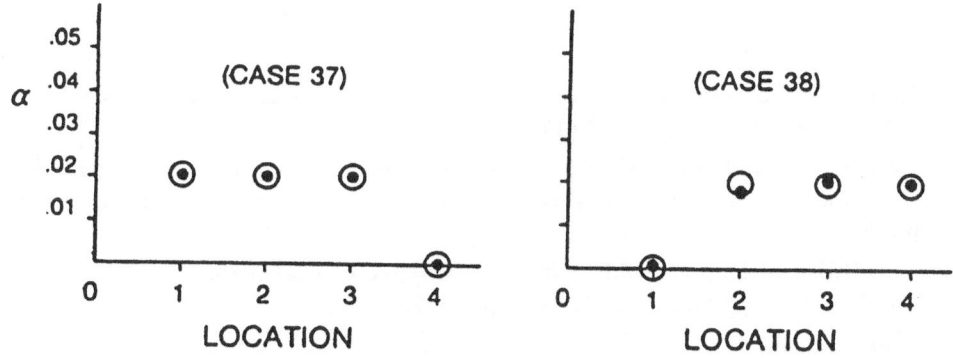

Figure 9: *Comparison between actual damage and predicted damage for cases 33-36*

Figure 10: *Comparison between actual damage and predicted damage for cases 37-38*

modelled to yield unique predictions. Furthermore, how the reliability of the prediction was affected by statistical uncertainty in material properties and dynamic measurements was addressed. The method was demonstrated using a numerical example of a multistory structure damaged at arbitrary locations and with arbitrary severities.

We concluded from this example that if the structure was damaged in one location, the proposed formulation accurately predicts the location of damage for magnitudes less than approximately one tenth the value of the original stiffness. Secondly, when the magnitude of damage was kept to below one tenth the value of the original stiffness in a member, the present formulation accurately predicts the location and magnitude of damage in a structure damaged at multiple locations. Finally, when the magnitude of damage in one location was greater than approximately one tenth the original stiffness of that location, the present formulation may have to be corrected.

Acknowledgement

This research was partially funded by the Center for Energy and Mineral Resources (Grant No. 18713) and the College of Architecture and Environmental Design (Grant No. 15250), Texas A & M University. In addition, the assistance of Charles S. Sikorsky in preparing this manuscript is greatly appreciated.

References

[1] Adams, R.D., Cawley, P., Pye, C.J., and Stone, B.J., "A Vibration Technique for Non-Destructively Assessing the Integrity of Structures," *J. Mechanics and Engineering Science*, Vol. 20, 1978, pp. 93-100.

[2] Cawley, P. and Adams, R.D., "The Location of Defects in Structures from Measurements of Natural Frequencies," *Journal of Strain Analysis*, Vol. 14, No. 2, 1979, pp. 4957.

[3] Craig, R.R., *Structural Dynamics: An Introduction to Computer Methods*, John Wiley and Sons, New York, 1981.

[4] Crohas, H. and Lepert, P., "Damage-Detection Monitoring Method for Offshore Platform is Field-Tested," *Oil and Gas Journal*, February 1982, pp. 94-103.

[5] Fox, R.L. and Kapoor, M.P., "Rates of Change of Eigenvalues and Eigenvectors," *AIAA J.*, Vol. 6, No. 2, 1968, pp. 2426.

[6] Gudmundson, P., "Eigenfrequencies Changes of Structures Due to Cracks, Notches or Other Geometrical Changes," *Journal of the Mechanics and Physics of Solids*, Vol. 30, 1983, pp. 339.

[7] Gudmundson, P., "The Dynamic Behavior of Slender Structures with Cross-Sectional Cracks," *Journal of the Mechanics and Physics of Solids*, Vol. 31, No. 4, 1983, pp. 329.

[8] Hasselman, T.K. and Hart, G.C., "Modal Analysis of Random Structural Systems,"*J. of the Engrg. Mech. Div.*, ASCE, June 1972, pp. 561.

[9] Kenley, R.M. and Dodds, C.J., "West Sole W.E. Platform: Detection of Damage by Structural Response Measurements," *OTC 3866*, 1980.

[10] Kreyszig, E., *Advanced Engineering Mathematics*, Third Edition, John Wiley and Sons, New York, 1972.

[11] Nataraja, R., "Structural Integrity Monitoring in Real Seas," *OTC paper 830514*, May 1983, pp. 221-228.

[12] O'Brien, T.K., "Stiffness Change as a Non-Destructive Damage Measurement," *Mechanics of Non-Destructive Testing*, ed. W.W. Stinchcomb, Plenum Press, New York, 1980, pp. 101-121.

[13] Penrose, R., "On the Best Approximate Solution of Linear Matrix Equations," *Proceedings of the Cambridge Philosophical Society*, Vol. 52, 1956, pp. 60-65.

[14] Spillers, W.R., *Automated Structural Analysis: An Introduction*, Pergamon Press, New York, 1972.

[15] Wittrick, W.H., " Rates of Change of Eigenvalues with Reference to Buckling and Vibration Problems," *J. of the Royal Aeronautical Society*, Vol. 66, September 1962, pp. 590.

[16] Wojnarowski, M.E., Stiansen, S.G., and Reddy, N.E., "Structural Integrity Evaluation of a Fixed Platform Using Vibration Criteria," *OTC 2909*, 1977.

[17] Yang, J.C.S., Chen, J., and Dagalakis, N.G., "Damage Detection in Offshore Structures by the Random Decrement Technique," *Journal of Energy Resources Technology*, ASME, Vol. 106, March 1984, pp. 38-42.

[18] Yang, J.C.S., Dagalakis, N.G., and Hirt, M., "Application of the Random Decrement Technique in the Detection of Induced Cracks on an Offshore Platform Model in Computational Methods for Offshore Structures," ASME, *AMD*, Vol. 37, 1980.

DISCONTINUOUS PULSE CONTROL
OF
LONGITUDINAL VIBRATIONS
WITHIN A TRAIN CONSIST

W. M. Sstrimbely, Ph.D., P.Eng.
Hatch Associates Ltd.
Toronto, Ontario, Canada

1. Introduction

The technological advances made over the past few decades have lead directly to the design and construction of larger and much more flexible fixed foundation structures, longer, more powerful and higher payload train ensembles. During the service life of each of these engineering systems excessive dynamic reactions to applied forces either man imposed or natural must be controlled to ensure the safety and integrity of the system. The control of the dynamic response of large engineering structures under seismic loading conditions has been one of the major areas of research by Civil Engineers in the past few years. Mechanical Engineers have concentrated on the control of large engineering vehicles such as the ocean going "Super Tankers", NASA's "Space Shuttle" or the one hundred car "Unit Trains" now used to carry material goods across North America. The Unit Train concept is used extensively throughout North America, whereby a train consist is made up of a number of locomotives and as many as one hundred identical rail cars of up to one hundred tons capacity each for carrying coal, or iron ore, etc. The Norfolk and Western Railroads have successfully tested the use of a five hundred car train consist between Williamson, West Virginia and Portsmouth, Ohio, as reported by Taylor [15]. The extreme lengths and payloads of the unit trains have lead to the development of new operating techniques throughout the railway industry. Trains cannot move without dynamic interactions between the wheels and the rails. The induced longitudinal oscillations of the cars results in longitudinal vibrations throughout the train consist. Excessive magnitudes of these vibrations as a consequence of increased train lengths, payloads and speeds must be reduced and/or eliminated in order to ensure stability and safety during operation.

The driving force developed between the locomotives' wheels and the rails, as well as the braking force developed between the wheels and rails are limited by the coefficients or friction present between the two interacting surfaces if all other variables remain constant. In starting a train from rest the usual practice is that of the locomotives backing up so as to bunch as many cars together as possible, and then to move forward starting one car at a time. The slack existing within the couplers between two adjoining cars allows this method to be used in order to maintain the drawbar force at the locomotives below the value where train pull-apart may occur or wheel slippage results.

Optimal control techniques have been used in railroad engineering whereby locomotives are distributed throughout the train consist and used to regulate coupler forces by applying driving or braking forces as required. This technique is known as "Locotrol" and has proved very useful when applied to the unit train. Examples are reported in the works of Levine [4], Martin [6], McLane [11], Peppard [12] and Wong [18]. Unfortunately the braking force as well as the driving force is limited by the coefficients of friction that exists between the locomotives wheels and the rails. In order to increase the effective braking force applied to a train, a means of applying controlled external forces in addition to the existing forces acting on the train is required. These external forces which are distributed throughout the train consist must be applied in such a manner as to be completely controllable by the engineer and adaptable to rapid changes in demand requirements. By definition we are lead to feedback control in order to accomplish this goal.

Through judicious application of the external control forces, the longitudinal vibrations throughout the train consist may be controlled and held below the level where coupler forces resulting from the excessive vibrations would lead to train pull-apart under tensile loads or train buckling and derailment under compressive loads. It is this longitudinal vibration control that the present work will address.

The specific problem under consideration is the longitudinal vibration control in a train consist during emerging braking. Under these circumstances the coupler forces within the train become excessive in both tension and compression and could very well lead to train derailment unless some control is introduced.

The discontinuous control strategy developed will when implemented with an active force generator result in:

1) increased braking effective force during emergency stopping,

2) reduction and/or elimination of shock wave propagation through the train consist,

3) elimination of excessive run-in and run-out coupler forces which lead to maintenance problems and in the extreme cases to train derailment.

A brief description of "Pulse Control" as applied to structural control will be presented followed by the theoretical development of a discontinuous control strategy for longitudinal vibration control within a train consist.

2. Discontinuous Pulse Control of Large Multidegree of Freedom Engineering Structures

The application of pulse control techniques to large engineering structures under seismic loading is given in the works of Masri [7-10] and Udwadia [16, 17], and to the control of offshore deep water structures by Soong [13]. The discontinuous pulse control strategy used by the first two authors is an "open loop adaptive control" technique which is incorporated to take into account the possible non-linear and time variant nature of the structures.

The pulse control strategy eliminates the need to apply large control forces for sustained periods of time and the open loop configuration reduces on-line computation by eliminating the need to solve high order Riccati equations resulting in the application of closed loop feedback optimal control techniques.

The pulse control strategy requires a continuous monitoring of system properties to determine if a threshold limit of displacement or velocity has been exceeded. The pulse duration is determined by the natural period of the system and must be applied to a direction to oppose the displacement and velocity so as to add complementary energy to the vibrating system. The pulse must be applied only at times when the response of the structure would exceed the threshold limits, in order to prevent the occurance of "Chatter" which may result under "Bang-Bang" control.

A detailed description of discontinuous control techniques is given by Flügge-Lotz [2,3] for both displacement dependent and velocity dependent control of engineering systems.

Train Braking

The forces effective in retarding the motion of a railway vehicle are 1) inherent resistance, 2) any incidental resistance effective at the time such as wind, rail curvature or ascending grade and 3) friction between the wheels and rails as the brakes are applied. The first two forces are neglected during emergency braking as they are approximately equal to 1% of the third force, Sztrimbely [14].

Brakes used on Northern American trains during emergency braking are operated by air pressure. A continuous pipe through the train called the "train line brake pipe" carries pressure from the compressor on the locomotives to a reservoir on each car through a control valve. Reduction of the air pressure in the brake line occurs as a result of manipulation of the brake valve control at the engineers station. The reduction in air pressure at the control valve of each car results in brake application. Once the process is initiated the "brake wave" travels rapidly through the train consist.

In order to prevent wheel slippage at the rails, the brake force must be maintained below the adhesion force limit defined as

$$f = \mu_f W \tag{1}$$

where μ_f = coefficient of wheel-rail adhesion,

and W = static weight upon the wheels.

When the vehicle is in motion, the actual or adhesive weight on the wheels is affected by weight transfer (moment transmitted to the trucks by the inertia of the car body through the truck centre plates) and verticle oscillations of body weight upon the truck springs. The value of μ_f remains constant at its static value at all speeds, being about 0.25 for a dry or sanded moist track. A value of 0.20 would provide reasonable assurance against sliding when used to include car body oscillation and weight transfer, Mark's [5].

During emergency braking the time delay that exists in the application of the brakes at each car leads to a "brake wave" that propagates through the consist. In order to reduce the delay times, the use of brake initialization cars have been introduced, where a reduction of brake line pressure occurs at these cars and the locomotives simultaneously, Martin [6]. By segmenting the train consist a more uniform application of the braking force results. The concept of brake utilization cars is shown in Figure 1. A unique problem that results from this method of emergency braking is that high compressive forces develop downstream from such a brake initialization car and high tensile forces develop upstream within the coupler assemblages. If either force exceeds a magnitude of ±250 KIPS, coupler failure possibly leading to train derailment may occur, Taylor [15].

By introducing discontinuous pulse control within the train consist, the excessive magnitudes of coupler forces resulting from the propagation of the brake wave through the train may be eliminated and a more even distribution of coupler forces achieved.

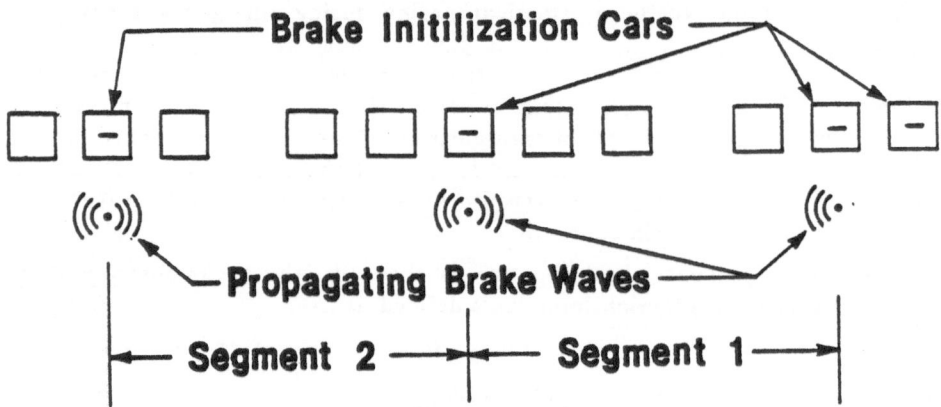

Figure 1: Brake Initialization Cars & Train Segments

Discrete Finite Difference Train Action Model

The physical make up of a train consist lends itself to analysis through a discrete train action model. The individual cars may be represented as point masses connected through a system of bilinear springs and dampers used to represent the coupler assemblages.

The analysis of the motion of a system of particles in particular the relative motion is made mathematically simpler through the introduction of the "Control Volume" as illustrated in Figure 2.

In Figure 2 the following definitions apply;

$X_c =$ displacement of centre of mass with respect to a non-accelerating frame of reference XOY, under applied external loading,

$x_i =$ displacement of ith car with respect to centre of mass,

$Z_i =$ the sum of the external forces applied to ith car within the control volume.

The centre of mass is the location at which the entire mass of the system may be assumed to be concentrated and to move under the influence of all the external forces summed by vector addition. We define the location of the centre of mass as

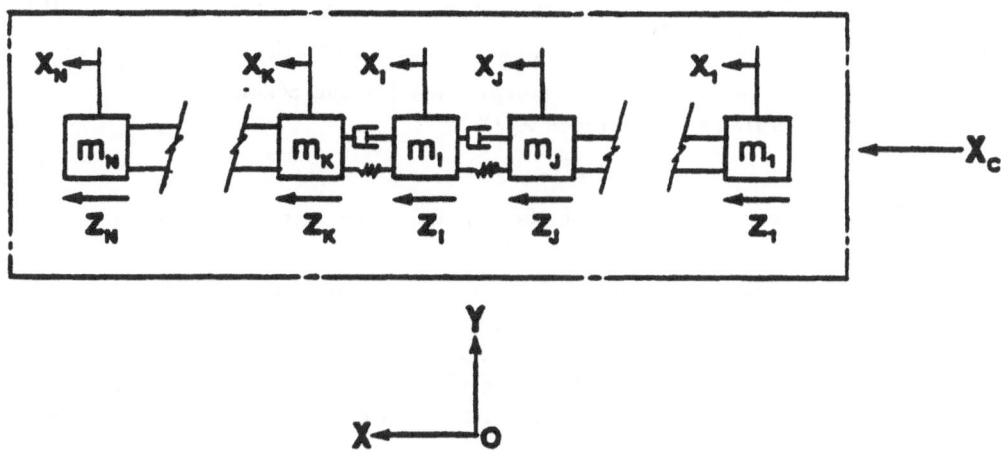

Figure 2: Control Volume of Train Consist

$$X_c = \frac{\sum\limits_{i=1}^{n} m_i r_i}{\sum\limits_{i=1}^{n} m_i}, \tag{2}$$

where r_i = position vector from the point "o" to the ith car.

Defining

$$M = \sum_{i=1}^{n} m_i \tag{3}$$

as the total mass of the system, we may express Newton's Second Law of Motion as

$$\sum_{i=1}^{n} Z_i = M \frac{d^2 X_c}{dt^2}, \tag{4}$$

The position of the ith car of the system with respect to the non-accelerating frame of reference is given as

$$x_i^o = X_c + x_i \tag{5}$$

where x_i = relative displacement of the ith car with respect to the centre of mass.

Taking the first derivative of (5) we obtain the expression for the velocity of the ith car with respect to the frame of reference XOY as,

$$\dot{x}_i^o = \dot{x}_c + \dot{x}_i \tag{6}$$

where $\dot{x}_i$ = relative velocity of the ith car with respect to the ith centre of mass

Differentiating (6) we obtain the expression for the acceleration of the car with respect to the frame of reference XOY as

$$\ddot{x}_i^o = \ddot{x}_c + \ddot{x}_i \tag{7}$$

where $\ddot{x}_i$ = relative acceleration of the ith car with respect to the centre of mass.

The overall aspects of train motion are modelled as shown in Figure 3 with coordinates of motion being assigned, one to each rigid body in the direction of the longitudinal car body axis. No coordinates are assigned to non-axial motion or to rotation about this axis. All forces acting due to external sources are assumed to act on the body mass centre. Forces on each car are summed and for each an equation of the form,

$$\sum F^i = 0 \tag{8}$$

is obtained.

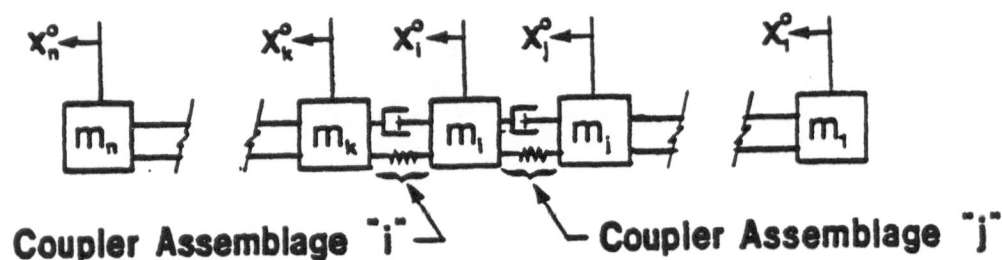

Figure 3: Coordinates of Motion Used in Train Action Model

The global free body diagram of the ith car of the consist is given in Figure 4. The assumed positive direction of motion and the corresponding direction of application of each force is indicated. The following definitions apply to Figure 4.

θ = angle of inclination of grade, positive if increasing towards front of train,

m_i = mass of ith car,

F_i^I = d'Alembert's inertia force,

F_i^s = spring coupling force of ith coupler assemblage,

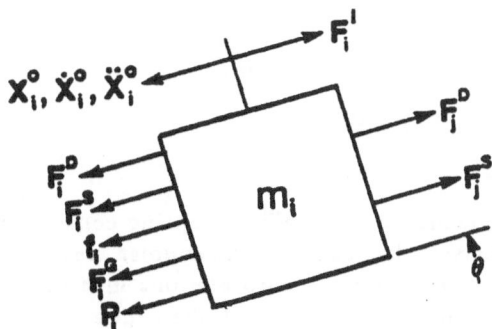

Figure 4: Global Free Body Diagram of ith Car of Consist

F_i^D = damping coupling force of ith coupler assemblage,

F_i^G = gravitational force on the ith car,

f_i = friction force on ith car,

P_i = control force applied at ith car,

x_i^o = displacement of ith car w.r.t. point "o",

$\dot{x}_i^o$ = velocity of ith car w.r.t. point "o",

$\ddot{x}_i^o$ = acceleration of ith car w.r.t. point "o",

Δx_i = relative displacement between ith and ith +1 cars measured from ith +1 car as reference,

$\Delta \dot{x}_i$ = relative velocity between the ith and ith +1 cars measured from ith +1 car as reference.

A description of each force will now be given.

Inertia Force

$$F_i^I = m_i \, \ddot{x}_i^o.$$

(9)

This force is applied according to d'Alembert's principal in the direction opposite to the assumed positive acceleration. By the introduction of (4) and (7) into (9) we obtain:

$$F_i^I = m_i \, \ddot{x}_i + m_i \, \frac{\sum\limits_{j=1}^{n} Z_j}{M}.$$

(10)

Gravitational Force

$$F_i^G = m_i g \sin\theta \tag{11}$$

where g = acceleration due to gravity.

Spring Coupling Force

The spring coupling is modelled using a bilinear spring constant in order to take into account the coupler slack that exists in the coupler assemblage. The magnitude of the coupler slack varies between zero and one inch per car coupler. The total slack present in the coupler assemblage is the sum existing in each coupler. We have used a value of one inch for total coupler slack in this work.

The bilinear spring model is shown in Figure 5 and the following definitions are used in determining the spring coupler Force, F_i^s.

$k_i \doteq$ stiffness coupling coefficient between ith and ith $+1$ cars; $k_i^{'} = 40$ [KIP/in.], Peppard [36] and $k_i = 1\%$ of $k_i^{'}$

$$\Delta x_i \doteq x_{i+1}^o - x_i^o \tag{12}$$

$\Delta x_i^{'} \doteq$ spring hardening value equal to half the total coupler slack existing in the coupler assemblage.

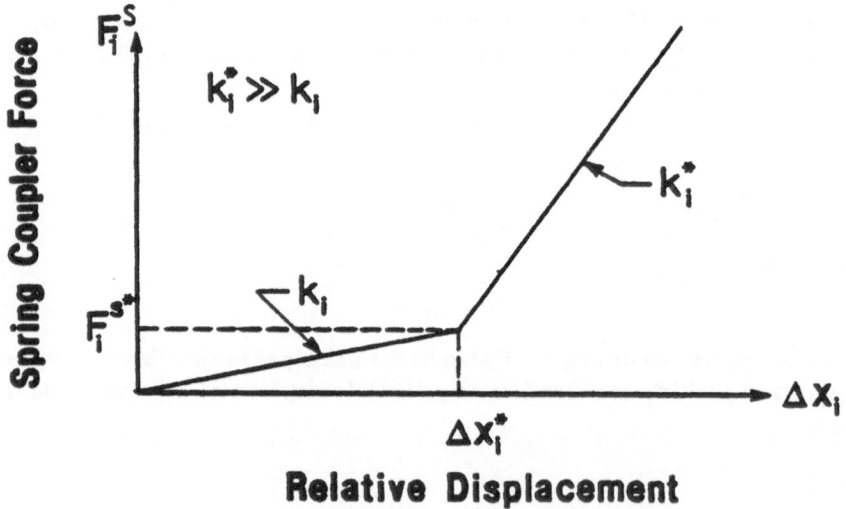

Figure 5: Bilinear Spring Force Model

The relative displacement between two cars as given in (12) may be rewritten with the introduction of (5) as

$$\Delta x_i = x_{i+1} - x_i,\tag{13}$$

From equation (13) we see that the value of Δx_i is determined from an algebraic equation and may have either a positive or a negative value, while $\Delta x_i^{\cdot}$ is taken as a constant positive value in the following development:

Case i)

$$0 \leq |\Delta x_i| \leq \Delta x_i^{\cdot}$$
$$F_i^s = k_i \Delta x_i.\tag{14}$$

Case ii)

$$|\Delta x_i| > \Delta x_i^{\cdot} \text{ in tension}$$
$$F_i^s = k_i \Delta x_i^{\cdot} + k_i^{\cdot}(\Delta x_i - \Delta x_i^{\cdot}).\tag{15}$$

Case iii)

$$|\Delta x_i| > \Delta x_i^{\cdot} \text{ in compression}$$
$$F_i^s = -k_i \Delta x_i^{\cdot} + k_i^{\cdot}(\Delta x_i + \Delta x_i^{\cdot}).\tag{16}$$

By defining

$$K_i = k_i - k_i^{\cdot}\tag{17}$$

and we may write (15) and (16) as

$$F_i^s = k_i^{\cdot}\Delta x_i + sign\,(\Delta x_i)K_i \Delta x_i^{\cdot}.\tag{18}$$

Damping Coupling Force

The damping coupling force is modelled as a bilinear force for the reasons stated for use of a bilinear spring model. The damping force is given as:

Case i)

$$0 \leq |\Delta x_i| \leq \Delta x_i^{\cdot}$$

$$F_i^D = c_i \Delta \dot{x}_i.\tag{19}$$

Case ii)

$$|\Delta x_i| > \Delta x_i^{\cdot}$$

$$\tag{20}$$

$$F_i^D = c_i^s \Delta \dot{x}_i \, ,$$

with

$$\Delta \dot{x}_i \dot{=} \dot{x}_{i+1}^o - \dot{x}_i^o \, , \tag{21}$$

and $\cdot c_i$ = damping coupling coefficient between ith and ith+1 cars with $c_i^{'}$ = 10% critical damping for a mass m_i and spring rate $k_i^{'}$ connected to a rigid support, Peppard [12], and c_i = 1% of $c_i^{'}$.

From the introduction of (6) into (21) we obtain the expression for relative velocity between two cars as

$$\Delta \dot{x}_i = \dot{x}_{i+1} - \dot{x}_i \, . \tag{22}$$

Friction Force

During emerging braking the major component of the total friction force acting on the car is due to the friction of the wheels on the rails. As stated earlier this force is limited to the magnitude of the adhesion force given in (1), and making use of this limit we define the variable of friction force as:

$$f_i = 0, \quad \text{brakes off},$$

$$\tag{23}$$

$$f_i = \mu_f m_i g \cos \theta \text{, brakes on},$$

with $\mu_f \dot{=}$ coefficient of adhesion, taken as 0.20.

The brake application model is illustrated in Figure 6, and is a modified version of the detailed brake model presented by Martin [6]. The various parameters of Figure 6 are defined as:

T_{del_i} = delay time between brake initialization and brake application at car "i".

T_{APPLY_i} = time required for full brake force to develop at car "i".

The *Steel Coil Train Test* reported in Martin [6] has been numerically simulated using the models and solution technique given in this work. The simulated results of car coupler forces and global train dynamics are in good agreement with the experimentally determined values. The comparison of values is reported in Sztrimbely, [14].

Control Force

The application of control force is in the form of a pulse, "P_i", required to limit the excessive coupler forces occurring as a result of brake wave induced relative displacements and velocities. The control strategy used for the application of the control pulse will be detailed in a later section.

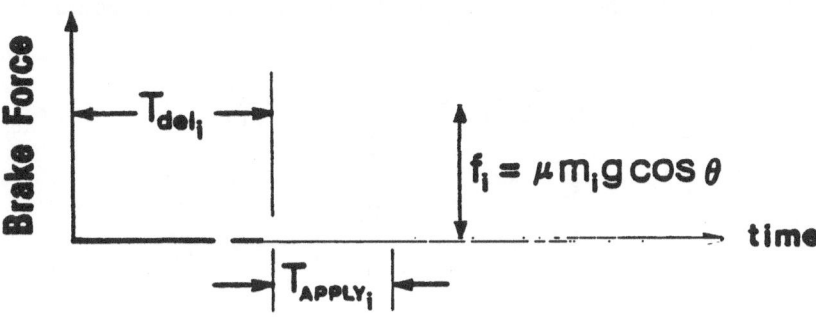

Figure 6: Brake Force Model

Finite Difference Equations of Longitudinal Motion

When the differential equations of motion cannot be integrated in closed form because, for example, the system is non-linear or is excited by force which is not expressible as a simple analytical function, numerical methods may be used to carry out the integration. The finite difference approximation used in this work is a first order backward difference scheme in which the velocity and acceleration of the ith car with respect to the centre of mass are given as

$$\dot{x}_I^j = \frac{x_I^j - x_I^{j-1}}{\Delta t} \tag{24}$$

and

$$\ddot{x}_I^j = \frac{x_I^j - 2x_I^{j-1} + x_I^{j-2}}{\Delta t^2} \tag{25}$$

respectively where

j = iteration number, superscript,

I = ith car, subscript.

The following variables are defined for use in the development of the finite difference equations of motion,

$J = I - 1$, jth car, subscript

$K = I + 1$, kth car, subscript, and

$$A_{cm} = \frac{\sum\limits_{i=1}^{n} Z_i}{M}, \tag{26}$$

as the acceleration of the centre of mass from (4).

The global free body diagram for the ith car of a train consist is given in Figure 7.

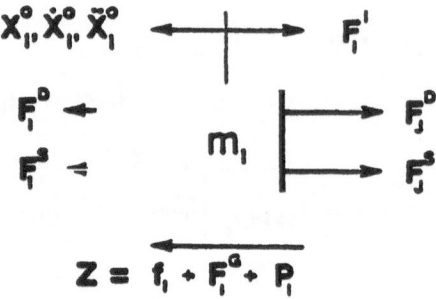

Figure 7: Global Free Body Diagram of ith Car

From Figure 7 we obtain the force equation (8) as

$$F_I^I + F_J^D + F_J^S = F_I^D + F_I^S + Z_I. \tag{27}$$

Substituting the definitions for each force shown in Figure 7 and representing each by a finite difference approximation leads to the following finite difference expressions for the longitudinal motion of the ith car.

a) $|\Delta x_I| \le \Delta \dot{x}_I$ & $|\Delta x_J| \le \Delta \dot{x}_J$ (28)

$$\left[m_I + c_J \Delta t + k_J \Delta t^2 + c_I \Delta t + k_I \Delta t^2 \right] x_i^j$$

$$= \left[c_J \Delta t + c_I \Delta t + 2m_I \right] x_i^{j-1} - m_I x_i^{j-2}$$

$$+ \left[c_I \Delta t + k_I \Delta t^2 \right] x_k^j - c_I \Delta t\, x_k^{j-1}$$

$$+ \left[c_J \Delta t + k_J \Delta t^2 \right] x_j^j - c_J \Delta t\, x_j^{j-1}$$

$$+ \left[Z_I - m_I A_{cm} \right] \Delta t^2 .$$

b) $|\Delta x_I| > \Delta \dot{x}_I$ & $|\Delta x_J| \le \Delta \dot{x}_J$ (29)

$$\left[m_I + c_J \Delta t + k_J \Delta t^2 + c_I' \Delta t + k_I' \Delta t^2 \right] x_i^j$$

$$= \left[c_J \Delta t + c_I' \Delta t + 2m_I \right] x_i^{j-1} - m_I x_i^{j-2}$$

$$+ \left[c_I' \Delta t + k_I' \Delta t^2 \right] x_k^j - c_I' \Delta t\, x_k^{j-1}$$

$$+ \left[c_J \Delta t + k_J \Delta t^2 \right] x_j^j - c_J \Delta t\, x_j^{j-1}$$

$$+ \left[Z_I - m_I A_{cm}\right]\Delta t^2 + \text{sign}(\Delta x_I)K_I \Delta x_I^{\cdot}\Delta t^2.$$

c) $|\Delta x_I| \leq \Delta x_I^{\cdot}$ & $|\Delta x_J| > \Delta x_J^{\cdot}$ (30)

$$\left[m_I + c_J^{\cdot}\Delta t + k_J^{\cdot}\Delta t^2 + c_I\Delta t + k_I\Delta t^2\right]x_I^j$$

$$= \left[c_J^{\cdot}\Delta t + c_I\Delta t + 2m_I\right]x_I^{j-1} - m_I x_I^{j-2}$$

$$+ \left[c_I\Delta t + k_I\Delta t^2\right]x_k^j - c_I\Delta t\, x_k^{j-1}$$

$$+ \left[c_J^{\cdot}\Delta t + k_J^{\cdot}\Delta t^2\right]x_j^j - c_J^{\cdot}\Delta t\, x_j^{j-1}$$

$$+ \left[Z_I - m_I A_{cm}\right]\Delta t^2 - \text{sign}(\Delta x_J)K_J \Delta x_J^{\cdot}\Delta t^2.$$

d) $|\Delta x_I| > \Delta x_I^{\cdot}$ & $|\Delta x_J| > \Delta x_J^{\cdot}$ (31)

$$\left[m_I + c_J^{\cdot}\Delta t + k_J^{\cdot}\Delta t^2 + c_I^{\cdot}\Delta t + k_I^{\cdot}\Delta t^2\right]x_I^j$$

$$= \left[c_J^{\cdot}\Delta t + c_I\Delta t + 2m_I\right]x_I^{j-1} - m_I x_I^{j-2}$$

$$+ \left[c_I^{\cdot}\Delta t + k_I\Delta t^2\right]x_k^j - c_I^{\cdot}\Delta t\, x_k^{j-1}$$

$$+ \left[c_J^{\cdot}\Delta t + k_J^{\cdot}\Delta t^2\right]x_j^j - c_J^{\cdot}\Delta t\, x_j^{j-1}$$

$$+ \left[Z_I - m_I A_{cm}\right]\Delta t^2 + \text{sign}(\Delta x_I)K_I \Delta x_I^{\cdot}\Delta t^2 - \text{sign}(\Delta x_J)K_J \Delta x_J^{\cdot}\Delta t^2.$$

Equations of the type (28) to (31) are developed for each car of the consist and integrated by means of a forward time stepping integration computer algorithm detailed in Sztrimbely [14].

Vibration Energy of the Train Consist

The vibrational energy of the mass-spring-damper system used to represent the train consist is composed of two components, the vibrational potential energy PE_v and the vibrational kinetic energy KE_v which are related by

$$E_{T_v} = KE_v + PE_v \tag{32}$$

where

$$E_{T_v} = \text{total vibrational energy.}$$

The vibrational potential energy of a coupler assemblage is equal to the area under the Spring Force vs. Displacement graph on Figure 5 and is obtained as

a) $|\Delta x_i| \le \Delta x_i^{\ast}$

$$PE_{v_i} = \frac{1}{2} k_i (\Delta x_i)^2$$

b) $|\Delta x_i| > \Delta x_i^{\ast}$

$$PE_{v_i} = \frac{1}{2} k_i (\Delta x_i)^2 + k_i \Delta x_i^{\ast} \left[|\Delta x_i| - \Delta x_i^{\ast} \right]$$

$$+ \frac{1}{2} k_i^{\ast} \left(|\Delta x_i| - \Delta x_i^{\ast} \right)^2 .$$

The total vibrational potential energy of the consist is given as

$$PE_v = \sum_{i=1}^{n} PE_{v_i} . \tag{35}$$

The total vibrational kinetic energy is defined by Sztrimbely [14] as

$$KE_v = \frac{1}{2} \sum_{i=1}^{n} m_i \dot{x}_i^2 . \tag{36}$$

Applied Total Impulse

The total impulse applied at each car is obtained from the integration of the applied pulse over time and is mathematically defined as

$$\text{Impulse}_i = \int_0^t P_i(t) \cdot dt . \tag{37}$$

The resulting total impulse applied to the train is the sum of the impulses applied at each car. In the discrete model if the pulse remains constant over a time step t the total impulse applied at each car may be obtained from

$$\text{Impulse}_i = \sum_{j=1}^{N} P_i^j \Delta t_j , \tag{38}$$

where $P_i^j \doteq$ magnitude of the pulse applied at the ith car during the jth time step Δt_j,

and

N = total number of time steps.

Control of Coupler Forces in a 33 Rail Car Train Segment Through Discontinuous Control

An example of the application of discontinuous pulse control to the longitudinal vibrations that develop in a train consist during emergency braking will now be presented. A thirty-three car train segment comprised of cars of 110 ton gross weight each are arranged as shown in segment 2 of Figure 1, with a brake initialization car located at each end of the segment. At time $t = 0$, the brakes are applied at these two cars and the brake application wave moves through the segment according to Figure 6. The brake waves cause longitudinal relative displacements and velocities through the segment and the resulting maximum force at each coupler location is shown in Figure 8. We see that at locations 12, 13, 20, 21 and 22 the coupler force exceeds the allowable safe limit of ±250 KIPS. The time of maximum coupler force at each location is shown in Figure 9 along with the time that brake application begins at each car and the time full brake force at each car is achieved. We see that at the location of excessive coupler force, the maximum force is experienced prior to full brake application at the two cars joined by the coupler. The total vibrational energy is plotted in Figure 10 as calculated from equation (32).

We now introduce the following control strategy assuming pulse generators are located at all cars within the segment, and the pulse applied is of constant magnitude equal to 75 KIPS. Based on the force conditions on each end of a car, the pulse will or will not be activated. These conditions are now given along with the control strategy.

The control strategy is to monitor the force on each end of a car at the couplers and apply the pulse so as to reduce upstream compression or downstream tension. In Figure 11 the control strategy is shown, where

F_J = Coupler force at location J, between cars J and I,

F_I = Coupler force at location I, between cars I and K,

P_I = Pulse control force applied at car I.

Condition 1 $F_J > 0$ and $F_I > 0$

$$P_I = \begin{cases} 75 \text{ [KIPS] if } F_J \leq 240 \text{ [KIPS] and } F_I > 100 \text{ [KIPS] and } \Delta \dot{x}_I > 0 \\ \quad\quad 0 \quad\quad\quad\quad\quad \text{other} \end{cases} \tag{39}$$

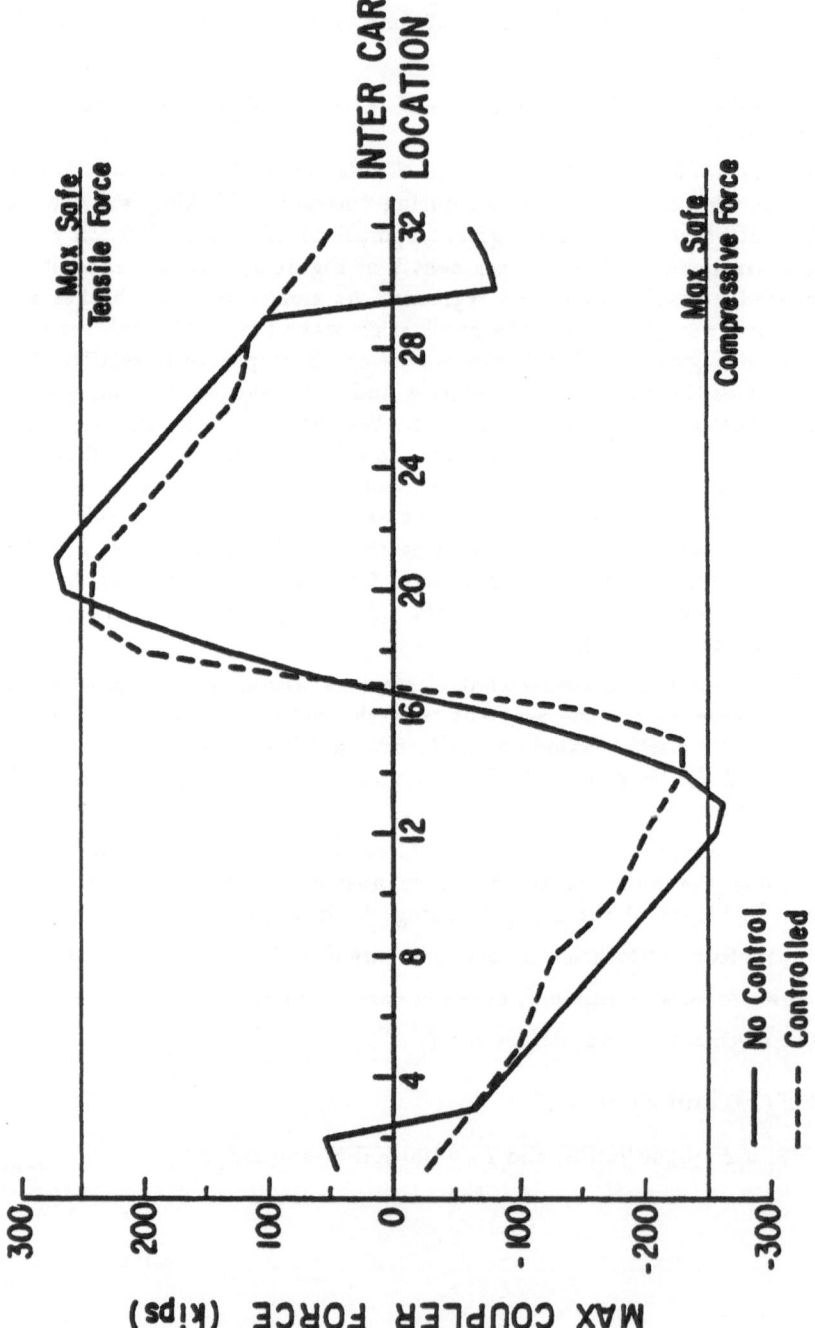

Figure 8: Max. Coupler Force Vs. Inter Car Location

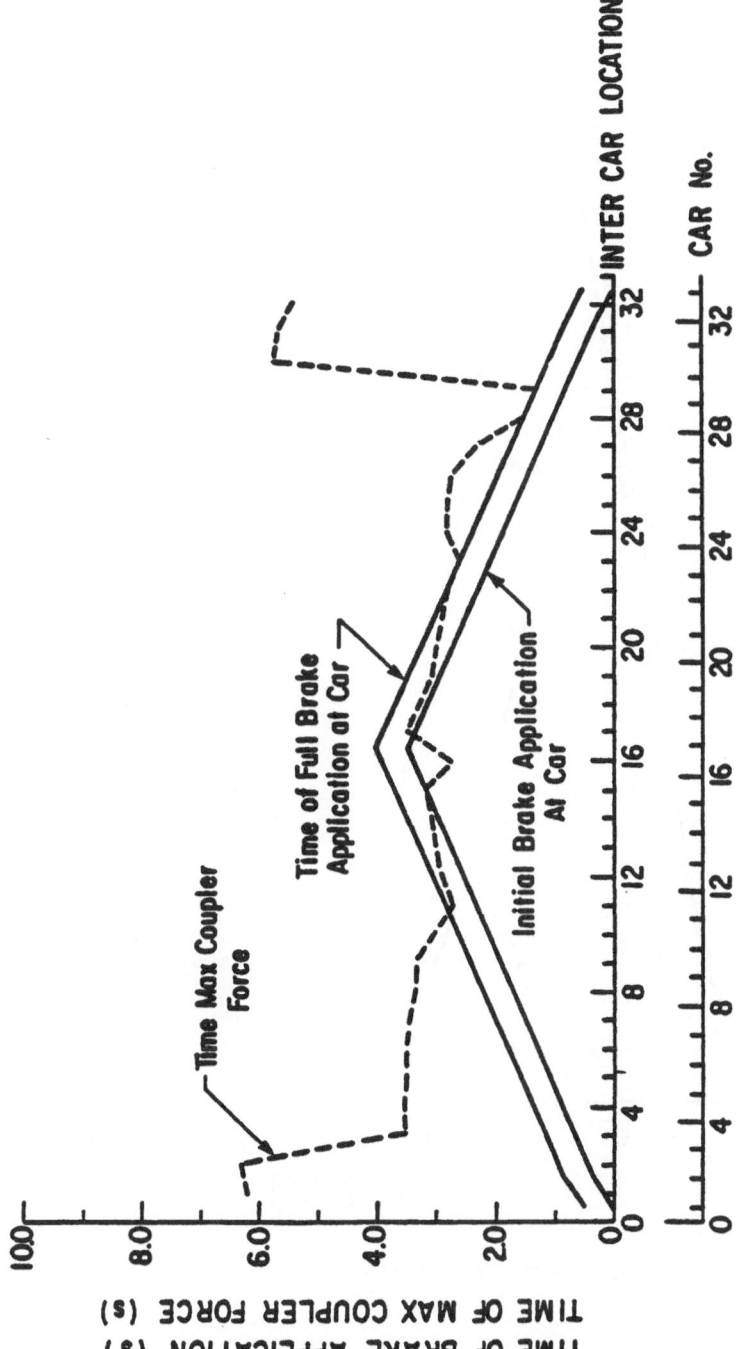

Figure 9: Time Max. Coupler Force Vs. Inter Car Location
Time of Brake Application Vs. Car No.

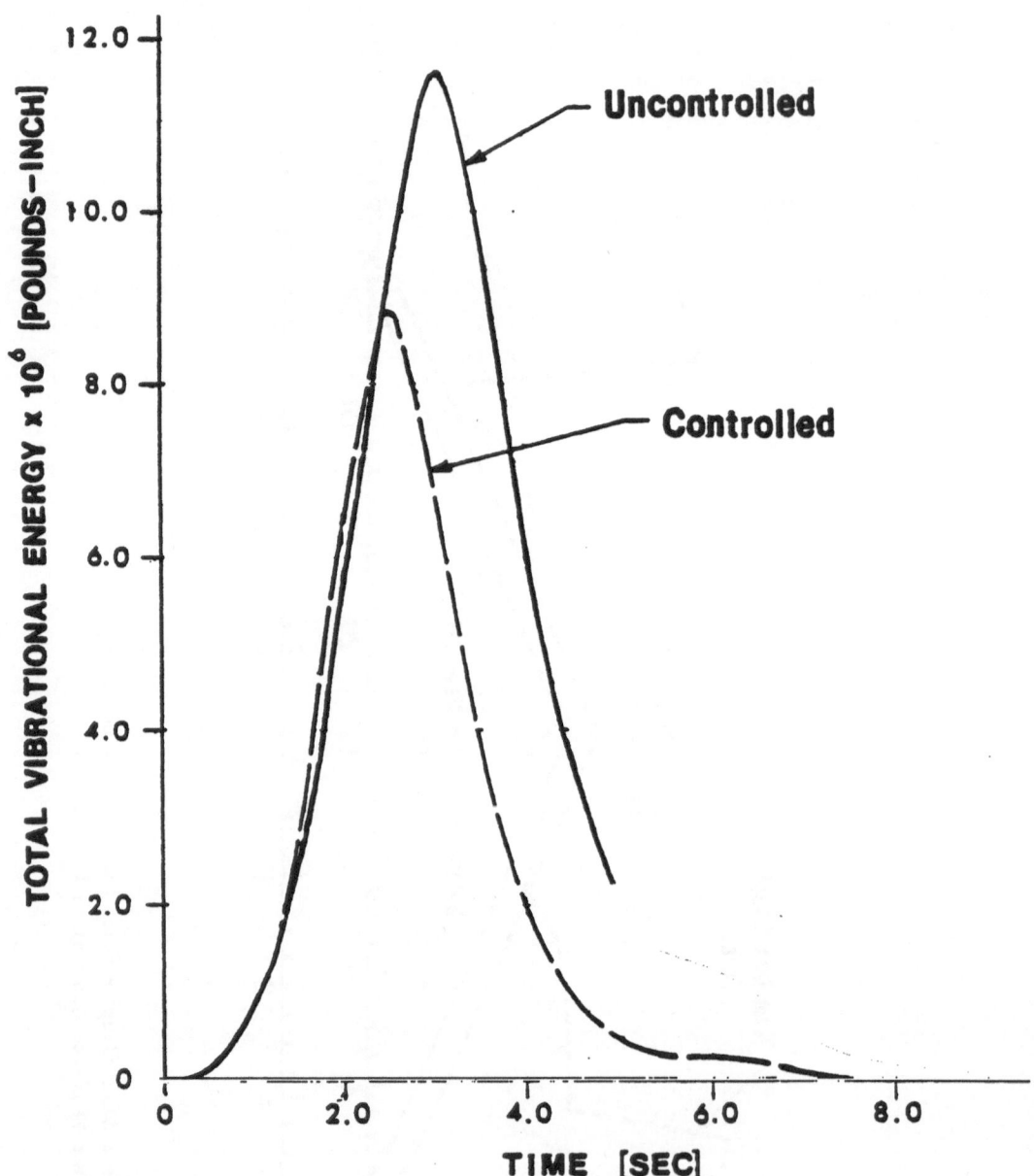

Figure 10:   Total Vibrational Energy Vs Time

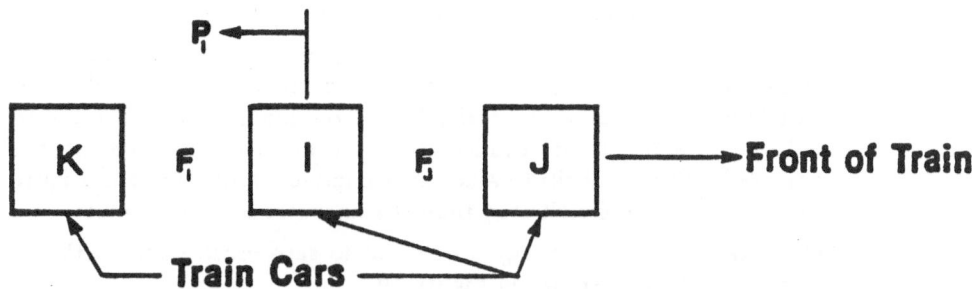

Figure 11: Control Strategy Set Up

Condition 2 $F_J < 0$ and $F_I < 0$

$$P_I = \begin{cases} 75 \text{ [KIPS]} & \text{if } F_J \leq -100 \text{[KIPS] and } F_I > -240 \text{[KIPS] and } \Delta \dot{x}_J < 0 \\ 0 & \text{other} \end{cases} \tag{40}$$

Condition 3 $F_J < 0$ and $F_I > 0$

$$P_I = \begin{cases} 75 \text{ [KIPS]} & \text{if } F_J < -100 \text{ [KIPS] and } \Delta \dot{x}_J < 0 \\ & \text{or } F_I > 100 \text{ [KIPS] and } \Delta \dot{x}_I > 0 \\ 0 & \text{other} \end{cases} \tag{41}$$

Condition 4 $F_J > 0$ and $F_I < 0$

$$P_I = 0. \tag{42}$$

The effect of the control pulse is to reduce the compressive force or increase the tensile force at location J and increase the compression force or reduce the tensile force at location I with respect to Figure 11.

The introduction of of the control strategy of equations (39) to (42) into the train action model results in new values of maximum coupler force at each location as shown in Figure 8 where we see that the excessive coupler forces have been reduced below the allowable limit. A plot of total vibrational energy is given in Figure 10.

The change in absolute magnitude of the maximum coupler forces divided by the absolute magnitude of the largest change in maximum coupler force is plotted in Figure 12. The total impulse applied at each car is shown in Figure 13 and the pulse train history at each car where a pulse was applied is given in Figure 14.

The following observations are made with respect to the results obtained in the previous example of discontinuous control.

1)   The effect of the applied pulses is to redistribute the maximum coupler force magnitudes to get a more balanced force distribution. From Figure 12 we see that the high force magnitudes are reduced and the low force magnitudes at the centre of the segment are increased. The controlled values of maximum coupler force in Figure 8 are more evenly distributed than the uncontrolled force values.

2)   The total vibrational energy is reduced to zero much faster with the control in effect as shown in Figure 10.

3)   The magnitude of applied impulse at a car is proportional to the magnitude of the coupler force as seen from Figures 13 and 8 respectively, and is very nearly symmetric in distribution about the centre of the segment, i.e. location 17.

Effects of Pulse Control In a Train Braking In a Curve

Applying the brakes during emergency stopping conditions as outlined earlier leads to the introduction of tensile forces between cars upstream from the brake initialization car and compressive forces downstream. On a curve these forces may lead to *draw-in* under the tensile force or *pop-out* under the compressive force if the force magnitude reaches the value where the lateral component of the braking induced force exceeds the design limit for safe traversing of the curve. The physical meaning of the terms draw-in and pop-out are illustrated in Figure 15.

Introducing a control strategy of the type investigated in this work, we see that by redistributing the coupler force as in Figures 8 and 12, the lateral components of these forces would be decreased at the locations of excessive force and the probability of *draw-in* or *pop-out* occurring is reduced. At the locations where due to the redistribution of force magnitude, there results an increase in force magnitude, no derailment occurs since the new force level is below the danger level.

3. General Comments and Conclusions

The Association of American Railroads [1] provides the design standards by which North American train cars are manufactured. In order to maintain structural integrity of rail cars, each car must be designed to:

a)   sustain a draft (tensile) of buff (compressive) drawbar and/or train action load of 350 KIPS applied along the nominal coupler centre lines. A load factor of 1.8 must be used in the design.

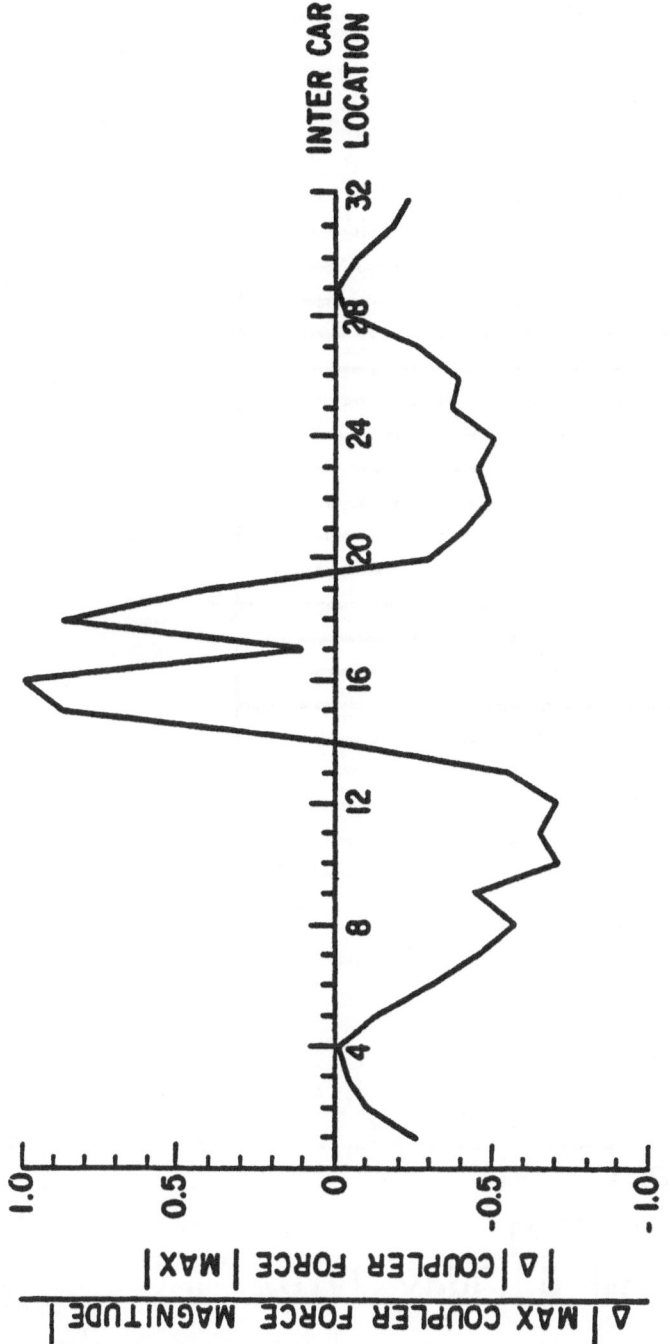

Figure 12: $\dfrac{\Delta|\text{Max. Coupler Force Magnitude}|}{|\Delta|\text{Coupler Force}|_{\text{Max.}}}$ *Vs Inter Car Location*

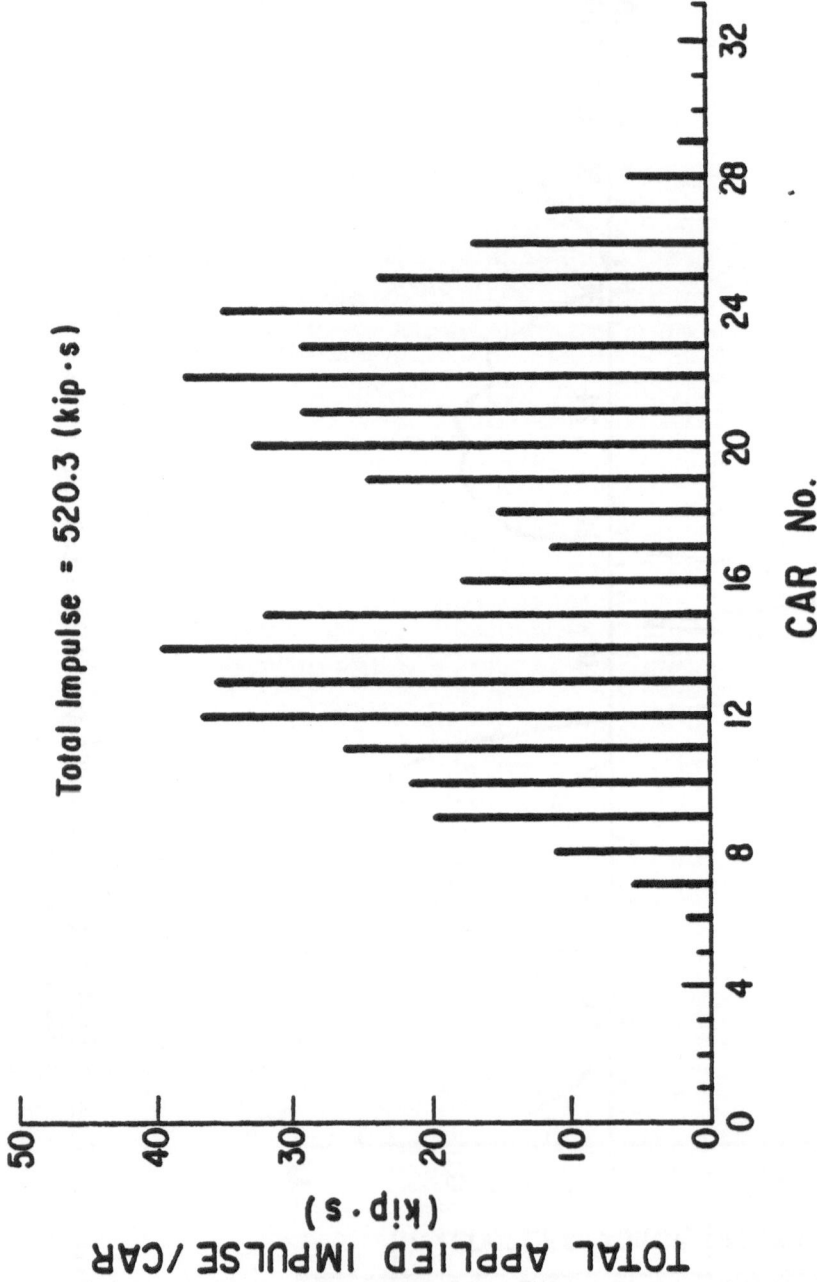

Figure 19: Total Applied Impulse/Car Vs. Car No.

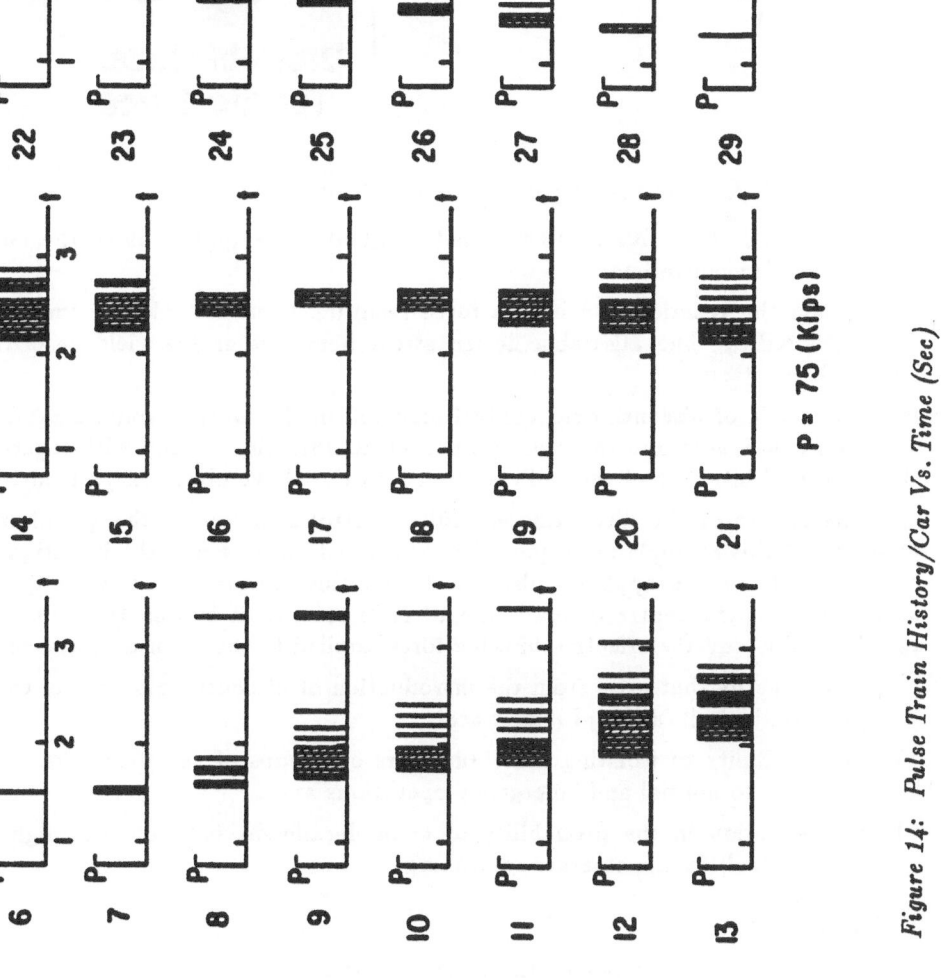

Figure 14: Pulse Train History/Car Vs. Time (Sec).

P = 75 (Kips)

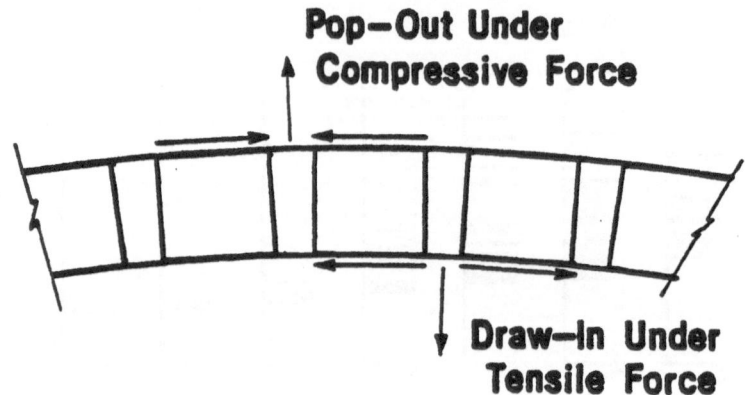

Figure 15:   Effects of Braking a Train on a Curve

b)   sustain a compressive columnar load of 1,000 KIPS applied along the nominal coupler centre line.

c)   sustain the reaction and inertia force resulting from a single end impact of 1,250 KIPS. The allowable design stress being either the yield or critical buckling stress.

In the example of discontinuous control reported in this work, a pulse magnitude of 75 KIPS was used and therefore we expect no difficulty arising with regard to the structural integrity of the rail cars based on the above given design loads.

The effects of the discontinuous pulse control strategy on the global train dynamics of this example are reported in Sztrimbely [14]. From the investigation of the effects of the applied pulse on the absolute displacement, velocity and acceleration of the centre of mass and of each rail car it is seen that the train remains stable and the effective braking force applied to the train is increased.

Two benefits that arise from the introduction of discontinuous control to the train longitudinal vibration problems are:

1)   the possibility of isolating individual cars or groups of cars from the vibrations due to normal and emergency operations and

2)   the reduction in the probability of train derailment both on a straight or curved track during emergency braking.

References

[1]   Association of American Railroads, *Manual of Standards and Recommended Practices, Section C - Part II, Specifications For Design, Fabrication and Construction of Freight Cars*, 1984, Vol. 1.

[2] Flügge-Lotz, I., *Discontinuous Automatic Control,* 1953, Princeton University Press.

[3] Flügge-Lotz, I., *Discontinuous and Optimal Control,* McGraw-Hill Book Company, 1968.

[4] Levine, W. S. and Athans, M., *On the Optimal Error Regulation of a String of Moving Vehicles,* IEEE Transactions on Automatic Control, Vol. AC-11, July 1966, pp. 355-361.

[5] *Mark's Standard Handbook for Mechanical Engineers* 8th Edition McGraw-Hill Book Company, 1978.

[6] Martin, G. C., and Tideman, H., *Detailed Longitudinal Train Action Model,* Association of American Railroads, R-221, 1977.

[7] Masri, S. F., Bekey, G. A. and Udwadian, F. E., *On-Line Pulse Control of Tall Buildings,* Structural Control Leipholz (editor), pp. 471-491, 1980, IUTAM.

[8] Masri, S. F., Bekey, G. A, Caughey, T. K., *Optimum Pulse Control of Flexible Structures,* University of Southern California, USC - CE - 8101, 1981.

[9] Masri, S. F., Bekey, G. A., Caughey, T. K., *On-Line Control of Non-Linear Flexible Structures,* University of Southern California, USC - CE - 8106, 1981.

[10] Masri, S. F., Miller, R. K., Bekey, G. A., Dehghanyar, T. J., and Caughey, T. K., *An Analytical and Experimental Study Into The Stability and Control of Non-Linear Flexible Structures,* 4th VPI and SU/AIAA Symposium on Dynamics and Control of Large Structures, June 1983, Virginia.

[11] McLane, P. J., *Control of Multi Locomotive Powered Trains,* Progress Report, Canadian Institute of Guided Ground Transport, Annual Report, 1972.

[12] Peppard, L. E., McLane, P. J., and Sundareswaran, K. K., *Localized Feedback Control for Multi Locomotive Powered Trains,* 1973, Proceedings of IEEE Conference on Decision and Control, San Diego paper TP 52 pp. 491-496.

[13] Soong, T. T., Prucz, Z., Reinhorn, A., *Pulse Control of Deep Water Offshore Structures,* National Science Foundation CEE 8010891.

[14] Sztrimbely, W. M., *Longitudinal Vibration Control by Discontinuous Feedback Control Methods,* 1985, Ph.D. Thesis, University of Waterloo, Dept. of Mech. Eng.

[15] Taylor, R. E., Tausch, R. G., Whitney, D. E., Kelly, J. J., *Hauling of Heavy Trains by U.S. Railroads,* Rail International August 1971, pp. 685-693.

[16] Udwadia, F. E. and Tabaie, S., *Pulse Control of Single Degree of Freedom System,* ASCE, Vol. 107, EM6, Dec. 1981, pp. 997-1009.

[17] Udwadia, F. E., and Tabaie, S., *Pulse Control of Structural and Mechanical Systems,* ASCE, Vol. 107, EM6, Dec. 1981, pp. 1011-1028.

[18] Wong, K. Y., Bayoumi, M. M., Ahmed, M. E., Oszoy, I. C., *Evaluation of Design Alternatives For a Freight Train Controller,* Report No. 83 -12 Transport Canada No. TP5114E, 1983 Canadian Institute of Guided Ground Transport.

RELIABLE STABILIZATION OF INTERCONNECTED SYSTEMS USING A SWITCHING MULTILAYER CONTROL STRUCTURE

W.Y. Yan
Institute of Systems Science
Academia Sinica
Beijing 100080, China

1. Introduction

The dynamic systems consisting of a number of interconnected subsystems, such as power systems, ecological systems, transportation systems, and etc., widespreadly appear in the natural world and man's life. One of the most striking characteristic of such a system is its interconnection structure, which may change while the system is operating, because of breakdown of circuits, or sudden change of loads, etc. Hence in the design of interconnected control systems, besides assuring the system to have satisfactory performance under the normal interconnection, the effects on the system, of interconnection perturbations, which may occur should also be considered, particularly for strongly interconnected systems. Siljak first recognized this point. In 1972, he presented the notion of connective stability for large-scale interconnected dynamic systems from an analytic point of view, and gave the sufficient conditions for an interconnected system to possess the connective exponential stability by using the method of vector Lyapunov functions. In 1976, from a design point of view, Siljak dealt with the problem of determining the conditions satisfied by perturbations of interconnection structure under which the perturbations do not violate the stability of the closed-loop system. In [3], Geromel, et al. considered the problem of stability of large-scale interconnected systems with a two-level control subjected to structural perturbations between two levels by means of Lyapunov vector functions. In a word, all these papers have concentrated on structural perturbations which may occur in interconnected control systems. Another striking characteristic of interconnected systems is the restriction of decentralized information and control structure. Therefore, one has to apply decentralized control in order to control such systems. The problem on the stabilization of decentralized control systems has been perfectly solved by Wang and Davison [4].

Based on the interconnected structure and decentralized structure of large-scale interconnected dynamic systems, we are forced to consider the following problem of decentralized reliable stabilization. Does there exist a decentralized controller for an interconnected system such that the resultant closed-loop system remains stable under arbitrary perturbations of the interconnected structure? This paper tries to solve the problem by establishing a switching multilayer control scheme where each layer's control is of the decentralized form. Clearly, this control structure is still decentralized, and what is particular consists in its having switching parameters attached.

The layout of the paper is as follows:

In Sec. 2, we formulate the problem considered hereafter. In Sec. 3, we give the necessary and sufficient conditions for the existence of a reliable stabilizing bilayer controller for an interconnected system with two subsystems, and the sequence of synthetic procedures of the controller.

2. Formulation

Consider a composite system consisting of two subsystems by linear interconnections, described by the following model:

$$\dot{x}_1 = A_1 x_1 + e_{12} A_{12} x_2 + B_1 u_1,$$
$$\dot{x}_2 = e_{21} A_{21} x_1 + A_2 x_2 + B_2 u_2,$$
$$y_i = C_i x_i, \quad i = 1,2, \tag{1}$$

where $x_i(t) \in R^{n_i}$, $u_i(t) \in R^{m_i}$ and $y_i(t) \in R^{r_i}$, $i = 1,2$ are the states, inputs and outputs of the ith subsystem S_i: (C_i, A_i, B_i); e_{ij} and A_{ij} are the interconnection coefficient and matrix of the jth subsystem to the ith one, $i \neq j$, $j = 1,2$.

We assume that the interconnection matrices remain invariant, but the interconnection coefficients are allowed to take values of 1 or 0 indicating that the interconnection of one subsystem to another may exist or not while the system works. Hence, a pair of the interconnection coefficients (e_{12}, e_{21}) solely determines the interconnection structure of the system (1), and visa versa. In addition, it does not lose generality to define $(e_{12}, e_{21}) = (1,1)$ as the normal interconnection structure of the system (1).

For the system (1), we propose a kind of switching bilayer decentralized control scheme as follows:

The first control layer is composed of local controllers of two subsystems,

$$\Sigma_{1i}: \quad \begin{array}{l} \dot{z}_{1i} = Q_{1i}z_{1i} + R_{1i}y_i + v_{1i}, \\ u_i = S_{1i}z_{1i} + K_{1i}y_i + v_{2i}, \end{array} \quad i = 1,2. \tag{2}$$

The second control layer is made up of the following decentralized dynamic compensator with interconnection coefficients as its switching parameters on the basis of the first.

$$\Sigma_{2i}: \quad \begin{array}{l} \dot{z}_{2i} = Q_{2i}z_{2i} + R_{2i}\begin{bmatrix} y_i \\ z_{1i} \end{bmatrix}, \\ \begin{bmatrix} v_{1i} \\ v_{2i} \end{bmatrix} = (1 - e_{12}e_{21})(S_{2i}z_{2i} + K_{2i}\begin{bmatrix} y_i \\ z_{1i} \end{bmatrix}), \quad i = 1,2. \end{array} \tag{3}$$

The feedback frame constituted by the system (1) and controller (2) - (3) can be explained by means of the following diagram.

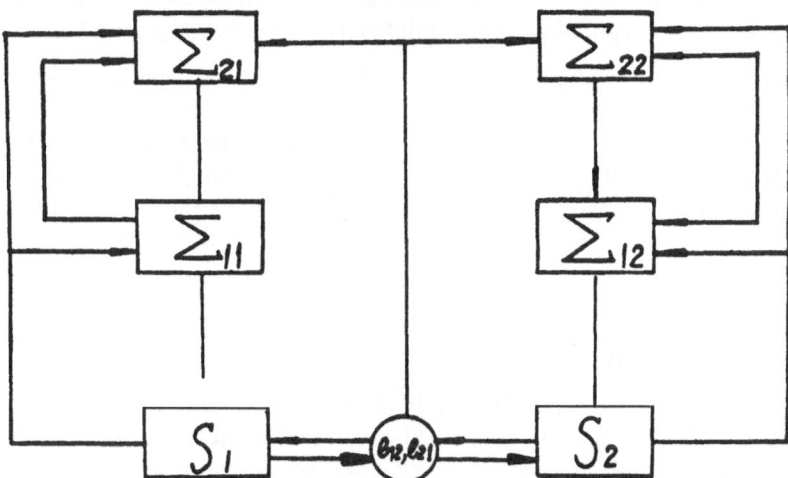

Figure 1: A switching two-layer control configuration

The problem considered in this paper is to determine under what conditions there exists a bilayer controller of the form (2)-(3) for the plant (1) such that the resultant closed-loop system is stable independently of any perturbation of the interconnection structure in (1). If such a controller exists, then it is said to stabilize reliably the system (1).

3. Existence and Synthesis of Reliably Stabilizing Bilayer Controllers

We have the following main result:

Theorem. Given an interconnected system (1). Then there exists a reliably stabilizing bilayer controller for (1) if and only if the system (1) as a two-channel decentralized system has no unstable fixed mode under the normal interconnection and each subsystem S_i $(i = 1,2)$ is jointly stabilizable and detectable.

The definition, characterizations and algorithm of fixed modes of decentralized systems can be found in [4, 5, 6]. In order to prove the above theorem, we have to draw support from the following auxiliary results.

Lemma 1 [4]. For a N-channel decentralized system $(C_i, A_i, B_i; i = 1,...,N)$, there exists a decentralized stabilizing compensator if and only if the system has no unstable fixed mode.

Lemma 2 [7]. For a linear time-invariant plant (C, A, B), there exists a stable stabilizing compensator if and only if the plant is jointly stabilizable and detectable, and its real zeros and poles in the right-half closed complex plane have the even interlacing property, i.e., if distinct finite real zeros of $C(sI - A)^{-1}B$ in Re $s \geq 0$ are denoted by $\sigma_1, \sigma_2, ..., \sigma_k$ and the total number of real poles of $C(sI - A)^{-1}B$ to the right of σ_i, each counted according to its Mcmillan degree, is denoted by ν_i $(i = 1,2,...,k)$, then the integers $\nu_1, \nu_2, ... , \nu_k$ are all even.

In [7], a plant for which there exists a stable stabilizing compensator is called strongly stabilizable.

Proof of Theorem. The necessary part of the theorem is easy to prove. Assume that there exists a bilayer controller of the form (2)-(3) reliably stabilizing the system (1). Clearly, when the system (1) is under the normal interconnection, i.e., $e_{12} = e_{21} = 1$, the second layer's compensator does not yield any effect on the first layer's controller and the system since

$$\begin{bmatrix} v_{11} \\ v_{21} \end{bmatrix} = \begin{bmatrix} v_{12} \\ v_{22} \end{bmatrix} = \begin{bmatrix} 0 \\ 0 \end{bmatrix}.$$

In this case, in order to make the whole closed-loop system stable, first, the first layer's controller (2) and the plant (1) must make up a stable system, which implies that (2) is a decentralized stabilizing compensator of (1); secondly, the second layer's controller must be also stable itself. On the other hand, when the system (1) is subjected to perturbations of interconnection structure, i.e., $e_{12}e_{21}$'s value changes from 1 into 0, two subsystems in (1) are in a state of triangular or complete decoupling. Hence, in order that the decentralized compensator made up of (2) and (3) stabilizes the system (1) perturbed structurally, each subsystem in (1) must be stabilized by its corresponding local controller, from which the necessary part of the theorem follows.

Sufficiency. By Lemma 1 and the assumption that the system (1) has no unstable decentralized fixed mode, a decentralized dynamic compensator of (1) can be designed as the first layer's controller (2) so that the following extended system, constituted by (1) and (2):

$$\begin{bmatrix} \dot{x}_1 \\ \dot{z}_{11} \end{bmatrix} = \begin{bmatrix} A_1 + B_1 K_{11} C_1 & B_1 S_{11} \\ R_1 C_1 & Q_{11} \end{bmatrix} \begin{bmatrix} x_1 \\ z_{11} \end{bmatrix} + e_{12} \begin{bmatrix} A_{12} & 0 \\ 0 & 0 \end{bmatrix} \begin{bmatrix} x_2 \\ z_{12} \end{bmatrix} + \begin{bmatrix} B_1 & 0 \\ 0 & I \end{bmatrix} \begin{bmatrix} v_{11} \\ v_{21} \end{bmatrix},$$

$$\begin{bmatrix} \dot{x}_2 \\ \dot{z}_{12} \end{bmatrix} = e_{21} \begin{bmatrix} A_{21} & 0 \\ 0 & 0 \end{bmatrix} \begin{bmatrix} x_1 \\ z_{11} \end{bmatrix} + \begin{bmatrix} A_2 + B_2 K_{12} C_2 & B_2 S_{12} \\ R_{12} C_2 & Q_{12} \end{bmatrix} \begin{bmatrix} x_2 \\ z_{12} \end{bmatrix} + \begin{bmatrix} B_2 & 0 \\ 0 & I \end{bmatrix} \begin{bmatrix} v_{12} \\ v_{22} \end{bmatrix},$$

$$\begin{bmatrix} y_i \\ z_{1i} \end{bmatrix} = \begin{bmatrix} C_i & 0 \\ 0 & I \end{bmatrix} \begin{bmatrix} x_i \\ z_{1i} \end{bmatrix}, \qquad i = 1,2, \tag{4}$$

is internally stable under the normal interconnection. Obviously, Eq. (4) is still an interconnected system composed of two extended subsystems. Now, we are to construct the second layer's controller so that under its action the extended system (4) remains stable when there occur interconnection faults. In view of the assumption that the subsystem S_i ($i = 1,2$) is jointly stabilizable and detectable, it is known that the extended subsystem S_i':

$$\left(\begin{bmatrix} C_i & 0 \\ 0 & I \end{bmatrix}, \begin{bmatrix} A_i & 0 \\ 0 & 0 \end{bmatrix}, \begin{bmatrix} B_i & 0 \\ 0 & I \end{bmatrix} \right), \qquad i = 1,2,$$

is also jointly stabilizable and detectable, and that its real zeros and poles have the even interlacing property since it has no zero. Therefore in terms of Lemma 2, for the extended subsystem S_i' ($i = 1,2$) we can design a stable stabilizing compensator denoted by (Q_{2i}, R_{2i}, S_{2i}, K_{2i}'). Set

$$K_{2i} \triangleq K_{2i}' - \begin{bmatrix} Q_{1i} & R_{1i} \\ S_{1i} & K_{1i} \end{bmatrix}, \qquad i = 1,2 .$$

Then the second layer's controller (3) can be constructed by ($Q_{2i}, R_{2i}, S_{2i}, K_{2i}$; $i = 1,2$). It is evident that each extended subsystem in (4) is stable under the feedback of its corresponding local controller in the second control layer, i.e., the decentralized compensator (3) makes the extended system (4) stable in the circumstance of the existence of interconnection faults. So the whole closed-loop system made up of (1), (2) and (3) is stable when there exist interconnection faults. Lastly, since the second control layer is stable itself, the whole closed-loop system is still stable under the normal interconnectin from the switching property of (3). To sum up, the so synthesized two-layer controller (2)-(3) must reliably stabilize the interconnected system (1). Thus the proof of the theorem is complete. Q.E.D.

From this theorem, the following corollary is easily obtained.

Corollary: The necessary conditions for the bilayer controller (2)-(3) to stabilize reliably the interconnected system (1) are that the first layer's controller makes (1) stable under the normal interconnection and that the second layer's is stable itself.

It can also be seen from the above theorem that the reliably stabilizing bilayer controller, if exists, can be synthesized in the following way. The sequence of procedures to synthesize a reliably stabilizing bilayer controller of (1):

Step 1. Find a decentralized stabilizing dynamic compensator for the normally interconnected system as the first layer's controller, denoted by $(Q_{1i}, R_{1i}, S_{1i}, K_{1i}; 1 = 1,2)$.

Step 2. Extend subsystem S_i into S_i':

$$\left[\begin{bmatrix} C_i & 0 \\ 0 & I \end{bmatrix}, \begin{bmatrix} A_i & 0 \\ 0 & 0 \end{bmatrix}, \begin{bmatrix} B_i & 0 \\ 0 & I \end{bmatrix} \right], \quad i = 1,2,$$

with its order equal to the sum of orders of S_i and,

Step 3. Construct a stable stabilizing compensator for the extended subsystem S_i', denoted by (Q, R, S, K') $(i = 1,2)$.

Step 4. Let:

$$K_{2i} = K_{2i}' - \begin{bmatrix} Q_{1i} & R_{1i} \\ S_{1i} & K_{1i} \end{bmatrix}, \quad i = 1,2 .$$

Then by putting $(Q_{2i}, R_{2i}, S_{2i}, K_{2i}; i = 1,2)$ into (3), the scond layer's controller is obtained. In this way, the synthesis of the reliably stabilizing bilayer controller is completed.

Finally, we give several evident concluding remarks.

Remark 1. If a subsystem in (1) is not strongly stabilizable, it is necessary to make its corresponding local controller in the first layer be dynamic in the synthesis, otherwise a stable stabilizing controller cannot be designed in the second layer.

Remark 2. It is not hard to see from the above theorem that for almost all interconnected systems of the form (1), there exists a reliably stabilizing bilayer controller of the form (2)-(3).

Remark 3. The prerequisite for an interconnected system of the form (1) to be stabilized by means of the switching two-layer control, is that whether when there occur interconnection faults in (1), they can be detected at any time during operation. Since the interconnection faults can frequently cause the instantaneous sudden change of the outputs of the subsystems, we can also identify whether or not there occur interconnection faults by observing the outputs continuously, instead of measuring the interconnection coefficients directly.

4. Conclusions

This paper has considered the problem on the reliable stabilization of large-scale dynamic systems consisting of two subsystems in terms of arbitrarily linear interconnections by means of the two-layer decentralized control with switching parameters, and has given the necessary and sufficient conditions for the existence of a reliably stabilizing bilayer controller and its concrete synthetic procedure as well. Nevertheless, how to extend the structure rationally so that it is suited for solving the same problem for the interconnected systems with a number of subsystems, awaits further probe. Eventually, it is necessary to point out that the control scheme presented in this paper can equally be adapted to solve the problem on the reliable servomechanism or regulation of interconnected systems, and it does not have any essential difficulty to do this.

Acknowledgement

The author expresses his gratitude to my advisor Professor E.P. Wang, Institute of Systems Science, Academia Sinica, Beijing, China, for his comments on this paper.

References

[1]   Siljak, D.D., "Stability of the Large Scale Systems Under Structural Perturbations," *IEEE Trans.*, SMC-2, 1972, pp. 657-663.

[2]   Siljak, D.D. and Sundareshan, M.K., "A Multilevel Optimization of Large Scale Dynamic Systems," *IEEE Trans.*, AC-21, 1976, pp. 79-84.

[3]   Geromel, J.G. and Bernussou, J., "Stability of Two-Layer Control Scheme Subjected to Structural Perturbations," *Int. J. Contr.*, Vol. 29, 1979, pp. 313-324.

[4]   Wang, S.H. and Davison, E.J., "On the Stabilization of Decentralized Control Systems," *IEEE Trans.*, AC-18, 1973, pp. 473-478.

[5]   Anderson, B.D.O. and Clements, D.J., "Algebraic Characterization of Fixed Modes in Decentralized Control," *Automatica*, Vol. 17, 1981, pp. 703-712.

[6]   Davison, E.J., "Decentralized Stabilization and Regulation in Large Multivariable Systems," in Ho, Y.C. and Mitter, S. (Eds), *Direction in Decentralized Control, Many-Person Optimization and Large Scale Systems*, Plenum Press, 1979, pp. 303-323.

[7]   Youla, D.C., Bongiorno, Jr. J.J., and Lu, C.N., "Single-Loop Feedback-Stabilization of linear Multivariable Dynamical Plants," *Automatica*, Vol. 10, 1974, pp. 159-173.

OPTIMAL CONTROL ALGORITHMS FOR EARTHQUAKE-EXCITED BUILDING STRUCTURES

J. N. Yang, A. Akbarpour and P. Ghaemmaghami
Department of Civil, Mechanical & Environmental Engineering
The George Washington University
Washington, D.C., U.S.A., 20052

1. Introduction

Considerable research efforts have been made recently to investigate the feasibility of the application of active control systems to building structures [e.g., 1-20]. The results of these studies indicate that some active control systems are promising for building structures under hostile environments, such as strong wind gusts, earthquakes, etc. The theoretical study, however, should be demonstrated experimentally. Recently, laboratory experiments for building structures subjected to seismic excitations have been undertaken at the State University of New York at Buffalo. In this connection, one important task, among many others, is the investigation of various optimal control algorithms for practical applications to building structures under earthquake excitations. It is the objective of this paper (i) to critically review various classical optimal control algorithms in relation to their effectiveness, practical implementation and the required on-line computations, and (ii) to propose an instantaneous optimal open-loop control algorithm. Because the base motions of the seismic-excited building can be measured, the implementation of such an open-loop algorithm is practical. A numerical example is given to demonstrate the effectiveness of various optimal control algorithms.

748

2. Classical Optimal Control

For simplicity, consider a one-dimensional building structure implemented by an active tendon control system as shown in Figure 1. It is idealized by an n degrees of freedom system and subjected to a one-dimensional earthquake ground acceleration $\ddot{X}_0(t)$. The matrix equation of motion can be expressed as:

$$\dot{Z}(t) = AZ(t) + BU(t) + W_1\ddot{X}_0(t) \tag{1}$$

with the initial condition $Z(0) = 0$. In equation (1), $Z(t) = 2n$ state vector, $U(t) = r$ dimensional control vector, $A = 2n \times 2n$ matrix, $B = a\ 2n \times r$ matrix specifying the locations of active controllers, and W_1 is an appropriate $2n$ vector.

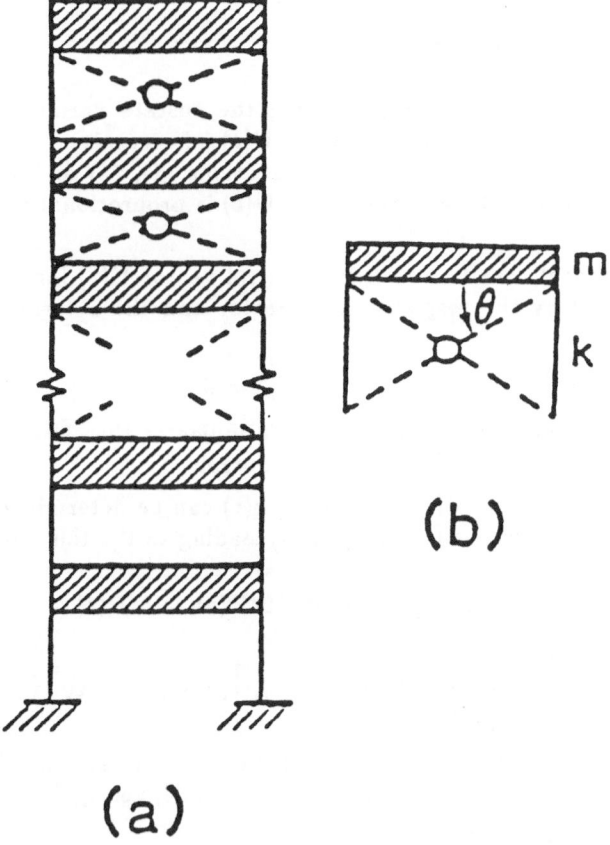

(b)

(a)

Figure 1: Structural model of a multi-story building with active tendon control system.

The standard quadratic performance index J given in the following is considered:

$$J = \int_0^{t_f} [\mathbf{Z}'(t)\mathbf{Q}\mathbf{Z}(t) + \mathbf{U}'(t)\mathbf{R}\mathbf{U}(t)]dt, \tag{2}$$

in which a prime denotes the transpose of a vector or matrix. In Equation (2), $\mathbf{Q}$ is a 2n × 2n positive semi-definite matrix, $\mathbf{R}$ is a r × r positive definite matrix and t_f is a duration defined to be longer than that of the earthquake.

To minimize the performance index J subjected to the constraint given by Equation (1), the necessary conditions can be shown as follows [e.g., Ref. 2]:

$$\dot{\lambda}(t) = \mathbf{A}'\lambda(t) - 2\mathbf{Q}\mathbf{Z}(t) \quad , \quad \lambda(t_f) = \mathbf{0}, \tag{3}$$

$$\mathbf{U}(t) = -\frac{1}{2}\mathbf{R}^{-1}\mathbf{B}'\lambda(t), \tag{4}$$

in which $\lambda(t)$ is a 2n vector representing the costate variables (or Lagrangian multipliers). The optimal control vector $\mathbf{U}(t)$, the costate vector $\lambda(t)$, and the state vector $\mathbf{Z}(t)$ can be solved using Equations (1), (3) and (4). It is noticed from Equation (4) that the control vector $\mathbf{U}(t)$ is proportional to the costate vector $\lambda(t)$.

For the general case in which the control vector $\mathbf{U}(t)$ (or the costate vector $\lambda(t)$) is regulated by the response state vector and the external excitations, one has

$$\lambda(t) = \mathbf{P}(t)\mathbf{Z}(t) + \mathbf{q}(t) \quad , \quad \lambda(t_f) = \mathbf{0}, \tag{5}$$

where the first term on the right hand side indicates the closed-loop control and the second term represents the open-loop control.

The unknown matrix $\mathbf{P}(t)$ and vector $\mathbf{q}(t)$ can be determined by substituting Equation (5) into Equations (1), (3) and (4) leading to the following expression:

$$\left[\dot{\mathbf{P}}(t) + \mathbf{P}(t)\mathbf{A} - \frac{1}{2}\mathbf{P}(t)\mathbf{B}\mathbf{R}^{-1}\mathbf{B}'\mathbf{P}(t) + \mathbf{A}'\mathbf{P}(t) + 2\mathbf{Q}\right]\mathbf{Z}(t)$$

$$+ \dot{\mathbf{q}}(t) - \left[\frac{1}{2}\mathbf{P}(t)\mathbf{B}\mathbf{R}^{-1}\mathbf{B}' - \mathbf{A}'\right]\mathbf{q}(t) + \mathbf{P}(t)\mathbf{W}_1\ddot{X}_0(t) = 0. \tag{6}$$

(1) *Optimal closed-loop control:* For the special case in which the control force is regulated by the response state vector alone, i.e., $\mathbf{q}(t) = 0$, one has

$$\lambda(t) = \mathbf{P}(t)\mathbf{Z}(t). \tag{7}$$

Then Equation (6) reduces to

$$\left[\dot{P}(t) + P(t)A - \frac{1}{2}P(t)BR^{-1}B'P(t) + A'P(t) + 2Q \right] Z(t)$$

$$+ P(t)W_1\ddot{X}_0(t) = 0 \quad , \quad P(t_f) = 0. \tag{8}$$

When the earthquake ground accelertion $\ddot{X}_0(t)$ is zero, Equation (8) becomes

$$\dot{P}(t) + P(t)A - \frac{1}{2}P(t)BR^{-1}B'P(t) + A'P(t) + 2Q = 0 \quad , \quad P(t_f) = 0. \tag{9}$$

Equation (9) is the classical Riccati equation and $P(t)$ is the Riccati matrix.

Strictly speaking, the Riccati matrix $P(t)$ obtained from Equation (9) does not result in the optimal closed-loop control, when a building structure is subjected to environmental loads, such as earthquakes or strong wind gusts. The optimal closed-loop control is achieved by the Riccati matrix only if the earthquake excitation is zero or a white noise random process [e.g. 13].

It is mentiond that the Riccati matrix, Equation (9), depends exclusively on the structural characteristics and the weighting matrices Q and R. For building structures, extensive experience indicates that the Riccati matrix $P(t)$ remains constant (i.e., each element of $P(t)$ remains constant) over the entire duration of earthquake excitation and it drops rapidly to zero near t_f. In other words, $P(t)$ establishes a stationary state in a very short period of time starting from t_f backwards. Typical elements of $P(t)$ for an eight story building are shown in Figure 2. As a result, the effectiveness of the control system is not affected when the Riccati matrix is approximated by a constant matrix, i.e., $P(t) = P$ or $\dot{P}(t) = 0$, as long as t_f is longer than the earthquake duration. Consequently, for building structures under earthquake excitations, the constant Riccati matrix P can be used, and Equation (9) becomes a matrix algebraic equation

$$PA = \frac{1}{2}PBR^{-1}B'P + A'P + 2Q = 0. \tag{10}$$

It is emphasized that the optimal closed-loop control requires the feedback measurements of the full state vector $Z(t)$, i.e., $2n$ sensors are needed.

2. Optimal Closed-Open-Loop Control: When the control vector $U(t)$ is expressed in the form of Equation (5), the control law is referred to as the optimal closed-open-loop control. In this case the control vector is determined by the measured state vector and the earthquake ground acceleration. The Riccati matrix P and the vector $q(t)$ are obtained from Equation (6) as follows:

$$PA - \frac{1}{2}PBR^{-1}B'P + A'P + 2Q = 0, \tag{11}$$

$$\dot{q}(t) - \left[\frac{1}{2}PBR^{-1}B' - A' \right]q(t) + PW_1\ddot{X}_0(t) = 0, \quad q(t_f) = 0, \tag{12}$$

in which the time dependent Riccati matrix $P(t)$ has been approximated by a constant matrix P, i.e., $P(t) = P$. The validity of such an approximation has

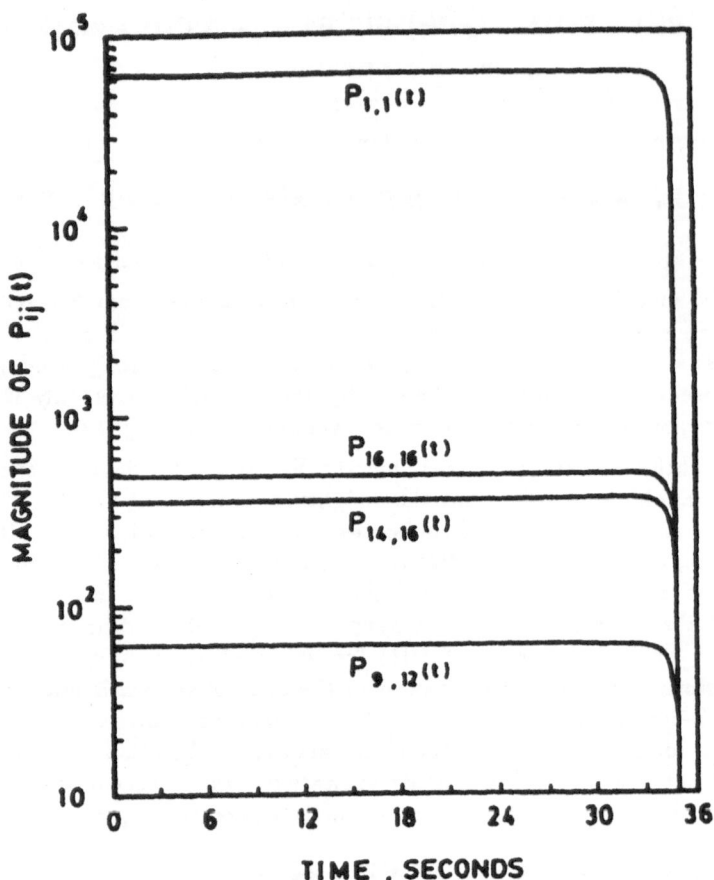

Figure 2: Some elements of Riccati matrix

been substantiated by extensive numerical results.

Unlike the optimal closed-loop control, in which the gain of the control vector is obtained independent of (or disregard with) the earthquake excitation, the optimal closed-open-loop control given by Equations (11) and (12) utilizes the information of earthquake excitations. Hence, it should be superior to the optimal closed-loop control.

Unfortunately, the optimal closed-open-loop control is not achievable for the earthquake excitation. This is because $q(t)$ in Equation (12) should be solved backwards from the terminal time t_f, indicating that the entire earthquake history $\ddot{X}_0(t)$ should be known a prior. Although the earthquake excitation $\ddot{X}_0(t)$ is measurable it is not known a prior.

The optimal closed-open-loop control is feasible only if Equation (12) can be solved forwardly starting from $t = 0$. One striking feature of Equation (12) is that the solution for $q(t)$ is always numerically unstable when solving forwardly from $t = 0$, because the real parts of the eigenvalues are positive. For instance, given the entire excitation history $\ddot{X}_0(t)$ a prior, $q(t)$ and $q(0)$ can be solved backwards from t_f. Then using the $q(0)$ obtained previously, not only is it impossible to reproduce $q(t)$ by solving Equation (12) forwardly but also the resulting numerical solution for $q(t)$ is always divergent. We have tried to resolve such a numerical instability problem without success.

Intuitively, the control vector $U(t)$ or $q(t)$ should be negligibly small at $t = 0$, because the earthquake excitation $\ddot{X}_0(0)$ and the response state vector $Z(0)$ are all zero. Such an assessment is also substantiated by many numerical examples when $q(t)$ is solved backwards from t_f. Unfortunately, $q(t)$ is numerically unstable when solved forwardly.

(3) Optimal Open-Loop Control: For the open-loop control, the control vector depends on the earthquake excitation, i.e., independent of the response state vector $Z(t)$. Thus Equation (5) becomes

$$\lambda(t) = q(t), \tag{13}$$

and Equation (6) reduces to

$$\dot{q}(t) = -A'q(t) - 2QA(t) \ , \ q(t_f) = 0. \tag{14}$$

which is identical to Equation (3). Substitution of Equations (4) and (13) into Equation (1) leads to the following expression

$$Z(t) = AZ(t) - \frac{1}{2}BR^{-1}B'q(t) + W_1\ddot{X}_0(t) \ ; \ Z(0) = 0. \tag{15}$$

Then, the state vector $Z(t)$ and $q(t)$ can be solved using Equations (14) and (15). Again, the optimal open-loop control cannot be implemented because $q(t)$ should be solved backwards from the terminal time t_f, and the earthquake excitation $\ddot{X}_0(t)$ should be known a prior. Furthermore, the solutions for $q(t)$ and $Z(t)$ are numerically unstable (divergent) when Equations (14) and (15) are solved forwardly from $t = 0$.

3. An Optimal Open-Loop Control Algorithm For Earthquake-Excited Building

As mentioned in the previous section, the optimal closed-loop control requires the complete measurements of the state vector $Z(t)$, which may not be possible for tall buildings with a very large number of degrees of freedom. For the classical optimal open-loop and closed-open-loop controls, the complete history of the earthquake excitation should be known a prior, and hence they are not feasible. Since the base excitations to tall buildings can be measured, an alternate form to the classical optimal open-loop control is suggested in this section.

The reason why it is not feasible to implement the optimal open-loop or closed-open-loop control algorithm stems from the definition of the performance index J. The performance index J given by Equation (2) is the integral of quadratic functions over the time interval $(0, t_f)$. Hence, to minimize the quantity J defined over the time interval $(0, t_f)$, the input excitation in that time interval should be known a prior. At any particular time t, however, the earthquake base excitation is measured only up to the time instant t. Therefore, for an open-loop control to be feasible a different performance index should be defined.

The system of equations of motion given in Equation (1) can be decoupled through the following transformation:

$$Z(t) = TY(t) \tag{16}$$

in which T is a $2n \times 2n$ modal matrix consisting of eigenvectors of A.

Substituting Equation (16) into Equation (1), one obtains the decoupled equations of motion as follows:

$$\dot{Y}(t) = \boldsymbol{\theta}\, Y(t) + F(t) \ , \ Y(0) = 0, \tag{17}$$

in which $\boldsymbol{\theta}$ is a $2n \times 2n$ diagonal matrix consisting of eigenvalues θ_j $(j = 1, 2, \ldots, 2n)$ of matrix A and

$$F(t) = T^{-1} B U(t) + T^{-1} W_1 \ddot{X}_0(t). \tag{18}$$

The solution of Equation (17) can be written as:

$$Y(t) = \int_0^t \exp[\boldsymbol{\theta}(t - \tau)] F(\tau) d\tau, \tag{19}$$

where $\exp[\boldsymbol{\theta}(t - \tau)]$ is a $2n \times 2n$ diagonal matrix with the jth diagonal element being $\exp[\theta_j(t - \tau)]$.

The solution for the response vector given by Equation (19) can be obtained numerically using the trapezoidal rule, i.e.,

$$Y(t) = \sum_{k=1}^{n-1} \exp[\boldsymbol{\theta}(n-k)\Delta t] F(k\Delta t)\Delta t + F(t)(\Delta t/2) \tag{20}$$

in which the conditions that the earthquake excitation $\ddot{X}_0(t)$ and the control force $U(t)$ are zero at $t = 0$ have been used. For simplicity, let

$$D(t - \Delta t) = \sum_{k=1}^{n-1} \exp[\boldsymbol{\theta}(n-k)\Delta t] F(k\Delta t)\Delta t \tag{21}$$

such that Equation (20) can be expressed as:

$$Y(t) = D(t - \Delta t) + F(t)(\Delta t/2). \tag{22}$$

The state vector $Z(t)$ follows from Equation (16) as:

$$\mathbf{Z}(t) = \mathbf{T}\mathbf{Y}(t) = \mathbf{T}[\mathbf{D}(t - \Delta t) + \mathbf{F}(t)(\Delta t/2)]. \tag{23}$$

It follows from Equation (21) that the vector $\mathbf{D}(t)$ can be expressed in a recurrent form,

$$\mathbf{D}(t) = \exp[(\boldsymbol{\theta}\,\Delta t)]\,[\mathbf{D}(t - \Delta t) + \mathbf{F}(t)\Delta t]. \tag{24}$$

The performance index to be optimized for the control system is defined by the time dependent quantity $J(t)$ as follows:

$$J(t) = \mathbf{Z}'(t)\mathbf{Q}\,\mathbf{Z}(t) + \mathbf{U}'(t)\mathbf{R}\mathbf{U}(t) \tag{25}$$

where $\mathbf{Q}$ and $\mathbf{R}$ are weighting matrices identical to those defined in Equation (2).

The optimal control vector $\mathbf{U}(t)$ will be determined such that the performance index $J(t)$ is minimum. To achieve such an objective, Equation (23) is substituted into Equation (25),

$$J(t) = [\mathbf{D}'(t-\Delta t)+\mathbf{F}'(t)(\Delta t/2)]\mathbf{T}'\mathbf{Q}\mathbf{T}[\mathbf{D}(t-\Delta t)+\mathbf{F}(t)(\Delta t/2)]+\mathbf{U}'(t)\mathbf{R}\mathbf{U}(t). \tag{26}$$

Taking the variation of both sides of Equation (26), one obtains

$$\begin{aligned}
\delta J(t) = {}& (\Delta t/2)[\delta\mathbf{F}'(t)]\mathbf{T}'\mathbf{Q}\mathbf{T}[\mathbf{D}(t - \Delta t) + \mathbf{F}(t)(\Delta t/2)] \\
& + [\mathbf{D}'(t - \Delta t) + \mathbf{F}'(t)(\Delta t/2)]\mathbf{T}'\mathbf{Q}\mathbf{T}[\delta\mathbf{F}(t)](\Delta t/2). \\
& + [\delta\mathbf{U}'(t)]\mathbf{R}\mathbf{U}(t) + \mathbf{U}'(t)\mathbf{R}[\delta\mathbf{U}(t)].
\end{aligned}$$

The variation of $\mathbf{F}(t)$ can be expressed in terms of the variation of $\mathbf{U}(t)$ using Equation (18) as follows:

$$\delta\mathbf{F}(t) = T^{-1}\mathbf{B}\delta\mathbf{U}(t). \tag{28}$$

Substituting Equation (28) into Equation (27) and setting the variation of $J(t)$ equal to zero, i.e., $\delta J(t) = 0$, one obtains the optimal control vector $\mathbf{U}(t)$ in the following:

$$\mathbf{U}(t) = \mathbf{L}\mathbf{G}(t) \tag{29}$$

in which

$$\mathbf{L} = [\mathbf{B}'\mathbf{Q}\mathbf{B}(\Delta t/2)^2 + \mathbf{R}]^{-1}, \tag{30}$$

$$\mathbf{G}(t) = -\mathbf{B}'\mathbf{Q}\mathbf{T}\mathbf{D}(t - \Delta t)(\Delta t/2) - \mathbf{B}'\mathbf{Q}\mathbf{W}_1(\Delta t/2)^2\ddot{X}_0(t). \tag{31}$$

Substitution of Equation (29) into Equations (18) and (23) leads to the response state vector $\mathbf{Z}(t)$ under optimal control as follows:

$$\mathbf{Z}(t) = \mathbf{T}\mathbf{D}(t - \Delta t) + \mathbf{B}\mathbf{U}(t)(\Delta t/2) + (\Delta t/2)\mathbf{W}_1\ddot{X}_0(t). \tag{32}$$

The control algorithm proposed herein is referred to as the instantaneous optimal open-loop control.

4. Numerical Example

An eight-story building in which every story is identically constructed is considered. The structural properties of each story are: m = floor mass = 345.6 tons; k = elastic stiffness = 3.404×10^5 KN/m and β = external damping coefficient = 100 tons/sec that corresponds to a 2% damping for the first vibrational mode. The internal damping is assumed to be zero. The computed natural frequencies are 5.78, 17.18, 27.98, 37.82, 46.38, 53.36, 58.52 and 61.68 rad/sec.

The earthquake ground acceleration $\ddot{X}_0(t)$ is assumed to be uniformly modulated random process, $\ddot{X}_0(t) = \psi(t)\ddot{X}(t)$, in which $\psi(t)$ is a deterministic modulating (or envelope) function and $X(t)$ is a stationary random process with a power spectral density, $\phi_{\ddot{X}\ddot{X}}(\omega) = [1 + 4\varsigma_g^2(\omega/\omega_g)^2]S^2 / \{[1 - (\omega/\omega_g)^2]^2 + 4\varsigma_g^2(\omega/\omega_g)^2\}$, where ω_g, ς_g and S^2 = parameters depending on the intensity and characteristics of earthquake in a particular geological location. The modulating function $\psi(t)$ used in the literature is considered herein; $\psi(t) = 0$, $t \leq 0$; $\psi(t) = (t/t_1)^2$, $0 \leq t \leq t_1$; $\psi(t) = 1$, $t_1 \leq t \leq t_2$; $\psi(t) = \exp[-c(t-t_2)]$, $t > t_2$.

In the present example, the following valus are used; $t_1 = 3$ sec., $t_2 = 13$ sec., $c = 0.26$, $\omega_g = 18.85$ rad/sec., $\varsigma_g = 0.65$ and $S^2 = 4.5 \times 10^{-4} m^2/sec^2$. With these numerical values, the earthquake intensity corresponds to the one associated with the Housner's response spectrum.

The active tendon controllers are installed in every story unit and the angle of inclination of the tendons with respect to the floor is 25°. Thus, the control force vector from the controllers is $U/\cos 25°$. Each element of the (16×16) $\mathbf{Q}$ matrix is zero except $Q_{11} = Q_{22} = Q_{33} = Q_{88} = Q = 10^5$. The (8×8) $\mathbf{R}$ matrix is chosen to be a diagonal matrix with $R_{11} = R_{22} = ... = R_{88} = 10^{-4}$.

The (16×16) time dependent Riccati matrix $\mathbf{P}(t)$ has been computed for the building with $t_f = 35$ sec. Some of the elements are presented in Figure 2. In fact all the elements of $\mathbf{P}(t)$ remain constant until they approach t_f. Hence a constant Riccati matrix $\mathbf{P}$, Equation (10), is adequate for applications.

With the stochastic model described above, a sample function of the earthquake ground acceleration $\ddot{X}_0(t)$ was simulated as shown in Figure 3. The relative displacement of the top floor with respect to the ground for the building without an active control system is shown in Figure 4(a). With the application of various control algorithms, including (i) optimal open-loop control, (ii) optimal closed-open-loop control, (iii) instantaneous optimal open-loop control, (iv) and optimal close-loop control, the same response quantity is displayed in Figures 4(b), (c), (d) and (e), respectively, for comparison. The reduction of the base shear force follows the same trend as that of the displacement response and hence it is not presented. The lowest controller requires a largest control force that is presented in Figure 5 when different control algorithms are used. In computing the results shown above, the incremental time step Δt is chosen to be 0.015 sec.

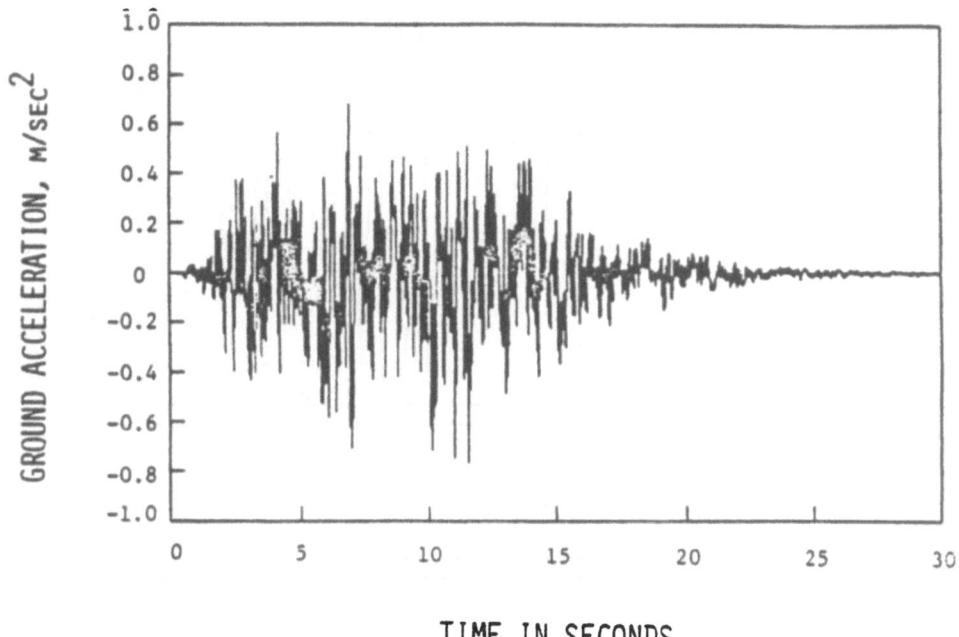

TIME IN SECONDS

Figure 3: Simulated ground acceleration

It is observed from Figure 5 that the maximum magnitude of the control force differs only slightly for various control algorithms. A comparison for the displacement response shown in Figure 4 indicates that the optimal closed-open-loop control, Figure 4(c), and the optimal open-loop control, Figure 4(b), are most efficient, in the sense that they achieve a best reduction for the structural response quantities. The difference between the results using these two control algorithms is extremely small. It should be emphasized, however, that these two control algorithms are not applicable to earthquake excitations, since they require a prior knowledge of the earthquake ground acceleration histroy $\ddot{X}_0(t)$. As mentioned previously, elements of $\lambda(0)$ are very small. However, the forward solution for $\lambda(t)$ starting from $\lambda(0)$ is numerically unstable due to error propagation.

It is further observed from Figures 4 and 5 that the instantaneous optimal open-loop control is slightly better than the optimal closed-loop control, in the sense that the building response and the required active control forces are slightly smaller.

Using the time dependent Riccati matrix $P(t)$, both the building response quantities and the required active control forces have been computed. The results are identical to those obtained using the constant Riccati matrix P, e.g., Figures 4 and 5.

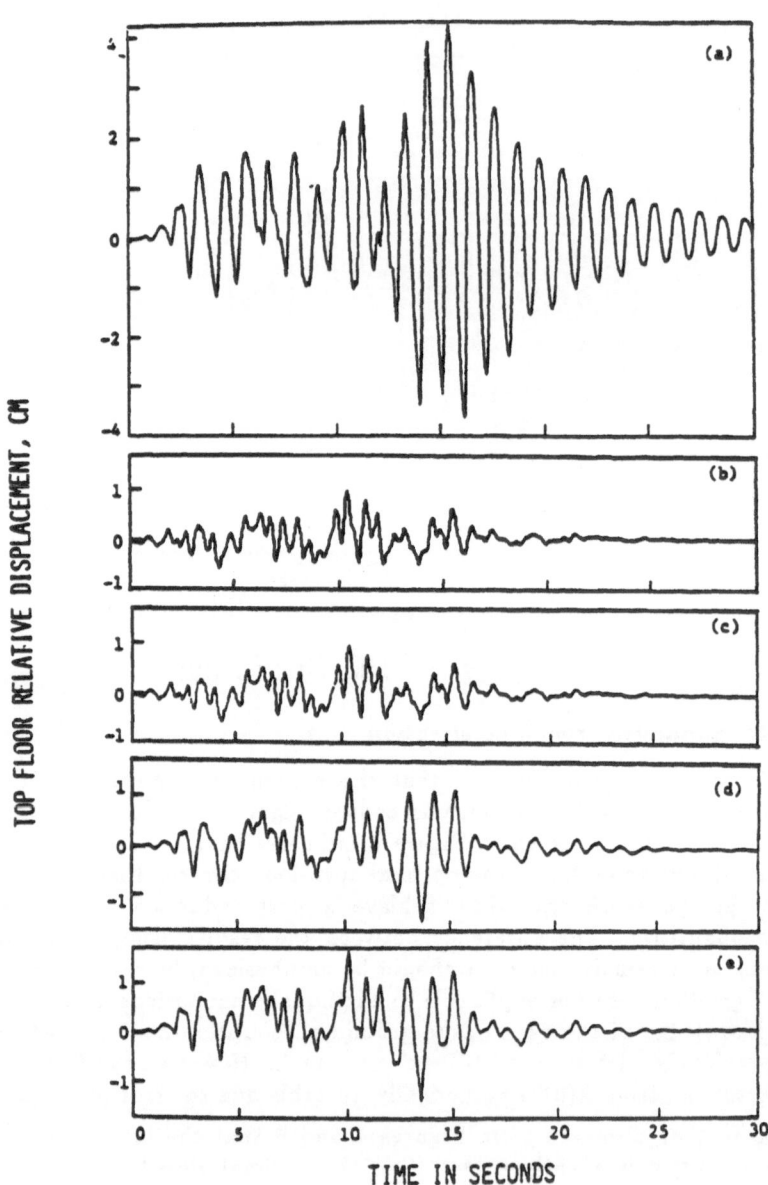

Figure 4: Relative displacement of top floor with respect to ground: (a) no control, (b) optimal open-loop control, (c) optimal closed-open-loop control, (d) instantaneous optimal open-loop control, and (e) optimal closed-loop control

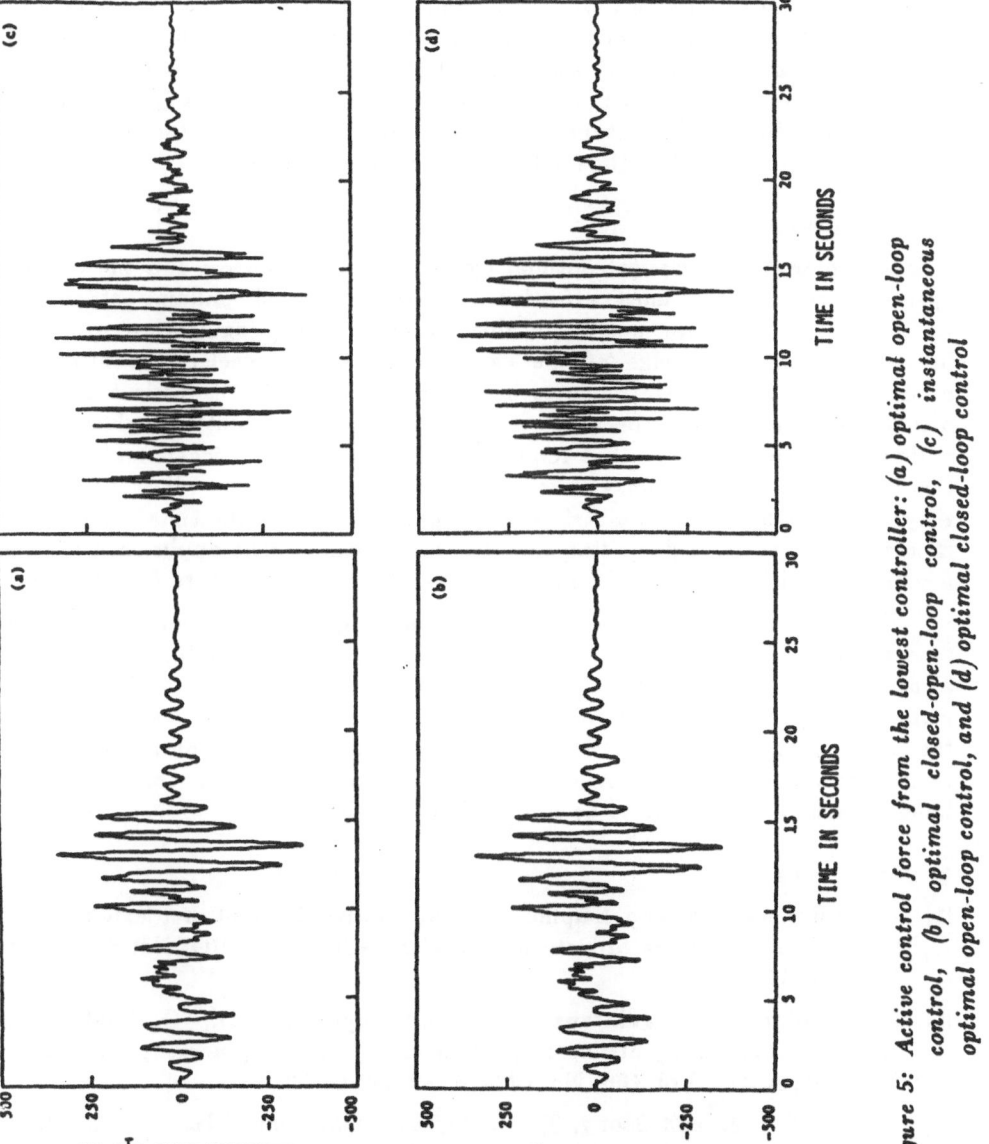

Figure 5: Active control force from the lowest controller: (a) optimal open-loop control, (b) optimal closed-open-loop control, (c) instantaneous optimal open-loop control, and (d) optimal closed-loop control

In the application of the instantaneous optimal open-loop control, the control force $U(t)$ at time t should be computed from the earthquake excitation $\ddot{X}_0(t)$ measured at t, see Equation (29). If the control system, including on-line computations, is not capable to react instantly, then the quantity $\ddot{X}_0(t)$ may be approximated by the measurement of $\ddot{X}_0(t - \Delta t/2)$ at time $T - (\Delta t/2)$. It is found that the results thus obtained do not differ from those presented in Figures 4 and 5. Consequently, for the instantaneous optimal open-loop control the efficiency is not sensitive to the possible time delay of the control system.

5. Conclusions

It is shown that the classical optimal closed-open loop control and the optimal open-loop control are most effective for seismic-excited building structures. Unfortunately, they are not applicable to earthquake engineering problems, because a prior knowledge of the earthquake ground acceleration history is needed. Attempt has been made to utilize these conrol algorithms in a different manner without success, because of the numerical instability involved. The optimal closed-loop control that requires the solution of the Riccati matrix equation is applicable to earthquake-excited buildings. In this regard it is shown that the time dependent Riccati matrix can be approximated by the time independent Riccati matrix without degrading the control efficiency. The application of such a control algorithm requires the measurement of the response state vector $Z(t)$.

Since the earthquake base motions of a building can be measured, an instantaneous optimal open-loop control algorithm is proposed herein. The control algorithm is shown to be slightly more efficient than the classical optimal closed-loop control. The application of such a control law requires the measurement of the building base motions only, and the associated real-time on-line computational effort is minimal. Hence, the implementation of such a control law for building structures appears to be practical and feasible.

6. References

[1] Abdel-Rohman, M. and Leipholz, H. H., "Active Control of Flexible Structures," *Journal of the Structural Division*, ASCE, Vol. 104, No. ST8, Aug., 1978, pp. 1251-1266.

[2] Abdel-Rohman, M., Quintana, V. H., and Leipholz, H. H., "Optimal Conrol of Civil Engineering Structures," *Journal of the Engineering Mechanics Division*, ASCE, Vol. 106, No. EM1, 1980, pp. 57-73.

[3] Chang, M. I. J., and Soong, T. T., "Optimal Controller Placement in Modal Control of Complex Systems," *Journal of Mathematical Analysis and Applications*, Vol. 75, No. 2, June, 1980, pp. 240-258.

[4] Chang, J. C. H., and Soong, T. T., "Structural Control Using Active Tuned Mass Damper," *Journal of the Engineering Mechanics Division*, ASCE, Vol. 106, No. EM6, Dec., 1980, pp. 1091-1098.

[5] Lund, R., "Active Damping of Large Structures in Winds," in Ref. 12, pp. 459-470.

[6] Masri, S. F. et al., "On-Line Pulse Control of Tall Buildings," in Ref. 12, pp. 471-492.

[7] Petersen, N. R., "Design of Large Scale Tuned Mass Dampers," in Ref. 12, pp. 581-596.

[8] Roorda, J., "Tendon Control in Tall Structures," *Journal of Structural Division*, ASCE, Vol. 101, No. ST3, Mar., 1975, pp. 505-521, 197.

[9] Roorda, J., "Experiments in Feedback Control of Structures," in Ref. 12, pp. 629-662.

[10] Soong, T. T., and Chang, M. I. J., "On Optimal Control Configuration in Theory of Modal Control," in Ref. 12, pp. 723-738.

[11] Samali, B., Yang, J. N., and Yeh, C. T., "Control of Lateral-Torsional Motion of Wind-Excited Buildings," *Journal of Engineering Mechanics*, ASCE, Vol. 111, No. 6, June, 1985, pp. 777-796.

[12] *Structural Control*, H. H. Leipholz, ed., North-Holland Publishing Company, Amsterdam, the Netherlands, 1980.

[13] Yang, J. N., "Application of Optimal Control Theory to Civil Engineering Structures," *Journal of the Engineering Mechanics Division*, ASCE, Vol. 101, No. EM6, Dec., 1975, pp. 818-838.

[14] Yang, J. N., and Giannopoulos, F., "Active Control and Stability of Cable-Stayed Bridge," *Journal of the Engineering Mechanics Division*, ASCE, Vol. 105, No. EM4, Aug., 1979, pp. 677-694.

[15] Yang, J. N., and Giannopoulos, F., "Active Control of Two Cable-Stayed Bridge," *Journal of the Engineering Mechanics Division*, ASCE, Vol. 105, No. EM5, Oct., 1979, pp. 795-810.

[16] Yang, J. N., "Control of Tall Buildings Under Earthquake Excitations," *Journal of the Engineering Mechanics Division*, ASCE, Vol. 108, No. EM5, Oct., 1982.

[17] Yao, J. T. P., "Concept of Structural Control," *Journal of the Structural Division*, ASCE, Vol. 98, No. ST7, July, 1972, pp. 1567-1574.

[18] Yao, J. T. P. and Sae-Ung, S., "Active Control of Building Structures," *Journal of the Engineering Mechanics Division*, ASCE, Vol. 104, No. EM2, Apr., 1978, pp. 335-350.

[19] Zuk, W., "Kinetic Structures," *Civil Engineering*, Vol. 39, No. 12, 1968, pp. 62-64.

[20] Zuk, W., "The Past and Future of Active Structural Control Systems," in Ref. 12, pp. 779-794.

RESEARCH TOPICS
FOR PRACTICAL IMPLEMENTATION
OF STRUCTURAL CONTROL

James T. P. Yao
Department of Civil Engineering
Purdue University
West Lafayette, IN U.S.A.
Mohamed Abdel-Rohman
Department of Civil Engineering
University of Kuwait, Kuwait

1. Introduction

To minimize the potential of catastrophic failures, many structures have been designed with redundant load paths. These structures with multiple load paths are said to be fail-safe. As an additional redundancy in safety considerations, the possible application of active control to civil engineering structures has been studied by many investigators during these past two decades.

To date, most studies on active control of structures deal with idealized mathematical models which include "perfect" sensing and control devices. As more experimental data becomes available, it is timely to examine several practical aspects of structural control prior to its implementation.

The objective of this paper is to outline and discuss important problems in the practical implementation of structural control. Several possible approaches for the solution of these problems are also presented.

2. State of the Art

In recent publications on structural control, it has been shown mathematically that it is possible to suppress undesirable vibrations in flexible civil engineering structures with the application of active control techniques. Various applications and techniques have been introduced including the use of active tendons or active tuned mass dampers to control the response of tall structures [1,3,10,11,12,16,17,19,20,25], flexible bridges [2,4,13,14,15,19,23], and flexible building frames [5,18] to unexpected excitations. Such unexpected disturbances included wind forces [2,25], moving loads [4,6], earthquakes [16], or uneveness in bridge decks [23]. Results of these investigations indicated that a significant amount of energy must be consumed in controlling the structure's response. Consequently, several researchers examined control techniques with less energy consumptions such as aerodynamic appendages [7,8,9,24] or a combination of active and passive techniques [26,27].

However, there remain several important problems for further study. In a recent attempt to solve such problems for tall buildings [27], it was found that active tuned mass dampers are not effective for controlling forced vibration response due to wind but very effective for controlling free vibration response [22,27]. The fact that active tuned mass dampers are not effective may be attributed to the change in the structural system parameters due to the large control force needed during wind storms. A new method for designing active tuned mass damper was proposed to improve its effectiveness by considering the change in structural parameters [21]. The ignorance of the change of structural parameters resulting from large feedback control forces may lead to erroneous conclusions.

Another important problem is the effect of time delay [28,29] on the controlled response. In reality, there is a timelag between taking measurements, processing them for estimation, generating the signal to activate the control force, and the application of the control force. In addition, it requires time to generate the required magnitude of the control force. The control forces are applied to the actual structural configuration which is different from the one designed. The reliability of both the control system and the structure may thus be affected. It has been found [27] that neglecting time delay effect can be counter productive and should be considered in the design of the control system.

It has been well known that any recorded data contains several types of noise, which must be treated. However, such procedures for data processing are not unique and the choice of a particular procedure may be arbitrary and/or subjective [30]. Results of using the same data with different procedures for data processing can also be different. Consequently, there is a question on how to interpret the feedback information which may be based on noise-polluted measurements.

3. Topics for Further Study

The time delay in using an actual control system should be considered for a controlled structure subjected to a typical disturbance. The delay effect for various control techniques should be studied experimentally [31]. The compensation of the delay effect and its relative cost should also be studied.

The design of any control system is based on the knowledge of the structural parameters. For an as-built structure, the structural parameters should be identified from a set of measurements. It is known that certain parameters of existing structures may change with time during strong earthquakes. Such measurements should be made in actual structures and considered in the design of control systems.

Feedback control introduces equivalent damping and/or stiffness. Any increase of frequencies causes an increase in oscillations with smaller amplitudes. It is desirable to determine the optimal active damping which does not cause extra fatigue problems or appreciable changes in the mathematical model of the structure. Moreover, various degrees maybe specified for control. For example, one may choose from completely passive, passive-active, and completely active techniques. The active technique should also be divided into levels according to the intensity of the disturbance and the strength of the structure.

Passive control systems have been used in practice to-date [20]. Combining them with active systems may decrease the control energy consumption. Moreover, passive systems may be used to compensate for such active control related problems as time delay and power failures.

Under certain circumstances, it may be desirable to combine the use of several types of control devices for various purposes. Such multi-purpose and multi-device structural control would require a set of strategies and decision-making processes in actual applications [32]. Moreover, the processing and interpretation of feedback information may require expertise in addition to mathematical modeling and analyses [33].

Recently, the theory of fuzzy sets has been applied to control systems. As an example, a fuzzy controller is used to interpret linguistic instructions in driving a model car which is equipped with ultrasonic sensors for the measurement of its position and direction [34]. A fuzzy control algorithm has also been applied to an automatic train operation system [35]. It seems desirable to study such fuzzy algorithms in structural control.

It is hoped that design guidelines for structural control may result from such studies. These guidelines may also be verified with available results of experimental investigations [31].

References

[1] Roorda, J., "Tendon Control in Tall Structures," *Journal of the Structural Division*, ASCE, Vol. 101, No. ST3, March 1975, pp. 505-521.

[2] Yang, J.N. and Giannopoulas, P., "Active Control of Two-Cable-Stayed Bridge," *Jounral of the Engineering Mechanics Division*, ASCE, Vol. 105, EM5, Oct. 1979, pp. 795-810.

[3] Yang, J.N. and Giannopoulas, P., "Active Tendon Control of Structures," *Jounral of the Engineering Mechanics Division*, ASCE, Vol. 194, EM3, June 1978, pp. 551-568.

[4] Abdel-Rohman, M. and Leipholz, H.H., "Automatic Active Control of Structures," *Journal of the Structural Division*, ASCE, Vol. 106, ST3, March 1980, pp. 663-877.

[5] Abdel-Rohman, M. and Leipholz, H.H., "General Approach to Active Structural Control," *Jounral of the Engineering Mechanics Division*, ASCE, Vol. 105, EM6, Dec. 1979, pp. 1007-1923.

[6] Abdel-Rohman, M. and Leipholz, H.H., "Structural Control by Pole Assignment Method," *Jounral of the Engineering Mechanics Division*, ASCE, Vol. 104, EM5, Oct. 1978, pp. 1159-1175.

[7] Klein, R.E., Cusano, C. and Slukel, J.V., "Investigation of Method to Stabilize Wind Induced Oscillations in Large Structures," presented at the 1972 ASME Winter Meeting, Paper No. 72-W+/AUT-11, New York, Nov. 1972.

[8] Chang, J.C. and Soong, T.T., "The Use of Aerodynamic Appendages for Tall Building Control," *Structural Control*, H. Leipholz (Editor), North-Holland Publishing Co. & SM Publications, IUTAM 1980, pp. 199-210.

[9] Soong, T.T. and Skinner, G.T., "Experimental Study of Active Structural Control," *Journal of the Engineering Mechanics Division*, ASCE, Vol. 107, EM6, Dec. 1981, pp. 1057-1068.

[10] Lund, R.A., "Active Daming of Large Structures in Winds," presented at the International IUTAM Symposium on Structural Control, University of Waterloo, Ontario, Canada, June 4-7, 1979, pp. 459-471.

[11] Peterson, N.R.M., "Design of Large Scale Tuned Mass Dampers," presented at the April 1979, ASCE Convention and Exposition, held at Boston, Mass., reprint 3578.

[12] Chang, J.C. and Soong, T.T., "Structural Control Using Active Tuned Mass Dampers," *Jounral of the Engineering Mechanics Division*, ASCE, Vol. 106, EM6, Dec. 1980, pp. 1091-1098.

[13] Abdel-Rohman, M., Quintana, V.H. and Leipholz, H.H., "Optimal Control of Civil Engineering Structures," *Journal of Engineering Mechanics Division*, ASCE, Vol. 106, EM1, Feb. 1980, pp. 57-73.

766   J. Yao, M. Abdel-Rohman

[14] Abdel-Rohman, M. and Leipholz, H.H., "Active Control of Flexible Structures," *Journal of the Structural Division*, ASCE, Vol. 104, ST8, Aug. 1978, pp. 1251-1266.

[15] Yang, J.N. and Giannopoulos, P., "Active Control and Stability of Cable-Stayed Bridge," *Journal of the Engineering Mechanics Division*, Vol. 105, EM4, Aug. 1979, pp. 677-694.

[16] Sae-Ung, S., and Yao, J.T.P., "Active Control of Building Structures," *Journal of the Engineering Mechanics Division, ASCE, Vol. 104, EM2, Apr. 1978, pp. 335-350.*

[17] Abdel-Rohman, M., and Leipholz, H.H., "Active Control of Tall Buildings," *Journal of the Structural Division," ASCE, Vol. 109, No. 3, March 1983, pp. 628-645.*

[18] Abdel-Rohman, M., "Active Control of Large Strucutes," *Journal of the Engineering Mechanics Division*, ASCE, Vol. 108, EM5, Oct. 1982, pp. 719-730.

[19] Abdel-Rohman, M., "Design of Optimal Observers for Structural Control," IEE Proceedings-D, *Journal of Control Theory and Applications*, England, Vol. 131, No. 4, July 1984, pp. 158-163.

[20] McNamara, R.J., "Tuned Mass Dampers for Buildings," *Journal of the Structural Division*, ASCE, Vol. 103, ST9, Sept. 1977, pp. 1785-1798.

[21] Abdel-Rohman, M., "Design of Active TMD for Tall Building Control," *Journal of Buildings and Environment*, Vol. 19, No. 3, Pergamon Press, England, 1984, pp. 191-195.

[22] Abdel-Rohman, M., "Effectivenss of Active TMD in Tall Buildings," *Transactions of Canadian Society of Mechanical Engineers*, Vol. 8, No. 4, Dec. 1984.

[23] Abdel-Rohman, M. and Leipholz, H.H., "Stochastic Control of Structures," *Journal of Structural Division*, ASCE, Vol. 107, No. ST7, July 1981, pp. 1313-1325.

[24] Abdel-Rohman, M., "Optimal Control of Tall Buildings by Appendages," *Journal of Structural Division*, ASCE, Vol. 110, May 1984, pp. 937-947.

[25] Abdel-Rohman, M., "Control of Tall Buildings Against Stochastic Wind Forces," *Journal of Wind Engineering and Industrial Aerodynamics*, Vol. 17, No. 2, August 1984.

[26] Hrovat, D., Barak, P. and Rabins, M., "Semi-Active Versus Passive or Active Tuned Mass Dampers for Structural Control," *Journal of Engineering Mechanics.*, Vol. 109, No. 3, June 1983, pp. 691-705.

[27] Abdel-Rohman, M., "Control of Dynamic Response of Tall Buildings Against Wind," Final Report, Kuwait Foundation for Advancement of Science, KFAS 81-08-02, 1984, Kuwait.

[28] Abdel-Rohman, M., "Structural Control Considering Time Delay Effect," Accepted for publication in *Transactions of Canadian Society of Mechanical Engineers,* Oct. 1984.

[29] Basharkhah, M. A. and Yao, J.T.P., "Reliability Aspects of Strucutral Control," *Civil Engineering Systems,* Vol. 1, June 1984, pp. 224-229.

[30] Stephens, J.E. and Yao, J.T.P., "Data Processing of Earthquake Acceleration Records, Technical Report No. CE-STR-85-5, School of Civil Engineering, Purdue University, West Lafayette, IN, 1985.

[31] Yao, J.T.P. and Soong, T.T., "Importance of Experimental Studies in Structural Control," Preprint 84-010, ASCE Atlanta Convention, 14-18 May 1984.

[32] Yao, J.T.P., "Identification and Control of Structural Damage," *Solid Mechanics Archives,* Vol. 5, Issue 3, August 1980, pp. 325-345.

[33] Furuta, H., Fu, K.S. and Yao, J.T.P., "Structural Engineering Applications of Expert Systems," Technical Report No. CE-STR-85-11, School of Civil Engineering, Purdue University, West Lafayette, IN, 1985.

[34] Sugeno, M. and Katayama, K., "Linguistic Control of Model Car," presented at the First Congress of International Fuzzy Systems Association, Palma de Mallorca, Spain, 1-6 July 1985.

[35] Yasunobu, S. and Miyamoto, S., "Predictive Fuzzy Control and Its Application to Transfer Function Model," presented at the First Congress of International Fuzzy Systems Association, Palma de Mallorca, Spain, 1-6, July 1985.

OPTIMAL DESIGN AND CONTROL
OF CENTRIFUGALLY TUNED
DYNAMIC VIBRATION ABSORBERS

M.I. Young
Professor of Mechanical and Aerospace Engineering
University of Delaware
Newark, Delaware 19716, U.S.A.

1. Introduction

Christian Huygens, Leonhard Euler, Johann Bernoulli, Adrien Legendre and others contributed to the concept of the Brachistochrone [4], the path requiring the least time of descent under gravity. Such paths are cycloids. It is a corollary that this time is indepedent of the starting point. If the path is made symmetrical about the vertical, the descending particle rises once again to its initial elevation. The rise and fall continues in a simple harmonic, oscillatory fashion. In effect, a pendulum type oscillator is created without the non-linear behavior of a pendulum where the period of the oscillation grows larger with the amplitude of the motion. Accordingly, a cycloidal path yields an isochronous oscillator, since the time of descent is constant as well as minimal. It is shown in the Analysis that in the case of the centrifugal force field a tuned centrifugal pendulum de-tunes as the amplitude grows. A formal search for an isochronous oscillator in this case yields higher plane curve paths which are cycloid-like, but not actually cycloids. This follows first from an approximative solution to the non-linear defferential equation of the path [5]. Then a numerical, digital computer solution method [6] is implemented to generate the needed ordinates of the higher plane curves in actual machine design, structural design applications [7].

Although amplitude of response with such advanced geometry paths would no longer de-tune such devices when used as dynamic vibration absorbers, practical motion limits are considered employing modern control science, quadratic minimization type techniques and criteria. The approach ensures that stress, acceleration and other practical constraints besides frequency are part of any design specification.

2. Analysis

In the case of a simple gravity pendulum where an ideal mass particle follows a circular path of length ℓ and the slope of the path with respect to the horizontal is the angle ϕ the familiar non-linear differential equation for the oscillations

$$\ddot{\phi} + (\frac{g}{\ell})\sin\phi = 0 \tag{1}$$

yields a period τ which increases with the amplitude of the motion as shown below [2]. ϕ_0 is the amplitude of the oscillation and τ_0 is the small oscillation period,

$$\tau_0 = 2\pi \left[\frac{\ell}{g}\right]^{1/2}, \tag{2}$$

(ϕ_0/π)	(τ/τ_0)
.1	1.0062
.2	1.0253
.3	1.0585
.4	1.1087
.5	1.1804
.6	1.2817
.7	1.4283
.8	1.6551
.9	2.0724
1.0	∞

If the path followed by the particle is modified to be cycloidal rather than circular [2], then it follows that the period is independent of the amplitude. Specifically defining the vertical displacement and ordinate η of the path and the abscissa of the path as ξ in terms of the parameter θ, the rolling angle of the circle of radius R generating the cycloid, then

$$\eta = R(1 - \cos\theta), \tag{3}$$

$$\xi = R(\theta + \sin\theta). \tag{4}$$

It follows at once that the governing differential equation in terms of arc length is

$$\ddot{s} + g\sin\phi = 0 \tag{5}$$

as for the circular path, but in the case of the cycloid

$$s = 4R\sin(\frac{\theta}{2}) = 4R\sin\phi. \tag{6}$$

Accordingly,

$$g\sin\phi = \frac{gs}{4R}, \tag{7}$$

and the constant period $\tau = 4\pi(\frac{R}{g})^{1/2}$ results.

Now consider the simple oscillator under the influence of a centrifugal force field as illustrated in Fig. 1.

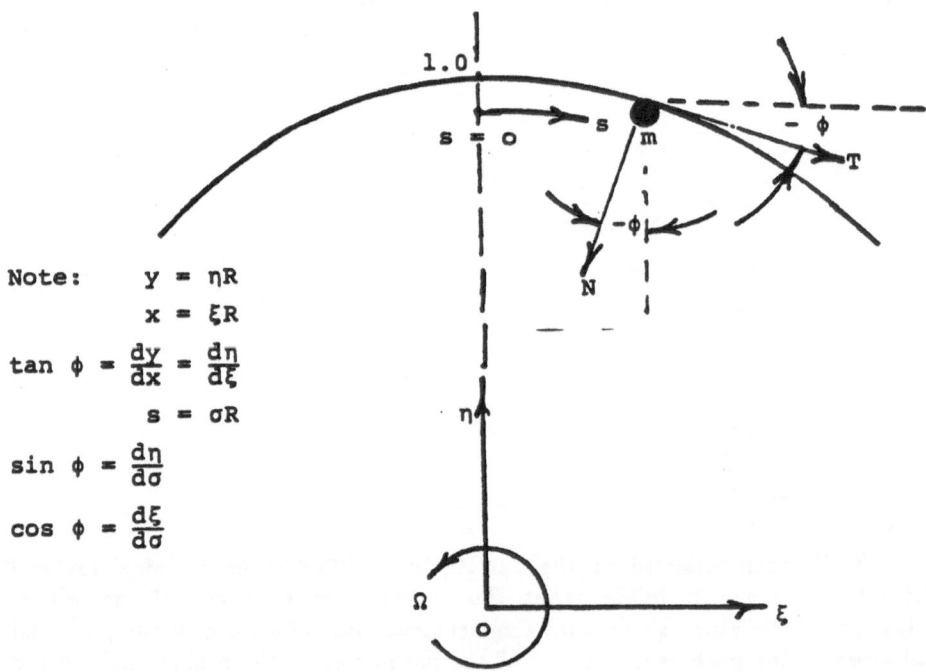

Figure 1

Two apparent forces now influence the motion of the ideal mass particle with respect to the rotating reference system. These are the centrifugal and the Coriolis forces [8]. A third apparent force, the Euler force [8], is absent under the assumed operating condition of no angular acceleration. That is, Ω is a constant. Resolving these apparent forces along the moving, local tangent and principal normal directions to the path in the rotating planar reference system,

$$F_T = mR\Omega^2 \left\{ [\xi + (\tfrac{\dot{\eta}}{\Omega})]\cos\phi + [\eta - (\tfrac{\dot{\xi}}{\Omega})]\sin\phi \right\}, \tag{8}$$

$$F_N = -mR\Omega^2 \left\{ -[\xi + (\tfrac{\dot{\eta}}{\Omega})]\sin\phi + [\eta - (\tfrac{\dot{\xi}}{\Omega})]\cos\phi \right\}. \tag{9}$$

As in the classical pendulum oscillator, it is assumed that the path is friction free. Accordingly, the component of force in the principal normal direction has no effect on the component of force in the tangential direction. Moreover, noting the geometric relationships between the slope of the path ϕ, the arc length s along the path and the tangential component of the Coriolis force, it follows that only the centrifugal force influences the motion. The equation of motion along the still to be specified path $\eta = \eta(\xi)$ follows as

$$\frac{d^2\sigma}{d\tau^2} = \xi\frac{d\xi}{d\sigma} + \eta\frac{d\eta}{d\sigma}, \tag{10}$$

where $\tau = \Omega t$, a dimensionless unit of time.

A key step in determining the appropriate path for an isochronous oscillator in a centrifugal force field now follows. It is desired that the natural frequency of oscillation ω_n remain constant in its relationship to the frequency of constant rotation Ω. That is, looking ahead to engineering type machine dynamics-system dynamics type application, it is required that

$$\left(\frac{\omega_n}{\Omega} \right) = n. \tag{11}$$

This results if, and only if,

$$\xi\frac{d\xi}{d\sigma} + \eta\frac{d\eta}{d\sigma} = -n^2\sigma, \tag{12}$$

since this yields the governing oscillator differential equation

$$\frac{d^2\sigma}{d\tau^2} + n^2\sigma = 0, \tag{13}$$

and the desired constant frequency ratio $(\frac{\omega_n}{\Omega}) = n$.

Noting that the differential equation for the path $\eta = \eta(\xi)$ consists, term by term, of perfect differentials, it is readily integrated with the result

$$\xi^2 + \eta^2 = 1 - n^2\sigma^2. \tag{14}$$

This equation is itself an integral equation, since the dimensionless arc length σ is given by:

$$\sigma = \int_o^\xi \left[1 + (\frac{d\eta}{d\xi})^2 \right]^{1/2} d\bar{\xi} = (n)^{-1} \left[1 - (\xi^2 + \eta^2) \right]^{1/2}, \tag{15}$$

where $\bar{\xi}$ is a dummy variable of integration. Rewriting the two foregoing equations, an implicit form of a first order, non-linear differential equation follows for $\eta = \eta(\xi,n)$, where the constant frequency ratio n is a parameter influencing the desired path geomatry.

$$\eta \frac{d\eta}{d\xi} + \xi = -n^2 \sigma \frac{d\sigma}{d\xi} = -n \left[1 - (\xi^2 + \eta^2) \right]^{1/2} \left[1 + (\frac{d\eta}{d\xi})^2 \right]^{1/2}. \tag{16}$$

3. Solving the Path Differential Equation

In the case of small oscillations, a circular path of dimensionless radius ρ, whose center is the dimensionless distance $(1 - \rho)$ from the axis of ratation yields the desired frequency ratio n. The governing differential equation is the non-linear equation

$$\frac{d^2\phi}{d\tau^2} + \left[\frac{1-\rho}{\rho} \right] \sin\phi = 0, \tag{17}$$

since

$$\xi = \rho\sin\phi \simeq \rho\phi, \tag{18}$$

$$\eta = (1 - \rho) + \rho\cos\phi \simeq 1 - \frac{1}{2}\rho\phi^2, \tag{19}$$

for small motions and the dimensionless arc length $\sigma = \rho\phi$, it follows that

$$\frac{d^2\phi}{d\tau^2} + \left[\frac{1-\rho}{\rho} \right] \sin\phi \simeq \frac{d^2\phi}{d\tau^2} + \left[\frac{1-\rho}{\rho} \right] \phi = 0, \tag{20}$$

and

$$\left[\frac{1-\rho}{\rho} \right] = n^2, \quad or \quad \rho = (n^2 + 1)^{-1}, \tag{21}$$

and

$$\xi \simeq (n^2 + 1)^{-1}\phi, \tag{22}$$

$$\eta \simeq 1 - \frac{1}{2}(n^2 + 1)\xi^2. \tag{23}$$

Thus a rough, first approximation to the ideal path is parabolic and a strong function of the frequency ratio n.

Employing a standard numerical integration routine and digital computation, the ordinates and figures of the ideal path for the large motions have been determined. These are cycloid like, but true higher plane curves [9].

Another case of practical interest where an isochronous oscillator is desirable in the presence of a centrifugal force field involves steady rotation about one of the two coordinate axes, rather than the foregoing case of steady rotation about an axis normal to the plane of rotation. In such a second case, taking the ξ-axis as the one about which the rotation occurs, yields the governing differential equation

$$\eta \frac{d\eta}{d\xi} = -n(1-\eta^2)^{1/2}\left[1+(\frac{d\eta}{d\xi})^2\right]^{1/2}. \tag{24}$$

Graphical results follow below without further elaboration.

CASE 1

Rotation About Axis Normal to the $\xi - \eta$ Plane
Note: n=1,2,2½,3,4,4½ From Right to Left

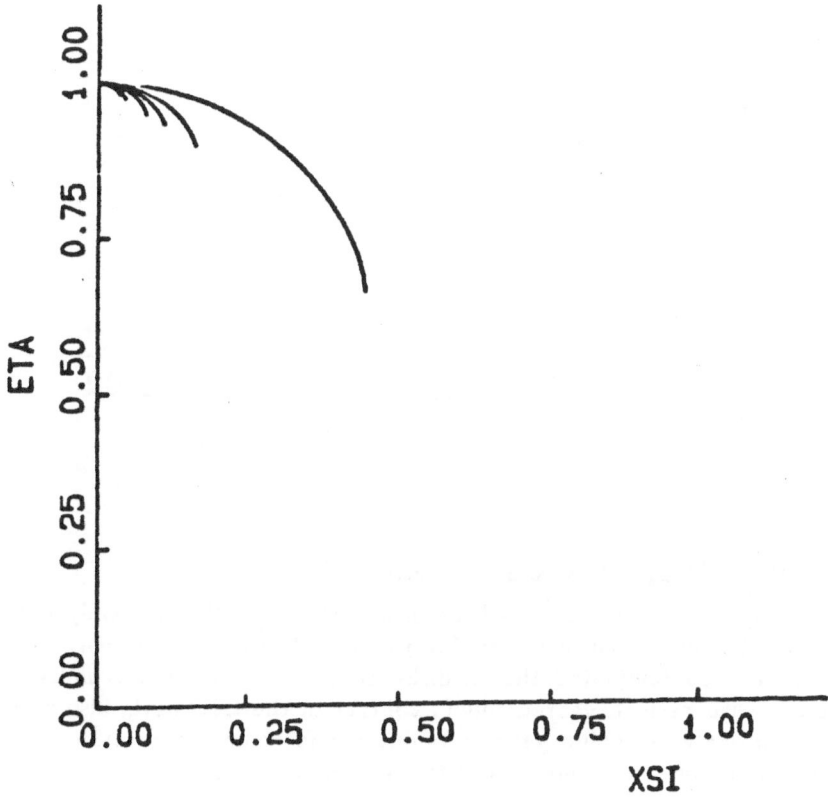

Figure 2

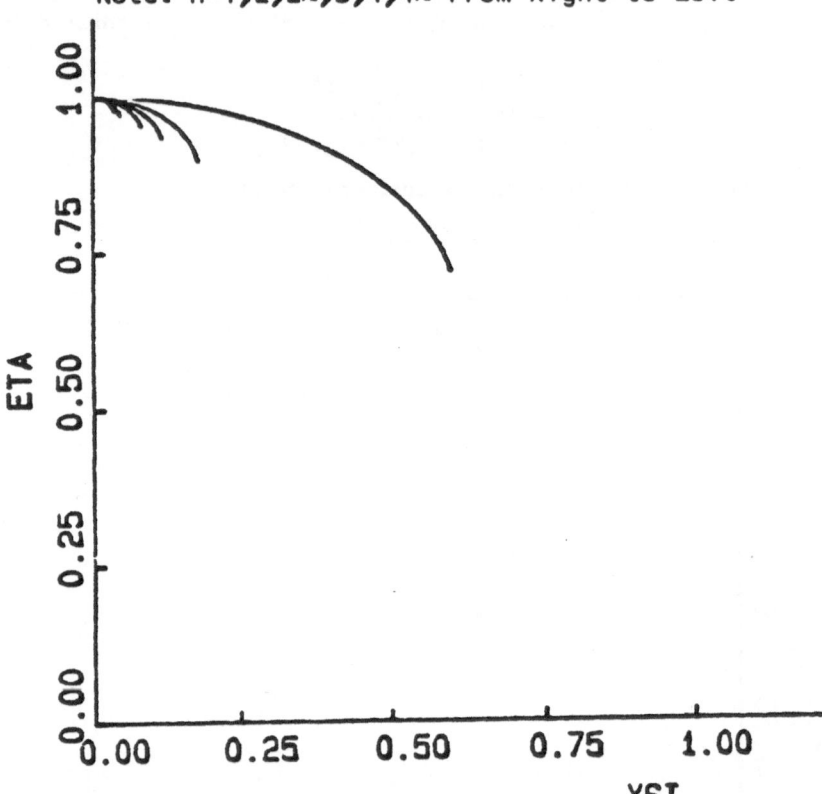

Figure 3

4. Some Rigid Body Dynamics Aspects

In applications such as dynamic vibration absorbers [1, 3], rigid body dynamics aspects must be considered in the tuning process. In the present case of cycloid-like paths, it is contemplated that a disk-like object such as a cylindrical gear would move with perfect, geared rolling contact in the trackpath, also with gear teeth. Accordingly, a rolling gear of radius r and radius of gyration νr would have the virtual mass m_v compared to the particle mass m,

$$m_v = m(1 + \nu^2). \tag{25}$$

In this case, the desired constant natural frequency ratio N is obtained by a path of different curvature than for a particle. That is,

$$N = \left(\frac{\omega_n}{\Omega}\right) = \left(\frac{m}{m_v}\right)^{1/2} n = \left[\frac{1}{1 + \nu^2}\right]^{1/2} n. \tag{26}$$

For example, in the case of a homogeneous cylinder $\nu^2 = 1/2$. Then for $N = 2$, $n = 2.45$. The path would then be computed for this equivalent n for an ideal mass particle.

5. Optimal Design and Control

First the appropriate, cycloid-like path for the centrifugally tuned dynamic vibration absorber is chosen. Then the device is employed to develop a cyclic counter-torque, for example, in the crankshafts of large internal combustion engines for marine, or for similar heavy duty industrial applications. In the classical application of such devices, no consideration is given to the behavior of the absorber itself in a balanced manner. Attention is directed to the primary system with a view towards suppressing ("absorbing") unwanted vibratory response to cyclic loadings. This frequently leads to excessive motion, stress and vibration of the absorber itself. Accordingly we now consider a modern systems science approach, where a performance index reflecting the behavior of both the primary system and the absorber is evaluated. The absorber system design parameters are then adjusted to make the performance index optimal. For example, it is shown in reference [3] that a performance index

$$P_o = \int_o^T x^2(t)\, dt, \tag{27}$$

where T is the period and x is the primary system response, yields the classical absorber solution and tuning when P_o is made a minimum by the choice of absorber system design parameters. Other indices involving a penalty factor P for absorber motion may be selected. An example of this is

$$P_1 = \int_o^T \left[x^2(t) + py^2(t)\right] dt, \tag{28}$$

where y is the absorber response which yields a new class of optimal solutions as p is varied [3].

In the present case, where no actual design experience exists and must be obtained, a general, quadratic performance index is proposed which incorporates absorber cyclic motion, cyclic acceleration and cyclic stress. This then takes the form

$$P = \int_o^T \left[x^2(t) + p_1 y^2(t) + p_2 \ddot{y}^2(t) + p_3 \sigma^2(t)\right] dt. \tag{29}$$

Quadratic minimization techniques [10] can then be applied to obtain the optimal design parameter for centrifugally tuned dynamic vibration absorber. As in the case of the linear, spring restrained type, the optimal tuning [3] will differ from the classical case.

6. Conclusions

In the foregoing analysis, it has been shown that a new class of cycloid-like paths result in isochronous oscillatory motions in the presence of a centrifugal force field. A representative class of these has been developed by numerical integration of the governing differential equation. Tables and Figures for these follow below. It has also been shown that an approximative numerical solution can be quickly obtained in terms of a parabolic path. In both cases, the detailed shape of the path depends on the offending frequency ratio n, the actual rigid body dynamics characteristics of the absorber and the modern performance index selected for choice of the optimal design. Tabular data and Fortran Program listing is available from the author upon request.

Acknowledgment

Digital computation and computer graphics assistance were ably provided by Mr. William Coleman, Graduate Student Department of Mechanical and Aerospace Engineering, University of Delaware.

References

[1] Ormandroyd, J. et al., "The Theory of the Dynamic Vibration Absorber," *Trans. ASME*, Paper APM-50-7, 1928.

[2] Young, M.I., "The Dynamics of Measuring Time by the Cycloidal Method," *Proceedings 6th International Symposium on Measurement and Control, IASTED*, Athens, Greece, September 1983.

[3] Young, M.I., "Systems Engineering, Optimization and Control of Dynamic Vibration Absorbers," *Proceedings 3rd International Conference in Systems Engineering*, Wright State University, Ohio, September 1984.

[4] Smith, D.E., *Source Book in Mathematics*, McGraw-Hill Book Co., pp. 644, 1929.

[5] Henrici, P., *Discreate Variable Methods in Ordinary Differential Equations*, John Wiley Co., pp. 108-163, 1962.

[6] Cheney, W. and Kinkaid, D., *Numerical Mathematics and Computing*, Brooks/Cole Publishing Co., pp. 179-200, 1980.

[7] Burr, A.H., *Mechanical Analysis and Design*, Elsevier North Holland, Inc., pp. 486-529, 1981.

[8] Lanczos, C., *The Variational Principles of Dynamics*, University of Toronto Press, Third Edition, 1966, pp. 100-103.

[9] Wilson, W.A. and Tracey, J.I., *Analytic Geometry*, Alternate Edition, D. C. Heath and Co., 1937, pp. 175-205.

[10] Kwakernak, H. and Sivan, R., *Linear Optimal Control Systems*, Wiley-Interscience, 1972, pp. 193-219.

CHAOTIC INSTABILITY
IN
MECHANICAL SYSTEMS

M. Zak

Applied Mechanics Technology Section
Jet Propulsion Laboratory, Caltech

1. Introduction

So far the random vibrations in mechanical systems, and particularly, in structures, are being considered only as a result of random loading. For instance, stationary random loadings are caused by the pressure from a wind gust, by the forces originated under the effect of waves on the structure of vehicles, by the pressure due to atmospheric turbulence, or acoustic radiation from operating jet engines. Non-stationary random loadings are caused by seismic and explosive effects. The problem of structural analysis in these cases is posed as follows: knowing the parameters of the statistical distribution of random loadings, find the corresponding parameters for quantities describing the behavior of the structure. This problem is well treated by using the usual deterministic structural models, i.e., a set of differential equation, with random inputs on the right-hand side.

Thus, the main properties of random vibrations are the following: 1) they can be excited only by random loadings, 2) they can be fully predicted if statistical properties of the random loadings are known.

However, in recent years, simple physical and mathematical simulation experiments, substantiated by some theoretical results, have demonstrated the existence of another type of random vibrations which are generated by fully deterministic systems *without* any random input. The main properties of these systems are the following: 1) they are necessarily non-linear, 2) they are

777

supersensitive to small changes in initial conditions.

The term "chaotic vibrations" is generally used to distinguish the latter type of vibrations (intrinsic stochasticity) from a true random process which is caused by random input.

It was proven by physical as well as numerical experiments Ref. [1], that chaotic vibrations are caused by such nonlinearities as bi-linear springs, unilateral constraints, i.e, by the jump-type nonlinearities. However, they also can occur in systems with smooth kinematic nonlinearities such as two-bar linkages (i.e., in deployment mechanisms), in systems with multiple equilibrium states (buckling beams and shells), in aeroelastic systems, and especially in structures including feedback control devices.

It is likely to expect chaotic instability in large flexible space structures such as reflector antennas, solar arrays, inflatable structures etc.

In constrast to random vibrations which occur as soon as random loadings are applied, the occurrence of the chaotic vibrations is much more difficult to predict because they are caused by the internal properties of the system, i.e., by its chaotic instability. Thus, even the prediction of only the occurrence of chaotic vibrations is strongly connected with the knowledge of the exact properties of the system itself, and particularly with the knowledge of the criteria of the chaotic instability. Obviously, the prediction of the characteristics of the chaotic vibrations as well as avoidance or suppression of chaos requires the full understanding of the post-instability behavior of the system.

Post-instability behavior of non-linear mechanical systems can appear in two different forms. The first one is characterized by an alternative stable state, or classical attractor, (for instance, buckled Euler's strut, or flutter of suspension bridge): after some transition period the system "finds" this state and its motion becomes stable again. The second form occurs if there is no alternative stable state, and consequently, the instability of the motion persists. However, because of the finiteness of the energy in all practical situations the parameters cannot grow without a limit, but they rather attain chaotic fluctuations. Thus, in contrast to the first (classical) form of instability, the chaotic instability is not necessarily accompained by large deviations of the controlled parameters. However, the modal representation of the motion becomes supersensitive to infinitesimal uncertainties in initial and boundary conditions, and therefore, its power spectrum attains noise. In other words, the motion becomes random and unpredictable without any random input. The importance of identification and understanding of this strongly non-linear phenomena in theory of structural control is obvious.

In this work analytical criteria for chaotic instability as well as a possible strategy of deterministic representation and control of chaotic motion with application to elastic structures are discussed.

2. Concept of Orbital Instability

The guiding principles for chaotic instability can be adopted from the orbital instability of an inertial motion of a particle M on a smooth surface S having a
constant negative Gaussian curvature G_o, Fig. 1. Remembering that trajectories
of inertial motions must be geodesic of S let us compare two different trajectories
assuming that initially they are parallel and that the distance between them, E_o
is very small. Then the distance between such geodesics will exponentially
increase with time t:

$$E = E_o \exp(\sqrt{-G_o})t, \quad G_o < 0. \tag{1}$$

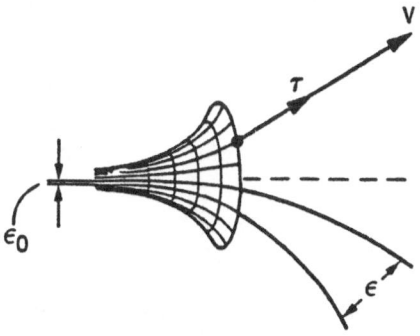

Figure 1:

If L is the accuracy to which E_o is known, then these two trajectories cannot be
distinguished for

$$t >> \frac{1}{|\sqrt{-G_o}|} \ell n \frac{L}{E_o} \tag{2}$$

and therefore, the real trajectories may "fill up" all the spacing between them
(Fig. 2). As shown by Anosov, Ref. [2], such a motion is characterized by positive
Liapunov exponents and can be qualified as chaotic.

The decomposition of the velocity

$$\mathbf{v} = v\mathbf{\tau} \tag{3}$$

where τ is the unit tangent to the trajectories, allows one to distinguish the classical (Liapunov) instability (which is associated with large deviations of the magnitude v, i.e., the total kinetic energy) from the orbital instability (which is
characterized by deviations of the trajectory directions τ). Indeed, the absence
of classical attractors in the course of Liapunov instability would mean the
unbounded growth of the velocity v, i.e., the total energy, which never can occur
in real situation. At the same time, the orbital instability leads only to the

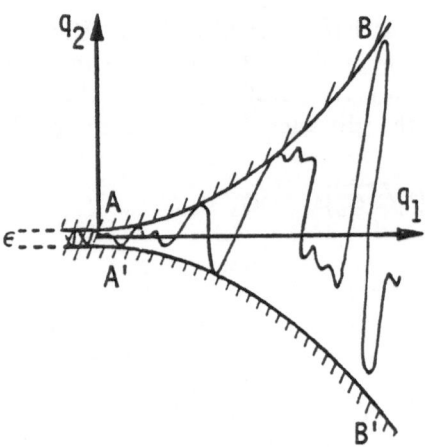

Figure 2

divergency of trajectories, i.e., to the energy redistributions between different coordinates, and therefore, the absence of classical attractors does not contradict the finiteness of the total energy of the particle. Thus, chaos is caused only by orbital instability, although it can be accompanied by the classical instability.

For non-inertial motions the trajectories will deviate from geodesics while the rate of their deviation is characterized by the geodesic curvature χ defined by the force field θ. As shown inRef. [3], the criterion of chaos in this case can be expressed as:

$$- G + \chi^2 + \nabla \cdot \chi > 0. \tag{4}$$

The results described above were related to motions of a particle on a surface. However, they can be easily generalized to motions of any finite-degree-of-freedom mechanical system by using the concept of configuration space, Refs. [3, 4].

3. Low of Motion in Configuration Space

Let us consider a mechanical system with N-degrees of freedom and the kinetic energy.

$$W = a_{ij}\dot{q}^i \dot{q}^j, \tag{5}$$

in which $\dot{q}^i$, and $\dot{q}^j$ are the generalized coordinates and velocities respectively, and introduce an N-dimensional (abstract) space with the metric:

$$ds^2 = a_{sk}\, dq^s dq^k, \qquad \dot{s}^2 = 2W. \tag{6}$$

Then the equations of motion

$$q^i = q^i(t) \tag{7}$$

satisfy the following differential equations:

$$\ddot{q}^\alpha + \Gamma^\alpha_{\beta\gamma}\dot{q}^\beta \dot{q}^\gamma = Q^\alpha, \tag{8}$$

where Q^α is the force vector, $\Gamma^\alpha_{\beta\gamma}$ are the Christoffel symbols:

$$\Gamma_{sk}{}^\ell = \frac{1}{2}a^{\ell p}\left[\frac{\partial a_{sp}}{\partial q^k} + \frac{\partial a_{kp}}{\partial q^s} - \frac{\partial a_{sk}}{\partial q^p}\right], \quad a^{\alpha\beta}a_{\beta\gamma} = \delta^\alpha_\gamma = \begin{cases} 0 & \text{if } \alpha \neq \gamma, \\ 1 & \text{if } \alpha = \gamma. \end{cases} \tag{9}$$

Equation (8) can be interpreted as a parametrical equation of the trajectory C of a representing point M with the contravariant coordinates q^α, Fig. 3. The unit tangent vector $\tau = \nu_o$ to this trajectory is defined as:

$$\tau^\alpha = \nu_o^\alpha = \frac{dq^\alpha}{ds} = \frac{1}{\sqrt{2W}}\dot{q}^\alpha, \quad a_{mn}\nu_o^m \nu_o^n = 1, \tag{10}$$

while the unit normals $\nu_1, \nu_2, \cdots \nu_{N-1}$ are given by the Frenet equations:

$$\frac{d\nu_p^i}{ds} + \Gamma^i_{kq}\nu_p^q \frac{dq^k}{ds} = -\chi_p \nu_{p-1}^i + \chi_{p+1}\nu_{p+1}^i, \tag{11}$$

where $\chi_1, \chi_2, ..., \chi_{N-1}$ are the curvatures of the trajectory, and S is the arc coordinate along this trajectory.

The principal normal ν_1 is coplanar with the tangent ν_o and the force vector Q. The rest curvatures as well as the directions of the rest normals are defined by Eq. (11).

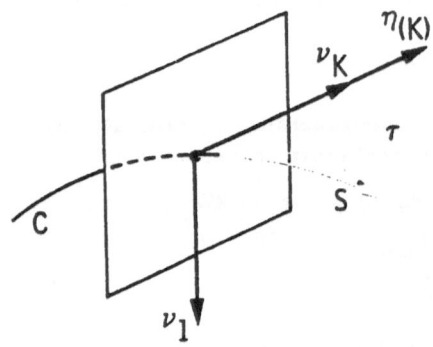

Figure 3

4. Low of Disturbed Motion in Configuration Space

Let ς^r be an infinitesimal disturbance vector. It characterizes the corresponding (simultaneous) point M^* of the disturbed trajectory C^*.

Substituting the vector $q^r + \varsigma^r$ in Eq. (8), after transformations similar to those performed by Syng, Ref. [4], one arrives at the differential equation with respect to the disturbance vector ς^r as a function of S(Ref. [11]):

$$\ddot{\varsigma}^r + \left(2\Gamma^r_{mn}\dot{q}^n - \frac{\partial Q^r}{\partial \dot{q}^s}\right)\dot{\varsigma}^s + \left(\dot{q}^m \dot{q}^n \frac{\partial \Gamma^r_{mn}}{\partial q^s} - \frac{\partial Q^r}{\partial q^s}\right)\varsigma^s = 0 \tag{12}$$

5. Criteria of Transition to Chaos

As shown in previous item, the behavior of disturbed trajectories of an N-degree-of-freedom mechanical system is described by $N-1$ $(p = 1,...N-1)$ second order differential Eqs. (12) which are linearized with respect to the undisturbed trajectories.

Let us consider now Eq. (12) in a small neighbourhood of a fixed point S_o on the undisturbed trajectories, and assume that at least one of its characteristic roots is bounded away from zero by the number b^2 in all configuration space. Then the exponential growth of ς^r would mean the orbital instability of the motion(Ref. [11]).

Now it can be concluded that the motion of the N-degree-of-freedom mechanical system will be chaotic if at least one of the roots λ • of the equation:

$$\det\|\lambda^2\delta_{rs} + (2\Gamma^r_{sn}\dot{q}^n - \frac{\partial Q^r}{\partial\dot{q}^s})\lambda + \dot{q}^m\dot{q}^n\frac{\partial\Gamma^r_{mn}}{\partial q^s} - \frac{\partial Q^r}{\partial q^s}\| = 0 \qquad (13)$$

has positive real part in all the configuration space.

Obviously, these criteria can be applied to continua which are approximated by finite-degree-of-freedom mechanical systems.

Example. For illustration we will consider a symmetric rigid body rotating about its center of gravity, Fig. 4. Determining its position by the Euler's angles.

$$\theta = q^1, \quad \psi = q^2, \quad \phi = q^3 \qquad (14)$$

one obtains the following expression for the kinetic energy:

$$W = \frac{A}{2}[\dot{\theta}^2 + \dot{\psi}^2\sin^2\theta + 2(\dot{\phi} + \dot{\psi}\cos\theta)^2] \qquad (15)$$

in which A and $C = 2A$ are the axial moments of inertia. Then:

$$a_{11} = A, \quad a_{12} = 0, \quad a_{13} = 0, \quad a_{22} = A(\sin^2\theta + 2\cos^2\theta),$$
$$a_{23} = 2A\cos\theta, \quad a_{33} = 2A, \qquad (16)$$

$$\Gamma^1_{22} = \sin\theta\cos\theta, \quad \Gamma^2_{21} = 0, \quad \Gamma^3_{21} = \frac{1}{2}, \quad \Gamma^1_{23} = \sin\theta,$$

$$\Gamma^2_{31} = -\frac{1}{\sin\theta}, \quad \Gamma^2_{31} = \cot\theta, \qquad (17)$$

and consequently, the Gaussian curvatures:

$$G_{(12)} = -2A\cos2\theta, \quad G_{(13)} = 0, \quad G_{(23)} = 0. \qquad (18)$$

Thus, any motion in the subspace θ, ψ will be chaotic if θ is bounded:

$$0 < \theta < \frac{\pi}{4}. \qquad (19)$$

The same type of chaotic instability was found in inertial motions of two-linkage bar, i.e., a system of two rigid rods AB and CD connected by an ideal hinge B and rotatings about a vertical axes normal to the plane ABC with the angular velocities ω_1 and ω_2, Fig. 6. It is proved in Ref. [3], that the motion will be chaotic if the angle β between the rods is bounded: $0 < \beta < \pi/2$ as shown in Fig. 5.

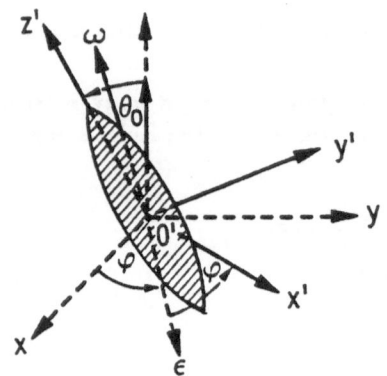

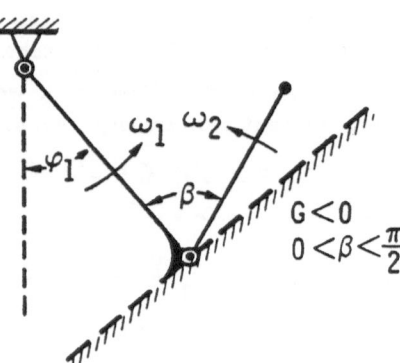

Figure 4   *Figure 5*

6. Deterministic Representation of Chaos

Thus, chaos in non-linear mechanics is caused by orbital instability. What is the mathematical meaning of this phenomenon?

It is a well established fact that the concept of stability or instability (as well as the concept of convergence or divergenve) must be related to a certain class of functions, or a type of space. Indeed, the same sequence of elements can converge in one space and diverge in another space, depending on how the distance between the elements is defined. A similar situation occurs with stability and instability: the same physical phenomena which "looks" as unstable from a certain "space of view" can be considered as stable from another "space of view". In other words, the concept of stability is an attribute of a mathematical model rather than of a physical phenomenon.

Now let us remember that the class of functions on which mechanics is based can be defined by the following: all functions are differentiable "as many times as necessary". Thus, it appears that the differentiability (or smoothness) is granted "for free". However, there is a price paid for it, and this price is instability and chaos in the models which are based on this assumptin.

Is such a strong assumption really necessary? Certainly not, because it follows neither from the principles of mechanics nor from the definition of continuity. Indeed, there are classical examples of continua which are non-differentiable in any point (Cantor discontinuum). On the other hand, the application of integral principles of mechanics do not require differentiability at all.

Thus, the assumption about differentiability serves only the purpose of mathematical convenience. Therefore, it is not surprising that in some situations the behavior of mechanical systems cannot be properly described in terms of smooth functions. In these cases, the classical models exhibit chaos, but it does not mean that such motins cannot be described in deterministic terms; it means only that the classical model should be modified by enlarging the class of function.

The first attempt to abandon the restriction about differentiability and enlarge the class of functions in order to describe chaos was reported in Ref. [5]. The guiding principles for the approach are the following. Let us return to Eqs. (1) and (2). As mentioned above, the real trajectory may "fill up" all the spacing between two exponentially diverging trajectories (Fig. 2). It implies that the solution should be sought in the class of functions having fractal dimension D (Ref. [6]):

$$1 < D < 2$$

The simplest analytical representation of such a function is given by a periodical function with finite amplitude and vanishingly small period

$$\Phi = f(\omega t), \quad \omega \to \infty. \tag{20}$$

Starting, for simplicity, with a conservative system described by the Lagrange equation

$$\frac{d}{dt}\left[\frac{\partial W}{\partial \dot{q}^s}\right] - \frac{\partial}{\partial q^s}(W - \Pi) = 0, \quad s = 1, 2. \tag{21}$$

let us decompose the velocity of each particle into smooth and fractal dimension components:

$$\overset{*}{\mathbf{v}} = \mathbf{v} + \tilde{\mathbf{v}} f(\omega t), \quad \omega \to \infty, \tag{22}$$

where $\mathbf{v}$ and $\tilde{\mathbf{v}}$ represent the mean and fluctuation velocities. Substituting Eq. (22) into Eq. (21) and averaging them over the vanishingly small period $2\pi/\omega$ yields:

$$\frac{d}{dt}\left[\frac{\partial W}{\partial \dot{q}^s}\right] - \frac{\partial}{\partial q^s}(W - \Pi) = \frac{\partial \tilde{\Pi}}{\partial q^s}, \tag{23}$$

in which, (Ref. [7, 8]):

$$\tilde{\Pi} = \frac{1}{2}a_{ij}\dot{\tilde{q}}^i \dot{\tilde{q}}^j. \tag{24}$$

The new governing Eq. (23) can be interpreted as a Lagrange equation written in a non-inertial frame of reference which oscillates with the transport velocity $\tilde{\mathbf{v}} f(\omega t), \omega \to \infty$. Therefore, the vector $\partial \tilde{\Pi}/\partial q^s$ represents the contribution of the corresponding inertia forces, while $\mathbf{v}$ in Eq. (22) is interpreted as a relative velocity.

The decomposition (22) as well as the transition to the new governing Eq. (23) will be usefull if solutions to Eq. (23) are orbitally stable. For this purpose $\tilde{v}$ should be appropriately coupled with v by a "feedback" which eliminate the origin chaotic instability. From physical standpoint it means that we are looking for such a frame of reference which provides the best "view" of the motion.

As follows from the conditions (13) the chaos in a conservative system is eliminated if

$$2WG + \frac{\partial^2 \Pi}{\partial q^m \partial q^n} \nu^m \nu^n < 0. \tag{25}$$

Now one can write the "feedback" in the following form:

$$\dot{\beta} = \left(\frac{\partial^2 \tilde{\Pi}}{\partial q^m \partial q^n} + \frac{\partial^2 \Pi}{\partial q^m \partial q^n} \right) \nu^m \nu^n + 2WG = 0, \tag{26}$$

in which the expression for $\tilde{\Pi}$ is presented by Eq. (24), while G is the Gaussian curvature of the configuration space.

Equations (23) and (26) form a coupled system the solution to which expresses velocities of the particles in the form (22). Such a solution gives a deterministic description of the originally chaotic motion because it is reproducible and predictable: small changes in initial condition lead to small changes in both mean and fluctuation velocities, i.e., in smooth and fractal dimansion components of the solution.

It is worth emphasizing that the feedback (26) is unique. Indeed, at first sight the condition (26) can be replaced by the inequality

$$\beta \geq 0. \tag{27}$$

However, it can be shown that such an "overstabilization" cannot be obtained by oscillations of the frame of reference because of the following reasons: the averaging procedure of Eq. (21) implies that the original motion is orbitally unstable, and therefore, it is represented by both smooth and fractal dimension velocity components. Otherwise, (i.e., when the original motion is smooth), the non-smooth relative and transport velocities must eliminate each other. That is why the inequality in Eq. (27) cannot be attained, and therefore, the feedback Eq. (26) is unique. The same strategy can be applied to nonconservative system, such as viscous fluids, Refs. [9, 10]. For instance, in theory of turbulence Eq. (23) can be considered as an analog to the Reynolds equations while the analog to Eq. (26) effects the closure problem.

7. Prediction, Avoidance and Suppression of Chaos in Controlled Structure

It was pointed out above that chaotic instability in structures leads to transition from periodic to non-periodic, or chaotic motion, i.e., to transition from a discrete to continuous (noisy) power spectrum. Because of such a transition the structure pratically falls into another category of controlled structures; it will require different control strategy since modal structural control will fail. That is why prediction, avoidance and suppression of transition to chaos in controlled structures is very important.

1. *Prediction of chaos* in controlled structures will help in the understanding in the decoding test results, and will lead to simplification of test programs. Analytical prediction of chaos can be performed in the course of design based on the criteria (13) in the same way as the classical stability analysis is performed.

2. *Avoidance of chaos* is especially important for structures with modal control strategy. Chaos can be avoided in the course of design by appropriate changes in parameters of the structure and/or of the control loop until the conditions (13) are eliminated.

3. *Suppression of Chaos* can be based upon an additional control loop with actuators generating high-frequency-excitation forces $\partial \bar{\Pi}/\partial q'$ (see Eq. (24)) and with a feedback given by Eq. (26). Indeed, as shown above, in this case the condition (13) is eliminated, and therefore, chaos is suppressed.

Acknowledgements

The research described in this paper was carried by the Jet Propulsion Laboratory, California Institute of Technology, under NASA Contract No. NAS 7-100.

References

[1] Holms, P.J. and Moon, F.C., "Strange Attractors and Chaos in Non-Linear Mechanics," *J. of Appl. Mechanics*, Vol. 50, No. 1021, Dec. 1983.

[2] Arnold, V.I., *Mathematical Methods in Classical Mechanics*, Springer-Verlag, New York, 1978, pp. 314.

[3] Zak, M., "Two Types of Chaos in Non-Linear Mechanics," *Int. J. of Non-Linear Mechanics*, Vol. 20, No. 4, 1985, pp. 297-308.

[4] Syng, J.L., "On the Geometry of Dynamics," *Phil. Transactions of the Roy. Soc. of London*, Ser. A, Vol. 226, 1926, pp. 31-106.

[5] Zak, M., "Deterministic Representation of Chaos in Classical Dynamics," *Physic Letters A*, Vol. 107, No. 3, pp. 125-128.

[6] Mandelbrot, B.B., *The Fractal Geometry of Nature*, W.H. Freeman and Company, New York, 1983.

[7] Zak, M., "Elastic Continua in High Frequency Excitation Field," *nt. J. of Non-Linear Mechanics*, Vol. 19, No. 5, pp. 479-487.

[8] Landau, L.D. and Lifshitz, E.M., *Course of Theoretical Physics, Vol. 1*, Pergamon Press, Oxford, 1965.

[9] Zak, M., "Deterministic Representation of Chaos in Turbulence," *Physica 18D*, 1986, pp. 486-487.

[10] Zak, M., "Deterministic Representation of Chaos with Application to Turbulence," Proceding of the Fifth International Conference on Mathematical Modelling, 29-31st July 1985, UC Berkeley.

[11] Zak, M., "Criteria of Chaos in Non-Linear Mechanics," *Int. J. of Non-Linear Mechanics*, Vol. 21, No. 3, 1986, pp. 175-182.